A SHORT COURSE IN
BACTERIAL GENETICS

A SHORT COURSE IN BACTERIAL GENETICS

A Laboratory Manual and Handbook
for *Escherichia coli* and Related Bacteria

JEFFREY H. MILLER

Department of Microbiology
and Molecular Genetics
and Molecular Biology Institute
University of California, Los Angeles

Cold Spring Harbor Laboratory Press
1992

A SHORT COURSE IN BACTERIAL GENETICS

A Laboratory Manual and Handbook
for *Escherichia coli* and Related Bacteria

9 8 7 6 5 4 3 2 1

Text design by Emily Harste
Cover design by Leon Bolognesi

Front cover: This strain of *Escherichia coli* with an altered *lacZ* gene reverts to Lac$^+$ only when certain transversion mutations occur. When grown on solid media containing a mix of β-galactosides, such mutants appear as blue Lac$^+$ papillae, or miniature colonies, growing out of a larger white colony of cells containing the still unreverted *lacZ* gene. (Photograph by J.H. Miller and his colleagues, University of California, Los Angeles.)

Back cover: Heterozygous colonies stained for the arabinose constitutive phenotype (*araC*) by the method of Lin et al. (*Biochim. Biophys. Acta*, vol. 60, pp. 422–424 [1962]). Stain: 1% solution of 2,3,5-triphenyltetrazolium chloride.

Library of Congress Cataloging-in-Publication Data

Miller, Jeffrey H.
  A short course in bacterial genetics: a laboratory manual and
  handbook for Escherichia coli and related bacteria / by Jeffrey H.
  Miller.
      p.   cm.
  Includes bibliographical references and index.
  ISBN 0-87969-349-5
  1. Bacterial genetics――Technique. 2. Escherichia coli――Genetics―
  ―Laboratory manuals. 3. Bacterial genetics――Laboratory manuals.
  I. Title.
  QH434.M54   1991
  589.9'015――dc20   91-3201
                                                     CIP

All Cold Spring Harbor Laboratory Press publications may be ordered directly from Cold Spring Harbor Laboratory Press, 10 Skyline Drive, Plainview, New York 11803. (Phone: 1-800-843-4388 in Continental U.S. and Canada. All other locations: (516) 349-1930. FAX: (516) 349-1946)

Handbook Contents

v

Laboratory Manual Contents

UNIT·3
THE *lac* SYSTEM

UNIT·4
MUTAGENESIS   81

UNIT·5
GENE TRANSFER   213

SECTION·1

Escherichia coli Genetic Map

The following map and tables of genetic symbols are reprinted, with permission, from Bachmann (1990, pp. 132–166). The references cited in the tables can be found in the original paper.

REFERENCE

Bachmann, B. 1990. Linkage map of *Escherichia coli* K-12, edition 8. *Microbiol. Rev.* **54:**130–197.

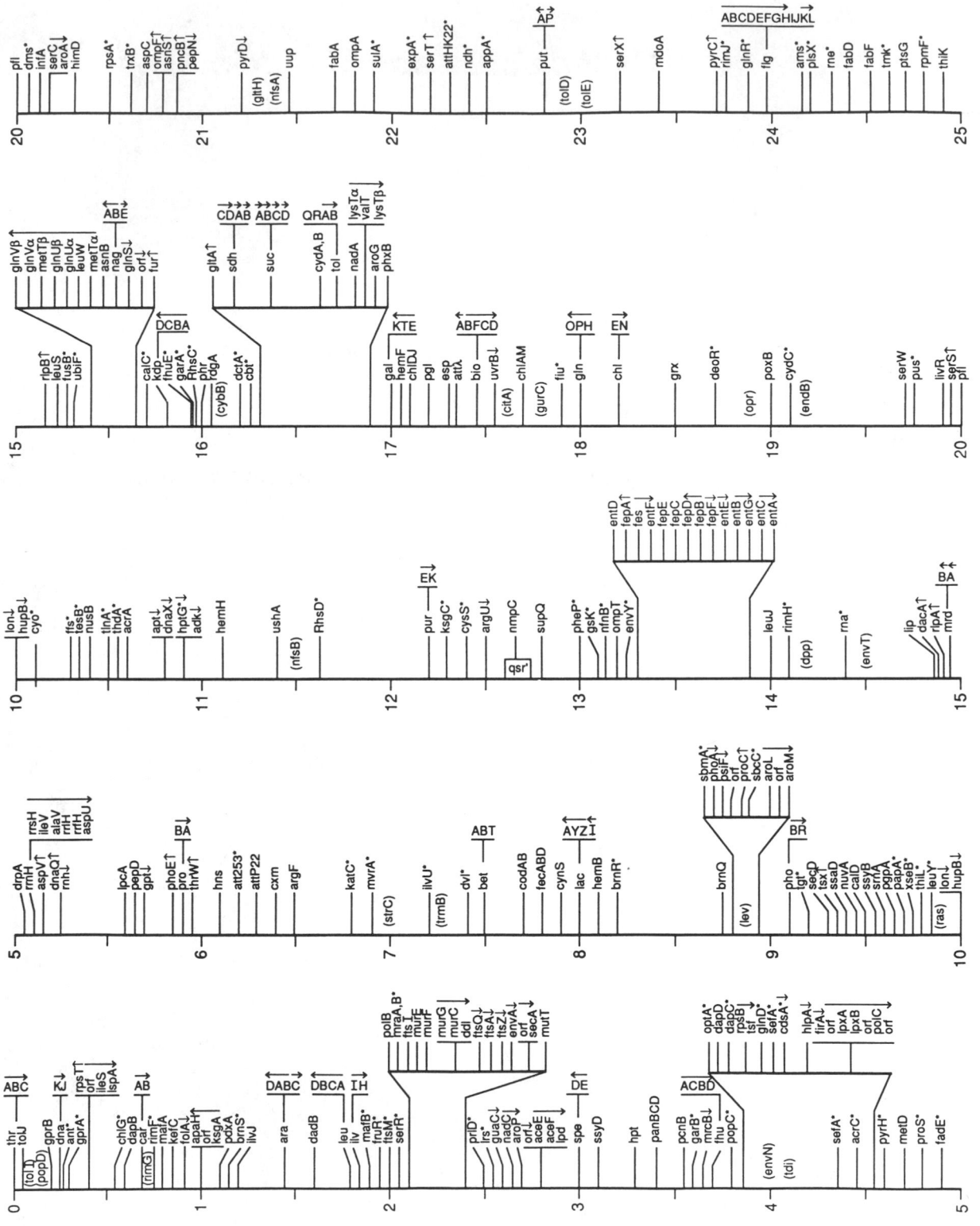

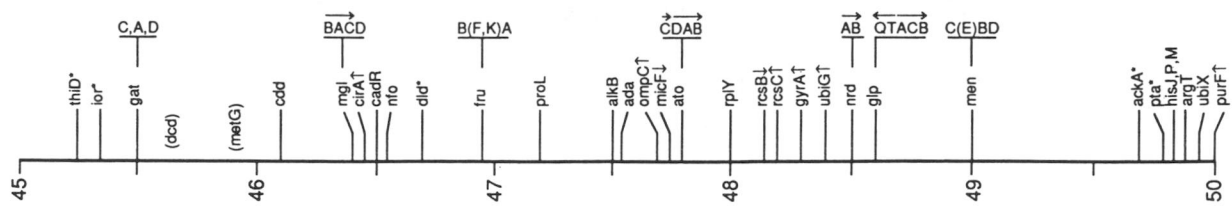

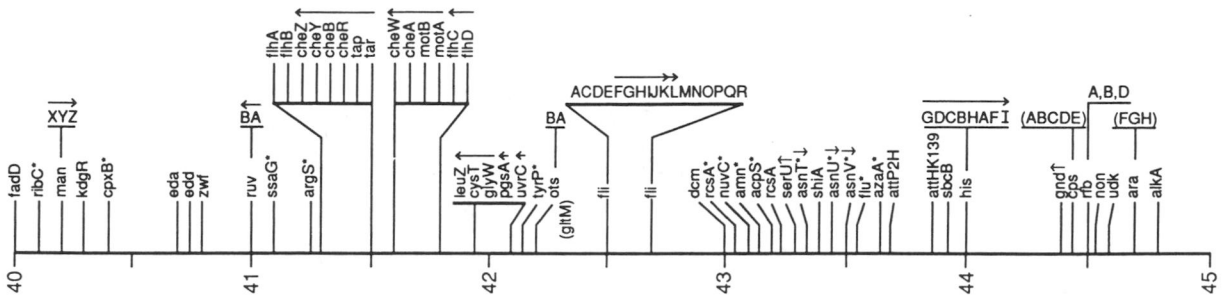

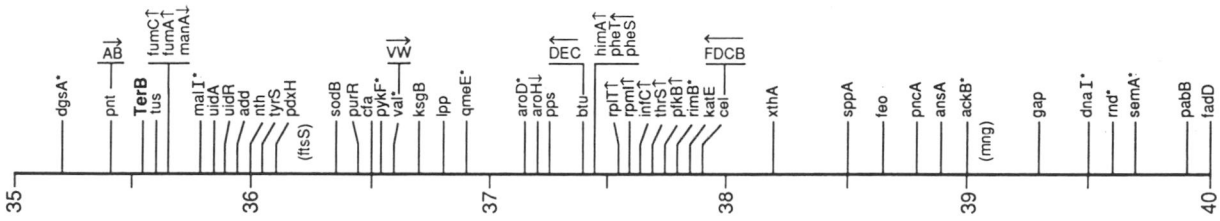

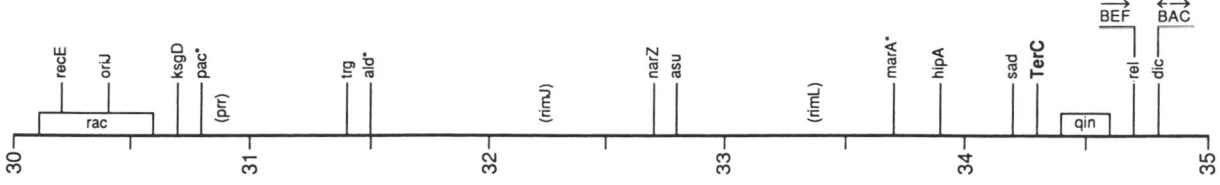

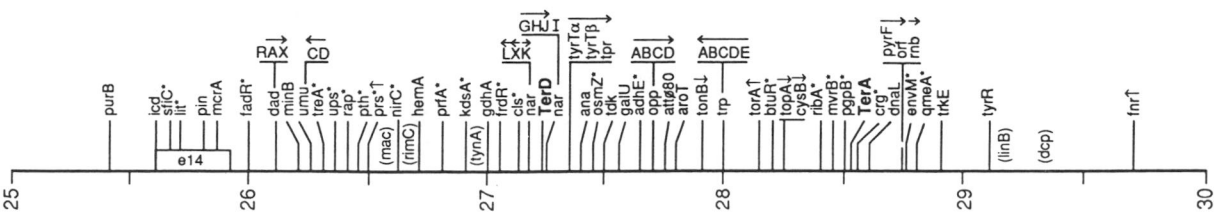

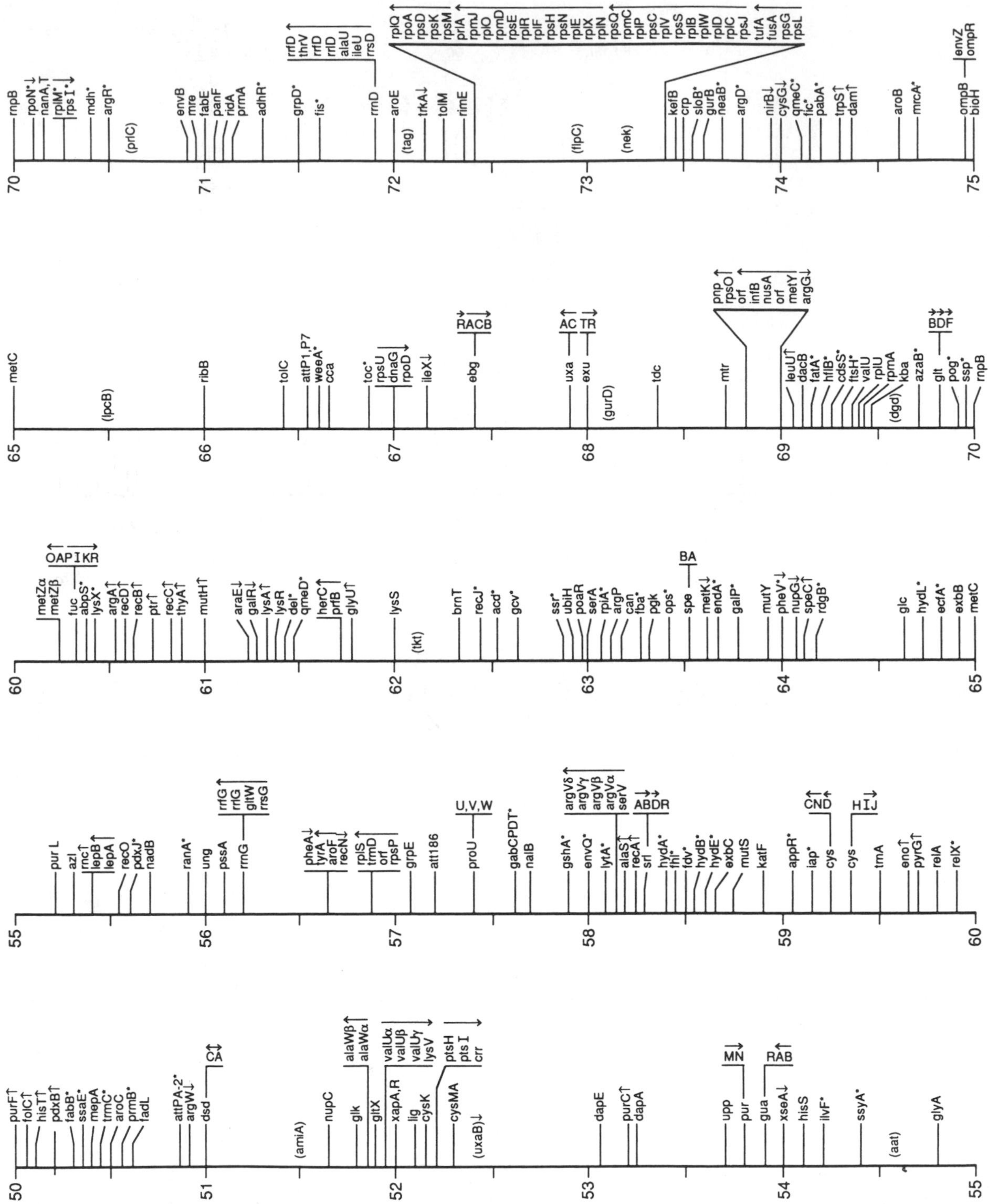
rnpB
rpoN*↓
nanA,T
rplM*
rpsI*↓
mdh*
argR*
(prlC)
envB
mre
fabE
panF
ridA
prmA
adhR*
grpD*
fis*
rrfD
thrV
rrlD
alaU
ileU
rrsD
rrnD
aroE
(tag)
trkA↓
tolM
rimE
rplQ
rpoA
rpsD
rpsK
rpsM
prlA
rpmJ
rplO
rpmD
rpsE
rplR
rplF
rpsH
rpsN
rplE
rplX
rplN
rpsQ
rpmC
rplP
rpsC
rplV
rpsS
rplB
rplW
rplD
rplC
rpsJ
tufA
fusA
rpsG
rpsL
(flpC)
(nek)
kefB
crp
sloB*
gurB
neaB*
argD*
nirB↓
cysG↓
qmeC*
fic*
pabA*
trpS↑
dam↑
aroB
mrcA*
envZ
ompR
ompB
bioH
70
71
72
73
74
75
metC
(lpcB)
ribB
tolC
attP1,P7
weeA*
cca
toc*
rpsU
dnaG
rpoD
ileX↓
RACB
ebg
AC
TR
uxa
exu
(gurD)
tdc
mtr
pnp
rpsO
orf
infB
nusA
orf
metY
argG↓
leuU↑
dacB
fatA*
hflB*
cdsS*
ftsH*
valU
rplU
rpmA
(dgd)
kba
azaB*
BDF
glt
pog*
ssp*
rnpB
65
66
67
68
69
70
OAPIKR
metZα
metZβ
fuc
abpS*
lysX*
argA↑
recD↑
recB↑
ptr↑
recC↑
thyA↑
mutH↑
araE↓
galR↓
lysA↑
lysR
del*
qmeD*
herC↑
prfB
glyU↑
lysS
(tkt)
brnT
recJ*
acd*
gcv*
ssr*
ubiH
poaR
serA
rpiA*
argP
can
fba*
pgk
ops*
BA
spe
metK↓
endA*
galP*
mutY
pheV*↓
nupG↓
speC↑
rdgB*
glc
hydL*
ecfA*
exbB
metC
60
61
62
63
64
65
pur L
azl
rnc↑
lepB↑
lepA
recO
pdxJ*
nadB
ranA*
ung
pssA
rrnG
rrfG
rrlG
gltW
rrsG
pheA↓
tyrA
aroF
recN↓
rplS
trmD
orf
rpsP
grpE
att186
U,V,W
proU
gabCPDT*
nalB
gshA*
envQ*
lytA*
argVδ
argVγ
argVβ
argVα
serV
alaS↑
recA↑
srl
ABDR
hydA*
fhl*
fdv*
hydB*
hydE*
exbC
mutS
katF
appR*
iap*
cys
CND
cys
HIJ
tmA
eno↑
pyrG↑
relA
relX*
55
56
57
58
59
60
purF↑
folC↑
hisT↑
pdxB↑
fabB*
ssaE*
mepA
trmC*
aroC
prmB*
fadL
attPA-2*
argW↓
dsd
CA
(amiA)
nupC
glk
alaWβ
alaWα
gltX
valUα
valUβ
valUγ
lysV
xapA,R
lig
cysK
ptsH
pts I
crr
cysMA
(uxaB)↓
dapE
purC↑
dapA
upp
MN
pur
gua
RAB
xseA↓
hisS
ilvF*
ssyA*
(aat)
glyA
50
51
52
53
54
55

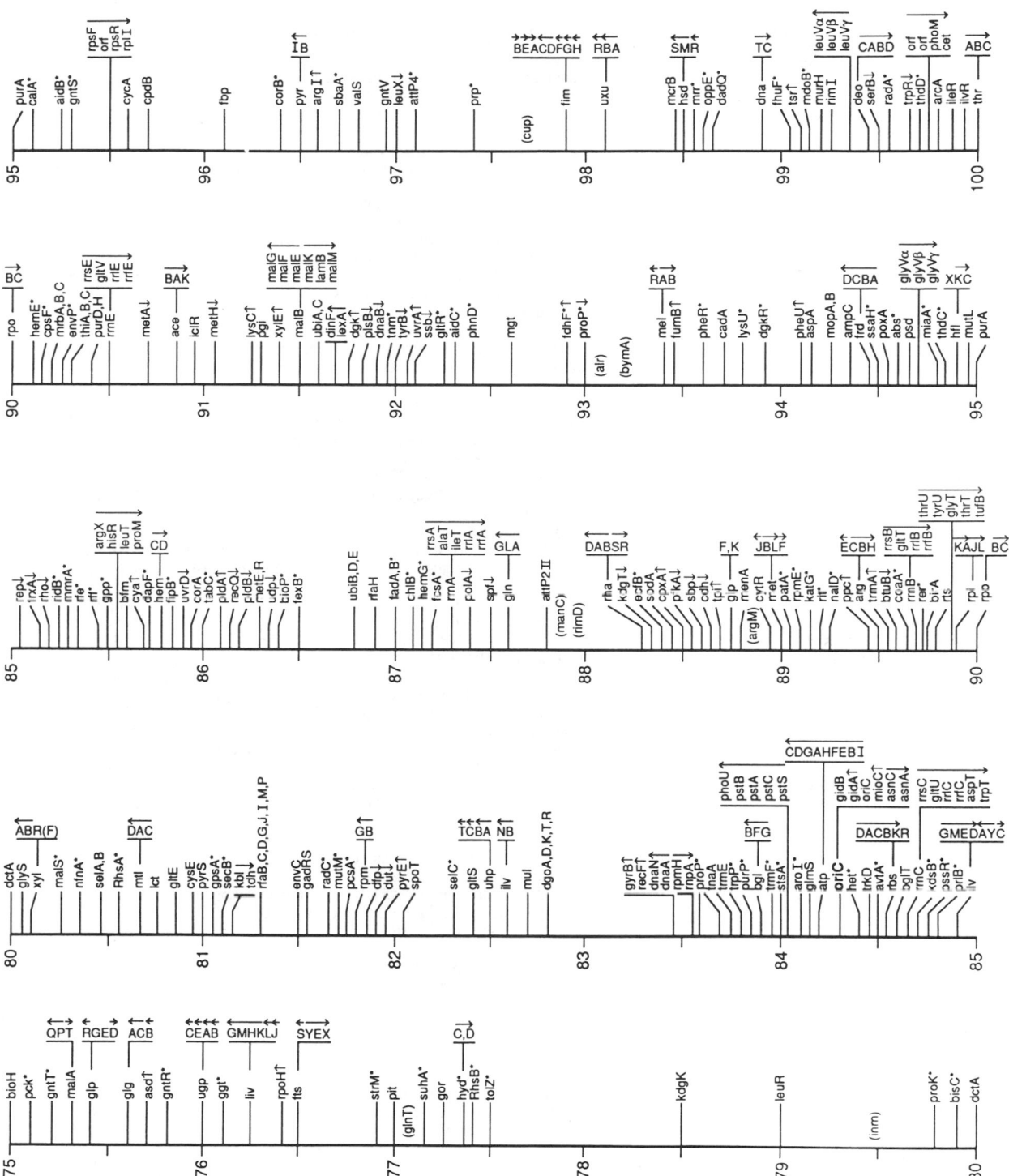

Table 1 Genetic Markers of *E. coli* K12

Gene symbol	Mnemonic	Map position (min)[a]	Alternate gene symbols; phenotypic trait affected[b]	Reference(s)[c]
aat		(55)	Aminoacyl-tRNA-protein-transferase (EC 2.3.2.6)	B
abpS		60	Low-affinity transport system for arginine and ornithine; periplasmic binding protein	180
abs		95	Sensitivity and permeability to antibiotics and dyes	195
acd		63	Acetaldehyde-CoA dehydrogenase	C
aceA	Acetate	91	*icl*; utilization of acetate; isocitrate lyase (EC 4.1.3.1)	A, C, 910
aceB	Acetate	91	*mas*; utilization of acetate; malate synthase A (EC 4.1.3.2)	A, C
aceE	Acetate	3	*aceE1*; acetate requirement; pyruvate dehydrogenase (decarboxylase component)	A, B, C, 422, 739, 1029
aceF	Acetate	3	*aceE2*; acetate requirement; pyruvate dehydrogenase (dihydrolipoyltransacetylase component)	A, B, C, 422, 1029, 1043
aceK		91	Isocitrate dehydrogenase kinase/phosphatase	194, 220, 576, 615, 616
ackA		50	Acetate kinase (EC 2.7.2.1) activity	B, C
ackB		39	Acetate kinase (EC 2.7.2.1) activity	C
acpS		43	CoA:apo-[acyl-carrier protein] pantetheinephosphotransferase (EC 2.7.8.7); holo-[acyl-carrier protein] synthase	C
acrA	Acridine	11	*lir*, *Mb*, *mbl*, *mtc*; sensitivity to acriflavine, phenethyl alcohol, and sodium dodecyl sulfate	A, C
acrC	Acridine	4	Sensitivity to acriflavine	C
ada		48	Inducible DNA repair system protecting against methylating and alkylating agents; O^6-methylguanine-DNA methyltransferase	C, 560, 635, 703, 744, 772, 1087, W
add		36	Adenosine deaminase (EC 3.5.4.4)	B
adhE		27	CoA-linked acetaldehyde dehydrogenase and alcohol dehydrogenase	B, C, 232
adhR		71	Regulatory gene for *acd* and *adhE*	196, 191, F
adk		11	*dnaW*; *plsA*; adenylate kinase (EC 2.7.4.3) activity; pleiotropic effects on glycerol-3-phosphate acyltransferase activity	A, B, C, 147
aidB		95	Induced by alkylating agents	1145, T
aidC		92	Induced by alkylating agents	1145
alaS	Alanine	58	*ala-act*; alanyl-tRNA synthetase (EC 6.1.1.7)	A, B, C
alaT	Alanine	87	*talA*; alanine tRNA 1B; in *rrnA* operon	B, 587
alaU	Alanine	72	*talD*; alanine tRNA 1B; in *rrnD* operon	B, 587
alaV	Alanine	5	Alanine tRNA 1B in *rrnH* operon	C, 587
alaW	Alanine	52	Alanine tRNA 2 (tandemly duplicated gene)	587
ald		31	Aldehyde dehydrogenase, NAD-linked	189
alkA	Alkylation	45	*aidA*; 3-methyl-adenine DNA glycosylase II, inducible	B, C, 199, 773, 774, 1087, 1145
alkB	Alkylation	47	*aidD*; DNA repair system specific for alkylated DNA	560, 561, 1145
alr		(93)	Alanine racemase (EC 5.1.1.1)	A
amiA		(51)	*N*-Acetylmuramyl-L-alanine amidase activity	1104
amn		43	AMP nucleosidase (EC 3.2.2.4)	636
ampC	Ampicillin	94	β-Lactamase; penicillin resistance	A, B, C, 525
ams		24	Alteration of mRNA stability	C, 183
ana		27	Alcohol dehydrogenase (EC 1.1.1.1) and acetaldehyde dehydrogenase (EC 1.2.1.10) activity	B, C
ansA		39	L-asparaginase I, cytoplasmic	C, 1033
ant		0	Na^+/H antiporter activity	381
apaH		1	Diadenosine tetraphosphatase	101, 729
appA		22	pH 2.5 acid phosphatase; exopolyphosphatase (EC 3.6.1.11)	117, 250
appR		59	Expression of pH 2.5 acid phosphatase	1110
apt		11	Adenine phosphoribosyltransferase (EC 2.4.2.7)	B, 457, 458
araA	Arabinose	1	L-Arabinose isomerase (EC 5.3.1.4)	A, B, 630
araB	Arabinose	1	Ribulokinase (EC 2.7.1.16)	A, B, C, 293, 630
araC	Arabinose	1	Regulatory gene; activator and repressor protein	A, B, C, 293
araD	Arabinose	1	L-Ribulosephosphate 4-epimerase (EC 5.1.3.4)	A, B, 630
araE	Arabinose	61	Low-affinity L-arabinose transport system; L-arabinose proton symport	A, C, 688, 1052
araF	Arabinose	45	L-Arabinose-binding protein	B, C, 982, 983
araG	Arabinose	45	High-affinity L-arabinose transport system	C, 983
araH	Arabinose	45	High-affinity L-arabinose transport system; membrane protein	983
arcA		100	*dye*, *fexA*, *msp*, *seg*, *sfrA*; negative regulatory gene of genes in aerobic pathways	A, B, C, 160, 161, 286, 519

Table 1 *Continued*

Gene symbol	Mnemonic	Map position (min)[a]	Alternate gene symbols; phenotypic trait affected[b]	Reference(s)[c]
argA	Arginine	61	*argB, Arg1, Arg2*; amino acid acetyltransferase; *N*-acetylglucosamine synthase (EC 2.3.1.1)	A, B, C, 142, 296, 796, 965
argB	Arginine	90	*argC*; acetylglutamate kinase (EC 2.7.2.8)	A, B, C
argC	Arginine	90	*argH, Arg2*; *N*-acetyl-γ-glutamyl-phosphate reductase (EC 1.2.1.38)	A, B, C, 88, 854
argD	Arginine	74	*argG, Arg1*; acetylornithine δ-aminotransferase (EC 2.6.1.11)	A, 911
argE	Arginine	90	*argA, Arg4*; acetylornithine deacetylase (EC 3.5.1.16)	A, B, C, 88, 854
argF	Arginine	7	*argD, Arg5*; ornithine carbamoyltransferase (EC 2.1.3.3) (duplicate gene)	A, B, C, 855, 1141
argG	Arginine	69	*argE, Arg6*; argininosuccinate synthetase (EC 6.3.4.5)	A
argH	Arginine	90	*argF, Arg7*; argininosuccinate lyase (EC 4.3.2.1)	A, B, C
argI	Arginine	97	Ornithine carbamoyltransferase (EC 2.1.3.3) (duplicate gene)	A, B, 86, 855, 921
argM	Arginine	(89)	Acetylornithine transaminase; cryptic gene; may be duplicate of *argD*	911
argP	Arginine	63	Transport of arginine, ornithine, and lysine	A
argR	Arginine	71	*Rarg*; regulatory gene; repressor of *arg* regulon	A, C, 651
argS	Arginine	(40)	Arginyl-tRNA synthetase (EC 6.1.1.19)	A, 793
argT	Arginine	50	Sequence homologous to *argT* of *S. typhimurium*, which codes for lysine-, arginine-, ornithine-binding protein	807
argU	Arginine	13	*dnaY, pin*; arginine tRNA 4	B, C, 341, 359, 373, 587
argV	Arginine	58	Arginine tRNA 2 (tandemly quadruplicated gene)	341
argW	Arginine	51	Arginine tRNA 5	587
argX	Arginine	85	Arginine tRNA 3	485, 587
aroA	Aromatic	20	3-*enol*-pyruvoylshikimate-5-phosphate synthase (EC 2.5.1.19)	A, 291, 292
aroB	Aromatic	75	Dehydroquinate synthase (EC 4.6.1.3)	A, 740
aroC	Aromatic	51	Chorismate synthase (EC 4.6.1.4)	A
aroD	Aromatic	37	3-Dehydroquinate dehydratase (EC 4.2.1.10)	A, C, 118
aroE	Aromatic	72	Dehydroshikimate reductase (EC 1.1.1.25)	A, 24, 730
aroF	Aromatic	57	DAHP synthetase (tyrosine repressible)	A, 488, 1008
aroG	Aromatic	17	DAHP synthetase (phenylalanine repressible)	A, C, 254
aroH	Aromatic	37	DAHP synthetase (tryptophan repressible)	A, C, 118, 290
aroI	Aromatic	84	Function unknown	A
aroL	Aromatic	9	Shikimate kinase II (EC 2.7.1.71)	B, 261, 262
aroM	Aromatic	9	Unknown function; regulated by *aroR*	261, 262
aroP	Aromatic	3	General aromatic amino acid transport	A, C, 422, 912
aroT	Aromatic	28	*aroR, trpR*; transport of aromatic amino acids, alanine, and glycine	A, B
asd		76	*dap + hom*; aspartate semialdehyde dehydrogenase (EC 1.2.1.11)	A, C
asnA	Asparagine	84	Asparagine synthetase A (EC 6.3.1.1)	A, B, C, 152
asnB	Asparagine	16	Asparagine synthetase B (EC 6.3.1.1)	B, C, 864
asnC	Asparagine	84	Regulatory gene	277
asnS	Asparagine	21	*lcs*; asparaginyl-tRNA synthetase (EC 6.1.1.22)	B, C, 506
asnT	Asparagine	43	Asparagine tRNA	B, C, 341, 587
asnU	Asparagine	43	Asparagine tRNA	587
asnV	Asparagine	43	Asparagine tRNA	587
aspA	Aspartate	94	ʟ-Aspartate ammonia-lyase (aspartase) (EC 4.3.1.1)	A, B, 423
aspC	Aspartate	21	Aspartate aminotransferase (EC 2.6.1.1)	B, 340, 607, 697
aspT	Aspartate	85	*tasC*; aspartate tRNA 1, triplicate gene; in *rrnC* operon	B, C, 587
aspU	Aspartate	5	Aspartate tRNA 1, triplicate gene; in *rrnH* operon	C, 587
aspV	Aspartate	5	Aspartate tRNA 1, triplicate gene	C, 477, 587
asu		33	Asparagine utilization, as sole nitrogen source	191
atoA	Acetoacetate	48	Acetyl-CoA:acetoacetyl-CoA transferase (EC 2.8.3.–) β-subunit	A, 588
atoB	Acetoacetate	48	Acetyl-CoA acetyltransferase (EC 2.3.1.9)	A, 528
atoC	Acetoacetate	48	Positive regulatory gene	A, 528, 529
atoD	Acetoacetate	48	Acetyl-CoA:acetoacetyl-CoA transferase (EC 2.8.3.–) β-subunit?	528
atpA	ATP	84	*papA, uncA*; membrane-bound ATP synthase (EC 3.6.1.3), F_1 sector, α-subunit	A, B, C, 578, 1152
atpB	ATP	84	*papD, uncB*; membrane-bound ATP synthase (EC 3.6.1.3), F_0 sector, subunit a	A, B, C, 801, 1146, 1152

Table 1   *Continued*

Gene symbol	Mnemonic	Map position (min)[a]	Alternate gene symbols; phenotypic trait affected[b]	Reference(s)[c]
atpC	ATP	84	*papG, uncC*; membrane-bound ATP synthase (EC 3.6.1.3), F_1 sector, ε-subunit	B, C, 1152
atpD	ATP	84	*papB, uncD*; membrane-bound ATP synthase (EC 3.6.1.3), F_1 sector, β-subunit	B, C, 1152
atpE	ATP	84	*papH, uncE*; membrane-bound ATP synthase (EC 3.6.1.3), F_0 sector, subunit c; DCCD-binding protein	B, C, 722, 1152
atpF	ATP	84	*uncF*; membrane-bound ATP synthase (EC 3.6.1.3), F_0 sector, subunit b	C, 1152
atpG	ATP	84	*papC, uncG*; membrane-bound ATP synthase (EC 3.6.1.3), F_1 sector, γ-subunit	B, C, 578, 1152
atpH	ATP	84	*papE, uncH*; membrane-bound ATP synthase (EC 3.6.1.3), F_1 sector, δ-subunit	C, 1152
atpI	ATP	84	*uncI*; membrane-bound ATP synthase (EC 3.6.1.3), subunit ?	148, 537, 801, 872, 1146, 1152
atte14	Attachment	26	Attachment site for element e14	C
attHK22	Attachment	22	*atthtt*; attachment site for phage HK022	C
attHK139	Attachment	44	Attachment site for phage HK139	C
attλ	Attachment	17	Integration site for prophages λ, 82, and 434	A, B, C
attP1,P7	Attachment	67	*loxB*; integration site for phages P1 and P7	C
attP2H	Attachment	44	Phage P2 integration site H	A
attP2II	Attachment	88	Phage P2 integration site II	A
attP4	Attachment	97	Integration site for phage P4	C
attP22	Attachment	6	*ata*; integration site for phage P22	A
attPA-2	Attachment	51	Integration site for phage PA-2	B
attϕ80	Attachment	28	Integration site for prophage ϕ80	A, B
att186	Attachment	57	Integration site for prophage 186	A
att253	Attachment	6	Integration site for phage 253	871
avtA		84	Alanine-α-ketoisovalerate transaminase, transaminase C	C, 1155
azaA	Azaserine	44	Resistance or sensitivity to azaserine	C
azaB	Azaserine	70	Resistance or sensitivity to azaserine	C
azl	Azaleucine	55	Regulation of *ilv* and *leu* genes; azaleucine resistance	A
betA	Betaine	7	Choline dehydrogenase	21, 1058
betB	Betaine	7	Betaine aldehyde dehydrogenase	21
betT	Betaine	7	High-affinity choline transport	21
bfm		86	Phage BF23 multiplication	A
bglA	β-Glucoside	84	*bglD*; phospho-β-glucosidase A	A, B
bglB	β-Glucoside	84	*blgA*; phospho-β-glucosidase B	A, B, C, 685, 973
bglF	β-Glucoside	84	*bglB, bglC*; β-glucoside transport	A, B, C, 128, 685
blgG	β-Glucoside	84	*bglC, bglS*; positive regulatory gene	A, 685, 904, 905, 973
bglT	β-Glucoside	85	*bglE*; regulatory gene for phospho-β-glucosidase A synthesis	A
bioA	Biotin	17	7,8-Diaminopelargonic acid synthetase	A, B, C, 1072
bioB	Biotin	17	Biotin synthetase	A, B, C, 1072
bioC	Biotin	17	Block prior to pimeloyl CoA	A, B, 1072
bioD	Biotin	17	Dethiobiotin synthetase	A, B, 1072
bioF	Biotin	17	7-Keto-8-aminopelargonic acid synthetase	A, B, 1072
bioH	Biotin	75	*bioB*; block prior to pimeloyl CoA	A, C
bioP	Biotin	86	*birB*; biotin transport	A, B, C
birA	Biotin retention	90	*bioR, dhbB*; biotin-[acetyl-CoA carboxylase] holoenzyme synthetase; biotin operon repressor	A, B, C, 483
bisC	Biotin sulfoxide	80	Biotin sulfoxide reductase, structural gene	B, C
brnQ	Branched chain	9	Transport system 1 for isoleucine, leucine, and valine	A, B
brnR	Branched chain	8	Component of transport systems 1 and 2 for isoleucine, leucine, and valine	A
brnS	Branched chain	1	Transport system for isoleucine, leucine, and valine	A
brnT	Branched chain	62	Low-affinity transport system for isoleucine	B
btuB	B_{12} uptake	90	*bfe, btuA, cer*; receptor for vitamin B_{12}, E colicins, and phage BF23	A, B, C, 40, 41, 455
btuC	B_{12} uptake	37	Vitamin B_{12} transport	B, 274, 276, 351
btuD	B_{12} uptake	37	Vitamin B_{12} transport, membrane-associated protein	274, 276, 351
btuE	B_{12} uptake	37	Vitamin B_{12} transport, periplasmic protein?	274, 351
btuR	B_{12} uptake	28	Regulatory gene affecting *btuB*	674
bymA		(93)	Bypass of maltose permease at *malB*	A
cadA	Cadaverine	94	Lysine decarboxylase (EC 4.1.1.18)	C
cadR	Cadaverine	46	*lysP?*; regulatory gene for lysine decarboxylase	C, W
calA	Calcium	95	Calcium transport	C

Table 1   *Continued*

Gene symbol	Mnemonic	Map position (min)[a]	Alternate gene symbols; phenotypic trait affected[b]	Reference(s)[c]
calC	Calcium	16	Calcium transport	C
calD	Calcium	10	Calcium transport	C
can	Canavanine	63	Canavanine resistance	A
carA		1	*arg + ura*, *cap*, *pyrA*; carbamoylphosphate synthase (EC 2.7.2.9), glutamine (light) subunit	A, C, 124, 856
carB		1	*arg + ura*, *cap*, *pyrA*; carbamoylphosphate synthase (EC 2.7.2.9), ammonia (heavy) subunit	A, C, 811
cbt		16	Dicarboxylate-binding protein production	A, C
cca		67	tRNA nucleotidyl transferase (EC 2.7.7.25)	A, 231
cdd		46	Deoxycytidine deaminase (EC 3.5.4.5)	A, B, C, 541, 736, W
cdh		89	CDP-diglyceride hydrolase	C, 11, 153, 456, 499
cdsA		4	CDP-diglyceride synthetase (CTP:phosphatidate cytidylyltransferase) (EC 2.7.7.41)	C, 501
cdsS		69	Stability of CDP-diglyceride synthetase activity	358
celB	Cellobiose	38	Transport of cellobiose, arbutin, and salicin	595, 596, 834, J
celC	Cellobiose	38	Transport of cellobiose, arbutin, and salicin	595, 596, 834, J
celD	Cellobiose	38	Negative regulatory gene	595, 596, 834, J
celF	Cellobiose	38	Phospho-β-glucosidase	595, 596, 834, J
cet	Colicin E2	100	*ref*, *refII*; tolerance to colicin E2	A, C, 287
cfa		36	Cyclopropane fatty acid synthase	413–415
cheA	Chemotaxis	42	Chemotactic response	A, B, C, 1016
cheB	Chemotaxis	41	Chemotactic response; protein methylesterase activity	A, B, C, 768, 1016
cheR	Chemotaxis	41	*cheX*; chemotactic response; protein methylesterase activity	B, C, 768, 1016
cheW	Chemotaxis	42	Chemotactic response	B, 768, 1016
cheY	Chemotaxis	41	Chemotactic response	B, C, 201, 714, 768, 1016
cheZ	Chemotaxis	41	Chemotactic response	B, C, 768, 1016
chlA	Chlorate	18	*bisA*, *narA*; biosynthesis of molybdopterin	A, B, C, 534, 901
chlB	Chlorate	87	*narB*; nitrate reductase, biosynthesis of molybdopterin	A, B, C, 901
chlD	Chlorate	17	*narD*; molybdenum uptake	A, B, C, 531
chlE	Chlorate	18	*bisB*, *narE*; biosynthesis of molybdopterin	A, B, C, 534, 803, 901
chlJ	Chlorate	17	Molybdenum transport?	531
chlM	Chlorate	18	Biosynthesis of molybdopterin	534, 1084
chlN	Chlorate	18	Biosynthesis of molybdopterin	534, 803, 1084
cirA	Colicine I resistance	43	*feuA*; production of colicin I receptor	A, B, C, 115, 408, 736, W
citA	Citrate	(18)	Cryptic gene of citrate transport system	429
citB	Citrate	(16)	Cryptic gene of citrate transport system	429
cls		27	Cardiolipin synthase activity	B, 817
cmlA	Chloramphenicol	19	Resistance or sensitivity to chloramphenicol	A, 1007
coaA		90	*panK*; pantothenate kinase	1134
codA		8	Cytosine deaminase (EC 3.5.4.1)	A
codB		8	Cytosine transport	A
corA	Cobalt resistance	86	Mg^{2+} transport, system I	B, C, 14
corB	Cobalt resistance	96	Mg^{2+} transport, system I	B
cpdB		96	2′,3′-Cyclic-nucleotide 2′-phosphodiesterase (EC 3.1.4.16)	C, 656
cpsA		44	Capsular polysaccharide synthesis	1113
cpsB		44	Capsular polysaccharide synthesis	1113
cpsC		44	Capsular polysaccharide synthesis	1113
		44	Capsular polysaccharide synthesis	1113
cpsD		44	Capsular polysaccharide synthesis	1113
cpsE		44	Capsular polysaccharide synthesis	1113
cpsF		90	Capsular polysaccharide synthesis	1113
cpxA		89	F-pilus formation, surface exclusion, conjugal donor activity	C, 11–13
cpxB		40	F-pilus formation, surface exclusion, conjugal donor activity	C
crg		29	Cold-resistant growth	565
crp		74	*cap*, *csm*; cyclic AMP receptor protein	A, C, 370, 821, 1019
crr		52	*gsr*, *iex*, *tgs*; glucose phosphotransferase system enzyme III^{Glc}	B, C, 134, 136, 270, 794, 835, 945
cup		98	Uptake of carbohydrates	686
cxm		6	*cxr*; methylglyoxal synthesis	A, B
cyaA		86	Adenylate cyclase (EC 4.6.1.1)	A, B, C, 6–8, 14, 59, 70, 578, 588, 930, 931

Table 1 *Continued*

Gene symbol	Mnemonic	Map position (min)[a]	Alternate gene symbols; phenotypic trait affected[b]	Reference(s)[c]
cybB		(16.5)	Cytochrome b_{561}	762
cycA	Cycloserine	96	*dagA*; resistance to D-cycloserine and D-serine; transport of D-alanine, D-serine, and glycine	A, B
cydA		17	Cytochrome *d* terminal oxidase, polypeptide subunit I	C, 400, 401
cydB		17	Cytochrome *d* terminal oxidase, polypeptide subunit II	A, B, 400–402
cydC		19	Cytochrome *d* terminal oxidase, possibly heme *d* component	371
cyn		8	*cnt*; cyanate aminohydrolase (EC 3.5.5.3), cyanase	424, 1063, 1064
cyo		10	Cytochrome *o* terminal oxidase complex	38, 39
cysA	Cysteine	52	Sulfate permease; chromate resistance	A, B, 945, 1012
cysB	Cysteine	28	Positive regulator for *cys* regulon	A, B, 825
cysC	Cysteine	59	Adenosine 5'-phosphosulfate kinase (EC 2.7.1.25)	A, 640
cysD	Cysteine	59	ATP sulfurylase (ATP:sulfate adenyltransferase) (EC 2.7.7.4)	A, 493, 640
cysE	Cysteine	81	Serine acetyltransferase (EC 2.3.1.30)	A, B, 269
cysG	Cysteine	74	Siroheeme synthesis	A, C, 522, 679, 680
cysH	Cysteine	59	Adenylylsulfate reductase (EC 1.8.99.2)	A, 493, 641
cysI	Cysteine	59	*cysQ*; sulfite reductase (EC 1.8.1.2), α subunit	A, 493, 641
cysJ	Cysteine	59	*cysP*; sulfite reductase (EC 1.8.1.2), β subunit	A, 493, 641
cysK	Cysteine	52	*cysZ*; *O*-Acetylserine sulfhydrolase A (EC 4.2.99.8)	B, C, 120, 134, 945, 1012
cysM	Cysteine	52	*O*-Acetylserine sulfhydrolase B (EC 4.2.99.8)	945, 1012
cysN	Cysteine	59	ATP-sulfurylase (ATP:sulfate adenylyltransferase) (EC 2.7.7.4), subunit	640
cysS	Cysteine	12	Cysteinyl-tRNA synthetase (EC 6.1.1.16)	C
cysT	Cysteine	42	Cysteine tRNA	341, 587
cytR		89	Regulatory gene for *deo* operon, *udp* and *cdd*	A, B, 63, 1130
dacA		15	D-Alanine carboxypeptidase, fraction A; penicillin-binding protein 5	A, B, C, 141, 832, 1048
dacB		69	D-Alanine carboxypeptidase, fraction B; penicillin-binding protein 4	B, C
dadA		26	*dadR*; D-amino acid dehydrogenase subunit	A, B, C, 1173
dadB		2	*alnA*; D-amino acid dehydrogenase subunit	A, B, C
dadQ		99	*alnR*; regulatory gene for *dad* regulon	A, C
dadR		26	Regulatory gene	1173
dadX		26	*msuA*?; alanine racemase (EC 5.1.1.1)	1173
dam		74	DNA adenine methylase	A, B, C, 32, 139
dapA	Diaminopimelate	53	Dihydrodipicolinate synthase (EC 4.2.1.52)	A, 833, 909
dapB	Diaminopimelate	1	Dihydrodipicolinate reductase (EC 1.3.1.26)	A, C, 125
dapC	Diaminopimelate	4	Tetrahydrodipicolinate succinylase	A, C
dapD	Diaminopimelate	4	Tetrahydrodipicolinate *N*-succinyltransferase	A, C, 908
dapE	Diaminopimelate	53	*dapB*; *N*-succinyl-diaminopimelate deacylase	A, 833, 909
dapF	Diaminopimelate	86	Diaminopimelate epimerase	907
dcd		(46)	*paxA*; 2'-deoxycytidine 5'-triphosphate deaminase (EC 3.5.4.–) activity	B
dcm		43	*mec*; DNA cytosine methylase	A, B, 90
dcp		(29)	Dipeptidyl carboxypeptidase	B
dctA		80	Uptake of C4-dicarboxylic acids	A
dctB		16	Uptake of C4-dicarboxylic acids	A
ddl		2	D-Alanine:D-alanine ligase	A, B, C, 914
del	Deletion	61	Frequency of IS*1*-mediated deletion	B
deoA	Deoxyribose	100	*tpp*, *TP*; thymidine phosphorylase (EC 2.4.2.4)	A, B, C, 1127, 1128
deoB	Deoxyribose	100	*drm*, *thyR*; phosphopentomutase (EC 2.7.5.6)	A, B, C, 1127
deoC	Deoxyribose	100	*dra*, *thyR*; deoxyribose-phosphate aldolase (EC 4.1.2.4)	A, B, C, 243, 1126
deoD	Deoxyribose	100	*pup*; purine-nucleoside phosphorylase (EC 2.4.2.1)	A, B, C, 617
deoR	Deoxyribose	19	*nucR*; regulatory gene for *deo* operon	A, B, 1007, 1129
dfp		82	*dnaS*, *dut*; flavoprotein affecting synthesis of DNA and pantothenate metabolism	1030, 1031
dgd		(70)	D-Galactose dehydrogenase production	B
dgkA	Diglyceride	92	Diglyceride kinase	B, C, 646
dgkR	Diglyceride	94	Level of diglyceride kinase	C
dgoA	D-Galactonate	82	2-Oxo-3-deoxygalactonate 6-phosphate aldolase (EC 4.1.2.21)	B
dgoD	D-Galactonate	82	Galactonate dehydratase (EC 4.2.1.6)	B
dgoK	D-Galactonate	82	2-Oxo-3-deoxygalactonate kinase (EC 2.7.1.58)	B
dgoR	D-Galactonate	82	Regulatory gene	B

Table 1 *Continued*

Gene symbol	Mnemonic	Map position (min)[a]	Alternate gene symbols; phenotypic trait affected[b]	Reference(s)[c]
dgoT	D-Galactonate	82	Galactonate transport	B
dgsA		35	Function of enzyme IIA/IIB of phosphotransferase system	C, 757
dicA		35	Regulatory gene	80, 81
dicB		35	Control of cell division	80, 81
dicC		35	Regulatory gene	80, 81
dinF		92	Locus induced by UV and mitomycin C; subject to *recA* and *lexA* regulation	C, 598
dld		(47)	D-Lactate dehydrogenase (EC 1.1.1.28)	C, 170, 932
dms		20	Dimethyl sulfoxide reductase	95
dnaA	DNA	83	DNA biosynthesis; initiation	A, B, C, 434, 578, 815
dnaB	DNA	92	*groP, grpA*; DNA biosynthesis; chain elongation	A, B, C, 783
dnaC	DNA	99	*dnaD*; DNA biosynthesis; initiation and chain elongation	A, B, 784, 927
dnaE	DNA	4	See *polC*	
dnaG	DNA	67	*dnaP, parB*; DNA biosynthesis; primase	A, B, C, 157, 677, 765, 784, 808, 927, 1018, 1086, 1182
dnaI	DNA	40	DNA biosynthesis	A
dnaJ	DNA	0	*groPAB, groPC*; DNA biosynthesis	B, 66, 814, 1196
dnaK	DNA	0	*groPAB, groPC, groPF, grpF*; DNA biosynthesis	B, 64, 951
dnaL	DNA	29	*dnaK*; DNA biosynthesis	B
dnaN	DNA	83	DNA biosynthesis; DNA polymerase III holoenzyme, β-subunit	C, 27, 434, 815, 894
dnaQ	DNA	5	*mutD*, DNA polymerase III holoenzyme, ε subunit, mutator activity	B, C, 225, 278, 298, 690, 707, 806, 971
dnaT	DNA	99	DNA biosynthesis; primosomal protein i	B, 709, 784
dnaX	DNA	11	*dnaZ*; DNA biosynthesis; DNA polymerase III holoenzyme, τ and γ subunits; DNA elongation factor III	B, C, 335, 583, 631, 761, 927, 1207
dpp	Dipeptide	(14)	Transport of dipeptides	A, B
drpA		5	DNA and RNA biosynthesis	626
dsdA	D-Serine	51	D-Serine deaminase	A, B, C, 119, 724
dsdC	D-Serine	51	Regulatory gene for *dsdA*	A, C, 119, 723, 830
dut	dUTPase	82	*dnaS, sof*; deoxyuridinetriphosphatase (EC 3.6.1.23)	A, B, C, 672, 673, 1031
dvl		7	Sensitivity to sodium dodecyl sulfate and toluidine blue plus light	1151
ebgA		67	Phospho-β-D-galactosidase, α subunit; cryptic gene	A, B, 430, 1050, 1051, J
ebgB		67	Possible homolog of *lacY*; in *ebg* operon	430, 1051, J
ebgC		67	Phospho-β-D-galactosidase, β subunit; cryptic gene	430, J
ebgR		67	Regulatory gene	B, 430, 1051, J
ecfA	Energy-coupling factor	65	Pleiotropic effects on active transport coupling to metabolic energy; may be *metC*	B
ecfB	Energy-coupling factor	88	*eup, ssd*; generalized resistance to aminoglycoside antibiotics; coupling of metabolic energy to active transport	B, C, 863
eda		41	*kdgA, kga*; 2-keto-3-deoxygluconate 6-phosphate aldolase (EC 4.1.2.14)	A
edd		41	Phosphogluconate dehydratase (EC 4.2.1.12)	A
endA		64	DNA-specific endonuclease I	B
eno		60	Enolase (EC 4.2.1.11)	A, B, C, 1169
entA	Enterochelin	14	2,3-Dihydro-2,3-dihydroxybenzoate dehydrogenase	A, B, C, 333, 770, 850
entB	Enterochelin	14	2,3-Dihydro-2,3-dihydroxybenzoate synthetase	A, B, C, 333, 770, 850
entC	Enterochelin	14	Isochorismate synthetase	A, B, C, 333, 770, 850
entD	Enterochelin	13	Enterochelin synthetase, component D	A, B, C, 205, 334
entE	Enterochelin	14	Enterochelin synthetase, component E	A, B, C, 333, 770, 850
entF	Enterochelin	14	Enterochelin synthetase, component F	A, B, C, 205, 333, 845
entG	Enterochelin	14	Enterochelin synthetase, component G	A, B, C, 333, 770, 850
envA	Envelope	2	Cell envelope and cell separation	A, B, C, 71, 535, 1060
envB	Envelope	71	*mon, rodY*; cell shape and sensitivity to antibiotics	A, B, 1147
envC	Envelope	81	Anomalous cell division; chain formation	A, 556
envM	Envelope	28	Osmotically remedial envelope defect	A
envN	Envelope	(4)	Osmotically remedial envelope defect	A
envP	Envelope	90	Osmotically remedial envelope defect	A
envQ	Envelope	58	Osmotically remedial envelope defect	A
envT	Envelope	(15)	Osmotically remedial envelope defect	A
envY	Envelope	13	Envelope protein; thermoregulation of porin synthesis	675

Table 1 *Continued*

Gene symbol	Mnemonic	Map position (min)[a]	Alternate gene symbols; phenotypic trait affected[b]	Reference(s)[c]
envZ	Envelope	75	*ompB*, *perA*, *tpo*; production of outer membrane proteins; regulatory gene	C, 179, 213, 363, 364, 748, 749, 933, 1191
esp		17	Site for efficient packaging of phage T1	B
exbB		65	Uptake of enterochelin; resistance or sensitivity to colicins	A, B, 302
exbC		59	Uptake of enterochelin; resistance or sensitivity to colicins	B
expA		22	Expression of a group of exported proteins	C
exuR		68	Negative regulatory gene for *exu* regulon, *exuT*, *uxaCA*, and *uxaB*	B, C, 489, 710
exuT		68	Transport of hexuronates	B, C, 102, 489, 710, 711
e14		26	Cryptic chromosomal element able to be excised; see *lit*, *mcrA*, *pin* and *sfiC*	137, 138, 553, 684, 861, 862, 887, 1139
fabA	Fatty acid biosynthesis	22	β-Hydroxydecanoyl thioester dehydrase (EC 4.2.1.60)	A, 228
fabB	Fatty acid biosynthesis	50	*fabC*; β-ketoacyl-[acyl-carrier protein] synthase I (EC 2.3.1.41)	A, B, C
fabD	Fatty acid biosynthesis	24	Malonyl-CoA-[acyl carrier protein] transacylase (EC 2.3.1.39)	A, C
fabE	Fatty acid biosynthesis	71	Acetyl-CoA carboxylase (EC 6.4.1.2)	B, 1147
fabF	Fatty acid biosynthesis	24	*cvc*, *vtrB*, β-ketoacyl-[acyl carrier protein] synthase II (EC 2.3.1.41)	B, C, 1123
fadA	Fatty acid degradation	87	*oldA*; thiolase I (EC 2.3.1.16)	A, C, 1032, 1202
fadB	Fatty acid degradation	87	*oldB*; 3-hydroxyacyl-CoA dehydrogenase (EC 1.1.1.35), 3-hydroxyacyl-CoA epimerase (EC 5.1.2.3), Δ^3-*C15*-Δ^2-*trans*-enoyl-CoA (EC 5.3.3.8) and enoyl-CoA-hydratase (crotonase) (EC 4.2.1.17)	A, 1032, 1201, 1202
fadD	Fatty acid degradation	40	*oldD*; acyl-CoA synthetase (EC 6.2.1.3)	A, B
fadE	Fatty acid degradation	5	Electron transport flavoprotein of β-oxidation	A, C
fadL	Fatty acid degradation	51	*ttr*; transport of long-chain fatty acids and sensitivity to phage T2	B, 98, 99, 755
fadR	Fatty acid degradation	26	*dec*, *oleR*, *thdB*; negative regulatory gene for *fad* regulon	B, C, 2, 279, F
fatA		69	Utilization of *trans*-unsaturated fatty acids	275
fba		63	*alc*, *fda*; fructose-bisphosphate aldolase (EC 4.1.2.13)	A
fbp		96	*fdp*; fructose-bisphosphatase (EC 3.1.3.11)	A, 578, 985
fcsA		86	Cell division; septation	B
fdhF		93	Formate dehydrogenase (formate hydrogen-lyase linked), selenopolypeptide	97, 840, 1188, 1189, 1220, 1221
fdv		59	*ant*?; formate dehydrogenase-2-activity	962, 1204
fecA	Iron	93	Citrate-dependent iron transport, outer membrane receptor	B, C, 882, 1218
fecB	Iron	8	Citrate-dependent iron transport, periplasmic protein	B, C, 882, 1218
fecD	Iron	8	Citrate-dependent iron transport, membrane-bound protein	882
feo	Iron	39	Ferrous iron transport system	438
fepA	Iron	13	*cbr*, *cbt*, *feuB*; receptor for ferric enterobactin (enterochelin) and colicins B and D	A, B, C, 205, 334, 676, 829, 853, 1185
fepB	Iron	13	Ferric enterobactin (enterochelin) uptake; periplasmic component	129, 829, 852, 853, 1185
fepC	Iron	13	Ferric enterobactin (enterochelin) uptake; cytoplasmic membrane component	829, 852
fepD	Iron	13	Ferric enterobactin (enterochelin) uptake	829
fepE	Iron	13	Ferric enterobactin (enterochelin) uptake	829
fepF	Iron	13	Ferric enterobactin (enterochelin) uptake	829
fes	Iron	13	Enterochelin esterase	A, B, C, 205, 334, 845
fexB		86	FexA phenotype affected	C
ffs		10	4.5S RNA	144, 486
fhl		58	Formate hydrogen-lyase activity, possibly electron transport system	962
fhuA	Ferric hydroxamate uptake	4	*tonA*, *T1*, *T5rec*; outer membrane receptor for ferrichrome, colicin M, and phages T1, T5, and φ80	129, 222, 223, 321
fhuB	Ferric hydroxamate uptake	4	Hydroxamate-dependent iron uptake, cytoplasmic membrane component	C, 129, 321, 591, 883
fhuC	Ferric hydroxamate uptake	4	Hydroxamate-dependent iron uptake, cytoplasmic membrane component	129, 154, 221, 321
fhuD	Ferric hydroxamate uptake	4	Hydroxamate-dependent iron uptake, cytoplasmic membrane component	129, 154, 221, 321

Table 1 *Continued*

Gene symbol	Mnemonic	Map position (min)[a]	Alternate gene symbols; phenotypic trait affected[b]	Reference(s)[c]
fhuE	Ferric hydroxamate up-take	16	Outer membrane receptor for ferric-rhodotorulic acid	436, 968
fhuF	Ferric hydroxamate up-take	99	Ferric hydroxamate transport	439
fic		74	Filamentation in presence of cyclic AMP in mutant	1125
fimA	Fimbriae	98	*fimD*, *pilA*; type 1 fimbrin (pilin), structural gene	A, B, C, 3, 304, 346, 347, 572, 575
fimB	Fimbriae	98	*pil*; regulatory gene for expression of *himA*	A, B, C, 304, 573, 575
fimC	Fimbriae	98	*pil*; biosynthesis of type 1 fimbriae	A, B, C, 575
fimD	Fimbriae	98	*pil*; biosynthesis of type 1 fimbriae	A, B, C, 575
fimE	Fimbriae	98	Regulatory gene for expression of *fimA*	304, 573
fimF	Fimbriae	98	Fimbrial morphology	574
fimG	Fimbriae	98	Fimbrial morphology	574
fimH	Fimbriae	98	Minor fimbrial subunit, adhesin	574
fipB		86	Morphogenesis of phage F1	665
fipC		(73)	Morphogenesis of phage F1	665
firA		4	Affects transcription	B, 1, 210, G
fis		72	Site-specific DNA inversion	581
fiu		18	Ferric iron uptake, outer membrane protein	436
flgA	Flagella	24	*flaU*; flagellar synthesis	B, C, 586
flgB	Flagella	24	*flbA*; flagellar synthesis	B, C
flgC	Flagella	24	*flaW*; flagellar synthesis, basal-body protein	B, C
flgD	Flagella	24	*flaV*; flagellar synthesis, basal-body rod modification	B
flgE	Flagella	24	*flaK*; flagellar synthesis, hook protein	B
flgF	Flagella	24	*flaX*; flagellar synthesis, basal-body rod protein	B, C
flgG	Flagella	24	*flaL*; flagellar synthesis, basal-body rod protein	B
flgH	Flagella	24	*flaY*; flagellar synthesis, basal-body L-ring protein	B, C, 536
flgI	Flagella	24	*flaM*; flagellar synthesis, basal-body P-ring protein	B, 536
flgJ	Flagella	24	*flaZ*; flagellar synthesis	B, C
flgK	Flagella	24	*flaS*; flagellar synthesis, hook-associated protein	B, C
flgL	Flagella	24	*flaT*; flagellar synthesis, hook-associated protein	B
flhA	Flagella	41	*flaH*; flagellar synthesis	A
flhB	Flagella	41	*flaG*; flagellar synthesis	A, C
flhC	Flagella	42	*flaI*; flagellar synthesis; regulatory gene	A, 68
flhD	Flagella	42	*flbB*; flagellar synthesis; regulatory gene; flagellum-specific σ factor	B, C, 68
fliA	Flagella	43	*flaD*; flagellar synthesis; regulation of late gene expression	A, B, C, 544, 586
fliC	Flagella	43	*flaF*, *hag*; flagellin; flagellar synthesis, filament structural protein	A, B, C, 612
fliD	Flagella	43	*flbC*; flagellar synthesis, hook-associated protein 2	C
fliE	Flagella	43	*flaN*; flagellar synthesis	A, B, C
fliF	Flagella	43	*flaBI*; flagellar synthesis, basal-body M-ring protein	A, B, C, 69
fliG	Flagella	43	*flaBII*; flagellar synthesis, motor switching and energizing	A, B, C, 69
fliH	Flagella	43	*flaBIII*; flagellar synthesis	A, B, C, 69
fliI	Flagella	43	*flaC*; flagellar synthesis	A, B, C, 69
fliJ	Flagella	43	*flaO*; flagellar synthesis	A, B, 69
fliK	Flagella	43	*flaE*; flagellar synthesis, hook length control	A, B, 69
fliL	Flagella	43	*flaAI*; flagellar synthesis	A, B, C, 69, 604, 696
fliM	Flagella	43	*flaAII*; flagellar synthesis, motor switching and energizing	A, B, C, 202, 604
fliN	Flagella	43	*motD*; flagellar synthesis, motor switching and energizing	696
fliO	Flagella	43	*flbD*; flagellar synthesis	696
fliP	Flagella	43	*flaR*; flagellar synthesis	A, B, C, 696
fliQ	Flagella	43	*flaQ*; flagellar synthesis	A, B, 696
fliR	Flagella	43	*flaP*; flagellar synthesis	A, B, 696
flu	Fluffing	44	Metastable gene affecting surface properties, piliation, and colonial morphology	B
fnr		30	*nirA*, *nirR*; regulatory gene for nitrite and nitrate reductases, hydrogenase, and fumarate reductase	A, B, C, 994, 995
folA	Folate	1	*tmrA*; dihydrofolate reductase (EC 1.5.1.3); trimethoprim resistance	A, B, C, 101
folC	Folate	50	Dihydrofolate:folylpolyglutamate synthetase	111, 323, 807
frdA		94	Fumarate reductase (EC 1.3.99.1), flavoprotein subunit	A, B, C, 538
frdB		94	Fumarate reductase (EC 1.3.99.1), iron-sulfur protein subunit	C, 538

Table 1 *Continued*

Gene symbol	Mnemonic	Map position (min)[a]	Alternate gene symbols; phenotypic trait affected[b]	Reference(s)[c]
frdC		94	Fumarate reductase (EC 1.3.99.1), membrane anchor polypeptide	C, 538
frdD		94	Fumarate reductase (EC 1.3.99.1), membrane anchor polypeptide	C, 538
frdR		27	Regulation of electron transport and fermentation-associated genes; may be *ana*	546
fruA	Fructose	47	*ptsF*; fructose phosphotransferase enzyme II	A, B, C, 367, W
fruB	Fructose	47	Fructose phosphotransferase enzyme III	367, W
fruF	Fructose	47	*fpr*; phosphohistidinoprotein-hexose phosphotransferase, fructose-specific	A, B, C, 281, 367, 589, W
fruK	Fructose	47	*fpk*; fructose-1-phosphate kinase (EC 2.7.1.3)	A, B, C, 115, 367, 736, W
fruR	Fructose	2	*fruC*; regulatory gene; possibly repressor of *fru* operon	367, 590
ftsA		2	*divA*; cell division	A, B, C, 78, 535, 913, 1059, 1206
ftsE		76	Cell division	B, 377, 954
ftsH		69	Cell division	B
ftsI		2	*pbp*, *sep*; peptidoglycan synthetase; septum formation; penicillin-binding protein 3	B, C, 78, 775
ftsM		2	Cell division	284
ftsQ		76	Cell division	C, 78, 913, 914, 1206
ftsS		76	Cell division	377, 954
ftsX		76	Cell division	377
ftsY		76	Cell division	377
ftsZ		2	*sfiB*, *sulB*; cell division	B, C, 78, 535, 1059, 1205, 1206
fucA	Frucose	60	*fucC*; L-fuculose-1-phosphate aldolase	A, B, 182, 190, 1013, 1217
fucI	Frucose	60	L-Fucose isomerase	A, B, 190, 1217
fucK	Frucose	60	L-Fuculose kinase (EC 2.7.1.51)	A, B, 190, 1217
fucO	Frucose	60	L-1,2-Propanediol oxidoreductase	A, B, 190, 1013, 1217
fucP	Frucose	60	Fucose permease	182, 190, 1013, 1217
fucR	Frucose	60	Positive regulatory protein	190, 1014, 1217
fumA	Fumarate	36	Regulatory gene?	421, 737, 738
fumB	Fumarate	93	Regulatory gene?	420, 421
fumC	Fumarate	36	Fumarase	420
fur		16	Ferric iron uptake; negative regulatory gene, repressor protein	51, 52, 266, 437, 969, K
fusA	Fusidic acid	73	*far*; protein chain elongation factor EF-G	A, B, C, 1215
fusB	Fusidic acid	15	Pleiotropic effects on RNA synthesis, ribosomes, and ribosomal protein S6	B
gabC	γ-Aminobutyrate	58	Regulatory gene for *gabPDT*	A, B, C
gabD	γ-Aminobutyrate	58	Succinate-semialdehyde dehydrogenase (EC 1.2.1.16), NADP-dependent, activity	B, C
gabP	γ-Aminobutyrate	58	Transport of γ-aminobutyrate	B, C
gabT	γ-Aminobutyrate	58	Aminobutyrate aminotransferase (EC 2.6.1.19) activity	A, B, C
gadR		82	Regulatory gene for *gadS*	A
gadS		82	Glutamate decarboxylase (EC 4.1.1.15)	A
galE	Galactose	17	*galD*; UDP-galactose 4-epimerase	A, B, C, 634
galK	Galactose	17	*galA*; galactokinase (EC 2.7.1.6)	A, 260
galP	Galactose	64	*Pgal*; galactose permease	B
galR	Galactose	61	*Rgal*; regulatory gene; repressor of *galETK* operon	A, C
galT	Galactose	17	*galB*; galactose-1-phosphate uridylyltransferase (EC 2.7.7.12)	A, 219, 634
galU	Galactose	28	Glucose-1-phosphate uridylyltransferase (EC 2.7.7.9)	A, B
gap		39	*gad*; glyceraldehyde 3-phosphate dehydrogenase (EC 1.2.1.12)	A, B
garA	Glucarate	16	Glucarate utilization	C
garB	Glucarate	4	Glucarate utilization	C
gatA	Galactitol	45	Galactitol-specific enzyme II of phosphotransferase system	B, C, 655, 736
gatC	Galactitol	45	Regulatory gene	B, C, 655, 736
gatD	Galactitol	45	Galactitol-1-phosphate dehydrogenase	B, C, 655, 736
gcv		63	Glycine cleavage pathway	858, 1040
gdhA		27	Glutamate dehydrogenase	B, C, 73, 726, 1135, 1136

Table 1 *Continued*

Gene symbol	Mnemonic	Map position (min)[a]	Alternate gene symbols; phenotypic trait affected[b]	Reference(s)[c]
ggt		76	γ-Glutamyltranspeptidase (EC 2.3.2.2)	1068, 1069
gidA		84	Glucose-inhibited division; chromosome replication?	152, 578, 1152
gidB		84	Glucose-inhibited division; chromosome replication?	578, 801, 1152
glc	Glycolate	65	Utilization of glycolate, malate synthase G (EC 4.1.3.2)	A
glgA	Glycogen	76	Glycogen synthase (EC 2.4.1.21)	A, B, C, 602
glgB	Glycogen	76	1,4-α-Glucan branching enzyme (EC 2.4.1.18)	A, B, C, 50
glgC	Glycogen	76	Glucose-1-phosphate adenylyltransferase (EC 1.7.7.27)	A, B, C, 49
glk		52	Glucokinase (EC 2.7.1.2)	A
glmS	Glucosamine	84	L-glutamine:D-fructose-6-phosphate aminotransferase (EC 2.6.1.16)	A, B, C, 1152
glnA	Glutamine	88	Glutamine synthetase (EC 6.3.1.2)	A, B, C, 212, 224, 742, 902, 915
glnD	Glutamine	4	Uridylyltransferase	B, C
glnF	Glutamine	70	See *rpoN*	B, C
glnG	Glutamine	88	*glnT, ntrC*; negative regulatory gene for *glnA*	C, 682, 742
glnH	Glutamine	18	Periplasmic glutamine-binding protein	804, 805
glnL	Glutamine	88	*glnR, ntrB*; negative regulatory gene for *glnA*	C, 682, 742, 805, 902, 915, 1122
glnP	Glutamine	18	Glutamine high-affinity transport system; membrane component	C, 805
glnQ		18	Glutamine high-affinity transport system	805
glnR	Glutamine	24	Level of glutaminyl-tRNA synthetase	C
glnS	Glutamine	16	Glutaminyl-tRNA synthetase (EC 6.1.1.18)	A, 864, 916, 1199
glnT	Glutamine	(75)	Levels of glutamine tRNA 1 and glutamine synthetase	B
glnU	Glutamine	15	*supB*; glutamine tRNA 1 (tandemly duplicated gene)	B, C, 341, 587, 864
glnV	Glutamine	15	*supE, Su2, suII*; glutamine tRNA 2 (tandemly duplicated gene)	B, C, 341, 587, 864
glpA	Glycerol phosphate	49	*sn*-Glycerol-3-phosphate dehydrogenase (anaerobic), large subunit	A, B, C, 208, 301
glpB	Glycerol phosphate	49	*sn*-Glycerol-3-phosphate dehydrogenase (anaerobic), membrane anchor subunit	208, 301
glpC	Glycerol phosphate	49	*sn*-Glycerol-3-phosphate dehydrogenase (anaerobic), small subunit	208, 301
glpD	Glycerol phosphate	75	*glyD*; *sn*-glycerol-3-phosphate dehydrogenase (aerobic)	A, 980, 981
glpE	Glycerol phosphate	75	Gene of *glp* regulon	981
glpF	Glycerol phosphate	89	Facilitated diffusion of glycerol	A
glpG	Glycerol phosphate	75	Gene of *glp* regulon	981
glpK	Glycerol phosphate	89	Glycerol kinase (EC 2.7.1.30)	A, 215, 844
glpQ	Glycerol phosphate	49	Glycerol-3-phosphate diesterase	C, 619
glpR	Glycerol phosphate	75	Regulatory gene	A, 979, 981
glpT	Glycerol phosphate	49	*sn*-Glycerol-3-phosphate permease	A, B, C, 301, 303
gltA	Glutamate	16	*glut*; citrate synthase (EC 4.1.3.7)	A, C, 110, 492, 795, 1174, 1184
gltB	Glutamate	70	*aspB*; glutamate synthase, large subunit	A, B, C, 178, 360, 824
gltD	Glutamate	70	*aspB*; glutamate synthase, small subunit	178, 360, 824
gltE	Glutamate	81	Glutamyl-tRNA synthetase; possible regulatory subunit	A
gltF	Glutamate	70	Regulatory gene?	178
gltH	Glutamate	(21)	Requirement	A
gltM	Glutamate	(43)	Level of glutamyl-tRNA synthetase activity	A
gltR	Glutamate	92	Regulatory gene for glutamate permease	A
gltS	Glutamate	82	*gltC*; glutamate permease	A
gltT	Glutamate	90	*tgtB*; glutamate tRNA 2; in *rrnB* operon	B, C, 587, 858, 885
gltU	Glutamate	85	*tgtC*; glutamate tRNA 2; in *rrnC* operon	B, 587
gltV	Glutamate	90	*tgtE*; glutamate tRNA 2; in *rrnE* operon	B, 587
gltW	Glutamate	56	Glutamate tRNA 2; in *rrnG* operon	C, 587
gltX	Glutamate	52	Catalytic subunit for glutamyl-tRNA synthetase (EC 6.1.1.17)	A, 131, 961, D
glyA	Glycine	55	Serine hydroxymethyltransferase (EC 2.1.2.1)	A, C, 859, 860, 1039
glyS	Glycine	80	*gly-act*; glycine-tRNA synthetase (EC 6.1.1.14)	A, 132, 566
glyT	Glycine	90	*supA36, sumA, sup15B*; glycine tRNA 2	A, B, C, 587
glyU	Glycine	62	*suA36, sufD, sumB, supT*; glycine tRNA 1	A, 587
glyV	Glycine	95	*suA58, suA78*; glycine tRNA 3 (duplicate gene, tandemly triplicated)	A, B, 341, 587
glyW	Glycine	42	*suA58, suA78*; glycine tRNA 3 (duplicate gene)	A, C, 587, 1117
gnd		44	Gluconate-6-phosphate dehydrogenase, (EC 1.1.1.44) decarboxylating	A, C, 788

Table 1   *Continued*

Gene symbol	Mnemonic	Map position (min)[a]	Alternate gene symbols; phenotypic trait affected[b]	Reference(s)[c]
gntR	Gluconate	76	Regulatory gene for *edd*; transport and phosphorylation of gluconate	A, B
gntS	Gluconate	95	*gntM*, *usgA*; second system for transport and possibly phosphorylation of gluconate	B
gntT	Gluconate	75	*gntM*, *usgA*; high-affinity transport of gluconate	A, B, 515
gntV	Gluconate	97	Glukokinase, thermosensitive	515
gor		77	Glutathione oxidoreductase (EC 1.6.4.2)	C, 308, 407
gpp		85	Guanosine pentaphosphatase activity	B, 14, 84
gprA		0	Replication of certain lambdoid phage	812
gprB		0	Replication of certain lambdoid phage	812
gpsA		81	*sn*-Glycerol-3-phosphate dehydrogenase [NAD(P)$^+$] (EC 1.1.1.94)	A, B
gpt		6	*gpp*, *gxu*; guanine-hypoxanthine phosphoribosyltransferase (EC 2.4.2.8)	A, B, 523, 810, 881, 906
grpD		71	Initiation of phage lambda DNA replication; host DNA synthesis	B
grpE		57	Phage lambda replication; host DNA synthesis	B
grx		19	Glutaredoxin	473, 594, 934
gshA		58	Glutathione synthetase (EC 6.3.2.2)	B, 404, 431
gsk		13	Guanosine kinase	B
guaA	Guanine	54	*gua*$_b$; GMP synthetase (EC 6.3.4.1)	A, B, 1097
guaB	Guanine	54	*gua*$_a$; IMP dehydrogenase (EC 1.2.1.14)	A, B, 1089, 1090, 1094
guuC	Guanine	3	GMP reductase (EC 1.6.6.8)	A, F, 751, 912
guaR	Guanine	54	Regulatory gene	1096
gurB		74	Utilization of methyl-β-D-glucuronide; possibly *crp*	A
gurC		(18)	Utilization of methyl-β-D-glucuronide	A
gurD		(68)	Utilization of methyl-β-D-glucuronide	A
gyrA	Gyrase	48	*nalA*; DNA gyrase (EC 5.99.1.3), subunit A; resistance or sensitivity to nalidixic acid	A, B, C, 497, 732, 733, 1208
gyrB	Gyrase	83	*acrB*, *cou*, *himB*, *nalC*, *pcbA*; DNA gyrase (EC 5.99.1.3), subunit B; resistance or sensitivity to coumermycin	B, C, 4, 5, 732, 733, 1197
hemA	Hemin	27	Glutamyl-tRNA dehydrogenase	A, 42
hemB	Hemin	8	*ncf*; 5-aminolevulinate dehydratase (EC 4.2.1.24) activity	A, B, 642
hemC	Hemin	86	*popE*; porphobilinogen deaminase (EC 4.3.1.18)	B, 539, 965, 966, 1091
hemD	Hemin	86	Uroporphyrinogen III synthase (EC 4.2.1.75)	B, 539, 540, 965, 966
hemE	Hemin	90	*hemC*; uroporphyrinogen decarboxylase (EC 4.1.1.37)	A
hemF	Hemin	17	*popB*, *sec*; coproporphyrinogen III oxidase (EC 1.3.3.3)	A
hemG	Hemin	87	Protoporphyrinogen oxidase activity	C
hemH	Hemin	11	*hemG*, *popA*; ferrochelatase (EC 4.99.1.1)	A
het		84	*cop*; binding of DNA sequences in *oriC* region to outer membrane; possibly structural gene for DNA-binding protein	A, C
herC		62	Suppressor of ColE1 mutation in primer RNA	564
hflB		69	Lysogeny and level of phage lambda *c*II protein	62
hflC		95	*hflA*; high frequency of lysogenization by phage lambda	61
hflK		95	*hflA*; high frequency of lysogenization by phage lambda	61
himA		37	*hid*; integration host factor (IHF), α subunit; site-specific recombination	C, 727, 728
himD		20	*hip*; integration host factor (IHF), β subunit; site-specific recombination	C, 332, 567
hipA		34	Frequency of persistence following inhibition of murein synthesis	759, 760
hisA	Histidine	44	N-(5′-phospho-L-ribosylformimino)-5-amino-1-(5′-phosphoribosyl)-4-imidazolecarboxamide isomerase (EC 5.3.1.16)	A
hisB	Histidine	44	Imidazoleglycerolphosphate dehydratase (EC 4.2.1.19) and histidinol phosphate phosphatase (EC 3.1.3.15)	A, 193, 410
hisC	Histidine	44	Histidinol phosphate aminotransferase (EC 2.6.1.9)	A, 409, 410
hisD	Histidine	44	L-Histidinol:NAD$^+$ oxidoreductase (EC 1.1.1.23)	A, C, 192
hisE	Histidine	44	See *hisI*	A
hisF	Histidine	44	Cyclase	A
hisG	Histidine	44	ATP phosphoribosyltransferase (EC 2.4.2.17)	A, B, C
hisH	Histidine	44	Amido transferase	A
hisI	Histidine	44	*hisE*; phosphoribosyl-AMP cyclohydrolase (EC 3.5.4.19), phosphoribosyl-ATP pyrophosphatase (EC 3.6.1.31)	A, 192

Table 1 *Continued*

Gene symbol	Mnemonic	Map position (min)[a]	Alternate gene symbols; phenotypic trait affected[b]	Reference(s)[c]
hisJ	Histidine	50	Histidine-binding protein of high-affinity histidine transport system	C
hisM	Histidine	50	Histidine transport	592
hisP	Histidine	50	Histidine permease	C, 592
hisR	Histidine	(86)	*hisT*; histidine tRNA	B, 485, 587
hisS	Histidine	54	Histidinyl-tRNA synthetase (EC 6.1.1.21)	B, C, 345
hisT	Histidine	50	*asuC, leuK*; pseudouridylate synthase I	B, C, 29–31, 708, 807, 984
hlpA		4	*skp*; histonelike protein HLP-I (BH1); DNA-binding nucleoid-associated protein	1, 210, 469, 621, G
hns		6	Histonelike protein HLP-II (HU, BH2, HD, NS); DNA-binding protein	870
hpt		3	Hypoxanthine phosphoribosyltransferase	A
hsdM	Host specificity	99	*hs, hsm, hsp, rm*; host modification; DNA methylase M	A, C, 660
hsdR	Host specificity	99	*hs, hsp, hsr, rm*; host restriction; endonuclease R	A, C, 660
hsdS	Host specificity	99	*hss*; specificity determinant for *hsdM* and *hsdR*	A, C, 391, 660, 767
htpG		11	Heat shock protein C 62.5	65
htpR			See *rpoH*	
hupB		10	Histonelike protein HU-1 (HU-β, NS1)	551, 552, 1053
hydA		58	Hydrogenase 1 activity	557, 629, 962
hydB		59	Hydrogenase activity	629, 962, 1161
hydC		77	Formate hydrogenlyase activity and fumarate-dependent H$_2$ uptake; possibly nickel metabolism	1187
hydD		77	Formate hydrogen-lyase activity and fumarate-dependent H$_2$ uptake	1187
hydE		59	Hydrogenase activity	188
hydL		65	*hup?*; hydrogenase L, possibly same as hydrogenase 3; hydrogen uptake	629, 1047
iap		59	Alkaline phosphatase isozyme conversion	B, 512, 780
icd		26	Isocitrate dehydrogenase, NADP$^+$ specific (EC 1.1.1.42)	B, C
iclR		91	Regulatory gene for *aceBA* operon	A, C
ileR		100	*avr, flrA?*; negative regulatory gene of *thr* and *ilv* operons	532, 533, 1164
ileS	Isoleucine	0	Isoleucyl-tRNA synthetase (EC 1.1.1.5)	A, B, 549, 741, 899, 1102, 1192, 1196
ileT	Isoleucine	87	*tilA*; isoleucine tRNA 1; in *rrnA* operon	B, 587
ileU	Isoleucine	72	*tilD*; isoleucine tRNA 1; in *rrnD* operon	B, 587
ileV	Isoleucine	5	Isoleucine tRNA 1; in *rrnH* operon	C, 587
ileX	Isoleucine	67	Isoleucine tRNA 2	766
ilvA	Isoleucine-valine	85	*ile*; threonine deaminase (EC 4.2.1.16)	A, B, C, 14, 285, 623
ilvB	Isoleucine-valine	83	Acetohydroxy acid synthase I (EC 4.1.3.18), valine-sensitive, large subunit	A, B, C, 349, 350, 448, 543, 993, 1167
ilvC	Isoleucine-valine	85	*ilvA*; ketol-acid reductoisomerase (EC 1.1.1.86)	A, B, C, 14, 70, 1165
ilvD	Isoleucine-valine	85	*ilvB*; dihydroxyacid dehydratase (EC 4.2.1.9)	A, B, C, 14, 285, 623
ilvE	Isoleucine-valine	85	*ilvC, ilvJ*; branched-chain amino-acid aminotransferase (EC 2.6.1.42)	A, B, C, 14, 168, 285, 606, 623, 664, 1166
ilvF	Isoleucine-valine	54	Acetohydroxy acid synthase (valine insensitive) activity	A
ilvG	Isoleucine-valine	85	Acetohydroxy acid synthase II (EC 4.1.3.18), valine-insensitive, large subunit	A, B, C, 14, 285, 623, 670
ilvH	Isoleucine-valine	2	Acetohydroxy acid synthase III (EC 4.1.3.18), valine-sensitive, small subunit	A, B, C, 1037
ilvI	Isoleucine-valine	2	Acetohydroxy acid synthase III (EC 4.1.3.18), valine-sensitive, large subunit	A, B, C, 447, 1037
ilvJ	Isoleucine-valine	2	Acetohydroxy acid synthase IV (EC 4.1.3.18), valine-insensitive	C
ilvM	Isoleucine-valine	85	Acetohydroxy acid synthase II (EC 4.1.3.18), valine-insensitive, small subunit	623, 664, 670, 1166
ilvN	Isoleucine-valine	83	Acetohydroxy acid synthase II (EC 4.1.3.18), valine-sensitive, small subunit	349, 1167
ilvR	Isoleucine-valine	100	Positive regulatory gene of *thr* and *ilv* operons	533
ilvU	Isoleucine-valine	7	Regulation of *ileS* and modification of isoleucine tRNA 2 and valine tRNA 2	C
ilvY	Isoleucine-valine	85	Positive regulatory locus for *ilvC*	B, C, 14, 1165
infA		20	Protein chain initiation factor 1	960
infB		69	*ssyG*; protein chain initiation factor 2	C, 510, 776, 865, 866, 868, 999–1001

Table 1　*Continued*

Gene symbol	Mnemonic	Map position (min)[a]	Alternate gene symbols; phenotypic trait affected[b]	Reference(s)[c]
infC		38	*fit*?; protein chain initiation factor 3	B, C, 158, 249, 306, 717, 878, 1170, 1190
inm		(79)	Susceptibility to mutagenesis by nitrosoguanidine	C
ior		45	Radiation sensitivity, particularly γ rays; recombination ability decreased	317
katC	Catalase	7	Catalase activity	B
katE	Catalase	38	Biosynthesis of catalase hydroperoxidase HPII (III)	661, 662
katF	Catalase	59	*nur*; biosynthesis of catalase hydroperoxidase HPII (III) and exonuclease III; regulatory gene	662, 950, 955
katG	Catalase	89	Catalase-peroxidase hydroperoxidase HPI (I), structural gene	663, 1112
kba		69	Ketose-bis-phosphate aldolase, temperature-sensitive enzyme, active on D-tagatose-1,6-diphosphate	B
kbl		81	2-Amino-3-ketobutyrate CoA ligase (EC 2.3.1.29) (glycine acetyltransferase)	28, 892
kdgK	Ketodeoxygluconate	78	Ketodeoxygluconokinase (EC 2.7.1.45)	A
kdgR	Ketodeoxygluconate	40	Regulatory gene for *kdgK*, *kdgT*, and *eda*	A
kdgT	Ketodeoxygluconate	88	2-Keto-3-deoxy-D-gluconate transport system	A, B, 698
kdpA	Potassium dependence	16	*kac*; high-affinity potassium transport system; probably K⁺-stimulated ATPase	A, B
kdpB	Potassium dependence	16	*kac*; high-affinity potassium transport system	A, B
kdpC	Potassium dependence	16	*kac*; high-affinity potassium transport system	A, B
kdpD	Potassium dependence	16	*kac*; high-affinity potassium transport system; regulatory gene	A, B
kdsA		27	3-Deoxy-D-*manno*-octulosonic acid 8-phosphate synthase	1180, 1181
kdsB		85	CTP:CMP-3-deoxy-D-*manno*-octulosonate cytidylyltransferase	382, 383, 848
kefB	K⁺ efflux	73	*trkB*; NEM-activable K⁺/H⁺ antiporter	A, 54
kefC	K⁺ efflux	1	*trkC*; NEM-activable K⁺/H⁺ antiporter	54
ksgA	Kasugamycin	1	S-Adenosylmethionine-6-N′,N′-adenosyl (rRNA) dimethyltransferase	A, C, 101, 1138
ksgB	Kasugamycin	37	Second step (high-level) resistance to kasugamycin	A, B, C, 415
ksgC	Kasugamycin	12	Kasugamycin resistance; affects ribosomal protein S2	B
ksgD	Kasugamycin	31	Kasugamycin resistance	C
lacA	Lactose	8	*a*, *lacAc*; galactoside acetyltransferase (EC 2.3.1.18)	A, B, 453
lacI	Lactose	8	*i*; regulatory gene; repressor protein of *lac* operon	A, B, C, 478
lacY	Lactose	8	*y*; galactoside permease (M protein)	A, B, C
lacZ	Lactose	8	*z*; β-D-galactosidase (EC 3.2.1.23)	A, B, C, 547
lamB	Lambda	92	*malB*; phage lambda receptor protein; maltose high-affinity uptake system	A, B, C
lct	Lactate	81	D-Lactate dehydrogenase (EC 1.1.1.27)	A, 653
lepA		55	GTP-binding membrane protein; function unknown	700, 701
lepB		55	Leader peptidase (signal peptidase I)	C, 700, 1183
leuA	Leucine	2	α-Isopropylmalate synthase (EC 4.1.3.12)	A, C, 368
leuB	Leucine	2	β-Isopropylmalate dehydrogenase (EC 1.1.1.85)	A, C
leuC	Leucine	2	α-Isopropylmalate isomerase subunit	A, C
leuD	Leucine	2	α-Isopropylmalate isomerase subunit	A, C
leuJ	Leucine	14	*flr*; regulation of *leu* and *ilv* operons	532
leuK	Leucine	19	See *hisT*	
leuR	Leucine	79	Level of leucyl-tRNA synthetase	B
leuS	Leucine	15	Leucyl-tRNA synthetase (EC 6.1.1.4)	A, B, C, 444
leuT	Leucine	(86)	Leucine tRNA 1 (duplicate gene)	B, 485, 587
leuU	Leucine	(69)	Leucine tRNA 2	B, 587, 1150
leuV	Leucine	(99)	Leucine tRNA 1 (duplicate gene, tandemly triplicated)	B, 289, 341, 587
leuW	Leucine	15	Leucine tRNA 3	B, C, 587, 864
leuX	Leucine	97	*Su-6*, *supP*; leucine tRNA 5	B, C, 587, 1092, 1210
leuY	Leucine	10	Level of leucyl-tRNA synthetase	B
leuZ	Leucine	42	Leucine tRNA 4	587
lev	Levallorphan	(9)	Resistance to levallorphan	B
lexA		92	*exrA*, *spr*, *tsl*, *umuA*; regulatory gene for SOS operon	A, B, C
lig	Ligase	52	*dnaL*, *pdeC*; DNA ligase	A, B, 513
linB	Lincomycin	(29)	High-level resistance to lincomycin	A
lip	Lipoate	15	Requirement	A, C
lit		25	Phage T4 late-gene expression; locus of element e14	B, 553, 554, L
livG	Leucine, isoleucine, and valine	76	High-affinity branched-chain amino acid transport system; membrane component	C, 791

Table 1 *Continued*

Gene symbol	Mnemonic	Map position (min)[a]	Alternate gene symbols; phenotypic trait affected[b]	Reference(s)[c]
livH	Leucine, isoleucine, and valine	76	High-affinity branched-chain amino acid transport system; membrane component	B, C, 790
livJ	Leucine, isoleucine, and valine	76	High-affinity branched-chain amino acid transport system; periplasmic binding protein for leucine, isoleucine, and valine	B, C, 613, 1098
livK	Leucine, isoleucine, and valine	76	High-affinity branched-chain amino acid transport system; leucine-specific periplasmic binding protein	B, C, 613, 790, 791, 1098
livL	Leucine, isoleucine, and valine	76	High-affinity branched-chain amino acid transport	P
livM	Leucine, isoleucine, and valine	76	High-affinity branched-chain amino acid transport	790, 791
livR	Leucine, isoleucine, and valine	20	*lss*, *lstR*; high-affinity branched-chain amino acid transport system; regulatory gene; repressor protein	B, 25, P
lon	Long form	10	*capR*, *deg*, *dir*, *muc*; DBA-binding, ATP-dependent protease La	A, B, C, 366
lpcA	Lipopolysaccharide core	6	*tfrA*; lipopolysaccharide core synthesis; resistance to phages T4, T7, and P1; deficiency in conjugation	A, B, Q
lpcB	Lipopolysaccharide core	(65)	*pon*; lipopolysaccharide core synthesis	A
lpd		3	*dhl*; lipoamide dehydrogenase (NADH) (EC 1.6.4.3)	A, B, C, 422, 1029, 1044
lpp	Lipoprotein	36	*mlpA*; murein lipoprotein structural gene	B, C, 415
lpxA		4	UDP-*N*-acetylglucosamine acetyltransferase	210, 229, 230
lpxB		4	*pgsB*; lipid A disaccharide synthase	210, 229, 230, 1103
lrs		3	Level of leucine tRNA	C
lspA		1	Prolipoprotein signal peptidase (SPaseII)	B, 505, 549, 741, 899, 1102, 1192, 1195, 1196, 1212
lysA	Lysine	61	Diaminopimelate decarboxylase (EC 4.1.1.20)	A, C, 1054, 1056
lysC	Lysine	91	*apk*; aspartokinase III	A, B, C, 175, 176
lysR	Lysine	61	Positive regulatory gene	1055, 1056
lysS	Lysine	62	Lysyl tRNA synthetase, constitutive	186, 310
lysT	Lysine	17	*suβ*, *supG*, *supL*; lysine tRNA (duplicated gene)	B, 341, 587, 880, 1211
lysU	Lysine	94	Lysyl tRNA synthetase, inducible	186, 1137
lysV	Lysine	52	*supN*; lysine tRNA (duplicated gene)	A, 587, 1121, D
lysX	Lysine	60	Lysine excretion	A
lytA		58	Tolerance to β-lactams; autolysis defective?	442, 1004
mac	Macrolide	(27)	Erythromycin growth dependence	A
mafA		1	Maintenance of F-like plasmids	A, B
mafB		2	Maintenance of F-like plasmids	B
malE	Maltose	92	*malB*; periplasmic maltose-binding protein; substrate recognition for transport and chemotaxis	A, B, C, 75, 76, 294, 816
malF	Maltose	92	*malB*; maltose transport; cytoplasmic membrane protein	A, B, 353
malG	Maltose	92	*malB*; active transport of maltose and maltodextrins	B, 251, 344
malI	Maltose	36	Production of oligosaccharide, probably glucose polymer	300
malK	Maltose	92	*malB*; maltose permeation	A, B, C, 75, 76, 379, 816
malM	Maltose	92	*molA*; periplasmic protein; function not known	380
malP	Maltose	75	*malA*; maltodextrin phosphorylase (EC 2.4.1.1)	A, B, C
malQ	Maltose	75	*malA*; amylomaltase (EC 2.4.1.25)	A
malS	Maltose	80	α-Amylase	348
malT	Maltose	75	*malA*; positive regulatory gene for *mal* regulon	A, B, C, 209
manA	Mannose	36	*pmi*; mannose-6-phosphate isomerase (EC 5.3.1.8)	A, C, 737, 738
manC	Mannose	(88)	*mni*; D-mannose isomerase regulation; utilization of D-lyxose	B
manX	Mannose	40	*gptB*, *mpt*, *ptsL*, *ptsM*, *ptsX*; mannose phosphotransferase system, protein II-A (III)	313, 315, 831, 1175
manY	Mannose	40	*pel*, *ptsP*, *ptsM*; mannose phosphotransferase system; Pel protein II-P; penetration of phage lambda	313, 315, 831, 1175
manZ	Mannose	40	*gptB*, *mpt*, *ptsM*, *ptsX*; mannose phosphotransferase system, enzyme IIB (II M)	831, 1175
marA		34	Multiple antibiotic resistance; tetracycline efflux system	369
mcrA		25	*rglA*; restriction of DNA at 5-methyl cytosine residues; locus of e14	887, 888, 890, L
mcrB		98	*rglB*; restriction of DNA at 5-methylcytosine residues	887, 888, 890, 924, 925
mdh		70	Malate dehydrogenase (EC 1.1.1.37)	A, 718, 1067, 1144
mdoA	Membrane-derived oligosaccharides	23	Membrane-localized component of glucosyl transferase system	112

Table 1   *Continued*

Gene symbol	Mnemonic	Map position (min)^a	Alternate gene symbols; phenotypic trait affected^b	Reference(s)^c
mdoB	Membrane-derived oligosaccharides	99	Phosphoglycerol transferase I activity	325, 521
melA	Melibiose	93	*mel-7*; α-galactosidase (EC 3.2.1.22)	A, 433, 648, 1003, 1162
melB	Melibiose	93	*mel-4*; thiomethylgalactoside permease II	A, 433, 1203
melR	Melibiose	93	Regulatory gene	1162
menA	Menaquinone	89	Conversion of 1,4-dihydroxy-2-naphthoate to demethylmenaquinone	A, B
menB	Menaquinone	49	1,4-Dihydroxy-2-naphthoate synthase	B, C, 996, 997
menC	Menaquinone	49	Conversion of chorismate to *o*-succinylbenzoate	B, C, 996, 997
menD	Menaquinone	49	Menaquinone biosynthesis	C, 996, 997
menE	Menaquinone	49	*o*-Succinylbenzoate-CoA synthase	996
mepA		50	Murein DD-endopeptidase	503
metA	Methionine	91	*met₃*; homoserine transsuccinylase (EC 2.3.1.46)	A, C, 734, N
metB	Methionine	89	*met-1*, *met₁*; cystationine γ-synthase (EC 4.2.99.9)	A, C, 288, 405, 406, 569, 647, 946
metC	Methionine	65	Cystathionine γ-lyase (EC 4.4.1.1)	A, 82, 735
metD	Methionine	5	High-affinity uptake of D- and L-methionine	A, B
metE	Methionine	86	*metB₁₂*; tetrahydropteroyltriglutamate methyltransferase (EC 2.1.1.14)	A, 14, 15, 822
metF	Methionine	89	*met-2*, *met₂*; 5,10-methylenetetrahydrofolate reductase (EC 1.1.1.68)	A, C, 405, 569, 946, 947, 1111
metG	Methionine	(46)	Methionyl-tRNA synthetase	A, B, 245
metH	Methionine	91	B₁₂-dependent homocysteine-N^5-methyltetrahydrofolate transmethylase	A, B, 822, N
metJ	Methionine	89	Regulatory gene; repressor of *metF*	A, B, C, 405, 569, 647, 946, 1020
metK	Methionine	64	Methionine adenosyltransferase (EC 2.5.1.6)	A, B, 127, 705
metL	Methionine	89	*metM*; aspartokinase II (EC 2.7.2.4), homoserine dehydrogenase II (EC 1.1.1.3)	A, C, 288, 405, 406, 569, 946, 1214
metR	Methionine	86	Regulatory gene for *metE* and *metH*	1124
metT	Methionine	15	Methionine tRNA$_m$ (duplicated gene)	B, C, 341, 587, 864
metY	Methionine	(69)	Methionine tRNA$_{f2}$ (tandemly duplicated gene)	B, 510, 511, 587
metZ	Methionine	(60)	Methionine tRNA$_{f1}$ (tandemly duplicated gene)	B, 587, 769
mglA	Methyl-galactoside	46	*mglP*; methyl-galactoside transport and galactose taxis, cytoplasmic membrane protein	A, B, C, 115, 440, 736, 975, W
mglB	Methyl-galactoside	46	*mglP*; galactose-binding protein; receptor for galactose taxis	A, B, C, 440, 736, 975, 982, W
mglC	Methyl-galactoside	46	*mglP*; methyl-galactoside transport and galactose taxis	A, B, C, 440, 736, W
mglD	Methyl-galactoside	45	Regulatory locus for methyl-galactoside transport	B, C
mglR	Methyl-galactoside	(16)	R-MG; regulatory gene	A
mgt	Magnesium transport	93	Mg²⁺ transport, system II	B
miaA		95	*trpX*; 2-methylthio-N^6-isopentyladenosine hypermodification	357
micF		48	*stc*; regulatory antisense RNA affecting *ompF* expression	20, 743, 747, 756
minB	Minicell	26	Formation of minute cells containing no DNA; complex locus; position of division septum	A, C, 259
mioC		84	Initiation of chromosome replication	659
mmrA		85	Recovery in rich medium following UV irradiation	990
mng	Manganese	(39)	Resistance or sensitivity to manganese	A
mopA	Morphogenesis of phages	94	*groE*, *groEL*, *hdh*, *tabB*; head assembly of phages T4 and lambda	A, B, C
mopB	Morphogenesis of phages	94	*groE*, *groES*, *hdh*, *tabB*; head assembly of phages T4 and lambda	A, B, C
motA	Motility	42	*flaJ*; flagellar paralysis	A, B, 258, 1016
motB	Motility	42	*flaJ*; flagellar paralysis	A, B, 1016, 1038
mraA	Murein	2	D-Alanine carboxypeptidase	A
mraB	Murein	2	D-Alanine requirement; cell wall peptidoglycan biosynthesis	A
mrbA	Murein	90	UDP-*N*-acetylglucosaminyl-3-enolpyruvate reductase activity	A
mrbB	Murein	90	D-Alanine requirement; cell wall peptidylglycan biosynthesis	A
mrbC	Murein	90	Cell wall peptidylglycan biosynthesis	A
mrcA	Murein	75	*ponA*; peptidoglycan synthetase; cell wall synthesis; penicillin-binding protein 1A	B, C, 140

Table 1 *Continued*

Gene symbol	Mnemonic	Map position (min)[a]	Alternate gene symbols; phenotypic trait affected[b]	Reference(s)[c]
mrcB	Murein	75	*ponB*; peptidoglycan synthetase; cell wall synthesis; penicillin-binding protein 1Bs	B, C, 140, 562, 917
mrdA		15	*pbpA*; cell shape; penicillin-binding protein 2	B, C, 35, 36, 78, 1048, 1049
mrdB		15	*rodA*; cell shape; sensitivity to radiation and drugs	A, B, C, 35, 36, 78, 1048, 1049
mre		71	Cell shape; sensitivity to antibiotics	1147
mrr		99	Restriction of methylated adenine	454
mtlA	Mannitol	81	Mannitol-specific enzyme II of phosphotransferase system	A, B, C, 628
mtlC	Mannitol	81	Regulatory locus	A, B
mtlD	Mannitol	81	Mannitol-1-phosphate dehydrogenase (EC 1.1.1.17)	A, B, C
mtr	Methyltryptophan	69	Resistance to 5-methyltryptophan	A, B
mul		83	Mutability of UV-irradiated phage λ	A
murC	Murein	2	L-Alanine-adding enzyme	A, B, C
murE	Murein	2	*meso*-Diaminopimelate-adding enzyme	A, B
murF	Murein	2	*mra*; D-alanyl:D-alanine-adding enzyme	A, B
murG	Murein	2	Murein or envelope biosynthesis	C
murH	Murein	99	Peptidoglycan synthesis, late stage	238
mutD	Mutator	5	See *dnaQ*	
mutH	Mutator	61	*mutR*, *prv*; methyl-directed mismatch repair	A, B, 398, 399
mutL	Mutator	95	*mut-25*; methyl-directed mismatch repair	A
mutM	Mutator	82	$G \cdot C \to T \cdot A$ transversions	165
mutS	Mutator	59	Methyl-directed mismatch repair	A, B
mutT	Mutator	2	$A \cdot T \to C \cdot G$ transversions; a nucleoside triphosphate	A, B, 9, 91
mutY	Mutator	64	$G \cdot C \to T \cdot A$ transversions	800
mvrA		7	Resistance or sensitivity to methyl viologen	753
nadA	NAD	17	*nicA*; quinolinate synthetase, A protein	A
nadB	NAD	56	*nicB*; quinolinate synthetase, B protein	A
nadC	NAD	3	Quinolinate phosphoribosyl transferase	A, C, 422, 912
nagA	N-Acetylglucosamine	16	N-Acetylglucosamine-6-phosphate deacetylase (EC 3.5.1.25)	A, 864
nagB	N-Acetylglucosamine	16	*glmD*; glucosamine-6-phosphate deaminase	A, 864, 916
nagE	N-Acetylglucosamine	16	*ptsN*; N-acetylglucosamine-specific enzyme II of phosphotransferase system	C, 864, 916
nalA	Nalidixic acid	48	See *gyrA*	
nalB	Nalidixic acid	58	Resistance or sensitivity to nalidixic acid	A
nalD	Nalidixic acid	89	Penetration of nalidixic acid through outer membrane	484
nanA		70	N-Acetylneuraminate lyase (aldolase) (EC 4.1.3.3)	1143
nanT		70	Sialic acid transport	1143
narG	Nitrate reductase	27	*chlC*, *narC*; nitrate reductase (EC 1.7.99.4), α subunit	A, B, C, 299, 427, 643, 644, 725, 920, 1024
narH	Nitrate reductase	27	*chlC*; nitrate reductase (EC 1.7.99.4), β subunit	A, B, C, 299, 1024
narI	Nitrate reductase	27	*chlI*; cytochrome b_{NR}, nitrate reductase (EC 1.7.99.4), γ subunit	A, B, C, 427, 920, 1024
narJ	Nitrate reductase	27	Nitrate reductase (EC 1.7.99.4), δ subunit	1024
narK	Nitrate reductase	27	Regulatory gene	1046
narL	Nitrate reductase	27	*narR*; regulatory gene	518, 1046
narX	Nitrate reductase	27	*narR*; regulatory gene	1046
narZ	Nitrate reductase	33	Cryptic gene(s) encoding a second nitrate reductase	114
ndh		22	Respiratory NADH dehydrogenase	B, C
neaB	Neamine	74	Resistance to neamine	A, B
nek		73	*amk*; resistance to neomycin, kanamycin, and other aminoglycoside antibiotics	A, B
nfnA		80	Sensitivity to nitrofurantoin	967
nfnB		13	Sensitivity to nitrofurantoin	967
nfo		47	Endonuclease IV	233, W
nfsA	Nitrofurazone sensitivity	(21)	Nitrofuran reductase I activity	B
nfsB	Nitrofurazone sensitivity	(11)	Nitrofuran reductase I activity	B
nirB	Nitrate reductase	74	*nirD*; NADH-nitrate oxidoreductase (EC 1.6.6.4) aproprotein, structural gene	522, 526, 679, 680
nirC	Nitrate reductase	26	NADH-nitrate reductase (EC 1.6.6.4) activity	B
nmpC	New membrane protein	13	Outer membrane porin protein; locus of qsr prophage	B, 462
non	Nonmucoid	45	Capsule formation	A
nrdA		49	*dnaF*; ribonucleoside diphosphate reductase (EC 1.17.4.1) subunit B1	A, B, C, 173, 802, 1118

Table 1   *Continued*

Gene symbol	Mnemonic	Map position (min)[a]	Alternate gene symbols; phenotypic trait affected[b]	Reference(s)[c]
nrdB		49	*ftsB*; ribonucleoside diphosphate reductase (EC 1.17.4.1) subunit B1	A, B, C, 173, 593, 1082
nth		36	"Endonuclease III"; a DNA glycosylase and phosphoric monoester lyase	53, 234, 1163
nupC		52	Transport of nucleosides, except guanosine	B, C
nupG		64	Transport of nucleosides	B, 800, 1171
nusA		69	Transcription termination; L factor	B, C, 407, 510, 511, 608, 776, 868, 949
nusB		10	*groNB*; transcription termination; L factor	B, C, 509, 1071
nuvA		9	Uridine thiolation factor A activity	B
nuvC		43	Thiazole biosynthesis; 4-thiouridine modification of tRNA; near-UV sensitivity and resistance	940
ompA	Outer membrane protein	22	*con*, *tolG*, *tut*; outer membrane protein 3a (II*;G;d), structural gene	A, B, C, 207, 403
ompC	Outer membrane protein	48	*meoA*, *par*; outer membrane protein 1b (Ib;c), structural gene	B, C, 745, 746
ompF	Outer membrane protein	21	*cmlB*, *coa*, *cry*, *tolF*; outer membrane protein 1a (Ia;b; F), structural gene	B, C, 239, 506, 507, 697, 1085
ompR	Outer membrane protein	75	*ompB*; positive regulatory gene for *ompC* and *ompF*	B, C, 213, 364, 748, 749, 785, 1191
ompT		13	Outer membrane protein 3b (a), a protease	387, 411, 412, 933
oppA		28	Oligopeptide transport; periplasmic binding protein	A, B, C, 22, 468
oppB		28	Oligopeptide transport	22, 468
oppC		28	Oligopeptide transport	22, 468
oppD		28	Oligopeptide transport	22, 468
oppE		28	Oligopeptide transport	22
opr		(19)	Rate of degradation of aberrant subunit proteins of RNA polymerase	988
ops		63	Level of exopolysaccharide production	1219
optA		4	Deoxyguanosine 5'-triphosphate triphosphohydrolase activity; phage T7 DNA metabolism	C, 72
oriC	Origin of replication	84	*poh*?; origin of replication of chromosome	B, C, 152, 542, 578, 712, 819, 820, 987, 1057, 1073, 1152
oriJ	Origin of replication	30	Origin function of *rac* prophage	C
osmZ		27	*bglY*, *cur*, *fimG*, *topX*; DNA supercoiling; expression of genes subject to osmotic regulation	232, 461, 1027
otsA		42	Trehalose phosphate synthase (EC 2.4.1.15) production	374
otsB		42	Trehalose phosphate synthase (EC 2.4.1.15) production	374
pabA	*p*-Aminobenzoate	74	*p*-Aminobenzoate synthetase, CoII	A, 555
pabB	*p*-Aminobenzoate	40	*p*-Aminobenzoate synthetase, CoII	A, B, 384
pac		31	Phenylacetate degradation	216
panB	Pantothenate	3	Ketopantoate hydroxymethyltransferase (EC 4.1.2.12)	A, B, C
panC	Pantothenate	3	Pantothenate synthetase (EC 6.3.2.1)	A, B, C
panD	Pantothenate	3	Aspartate 1-decarboxylase (EC 4.1.1.11)	A, B, C
panF	Pantothenate	71	Pantothenate permease	1133
pat		89	Putrescine aminotransferase activity	989
pbpA			See *mrdA*	B, C, 149
pbpB			See *ftsI*	
pck		75	Phosphoenolpyruvate carboxykinase (EC 4.1.1.49)	C
pcnB		4	Plasmid copy number	666
pcsA		82	Cell division; chromosome segregation	B
pdxA	Pyridoxine	1	Requirement	A, B, C
pdxB	Pyridoxine	50	Placement of 5, 5', and 6' carbons into pyridine ring of pyridoxine	A, B, 30, 31
pdxC			See *serC*	
pdxH	Pyridoxine	36	Pyridoxinephosphate oxidase	A, B, 415
pdxJ	Pyridoxine	56	Requirement	A, B
pepD	Peptides	6	*pepH*; peptidase D, a dipeptidase	A, B, 145, 571
pepN	Peptides	21	Aminopeptidase N	B, C, 55–57, 145, 336, 719–721
pfkA		89	6-Phosphofructokinase I (EC 2.7.1.11)	A, B, 240, 456
pfkB		38	Level of 6-phosphofructokinase II; suppressor of *pfkA*	A, B, C, 241
pfl		20	Pyruvate formate-lyase	B, 839
pgi		91	Glucosephosphate isomerase (EC 5.3.1.9)	A, N
pgk		63	Phosphoglycerate kinase (EC 2.7.2.3)	A, B

Table 1 *Continued*

Gene symbol	Mnemonic	Map position (min)[a]	Alternate gene symbols; phenotypic trait affected[b]	Reference(s)[c]
pgl		17	*blu*; 6'-phosphogluconolactonase (EC 3.1.1.31)	A
pgm		(15)	Phosphoglucomutase (EC 2.7.5.1)	A
pgpA		10	Phosphatidylglycerophosphate phosphatase, membrane bound	500
pgpB		28	Phosphatidylglycerophosphate phosphatase, membrane bound	500
pgsA		42	Phosphatidylglycerophosphate synthetase (EC 2.7.8.5)	A, B, C, 386
pgsB			See *lpxB*	
pheA	Phenylalanine	57	Chorismate mutase-P-prephenate dehydrogenase	A, B, C, 488
pheP	Phenylalanine	13	Phenylalanine-specific transport system	C
pheR	Phenylalanine	94	Regulatory gene for *pheA*	C, 395
pheS	Phenylalanine	37	*phe-act*; phenylalanyl-tRNA synthetase (EC 6.1.1.20), α subunit	A, B, C, 306, 320, 727, 728, 867, 1035, 1036, 1190
pheT	Phenylalanine	37	*pheS*; phenylalanyl-tRNA synthetase (EC 6.1.1.20), β subunit	A, B, C, 306, 320, 727, 728, 867, 1035, 1190
pheU	Phenylalanine	94	Phenylalanyl-tRNA (duplicate gene)	357, 587, 976
pheV	Phenylalanine	64	Phenylalanyl-tRNA (duplicate gene)	166, 587, 1179
phnD		92	*psiD*; carbon-phosphorus lyase	1148, U
phoA	Phosphate	9	Alkaline phosphatase (EC 3.1.3.1)	A, B, C, 184, 1011
phoB	Phosphate	9	*phoRc*, *phoT*; positive regulatory gene for *pho* regulon	A, B, C, 691, 693, 695, 1006, 1157
phoE	Phosphate	6	*ompE*; outer membrane pore protein e (E,Ic,NmpAB), structural gene	C, 450, 810, 827, 1106
phoM	Phosphate	100	Positive regulatory gene for *pho* regulon	C, 17, 671, 694, 695, 1105, 1156, 1159
phoR	Phosphate	9	*nmpB*, *phoR1*, *R1pho*; positive and negative regulatory gene for *pho* regulon	A, B, C, 692, 695, 1006, 1157
phoS	Phosphate	84	See *pstS*	
phoT	Phosphate	84	See *pstA*, *pstB*, and *phoU*	
phoU	Phosphate	84	*phoT*; high-affinity phosphate-specific transport system, regulatory gene	C, 16, 18, 779, 1065
phr	Photoreactivation	16	Deoxyribodipyrimidine photolyase (EC 4.1.99.3)	A, B, 495, 957, 958
phxB	Phi-X	17	Adsorption of φX174	B
pil	Pili	98	See *fim*	
pin		26	Inversion of adjacent DNA; locus of element e14	312, 563, 861, 862, 1139, L
pit	P$_i$ transport	77	Low-affinity P$_i$ transport	A, 308, 309
pldA		86	Detergent-resistant phospholipase A	A, 14, 263, 264, 471, 472, 579
pldB		86	Lysophospholipase L$_2$	14, 472, 579, 580
plsA	Phospholipid synthesis	11	See *adk*	
plsB	Phospholipid synthesis	92	Glycerolphosphate acyltransferase activity	A, B, C, 646
plsX	Phospholipid synthesis	24	Glycerolphosphate auxotrophy in *plsB* background	620
pncA	Pyridine nucleotide cycle	39	*nam*; nicotinamide deamidase (EC 3.5.1.19)	A, B
pncB	Pyridine nucleotide cycle	(21)	Nicotinate phosphoribosyltransferase (EC 2.4.2.11)	C
pnp		69	Polynucleotide phosphorylase (EC 2.7.7.8)	A, C, 227, 318, 470, 873, 897, 898, 1078
pntA		35	Pyridine nucleotide transhydrogenase (EC 1.6.1.1), α subunit	B, 197, 198
pntB		35	Pyridine nucleotide transhydrogenase (EC 1.6.1.1), β subunit	197, 198
poaR		63	Regulation of proline oxidase production	A
pog		70	Growth of phage P1	C
polA	Polymerase	87	*resA*; DNA polymerase I (EC 2.7.7.7)	A, B, C
polB	Polymerase	2	DNA polymerase II (EC 2.7.7.7)	A
polC	Polymerase	4	*dnaE*; DNA biosynthesis; DNA polymerase III α subunit	A, B, C, 998, 1103, 1168
popC	Porphyrin	4	Synthesis of δ-aminolevulinate	A
popD	Porphyrin	(1)	Level of 5-aminolevulinate dehydratase (EC 4.2.1.24) activity	A, B
poxA		95	Regulatory gene for *poxB*	C
poxB		19	Pyruvate oxidase (EC 1.2.2.2), structural gene	C, 185, 396, 397
ppc	Phosphoenolpyruvate	89	*glu*, *asp*; phosphoenolpyruvate carboxylase (EC 4.1.1.31)	A, B, C, 354, 943
pps	Phosphoenolpyruvate	37	Phosphoenolpyruvate synthase	A
prfA		27	*asuA?*, *sueB*, *uar*, *ups?*; protein release factor 1	174, 941, 942
prfB		62	*supK*; protein release factor 2	564

Table 1 *Continued*

Gene symbol	Mnemonic	Map position (min)[a]	Alternate gene symbols; phenotypic trait affected[b]	Reference(s)[c]
prlA		73	*secY*; protein export; membrane protein	10, 181, 311, 516, 517, 1002, 1009
prlB		85	Protein export	311
prlC		71	Protein export	311
prlD		2	Protein export	60
prmA		71	*prm-1*; methylation of 50S ribosomal subunit protein L11	B
prmB		51	*prm-2*; methylation of 50S ribosomal subunit protein L3	B
proA	Proline	6	*pro1*; γ-glutamyl phosphate reductase (EC 1.2.1.41)	A, C, 272, 450, 687, 797
proB	Proline	6	*pro2*; γ-glutamyl kinase (EC 2.7.2.11)	A, C, 272, 450, 687, 797
proC	Proline	9	*pro3*, *pro2*; pyrroline-5-carboxylate reductase (EC 1.5.1.2)	A, C, 262, 273
proK	Proline	80	*proV*; proline tRNA 1	341, 587, 599
proL	Proline	47	*proW*; proline tRNA 2	587
proM	Proline	86	*proU*; proline tRNA 3	587
proP	Proline	93	Low-affinity transport system for glycine betaine and proline; proline permease II	C, 392, 393, 716
proS	Proline	5	Prolyl-tRNA synthetase (EC 1.1.1.15)	C
proT	Proline	83	Proline transport	B, C
proU	Proline	57	*osrA*; high-affinity transport system for glycine betaine and proline	319, 392, 394, 418, 716, I
proV	Proline	57	High-affinity transport system for glycine betaine and proline; glycine betaine-binding protein	319, 341, 394, I
proW	Proline	57	High-affinity transport system for glycine betaine and proline	I
prp	Propionate	97	Propionate metabolism	C
prr		31	γ-Aminobutyraldehyde (pyrroline) dehydrogenase activity	989
prs		26	Phosphoribosylpyrophosphate synthetase (EC 2.7.6.1)	C, 480–482
psd		95	Phosphatidylserine decarboxylase	A, B
psiF		9	Induced by phosphate starvation	184, 1158, U
pssA		56	Phosphatidylserine synthetase (EC 2.7.8.8)	B, C
pssR		85	Regulatory gene	1026
pstA		84	*phoR2b*, *phoT*, *R2pho*; high-affinity phosphate-specific transport system	A, B, C, 16, 18, 922, 1066
pstB		84	*phoT*; high-affinity phosphate-specific transport system, cytoplasmic membrane protein?	A, B, C, 16, 18, 638, 922, 1066
pstC		84	*phoW*; high-affinity phosphate-specific transport system, cytoplasmic membrane component	A, B, C, 16, 18, 922, 1066
pstS		84	*nmpA*, *phoR2a*, *phoS*, *R2pho*; high-affinity phosphate-specific transport system; periplasmic phosphate-binding protein	A, B, C, 16, 18, 146, 520, 639, 683, 754, 922, 1065, 1066
pta		50	Phosphotransacetylase (EC 2.3.1.8) activity	B, C
pth		26	Peptidyl-tRNA hydrolase	A
ptr		61	Protease III	C, 61, 200, 295, 296, 330
ptsG	Phosphotransferase system	25	*car*, *CR*, *gpt*, *gptA*, *tgl*, *umg*; glucosephosphotransferase enzyme II	A, B, 123, 314
ptsH	Phosphotransferase system	52	*ctr*, *Hpr*; phosphohistidinoprotein-hexose phosphotransferase (EC 2.7.1.69)	A, B, 134–136, 270, 271, 835, 945
ptsI	Phosphotransferase system	52	*ctr*; phosphotransferase system enzyme I	A, B, 134–136, 270, 271, 835, 945
purA	Purine	95	ade_k; Ad_4; adenylosuccinate synthetase (EC 6.3.4.4)	A
purB	Purine	25	ade_h; adenylosuccinate lyase (EC 4.3.2.2)	A, L
purC	Purine	53	ade_g; phosphoribosylaminoimidazole-succinocarboxamide synthetase (EC 6.3.2.6)	A, 833
purD	Purine	90	$adth_a$; phosphoribosylglycineamide synthetase (EC 6.3.4.13)	A
purE	Purine	12	ade_3, ade_f, Pur_2; phosphoribosylaminoimidazole carboxylase (EC 4.1.1.21), catalytic subunit	A, C, 548, 1095
purF	Purine	50	$ade_{u,b}$, *purC*; amidophosphoribosyl transferase (EC 2.4.2.14)	A, C, 689, 807
purG	Purine		See *purM*	
purH	Purine	90	ade_i, *purJ*; phosphoribosylaminoimidazolecarboxamide formyltransferase (EC 2.1.2.3)	A
purI	Purine		See *purL*	
purK	Purine	12	$purE_2$; phosphoribosylaminoimidazole carboxylase (EC 4.1.1.21), CO_2-fixing subunit	548, 1095

Table 1 *Continued*

Gene symbol	Mnemonic	Map position (min)[a]	Alternate gene symbols; phenotypic trait affected[b]	Reference(s)[c]
purL	Purine	55	*purI*; phosphoribosylformylglycineamide synthetase (EC 6.3.5.3); homologous to *purG* of *S. typhimurium*	A, C, 479
purM	Purine		*purG*; phosphoribosylaminoimidazole synthetase (EC 6.3.3.1) homologous to *purI* of *S. typhimurium*	A, B, C, 479, 1022, 1023
purN	Purine	54	*ade*_c; 5'-phosphoribosyglycinamide transformylase (EC 2.1.2.2)	1023
purP	Purine	84	High-affinity adenine transport	156, E
purR	Purine	36	Regulatory gene for *pur* regulon	568, 919, S
pus		20	Effect of suppressors on *relB* mutations	C
putA	Proline utilization	23	*poaA*; proline dehydrogenase (EC 1.5.99.8)	A, C, 752, 778
putP	Proline utilization	23	Proline utilization; major proline permease	C, 752, 777, 778
pykF		37	Pyruvate kinase F	365
pyrA	Pyrimidine	1	See *car*	
pyrB	Pyrimidine	97	Aspartate carbamoyltransferase (EC 2.1.3.2) catalytic subunit	A, B, C, 474, 578, 637, 789, 838, 921, 1119
pyrC	Pyrimidine	24	Dihydroorotase (EC 3.5.2.3)	A, 48, 530, 1178
pyrD	Pyrimidine	21	Dihydroorotate oxidase (EC 1.3.3.1)	A, 530, 618
pyrE	Pyrimidine	82	Orotate phosphoribosyltransferase (EC 2.4.2.10)	A, C, 876, 877
pyrF	Pyrimidine	28	Orotidine-5'-phosphate decarboxylase (EC 4.1.1.23)	A, 282, 530, 1120
pyrG	Pyrimidine	60	CTP synthetase (EC 6.3.4.2)	B, C, 1169
pyrH	Pyrimidine	5	UMP kinase	A
pyrI	Pyrimidine	97	Aspartate carbamoyltransferase (EC 2.1.3.2) regulatory subunit	C, 474, 838, 921
pyrS	Pyrimidine	81	Regulatory gene	809
qin		35	kim; cryptic lambdoid phage	122, 316
qmeA		29	*gts*; unspecified membrane defect	A
qmeC		74	Unspecified membrane defect; tolerance to glycine, penicillin sensitivity	A
qmeD		61	Unspecified membrane defect; tolerance to glycine, penicillin sensitivity	A
qmeE		37	Unspecified membrane defect	A
qsr'		13	Cryptic lambdoid phage	23, 109, 462, 545
rac		30	Defective prophage rac; see *recE* and *oriJ*	A, C, 1176, 1177
radA		100	Sensitivity to γ and UV radiation and methyl methanesulfonate	280
radC		82	Sensitivity to radiation	A, 322
ranA		56	RNA metabolism	A
rap		26	Growth of phage lambda	419
ras	Radiation sensitivity	(10)	Sensitivity to UV and X-rays	A
rbsA	Ribose	84	*rbsP*, *rbsT*; D-ribose high-affinity transport system; membrane-associated protein	85, 502, 667
rbsB	Ribose	84	*rbsP*; D-ribose periplasmic binding protein	502, 667
rbsC	Ribose	84	*rbsP*, *rbsT*; D-ribose high-affinity transport system; membrane-associated protein	85, 502, 667
rbsD	Ribose	84	*rbsP*; D-ribose high-affinity transport system; membrane-associated protein	85
rbsK	Ribose	84	Ribokinase (EC 2.7.1.15)	A, B, C, 475, 502, 667
rbsR	Ribose	84	Regulatory gene	667
rcsA		43	Positive regulatory gene for capsule synthesis	390, 1107
rcsB		48	Positive regulatory gene for capsule synthesis	133, 390
rcsC		48	Negative regulatory gene for capsule synthesis	133, 390
rdgA		16	Dependence of growth upon *recA* gene product	C
rdgB		64	Dependence of growth and viability upon *recA* function	204
recA	Recombination	58	*lexB*, *recH*, *rnmB*, *tif*, *umuB*, *zab*; general recombination, DNA repair and induction of phage lambda	A, B, C
recB	Recombination	61	*rorA*; recombination and DNA repair; exonuclease V (EC 3.1.11.5)	A, B, C, 296, 329, 965
recC	Recombination	61	Recombination and DNA repair; exonuclease V (EC 3.1.11.5)	A, B, C, 296, 331, 333, 965
recD	Recombination	61	Recombination and DNA repair; exonuclease V (EC 3.1.11.5), α subunit	19, 94, 328
recE	Recombination	30	rac; locus of rac prophage; recombination and DNA repair; exonuclease VIII	A, B, C, 342, 1177
recF	Recombination	83	*uvrF*; recombination and DNA repair	A, B, C, 4, 27, 100, 894
recJ	Recombination	62	Recombination and DNA repair	668, 669
recN	Recombination	57	*radB*; recombination and DNA repair	658, 851, 926, 963, 964

Table 1   *Continued*

Gene symbol	Mnemonic	Map position (min)^a	Alternate gene symbols; phenotypic trait affected^b	Reference(s)^c
recO	Recombination	56	Conjugational recombination and DNA repair	585
recQ	Recombination	86	Conjugational recombination and DNA repair	14, 508, 781, 782
relA	Relaxed	60	*RC*; regulation of RNA synthesis; stringent factor; ATP:GTP 3′-pyrophosphotransferase	A, B
relB	Relaxed	35	Regulation of RNA synthesis	B, C, 74
relE	Relaxed	35	Locus in *relB* operon; function unknown	74, 372
relF	Relaxed	35	Locus in *relB* operon; function unknown	74, 372
relX	Relaxed	60	Control of synthesis of guanosine-5′-diphosphate-3′-diphosphate	B
rep		85	*dasC*?, *mmrA*?; Rep helicase, a single-stranded DNA dependent ATPase	A, B, C, 14, 70, 92, 93, 285, 376
rer		90	Resistance to UV and γ-radiation	B
rfa	Rough	81	*con*, *lpsA*, *phx*; cluster of genes coding for enzymes involved in lipopolysaccharide core biosynthesis	A, B
rfaB	Rough	81	UDP-D-galactose:(glucosyl)lipopolysaccharide-1,6-D-galactosyltransferase	226
rfaC	Rough	81	Lipopolysaccharide core biosynthesis; proximal hexose	C
rfaD	Rough	81	ADP-L-Glycero-D-mannoheptose-6-epimerase	C, 211
rfaG	Rough	81	Lipopolysaccharide core biosynthesis; glucosyltransferase I	226
rfaH	Rough	87	*sfrB*; lipopolysaccharide core biosynthesis; positive regulation of production of glucosyltransferase; expression of *tra* operon of F factor; antiterminator	226, 900
rfaI	Rough	81	UDP-D-galactose:(glucosyl)lipopolysaccharide-α-1,3-D-galactosyltransferase	226
rfaJ	Rough	81	UDP-D-glucose:(galactosyl)lipopolysaccharide glucosyltransferase	226
rfaM	Rough	81	Lipopolysaccharide core biosynthesis; glucosyltransferase II	226
rfaP	Rough	81	Lipopolysaccharide core biosynthesis; phosphorylation of core heptose	C
rfbA	Rough	45	TDP-glucose pyrophosphorylase	A
rfbB	Rough	45	TDP-glucose oxidoreductase	A
rfbD	Rough	45	TDP-rhamnose synthetase	A
rfe	Rough	(85)	Synthesis of enterobacterial common antigen and O antigen	B, 731
rff	Rough	(85)	Synthesis of enterobacterial common antigen	B, 731
rhaA	Rhamnose	88	L-Rhamnose isomerase (EC 5.3.1.14)	A, 1101
rhaB	Rhamnose	88	Rhamnulokinase (EC 2.7.1.5)	A, 1101
rhaD	Rhamnose	88	Rhamnulose-1-phosphate aldolase (EC 4.1.2.19)	A, 1101
rhaR	Rhamnose	88	*rhaC*; positive regulatory gene	1101
rhaS	Rhamnose	88	*rhaC*; positive regulatory gene	1101
rho		85	*nitA*, *psu*, *rnsC*, *SuA*, *sun*, *tsu*; transcription termination factor rho; polarity suppressor	A, B, C, 14, 70, 143, 167, 650, 713, 857
RhsA		81	Repetitive sequence responsible for duplications within chromosome	652, 944
RhsB		77	Repetitive sequence responsible for duplications within chromosome	652, 944
RhsC		16	Repetitive sequence responsible for duplications within chromosome	944
RhsD		12	Repetitive sequence responsible for duplications within chromosome	944
ribA	Riboflavin	28	GTP cyclohydrolase II	C, 58, 1088
ribB	Riboflavin	66	Block before 6,7-dimethyl-8-ribityllumazine	C, 58, 1088
ribC	Riboflavin	40	Riboflavin synthase	58, 1088, R
ridA		71	Transcription and translation; rifampin and kasugamycin dependence	C
ridB		85	Transcription and translation; rifampin dependence	236
rimB	Ribosomal modification	38	Maturation of 50S ribosomal subunit	A, 415
rimC	Ribosomal modification	(26)	Maturation of 50S ribosomal subunit	A
rimD	Ribosomal modification	(88)	Maturation of 50S ribosomal subunit	A
rimE	Ribosomal modification	72	Modification of ribosomal proteins	B, 730
rimF	Ribosomal modification	1	*res*; ribosomal modification	A
rimG	Ribosomal modification	(1)	*ramB*; modification of 30S ribosomal subunit protein S4	A
rimH	Ribosomal modification	14	*stsB*; ribosomal modification	A, B

Table 1 *Continued*

Gene symbol	Mnemonic	Map position (min)[a]	Alternate gene symbols; phenotypic trait affected[b]	Reference(s)[c]
rimI	Ribosomal modification	99	Modification of 30S ribosomal subunit protein S18; acetylation of N-terminal alanine	B, C, 1209
rimJ	Ribosomal modification	(32)	Modification of 30S ribosomal subunit protein S5; acetylation of N-terminal alanine	B, 524, 1209
rimL	Ribosomal modification	(33)	Modification of 30S ribosomal subunit protein L7; acetylation of N-terminal serine	C
rit		89	Affects thermolability of 50S ribosomal subunit	B
rlpA		15	A minor lipoprotein	1076
rlpB		15	A minor lipoprotein	1076
rna	RNase	14	*rns, rnsA*; RNase I	A
rnb	RNase	29	RNase II	B, C
rnc	RNase	55	RNase III	A, B, 699, 787, 1160
rnd	RNase	40	RNase D	C, 1216
rne	RNase	24	RNase E activity	B, C, 893
rnh	RNase	5	*dasF, herA, sdrA, sin*; RNase H (EC 3.1.26.4)	C, 225, 550, 690, 771, 806, 813, 1108
rnpA	RNase	83	RNase P, protein component	435
rnpB	RNase	70	RNase P, RNA subunit, M1 RNA	B, C, 758, 895, 896, 952
rodA		15	See *mrdB*	
rpiA		63	Ribose phosphate isomerase (EC 5.3.1.6), constitutive	A
rplA	Ribosomal protein, large	90	50S ribosomal subunit protein L1	A, B, C, 283, 491, 889
rplB	Ribosomal protein, large	73	50S ribosomal subunit protein L2	A, B, 1222
rplC	Ribosomal protein, large	73	50S ribosomal subunit protein L3	A, B, 1222
rplD	Ribosomal protein, large	73	*eryA*; 50S ribosomal subunit protein L4	A, B, C, 1222
rplE	Ribosomal protein, large	73	50S ribosomal subunit protein L5	A, B, 181
rplF	Ribosomal protein, large	73	50S ribosomal subunit protein L6	A, B, 181
rplI	Ribosomal protein, large	96	50S ribosomal subunit protein L9	B, 974
rplJ	Ribosomal protein, large	90	50S ribosomal subunit protein L10	A, B, C, 203, 283, 491, 889
rplK	Ribosomal protein, large	90	*relC*; 50S ribosomal subunit protein L11	A, B, C, 283, 491, 889
rplL	Ribosomal protein, large	90	50S ribosomal subunit protein L7/L12	A, B, C, 283, 491, 889
rplM	Ribosomal protein, large	70	50S ribosomal subunit protein L13	C, 514
rplN	Ribosomal protein, large	73	50S ribosomal subunit protein L14	A, B, 181
rplO	Ribosomal protein, large	73	50S ribosomal subunit protein L15	A, B, 181, 516
rplP	Ribosomal protein, large	73	50S ribosomal subunit protein L16	A, B, 1222
rplQ	Ribosomal protein, large	73	50S ribosomal subunit protein L17	A, B, C, 77, 181, 730
rplR	Ribosomal protein, large	73	50S ribosomal subunit protein L18	A, B, 181
rplS	Ribosomal protein, large	57	50S ribosomal subunit protein L19	B, 164
rplT	Ribosomal protein, large	38	*pdzA*; 50S ribosomal subunit protein L20	320
rplU	Ribosomal protein, large	69	50S ribosomal subunit protein L21	B
rplV	Ribosomal protein, large	73	50S ribosomal subunit protein L22	A, B, 1222
rplW	Ribosomal protein, large	73	50S ribosomal subunit protein L23	B, 1222
rplX	Ribosomal protein, large	73	50S ribosomal subunit protein L24	A, B, 181, 235
rplY	Ribosomal protein, large	48	50S ribosomal subunit protein L25	B
rpmA	Ribosomal protein, large	69	50S ribosomal subunit protein L27	B
rpmB	Ribosomal protein, large	82	50S ribosomal subunit protein L28	B, C, 1031
rpmC	Ribosomal protein, large	73	50S ribosomal subunit protein L29	A, B, 1222
rpmD	Ribosomal protein, large	73	50S ribosomal subunit protein L30	A, B, 181
rpmE	Ribosomal protein, large	89	50S ribosomal subunit protein L31	C
rpmF	Ribosomal protein, large	24	50S ribosomal subunit protein L32	524
rpmG	Ribosomal protein, large	82	50S ribosomal subunit protein L33	B, C, 1031
rpmH	Ribosomal protein, large	83	*rimA, ssaF*; 50S ribosomal subunit protein L34	C, 823, O
rpmI	Ribosomal protein, large	38	50S ribosomal subunit protein A	1149
rpmJ	Ribosomal protein, large	73	50S ribosomal subunit protein X	181, 1149
rpoA	RNA polymerase	73	RNA polymerase (EC 2.7.7.6), α subunit	A, B, C, 77, 181, 730, 928, 929
rpoB	RNA polymerase	90	*groN, nitB, rif, ron, stl, stv, tabD*; RNA polymerase (EC 2.7.7.6), β subunit	A, B, C, 283, 491, 889
rpoC	RNA polymerase	90	*tabD*; RNA polymerase (EC 2.7.7.6), β subunit	A, B, C, 283, 889
rpoD	RNA polymerase	67	*alt*; RNA polymerase (EC 2.7.7.6), σ^{70} subunit	B, C, 157, 677, 1086
rpoH	RNA polymerase	76	*fam, hin, htpR*; RNA polymerase (EC 2.7.7.6), σ^{32} subunit; regulatory gene for proteins induced at high temperatures	C, 416, 417, 614, 792, 1098, 1116, 1213
rpoN	RNA polymerase	70	*glnF, ntrA*; RNA polymerase (EC 2.7.7.6), σ^{60} subunit	B, C, 70, 177, 467, 494
rpsA	Ribosomal protein, small	21	*ssyF*; 30S ribosomal subunit protein S1	B, C, 291, 841, 1001
rpsB	Ribosomal protein, small	4	30S ribosomal subunit protein S2	A, B, C

Table 1 *Continued*

Gene symbol	Mnemonic	Map position (min)[a]	Alternate gene symbols; phenotypic trait affected[b]	Reference(s)[c]
rpsC	Ribosomal protein, small	73	30S ribosomal subunit protein S3	A, B, 1222
rpsD	Ribosomal protein, small	73	*ramA*, *sud2*; 30S ribosomal subunit protein S4	A, B, C, 77, 181
rpsE	Ribosomal protein, small	73	*eps*, *spcA*, *spc*; 30S ribosomal subunit protein S5	A, B, 181
rpsF	Ribosomal protein, small	95	30S ribosomal subunit protein S6	A, B, 974
rpsG	Ribosomal protein, small	73	*K12*; 30S ribosomal subunit protein S7	A, B, C
rpsH	Ribosomal protein, small	73	30S ribosomal subunit protein S8	A, B, 181
rpsI	Ribosomal protein, small	70	30S ribosomal subunit protein S9	237, 514
rpsJ	Ribosomal protein, small	73	30S ribosomal subunit protein S10	A, B, C
rpsK	Ribosomal protein, small	73	30S ribosomal subunit protein S11	A, B, C, 77, 181
rpsL	Ribosomal protein, small	73	*strA*; 30S ribosomal subunit protein S12	A, B, C
rpsM	Ribosomal protein, small	73	30S ribosomal subunit protein S13	A, B, C, 77, 181
rpsN	Ribosomal protein, small	73	30S ribosomal subunit protein S14	A, B, 181
rpsO	Ribosomal protein, small	69	*secC*; 30S ribosomal subunit protein S15	B, C, 318, 324, 873, 898, 903, 1077, 1078, H
rpsP	Ribosomal protein, small	57	30S ribosomal subunit protein S16	B, 164
rpsQ	Ribosomal protein, small	73	*neaA*; 30S ribosomal subunit protein S17	A, B, 1222
rpsR	Ribosomal protein, small	96	30S ribosomal subunit protein S18	A, B, 974
rpsS	Ribosomal protein, small	73	30S ribosomal subunit protein S19	A, B, 1222
rpsT	Ribosomal protein, small	0	*supS20*; 30S ribosomal subunit protein S20	A, B, 681, 1196
rpsU	Ribosomal protein, small	67	30S ribosomal subunit protein S21	B, C, 157, 677
rrfA	rRNA 5S	87	5S rRNA gene of *rrnA* operon	B, C
rrfB	rRNA 5S	90	5S rRNA gene of *rrnB* operon	B, C
rrfC	rRNA 5S	85	5S rRNA gene of *rrnC* operon	B, C
rrfD	rRNA 5S	72	5S rRNA gene of *rrnD* operon	C
rrfE	rRNA 5S	90	5S rRNA gene of *rrnE* operon	B, C
rrfG	rRNA 5S	56	5S rRNA gene of *rrnG* operon	C
rrfH	rRNA 5S	5	5S rRNA gene of *rrnH* operon	C
rrlA	rRNA, 23S	87	23S rRNA gene of *rrnA* operon	A, B, C
rrlB	rRNA, 23S	90	23S rRNA gene of *rrnB* operon	B, C
rrlC	rRNA, 23S	85	23S rRNA gene of *rrnC* operon	B
rrlD	rRNA, 23S	72	23S rRNA gene of *rrnD* operon	B, C
rrlE	rRNA, 23S	90	23S rRNA gene of *rrnE* operon	B, C
rrlG	rRNA, 23S	56	23S rRNA gene of *rrnG* operon	B, C
rrlH	rRNA, 23S	5	23S rRNA gene of *rrnH* operon	C
rrnA	rRNA	87	*cqsA*; rRNA operon	A, B, C, 578
rrnB	rRNA	90	*cqsE*, *rrnB1*; rRNA operon	A, B, C, 89, 446
rrnC	rRNA	85	*cqsb*, *rrnB*, *rrnB2*; rRNA operon	A, B, C, 446, 578
rrnD	rRNA	72	*cqsD*; rRNA operon	B, C
rrnE	rRNA	90	*rrnD1*; rRNA operon	B, C, 645
rrnG	rRNA	56	rRNA operon	B, C, 446
rrnH	rRNA	5	rRNA operon	C
rrsA	rRNA, 16S	87	16S rRNA gene of *rrnA* operon	B
rrsB	rRNA, 16S	90	16S rRNA gene of *rrnB* operon	B, C, 89, 121
rrsC	rRNA, 16S	85	16S rRNA gene of *rrnC* operon	B
rrsD	rRNA, 16S	72	16S rRNA gene of *rrnD* operon	B
rrsE	rRNA, 16S	90	16S rRNA gene of *rrnE* operon	B
rrsG	rRNA, 16S	56	16S rRNA gene of *rrnG* operon	B, C
rrsH	rRNA, 16S	5	16S rRNA gene of *rrnH* operon	C
rts		90	*ts-9*; uncharacterized growth defect	A, B, C, 1079
ruvA		41	Filament formation and sensitivity to UV radiation	A, 37, 87, 1010
ruvB		41	Filament formation and sensitivity to UV radiation	87
sad		34	Succinate-semialdehyde dehydrogenase (EC 1.2.1.16), NAD dependent	C, 702
sbaA		97	Regulation of serine and branched-chain amino acid metabolism	C
sbcB		44	*xonA*; exonuclease I; suppression of *recB recC* mutations	A, 846, 847, 879
sbcC		9	Suppression of *recB recC* mutations	657
sbp		89	Periplasmic sulfate-binding protein	456
sbmA		9	Sensitivity to microcin B17	622
sdhA		16	Succinate dehydrogenase (EC 1.3.99.1), flavoprotein subunit	A, C, 247, 1174, 1184
sdhB		16	Succinate dehydrogenase (EC 1.3.99.1), iron sulfur protein	A, C, 247, 1174, 1184
sdhC		16	*cybA*; succinate dehydrogenase (EC 1.3.99.1), cytochrome b_{556}	A, C, 763, 764

Table 1 *Continued*

Gene symbol	Mnemonic	Map position (min)[a]	Alternate gene symbols; phenotypic trait affected[b]	Reference(s)[c]
sdhD		16	Succinate dehydrogenase (EC 1.3.99.1), hydrophobic subunit	A, C, 1174, 1184
secA		2	*azi, pea*; secretion of envelope proteins	C, 71, 972, O
secB		81	Protein export	600, 601
secD		9	Protein export	361
sefA		4	Septum formation	B
selA	Selenium	81	*fdhA*; selenium metabolism; biosynthesis or incorporation of selenocystein	632
selB	Selenium	81	*fdhA*; selenium metabolism; biosynthesis or incorporation of selenocystein	632
selC	Selenium	82	*fdhC*; selenium metabolism; selenocysteine tRNA	587, 632, 633
semA		40	Sensitivity to microcin E492	884
serA	Serine	63	D-3-phosphoglycerate dehydrogenase (EC 1.1.1.95)	A, 1100
serB	Serine	100	Phosphoserine phosphatase (EC 3.1.3.3)	A, C, 362, 799
serC	Serine	20	*pdxC*; 3-phosphoserine aminotransferase (EC 2.6.1.52)	A, B, 291
serR	Serine	2	Level of seryl-tRNA synthetase	B
serS	Serine	20	Seryl-tRNA synthetase (EC 6.1.1.11)	A, 445
serT	Serine	(22)	*divE*; serine tRNA 1	B, 587, 1081
serU	Serine	43	*supD, supH, Su-1, suI*; serine tRNA 2	A, C, 587, 1041, 1093
serV	Serine	(58)	Serine tRNA 3	B, 587
serW	Serine	20	Serine tRNA 5 (duplicate gene)	587, 960
serX	Serine	23	Serine tRNA 5 (duplicate gene)	587
sfiC		26	Cell division inhibition; locus of element e14	246, 684, L
shiA	Shikimate	43	Shikimate and dehydroshikimate permease	A
sloB	Slow growth	74	Low growth rate; tolerance to amidinopenicillin and nalidixic acid	B
sodA		88	Superoxide dismutase, manganese	172, 1080, 1109
sodB		36	Superoxide dismutase, iron	171, 172, 415, 798
speA	Spermidine	64	Arginine decarboxylase (EC 4.1.1.19)	A, 127
speB	Spermidine	64	Agmatinase (EC 3.5.3.11)	A, 127
speC	Spermidine	64	Ornithine decarboxylase (EC 4.1.1.17)	A, B, 127
speD	Spermidine	3	S-Adenosylmethionine decarboxylase (EC 4.1.1.50)	B, 1074, 1075
speE	Spermidine	3	Spermidine synthase (putrescine aminopropyltransferase) (EC 2.5.1.16)	1074, 1075
spf		87	"Spot 42" RNA	C, 869
spoT		82	Guanosine 3′,5′-bis(diphosphate) 3′-pyrophosphatase	A, B
sppA		39	Protease IV, a signal peptide peptidase	498, 1070
srlA	Sorbitol	58	*gutA, sbl*; D-glucitol-specific enzyme II of phosphotransferase system	A, B, C, 1193, 1194
srlB	Sorbitol	58	*gutB*; D-glucitol (sorbitol)-specific enzyme III of phosphotransferase system	1193, 1194
srlD	Sorbitol	58	*gutD, sbl*; glucitol (sorbitol)-6-phosphate dehydrogenase (EC 1.1.1.140)	A, B, C, 1193, 1194
srlR	Sorbitol	58	Regulatory gene	B, C
srnA		10	Degradation of stable RNA	A
ssaD		9	Suppression of *secA* mutation	361, 823
ssaE		50	Suppression of *secA* mutation	361
ssaG		41	Suppression of *secA* mutation	361
ssaH		94	Suppression of *secA* mutation	361
ssb	Single-strand binding	92	*exrB, lexC*; single-strand DNA-binding protein	B, C
ssp		70	Stringent starvation protein	355, 356, 986
ssr		63	Stable 6S RNA	487, 627
ssyA		54	Suppressor of *secY* mutation	1000
ssyB		10	Suppressor of *secY* mutation	1001
ssyD		3	Suppressor of *secY* mutation	1001
strC	Streptomycin	7	*strB*; low-level streptomycin resistance	A
strM	Streptomycin	77	Control of ribosomal ambiguity	A
stsA		84	Altered RNase activity	A
sucA	Succinate	16	*lys + met*; succinate requirement; α-ketoglutarate dehydrogenase (decarboxylase component)	A, C, 151, 247, 248, 1174, 1184
sucB	Succinate	16	*lys + met*; succinate requirement; α-ketoglutarate dehydrogenase (dihydrolipoyltranssuccinase component)	A, C, 151, 1028, 1029
sucC	Succinate	16	Succinyl-CoA synthetase (EC 6.2.1.5), β subunit	150, 151, 1028, 1029
sucD	Succinate	16	Succinyl-CoA synthetase (EC 6.2.1.5), α subunit	150, 151
suhA		77	Induction of heat shock genes	1099

Table 1   *Continued*

Gene symbol	Mnemonic	Map position (min)[a]	Alternate gene symbols; phenotypic trait affected[b]	Reference(s)[c]
sulA		22	*sfiA*, *suf*; suppressor of *lon*	A, B, 206, 750
sulB		2	See *ftsZ*	
supB	Suppressor	16	*su$_B$*; suppressor of ochre (UAA) and amber (UAG) mutations; see *glnU*	
supC	Suppressor	27	*su$_C$*, *Su-4*; suppressor of ochre (UAA) and amber (UAG) mutations; see *tyrT*	
supD	Suppressor	43	*su$_I$*, *Su-1*; suppressor of ochre (UAG) mutations; see *serU*	
supE	Suppressor	15	*su$_{II}$*, *Su-2*; suppressor of ochre (UAG) mutations; see *glnV*	
supF	Suppressor	27	*su$_{III}$*, *Su-3*; suppressor of amber (UAG) mutations; see *tyrT*	
supG	Suppressor	17	*su-5*; suppressor of ochre (UAA) and amber (UAG) mutations; see *lysT*	
supH	Suppressor	43	Suppressor; see *serU*	
supK	Suppressor	62	Suppressor of opal (UGA) mutations; see *prfB*	
supL	Suppressor	17	*su$_B$*; suppressor of ochre (UAA) and amber (UAG) mutations; see *lysT*	
supM	Suppressor	90	Suppressor of ochre (UAA) and amber (UAG) mutations, see *tyrU*	
supN	Suppressor	52	Suppressor of ochre (UAA) and amber (UAG) mutations; see *lysV*	
supO	Suppressor	27	Suppressor of ochre (UAA) and amber (UAG) mutations; see *tyrT*	
supP	Suppressor	97	*Su-6*; suppressor of amber (UAG) mutations; see *leuX*	
supQ	Suppressor	13	Suppressor	A
tabC		86	Development of phage T4	B
tag		(72)	3-Methyladenine DNA glycosylase I, constitutive	B, C, 199, 953, 1042
tap		42	Methyl-accepting chemotaxis protein IV	C, 597, 1016, 1017
tar		42	*cheM*; methyl-accepting chemotaxis protein II	B, C, 597, 1016, 1017
tdc		68	Threonine dehydratase (EC 4.2.1.16)	252, 388, 389
tdh		81	Threonine dehydrogenase (EC 1.1.1.103)	891, 892
tdi		(4)	Transduction, transformation and rates of mutation	B
tdk		27	Thymidine kinase (EC 2.7.1.75)	A, B
TerA	Terminus	28	DNA replication fork inhibition	267, 343, 460, 463–465, 842, M
TerB	Terminus	36	DNA replication fork inhibition	267, 343, 460, 463–465, 842, M
TerC	Terminus	34	DNA replication fork inhibition	343, 460, M
TerD	Terminus	27	DNA replication fork inhibition	343, 460, M
tesB		10	Thioesterase II	786
tgt		9	tRNA-guanine transglycosylase	C
thdA	Thiophene degradation	11	Utilization of furans and thiopenes; may be *tln*	2, F
thdC	Thiophene degradation	95	Utilization of furans and thiopenes	2, F
thdD	Thiophene degradation	100	Utilization of furans and thiopenes	2, F
thiA	Thiamin	90	Thiamin thiazole requirement	A
thiB	Thiamin	90	Thiaminphosphate pyrophosphorylase (EC 2.5.1.3)	A
thiC	Thiamin	90	Thiamin pyrimidine requirement	A
thiD	Thiamin	45	Phosphomethylpyrimidine kinase activity	C
thiK	Thiamin	25	Thiamin kinase	C
thiL	Thiamin	10	Thiamin monophosphate kinase	C
thrA	Threonin	0	*HS*, *thrD*; aspartokinase I–homoserine dehydrogenase I (EC 2.3.2.4–EC 1.1.1.3)	A, B, C, 244, 678
thrB	Threonine	0	Homoserine kinase (EC 2.7.1.39)	A, C, 948
thrC	Threonine	0	Threonine synthase (EC 4.2.99.2)	A, C, 836
thrS	Threonine	38	Threonyl-tRNA synthetase (EC 6.1.1.3)	B, C, 717, 867, 878, 1034, 1170, 1190
thrT	Threonine	90	Threonine tRNA 3	A, B, C, 587
thrU	Threonine	90	Threonine tRNA 4	B, C, 587
thrV	Threonine	72	Threonine tRNA 1; at distal end of *rrnD* operon	C, 214, 587
thrW	Threonine	6	Threonine tRNA 2	214, 242, 587
thyA	Thymine	61	Thymidylate synthetase (EC 2.1.1.45)	A, C, 83, 296, 965
tkt		(62)	Transketolase (EC 2.2.1.1)	A
tlnA		11	*tlnI*; resistance or sensitivity to thiolutin	C
tmk		25	Deoxythymidine kinase	96, 257
tnaA		84	*ind*, *tnaR*; tryptophanase (EC 4.1.99.1)	A, B, C, 1045

Table 1 *Continued*

Gene symbol	Mnemonic	Map position (min)[a]	Alternate gene symbols; phenotypic trait affected[b]	Reference(s)[c]
tnm		92	Transposition of Tn9 and other transposons; development of phage Mu	C
toc		67	Compensation for loss of DNA topoisomerase I	886
tolA	Tolerance	17	*cim, excC, lky, tol-2*; tolerance to group A colicins and single-stranded filamentous DNA phage	A, 337, 1061, 1062, V
tolB	Tolerance	17	*tol-3*; tolerance to colicins E2, E3, A, and K	A, 337, 624, 625, 1061, 1062, V
tolC	Tolerance	66	*colE1-i, mtcB, refI, tol-8*; specific tolerance to colicin E1, expression of outer membrane proteins	A, C, 425, 426, 826
tolD	Tolerance	(23)	Tolerance to colicins E2 and E3; ampicillin resistance	A
tolE	Tolerance	(23)	Tolerance to colicins E2 and E3; ampicillin resistance	A
tolI	Tolerance	(0)	Tolerance to colicins Ia and Ib	A
tolJ	Tolerance	0	Resistance to colicins L, A, and S4; partial resistance to colicins E and K	B
tolM	Tolerance	72	*cmt*; high-level tolerance to colicin M	C, 432, 730
tolQ	Tolerance	17	*fii, tolP?*; tolerance to group A colicins and single-stranded filamentous DNA phage	1061, 1062, V
tolR	Tolerance	17	Tolerance to group A colicins and single-stranded filamentous DNA phage	1062, V
tolZ	Tolerance	77	Tolerance to colicins E2, E3, D, Ia, and Ib; generation of chemical proton gradient	715
tonA	T-one	4	See *fhuA*	
tonB	T-one	28	*exbA, T1rec*; uptake of chelated iron and cyanobalimin; sensitivity to phages T1 and ϕ80 and colicins	A, B, C, 874, 875
topA	Topoisomerase	28	*supX*; DNA topoisomerase I, ω protein	C, 704, 825, 1115, 1154
torA		28	Trimethylamine *N*-oxide reductase	837
tpiA		89	Triosephosphate isomerase (EC 5.3.1.1)	A, B, 456, 849
tpr		27	A protaminelike protein	C
treA	Trehalose	26	Trehalase, periplasmic	B, 116, 730
trg		31	Methyl-accepting chemotaxis protein III	B, C, 113, 441
trkA		72	Transport of potassium	A, 432, 730
trkB		73	See *kefB*	
trkC		1	See *kefB*	
trkD		84	Transport of potassium	A, B, C
trkE		29	Transport of potassium	A
trmA	tRNA methyltransferase	90	tRNA (uracil-5)-methyltransferase (EC 2.1.1.35)	A, B, C, 654
trmB	tRNA methyltransferase	(7)	tRNA (guanine-7)-methyltransferase (EC 2.1.1.33)	A
trmC	tRNA methyltransferase	(56)	5-Methylaminoethyl-2-thiouridine in tRNA	A, B, 428
trmD	tRNA methyltransferase	(57)	tRNA (guanine-7)-methyltransferase (EC 2.1.1.31)	B, 162–164
trmE	tRNA methyltransferase	84	*asuE?*; 5-methylaminoethyl-2-thiouridine biosynthesis	307
trmF	tRNA methyltransferase	84	5-methylaminoethyl-2-thiouridine biosynthesis	307
trnA		59	*glnU*; level of several tRNAs	C
trpA	Tryptophan	28	*tryp-2*; tryptophan synthase (EC 4.2.1.20), A protein	A, B, C
trpB	Tryptophan	28	*tryp-1*; tryptophan synthase (EC 4.2.1.20), B protein	A, C
trpC	Tryptophan	28	*tryp-3*; *N*-(5-phosphoribosyl)anthranilate isomerase indole-3-glycerolphosphate synthetase	A, B, C
trpD	Tryptophan	28	*tryE*; glutamine amidotransferase-phosphoribosyl anthranilate transferase	A, C
trpE	Tryptophan	28	*anth, tryp-4, tryD*; anthranilate synthase (EC 4.1.3.27)	A, B, C
trpP	Tryptophan	84	Low-affinity tryptophan-specific permease	C
trpR	Tryptophan	100	*Rtry*; regulation of *trp* operon and *aroH*; *trp* aporepressor	A, B, C
trpS	Tryptophan	74	Tryptophanyl-tRNA synthetase (EC 6.1.1.2)	A, B, C
trpT	Tryptophan	85	*su7, su8, su9, supU, supV*; tryptophan tRNA gene at distal end of *rrnC* operon	A, B, C, 587
trxA	Thioredoxin	86	*fip, tsnC*; thioredoxin	B, 14, 649, 650, 713, 935, 936, 938, 1153
trxB	Thioredoxin	21	Thioredoxin reductase	431, 937, 939
tsf		4	Protein chain elongation factor, EF-Ts	B, C
tsr		99	*cheD*; methyl-accepting chemotaxis protein I	B, C, 126, 169
tsx	T-six	9	*nupA, T6rec*; nucleoside uptake; receptor for phage T6 and colicin K	A, B, 130
tufA		74	Protein chain elongation factor; EF-Tu (duplicate gene)	A, B, C
tufB		90	Protein chain elongation factor; EF-Tu (duplicate gene)	A, B, C, 1079
tus		36	DNA-binding protein; inhibition of replication at *Ter* sites	C, 464, 466, 842, M
tynA		(27)	Tyramine oxidase (EC 1.4.3.4)	B

Table 1 *Continued*

Gene symbol	Mnemonic	Map position (min)[a]	Alternate gene symbols; phenotypic trait affected[b]	Reference(s)[c]
tyrA	Tyrosine	57	Chorismate mutase T (EC 5.4.99.5)–prephenate dehydrogenase (EC 1.3.1.12)	A, 488
tyrB	Tyrosine	92	Tyrosine aminotransferase (EC 2.6.1.5), tyrosine repressible	B, 340, 605, 1200
tyrP	Tyrosine	42	Tyrosine-specific transport system	C, 558, 559, 1186
tyrR	Tyrosine	29	Regulation of *aroF*, *aroG*, and *tyrA* and aromatic amino acid transport systems	A, B, C, 217, 218
tyrS	Tyrosine	36	Tyrosyl-tRNA synthetase (EC 6.1.1.1)	A, B, C, 67
tyrT	Tyrosine	27	*sul$_{III}$*, *Su-3*, *su$_c$*, *Su-4*, *supF*, *supE*, *tyrV*; tyrosine tRNA 1 (tandemly duplicated gene)	A, B, C, 587
tyrU	Tyrosine	90	*supM*; tyrosine tRNA 2	A, B, C, 587
ubiA	Ubiquinone	92	4-Hydroxybenzoate → 3-octaprenyl 4-hydroxybenzoate	A, C
ubiB	Ubiquinone	87	2-Octaprenylphenol → 2-octaprenyl-6-methoxyphenol	A
ubiC	Ubiquinone	92	Chorismate lyase	A
ubiD	Ubiquinone	87	3-Octaprenyl-4-hydroxybenzoate → 2-octaprenylphenol	A
ubiE	Ubiquinone	87	2-Octaprenyl-6-methoxy-1,4-benzoquinone → 2-octaprenyl-3-methyl-6-methoxy-1,4-benzoquinone	A
ubiF	Ubiquinone	15	2-Octaprenyl-3-methyl-6-methoxy-1,4-benzoquinone → 2-octaprenyl-3-methyl-5-hydroxy-6-methoxy-1,4-benzo-uinone	A
ubiG	Ubiquinone	48	2-Octaprenyl-3-methyl-5-hydroxy-6-methoxy-1,4-benzo-uinone → ubiquinone 8	A, B, C, 375
ubiH	Ubiquinone	63	2-Octaprenyl-6-methoxyphenol → 2-octaprenyl-6-methoxy-1,4-benzoquinone	A
ubiX	Ubiquinone	50	Sequence homologous to *ubiX* of *S. typhimurium*, which codes for polyprenyl *p*-hydroxybenzoate carboxylase	807
udk		45	Uridine kinase (EC 2.7.1.48)	A, C
udp		86	Uridine phosphorylase (EC 2.4.2.3)	A, 14, 15
ugpA		76	*sn*-Glycerol-3-phosphate transport system	A, 828, 977, 978
ugpB		76	*sn*-Glycerol-3-phosphate transport system; periplasmic binding protein	C, 828, 977, 978
ugpC		76	*sn*-Glycerol-3-phosphate transport system	828, 977, 978
ugpE		76	*sn*-Glycerol-3-phosphate transport system, membrane protein	828, 977, 978
uhpA		82	Positive activator of *uhpt* transcription	352, 543, 993, 1172
uhpB		82	Regulatory gene	352, 1172
uhpC		82	Regulatory gene	A, B, C, 352, 543, 993, 1172
uhpR		82	Regulation of hexose phosphate transport; possibly outer membrane receptor for glucose 6-phosphate	A, B, C
uhpT		82	Transport of hexose phosphates, transport protein	A, B, C, 352, 543, 993, 1172
uidA		36	*gurA*; β-D-glucuronidase (EC 3.2.1.31)	A, B, C, 107, 527
uidR		36	Regulatory gene	A, B, C, 105, 107, 108
umuC		26	*uvm*, induction of mutations by UV; error-prone repair	B, 137, 259, 305, 570, 706, 843, 1005
umuD		26	*uvm*; inducible mutagenesis; error-prone repair	C, 259, 305, 570, 706, 843, 1005
unc			See *atp*	
ung		56	Uracil-DNA-glycosylase	B, 1142
upp		54	*uraP*; uracil phosphoribosyltransferase (EC 2.4.2.9)	A, B, 476
ups		26	Efficiency of nonsense suppressors	B, 941
ushA		11	UDP-glucose hydrolase (F′-nucleotidase)	A, B, C, 155
uup		21	Precise excision of insertion elements	476
uvrA	Ultraviolet	92	*dar*; repair of UV damage to DNA; excision nuclease	A, B, C, 496
uvrB	Ultraviolet	18	*dar-1,6*; DNA repair; excision nuclease	A, B, C, 26, 47, 1072
uvrC	Ultraviolet	42	*dar-4,5*; repair of UV damage to DNA; excision nuclease	A, C, 338, 339, 956, 991, 992, 1140
uvrD	Ultraviolet	86	*dar-2*, *dda*, *mutU*, *pdeB*, *recL*, *uvrE*, *uvr502*; DNA-dependent ATPase I and DNA helicase II	A, B, C, 14, 33, 297, 326, 327, 459, 603, 1083, 1198
uxaA		68	Altronate hydrolase (EC 4.2.1.7)	A, B, C
uxaB		(52)	Altronate oxidoreductase (EC 1.1.1.58)	A, B, 103, 104, 490, 711
uxaC		68	Uronate isomerase (EC 5.3.1.12)	A, B, C, 102, 710, 711
uxuA		98	Mannonate hydrolase (EC 4.2.1.8)	A, B, C, 106, 347
uxuB		98	Mannonate oxidoreductase (EC 1.1.1.57)	A, B, C, 106, 347

Table 1   *Continued*

Gene symbol	Mnemonic	Map position (min)[a]	Alternate gene symbols; phenotypic trait affected[b]	Reference(s)[c]
uxuR		98	Regulatory gene for *uxuBA* operon	B, C
valS	Valine	97	*val-act*; valyl-tRNA synthetase (EC 6.1.1.9)	A, B, C, 443, 451, 452, 1015
valT	Valine	17	Valine tRNA 1	B, C, 587, 1211
valU		52	Valine tRNA 1 (tandemly triplicated gene)	587, D
valV		37	Valine tRNA 2B	587
valW		37	Valine tRNA 2A	587
weeA		67	Cell elongation	265
xapA		52	*pndA*; xanthosine phosphorylase	C, 582
xapR		52	*pndR*; regulatory gene	C, 582
xthA		38	Exonuclease III	A, B, C
xseA		54	Exonuclease VII, large subunit	B, 187, 1131
xseB		10	Exonuclease VII, small subunit	1131, 1132
xylA	Xylose	80	D-Xylose isomerase (EC 5.3.1.5)	A, C, 132, 610, 923, 970
xylB	Xylose	80	Xylulokinase (EC 2.7.1.17)	A, C, 132, 923
xylE	Xylose	91	Xylose-proton symport	255, 256, 344, 609
xylF	Xylose	80	*xylT*; xylose binding protein transport system	256, 610, 923
xylR	Xylose	80	Regulatory gene	A, C, 923
zwf	Zwischenferment	41	Glucose-6-phosphate dehydrogenase (EC 1.1.1.49)	A

[a] Numbers refer to the time scale shown in Fig. 1. Parentheses indicate approximate map locations.
[b] Abbreviations: CoA, coenzyme A; DCCD, *N,N'*-dicyclohexylcarbodiimide; NEM, *N*-ethylmaleimide.
[c] Numbers refer to Literature Cited. Letters refer to: (A) literature cited in Table 2 of reference 46; (B) literature cited in Table 1 of reference 45; (C) literature cited in Table 1 of reference 43; and personal communications from (D) Y. Brun, (E) K. Burton, (F) D. Clark, (G) J. Coleman, (H) S. Ferro-Novick, J. Sands, and J. Beckwith, (I) J. Gowrishankar, (J) B. G. Hall, (K) K. Hantke, (L) C. W. Hill, (M) P. Kuempel, (N) I. G. Old, (O) D. B. Oliver, (P) D. Oxender, (Q) R. Plapp, (R) P. Rabinovich, (S) J. M. Smith, (T) M. Volkert, (U) B. Wanner, (V) R. Webster, and (W) B. Weiss

Table 2   Alternate Gene Symbols

Alternate symbol	Symbol in Table 1	Alternate symbol	Symbol in Table 1
acrB	*gyrB*	*bisA*	*chlA*
ade	*pur*	*bisB*	*chlE*
adth_a	*purD*	*bisD*	*narG*
adth_b	*purG*	*blu*	*pgl, pgm, malP*
aidA	*alkA*	*brnP*	*ilvH*
aidD	*alkB*	*cap*	*car, crp*
ala-act	*alaS*	*capR*	*lon*
ald	*fba*	*car*	*ptsG*
alnA	*dadB*	*cat*	*ptsG*
alnR	*dadQ*	*cbr*	*fep*
alt	*rpoD*	*cbt*	*fep*
amk	*nek*	*cer*	*btuB*
ampA	*ampC*	*cheC*	*flaA*
anth	*trpE*	*cheD*	*tsr*
apk	*lysC*	*cheM*	*tar*
arg + ura	*car*	*cheX*	*cheR*
aroR	*aroT*	*chlC*	*narG, narH*
asp	*ppc*	*chlG*	*narG*
aspB	*gltB*	*chlI*	*narI*
asuC	*hisT*	*cim*	*tolA*
asuD	*lysS*	*cmlB*	*ompF*
ata	*attP22*	*cmt*	*tolM*
att82, att434	*attγ*	*coa*	*ompF*
azi	*secA*	*colE1-i*	*tolC*
bfe	*btuB*	*con*	*ompA, rfa*
bglY	*osmZ*	*cop*	*het*
bioR	*birA*	*Cou*	*gyrB*
birB (bir)	*bioP*	*cqsA*	*rrnA*

Table 2   *Continued*

Alternate symbol	Symbol in Table 1	Alternate symbol	Symbol in Table 1
mtcA	*acrA*	*psuA*	*rho*
mtcB	*tolC*	*ptsF*	*fruA*
muc	*lon*	*ptsL*	*manX*
mutD	*dnaQ*	*ptsM*	*manX,Y,Z*
mutR	*mutH*	*ptsN*	*nagE*
mutU	*uvrD*	*ptsP*	*manY*
nalA	*gyrA*	*ptsX*	*manX,Y,Z*
nalC,D	*gyrB*	*pup*	*deoD*
nam	*pncA*	*pyrA*	*car*
ncf	*hemB*	*rad*	*uvrD*
neaA	*rpsQ*	*radB*	*recN*
nic	*nad*	*ramA*	*rpsD*
nirA	*fnr*	*ramB*	*rimG*
nirR	*fnr*	*RC*	*rel*
nitA	*rho*	*recL*	*uvrD*
nitB	*rpoB*	*refI*	*tolC*
nmpA	*pst, phoS,T*	*refII*	*cet*
nmpB	*phoR*	*relC*	*rplK*
ntrA	*rpoN*	*res*	*rimF*
ntrB	*glnL*	*resA*	*polA*
ntrC	*glnG*	*rglA*	*mcrA*
nucR	*deoR*	*rglB*	*mcrB*
nupA	*tsx*	*rif*	*rpoB*
nusE	*rpsJ*	*rimA*	*rpmH*
old	*fad*	*rm*	*hsd*
ole	*fadR*	*rnsA*	*rna*
ompB	*envZ, ompR*	*rnsC*	*rho*
ompE	*phoE*	*rodA*	*mrdB*
osrA	*proU*	*rodY*	*envB*
par	*ompC*	*ron*	*rpoB*
paxA	*dcd*	*rorA*	*recB*
pbpA	*mrdA*	*rpx*	*rps*
pbpB	*ftsI*	*rpy*	*rpl*
pbpF	*mrcB*	*rpz*	*rpm*
pcbA	*gyrB*	*sbcA*	*rac*
pdeB	*uvrD*	*sbl*	*srl*
pdeC	*lig*	*sdrA*	*rnh*
pdxC	*serC*	*sec*	*hemF*
pdxF	*serC*	*secC*	*rpsO*
pdzA	*rplT*	*secY*	*prlA*
pea	*secA*	*seg*	*arcA*
pel	*manY*	*sep*	*ftsI*
perA	*envZ*	*sez*	*rpoA*
pfv	*dacA*	*sfiA*	*sulA*
pgsB	*lpxB*	*sfiB*	*ftsZ*
phe-act	*pheS*	*sfrA*	*arcA*
phoS	*pstS*	*sfrB*	*rfaH*
phoT	*pstA,B, phoU*	*sin*	*rnh*
phoW	*pstC*	*skp*	*hlpA*
phs	*rpoA*	*sof*	*dut*
phx	*rfa*	*som*	*rfb*
pil	*fim*	*spcA*	*rpsE*
plsA	*adk*	*spr*	*lexA*
PMG	*mgl*	*ssd*	*ecfB*
pmi	*manA*	*ssyF*	*rpsA*
pndA	*xapA*	*ssyG*	*infB*
pndR	*xapR*	*stc*	*micF*
poaA	*putA*	*stl*	*rpoB*
poh	*oriC*	*strA*	*rpsL*
polC	*dnaE*	*stsB*	*rimH*
pon	*lpcB, mrc*	*stv*	*rpoB*
popA	*hemH*	*Su, su*	*sup*
popB	*hemF*	*suβ*	*lysTβ*
popE	*hemC*	*sud2*	*rpsD*
prd	*fuc*	*sueB*	*prfA*
prv	*mutH*	*sufD*	*glyU*
psiB	*ugpA,B*	*sulB*	*ftsZ*
psiC	*ugpA,B*	*sumA*	*glyT*
psiD	*phnD*	*sumB*	*glyU*

Table 2 *Continued*

Alternate symbol	Symbol in Table 1	Alternate symbol	Symbol in Table 1
cqsB	*rrnC*	*gpp*	*gpt*
cqsD	*rrnD*	*gpt*	*ptsG*
CR	*ptsG*	*gptB*	*manX,Z*
cru	*nupC*	*groE*	*mop*
cry	*ompR, ompF*	*groN*	*rpoB*
csm	*crp*	*groP*	*dnaB, dnaJ, dnaK*
ctr	*ptsH, ptsI*	*grpA*	*dnaB*
cur	*osmZ*	*grpC*	*dnaJ, dnaK*
cxr	*cxm*	*grpF*	*dnaK*
cybA	*sdhC*	*gts*	*qmeA*
dagA	*cycA*	*gurA*	*uidA*
dap + hom	*asd*	*gut*	*srl*
dar	*uvr*	*gxu*	*gpt*
dasC	*rep?*	*H*	*fliC*
dasF	*rnh*	*hag*	*fliC*
dda	*uvrD*	*hdh*	*mop*
deg	*lon*	*herA*	*rnh*
dhbB	*bioR*	*hid*	*himA*
dhl	*lpd*	*himB*	*gyrB*
dir	*lon*	*hin*	*rpoH*
divA	*ftsA*	*hip*	*himD*
divE	*serT*	*Hpr*	*ptsH*
dnaF	*nrdA*	*hrbA*	*brnQ*
dnaL	*lig*	*hrbB,C,D*	*livG,H,J,K*
dnaS	*dut*	*hs*	*hsd*
dnaW	*adk*	*Hs*	*thrA*
dnaY	*argU*	*hsm*	*hsdM*
dra	*deoC*	*hsp*	*hsd*
drm	*deoB*	*hsr*	*hsdR*
dye	*arcA*	*hss*	*hsdS*
eps	*rpsE*	*htpR*	*rpoH*
eryA	*rplD*	*icl*	*aceA*
eryB	*rplV*	*ile*	*ilvA*
eup	*ecfB*	*ind*	*tnaA*
exbA	*tonB*	*ins*	*glyV, glyW*
excC	*tolA*	*K12*	*rpsG*
exrA	*lexA*	*kac*	*kdp*
exrB	*ssb*	*kdgA*	*eda*
fabC	*fabB*	*kga*	*eda*
fam	*rpoH*	*kim*	*qin*
far	*fusA*	*kmt*	*ompB*
fda	*fba*	*lcs*	*asnS*
fdhA	*selA, selB*	*ldh*	*dld*
fdhB	*selD*	*leuK*	*hisT*
fdhC	*selC*	*leuX*	*leuSo, leuSp*
fdp	*fbp*	*lexB*	*recA*
feuA	*cir*	*lexC*	*ssb*
feuB	*fep*	*lir*	*acrA*
fexA	*arcA*	*lky*	*tolB*
fii	*tolQ, tolR*	*lop*	*ligA*
fipA	*trxA*	*loxB*	*attP1,P7*
fit	*infC*	*lps*	*rfa*
fla	*flg, flh, fli*	*lss*	*livR*
flaJ	*motA, motB*	*lstR*	*livR*
flaF	*fliC*	*lys + met*	*sucA, sucB*
flb	*flg, flh, fli*	*lysP*	*cadR*
flrA	*ileR?*	*mas*	*aceB*
fpk	*fruK*	*Mb*	*acrA*
fpr	*fruF*	*mbl*	*acrA*
frdB	*fnr*	*mec*	*dcm*
ftsB	*nrdB*	*meoA*	*ompC*
gad	*gap*	*mlpA*	*lpp*
glmD	*nagB*	*mni*	*manC*
glnF	*rpoN*	*molA*	*malM*
gltC	*gltSo*	*mon*	*envB*
glu	*ppc*	*motD*	*fliN*
glut	*gltA*	*mpt*	*manX,Z*
gly-act	*glyS*	*mra*	*murF*
glyD	*gpt*	*msp*	*arcA*

Table 2 *Continued*

Alternate symbol	Symbol in Table 1	Alternate symbol	Symbol in Table 1
sun	*rho*	*tpp*	*deoA*
sup$_{s20}$	*rpsT*	*tre*	*Ter*
supK	*prfB*	*trkB*	*kefB*
T1rec	*tonB*	*trkC*	*kefC*
T1, T5rec	*fhuA*	*trpP*	*aroT*
T6rec	*tsx*	*try*	*trp*
tabB	*mop*	*tryp*	*trp*
tabD	*rpoB, rpoC*	*ts-9*	*rts*
talA	*alaT*	*tsl*	*lexA*
talD	*alaU*	*tsnC*	*trxA*
tasC	*aspT*	*tsu*	*rho*
tfrA	*lpcA*	*tss*	*asnS*
tgl	*ptsG*	*ttr*	*fadL*
tgs	*crr?*	*tut*	*ompA*
tgtB	*gltT*	*uar*	*prfA*
tgtC	*gltU*	*umg*	*ptsG*
tgtE	*gltV*	*umuA*	*lexA*
thdB	*fadR*	*umuB*	*recA*
thyR	*deoB, deoC*	*unc*	*atp*
tif	*recA*	*uraP*	*upp*
tmr	*fol*	*usgA*	*gntT*
tolF	*ompF*	*uvm*	*umu*
tolG	*ompA*	*uvrF*	*recF*
tolP	*tolQ*	*val-act*	*valS*
tonA	*fhuA*	*vtr*	*fabF*
TP	*deoA*	*xonA*	*sbcB*
tpo	*envZ*	*zab*	*recA*

SECTION·2

Physical Maps of *Escherichia coli* and Their Correlation with the Genetic Map

The entries in this section are:

A. Alignment of *E. coli* DNA sequences to a revised, integrated genomic restriction map

B. Location of cosmid clones on the genetic map of *E. coli*

C. Restriction maps of *E. coli* K12

D. *E. coli* proteins identified on two-dimensional gels and a plasmid map of *E. coli*

E. Alignment of genetic maps of *E. coli* and *S. typhimurium*

F. Sizes of bacterial chromosomes

Physical maps of the *E. coli* chromosome have been generated by different methods. The most prominent map was constructed by Kohara et al. (1987) and involved a 4700-kb long integrated restriction map for eight six-base-recognition enzymes. This same group has constructed an ordered cosmid library representing 70% of the *E. coli* genome (Tabata et al. 1989). Also, Smith et al. (1987) have used pulse field gel electrophoresis to order *Not*I restriction fragments of the *E. coli* chromosome. A nice use of some of these maps to localize genes by physical approaches has been published by Lee et al. (1988). An ultimate goal is the alignment of the genetic map with the DNA sequence of the chromosome and with a genomic restriction map. Rudd et al. (1990), Kröger et al. (1990), and Kunisawa et al. (1990) have compiled the DNA sequences of *E. coli*, and Rudd has updated this compilation and correlated it with the genetic map and with a revised version of the genomic restriction map of Kohara and co-workers. These data are presented below. The maps of Tobata et al. (1989) and Smith et al. (1987) are also reproduced below. The reader is also directed to an analysis of chromosomal location and transcriptional direction of *E. coli* genes by Watanabe and Kunisawa (1990), who charted the transcriptional direction of more than 500 genes.

Neidhardt and co-workers (Phillips et al. 1987; VanBogelen et al. 1990; VanBogelen and Neidhardt 1991) have compiled a gene-protein index that cross-references many of the known genes and proteins of *E. coli* K12 and many of the hybrid plasmids that contain specific fragments of the genome. The latest edition (VanBogelen and Neidhardt 1991) allows the location of genes for newly recognized proteins and contains several valuable tables and figures. Here we reproduce a table of *E. coli* proteins identified on two-dimensional gels and a detailed plasmid map (Nishimura et al. 1992) that tells whether a particular region of interest has been cloned on plasmids in the Clarke-Carbon library (see Sections 2D and 7 of this Handbook).

The alignment of the genetic maps of *E. coli* and *S. typhimurium* is based on previous *E. coli* and *S. typhimurium* maps, not on the most recent ones. No new alignments of these two maps have yet been generated. However, these maps are still useful, and so they are included here.

REFERENCES*

Clarke, L. and J. Carbon. 1976. A colony bank containing synthetic col E1 hybrid plasmids representative of the entire *E. coli* genome. *Cell* **9:** 91–99.

Daniels, D. 1990. The complete *Avr*II restriction map of the *Escherichia coli* genome and comparisons of several laboratory strains. *Nucleic Acids Res.* **18:** 2649–2651.

Heath, J.D., J.D. Perkins, B.R. Sharma, and G.M. Weinstock. 1992a. *Not*I genomic cleavage map of *Escherichia coli* K-12 strain MG1655. *J. Bacteriol.* **174:** 558–567.

Heath, J.D., J.D. Perkins, B.R. Sharma, and G.M. Weinstock. 1992b. *Xba*I genomic cleavage map of *Escherichia coli* K-12 strain MG1655 and comparative analysis of other strains. *J. Mol. Biol.* (in press).

Kohara, Y., K. Akiyama, and K. Isono. 1987. The physical map of the whole *E. coli* chromosome: Application of a new strategy for rapid analysis and sorting of a large genomic library. *Cell* **50:** 495–508.

Krawiec, S. and M. Riley. 1990. Organization of the bacterial chromosome. *Microbiol. Rev.* **54:** 502–539.

Kröger, M., R. Wahl, and P. Rice. 1990. Compilation of DNA sequence of *Escherichia coli* (update 1990). *Nucleic Acids Res.* **18:** 2549–2587.

Kunisawa, T., M. Nakamura, H. Watanabe, J. Otsuka, A. Tsugita, L.-S.L. Yeh, D.G. George, and W.C. Barker. 1990. *Escherichia coli* K12 genomic database. *Protein Seq. Data Anal.* **3:** 157–162.

Lee, C.C., Y. Kohara, K. Akiyama, C.L. Smith, W.J. Craigen, and T. Caskey. 1988. Rapid and precise mapping of the *Escherichia coli* release factor genes by two physical approaches. *J. Bacteriol.* **170:** 4537–4541.

Neidhardt, F.C., V. Vaughn, T.A. Phillips, and P.L. Bloch. 1983. *Sfi*I genomic cleavage map of *Escherichia coli* K-12 strain MG1655. *Microbiol. Rev.* **47:** 231–284.

Nishimura, A., K. Okiyama, Y. Kohara, and K. Horiuchi. 1992. Correlation of a subset of the pLC plasmids to the physical map of *Escherichia coli* K-12. *Microbiol. Rev.* **56:** 137–151.

Perkins, J.D., J.D. Heath, B.R. Sharma, and G.M. Weinstock. 1992. Gene-protein index of *Escherichia coli* K-12. *Nucleic Acids Res.* **20:** 1129–1137.

Phillips, T.A., V. Vaughn, P.L. Bloch, and F.C. Neidhardt. 1987. Gene-protein index of *Escherichia coli* K-12, edition 2. In Escherichia coli *and* Salmonella typhimurium. *Cellular and Molecular Biology* (ed. F.C. Neidhardt), pp. 919–967. American Society for Microbiology, Washington, D.C.

Riley, M. and S. Krawiec. 1987. Genome organization. In Escherichia coli *and* Salmonella typhimurium. *Cellular and Molecular Biology* (ed. F.C. Neidhardt), pp. 967–981. American Society for Microbiology, Washington, D.C.

Riley, M. and K.E. Sanderson. 1990. Comparative genetics of *Escherichia coli* and *Salmonella typhimurium*. In *The Bacterial Chromosome* (ed. K. Drlica and M. Riley), pp. 85–95. American Society for Microbiology. Washington, D.C.

Rudd, K.E., W. Miller, J. Ostell, and D.A. Benson. 1990. Alignment of *Escherichia coli* K12 DNA sequences to a genomic restriction map. *Nucleic Acids Res.* **18:** 313–321.

Smith, C.L., J.G. Econome, A. Schutt, S. Klco, and C.R. Cantor. 1987. A physical map of the *Escherichia coli* K12 genome. *Science* **236:** 1448–1453.

Tobata, S., A. Higashitani, M. Takanami, K. Akiyama, Y. Kohara, Y. Nishimura, A. Nishimura, S. Yasuda, and Y. Hirota. 1989. Construction of an ordered cosmid collection of the *Escherichia coli* K-12 W3110 chromosome. *J. Bacteriol.* **171:** 1214–1218.

VanBogelen, R.A. and F.C. Neidhardt. 1991. The gene-protein database of *Escherichia coli*: Edition 4. *Electrophoresis* **12:** 955–994.

VanBogelen, R.A., M.E. Hutton, and F.C. Neidhardt. 1990. Gene-protein database of *Escherichia coli* K-12: Edition 3. *Electrophoresis* **11:** 1131–1166.

Watanabe, H. and T. Kunisawa. 1990. Computer-assisted analysis of chromosomal locations and transcriptional directions of *Escherichia coli* genes. *Protein Seq. Data Anal.* **3:** 149–156.

The references cited in Section 2A are listed on p. 2.8.

SECTION 2

A. Alignment of *E. coli* DNA Sequences to a Revised, Integrated Genomic Restriction Map

Contributed by

Kenneth E. Rudd

National Center for Biotechnology Information, National Library of Medicine, National Institutes of Health, Bethesda, MD 20894

The revised, integrated *E. coli* genomic restriction map given here can be used to correlate *E. coli* DNA sequences to the DNA inserts in the Kohara bacteriophage λ clone miniset collection (a subset of the 3400 clones used to generate the genomic restriction map) (Kohara et al. 1987). Table 1 and Figure 1 can be used to identify miniset clones that are likely to contain a particular gene. Specifically, Table 1 allows one to look up a particular genetic map position (in minutes) and find the corresponding physical genomic map coordinates (in kb). Inspection of that region of Figure 1 then identifies miniset clones (depicted directly below the restriction map) from that region of the chromosome. Since the miniset clone collection is freely distributed by Yuji Kohara, researchers can quickly obtain DNA fragments containing particular *E. coli* genes. For example, a researcher determines that his new mutation *beeR* maps very close to the *nag* operon at 16 minutes. Table 1 directs him to the 715-kb region of Figure 1. There he finds that miniset clones [171]16A8 or [172]3A6 (clone serial numbers are in brackets) may contain the *beeR* gene. The revised restriction map in this region assists him in subcloning the *beeR* gene. Another researcher isolates a plasmid clone that complements her newly isolated, but unmapped, mutation *smiL*. She hybridizes her clone to a filter containing plaque lifts of the Kohara miniset and finds that her plasmid hybridizes to clones [351]21H10 and [352]1C5. Since the miniset clones are numbered sequentially (note the inversion region described below), she quickly locates these clones in Figure 1. She sees that *smiL* may be linked to the *his* operon and quickly confirms this by demonstrating genetic linkage of *smiL* to a *his* mutation.

The revised, integrated restriction map presented here is substantially different from the original restriction map published by Kohara et al. (1987). To define those differences clearly, the process used to derive the new map is described in detail below. It is expected that this map will be of general utility to *E. coli* research geneticists, and therefore it is important to appreciate exactly what it represents.

We begin with an outline of our approach. Genomic map revision and subsequent integration with DNA sequence data were used to improve the accuracy and utility of the original genomic restriction map of Kohara et al. (1987). First, the original eight-enzyme map (*Bam*HI, *Hind*III, *Eco*RI, *Eco*RV, *Bgl*I, *Kpn*I, *Pst*I, and *Pvu*II) was revised to accommodate several subsequent corrections and to portray a consensus of *E. coli* K12 strains. Then, 693 individual *E. coli* DNA sequences were aligned to the revised map using the computer program MapSearch (Miller et al. 1990, 1991; Rudd et al. 1990, 1991), and the restriction map information derived from these DNA sequences was substituted for the corresponding segments of the revised map using a new program, AlterMap (Rudd et al. 1991). AlterMap integrates DNA sequence and physical mapping data into a single, composite, genomic restriction map and automatically updates a file containing the genomic map positions for the endpoints of the chromosomal DNA present in the miniset of phage λ clones (Kohara et al. 1987). An additional 175 individual sequences were then aligned to the map without the benefit of a MapSearch alignment for the sake of completeness. The alignment and integration of these additional sequences is discussed in more detail below. The revised, integrated genomic map, cloned DNA segments, and DNA sequence alignments are depicted in Figure 1. Figure 1 is a laser printer output of a new program, PrintMap (Rudd et al. 1991), that uses the Plasmid Description Language (C. Werner, in prep.) to draw genomic restriction maps using Postscript output devices.

We now turn to a more detailed description of the process of revising and updating the map. Before a comprehensive computerized alignment of *E. coli* sequences to the genomic restriction map could be attempted, it was necessary to compile a dataset of non-overlapping *E. coli* DNA sequences, designated EcoSeq. In the current version of EcoSeq (EcoSeq5), a total of 603 individual DNA sequences were condensed into 177 entries on the basis of DNA sequence overlaps. This process is called DNA sequence melding. Melded entries are designated with a gene name followed by the suffix "ecoM" (see Table 1). The EcoSeq5 dataset also contains 282 individual DNA sequence entries that did not overlap with other DNA sequence entries. Individual sequence entries include ones from the GenBank (Release 69) and EMBL (Release 28) databases (and daily updates), ones transcribed directly from publications, and ones received as personal communications. DNA sequences not taken verbatim from GenBank or EMBL end with the suffixes "eco" or "ecoM." A number of genes have been sequenced more than once. The longest and most accurate (if known) sequence of multiply sequenced entries was used in the EcoSeq database. In the case of the IS elements, the *rhs* genes, and the rRNA genes, prototype DNA sequences were used as analogs to represent various copies of these duplicated genes even though the unsequenced copies may differ slightly from the prototype. The rpsGeco entry (73.375′) was derived from the RpsG protein sequence (Reinbolt et al. 1979) using a computer-generated back translation and was used to fill in the central 195 base pairs of the *rpsG* gene that has not yet been DNA sequenced.

The EcoSeq database contains several individual sequences that were modified from GenBank or EMBL entries to eliminate vector DNA sequences, to complete terminal restriction sites, or to correct sequence errors. These include the Table 1 entries aroHeco (derived from ECOAROH, J04221), aspAeco (ECOASPA, X53863), bioHeco (ECOBIOH, X15587), cysPeco (ECOCYS, M32101), envCeco (ECOENVCD, X57948), deoDeco (ECOPNP, M60917), fadReco (ECOFADR, X08087), fecAeco (ECOFEC, M20981), fdhFeco (ECOFDHF, M18632), greAeco (ECOGREAG, X54718), hemBeco (M24488), hemHeco (ECOADKVIS, D90259), hipAeco (ECOHIPO, M61242), hisSeco (ECOHISS, M11843), hydGeco (ECOHYDHGA, M28369), ilvIeco (ECOILVIH, X01609), malGeco (ECOMALB, J01648), nlpAeco (ECLP28, M12163), ntrLeco (ECONTRLA, M15328), nusAeco (ECONUSA, X00513), ompAeco (ECOOMPA, J01654), pckAeco (ECOPCKA, M59823), pepQeco (ECOPEPQ, X54687), pssAeco (ECOPSS, M58699), pth-eco (ECPTHG, X61941), rfaDeco (ECORFADA, M33577), smpAeco (ECOSPROT, X52620), thrSeco (ECOTHRINF, V00291), tyrBeco (ECOTYBA, M12047), and udp-eco (ECOUDP, X15689). Vector DNA sequences were also eliminated during the melding process.

A total of 1,774,683 base pairs (contained in 459 non-overlapping entries) of *E. coli* DNA sequence is present in EcoSeq5. Using the revised, integrated genomic restriction map (see below) to estimate the length of the *E. coli* chromosome as 4673.6 kb, 38.0% of the *E. coli* genome is represented in EcoSeq5, and 87.1% (1,545,422 base pairs) of the DNA sequence present in EcoSeq5 was aligned to the revised genomic restriction map using MapSearch. In a few cases of uncertain alignment, the 1990 *E. coli* genetic map (Bachmann 1990) was used to help choose the correct MapSearch alignment. Additional DNA sequences can be positioned using other methods. The Kohara miniset clones have been used by many researchers to identify gene locations on the genomic restriction map by using Southern blot, Western blot, genetic complementation, and marker rescue techniques. DNA sequences can also be positioned using DNA sequence data from closely related organisms to bridge a DNA sequencing gap. For example, the ECODNAB (91.900′) was positioned using the STYALR (M12847) DNA sequence entry to bridge the gap to the *tyrB* gene. The flushmeld (no sequence overlap depicted [see

Rudd et al. 1991 for more discussion of flushmelds]) of ECOPABB (39.900') and ECOSDAA (41.000') was detected using the KPNPABB (M22078) DNA sequence, which bridges these two sequences together. This article presents the first precise mapping of the *sdaA* (and many other) genes. As with many of the other additional non-MapSearch alignments, ECOPABB was positioned using published gel-derived restriction maps (instead of DNA-sequence-generated restriction maps) as MapSearch probes (see Miller et al. 1990 for details of this process). The previously described (Miller et al. 1991) flushmeld of ECOPRS (26.500') and ECOHEMA (26.700') has recently been confirmed by the GenBank entry ECOPRSA (M77237). These additional DNA sequence positions are included in Table 1 and Figure 1 and are noted with "ND" in both the rank and alignment columns since they contain insufficient restriction site information to permit processing with MapSearch. Most, but not all, of the gene-to-miniset clone assignment data can be found in the DNA sequence citations in Table 1. A more detailed description of these DNA sequence assignments will be published elsewhere (K. Rudd and W. Miller, in prep.). Anyone wishing to cite positions of the MapSearch-independent ("ND ND" in Table 1) alignments should contact us for the proper citations and methods employed for any particular alignment. The tRNA gene sequences were aligned using the data of Komine et al. (1990). The IS insertion elements were placed using published Kohara miniset clone hybridization data (Birkenbihl and Vielmetter 1989, 1991; Umeda and Ohtsubo 1989, 1990a,b). If an additional DNA sequence that was aligned to the genomic restriction map contained more than one restriction site, these sites were integrated into the genomic restriction map using a new program, AlignMaker, to create pseudo-MapSearch alignments to use as AlterMap input files (Miller et al. 1990; K. Rudd and W. Miller, in prep.). Using all these additional methods, an extra 200,091 base pairs (175 individual sequences) were aligned to the revised, integrated genomic restriction map. Some of these are aligned in the positive orientation by default, since the true orientation has not been accurately determined. These are noted with a "?" in the orientation column of Table 1. Only 17 EcoSeq5 entries (27,389 bp) were not assigned genomic restriction map positions. These are listed, along with their genetic map locations, at the end of Table 1.

Table 1 describes the DNA sequences that were aligned to the genomic map, as depicted in Figure 1. Altogether, 603 entries (which are the constituents of 177 melded entries) and 265 unmelded sequence entries are aligned, for a total of 868 individual sequence alignments. The 603 entries used to form the melded entries are depicted as individual alignments in Figure 1 and are listed in Table 1. However, these sequences were aligned only as part of the larger melds since many of them were too small to be aligned by themselves. Thus, the rank and *p* (probability) value listed in Table 1 for each of these 603 constituent entries are the same as those of the encompassing melded entry.

The revised, integrated genomic restriction map presented here was produced using a two-step process. In the first step, a digital version of the original map of Kohara et al. (1987) was revised according to certain additional information. The original map contained eight gaps that were either arbitrarily set at 2 kb apiece or filled using published DNA sequence information (Kohara et al. 1987). DNA covering the gap regions has been cloned, gap lengths have been estimated, and restriction sites for three enzymes (*Eco*RI, *Bam*HI, and *Hin*dIII) in the gap regions have been determined (Knott et al. 1988, 1989). We used this information to revise the original genomic map data. Recent restriction map information for the gap at 40' has also been incorporated, allowing the alignment of the *edd*, *eda*, and *zwf* gene sequences (T. Conway, pers. comm.; see Table 1). We also incorporated Yuji Kohara's corrections to the region around the *htrA* gene (181.1 kb) (cited and published in Lipinska et al. 1989). The *E. coli* W3110 strain that was used to generate the genomic restriction map differs from other *E. coli* K12 strains in that it has five copies of a region containing the *tdc* operon (Schweizer and Datta 1990; Umeda and Ohtsubo 1990a; Rudd et al. 1991) and the *rnpB* gene (Komine and Inokuchi 1991). The pre-duplication position of the *tdc* operon is at 3282.2 kb (Table 1, Figure 1). The other four copies of this duplicated region have been removed from the revised genomic restriction map. Removal of the copies eliminated or shortened the depictions of several miniset clones and shortened the estimated length of the *E. coli* chromosome by 51.3 kb. The sections removed were 3318.0 kb to 3329.0 kb and 3144.5 kb to 3184.8 kb of the original genomic restriction map. A large inverted segment between the *rrnE* and *rrnD* operons present in W3110 has been returned to the wild-type configuration (3444.6 kb to 4241.0 kb in Figure 1) as previously described (Rudd et al. 1990). This process interrupted the depiction of five λ clones; the separated clone segments are denoted with the suffixes "A" and "B" in Figure 1. In this first step, the original digitized map (named Ecoli1.map) 4,719,600 bp in length (containing 7112 restriction sites [see Rudd et al. 1990]) was converted to a map (named Ecoli2.map) 4,675,800 bp in length (containing 7035 sites).

In the second step, an integrated genomic map was produced. The program AlterMap was used to replace regions of the genomic restriction map with DNA-sequence-derived restriction maps that had been aligned using MapSearch (Rudd et al. 1990) or the alternative procedure described above. **All of the restriction sites depicted above aligned DNA sequence entries in Figure 1 were derived from DNA sequences.** The remainder of the map was taken from the map that resulted from the first step, Ecoli2.map. A restriction map derived from the DNA sequence should be more accurate than one determined by physical mapping techniques, although some sequence-derived sites may not be present in the miniset clones (and vice versa) due to restriction site polymorphism or sequencing errors. This second step produced a genomic restriction map (named Ecoli5.map) 4,673,600 bp in length (containing 7845 restriction sites). (Ecoli3.map and Ecoli4.map were interim maps released along with EcoSeq1 and EcoSeq2, respectively.) A total of 37.3% of the revised, integrated *E. coli* genomic map (Ecoli5.map) was derived from DNA sequence. Largely because closely spaced sites for the same enzyme were not resolved by gel electrophoresis (discussed in Rudd et al. 1990), the AlterMap replacements increased the number of genomic restriction sites by 810. The estimated size of the *E. coli* genome was changed by only 2200 base pairs in the second step (0.13% of the entire replaced regions), indicating an unexpectedly high accuracy of the genome-length estimate produced by physical mapping techniques.

The 1990 *E. coli* genetic map (Bachmann 1990) position of a representative gene for each of the DNA sequence entries is presented in Table 1. These positions are generally consistent with the MapSearch alignments. The relative order of some linked genes differs from the order presented in the 1990 genetic map (Bachmann 1990). In several cases, new gene positions have been determined by detection of DNA sequence overlaps with other mapped genes and by highly significant MapSearch alignments of the melded DNA sequences. These include *ddlA* (8.800′), *mvrC* (12.550), *pal* (16.775′), *dacC* (18.650′), *fhuE* (24.750′), *tag* (79.875′), and *hydGH* (90.400′) (see above and Table 1). The *fec* operon DNA sequence has been relocated to 97.300′, as previously predicted and confirmed (Rudd et al. 1991). Genes whose minute values were not taken from the 1990 genetic map but from MapSearch alignments (and extrapolated to 1990 genetic map positions) are marked with a # in the "Gene" column of Table 1. Uncertainties in 1990 genetic map positions are indicated with an * in the "Gene" column, as was done in the 1990 genetic map. A new genomic distance scale has been calculated using the value of 46,736 bp/unit and is depicted immediately below the restriction sites depicted in Figure 1. These new units are equal to 1% of the length of the chromosome and are designated as centisomes. Although centisome values are similar to the minute values taken from the 1990 genetic map (Bachmann 1990) and listed in Table 1, the deviation from strict collinearity of these two scales previously reported (Rudd et al. 1990) has persisted despite the removal of 51.3 kb from the physical map (Rudd et al. 1991 and see above) in the region of maximal deviation. (K. Rudd and W. Miller, in prep.). Several novel DNA sequence identifications are presented in Table 1: the *rpmE* gene (89.100′) is upstream of the *priA* gene on the opposite DNA strand; *drpA* (5.050′) and *proS* (Eriani et al. 1990) are the same gene (although the published sequences differ slightly); *sohA* (68.500′) and *prlF* are the same gene; *shl* (ECOSHL, M35034) and *fruR* (1.950′) are identical. *hlpA* (*skp*) (4.250′) and *ompH* have previously been shown to be the same gene (Hirvas et al. 1990).

All of the DNA sequences in Table 1 were determined using *E. coli* K12 DNA with a few exceptions (see Table 1 for citations). The *gshA* (57.900′) and *gshB* (63.700′) DNA sequences were determined using *E. coli* B DNA, but they are likely to be very close to their K12 counterparts since all of the restriction sites in the DNA sequences could be found in the genomic restriction map. Likewise, part of the *phn* operon (92.400′, ECOPHNAQ) DNA sequence was derived from *E. coli* B. The *cysJ* and *cysI* genes (59.350′, ECOCYSJIHA) were sequenced using *E. coli* B DNA. The *araBA* gene sequences (1.450′, ECOARAABD) were determined using *E. coli* B/rDNA. The *ddlA* and *phoA* genes (8.825′, ECOPHOAJ) were linked using 108 base pairs of *Escherichia fergusonii phoA* DNA.

All of the DNA sequences larger than approximately 2500 bp could be aligned using MapSearch, with two exceptions that may indicate errors in the genomic restriction map. The *pheA*, *tyrA*, and *aroF genes* (56.700′) (ECOPHEAB, M10432) could not be aligned with MapSearch, probably due to missing sites in the 2753-kb to 2760-kb region of the genomic map. The *cysK*, *ptsHI*, and *crr* genes (52.200′) (ECOCYSK, X12615; ECOPHOSYS, M21994) probably lie between *lig* (52.125′) and *cysP-M* (52.300′), but they could not be aligned using MapSearch, possibly due to a deletion of DNA in the corresponding Kohara clones. Other sequences did not align well by themselves but were aligned as part of a meld. For example, the second half of the entry entDecoM (13.300′) and the first half of the entry rplQecoM (72.425′) do not correspond well to the Kohara map.

In the original genomic restriction map of Kohara et al. (1987), 40 regions of the genome were depicted as lacking information about *Eco*RV cleavage sites. Table 2 presents the revised, integrated genomic map addresses of these regions. The DNA-sequence-derived segments of the revised, integrated genomic restriction map do depict *Eco*RV sites in some of these 40 regions (Figure 1).

The miniset of *E. coli* phage λ clones can be obtained from Dr. Yuji Kohara, DNA Research Center, National Institute of Genetics, Mishima 411, Japan. DNA hybridization filters with dots of each of the miniset clone DNAs can be obtained from Biotechnology Research Laboratories, Takara Shuzo Co., Ltd., Seta 3-4-1, Otsu, Shiga, 520-21, Japan.

Additional information about *E. coli* DNA sequences can be obtained from a compilation prepared by Kröger et al. (1991). The reader should also refer to the 1990 *E. coli* genetic map (Bachmann 1990) for more information about the *E. coli* genome, including many additional literature citations. The GenBank and EMBL databases contain further information about *E. coli* DNA sequences, including information about the various genes sequenced within each entry. For example, the GenBank entry ECOAPTADK (10.800′) contains DNA sequences for the genes *apt*, *dnaX*, *htpG*, *recR*, and *adk*. The reader should also refer to the section on *E. coli* translation initiation sites by K. Rudd and T. Schneider in this Handbook (Section 17C) to obtain gene names, gene positions in centisomes, and SWISSPROT accession numbers for the genes present in EcoSeq5. The literature survey and database (GenBank and EMBL) daily update surveys for *E. coli* sequences included in EcoSeq5 and Table 1 were terminated on December 1, 1991.

ASCII files of EcoSeq5 database entries, digital genomic restriction maps, DNA-sequence-derived restriction maps, a meld description file, and C language source code for MapSearch, AlterMap, and PrintMap can be obtained from the author or downloaded directly from an anonymous ftp site, ncbi.nlm.nih.gov (in the "repository/ Eco" directory). The author can be contacted directly: electronic mail (Internet) address = rudd ncbi.nlm.nih.gov; FAX = 301-480-4941; telephone = 301-496-2475.

Individuals using novel information obtained from Table 1 or Figure 1 should cite Rudd et al. (1991) and this volume.

This project could not have been completed without the computer programming skill and patience of Webb Miller (MapSearch, AlterMap) and Craig Werner (PrintMap). I also thank the individuals who have sent me unpublished DNA sequences. Finally, I would like to acknowledge help received from B. Baum, D. Benson, G. Bouffard, T. Conway, D. Johnson, S. Karlin, T. Kazic, D. Lipman, J. Miller, J. Ostell, R. Overbeek, G. Redfern, M. Riley, S. Satterfield, T. Schneider, and C. Tolstoshev.

REFERENCES

Bachmann, B.J. 1990. Linkage map of *Escherichia coli* K12, edition 8. *Microbiol. Rev.* **54:** 130–197.

Birkenbihl, R.P. and W. Vielmetter. 1989. Complete maps of IS*1*, IS2, IS3, IS4, IS5, IS30, and IS*150* locations in *Escherichia coli* K12. *Mol. Gen. Genet.* **220:** 147–153.

Birkenbihl, R.P. and W. Vielmetter. 1991. Completion of the IS map in *E. coli*: IS*186* positions on the *E. coli* chromosome. *Mol. Gen. Genet.* **226:** 318–328.

Eriani, G., M. Delarue, O. Poch, J. Gangloff, and D. Moras. 1990. Partition of tRNA synthetases into two classes based on mutually exclusive sets of sequence motifs. *Nature* **347:** 203–206.

Hirvas, L., J. Coleman, P. Koski, and M. Vaara. 1990. Bacterial "histone-like protein I" (HLP-I) is an outer membrane constituent? *FEBS Lett.* **262:** 123–126.

Knott, V., D.J. Blake, and G.G. Brownlee. 1989. Completion of the detailed restriction map of the *E. coli* genome by the isolation of overlapping cosmid clones. *Nucleic Acids Res.* **17:** 5901–5912.

Knott, V., D.J.G. Rees, Z. Cheng, and G.G. Brownlee. 1988. Randomly picked cosmid clones overlap the *pyrB* amd *oriC* gap in the physical map of the *E. coli* chromosome. *Nucleic Acids Res.* **16:** 2601–2612.

Kohara, Y., K. Akiyama, and K. Isono. 1987. The physical map of the whole *E. coli* chromosome: Application of a new strategy for rapid analysis and sorting of a large genomic library. *Cell* **50:** 495–508.

Komine, Y. and H. Inokuchi. 1991. Precise mapping of the *rnpB* gene encoding the RNA component of RNase P in *Escherichia coli* K-12. *J. Bacteriol.* **173:** 1813–1816.

Komine, Y., T. Adachi, H. Inokuchi, and H. Ozeki. 1990. Genomic organization and physical mapping of the transfer RNA genes in *Escherichia coli* K12. *J. Mol. Biol.* **212:** 579–598.

Kröger, M., R. Wahl, and P. Rice. 1991. Compilation of DNA sequences of *Escherichia coli* (update 1991). *Nucleic Acids Res.* (Suppl.) **19:** 2023–2043.

Lipinska, B., O. Fayet, L. Baird, and C. Georgopoulos. 1989. Identification, characterization, and mapping of the *Escherichia coli htrA* gene, whose product is essential for bacterial growth only at elevated temperatures. *J. Bacteriol.* **171:** 1574–1584.

Miller, W., J. Ostell, and K.E. Rudd. 1990. An algorithm for searching restriction maps. *Comput. Applic. Biosci.* **6:** 247–252.

Miller, W., J. Barr, and K.E. Rudd. 1991. Improved algorithms for searching restriction maps. *Comput. Applic. Biosci.* **7:** 447–456.

Reinbolt, J., D. Tritsch, and B. Wittmann-Liebold. 1979. The primary structure of ribosomal protein S7 from *E. coli* strains K and B. *Biochimie* **61:** 501–522.

Rudd, K.E., W. Miller, J. Ostell, and D.A. Benson. 1990. Alignment of *Escherichia coli* K12 DNA sequences to a genomic restriction map. *Nucleic Acids Res.* **18:** 313–321.

Rudd, K.E., W. Miller, C. Werner, J. Ostell, C. Tolstoshev, and S.G. Satterfield. 1991. Mapping sequenced *E. coli* genes by computer: Software, strategies and examples. *Nucleic Acids Res.* **19:** 637–647.

Schweizer, H.P. and P. Datta. 1990. Physical map location of the *tdc* operon of *Escherichia coli*. *J. Bacteriol.* **172:** 2825.

Umeda, M. and E. Ohtsubo. 1989. Mapping of insertion elements IS*1*, IS2 and IS3 on the *Escherichia coli* K-12 chromosome. Role of the insertion elements in formation of Hfrs and F′ factors and in rearrangement of bacterial chromosomes. *J. Mol. Biol.* **208:** 601–614.

Umeda, M. and E. Ohtsubo. 1990a. Mapping of insertion element IS5 in the *Escherichia coli* K-12 chromosome. Chromosomal rearrangements mediated by IS5. *J. Mol. Biol.* **213:** 229–237.

Umeda, M. and E. Ohtsubo. 1990b. Mapping of insertion element IS30 in the *Escherichia coli* K12 chromosome. *Mol. Gen. Genet.* **222:** 317–322.

Table 1   Alignment of DNA Sequences to a Revised, Integrated *E. coli* Genomic Restriction Map

Gene[a]	Ori[b]	Min[c]	Rnk[d]	p val[e]	Locus[f]	Access[g]	Bp[h]	Sites[i]	Addr[j]	Reference[k]	
thrA	+	0.000	1	0.001	ECOTHR	J01706	5922	11	0.0	NAR	11:7331-7345(1983)
thrA		0.000	1	0.001	thrAecoM	ES1216	6045	11	0.1		
thrA	+	0.000	1	0.001	thrAeco	ES3161	349	0	0.1	JBC	257:3896-3904(1991)
thrA	+	0.000	1	0.001	ECOTHR	J01706	5922	11	0.2	NAR	11:7331-7345(1983)
dnaK		0.250	1	0.000	dnaKecoM	ES1244	22431	44	12.3		
dnaK	+	0.250	1	0.000	ECODNAKA	M10420	182	0	12.3	PNAS	82:2679-2683(1985)
dnaK	+	0.250	1	0.000	ECODNAK	K01298	1917	7	12.5	PNAS	81:848-852(1984)
dnaJ	+	0.250	1	0.000	ECODNAJK	M12544	1623	5	14.2	JBC	261:1778-1781(1986)
is186A	+	0.275	1	0.000	IS186A	ES3185	1336	2	15.7	FEBS	192:47-52(1985)
gef#	-	0.275	1	0.000	gef-eco	ES3035	780	2	16.8	MM	3:1463-1472(1989)
nhaA#	+	0.275	1	0.000	nhaAeco	ES3221	315	0	17.6	JBC	266:21753-21759(1991)
nhaA#	+	0.275	1	0.000	ECOANTAPA	J03879	1349	4	17.7	JBC	263:10408-10414(1988)
rpsT	-	0.375	1	0.000	ECORPST	J01683	2881	7	18.8	NAR	14:6965-6981(1986)
ileS	+	0.400	1	0.000	ECORPSTA	M10428	1806	3	21.1	JBC	260:5616-5620(1985)
ileS	+	0.400	1	0.000	ileSeco	ES3048	2817	5	22.7	T.Webster	
lspA	+	0.425	1	0.000	ECOLSP	K01990	1135	0	25.1	PNAS	81:3708-3712(1984)
lspA	+	0.425	1	0.000	ECOLSPDAP	X54945	2696	4	26.0	NAR	19:180-180(1991)
dapB	+	0.600	1	0.000	ECODAPB	M10611	1281	3	28.6	JBC	259:14829-14834(1984)
carA	+	0.700	1	0.000	ECOCARAB	J01597	5227	8	29.5	PNAS	81:4139-4143(1984)
kefC		0.850	1	0.000	kefCecoM	ES1186	7552	17	48.4		
kefC	+	0.850	1	0.000	ECOKEFC	X56742	2253	5	48.4	MM	5:607-616(1991)
folA	+	0.900	1	0.000	ECODHFOLG	X05108	1200	3	50.1	EJB	162:473-476(1987)
apaH	-	1.000	1	0.000	ECOAPAH	X04711	2396	3	51.1	MGG	205:515-522(1986)
pdxA	-	1.100	1	0.000	ECOPDXA	M68521	1469	4	53.1	JBact	171:4767-4777(1989)
surA#	-	1.125	1	0.000	surAeco	ES3143	1666	6	54.3	M.Almiron/R.Kolter	
polB		2.100	1	0.000	polBecoM	ES1171	9396	18	62.6		
polB	-	2.100	1	0.000	ECPOLB	X54847	4081	7	62.6	MGG	226:24-33(1991)
polB	-	2.100	1	0.000	ECOPOLBDA	M35371	4666	10	63.0	DCB	9:631-635(1990)
araD	-	1.450	1	0.000	ECOARAABD	M15263	4478	10	66.3	Gene	47:231-244(1986)
araB	-	1.450	1	0.000	ECOARABOP	J01641	1335	2	70.6	JBC	253:6931-6933(1978)
araC	+	1.450	1	0.000	ECOARACK	V00259	1172	1	70.8	JMB	154:649-652(1982)
leuA		1.800	1	0.000	leuAecoM	ES1238	28278	77	83.7		
leuA	-	1.800	1	0.000	ECOLEUA	J01642	306	0	83.7	JMB	149:579-597(1981)
leuA	-	1.800	1	0.000	ECOLEUP	M12891	855	0	84.0	JBact	166:1113-1117(1986)
leuO	+	1.800	1	0.000	ECOLEUO	M21150	960	2	84.6	PNAS	85:6602-6606(1988)
ilvI	+	1.850	1	0.000	ECOILVIHP	M10738	389	0	85.5	JBact	163:186-198(1985)
ilvI	+	1.850	1	0.000	ilvIeco	ES3128	2332	4	85.8	MGG	226:332-336(1991)
fruR*	+	1.950	1	0.000	ECOFRURG	X55457	1820	5	88.0	MGG	226:332-336(1991)
ftsI	+	2.150	1	0.000	ECOPBPBRR	X52063	1500	6	89.8	NAR	18:2813-2813(1990)
ftsI	+	2.150	1	0.000	ECOPBPB	K00137	2758	4	91.0	MGG	191:1-9(1983)
murE	+	2.175	1	0.000	murEeco	ES3065	1656	3	93.3	CJM	35:1051-1054(1989)
murF	+	2.200	1	0.000	ECOMURF	X15432	1491	4	94.9	NAR	17:5379-5379(1989)
murD#	+	2.200	1	0.000	ECOMUROY	X51584	2608	10	96.2	NAR	18:1058-1058(1990)
ftsW#	+	2.250	1	0.000	ECOFTSW	M30807	1497	3	98.5	JBact	171:6375-6378(1989)
murG	+	2.275	1	0.000	ECOMURGC	X52644	2793	6	99.8	NAR	18:4014-4014(1990)
ddlB	+	2.300	1	0.000	ECOFTSQA	K02668	3333	17	102.4	JBact	160:546-555(1984)
ftsZ	+	2.325	1	0.000	ECOFTSQAB	M10429	1870	5	105.2	Gene	36:241-247(1985)
envA	+	2.350	1	0.000	ECOENVAA	M19211	2048	8	106.7	JBact	169:5408-5415(1987)
secA	+	2.375	1	0.000	ECOSECA	M20791	3811	14	107.7	JBact	170:3404-3414(1988)
mutT	+	2.400	1	0.000	ECOMUTT	X04831	867	2	111.1	MGG	206:9-16(1987)
guaC	+	2.600	ND	ND	ECOGUAC	M33020	1991	3	113.5	BJ	255:35-43(1988)
ampD		2.675	1	0.000	ampDecoM	ES1003	11345	34	118.8		
ampD	+	2.675	1	0.000	ECOAMPDE	X15237	1983	6	118.8	MM	3:1091-1102(1989)
aroP	-	2.700	1	0.000	ECOAROP	X17333	1440	2	120.7	NAR	18:653-653(1990)
aroP	-	2.700	1	0.000	ECOAROPRA	M15312	585	0	122.0	JBact	169:386-393(1987)

Table 1 *Continued*

Gene[a]Ori[b]		Min[c]	Rnk[d]	*p* val[e]	Locus[f]	Access[g]	Bp[h]	Sites[i]	Addr[j]	Reference[k]	
aceE	+	2.800	1	0.000	ECOACE	V01498	7740	27	122.4	EJB	135:519-527(1983)
speD	-	3.000	3F	0.769	ECOSPDE	J02804	2792	5	135.2	JBact	171:4457-4465(1989)
gcd#	+	3.050	ND	ND	ECOGCD	X51323	2634	2	140.1	JBact	172:6308-6315(1990)
pcnB	-	3.550	1	0.031	ECOPCNB	M20574	1542	6	157.7	JBact	171:1254-1261(1988)
dksA#		3.600	ND	ND	dksAecoM	ES1164	1780	5	160.0		
dksA#	-	3.600	ND	ND	dksAeco	ES3123	1276	3	160.0	JBact	172:2055-2064(1990)
sfsA#	-	3.603	ND	ND	ECOSFS1A	M60726	1230	2	160.6	JBact	173:2644-2648(1991)
mrcB		3.650	1	0.000	mrcBecoM	ES1161	10775	19	164.8		
mrcB	+	3.650	1	0.000	ECOPONB	X02163	2765	7	164.8	EJB	147:437-446(1985)
fhuA	+	3.700	1	0.000	ECOFHUACD	M12486	4607	7	167.1	JBact	169:3844-3849(1987)
fhuB	+	3.725	1	0.000	ECOFHUB	X04319	2563	6	171.4	MGG	204:435-442(1986)
hemL#	-	3.800	1	0.000	ECOGSAG	X53696	1920	4	173.7	MGG	225:1-10(1991)
dgt#		3.750	1	0.001	dgt-ecoM	ES1036	4369	12	178.6		
dgt#	+	3.750	1	0.001	ECODGTP	M31772	2760	8	178.6	PNAS	87:2740-2744(1990)
htrA#	+	3.775	1	0.001	ECOHTRA	X12457	1855	4	181.1	NAR	16:10053-67(1988)
dapD	-	3.900	ND	ND	ECODAPD	K02970	1182	1	190.8	JBC	259:14824-14828(1984)
map#		3.900	24	1.000	map-ecoM	ES1073	3255	9	196.0		
map#	-	3.900	24	1.000	ECOMAP	M15106	1197	2	196.0	JBact	169:751-757(1987)
rpsB	+	4.000	24	1.000	ECORPSBTS	J01684	2192	7	197.1	NAR	9:4163-4172(1981)
frr#		4.100	ND	ND	frr-ecoM	ES1207	893	3	199.6		
frr#	-	4.100	ND	ND	ECORRFX	J05113	831	2	199.6	JBC	264:20054-20059(1989)
frr#	-	4.100	ND	ND	ECOFRRPR	M69029	153	2	200.4	JBact	173:5181-5187(1991)
cdsA*	+	4.150	5	1.000	ECOCDS	M11330	1259	3	203.2	JBC	260:12078-12083(1985)
hlpA		4.250	1	0.000	hlpAecoM	ES1195	8487	20	208.0		
hlpA	+	4.250	1	0.000	ompHeco	ES3152	780	6	208.0	JBact	173:1223-1229(1991)
firA	+	4.275	1	0.000	ECOFIRA	X54797	1459	3	208.5	JBact	173:334-344(1991)
lpxA	+	4.500	1	0.000	ECOLPXA	M19334	6627	13	209.9	JBact	170:1268-1274(1988)
proS*	+	4.800	1	0.000	ECODRPA	M32357	2791	15	224.4	JBact	172:281-286(1990)
rrsH		5.100	4	0.011	rrnHecoM	ES1137	5878	8	231.5		
rrsH	+	5.100	4	0.011	ECORGNX1	J01700	478	0	231.5	Cell	17:225-234(1979)
rrsH	+	5.100	4	0.011	rrsHeco	ES3098	1542	4	231.9	PNAS	75:4801-4805(1978)
ileV	+	5.100	4	0.011	ECORGNX2	K00763	545	1	233.5	JBC	254:3264-3271(1979)
rrlH	+	5.100	4	0.011	rrlHeco	ES3092	2904	2	233.9	PNAS	77:201-204(1980)
rrfH	+	5.100	4	0.011	ECORGNF	J01698	600	1	236.8	NAR	8:3809-3827(1980)
dniR#	-	5.200	1	0.014	ECODNIR	X60739	2596	6	239.5	FEMS	83:205-212(1991)
rnhA		5.250	1	0.019	rnhAecoM	ES1180	1594	6	244.0		
rnhA	-	5.250	1	0.019	ECORNHQ	K00985	1592	6	244.0	MGG	205:9-13(1986)
aspV	+	5.150	1	0.019	ECOASPV	X14007	284	1	245.3	MGG	206:356-357(1987)
pepD		5.650	1	0.000	pepDecoM	ES1090	8710	15	261.7		
pepD	-	5.650	1	0.000	ECOPEPD	M34034	2311	2	261.7	JBact	172:4641-4651
gpt	+	5.700	1	0.000	ECOGPTA	M13422	2253	3	263.8	Gene	32:243-249(1984)
phoE	-	5.850	1	0.000	ECOPHOE	J01662	1980	3	265.9	JMB	163:513-532(1983)
proB	+	5.900	1	0.000	ECOPHOEA	X00786	3041	9	267.4	NAR	12:6337-6355(1984)
is30A	+	6.100	ND	ND	IS30A	ES3177	1221	1	277.9	EMBO	3:2145-2149(1984)
is5A	+	6.200	ND	ND	IS5A	ES3198	1195	2	282.0	Gene	14:165-174(1981)
is1B	-	6.300	ND	ND	ECOIS1B	X17345	1001	2	287.0	MGG	222:317-322(1990)
is30B	+	6.300	ND	ND	IS30B	ES3178	181	0	287.8	MGG	222:317-322(1990)
argF	-	6.500	1	0.555	ECOARGF	X00759	1405	5	296.8	NAR	12:6277-6289(1984)
is1C	-	6.500	ND	ND	ECOIS1C	X52535	880	2	298.6	Gene	98:1-5(1991)
is3A	+	7.000	4	0.783	INS3INVA	M55510	1258	4	321.5	JBact	173:906-909(1991)
betA	-	7.500	1	0.004	ECOBET	X52905	7412	18	331.7	MM	5:1049-1064(1991)
site	+	7.800	ND	ND	lacTeco	ES3052	1433	2	362.7	JBact	159:618-623(1984)
cynS		7.900	1	0.000	cynTecoM	ES1026	8904	14	365.5		
cynS	+	7.900	1	0.000	M23219	M23219	2633	4	365.5	JBC	263:14769-14775(1988)
lacA	+	8.000	1	0.000	ECOLAC	J01636	7477	11	366.9	PNAS	82:6414-6418(1985)
is30E	+	8.050	ND	ND	IS30E	ES3181	1150	1	380.0	EMBO	3:2145-2149(1984)
is2A	-	8.400	11	0.975	IS2A	ES3188	1331	3	388.4	Gene	59:291-296(1987)
hemB		8.100	12	0.999	hemBecoM	ES1062	2051	4	395.6		
hemB	-	8.100	12	0.999	M24488	M24488	1907	4	395.6	Gene	75:177-184(1989)
hemB	-	8.100	12	0.999	hemBeco	ES3042	175	0	397.5	JBact	173:94-100(1991)

Table 1 *Continued*

Gene[a]	Ori[b]	Min[c]	Rnk[d]	p val[e]	Locus[f]	Access[g]	Bp[h]	Sites[i]	Addr[j]	Reference[k]	
is3B	-	8.600	25	1.000	INS3INVB	M55511	1258	4	398.3	JBact	173:906-909(1991)
sbmA*	?	8.800	ND	ND	ECOSBMA	X54153	1992	0	402.8	F.Moreno	
ddlA#		8.800	1	0.000	ddlAecoM	ES1152	4512	7	406.5		
ddlA#	-	8.800	1	0.000	ECODDLA	M58467	1687	1	406.5	Bioch	30:1673-1682(1991)
phoA	+	8.825	1	0.000	ECOPHOAJ	M33966	3133	4	408.0	MBE	7:547-577(1990)
phoA	+	8.825	1	0.000	ECOPHOAA	M13345	2715	6	408.3	Gene	44:121-125(1986)
proC	-	8.950	6	1.000	ECOPROC	J01665	968	3	411.6	NAR	10:7701-7714(1982)
aroL		8.900	6	1.000	aroLecoM	ES1219	2077	2	412.9		
aroL	+	8.900	6	1.000	ECOAROL	X04064	810	0	412.9	BJ	237:427-437(1986)
aroL	+	8.900	6	1.000	ECOAROLM	M13045	2021	2	412.9	JBact	165:233-239(1986)
araJ#		9.000	1	0.001	araJecoM	ES1166	9532	19	417.0		
araJ#	-	9.000	1	0.001	ECOARAJ	M64787	4355	9	417.0	JBact	173:7765-7771(1991)
sbcC*	-	8.875	1	0.001	ECOSBCC	X15981	5125	13	418.6	NAR	17:8033-8046(1989)
phoB	+	9.100	1	0.001	ECOPHOB	X04026	976	3	423.6	JMB	190:37-44(1986)
phoR	+	9.100	1	0.001	ECOPHORG	X04704	1972	3	424.5	JMB	192:549-556(1986)
malZ#		9.125	1	0.000	malZecoM	ES1229	8762	16	428.9		
malZ#	+	9.125	1	0.000	ECOMALZ	X59839	2345	6	428.9	JBC	266:19450-19458(1991)
queA#	+	9.150	1	0.000	ECOQBIO	M37702	3013	5	429.7	JBact	173:2256-2264(1991)
tgt*	+	9.200	1	0.000	ECOTGT	M63939	1823	3	432.6	JBact	173:2256-2264(1991)
secD	+	9.300	1	0.000	ECOSECDF	X56175	4435	8	433.2	EMBO	9:3209-3216(1990)
tsx	-	9.350	ND	ND	ECOTSX	M57685	1477	4	437.8	Gene	96:59-65(1991)
ispA#	-	9.600	ND	ND	ECOISPA	D00694	1452	2	447.4	JBioch	108:995-1000(1990)
cyoE*	-	10.100	1	0.000	ECOCYOA	J05492	5819	8	453.6	JBC	265:11185-11192(1990)
bolA#		9.900	1	0.000	bolAecoM	ES1021	7719	15	461.0		
bolA#	+	9.900	1	0.000	ECOBOLA	X17642	1596	2	461.0	EMBO	8:3923-3931(1989)
tig#	+	9.920	1	0.000	tig-eco	ES3106	1809	2	461.9	JBact	172:5555-5562(1990)
clpP#	+	9.950	1	0.000	ECOCLPPA	J05534	1236	4	463.2	JBC	265:12536-12545(1990)
clpX#	+	9.975	1	0.000	clpXeco	ES3012	1245	2	464.4	W.Clark/S.Gottesman	
lon	+	10.000	1	0.000	ECOLON	J03896	3002	4	465.4	JBC	263:11718-11728(1988)
hupB	+	10.000	1	0.000	ECOHUPB	X16540	460	2	468.3	MGG	201:360-362(1985)
nusB	?	10.400	ND	ND	ECONUSAA	M26839	1012	5	471.0	AdvB	21:175-192(1986)
ffs*	+	10.300	ND	ND	ECOSGK	X01074	764	1	477.7	JMB	178:509-531(1984)
tesB*	-	10.350	1	0.032	ECOTESB	M63308	1455	4	481.2	JBC	266:11044-11050(1991)
hha#	?	10.575	ND	ND	ECHHAE	X57977	707	0	486.5	MM	5:1285-1293(1991)
priC#		10.200	1	0.000	priCecoM	ES1247	11068	20	496.7		
priC#	-	10.200	1	0.000	ECOPRIC	D00727	1225	5	496.7	JBC	266:13988-13995(1991)
apt	+	10.800	1	0.000	ECOAPTADK	M38777	6820	11	497.9	Gene	43:287-293(1986)
hemH	+	11.100	1	0.000	hemHeco	ES3230	3576	6	504.2	JMB	219:393-398(1991)
ushA	+	11.400	6F	0.845	ECOUSHA	X03895	1819	7	511.4	NAR	14:4325-4342(1986)
rhsD*	+	11.600	1	0.006	ECORHSDG	X60999	4846	10	529.5	A.Sadosky/C.Hill	
purK	-	12.200	1	0.287	ECOPUREK	M19657	2449	2	557.4	JBact	171:198-204(1989)
ppiB#		12.375	2	0.905	ppiBecoM	ES1206	2745	7	560.2		
ppiB#	+	12.375	2	0.905	ECOAPPIB	M55430	949	2	560.2	Bioch	30:3041-3048(1991)
cysS*	+	12.400	2	0.905	ECOCYSSG	X56234	2194	7	560.7	NAR	19:265-269(1991)
folD#	?	12.450	ND	ND	ECOFOLD	M74789	1044	0	564.4	JBC	266:23953-23958(1991)
argU		12.500	3	0.913	argUecoM	ES1199	5391	8	570.1		
argU	+	12.500	3	0.913	argUeco	ES3232	2520	4	570.1	MGG	220:325-328(1990)
argU	+	12.500	3	0.913	ECOINTDLP	M27155	2269	5	571.1	JBact	171:6197-205(1989)
is3C	+	12.525	3	0.913	IS3C	ES3182	1258	4	573.2	NAR	13:2127-2139(1985)
mvrC#	+	12.550	3	0.913	ECOMVRC	M62732	1002	1	574.2	M.Morimyo	
mvrC#	+	12.550	3	0.913	ebrAeco	ES3155	1244	1	574.2	FEMS	82:229-232(1991)
is5B	+	12.625	1	0.319	IS5B	ES3138	1195	2	580.9	Gene	14:165-174(1981)
nmpC		12.650	1	0.319	nmpCecoM	ES1086	2667	5	580.9		
nmpC	-	12.650	1	0.319	ECOEPNMPC	M13457	1611	4	582.0	JBC	261:12723-12732(1986)
appY#		13.150	1F	0.006	appYecoM	ES1007	4262	11	589.2		
appY#	+	13.150	1F	0.006	ECOAPPYAA	M24530	1874	4	589.2	JBact	171:1683-1691(1989)
ompT	-	13.175	1F	0.006	ECOOMPT1	X06903	2035	6	590.5	NAR	16:1209-1209(1988)
envY*	-	13.200	1F	0.006	ECOENVY	X13548	945	3	592.5	NAR	17:800-800(1989)
pheP*	+	13.000	1	0.240	ECOPHEPA	M58000	2256	3	608.7	JBact	173:3622-3629(1991)
entD		13.300	1	0.000	entDecoM	ES1176	24611	40	615.3		

Table 1 *Continued*

Gene[a]	Ori[b]	Min[c]	Rnk[d]	*p* val[e]	Locus[f]	Access[g]	Bp[h]	Sites[i]	Addr[j]	Reference[k]	
is186B	+	13.275	1	0.000	IS186B	ES3186	1336	2	615.3	FEBS	192:47-52(1985)
entD	-	13.300	1	0.000	entDeco	ES3021	1137	1	616.5	MM	3:757-766(1989)
fepA	-	13.350	1	0.000	ECOFEPAA	M13748	2624	4	617.5	JBC	261:10797-10801(1986)
fes	+	13.400	1	0.000	ECOFE6	J04216	1997	2	619.6	JBC	263:18857-18863(1988)
entF	+	13.450	1	0.000	ECOENTF	M60177	4756	13	621.1	Bioch	30:2916-2927(1991)
fepE	+	13.500	1	0.000	fepEeco	ES3030	1732	5	625.3	M.McIntosh	
fepC	-	13.550	1	0.000	ECOFEPCDG	X57471	2961	4	626.6	MM	5:1415-1428(1991)
fepD	-	13.650	1	0.000	ECOP43MP	X57470	1391	0	629.5	MM	5:1415-1428(1991)
fepB	-	13.700	1	0.000	ECOFEPB	M29730	2177	2	630.1	JBact	171:5443-5451(1989)
entC	+	13.750	1	0.000	M24142	M24142	1659	5	631.8	JBact	171:775-783(1989)
entE	+	13.800	1	0.000	ECOENTB	X15058	1655	1	633.4	FEMS	50:15-19(1989)
entA	+	13.900	1	0.000	M24148	M24148	3249	2	634.6	JBact	171:791-798(1989)
cstA#	+	13.925	1	0.000	ECOCASTGE	X52904	3253	5	636.7	JMB	218:129-140(1991)
bnt104	?	13.700	ND	ND	BENT104	ES3168	279	1	643.4	MGG	226:367-376(1991)
rna*	-	14.400	13	ND	ECORIBI34	M55687	1206	2	651.6	Gene	95:1-7(1990)
is5C	-	14.700	ND	ND	IS5C	ES3199	1195	2	662.8	Gene	14:165-174(1981)
dacA		14.875	1	0.003	dacAecoM	ES1031	6643	12	672.3		
dacA	-	14.875	1	0.003	ECODACA	X06479	1597	4	672.3	NAR	16:1617-1617(1988)
rlpA	-	14.900	1	0.003	ECORLPA	M18276	1408	5	673.5	JBact	169:5692-5699(1987)
mrdB	-	14.950	1	0.003	ECORODA	M22857	1260	0	674.8	JBact	171:558-560(1989)
mrdA	-	14.950	1	0.003	ECOPBPA	D00001	2936	4	676.0	EJB	160:231-238(1986)
rlpB		15.200	1	0.001	rlpBecoM	ES1108	3898	13	681.6		
rlpB	-	15.200	1	0.001	ECORLPB	M18277	938	3	681.6	JBact	169:5692-5699(1987)
leuS	-	15.225	1	0.001	ECOLEUS	X06331	3618	13	681.9	NAR	15:10199-10210(1987)
cutE#		15.525	1	0.007	cutEecoM	ES1234	3424	7	697.8		
is5D	+	15.500	1	0.007	IS5D	ES3200	1195	2	697.8	Gene	14:165-174(1981)
cutE#	-	15.525	1	0.007	ECOCUTE	X58070	2333	6	698.9	JBact	173:6742-6748(1991)
glnV		15.400	1	0.000	glnVecoM	ES1054	12152	21	706.5		
glnV	-	15.400	1	0.000	ECOTGOP	J01713	1100	2	706.5	Cell	23:239-249(1981)
asnB	-	15.550	1	0.000	ECOASNB	J05554	3080	6	707.3	JBC	265:12895-12902(1990)
nagA	-	15.575	1	0.000	ECONAGACD	X14135	3619	8	709.4	MM	3:505-515(1989)
nagB	+	15.600	1	0.000	ECONAGBE	M19284	3396	7	712.8	Gene	62:197-207(1988)
glnS	+	15.625	1	0.000	ECOGLNSA	M10187	2586	1	716.0	JBC	257:11639-11643(1982)
fur		15.650	1	1.000	fur-ecoM	ES1151	1799	2	720.5		
fur	-	15.650	9	1.000	ECOFUR	X02589	868	0	720.5	MGG	200:110-113(1985)
fldA#	-	15.675	9	1.000	ECOFLDA	M59426	1164	2	721.2	JBact	173:1729-1737(1991)
potE#		15.700	1	0.000	potEecoM	ES1248	12316	19	727.4		
potE#	-	15.700	1	0.000	ECOPOTESPE	M64495	4341	4	727.4	JBC	266:20922-20927(1991)
kdpD	-	15.800	1	0.000	ECOKDPDE	M36066	3600	8	731.6	W.Epstein	
kdpC	-	15.800	1	0.000	ECOKDPABC	K02670	4933	9	734.6	PNAS	81:4746-4750(1984)
bnt102	+	15.825	1	0.000	BENT102	ES3166	243	1	739.5	MGG	226:367-376(1991)
rhsC*		15.950	2	0.003	rhsCecoM	ES1183	4294	13	740.2		
rhsC*	+	15.950	2	0.003	ECORHSC1	M21763	400	1	740.2	JBact	171:636-642(1989)
rhsC*	+	15.950	2	0.003	rhsCeco	ES3142	4134	12	740.3	JBact	172:446-456(1990)
rhsC*	+	15.950	2	0.003	ECORHSC	M29718	538	1	744.0	JBact	172:446-456(1990)
phr	+	16.000	12	1.000	ECOPHRORF	K01299	2039	3	749.7	JBC	259:6033-6038(1984)
gltA		16.300	1	0.000	gltAecoM	ES1058	13881	28	762.3		
gltA	-	16.300	1	0.000	ECOGLTA	J01619	13063	28	762.3	Bioch	24:6245-6252(1985)
sucD	+	16.500	1	0.000	ECOG30SUC	X15790	1285	2	774.9	BJ	260:737-747(1989)
cydA	+	16.650	1	0.027	ECOCYD	J03939	3845	5	781.0	JBC	263:13138-13143(1988)
tolQ		16.750	1	0.000	tolQecoM	ES1127	5037	25	785.0		
tolQ	+	16.750	1	0.000	ECOTOLQRA	M16489	1855	2	785.0	JBact	169:2667-2674(1987)
tolA	+	16.750	1	0.000	ECOTOLAB	M28232	2900	18	786.6	JBact	171:6600-6609(1989)
pal#	+	16.775	1	0.000	ECOPAL	X05123	713	6	789.3	EJB	163:73-77(1987)
nadA	?	16.800	ND	ND	ECONADA	X12713	1470	1	792.1	EJB	175:221-228(1988)
lysT	+	16.800	ND	ND	ECOTRNKV	X04171	730	2	793.7	JMB	177:609-625(1984)
aroG	+	16.850	ND	ND	ECOAROG	J01591	2107	2	794.4	NAR	10:4045-4058(1982)
galE	-	17.000	1	0.156	galEeco	ES3034	4234	6	797.0	S.Adhya	
chlD	+	17.100	ND	ND	ECOCHLJD	M16182	1609	0	806.8	JBact	169:1911-1916(1987)
attL	+	17.350	1	0.000	ECOLAMATT	J01638	260	1	816.4	Science	197:1147-1160(1977)

Table 1 *Continued*

Gene[a]	Ori[b]	Min[c]	Rnk[d]	p val[e]	Locus[f]	Access[g]	Bp[h]	Sites[i]	Addr[j]	Reference[k]	
bioA		17.450	1	0.000	bioAecoM	ES1203	8634	22	816.4		
bioA	-	17.450	1	0.000	ECOBIO	J04423	5793	15	816.6	JBC	263:19577-19585(1988)
uvrB	+	17.550	1	0.000	ECOUVRB	J01722	2605	6	822.3	NAR	14:2637-2650(1986)
uvrB	+	17.550	1	0.000	ECOUVRB2	X03722	2400	6	822.6	NAR	14:2877-2890(1986)
rhlE#	?	17.800	ND	ND	ECORHLE	X56307	858	1	846.3	NB	3:886-895(1991)
glnH	-	18.000	1	0.017	ECOGLNHPQ	X14180	3436	8	854.1	MGG	205:260-269(1986)
chlE	-	18.200	1	0.063	ECOCHLEN	M21151	2492	7	873.1	JBact	170:4097-4102(1988)
grx	+	18.500	ND	ND	ECOGRX	M13449	1147	2	879.9	Gene	43:13-21(1986)
dacC#		18.650	1	0.001	dacCecoM	ES1032	2304	6	889.7		
dacC#	+	18.650	1	0.001	ECODACC	X06480	1505	4	889.7	NAR	16:1617-1617(1988)
deoR*	-	18.700	1	0.001	ECODEOR	X02387	986	2	891.0	NAR	13:5927-5936(1985)
rimK#	+	18.700	ND	ND	ECORIMK	X15859	1559	1	900.7	MGG	217:281-288(1989)
poxB	?	19.000	ND	ND	ECOPOXB	M13947	1974	2	908.3	NAR	14:5449-5460(1986)
clpA#		19.000	1	0.000	clpAecoM	ES1189	5585	17	931.6		
clpA#	+	19.000	1	0.000	ECOCLPAA	M31045	3380	10	931.6	PNAS	87:3513-3517(1990)
serW	-	19.700	1	0.000	ECOINFSERW	M63145	3060	10	934.1	JBC	266:16491-16498(1991)
lrp#	+	19.800	ND	ND	lrp-eco	ES3055	1225	1	940.8	S.Short	
serS		19.950	1	0.000	serSecoM	ES1117	7578	20	947.9		
serS	+	19.950	1	0.000	ECOSERS	X05017	1854	11	947.9	NAR	15:1005-1017(1987)
dmsA*	+	20.050	1	0.000	ECODMS	J03412	6492	12	949.0	MM	2:785-795(1988)
pfl		20.000	1	0.000	pfl-ecoM	ES1221	4884	9	958.6		
pfl	-	20.000	1	0.000	ECOPFL	X08035	3592	6	958.6	EJB	177:153-158(1988)
pfl	-	20.000	1	0.000	M26413	M26413	1431	3	962.0	JBact	171:2485-2498(1989)
serC	+	20.175	2	0.401	ECOAROA	X00557	2983	6	965.6	BJ	234:49-57(1986)
rpsA*		20.500	1	0.000	rpsAecoM	ES1110	2762	13	969.4		
rpsA*	+	20.500	1	0.000	ECORPSA	J01682	2412	13	969.4	NAR	10:1857-1865(1982)
himD	+	20.300	1	0.000	ECOHIP	X04864	600	1	971.7	JMB	183:117-128(1985)
msbA#	+	20.400	1	0.026	msbAeco	ES3064	3470	7	973.8	M.Karow/C.Georgopoulos	
kdsB*	+	84.700	ND	ND	ECOKDSB	J02614	1308	2	977.7	JBC	261:15831-15835(1986)
bnt103	?	20.700	ND	ND	BENT103	ES3167	238	0	980.5	MGG	226:367-376(1991)
mukB#	+	20.550	1	0.000	ECOMUKB	M63930	5353	16	982.6	EMBO	10:183-193(1991)
aspC		20.700	1	0.001	aspCecoM	ES1018	9132	15	991.1		
aspC	-	20.700	1	0.000	ECOASPC	X03629	1415	2	991.1	BJ	234:593-604(1986)
ompF	-	20.700	1	0.000	ECOOMPF	J01655	1808	3	992.3	NAR	10:6957-6968(1982)
asnS	-	20.800	1	0.000	ECOTGASNS	M33145	2040	3	994.0	Gene	84:481-485(1989)
pncB	-	20.875	1	0.000	ECOPNCB	J05568	1490	3	995.8	JBC	265:17665-17672(1990)
pepN	+	20.900	1	0.000	ECOPEPN	M15273	3409	6	996.8	Gene	48:145-153(1986)
pyrD	+	21.200	1	0.042	ECOPYRD	X02826	1357	4	1011.7	EJB	151:59-65(1985)
fabA	-	21.700	ND	ND	ECOFABAA	J03186	979	2	1023.1	JBC	263:4641-4646(1988)
ompA	-	21.800	1	0.001	ompAeco	ES3070	2271	7	1026.6	NAR	8:3011-3024(1980)
helD#	+	22.000	3	0.744	ECOHELIV	J04726	2821	4	1031.6	JBC	264:8297-8303(1989)
serT	-	22.200	ND	ND	ECOTGS	X00547	1344	3	1037.9	EMBO	3:1103-1107(1984)
hyaA#	+	21.100	1	0.000	ECOHYA	M34825	6023	8	1039.7	JBact	172:1969-1977(1990)
appA*		22.500	14	1.000	appAecoM	ES1006	2076	3	1048.4		
appA*	+	22.500	14	1.000	ECOAPPA	X05471	700	1	1048.4	Bchim	69:215-221(1987)
appA*	+	22.500	14	1.000	ECOAPPAA	M58708	1901	2	1048.5	JBact	172:5497-5500(1990)
is1D	+	22.300	5	0.865	ECOIS1D	X52536	808	4	1058.2	Gene	98:1-5(1991)
agp#	+	22.500	1	0.183	ECOAGPA	M33807	1675	3	1074.0	JBact	172:802-807(1990)
putP	+	22.800	6	0.955	ECPUTP	X05653	2212	2	1087.4	MGG	208:70-75(1987)
TerE#	+	22.850	ND	ND	terEeco	ES3215	22	0	1091.3	JBact	173:391-393(1991)
is3D	-	23.100	3	0.772	IS3D	ES3183	1258	4	1102.7	NAR	13:2127-2139(1985)
serX	-	23.200	ND	ND	serXeco	ES3102	1000	0	1109.6	JMB	212:579-598(1990)
is2B	+	23.300	2	0.385	IS2B	ES3189	1331	3	1114.1	Gene	59:291-296(1987)
htrB#	-	23.600	1	0.158	ECOHTRB	X61000	3129	4	1125.0	MM	5:2285-2292(1991)
pyrC	-	23.725	1	0.035	ECOPYRC	D00002	2046	4	1130.7	EJB	160:77-82(1986)
rimJ	+	23.750	ND	ND	ECORIMJ	X06118	1220	2	1135.2	MGG	209:481-488(1987)
ams*	-	24.150	3	1.000	ECOAMSG	M62747	3959	8	1151.0	JBC	266:2843-2851(1991)
rpmF*	-	24.800	4	1.000	ECORPMFA	M29698	1191	3	1157.1	JBact	171:5707-5712(1989)
ptsG		24.700	1	0.000	ptsGecoM	ES1078	4102	7	1167.9		
ptsG	+	24.700	1	0.000	ECOPTSG	J02618	1523	4	1167.9	JBC	261:16398-16403(1986)

Table 1   *Continued*

Gene[a]	Ori[b]	Min[c]	Rnk[d]	*p* val[e]	Locus[f]	Access[g]	Bp[h]	Sites[i]	Addr[j]	Reference[k]	
fhuE#	-	24.750	1	0.000	ECOFHUE1	X17615	2900	4	1169.1	MM	4:427-437(1990)
ndh*	+	22.400	2	0.130	ECONDH	J01653	2057	6	1176.3	EJB	116:165-170(1981)
phoP#	-	25.500	ND	ND	ECOPHOPQ	D90393	4097	2	1192.4	JBact	InPress
icdE		25.600	ND	ND	icdEecoM	ES1201	1659	2	1199.9		
icdE	+	25.600	ND	ND	ECOICD	J02799	1568	2	1199.9	JBC	262:10422-10425(1987)
icdE	+	25.950	ND	ND	REIATTE	M19693	439	0	1201.1	JBact	170:2040-2044(1988)
lit*	?	25.700	ND	ND	ECOLIT	M19634	1772	2	1203.5	JBact	170:2056-2062(1988)
pin	+	25.800	2	1.000	ECOPINP	K03521	2614	2	1213.6	EMBO	4:237-242(1985)
mcrA	?	25.875	ND	ND	ECOMCRA	M76667	853	1	1216.6	S.Sedgwick/K.Hiom	
icdC#		25.950	ND	ND	icdCecoM	ES1202	439	0	1217.8		
icdC#	+	25.950	ND	ND	ECOATTE	M19683	439	0	1217.8	JBact	170:2040-2044(1988)
icdC#	+	25.600	ND	ND	REIATTE	M19693	439	0	1217.8	JBact	170:2040-2044(1988)
minB	-	26.200	2	0.326	ECOMINB	J03153	2400	3	1230.0	Cell	56:641-649(1989)
umuC	+	26.250	1	0.068	ECOUMUCD	M10107	2454	7	1236.6	PNAS	82:4336-4340(1985)
umuC	-	26.250	ND	ND	ECOUMUDCA	M29542	363	1	1247.8	JBact	170:1610-1616(1988)
treA	-	26.325	11	1.000	ECOTREA	X15868	2538	3	1251.7	MGG	217:347-354(1989)
pth*	-	26.450	8	1.000	pth-eco	ES3213	1628	3	1265.0	EMBO	10:3549-3555(1991)
prs*		26.500	1	0.000	prs-ecoM	ES1077	5494	12	1267.9		
prs*	-	26.500	1	0.000	ECOPRS	M13174	1785	6	1267.9	JBC	261:6765-6771(1986)
hemA	+	26.700	1	0.000	ECOHEMA	M30785	2924	9	1269.7	Gene	82:209-217(1989)
prfA*	+	26.825	1	0.000	ECORF1X	M11519	1441	1	1272.0	JBact	170:4537-4541(1988)
kdsA*	+	26.925	2	0.069	ECOKDSA	X05552	1504	6	1275.2	MGG	207:369-373(1987)
narL		27.175	1	0.000	narLecoM	ES1175	13341	38	1282.4		
narL	-	27.175	1	0.000	ECONARXL	X13360	3211	10	1282.4	NAR	17:2947-2957(1989)
narK	+	27.200	1	0.000	ECONARK	X15996	2482	3	1285.0	FEBS	252:139-143(1989)
narG	+	27.225	1	0.000	narGeco	ES3131	7301	20	1286.9	J.Cock/J.Wootton	
tyrT	-	27.350	1	0.000	ECOTGY1	K01197	1949	6	1293.8	JBC	253:3607-3622(1978)
hns#		27.500	1	0.443	hns-ecoM	ES1144	1903	4	1300.8		
hns#	-	27.450	1	0.443	ECOHNS	X07688	1380	4	1300.8	MGG	224:81-90(1990)
tdk	+	27.500	1	0.443	ECOTDKG	X53733	1095	1	1301.6	MM	5:373-379(1991)
adhE		27.650	1	0.743	adhEecoM	ES1181	10517	20	1303.9		
adhE*	-	27.650	1	0.743	ana-eco	ES3139	76	0	1303.9	FEBS	281:59-63(1991)
adhE*	-	27.650	1	0.743	ECOADHEX	M33504	2759	10	1303.9	Gene	85:209-214(1989)
adhE*	-	27.650	1	0.743	adhEeco	ES3140	2023	5	1306.6	FEBS	281:59-63(1991)
is2C	-	27.400	4	0.888	IS2C	ES3190	-1331	3	1306.8	Gene	59:291-296(1987)
oppA	+	27.700	1	0.743	oppAeco	ES3072	7005	9	1307.4	S.Short	
oppA	+	27.700	1	0.743	ECOOPPA	J05433	1920	4	1308.3	JBC	265:8387-8391(1990)
tonB		27.925	1	0.000	tonBecoM	ES1128	13504	22	1318.3		
tonB	+	27.925	1	0.000	ECOTONB	K00431	1697	0	1318.3	Cell	41:577-585(1985)
tonB	+	27.900	1	0.000	ECOTRTOI	X13583	3779	8	1319.8	Genet	120:345-358(1988)
trpA	-	28.000	1	0.000	ECOTGP	J01714	7539	13	1323.4	JMB	156:245-256(1982)
trpA	-	28.000	1	0.000	ECOTRPZ	M38366	1202	3	1330.6	BioKh	10:415-417(1984)
btuR*	-	28.200	ND	ND	ECOBTUR	M21528	958	2	1334.9	JBact	171:154-161(1989)
sohB#		28.225	1	0.000	sohBecoM	ES1211	6052	15	1336.7		
sohB#	+	28.225	1	0.000	ECOSOHB	M73320	1588	5	1336.7	JBact	173:5763-5770(1991)
topA	+	28.250	1	0.000	ECTOPA	X04475	4074	10	1337.5	JMB	191:321-331(1986)
cysB	+	28.275	1	0.000	ECOCYSB	M15041	1840	6	1340.9	JBC	262:5999-6005(1987)
pyrF		28.750	1	0.001	pyrFecoM	ES1239	2155	9	1350.5		
TerA	+	28.525	1	0.001	ECOTERC1A	M23250	488	1	1350.5	Cell	55:467-475(1988)
pyrF	+	28.750	1	0.001	ECOPYRF	J02768	1549	9	1350.6	JBC	262:10239-10245(1987)
osmB#	-	28.775	1	0.001	ECOOSMB	M22859	652	1	1352.0	JBact	171:511-520(1989)
dcp*	+	29.400	3	0.552	ECODCPG	X57947	2906	9	1355.6	S.Becker/R.Plapp	
aldH#	+	29.300	1	0.074	ECOALDHQ3	M38433	2883	11	1370.7	Gene	99:15-23(1991)
pspA#	?	29.000	ND	ND	ECOPSP	X57560	2424	3	1376.2	JMB	220:35-48(1991)
is5F	-	29.400	ND	ND	IS5F	ES3201	1195	2	1405.2	Gene	14:165-174(1981)
fnr		29.700	1	0.016	fnr-ecoM	ES1226	2019	8	1407.3		
fnr	-	29.700	1	0.016	ECONIRR	J01608	1641	7	1407.3	NAR	10:6119-6130(1982)
ogt#	-	29.725	1	0.016	ECOOGT	Y00495	719	4	1408.6	NAR	15:9177-9193(1987)
bnt108	?	29.800	ND	ND	BENT108	ES3171	315	0	1416.7	MGG	226:367-376(1991)
dbpA#	?	30.000	ND	ND	ECODEADA	X52647	1668	2	1417.8	NAR	18:5413-5417(1990)

Table 1 *Continued*

Gene[a]	Ori[b]	Min[c]	Rnk[d]	p val[e]	Locus[f]	Access[g]	Bp[h]	Sites[i]	Addr[j]	Reference[k]
recE	-	30.200	3	0.453	M24905	M24905	2475	6	1423.5	JBact	171:2101-2109(1989)
trkG#	+	30.400	ND	ND	ECTRKG	X56783	1817	1	1432.4	JBact	173:3170-3176(1991)
is2D	+	31.000	3	0.432	IS2D	ES3191	1331	3	1477.5	Gene	59:291-296(1987)
is30C	+	31.000	ND	ND	IS30C	ES3179	1221	1	1478.9	EMBO	3:2145-2149(1984)
ald*	+	31.500	2	0.394	ald-eco	ES3162	1764	5	1497.8	JBact	173:6118-6123(1991)
cybB*	+	16.100	ND	ND	ECOCYBB	X07569	1439	1	1499.8	MGG	212:1-5(1988)
trg	+	31.400	ND	ND	ECOTRG	K02073	1722	2	1501.3	PNAS	81:3287-3291(1984)
rimL*	?	33.375	2	0.942	ECORIML	X15830	1248	3	1507.1	MGG	217:289-293(1989)
rhsE#	+	32.500	12	1.000	ECORHSEG	X60998	2440	2	1536.7	A.Sadosky/C.Hill	
narZ	-	32.700	1	0.000	ECNARZYW	X17110	7080	15	1544.5	MGG	222:104-111(1990)
fdnG#	+	32.900	1	0.000	ECOFDNGHI	M75029	4981	7	1555.8	JBC	266:22380-22385(1991)
sfcA#		34.000	4	0.884	sfcAecoM	ES1205	2757	6	1563.4		
sfcA#	-	34.000	4	0.884	sfcAeco	ES3103	1765	5	1563.3	Genet	125:261-273(1990)
osmC#	+	34.100	4	0.884	ECOOSMC	X57433	998	2	1565.0	JMB	220:959-973(1991)
hipA	-	33.900	1	0.061	hipAeco	ES3116	2234	3	1599.9	JBact	173:5732-5739(1991)
TerC	-	34.300	ND	ND	ECOTERC3A	M23252	454	2	1618.0	Cell	55:467-475(1988)
relB	-	34.700	6	1.000	ECORELB	X02405	2142	2	1652.4	EMBO	4:1059-1066(1985)
dicA	+	34.800	1	0.037	ECOICABC	X07465	4440	5	1655.9	NAR	16:10388-10388(1988)
pntA	-	35.400	1	0.007	ECOPNTAB	X04195	3240	8	1684.1	EJB	158:647-653(1986)
tus		35.600	1	0.000	tus-ecoM	ES1132	6154	14	1692.3		
tus	+	35.600	1	0.000	tus-eco	ES3108	2416	8	1692.3	PNAS	86:1593-7(1989)
fumC	-	35.625	1	0.000	ECOFUMC	X04065	2250	6	1693.2	BJ	237:547-557(1986)
fumA	-	35.650	1	0.000	ECOFUMA	X00522	2409	3	1695.0	NAR	12:3631-3642(1984)
manA	+	35.650	1	0.000	ECOMANAA	M15380	1604	3	1696.8	Gene	32:41-48(1984)
uidA	-	35.850	ND	ND	ECOUIDAA	M14641	2439	4	1703.1	PNAS	83:8447-8451(1986)
hdhA#	-	35.750	1	0.000	hdhAeco	ES3124	1785	3	1705.8	JBact	173:2173-2179(1991)
malI*		35.800	1	0.000	malIecoM	ES1168	6319	16	1705.8		
malI*	-	35.800	1	0.000	ECOMALAA	M60722	4202	9	1706.7	JBact	173:4862-4876(1991)
add	+	35.950	1	0.000	ECOADD	M59033	1200	5	1710.9	Bioch	30:2273-2280(1991)
nth	+	36.000	ND	ND	ECONTH	J02857	780	3	1720.7	Bioch	28:4444-4449(1989)
tyrS	-	36.050	11	ND	ECOTYRS	J01719	1275	2	1725.4	FEBS	150:419-423(1982)
sodB	+	36.350	6	ND	ECOSODB	J03511	970	2	1745.0	JBC	263:1555-1562(1988)
purR	+	36.450	13	ND	ECOPURRRP	X51368	2041	2	1747.3	EJB	187:373-379(1990)
valV*	+	36.600	ND	ND	valVeco	ES3111	300	0	1756.4	JMB	212:579-598(1990)
pykF*		36.550	ND	ND	pykFecoM	ES1100	2485	4	1764.6		
pykF*	+	36.550	ND	ND	ECOPK1	M24636	1830	2	1764.6	PNAS	86:6883-6887(1989)
lpp	+	36.800	ND	ND	ECOLPP	J01645	814	4	1766.3	Cell	18:1109-1117(1979)
aroD*	+	37.150	7	1.000	ECOAROD	X04306	1798	6	1783.6	BJ	238:475-483(1986)
ppsA		37.500	1	0.662	ppsAecoM	ES1174	5711	10	1792.5		
ppsA	-	37.500	1	0.662	ECOPEPSYN	M69116	3662	5	1792.5	D.Holzschu/A.Berry	
aroH	+	37.200	1	0.662	ECOAROHA	M38266	2054	5	1796.2	Gene	102:87-91(1991)
btuD		37.400	1	0.000	btuDecoM	ES1242	10766	20	1800.8		
btuD	-	37.400	1	0.000	ECOBTUCED	M14031	3169	5	1800.8	JBact	167:928-934(1986)
himA	-	37.450	1	0.000	thrSeco	ES3218	7783	15	1803.8	JBact	163:787-791(1985)
pfkB	+	37.750	16	1.000	ECOPFKBK	K02500	1249	2	1814.3	Gene	28:337-342(1984)
katE	+	37.850	2	0.614	ECOKATE	M55161	3466	9	1821.4	JBact	173:514-520(1991)
celF		37.900	1	0.000	celFecoM	ES1177	6706	7	1825.6		
celF	-	37.900	1	0.000	ECOCELA	M64438	4989	5	1825.6	Genet	124:455-471(1990)
anr#	-	37.950	1	0.000	ECANRG	X60186	520	2	1830.5	S.Kim/D.Wulff	
ntrL#	+	38.000	1	0.000	ntrLeco	ES3068	1379	1	1830.9	JBact	172:6619-6619(1990)
bnt106	?	38.200	ND	ND	BENT106	ES3169	392	0	1836.8	MGG	226:367-376(1991)
xthA	+	38.200	ND	ND	ECOXTHA	X13002	1246	3	1841.8	H.Wurst/F.Pohl	
gdhA	+	27.000	ND	ND	ECOGDHAK	K02499	1937	5	1852.7	Gene	27:193-199(1984)
topB#		38.700	1	0.000	topBecoM	ES1208	8138	15	1854.9		
topB#	-	38.700	1	0.000	ECOTOPB	J05076	2540	3	1854.9	JBC	264:17924-17930(1989)
selD#	-	38.725	1	0.000	ECOSELD	M30184	1202	2	1857.0	PNAS	87:543-547(1990)
selD#	-	38.725	1	0.000	ECOORF183	M68961	780	2	1858.2	JBact	173:4983-4993(1991)
sppA	+	38.500	1	0.000	ECOSPPA	M13359	2252	6	1858.7	JBC	261:9405-9411(1986)
ansA	+	38.900	1	0.000	ECOANSORA	M26934	2156	4	1860.9	Gene	78:37-46(1989)
gapA	+	39.300	ND	ND	ECOGAP	X02662	1523	1	1873.0	EJB	150:61-66(1985)

Table 1   *Continued*

Gene[a]	Ori[b]	Min[c]	Rnk[d]	*p* val[e]	Locus[f]	Access[g]	Bp[h]	Sites[i]	Addr[j]	Reference[k]	
rnd*	-	39.600	1	0.184	ECORND	X07055	1354	4	1897.8	NAR	16:6265-6278(1988)
pabB		39.900	ND	ND	pabBecoM	ES1213	4222	8	1905.5		
pabB	+	39.900	ND	ND	ECOPABB	K02673	1623	4	1905.5	JBact	159:57-62(1984)
sdaA#	+	41.000	ND	ND	ECOSDAA	M28695	2605	4	1907.1	JBact	171:5095-5102(1989)
manX	+	40.200	1	0.000	ECOPTSLPM	J02699	3188	15	1912.9	JBC	262:5238-5247(1987)
htpX#		40.400	1	0.000	htpXecoM	ES1190	3466	11	1923.4		
htpX#	-	40.400	1	0.000	ECOHTPX	M58470	1224	5	1923.4	JBact	173:2944-2953(1991)
prc#	-	40.425	1	0.000	ECOPRC	D00674	3178	9	1923.7	JBact	173:4799-4813(1991)
eda		40.700	1	0.000	eda-ecoM	ES1246	5194	11	1942.5		
eda	-	40.700	1	0.000	edd-eco	ES3237	3017	6	1942.5	T.Conway	
zwf	-	40.800	1	0.000	ECOZWF	M50055	2330	5	1945.3	JBact	173:968-977(1991)
pykA#		40.850	ND	ND	pykAecoM	ES1212	3694	8	1948.2		
pykA#	+	40.850	ND	ND	ECOPYKAA	M63703	1640	3	1948.2	S.Bledig/M.Hunter	
msbB#	-	40.900	ND	ND	ECOMSBBA	M77039	2589	6	1949.3	JBact	Inpress
ruvB		41.000	ND	ND	ruvBecoM	ES1204	8243	12	1956.2		
ruvA	-	41.000	ND	ND	ECORUVABA	M21298	2685	6	1956.2	JBact	170:4322-4329(1988)
ruvC#	-	41.150	ND	ND	ruvCeco	ES3158	3255	3	1958.5	JBact	173:5747-5753(1991)
aspS#	-	41.200	ND	ND	ECOASPS	X53863	3868	5	1960.6	NAR	18:7109-7118(1990)
argS	+	41.250	1	0.012	ECOARGS	X15320	2372	6	1972.4	NAR	17:5725-5736(1989)
cheZ		41.300	1	0.000	cheZecoM	ES1024	12141	25	1979.3		
cheZ	-	41.300	1	0.000	ECOCHE3	M13463	3063	9	1979.3	JBact	165:161-166(1986)
tap	-	41.475	1	0.000	ECOCHE2	J01705	3465	7	1982.4	Cell	33:615-622(1983)
cheW	-	41.600	1	0.000	ECOCHE1	M13462	1360	3	1985.7	JBact	165:161-166(1986)
cheA	-	41.650	1	0.000	ECOCHEA	M34669	2190	3	1986.2	JBact	173:2116-2119(1991)
motB	-	41.675	1	0.000	ECOMOTAB	J01652	2005	2	1988.3	JBact	166:244-252(1986)
flhC	-	41.675	1	0.000	ECOFLBA	M19439	1301	3	1990.2	JBact	170:1575-1581(1988)
is5G	+	41.800	ND	ND	IS5G	ES3202	1195	2	1991.7	Gene	14:165-174(1981)
araF		44.700	1	0.003	araFecoM	ES1008	4208	8	1996.2		
araF	-	44.700	1	0.003	ECOARAFGH	X06091	4200	7	1996.2	JMB	197:37-46(1987)
araF	-	44.700	1	0.003	araFeco	ES3002	507	2	1999.9	JMB	215:497-510(1990)
rsgA#	+	42.125	ND	ND	ECORSGA	X53513	823	1	2002.3	MGG	225:510-513(1991)
tyrP*		42.150	1	0.000	tyrPecoM	ES1133	7881	16	2003.3		
tyrP*	+	42.150	1	0.000	ECOTYRPA	M23240	1947	3	2003.3	JBact	170:4946-4949(1988)
leuZ	-	41.900	1	0.000	ECOTGGA	M14391	890	2	2005.2	Bchim	67:1053-1057(1985)
glyW	-	42.000	1	0.000	ECOGLYWA	M12299	955	1	2005.8	JBC	261:1329-1338(1986)
uvrC	-	42.100	1	0.000	ECOUVRC	X03691	4549	11	2006.6	NAR	14:2301-2318(1986)
fliC		42.525	3	0.966	fliCecoM	ES1047	1889	5	2016.0		
fliC	-	42.525	3	0.966	ECOHAG	M14358	1667	5	2016.0	JBact	168:1479-1483(1986)
fliC	-	42.525	3	0.966	ECOHAGFLG	J01607	351	0	2017.5	JBact	155:74-81(1983)
fliL		42.700	1	0.140	fliLecoM	ES1048	2211	9	2032.8		
fliL	+	42.700	1	0.140	ECOFLAA	M12784	1763	6	2032.8	JBact	166:1007-1012(1986)
fliN	+	42.700	1	0.140	M26294	M26294	499	3	2034.5	JBact	171:2728-2734(1989)
rcsA*	+	43.050	ND	ND	ECORCSA	M58003	934	1	2036.5	JBact	173:1738-1747(1991)
dcm	-	43.000	1	0.000	ECODCM	X13330	3655	12	2042.2	NAR	17:5844-5844(1989)
serU	-	43.300	ND	ND	serUeco	ES3099	287	1	2056.2	JMB	212:579-598(1990)
asnT*	+	43.350	ND	ND	asnTeco	ES3010	200	1	2058.1	JMB	212:579-598(1990)
amn*	+	43.150	1	0.086	ECOAMN	M30469	1803	5	2067.3	Bioch	28:8726-8733(1989)
asnU*	+	43.450	ND	ND	asnUeco	ES3008	200	0	2073.5	JMB	212:579-598(1990)
asnV*	+	43.500	ND	ND	asnVeco	ES3009	200	0	2075.4	JMB	212:579-598(1990)
srmB#	+	43.200	ND	ND	ECOSRMB	X14152	1973	8	2076.1	Nature	336:496-498(1988)
is5H	+	43.800	ND	ND	IS5H	ES3203	1195	2	2078.9	Gene	14:165-174(1981)
is2F	+	43.400	13	1.000	IS2F	ES3192	1331	3	2081.5	Gene	59:291-296(1987)
sbcB	+	43.925	2	0.283	ECOSBCB	J02641	1927	3	2094.4	JBC	262:455-459(1987)
hisG		44.000	1	0.012	hisGecoM	ES1222	7789	13	2101.5		
hisG	+	44.000	1	0.012	ECOHIS1	V00284	734	0	2101.5	NAR	9:2075-2086(1981)
hisG	+	44.000	1	0.012	ECOHISOPA	X13462	7390	13	2101.9	JMB	203:585-606(1988)
gnd		44.400	ND	ND	gnd-ecoM	ES1059	3196	8	2111.0		
gnd	-	44.400	ND	ND	ECOGND	K02072	1887	6	2111.0	Gene	27:253-264(1984)
gnd	-	44.400	ND	ND	ECOGNDF	M23181	1013	1	2112.1	MBE	5:691-703(1988)
is5I#	+	44.400	ND	ND	IS5I	ES3176	1195	2	2113.0	Gene	14:155-163(1981)

Table 1 *Continued*

Gene[a]	Ori[b]	Min[c]	Rnk[d]	p val[e]	Locus[f]	Access[g]	Bp[h]	Sites[i]	Addr[j]	Reference[k]	
alkA	-	44.800	ND	ND	ECOALKA	K02498	1954	4	2157.8	JBC	259:13730-13736(1984)
is3E	+	44.300	1	0.227	IS3E	ES3184	1258	4	2179.0	NAR	13:2127-2139(1985)
is5J	-	44.400	ND	ND	IS5J	ES3204	1195	2	2184.1	Gene	14:165-174(1981)
metG*	+	45.925	1	0.000	ECOMRPMET	X55791	3714	13	2203.6	MGG	223:121-133(1990)
dld*	+	46.700	1	0.252	ECODLD	X01067	2340	6	2232.6	EJB	144:367-373(1984)
cdd		46.100	ND	ND	cdd-ecoM	ES1023	1453	1	2241.6		
cdd	+	46.100	ND	ND	ECOCCDP	X16419	485	1	2241.6	MM	3:1385-1890(1989)
cdd	+	46.100	ND	ND	cdd-eco	ES3011	1291	1	2241.7	S.Short	
mglB		46.350	1	0.000	mglBecoM	ES1139	6243	9	2247.0		
mglB	-	46.400	1	0.000	ECOMGLABCO	M59444	4509	6	2247.0	MGG	229:453-459(1991)
galS#	-	46.450	1	0.000	galSeco	ES3112	1990	3	2251.2	M.Weickert/S.Adhya	
cirA	-	46.450	1	0.019	ECOCIR	J04229	2499	5	2255.9	JBact	171:1041-1047(1989)
nfo	+	46.550	4	1.000	ECONFO	M22591	1020	3	2262.2	JBact	170:5141-5145(1988)
fruA		46.950	1	0.020	fruAecoM	ES1163	3748	10	2270.3		
fruA	-	46.950	1	0.020	M23196	M23196	2600	6	2270.3	JGM	134:2757-2768(1988)
fruK	-	46.950	1	0.020	ECOFRUK	X53948	1491	5	2272.6	PRSL	242:87-90(1990)
rplY	+	48.000	2	0.119	rplYeco	ES3130	1116	7	2293.8	MGG	226:341-344(1991)
proL	+	47.200	ND	ND	proLeco	ES3078	300	0	2296.3	JMB	212:579-598(1990)
is5K	+	46.600	ND	ND	IS5K	ES3205	1195	2	2299.8	Gene	14:165-174(1981)
alkB		47.550	ND	ND	alkBecoM	ES1194	1997	3	2320.6		
alkB	+	47.500	ND	ND	ECOADAB	J02607	959	1	2320.6	JBC	261:15772-15777(1986)
ada	+	47.550	ND	ND	ECOADA	M10211	1324	4	2321.2	JBC	260:7281-7288(1985)
ada	+	47.550	ND	ND	ECOADAPA	M13155	267	3	2322.3	Cell	45:315-324(1986)
ompC	-	47.700	8	1.000	ECOOMPC	K00541	1713	5	2323.8	JBC	258:6932-6940(1983)
rcsC	-	48.200	6	0.999	ECORCSBC	M28242	5246	11	2327.4	JBact	172:659-669(1990)
is2G	+	47.500	9	0.959	IS2G	ES3193	-1331	3	2331.9	Gene	59:291-296(1987)
gyrA	-	48.300	1	0.002	ECOGYRAAM	Y00544	4729	5	2349.9	MM	1:259-273(1987)
nrdA		48.500	1	0.000	nrdAecoM	ES1162	15795	23	2355.9		
nrdA	+	48.500	1	0.000	ECONRDA	K02672	8554	13	2355.9	PNAS	81:4294-4297(1984)
nrdA	+	48.500	1	0.000	ECNRDAB1	X06999	2286	3	2359.4	NAR	16:4174-4174(1988)
glpQ	-	48.600	1	0.000	ECOGLPQ	X56907	1118	2	2364.5	MGG	226:321-327(1991)
glpT	-	48.600	1	0.000	ECOGLPT	Y00536	1560	1	2365.5	MM	1:251-258(1987)
glpA	+	48.600	1	0.000	ECOGLPA	M20938	4739	8	2367.0	JBact	170:2448-2456(1988)
menD	-	49.000	2	0.506	ECOMEND	M21787	2345	7	2392.1	JBact	171:4349-4354(1989)
ackA*	+	49.700	1	0.004	ECOACKA	M22956	1758	8	2429.2	JBact	171:577-580(1989)
hisP	-	49.850	14	1.000	ECOHISMP	Y00455	1332	2	2439.9	NAR	15:8568-8568(1987)
argT		49.875	1	0.000	argTecoM	ES1245	10721	28	2445.5		
purF	-	50.000	1	0.000	ECOHISPUR2	M68935	1440	5	2445.5	JBC	262:12209-12217(1987)
purF	-	50.000	1	0.000	ECOPURF1	X12423	1518	5	2446.9	NAR	16:8717-8717(1988)
argT	-	49.875	1	0.000	ECOHISPUR1	M68934	4732	7	2448.3	JBC	262:12209-12217(1987)
hisT	-	50.125	1	0.000	ECOHIST1	X02743	2323	9	2452.9	NAR	13:5297-5315(1985)
pdxB	-	50.200	1	0.000	ECOPDXB	M29962	1500	4	2454.7	JBact	171:6084-6092(1989)
fabB*	-	50.300	ND	ND	ECOFABB	M24427	1748	5	2458.2	CRC	53:357-370(1988)
mepA		50.400	1F	0.205	mepAecoM	ES1097	3230	8	2462.8		
mepA	-	50.400	1F	0.205	ECOMEPAMR	X16909	1765	4	2462.8	MM	4:209-219(1990)
aroC	-	50.500	1F	0.205	ECOAROCX	M33021	1690	6	2464.4	BJ	251:313-322(1988)
fadL	+	50.625	1	0.283	ECOFADLA	M60607	2197	6	2479.2	JBact	173:435-442(1991)
argW	+	50.900	ND	ND	argWeco	ES3005	300	1	2483.6	JMB	212:579-598(1990)
dsdC		51.000	ND	ND	dsdCecoM	ES1038	3155	2	2497.2		
dsdC	-	51.000	ND	ND	dsdCeco	ES3019	1466	2	2497.2	JBact	170:330-334(1988)
dsdA	+	51.000	ND	ND	ECODSDA	J01603	1989	0	2498.4	JBC	263:16926-16933(1988)
is186C	+	51.800	9	ND	IS186C	ES3187	1336	2	2532.1	FEBS	192:47-52(1985)
alaW		51.850	1	0.000	alaWecoM	ES1002	5152	18	2535.3		
alaW	-	51.850	1	0.000	alaWeco	ES3001	1862	7	2535.3	JMB	214:845-864(1990)
gltX	-	51.900	1	0.000	ECOGLTX	M13687	1514	3	2537.1	JBC	261:10610-10617(1986)
valU	-	51.950	1	0.000	valUeco	ES3110	1880	9	2538.6	JMB	214:845-864(1990)
lig	-	52.125	1	0.004	ECOLIG	M24278	3195	12	2545.5	M.O'Connor/K.Backman	
cysK		52.150	ND	ND	cysKecoM	ES1029	4859	11	2549.2		
cysK	+	52.150	ND	ND	ECOCYSK	X12615	1923	3	2549.2	MM	2:777-783(1988)
ptsH	+	52.225	ND	ND	ECOPHOSYS	M21994	3144	10	2550.9	JBact	170:3827-3837(1988)

Table 1   *Continued*

Gene[a]	Ori[b]	Min[c]	Rnk[d]	p val[e]	Locus[f]	Access[g]	Bp[h]	Sites[i]	Addr[j]	Reference[k]	
cysA	-	52.150	1	0.020	cysPeco	ES3014	5636	8	2555.8	JBact	172:3351-3357(1990)
bnt111	?	56.700	ND	ND	BENT111	ES3172	361	0	2570.1	MGG	226:367-376(1991)
dapE	+	53.075	1	0.133	ECODAPE	X57403	2270	4	2598.9	P.Stragier	
purC		53.200	1	0.000	purCecoM	ES1154	5082	14	2605.3		
purC	-	53.200	1	0.000	ECOPURCA	M33928	2060	8	2605.3	JBact	172:6035-6041(1990)
nlpB	-	53.225	1	0.000	ECONLB34	X57402	1366	5	2606.0	JBact	173:5523-5531(1991)
dapA	-	53.250	1	0.000	ECODAPA	M12844	1197	0	2607.3	JBact	166:297-300(1986)
bcp#	+	53.300	1	0.000	ECOORF123	M63654	1985	6	2608.4	JGM	137:361-367(1991)
uxaB*	+	52.400	ND	ND	ECOUXAB	M15737	286	1	2623.2	JGM	132:697-705(1986)
purM	+	53.800	1	0.015	ECOPURMN	M13747	2899	6	2628.4	JBC	262:10565-10569(1987)
guaA		53.900	1	0.000	guaAecoM	ES1060	4825	9	2638.8		
guaA	-	53.900	1	0.000	ECOGUABA	M10101	3531	7	2638.8	JBC	260:8676-8679(1985)
xseA	+	54.000	1	0.000	ECOXSEA	J02599	1616	2	2642.0	JBC	261:14929-14935(1986)
hisS		54.100	1	0.014	hisSecoM	ES1187	3239	6	2647.1		
hisS	-	54.100	1	0.014	hisSeco	ES3132	1679	3	2647.1	JBC	260:10063-10068(1985)
gcpE#	-	54.125	1	0.014	gcpEeco	ES3144	1697	5	2648.7	J.Parker	
ndk#	-	54.150	1	0.079	ndk-eco	ES3121	724	3	2652.5	Gene	105:31-36(1991)
suhB#	+	54.600	ND	ND	ECOSUHBA	M34828	1017	4	2672.0	JBact	172:2124-2130(1990)
glyA		54.800	2	0.396	glyAecoM	ES1173	3634	4	2693.5		
glyA	-	54.800	2	0.396	ECOGLYA	J01620	1902	2	2693.5	NAR	11:2065-2075(1983)
hmp#	-	55.000	2	0.396	hmp-eco	ES3129	2054	2	2695.1	MGG	226:49-58(1991)
purL	-	55.200	1	0.000	ECOPURLA	M19501	5865	13	2700.9	Bioch	28:2459-2471(1989)
pdxJ*		55.600	1	0.000	pdxJecoM	ES1198	6881	23	2710.2		
pdxJ*	-	55.600	1	0.000	ECORECOPDX	M76470	1444	6	2710.2	H.Takiff/D.Court	
recO	-	55.550	1	0.000	ECORECO	M27251	1590	5	2711.4	JBact	171:2581-2590(1989)
era#	-	55.375	1	0.000	ECOERA	M14658	1020	4	2712.1	PNAS	83:8849-8853(1986)
rnc	-	55.425	1	0.000	ECORNC1	X02673	1076	3	2712.9	NAR	13:4677-4685(1985)
lepB	-	55.400	1	0.000	ECOLEP	K00426	3131	12	2714.0	JBC	260:7206-7213(1985)
nadB	+	55.700	3	0.193	ECONADB	X12714	1724	5	2720.8	EJB	175:221-228(1988)
ung	+	56.000	ND	ND	ECOUNG	J03725	1532	2	2726.8	JBC	263:7776-7784(1988)
pssA	+	56.100	ND	ND	pssAeco	ES3136	1950	1	2732.7	JBC	266:5323-5332(1991)
kgtP#		56.150	1	0.000	kgtPecoM	ES1196	11651	16	2735.0		
kgtP#	-	56.150	1	0.000	kgtPeco	ES3120	1873	3	2735.0	PNAS	88:3802-3806(1991)
rrfG	-	56.200	1	0.000	ECRRFG	X52363	480	0	2736.5	NAR	18:3056-3056(1990)
rrlG	-	56.200	1	0.000	rrlGeco	ES3091	3000	2	2737.0	PNAS	77:201-204(1980)
gltW	-	56.200	1	0.000	M20397	M20397	431	0	2740.0	JBact	170:1235-1238(1988)
rrsG	-	56.200	1	0.000	rrsGeco	ES3097	1542	4	2740.4	PNAS	75:4801-4805(1978)
clpB#	-	56.300	1	0.000	ECOPROT	M29364	4248	6	2741.3	PNAS	87:3513-3517(1990)
clpB#	-	56.300	1	0.000	ECCLPB	X57620	3503	6	2743.2	JBact	173:4247-4253(1991)
pheA	+	56.650	ND	ND	ECOPHEAB	M10431	4509	5	2750.2	JMB	180:1023-1051(1984)
rplS	-	56.875	1	0.007	ECOTRMD	X01818	4586	12	2755.8	EMBO	2:899-905(1983)
grpE	-	57.100	ND	ND	ECOGRPE	X07863	1582	6	2761.6	NAR	16:7545-7562(1988)
recN		56.650	15	1.000	recNecoM	ES1169	2750	7	2763.6		
recN	+	56.650	15	1.000	ECORECN	Y00357	2224	6	2763.6	NAR	15:5041-5049(1987)
smpA#	+	56.700	15	1.000	smpAeco	ES3125	611	3	2765.7	JBact	173:3271-3272(1991)
ssrA#	+	57.500	ND	ND	ECOSSRA	X16382	648	2	2767.3	MM	3:1481-1485(1989)
gabT*	+	57.600	ND	ND	ECOGABT	M38417	1608	3	2810.8	JBact	172:7035-7042(1990)
proU	+	57.400	1	0.045	ECOPROU	M24856	4361	5	2822.1	JBact	171:1923-1931(1989)
mprA#	+	57.550	ND	ND	ECOMPRA	X54151	742	1	2828.6	JBact	173:3914-3929(1991)
gshA*	-	57.900	1	0.017	ECOGSHI	X03954	2098	4	2833.2	NAR	14:4393-4400(1986)
serV	-	58.150	4	0.993	serVeco	ES3100	1200	5	2836.1	JMB	212:579-598(1990)
alaS	-	58.200	2	0.315	ECOALAS	J01581	2767	6	2837.9	Science	213:1497-1501(1981)
recA	-	58.250	ND	ND	ECORECE	V00328	1391	4	2841.5	MGG	193:288-292(1984)
srlA		58.300	1	0.079	srlAecoM	ES1122	5454	4	2843.5		
srlA	+	58.300	1	0.079	ECOGUT	J02708	4770	2	2843.5	JBC	262:5455-5463(1987)
srlQ	+	58.300	1	0.079	ECOGUTQ	X51361	935	2	2848.0	DNAS	1:141-145(1990)
ascG#	-	58.600	1	0.029	ascGeco	ES3119	4286	8	2856.2	MBE	Inpress
hycH#		58.700	1	0.000	hycHecoM	ES1158	16907	30	2860.7		
hycH#	-	58.350	1	0.000	ECOHEVOP	X17506	7829	15	2860.7	MM	4:231-243(1990)
hyp#	+	58.400	1	0.000	ECOHYP	X54543	3793	7	2868.4	MM	5:123-135(1991)

Table 1 *Continued*

Gene[a]	Ori[b]	Min[c]	Rnk[d]	p val[e]	Locus[f]	Access[g]	Bp[h]	Sites[i]	Addr[j]	Reference[k]
fhlA*	+	58.450	1	0.000	ECOFHLA	M58504	2443	5	2872.0	JBact 172:4798-4806(1990)
mutS	+	58.750	1	0.000	ECOMUTS	M64730	3327	4	2874.3	V.Schlensog/A.Boeck
katF	-	58.900	ND	ND	ECOKATF	X16400	1483	1	2884.8	MM 5:49-59(1991)
pcm#	-	58.950	5	1.000	ECOPCM	M63493	1146	3	2887.1	JBC 266:14562-14572(1991)
cysC		59.250	1	0.000	cysCecoM	ES1209	5987	10	2891.5	
cysC	-	59.250	1	0.000	ECOCYSDNC	M74586	3821	7	2891.5	T.Leyh/T.Vogt
iap*	+	59.150	1	0.000	ECOIAP	M18270	1664	3	2894.7	JBact 169:5429-5433(1987)
repeats	+	59.175	1	0.000	M27059	M27059	1168	1	2896.3	JBact 171:3553-3556(1989)
cysH		59.350	24	1.000	cysHecoM	ES1155	5973	7	2905.5	
cysH	-	59.350	24	1.000	ECOCYSH	Y07525	2866	3	2905.5	MGG 225:314-319(1991)
cysH	-	59.350	24	1.000	ECOCYSJIHA	M23008	5720	5	2905.8	JBC 264:15796-15808(1989)
rpts	+	59.600	ND	ND	M27060	M27060	1531	3	2922.7	171:3553-3556(1989)
pyrG		59.700	1	0.000	pyrGecoM	ES1243	8647	14	2926.3	
pyrG	-	59.700	1	0.000	ECOPYRG	M12843	2442	5	2926.3	JBC 261:5568-5574(1986)
mazG	-	59.700	1	0.000	mazGeco	ES3057	1059	3	2928.8	G.Glaser/M.Cashel
relA	-	59.800	1	0.000	ECORELA	J04039	2858	6	2929.8	JBC 263:15699-15704(1988)
relA	-	59.800	1	0.000	relAeco	ES3220	2294	0	2932.7	G.Glaser/M.Cashel
fucO	-	60.325	1	0.002	ECOFUCOSE	X15025	8901	17	2950.4	NAR 17:4883-4884(1989)
metZ	?	60.250	ND	ND	ECOTGMZY	M21681	344	1	2960.8	Gene 67:49-57(1988)
argA		60.550	1	0.000	argAecoM	ES1010	16371	34	2967.2	
argA	+	60.550	1	0.000	ECOARGA	Y00492	1575	7	2967.2	NAR 15:10586-10586(1987)
recD	-	60.575	1	0.000	ECORECD	X04582	2160	3	2968.8	NAR 14:8583-8594(1986)
recB	-	60.625	1	0.000	ECORECB	X04581	3960	10	2970.6	NAR 14:8573-8582(1986)
ptr	-	60.725	1	0.000	ECOPTR	X06227	3120	5	2974.2	NAR 14:7695-7703(1986)
recC	-	60.825	1	0.000	ECORECC	X03966	6000	9	2977.2	NAR 14:4437-4451(1986)
thyA	-	60.875	1	0.000	ECOTHYA	J01710	1163	4	2982.4	PNAS 80:4914-4918(1983)
bnt107	?	61.100	ND	ND	BENT107	ES3170	316	1	2986.9	MGG 226:367-376(1991)
mutH	+	61.000	ND	ND	ECOMUTH	Y00113	790	0	2987.9	NAR 15:3073-3084(1987)
galR		61.275	1	0.000	galRecoM	ES1052	6097	12	2994.6	
galR	+	61.275	1	0.000	ECOGALLYS	J01614	4295	10	2994.6	JMB 168:321-331(1983)
araE	-	61.250	1	0.000	ECOARAEA	J03732	2866	3	2997.8	JBC 263:8003-8010(1988)
is2H	+	61.500	12	0.951	IS2H	ES3194	1331	3	3013.3	Gene 59:291-296(1987)
glyU	-	61.750	ND	ND	glyUeco	ES3038	300	0	3015.5	JMB 212:579-598(1990)
lysS		62.000	1	0.000	lysSecoM	ES1071	6279	11	3051.4	
lysS	-	62.000	1	0.000	ECOHERC	J03795	1832	4	3051.4	PNAS 85:5620-5624(1988)
prfB	-	61.725	1	0.000	ECORF2X	M11520	1441	3	3053.0	JBact 170:4537-4541(1988)
recJ*	-	62.450	1	0.000	ECORECJXPR	M54884	3772	6	3053.9	JBact 173:353-364(1991)
ssr*	+	62.850	ND	ND	ECOSSR	M12965	954	1	3072.8	JBact 161:1162-1170(1985)
serA		63.000	4	0.459	serAecoM	ES1214	1690	5	3075.1	
serA	-	63.000	4	0.459	ECOSERA	N00029	1233	5	3075.1	JBC 261:12179-12183(1986)
serA	-	63.000	4	0.459	ECOSERAP	M64630	545	0	3076.2	JBact 173:5944-5953(1991)
iciA#	+	63.050	ND	ND	ECOICIA	M62865	1815	0	3077.5	PNAS 88:406-4070(1991)
fba*	-	63.275	1	0.000	ECOFDAPGK	X14436	8029	23	3085.3	MM 3:723-732(1989)
speB		63.525	1	0.000	speBecoM	ES1119	6530	15	3100.0	
speB	-	63.525	1	0.000	ECOSPEAA	M32363	2458	5	3100.0	JBact 172:538-547(1990)
speA	-	63.525	1	0.000	ECOSPEA	M31770	3236	9	3102.2	JBact 172:4631-4640(1990)
metK	+	63.625	1	0.000	ECOMETK	K02129	1462	4	3105.1	JBC 259:14505-14507(1984)
gshB#	+	63.700	1	0.007	ECOGSHII	X01666	1477	6	3110.4	NAR 12:9299-9307(1984)
ansB#	+	63.800	1	0.005	ansBecoM	ES1004	1673	4	3118.4	
ansB#	-	63.800	1	0.005	ECOANSBA	M34234	1643	4	3118.4	Gene 91:101-105(1990)
ansB#	-	63.800	1	0.005	ECOLASNII	M34277	1530	3	3118.5	JBact 172:1491-8(1990)
mutY	+	63.925	1	0.002	ECOMICA	M59471	2293	6	3120.8	JBact 173:1902-1910(1991)
nupG		64.075	1	0.000	nupGecoM	ES1087	3740	12	3124.5	
nupG	+	64.075	1	0.000	ECONUPG	X06174	1486	8	3124.5	EJB 168:385-391(1987)
speC	-	64.100	1	0.000	ECOSPEC	M33766	2330	4	3125.9	S.Boyle
pheV*	+	64.000	ND	ND	ECOPHEV	X02480	487	2	3129.1	NAR 13:3699-3710(1985)
is5LO	-	64.400	ND	ND	IS5LO	ES3206	1195	2	3149.6	Gene 14:165-174(1981)
exbD		64.925	2	0.206	exbDecoM	ES1042	3212	10	3169.8	
exbD	-	64.925	2	0.206	ECOEXBBD	M28819	2195	9	3169.8	JBact 171:5117-5126(1989)
metC	+	65.000	2	0.206	ECOMETC	M12858	1880	8	3171.2	PNAS 83:867-871(1986)

Table 1 *Continued*

Gene[a]	Ori[b]	Min[c]	Rnk[d]	*p* val[e]	Locus[f]	Access[g]	Bp[h]	Sites[i]	Addr[j]	Reference[k]	
parC#	-	65.500	6	0.963	ECOPARC	M58408	2284	5	3183.4	Cell	63:393-404(1990)
parE#	-	66.000	2	0.075	ECOPARE	M58409	1870	6	3193.2	Cell	63:393-404(1990)
tolC		66.400	1	0.000	tolCecoM	ES1210	7806	16	3197.2		
tolC	+	66.400	1	0.000	ECOTOLCMP	X54049	2096	5	3197.2	NAR	18:5547-5547(1990)
htrP#	-	66.450	1	0.000	ECOLUXH	M77129	5058	11	3199.0	T.-P.Yang/R.Depew	
htrP#	-	66.450	1	0.000	ECOHTRP	M64472	1993	3	3203.0	JBact	173:5999-6008(1991)
is2I	-	66.500	15	1.000	IS2I	ES3195	1331	3	3205.6	Gene	59:291-296(1987)
cca	+	66.675	1	0.096	ECOCCA	M12788	2257	4	3221.4	JBC	261:6444-6449(1986)
rpsU		67.000	1	0.000	rpsUecoM	ES1111	8989	15	3226.1		
rpsU	+	67.000	1	0.000	ECORPSU	M16194	4653	7	3226.1	Gene	51:149-161(1987)
rpsU	+	67.000	1	0.000	ECORPSRPO	J01687	5059	8	3230.0	MGG	189:193-198(1983)
ileX	+	67.200	ND	ND	ileXeco	ES3049	200	0	3235.8	JMB	212:579-598(1990)
ebgR	+	67.425	1	0.001	ECOEBGRA	M64441	4983	11	3241.7	Genet	123:635-648(1989)
uxaC	-	67.900	ND	ND	ECOUXEX	M35280	318	1	3260.7	FEMS	33:205-209(1986)
tdcC		68.375	1	0.000	tdcCecoM	ES1125	10619	18	3282.2		
tdcC	-	68.375	1	0.000	ECTDCRAB	X14430	6295	9	3282.2	NAR	17:3994-3994(1989)
rnpB	-	70.000	1	0.000	ECORNPBW	D90212	4434	11	3288.4	JBact	173:1813-1819(1991)
sohA		68.500	2	0.268	sohAecoM	ES1118	1159	5	3295.7		
sohA#	+	68.500	2	0.268	ECOSOHA	M30178	828	5	3295.7	JBact	172:1587-1594(1990)
sohA#	+	68.500	2	0.268	ECOPRLF	M32358	927	3	3295.9	JBact	172:185-192(1990)
mtr		68.700	1	0.000	mtr-ecoM	ES1184	16052	45	3323.0		
mtr	-	68.700	1	0.000	ECOMTRA	M58338	2594	5	3323.0	JBact	173:4133-4143(1991)
mtr	-	68.700	1	0.000	ECOMTR	M59862	1867	1	3323.5	JBact	173:108-115(1991)
deaD#	-	68.775	1	0.000	ECODEAD	M63288	2982	7	3324.9	JBact	173:3291-3302(1991)
pnp	-	68.825	1	0.000	ECORPSOP	J02638	3030	10	3327.9	JBC	262:63-68(1987)
rpsO	-	68.875	1	0.000	ECOP15B35	X13270	1861	6	3330.6	NAR	16:10803-16(1988)
infB	-	68.900	1	0.000	nusAeco	ES3069	5426	18	3332.0	JBact	173:1485-1491(1991)
metY	-	68.975	1	0.000	ECOTGMETY	M28401	800	2	3337.2	JBact	172:2336-2342(1990)
argG	+	69.000	1	0.000	ECOARGGA	M35236	1344	4	3337.7	Gene	95:99-104(1990)
leuU		69.050	ND	ND	leuUecoM	ES1069	368	3	3341.7		
leuU	-	69.050	ND	ND	ECOTGL3A	M29082	300	3	3341.7	Gene	81:193-194(1989)
leuU	-	69.050	ND	ND	leuUeco	ES3053	300	3	3341.7	JMB	212:579-598(1990)
greA#		69.125	2	0.060	greAecoM	ES1156	3031	6	3346.9		
greA#	-	69.125	2	0.060	greAeco	ES3041	1745	5	3346.9	NAR	18:6443-6443(1990)
dacB	+	69.100	2	0.060	dacBeco	ES3118	1884	2	3348.0	FEMS	78:213-220(1991)
nlp#	+	69.300	ND	ND	nlp-eco	ES3066	1342	2	3353.5	JBact	171:5222-5225(1989)
rpoN*	+	70.100	2	0.044	ECOSIG54	M57612	1434	5	3363.5	Cell	62:945-954(1990)
arcB#	-	69.800	2	0.433	ECOARCB	X53315	2534	4	3369.1	MM	4:715-727(1990)
gltB	+	69.825	1	0.000	ECOGLTB	M18747	6292	22	3373.1	Gene	60:1-11(1987)
is5R	+	70.000	ND	ND	IS5R	ES3207	1195	2	3384.4	Gene	14:165-174(1981)
nanA	-	70.175	1	0.301	ECONANA	M20207	1243	5	3391.2	ABC	50:2155-2158(1986)
sspA*		69.950	ND	ND	sspBecoM	ES1200	2623	3	3395.5		
sspB#	-	69.925	ND	ND	sspBeco	ES3156	534	2	3395.5	Gene	Inpress
sspA*	-	69.950	ND	ND	ECOSSPG	X05088	1616	2	3395.6	NAR	15:1153-1163(1987)
rpsI*	-	70.275	ND	ND	ECORPSI	X02130	1184	2	3396.9	MGG	198:279-282(1985)
mdh*		70.400	1	0.005	mdh-ecoM	ES1075	2909	11	3401.7		
mdh*	-	70.400	1	0.005	M24777	M24777	2470	10	3401.7	AM	149:36-42(1987)
argR*	+	70.500	1	0.005	ECOARGR	M17532	806	2	3403.8	PNAS	84:6697-6701(1987)
mreD		70.950	1	0.000	mreDecoM	ES1082	3587	9	3417.8		
mreD	-	70.950	1	0.000	ECOMERBCD	M31792	1966	4	3417.8	JBact	171:6511-6516(1989)
mreB	-	70.950	1	0.000	ECOMREB	M22055	2105	5	3419.3	JBact	170:4619-4624(1988)
fabE		71.000	1	0.004	fabEecoM	ES1237	4160	12	3424.8		
fabE	-	71.000	1	0.004	ECOBENT18	X05966	353	0	3424.8	NAR	17:3982-3982(1989)
fabE	+	71.000	1	0.004	ECOFABEA	M32214	1229	5	3425.0	DNA	8:779-789(1989)
fabG#	+	71.025	1	0.004	ECOFABG	M79446	1750	6	3425.7	PNAS	88:9730-9733(1991)
panF	+	71.050	2	0.526	ECOPANF	M30953	1904	5	3427.0	JBact	172:3842-3848(1990)
fis*	+	71.600	ND	ND	ECOFISA	J03816	536	2	3430.6	PNAS	85:4237-4241(1988)
rrfD		71.900	1	0.000	rrnDecoM	ES1113	5841	8	3443.1		
rrfD	-	71.900	1	0.000	ECORGNDIS	J01693	667	0	3443.1	NAR	8:3793-3807(1980)
rrlD	-	71.900	1	0.000	rrlDeco	ES3089	2904	2	3443.7	PNAS	77:201-204(1980)

Table 1 *Continued*

Gene[a]	Ori[b]	Min[c]	Rnk[d]	p val[e]	Locus[f]	Access[g]	Bp[h]	Sites[i]	Addr[j]	Reference[k]
alaU	-	71.900	1	0.000	ECORGNDS2	J01702	706	1	3446.5	JBC 254:3264-3271(1979)
rrsD	-	71.900	1	0.000	rrsDeco	ES3095	1542	4	3447.0	PNAS 75:4801-4805(1978)
rrsD	-	71.900	1	0.000	ECORGNDS1	J01692	472	1	3448.5	Cell 17:225-234(1979)
aroE	?	72.000	ND	ND	ECOAROE1	Y00710	819	0	3452.9	BJ 249:319-326(1988)
aroE	?	72.000	ND	ND	ECOAROI	X00767	242	2	3454.5	NAR 12:5813-5821(1984)
trkA	+	72.150	23	1.000	ECOTRKAG	X52114	1788	4	3456.3	A.Hamann/E.Bakker.
rplQ		72.425	1	0.005	rplQecoM	ES1109	14392	42	3460.6	
rplQ	-	72.425	1	0.005	ECORPA	X02543	3154	13	3460.6	NAR 13:3891-3903(1985)
rpsM	-	72.500	1	0.005	ECORPLN	X01563	5922	14	3463.5	NAR 11:2599-2616(1983)
rpsQ	-	73.000	1	0.005	ECORPOS10	X02613	5422	15	3469.2	NAR 13:4521-4526(1985)
rpsJ	-	73.300	1	0.005	ECORPSJ	J01680	1241	0	3473.7	Cell 26:205-211(1981)
bfr#	-	73.300	ND	ND	ECOBFR	M27176	1350	3	3487.1	JBact 171:3940-3947(1989)
tufA	-	73.325	ND	ND	ECOSTR4	J01691	200	0	3490.6	Gene 12:25-31(1980)
tufA		73.325	1	0.000	tufAecoM	ES1191	4757	17	3490.9	
tufA	-	73.325	1	0.000	ECOSTR3	J01690	1374	3	3490.9	Gene 12:25-31(1980)
fusA	-	73.350	1	0.000	ECOSTRA	X00415	2076	9	3492.3	NAR 12:2181-2192(1984)
fusA	-	73.350	1	0.000	ECOSTR2	J01689	405	0	3494.0	JBC 255:4660-4666(1980)
rpsG	-	73.375	1	0.000	rpsGeco	ES3149	547	0	3494.3	Bchim 61:501-522(1979)
rpsL	-	73.400	1	0.000	ECOSTR1	J01688	1016	3	3494.6	JBC 255:4660-4666(1980)
crp		73.500	1	0.035	crp-ecoM	ES1236	1709	4	3505.8	
crp	+	73.500	1	0.035	ECOCRPDDNA	X61919	783	2	3505.8	R.Bhasin/M.Freundlich
crp	+	73.500	1	0.035	ECOCRP	J01598	1127	2	3506.4	NAR 10:1363-1378(1982)
argD*		74.150	2	0.129	argDecoM	ES1012	3923	10	3509.7	
argD*	-	73.800	2	0.129	ECOARGD	M32796	1221	3	3509.7	Gene 90:69-78(1990)
pabA*	-	74.200	2	0.129	ECOPABAA	M32354	2059	7	3510.9	JBact 172:397-410(1990)
fic*	-	74.150	2	0.129	ECOFIC1	M28363	2496	7	3511.2	JBact 171:4525-4529(1989)
nirB	+	73.950	1	0.082	ECNIRBC	X14202	5618	11	3514.5	EJB 191:315-323(1990)
trpS		74.300	ND	ND	trpSecoM	ES1223	1132	3	3534.1	
trpS	-	74.300	ND	ND	ECOTRPS	J01716	1005	2	3534.1	JBact 148:941-949(1981)
trpS	-	74.300	ND	ND	trpSeco	ES3164	127	0	3535.1	JBact 148:941-949(1981)
dam		74.350	1	0.000	dam-ecoM	ES1197	5959	12	3536.4	
dam	-	74.350	1	0.000	ECODAM	J01600	1134	4	3536.4	NAR 11:837-851(1983)
dam	-	74.350	1	0.000	ECOURF743	X15162	1560	6	3537.3	MGG 217:85-96(1989)
aroB	-	74.600	1	0.000	ECOAROB	X03867	1644	2	3538.7	FEBS 200:11-17(1986)
aroK#	-	74.625	1	0.000	ECOAROK	M76389	2225	3	3540.1	A.Lobner-Olesen/M.Marinus
mrcA*	+	74.700	4	0.815	ECOPONA	X02164	2764	8	3544.5	EJB 147:437-446(1985)
pckA*		75.100	1	0.004	pckAecoM	ES1089	4289	8	3554.5	
pckA*	+	75.100	1	0.004	pckAeco	ES3073	1895	6	3554.5	JBact 172:7151-7156(1990)
ompR	-	74.950	1	0.004	ECOOMPB	J01656	2703	3	3556.0	MGG 202:194-199(1986)
bioH	-	75.000	1	0.029	bioHeco	ES3007	2537	4	3564.5	NAR 17:8004-8004(1989)
malQ		75.325	1	0.000	malQecoM	ES1072	8703	22	3569.9	
malQ	-	75.325	1	0.000	ECOMALQP	M32793	2866	9	3569.9	MM 2:473-479(1988)
malP	-	75.325	1	0.000	ECOMALP	X02003	2600	9	3572.1	Nature 313:500-502(1985)
malP	-	75.325	1	0.000	ECOMALAR	J01647	824	0	3574.5	JMB 163:395-408(1983)
malT	+	75.325	1	0.000	ECOMALT	M13585	3508	8	3575.1	Gene 42:201-208(1986)
glpR		75.400	1	0.000	glpRecoM	ES1056	15820	40	3581.7	
glpR	-	75.400	1	0.000	ECOGLPREG	X07520	2913	5	3581.7	NAR 16:7732-7732(1988)
glpD	+	75.400	1	0.000	ECOSNGLPD	M55989	2812	4	3584.3	JBact 173:101-107(1991)
glgA	-	75.550	1	0.000	ECOGLGPA	J03966	2580	6	3586.5	JBC 263:13706-13711(1988)
glgA	-	75.575	1	0.000	ECOGLGA	J02616	1601	4	3589.0	JBC 261:16256-16259(1986)
glgC	-	75.600	1	0.000	ECOGLG	J01616	3354	12	3590.4	Gene 70:363-376(1988)
glgB	-	75.625	1	0.000	ECOGLGBA	M13751	2559	5	3593.6	JBC 261:8738-8743(1986)
asd	-	75.700	1	0.000	ECOASD	V00262	1674	6	3595.9	JBact 169:386-393(1987)
is1E	+	75.600	ND	ND	ECOIS1E	X52537	920	2	3605.7	Gene 98:1-5(1991)
ggt*		76.125	1	0.000	ggt-ecoM	ES1160	20574	37	3607.1	
ggt*	-	76.125	1	0.022	M28722	M28722	2148	6	3607.1	JBact 171:5169-5172(1989)
ugpQ#	-	76.000	1	0.000	ECOUGPQQ	X56908	1222	1	3609.2	MGG 226:321-327(1991)
ugpC	-	76.000	1	0.000	ECOUGP	X13141	4717	6	3610.3	MM 2:767-775(1988)
livG	-	76.500	1	0.000	ECOLIVHMGF	J05516	8703	15	3614.7	JBC 265:11436-11443(1990)
ftsX	-	76.500	1	0.000	ECOFTSYEX	X04398	4480	11	3623.2	MGG 205:134-145(1986)

Table 1   *Continued*

Gene[a]	Ori[b]	Min[c]	Rnk[d]	*p* val[e]	Locus[f]	Access[g]	Bp[h]	Sites[i]	Addr[j]	Reference[k]	
rhsB*		77.400	2	0.012	rhsBecoM	ES1182	4336	15	3641.9		
rhsB*	+	77.400	1	0.012	ECORHSB	M21762	400	0	3641.9	JBact	171:636-642(1989)
rhsB*	+	77.400	2	0.012	rhsBeco	ES3141	4134	12	3642.0	JBact	172:446-456(1990)
rhsB*	+	77.400	1	0.012	ECORHSBA	M29717	577	4	3645.6	JBact	172:446-456(1990)
gor	?	77.250	ND	ND	ECOGOR	M13141	1500	1	3669.2	Bioch	25:2736-2742(1986)
is5T	+	77.500	ND	ND	IS5T	ES3208	1195	2	3675.5	Gene	14:165-174(1981)
dppA#	-	79.500	1	0.000	ECODPPA	M35045	1950	7	3728.5	JBact	173:234-244(1990)
proK*	-	79.800	ND	ND	ECOTGPRO	X02434	1085	4	3733.5	NAR	13:3213-3220(1985)
tag*		72.100	1	0.024	tag-ecoM	ES1124	3894	6	3735.1		
tag*	+	72.100	1	0.024	ECOTAG	J02606	869	1	3735.1	JBC	261:15761-15766(1986)
bisC*	-	79.900	1	0.024	ECOBISCASD	M34827	3337	5	3735.6	JBact	172:2194-2198(1990)
cspA#		80.000	ND	ND	cspAecoM	ES1228	2637	5	3741.8		
cspA#	+	80.000	ND	ND	ECOCSPAA	M30139	1205	3	3741.8	PNAS	87:283-287(1990)
is150	+	79.400	ND	ND	INS150CG	X07037	1443	2	3743.0	NAR	16:6789-6802(1988)
glyS	-	80.050	3	1.000	ECOGLYS	J01622	3333	4	3744.5	JBC	258:10637-10641(1983)
xylA	-	80.100	1	0.089	ECOXYLABA	X00772	4176	8	3750.2	AEM	47:15-21(1984)
avtA*	-	84.500	ND	ND	ECOAVT	Y00490	1752	5	3763.0	NAR	15:9461-9469(1987)
selB		80.450	1	0.000	selBecoM	ES1192	3385	4	3781.3		
selB	-	80.450	1	0.627	ECOSELB	X16644	2000	3	3781.3	Nature	342:453-456(1989)
selA	-	80.475	1	0.627	ECOSELA	M64177	1451	1	3783.2	JBC	266:6318-6323(1991)
rhsA*	+	80.550	1	0.000	ECORHSA	M29716	9234	28	3784.8	JBact	172:446-456(1990)
mtlA		80.650	1	0.005	mtlAecoM	ES1083	3800	6	3795.4		
mtlA	+	80.650	1	0.005	ECMTLOP1	X51360	299	1	3795.4	MM	4:2003-2006(1990)
mtlA	+	80.675	1	0.005	ECOMTLA	K00051	2162	2	3795.6	JBC	258:10761-10767(1983)
mtlD	+	80.650	1	0.005	ECOMTLD1	X51359	1436	3	3797.8	MM	4:2003-2006(1990)
cysE	-	80.950	ND	ND	ECOCYSXE	M34333	1396	2	3805.1	BBRC	167:948-955(1990)
secB	-	81.100	ND	ND	ECOSECB	M24489	593	1	3807.1	Gene	75:167-175(1989)
tdh		81.150	1	0.000	tdh-ecoM	ES1231	8815	7	3813.2		
tdh	-	81.150	1	0.000	ECOKBLTDH	X06690	3563	3	3813.2	NAR	16:3586-3586(1988)
rfaD	+	81.300	1	0.000	rfaDeco	ES3084	2040	2	3816.8	JBact	172:4652-4660(1990)
rfaD	+	81.300	1	0.000	ECOHTRMG	X54492	1654	1	3817.5	NAR	19:3811-3819(1991)
rfaD	+	81.300	1	0.000	ECORFA2	X62530	3219	3	3818.8	L.Chen/W.Coleman	
kdtA#		81.750	1	0.002	kdtAecoM	ES1147	4147	9	3830.8		
kdtA#	+	81.750	1	0.002	ECOKDTA	M60670	2864	8	3830.8	JBC	266:9687-9696(1991)
fpg#	-	81.775	1	0.002	ECOFPG	X06036	1093	3	3833.2	EMBO	6:3177-3183(1987)
rpmG	-	81.800	1	0.002	ECORPMBG	J01677	764	0	3834.2	MGG	184:218-223(1981)
dut		81.900	1	0.001	dut-ecoM	ES1039	5350	9	3836.7		
dut	+	81.900	1	0.001	ECODUTPYR	V01578	2568	6	3836.7	EMBO	2:967-971(1983)
pyrE	-	81.950	1	0.001	ECPYRE	X00781	1872	3	3838.2	EMBO	3:1783-1790(1984)
pyrE	-	81.950	1	0.001	ECORFPYRE	X14235	2375	3	3839.7	MM	3:393-404(1989)
spoT		82.050	1	0.000	spoTecoM	ES1232	10669	18	3843.8		
spoR#	+	82.000	1	0.000	spoReco	ES3104	1531	3	3843.8	C.Bengra/M.Cashel	
spoT	+	82.050	1	0.000	ECOSPOT	M24503	3171	6	3845.3	JBC	264:15074-82(1989)
recG#	+	82.300	1	0.000	ECORCG	X59550	3041	4	3848.1	JBact	173:6837-6843(1991)
gltS	-	82.400	1	0.000	ECOGLTS	X17499	1881	3	3850.4	MGG	225:379-386(1991)
gltS	-	82.400	1	0.000	gltSeco	ES3127	2225	5	3852.2	M.Kalman/M.Cashel	
selC*	+	82.300	ND	ND	selCeco	ES3160	1445	2	3859.4	JBact	173:4171-4181(1991)
nlpA#	-	77.000	ND	ND	nlpAeco	ES3175	1684	3	3862.2	JBC	261:2284-2288(1986)
uhpT		82.500	1	0.104	uhpTecoM	ES1135	7651	14	3869.7		
uhpT	-	82.500	1	0.104	ECOUHP	M17102	5403	9	3869.7	JBact	169:3556-3563(1987)
ilvN	-	82.600	1	0.104	ECOILVBPR	J01633	2470	6	3874.9	NAR	13:3995-4010(1985)
gyrB		83.450	1	0.000	gyrBecoM	ES1230	15535	32	3900.8		
gyrB	-	83.450	1	0.000	ECORECFA	X04341	4931	6	3900.8	NAR	15:771-784(1987)
dnaN	-	83.400	1	0.000	ECODNAAOP	J01602	3873	9	3905.2	Gene	28:159-170(1984)
rpmH	+	83.550	1	0.000	ECORNPA	M11056	1069	3	3908.4	Gene	38:85-93(1985)
rnpA	+	83.550	1	0.000	rnpAeco	ES3214	714	3	3909.5	Gene	93:27-34(1990)
tdhF#	+	83.600	1	0.015	thdFeco	ES3163	3777	12	3909.9	JBact	173:6018-6024(1991)
tnaA	+	83.700	1	0.015	ECOTNAA	K00032	2083	2	3912.7	JBact	147:787-796(1981)
is5U	+	83.000	ND	ND	IS5U	ES3209	-1195	2	3912.9	Gene	14:165-174(1981)
tnaB*	+	83.725	1	0.015	ECOTNAB	M59914	1568	3	3914.8	JBact	173:3231-3234(1991)

Table 1 *Continued*

Gene[a]	Ori[b]	Min[c]	Rnk[d]	p val[e]	Locus[f]	Access[g]	Bp[h]	Sites[i]	Addr[j]	Reference[k]	
is5V	+	83.100	ND	ND	IS5V	ES3210	1195	2	3916.4	Gene	14:165-174(1981)
is2J	+	83.200	18	1.000	IS2J	ES3196	1331	3	3918.8	Gene	59:291-296(1987)
bglB		83.900	1	0.000	bglBecoM	ES1019	27175	58	3929.2		
bglB	-	83.900	1	0.000	ECOBGLO	M16487	5270	8	3929.2	JBact	169:2579-2590(1987)
phoU	-	84.050	1	0.000	ECOPHOS	K01992	5032	5	3934.1	JBact	161:189-198(1985)
glmS	-	84.150	1	0.000	ECOUNCC	X01631	14526	36	3938.9	BJ	224:799-815(1984)
asnA	+	84.300	1	0.000	ECOORIASN	K00826	4012	14	3952.4	Gene	24:265-279(1983)
rbsD		84.550	1	0.001	rbsDecoM	ES1104	6197	12	3960.4		
rbsD	+	84.550	1	0.001	ECORBS	M13169	5820	11	3960.4	JBC	261:7652-7658(1986)
rbsR	+	84.550	1	0.001	rbsReco	ES3081	1148	5	3965.5	C.Mauzy/M.Hermodson	
rrsC		84.650	1	0.000	rrnCecoM	ES1227	18507	45	3969.1		
rrsC	+	84.650	1	0.000	ECORGNC	M10739	682	0	3969.7	PNAS	82:5073-5077(1985)
rrsC	+	84.650	1	0.000	rrsCeco	ES3094	1542	4	3969.7	PNAS	75:4801-4805(1978)
gltU	+	84.650	1	0.000	ECO16S23S	X12420	535	0	3971.1	E.Morgan	
rrlC	+	84.650	1	0.000	rrlCeco	ES3088	2904	2	3971.6	PNAS	77:201-204(1980)
rrfC	+	84.650	1	0.000	ECORGNX3	J01696	725	1	3974.4	Cell	19:393-401(1980)
rrfC	+	84.650	1	0.000	ECORRNILV	M37337	4900	12	3974.8	Gene	97:21-27(1991)
ilvG	+	84.900	1	0.000	ECOILVGMED	M32253	7203	20	3977.7	NAR	15:2137-2155(1987)
ilvG	+	84.900	1	0.000	ECOILVGE	M10313	9456	32	3978.1	Gene	56:185-198(1987)
rep		85.025	1	0.000	rep-ecoM	ES1106	9287	20	3987.9		
rep	+	85.025	1	0.000	ECOREPHEL	X04794	2671	3	3987.9	NAR	15:465-475(1987)
gppA*	-	85.500	1	0.000	gppAeco	ES3174	1329	1	3990.6	M.Kalman/M.Cashel	
rhlB#	-	85.100	1	0.000	ECORHLB	X56310	1536	4	3991.9	NB	3:886-895(1991)
trxA	+	85.125	1	0.000	ECORHOB	K02845	842	0	3993.2	Gene	32:399-408(1984)
rho	+	85.175	1	0.000	ECORHO	J01673	1880	6	3993.6	NAR	11:3531-3545(1983)
rfe*	+	85.350	1	0.000	rfe-eco	ES3159	1939	8	3995.3	MM	5:1853-1862(1991)
argX		85.050	1	0.000	argXecoM	ES1016	11953	25	4010.4		
argX	+	85.050	1	0.000	ECOTGRHLP	K01994	646	1	4010.4	JBact	158:934-942(1984)
aslB	+	85.050	1	0.000	aslBeco	ES3212	3734	3	4011.0	H.Murphy/M.Cashel	
hemC	-	85.775	1	0.000	ECOHEMCD	X12614	4260	14	4014.7	NAR	16:9871-9871(1988)
cyaA	+	85.700	1	0.000	ECOCYAG	K02969	3699	12	4018.6	JBC	260:3063-3070(1985)
dapF		85.725	1	0.000	dapFecoM	ES1034	6000	14	4022.7		
dapF	+	85.725	1	0.000	ECODAPF	X12968	1308	5	4022.7	NAR	16:10367-10367(1988)
xerC#	+	85.800	1	0.000	ECOXERC	M38257	2500	5	4023.8	JBact	172:6973-6980(1990)
uvrD	+	85.900	1	0.000	ECOUVRD	X00738	2869	5	4025.8	NAR	12:5789-5799(1984)
pldA		86.100	1	0.001	pldAecoM	ES1093	3504	10	4033.1		
pldA	+	86.100	1	0.001	ECOPLDAA	X02143	1319	5	4033.1	Bchim	96:1655-1664(1984)
recQ	+	86.150	1	0.001	ECORECQ	M30198	2695	7	4033.9	MGG	205:298-304(1986)
pldB	+	86.200	ND	ND	ECOPLDB	X03155	1576	2	4037.3	JBioch	98:1017-1025(1985)
metR		86.300	ND	ND	metRecoM	ES1235	1275	2	4042.8		
metR	-	86.300	ND	ND	ECOMETR	M37630	1013	2	4042.8	PNAS	87:7076-7079(1990)
metE	+	86.300	ND	ND	ECOMETER	J04155	360	0	4043.7	PNAS	86:85-89(1989)
udp		86.350	1	0.000	udp-ecoM	ES1225	24328	44	4045.6		
udp	+	86.350	1	0.000	ECOUDP	X15689	2479	6	4045.6	NAR	17:6741-6741(1989)
rfaH	-	86.900	1	0.000	rfaHeco	ES3165	7661	14	4048.0	D.Daniels/F.Blattner	
fre#	+	86.950	1	0.000	ECOFLRDA	M61182	894	3	4055.7	JBact	173:3673-3679(1991)
fadA*	-	87.000	1	0.000	ECOFADAB	M59368	4104	9	4056.5	JBact	72:6459-6468(1990)
pepQ#	+	87.100	1	0.000	pepQeco	ES3135	4150	7	4060.4	NAR	18:6439-6439(1990)
rrsA	+	87.300	1	0.000	ECORGNA	J01694	463	1	4064.4	Cell	17:201-209(1979)
rrsA	+	87.300	1	0.000	rrsAeco	ES3093	1542	4	4064.8	PNAS	75:4801-4805(1978)
ileT	+	87.300	1	0.000	ECORGNDS2	J01702	706	1	4066.2	JBC	254:3264-3271(1979)
rrlA	+	87.300	1	0.000	rrlAeco	ES3087	2904	2	4066.8	PNAS	77:201-204(1980)
rrlA	+	87.300	1	0.000	ECORGNDS3	K00766	247	0	4069.6	Cell	19:393-401(1980)
rrfA	+	87.320	1	0.000	ECORRAA	K00609	120	0	4069.8	JMB	3:379-412(1968)
polA	+	87.400	1	0.110	ECOPOLA	J01663	4127	6	4075.9	JBC	257:1958-1964(1982)
glnA		87.600	1	0.696	glnAecoM	ES1053	4969	11	4083.2		
glnA	-	87.600	1	0.696	ECOGLN	X05173	4311	11	4083.2	NAR	15:2757-2770(1987)
glnA	-	87.600	1	0.696	ECOGLNACR	J01618	813	1	4087.3	JBact	164:1032-1038(1985)
fdhE#	-	88.000	ND	ND	ECOFDHE	X16016	1200	2	4110.1	Gene	97:147-148(1991)
rhaB		88.300	10	1.000	rhaBecoM	ES1107	4260	5	4125.7		

Table 1 *Continued*

Gene[a]	Ori[b]	Min[c]	Rnk[d]	*p* val[e]	Locus[f]	Access[g]	Bp[h]	Sites[i]	Addr[j]	Reference[k]	
rhaB	-	88.300	10	1.000	rhaBeco	ES3085	2065	3	4125.7	J.Tobin/R.Schleif	
rhaS	+	88.300	10	1.000	ECORHAC	X06058	2201	3	4127.8	JMB	196:789-799(1987)
sodA	+	88.450	ND	ND	ECOSOD	X03951	1053	3	4130.7	NAR	14:4577-4589(1986)
cpxA	-	88.500	16	1.000	ECOCPXA	M36795	1841	3	4133.2	JMB	203:467-478(1988)
bnt5	-	88.550	1	0.000	ECOBENT5	X05960	188	0	4136.6	NAR	15:6827-6841(1987)
pfkA		88.550	1	0.000	pfkAecoM	ES1143	4425	11	4136.6		
pfkA	+	88.550	1	0.000	ECOCDHA	X02519	3304	8	4136.8	EJB	149:363-373(1985)
tpiA	-	88.700	1	0.000	ECOTPIA	X00617	1338	5	4139.7	MGG	195:314-320(1984)
glpK		88.750	1	0.000	glpKecoM	ES1055	2781	9	4144.8		
glpK	-	88.750	1	0.000	ECOGLYK	M18393	2028	4	4144.5	JBC	263:135-139(1988)
glpF	-	88.750	1	0.000	ECOGLPF	X15054	1170	6	4146.4	NAR	17:4378-4378(1989)
cytR		88.950	1	0.168	cytRecoM	ES1030	3843	6	4152.7		
cytR	-	88.950	1	0.168	ECOCYTR	X03683	1384	6	4152.7	NAR	14:2215-2228(1986)
priA#	-	88.975	1	0.168	ECOPRIA	M33293	2658	3	4153.6	PNAS	87:4620-4624(1990)
rpmE*	+	89.100	1	0.168	ECOPRIAY	D00616	2907	2	4153.7	PNAS	87:4615-4619(1990)
metJ		89.000	1	0.001	metJecoM	ES1098	8542	13	4157.1		
metJ	-	89.000	1	0.001	ECOMETJA	M12869	729	0	4157.1	JBC	259:14282-14285(1984)
metB	+	89.000	1	0.001	ECOMETLB1	K01546	1411	1	4157.6	JBC	258:14868-14871(1983)
metL	+	89.000	1	0.001	ECOMETL	J01651	2433	3	4159.0	JBC	258:3028-3031(1983)
metL	+	89.000	1	0.001	ECOMETLB2	K01547	240	0	4161.4	JBC	258:14868-14871(1983)
metF	+	89.000	1	0.001	ECOMETF	V01502	1238	3	4161.7	NAR	11:6723-6732(1983)
katG	+	89.150	1	0.001	ECOKATGA	M21516	2805	7	4162.9	JBact	170:4415-4419(1988)
ppc	-	89.450	1	0.000	ECOPPCG	X05903	3106	10	4179.6	Bchim	95:909-916(1984)
argC		89.500	1	0.122	argCecoM	ES1011	2201	6	4183.9		
argC	+	89.500	1	0.122	ECOARGOP1	J01587	344	0	4183.9	NAR	NAR
argC	+	89.500	1	0.122	ECOARGBCH	M21446	2117	6	4184.0	Gene	68:275-283(1988)
oxyR#		89.525	1	0.000	oxyRecoM	ES1145	32329	87	4187.5		
oxyR#	+	89.525	1	0.000	ECOOXYS	X16531	1471	6	4187.5	MGG	218:371-376(1989)
oxyR#	+	89.525	1	0.000	mor-eco	ES3061	811	5	4189.0	S.Warne	
trmA	-	89.550	1	0.000	trmAeco	ES3115	1387	3	4189.8	C.Gustafsson/G.Bjork	
trmA	-	89.550	1	0.000	ECOTRMA	M57568	1684	4	4191.2	JBact	173:1757-1764(1991)
btuB	+	89.600	1	0.000	ECOBTUB	M10112	2220	7	4192.5	JBact	161:904-908(1985)
rrsB	+	89.700	1	0.000	ECORGNB	J01695	7508	10	4194.4	PNAS	148:107-127(1981)
birA	+	89.750	1	0.000	ECOBIRA	M10123	2491	7	4201.0	Gene	35:321-331(1985)
rts	-	89.800	1	0.000	ECORTSA	M36321	1407	1	4203.2	Gene	125:261-273(1990)
tufB	+	89.875	1	0.000	ECOTGTUFB	J01717	1973	3	4204.4	Gene	12:33-39(1980)
secE#	+	89.900	1	0.000	ECOSECE	M30610	1380	5	4206.3	JBact	172:1621-1627(1990)
rplK	+	89.900	1	0.000	ECORPLRPO	J01678	12337	46	4207.5	MGG	190:344-348(1983)
hupA#	+	90.100	ND	ND	hupAeco	ES3046	584	1	4229.9	JMB	213:27-36(1990)
hydG#		90.400	1	0.000	hydGecoM	ES1148	24697	42	4233.1		
hydG#	+	90.400	1	0.000	hydGeco	ES3047	1745	4	4233.1	JBact	171:4448-4456(1989)
purD	-	90.425	1	0.000	ECOPURHD	J05126	3535	10	4234.5	JBC	264:21239-21246(1989)
rrsE	+	90.500	1	0.000	ECORGNE	J01697	410	0	4238.0	Cell	17:201-209(1979)
rrsE	+	90.500	1	0.000	rrsEeco	ES3096	1542	4	4238.3	PNAS	75:4801-4805(1978)
gltV	+	90.500	1	0.000	gltVeco	ES3037	354	0	4239.9	JBact	170:1235-1238(1988)
rrlE	+	90.500	1	0.000	rrlEeco	ES3090	2904	2	4240.2	PNAS	77:201-204(1980)
rrlE	+	90.500	1	0.000	ECORGNDS3	K00766	247	0	4243.0	Cell	19:393-401(1980)
rrfE	+	90.500	1	0.000	ECORRNE	X02800	1444	0	4243.2	NAR	13:5515-5525(1985)
metA	+	90.700	1	0.000	ECOMETAG	X14501	973	1	4244.4	NAR	17:2856-2856(1989)
aceB	+	90.850	1	0.000	ECACEB	X12431	3720	7	4245.1	NAR	16:10924-10924(1988)
aceK	+	90.850	1	0.000	ECOIDHKPA	M20714	2358	2	4248.2	JBact	170:2763-2769(1988)
aceK	+	90.850	1	0.000	ECOINTER	M63497	2509	7	4250.4	Gene	97:149-150(1991)
iclR	-	90.950	1	0.000	ECOICLR	M31761	1166	0	4252.8	JBact	172:2642-2649(1990)
metH	+	91.075	1	0.000	ECOMTHM	X16584	4098	8	4253.7	Gene	87:15-21(1990)
lysC		91.250	1	0.000	lysCecoM	ES1070	4157	15	4261.2		
lysC	-	91.250	1	0.000	ECOLYSC	M11812	1587	11	4261.2	JBC	261:1052-1057(1986)
pgi	+	91.300	1	0.000	ECOPGI	X15196	2573	5	4262.8	MGG	217:126-131(1989)
xylE		91.400	1	0.000	xylEecoM	ES1241	10633	13	4269.1		
xylE	-	91.400	1	0.000	ECOXYLE	J02812	2842	4	4269.0	JBC	262:13928-13932(1987)
malG	-	91.500	1	0.000	ECOMALG	X02871	1100	0	4271.7	EMBO	4:2287-2293(1985)

Table 1 *Continued*

Gene[a]	Ori[b]	Min[c]	Rnk[d]	p val[e]	Locus[f]	Access[g]	Bp[h]	Sites[i]	Addr[j]	Reference[k]	
malG	-	91.500	1	0.000	malGeco	ES3217	6548	7	4272.7	JBC	259:10606-10613(1984)
malM	+	91.500	1	0.000	ECOMALM	X04477	1313	2	4278.5	JMB	191:303-311(1986)
plsB		91.825	1	0.000	plsBecoM	ES1094	4811	9	4282.5		
plsB	-	91.825	1	0.000	ECOPLSB	K00127	3865	8	4282.5	JBC	258:10856-10861(1983)
lexA	+	91.675	1	0.000	ECOLEXA	J01643	952	2	4286.4	NAR	9:4149-4161(1981)
dnaB	+	91.900	ND	ND	ECODNAB	K01174	1661	0	4293.9	JBC	259:97-101(1984)
tyrB	+	92.000	1	0.000	tyrBeco	ES3153	1734	6	4296.5	BBRC	133:134-139(1985)
uvrA		92.075	1	0.005	uvrAecoM	ES1136	3850	8	4300.3		
uvrA	-	92.075	1	0.005	ECOUVRAA	M13495	3205	7	4300.3	JBC	261:4895-4901(1986)
ssb	+	92.100	1	0.005	ECOSSB	J01704	764	1	4303.4	PNAS	78:4274-4278(1981)
soxR*	+	92.150	1	0.991	ECOSOXRS	M60111	1099	4	4306.5	JBact	173:2864-2871(1991)
gltP#	+	92.500	ND	ND	ECOGACAR	M32488	1630	4	4323.6	JBact	172:3214-3220(1990)
fdhF*	-	92.900	1	0.051	fdhFeco	ES3025	2871	4	4326.1	AM	148:44-51(1987)
is5W	-	92.700	ND	ND	IS5W	ES3211	1195	2	4335.9	Gene	14:165-174(1981)
phnD*		92.400	1	0.000	phnQecoM	ES1170	15916	31	4343.6		
phnD*	-	92.400	1	0.000	ECOPHN	D90227	11672	25	4343.6	JBact	173:2665-2672(1991)
phnD*	-	92.400	1	0.000	ECOPHNAQ	J05260	15611	30	4343.9	JBC	265:4461-4471(1990)
melA		93.400	1	0.000	melAecoM	ES1076	4607	11	4370.4		
melA	+	93.400	1	0.000	ECOMELOPA	M18425	1628	4	4370.4	Gene	59:253-263(1987)
melA	+	93.400	1	0.000	ECOMELA	X04894	1835	2	4371.7	NAR	15:2213-2220(1987)
melB	+	93.400	1	0.000	ECOMELB	K01991	1575	6	4373.4	JBC	259:4320-4326(1984)
fumB	-	93.450	1	0.001	ECOFUMB	M27058	3162	6	4375.7	JBact	171:3494-3503(1989)
lysU*	-	93.800	1	0.016	ECOLYSU	M30630	3034	9	4382.3	JBact	172:3237-3243(1990)
cadA		93.700	1	0.000	cadAecoM	ES1217	6625	14	4387.3		
cadA	-	93.700	1	0.000	ECOCADABC	M76411	4349	10	4387.3	S.Meng/G.Bennett	
cadA	-	93.700	1	0.000	ECOCADAB	M67452	5028	11	4388.9	E.Olson/D.Dunyak	
aspA		94.150	1	0.001	aspAecoM	ES1240	3366	10	4396.1		
aspA	-	94.150	1	0.001	aspAeco	ES3216	2903	10	4396.1	NAR	13:2063-2074(1985)
aspA	-	94.150	1	0.001	ECOASPAG	X04066	2921	10	4396.5	BJ	237:547-557(1986)
mopA	+	94.500	1	0.001	ECOGROESL	X07850	2267	14	4401.8	Nature	333:330-334(1988)
ampC	-	94.350	1	0.000	ECOAMPCFR	J01611	5482	15	4409.4	EJB	122:479-484(1982)
psd	-	94.650	1	0.030	ECOPSD	J03916	1350	4	4420.9	JBC	263:11516-11522(1988)
glyV	+	94.700	ND	ND	glyVeco	ES3039	423	0	4424.0	JMB	212:579-598(1990)
miaA*		94.800	3	0.932	miaAecoM	ES1080	7712	19	4430.4		
miaA*	+	94.800	3	0.932	ECOMIAA	M63655	1328	5	4430.4	JBact	173:1711-1721(1991)
hflX	+	94.900	3	0.932	hflXeco	ES3043	4843	8	4430.8	F.Banuett/I.Herskowitz	
hfq#	+	94.875	3	0.932	hfq-eco	ES3150	1290	2	4430.8	NAR	19:1063-1066(1991)
purA	+	95.000	3	0.932	ECOPURAA	J04199	2726	9	4435.4	JBC	263:19147-19153(1988)
rpsF	+	95.500	1	0.001	ECORPSFRI	X04022	1979	12	4455.9	MGG	204:126-132(1986)
cpdB		95.700	1F	0.081	cpdBecoM	ES1025	3278	8	4465.6		
cpdB	-	95.700	1F	0.081	ECOCPDB	M13464	2198	5	4465.6	JBact	165:1002-1010(1986)
amtA#	+	95.800	1F	0.081	ECOAMTA	M55170	2121	5	4466.7	JGM	137:983-989(1991)
ppa#		99.900	ND	ND	ppa-ecoM	ES1096	2051	3	4479.4		
ppa#	-	99.900	ND	ND	ECOILER	M14018	890	1	4479.4	JBC	261:9966-9971(1986)
ppa#	-	99.900	ND	ND	M23550	M23550	1195	2	4480.3	JBact	170:5901-5907(1988)
fbp	-	96.125	ND	ND	ECOFBPASE	X12545	1611	6	4485.7	NAR	16:8707-8707(1988)
pmbA#	+	96.200	ND	ND	ECOPMBA	X54152	1682	5	4488.4	MM	4:1921-1932(1990)
pyrI		96.500	ND	ND	pyrIecoM	ES1179	2028	4	4500.9		
pyrI	-	96.500	ND	ND	ECOPYRBIA	K01472	1410	3	4500.9	PNAS	81:115-119(1984)
pyrB	-	96.500	ND	ND	ECOPYRBI	J01670	1593	4	4501.3	PNAS	80:2462-2466(1983)
argI		96.700	ND	ND	argIecoM	ES1013	1222	5	4507.0		
argI	-	96.700	ND	ND	ECOARGI	X00210	1085	5	4507.0	NAR	11:8509-8518(1983)
argI	-	96.700	ND	ND	ECOARGIPRM	M24186	300	2	4507.9	EMBO	1:853-857(1982)
valS		96.800	1	0.000	valSecoM	ES1115	5325	21	4510.4		
valS	-	96.800	1	0.000	ECOVALS	X05891	3293	12	4510.4	NAR	15:9081-9082(1987)
pepA#	-	96.850	1	0.000	ECOXERB	X15130	2038	10	4513.7	EMBO	8:1623-1627(1989)
leuX	+	97.000	ND	ND	leuXeco	ES3054	510	2	4527.0	JMB	212:579-598(1990)
is2K	+	97.050	39	1.000	IS2K	ES3197	1331	3	4528.8	Gene	59:291-296(1987)
is4	+	97.100	ND	ND	INS4ECO	J01733	1426	2	4533.3	MGG	181:169-175(1981)
is30D	-	97.100	ND	ND	IS30D	ES3180	1221	1	4538.1	EMBO	3:2145-2149(1984)

Table 1   *Continued*

Gene[a]	Ori[b]	Min[c]	Rnk[d]	p val[e]	Locus[f]	Access[g]	Bp[h]	Sites[i]	Addr[j]	Reference[k]	
fecD		7.800	1	0.000	fecEecoM	ES1033	9821	15	4540.1		
fecD	-	7.800	1	0.000	M26397	M26397	4842	6	4540.1	JBact	171:2626-2633(1989)
fecA	-	7.800	1	0.000	fecAeco	ES3026	2645	7	4544.8	JBact	172:6749-6758(1990)
fecI#	-	97.200	1	0.000	ECOFECIR	M63115	4074	5	4545.8	JBact	172:6749-6758(1990)
fimB		97.875	1	0.000	fimBecoM	ES1046	9679	15	4571.5		
fimB	+	97.875	1	0.000	ECOFIMBE	X03923	3050	0	4571.5	EMBO	5:1389-1393(1986)
fimA	+	97.900	1	0.000	ECOFIMA	X00981	1450	3	4574.0	EJB	143:395-399(1984)
fimD	+	97.900	1	0.000	fimDeco	ES3033	3740	8	4575.4	P.Klemm	
fimF	+	97.925	1	0.000	ECOFIMFGH	X05672	2050	4	4579.2	MGG	208:439-445(1987)
mcrB		98.450	1	0.000	mcrCecoM	ES1074	11970	23	4608.6		
mcrB	-	98.450	1	0.000	ECOMCRBC	M58752	2705	3	4608.6	JBact	172:4888-4900(1990)
hsdS	-	98.500	1	0.000	ECOHSDSK	J01632	2528	5	4611.0	JMB	166:1-19(1983)
hsdM	-	98.500	1	0.000	ECOHSDRM	X06545	5591	13	4613.3	JMB	198:159-170(1987)
mrr*	+	98.550	1	0.000	ECOMRR	X54198	2017	6	4618.6	JBact	173:5207-5219(1991)
tsr	+	99.100	ND	ND	ECOTSR	J01718	1788	7	4623.5	Nature	301:623-626(1983)
dnaC	-	98.900	2	0.439	ECODNATC	J04030	2554	2	4631.6	JBC	263:15083-15093(1988)
leuV	-	99.350	ND	ND	ECOTGLEUV	J01712	699	1	4637.8	NAR	9:2121-2139(1981)
rimI	-	99.250	ND	ND	ECORIMI	X06117	1423	1	4640.6	MGG	209:481-488(1987)
deoC		99.450	1	0.059	deoCecoM	ES1035	3516	2	4647.9		
deoC	+	99.450	1	0.059	ECODEOP1	X04151	480	2	4647.9	EMBO	5:2015-2021(1986)
deoC	+	99.450	1	0.059	ECODEOC	J01601	1538	1	4648.3	EJB	125:561-566(1982)
deoC	+	99.450	1	0.059	ECODEOCA	X00314	204	0	4649.8	EMBO	3:179-183(1984)
deoA	+	99.450	1	0.059	deoAeco	ES3017	1573	0	4649.8	S.Short	
deoD	+	99.450	ND	ND	deoDeco	ES3018	1157	1	4652.2	PNAS	88:7185-7189(1991)
serB	+	99.500	3	1.000	ECOSERB	X03046	2011	3	4655.8	NAR	13:7025-7039(1985)
slt#		99.625	1	0.015	slt-ecoM	ES1233	3381	6	4662.2		
slt#	+	99.625	1	0.015	ECOSLTY	M69185	2548	4	4662.2	JBact	173:6773-6782(1991)
trpR	+	99.650	1	0.015	ECOTRPR	J01715	1289	3	4664.3	NAR	8:1551-1560(1980)
phoM		99.750	1	0.000	phoMecoM	ES1092	5462	10	4666.7		
phoM	+	99.750	1	0.000	ECOPHOM	M13608	4658	8	4666.7	JBact	168:294-302(1986)
arcA	-	99.700	1	0.000	ECODYE	M10044	1468	3	4670.7	JBC	260:4236-4242(1985)
mvrA*	?	6.900	ND	ND	ECOMVRA	M19644	981	3		JBact	170:2136-2142(1988)
pgpA	?	9.650	ND	ND	M23546	M23546	711	0		JBact	170:5110-5116(1988)
potA#	?	15.350	ND	ND	ECOPOTABCD	M64519	4385	9		JBC	266:20928-20933(1991)
trxB*	?	20.600	ND	ND	ECOTRXB	J03762	1202	1		JBC	263:9015-9019(1988)
fadR*	?	26.000	ND	ND	fadReco	ES3024	1064	1		NAR	16:7995-8009(1988)
pgpB*	?	28.500	ND	ND	M23628	M23628	1113	3		JBact	170:5117-5124(1988)
tyrR	?	29.100	ND	ND	ECOTYRR	M12114	1965	1		JBC	261:403-410(1986)
cpsB#	?	44.450	ND	ND	ECOPGMPMI	M77127	1945	3	~2132	R.Tal/H.Wong	
gcvH*	?	62.650	ND	ND	ECOGCVH	M57690	516	0		DNAS	2:13-17(1991)
pepP#	?	63.100	ND	ND	ECOAPP2	D00398	1851	1		JBioch	105:412-416(1989)
envC	?	81.500	ND	ND	envCeco	ES3147	5465	9		J.Klein/R.Plapp	
uxuA	?	98.100	ND	ND	ECOUXU1	X03411	434	1		MGG	202:112-119(1986)
eco#	?	?	ND	ND	ECOECOA	M60876	1053	3		JBC	266:6620-6625(1991)
fabH#	?	?	ND	ND	ECOFABH	M77744	1273	3		J.Tsay/C.Rock	
orf	?	?	ND	ND	ECOLIVRA	M36020	729	0		Prot	1:125-133(1986)
tlp#	?	?	ND	ND	tlp-eco	ES3157	2339	4		JBioch	110:315-320(1991)
xylU#	?	?	ND	ND	ECOXYLUP1	X04387	363	0		NAR	14:7115-7123(1986)

Notes to Table 1

(a) Gene (usually the most clockwise) contained in DNA sequence. Genes marked with an asterisk were noted in the 1990 genetic map as being less accurately mapped than genes without an asterisk. Genes marked with # were not present in the 1990 genetic map a their minute values are assigned in order to be consistent with the 1990 genetic map and the MapSearch alignments.(b) Orientations of the aligned genes as determined by the MapSearch program. A plus (+) sign indicates that genes are transcribed in the direction of increasing genomic map coordinates (clockwise); minus (-) sign indicates counterclockwise transcription. (c) The map position (in minutes). The positions are approximated from the 1990 *E. coli* genetic map. We realize that the genetic map positions were not originally determined to this level of accuracy but we imposed a resolution of 0.025 minutes to preserve map order information. (d) The rank and (e) *p* (probability) value calculated for the MapSearch alignment (based on 100 map shuffles). 1F, 2F, etc. indicate that the enzyme EcoRV(F) is ignored in the MapSearch alignment. Sequences which are part of a larger meld are assigned the *p* value and rank of the meld (ecoM) entry. A *p* value of 0.000 means <0.001. (f) The locus name of the database entry. Entries beginning with ECO are from GenBank; entries beginning with EC are from EMBL; entries ending with eco are from the EcoSeq database. All entries without locus names are GenBank entries. An M at the end of an EcoSeq database entry name denotes it is a meld of one or more database entries. The locus name of the EcoSeq meld entry is derived from the most clockwise gene in the meld. The various rrs-eco and rrl-eco entries are derived from the ECORGNB sequence. (g) Database entry accession numbers. (h) The number of DNA basepairs in the entry. (i) The number of genomic map restriction sites in the entry. (j) The revised genomic map address (in kilobase pairs) of the clockwise end of the DNA sequences aligned. (k) One reference for DNA sequence data. Journal abbreviations: ABC, Agric. Biol. Chem.; AdvB, Adv. Biophys.; AEM, Appl. Environ. Microbiol.; AM, Arch. Microbiol.; BBRC, Biochem. Biophys. Res. Commun.; Bchim, Biochimie; Bioch, Biochemistry; BioKh, Bioorg. Khim.; BJ, Biochem. J.; CJM, Can.J. Microbiol.; CRC, Carlsberg Res. Commun.; DCB, DNA Cell Biol.; DNAS, DNA Sequence; EJB, European Journal of Biochemistry; EMBO, EMBO J.; FEBS, FEBS Lett.; FEMS, FEMS Microbiol. Lett.; Genet, Genetics; JBact, J. Bacteriol.; JBioch, J. Biochem.; JGM, J. Gen. Microbiol.; JMB, J. Mol. Biol.; MBE, Mol. Biol. Evol.; MGG, Mol. Gen. Genet.; MM, Molecular Microbiology; NAR, Nucleic Acids Res.; NB, New Biol.; PNAS, Proc. Natl. Acad. Sci.; PRSL, Proc. R. Soc. Lond.; Prot, Proteins; ND, no data.

Table 2   Regions of the Revised, Integrated Map Lacking *Eco*RV Site Information

Start (kb)	Stop (kb)	Region	Start (kb)	Stop (kb)	Region
23.6	38.4	noF1	1956.6	1962.6	noF21
23.9	38.8	noF1	1955.3	1961.3	noF21
97.1	118.7	noF2	2290.8	2301.7	noF22
132.8	144.8	noF3	2449.2	2477.3	noF23
226.0	238.3	noF4	2659.8	2665.4	noF24
257.3	265.1	noF5	2801.8	2814.4	noF25
269.4	291.4	noF6	3343.1	3350.8	noF26
367.7	371.9	noF7	3512.5	3524.4	noF27
372.2	389.1	noF8	3639.2	3650.7	noF28
415.0	434.6	noF9	3743.7	3766.9	noF29
509.1	526.3	noF10	3794.4	3802.3	noF30
585.4	601.8	noF11	3845.6	3883.0	noF31
741.2	750.5	noF12	3905.1	3909.0	noF32
1143.4	1161.7	noF13	3941.3	3951.3	noF33
1291.8	1309.9	noF14	4154.3	4178.3	noF34
1415.9	1443.6	noF15	4219.4	4240.4	noF35
1523.1	1531.6	noF16	4422.6	4432.9	noF36
1617.4	1665.8	noF17	4449.3	4467.7	noF37
1731.5	1738.4	noF18	4618.7	4627.9	noF38
1766.3	1774.2	noF19	4630.8	4634.8	noF39
1837.8	1863.7	noF20	4647.9	4661.0	noF40

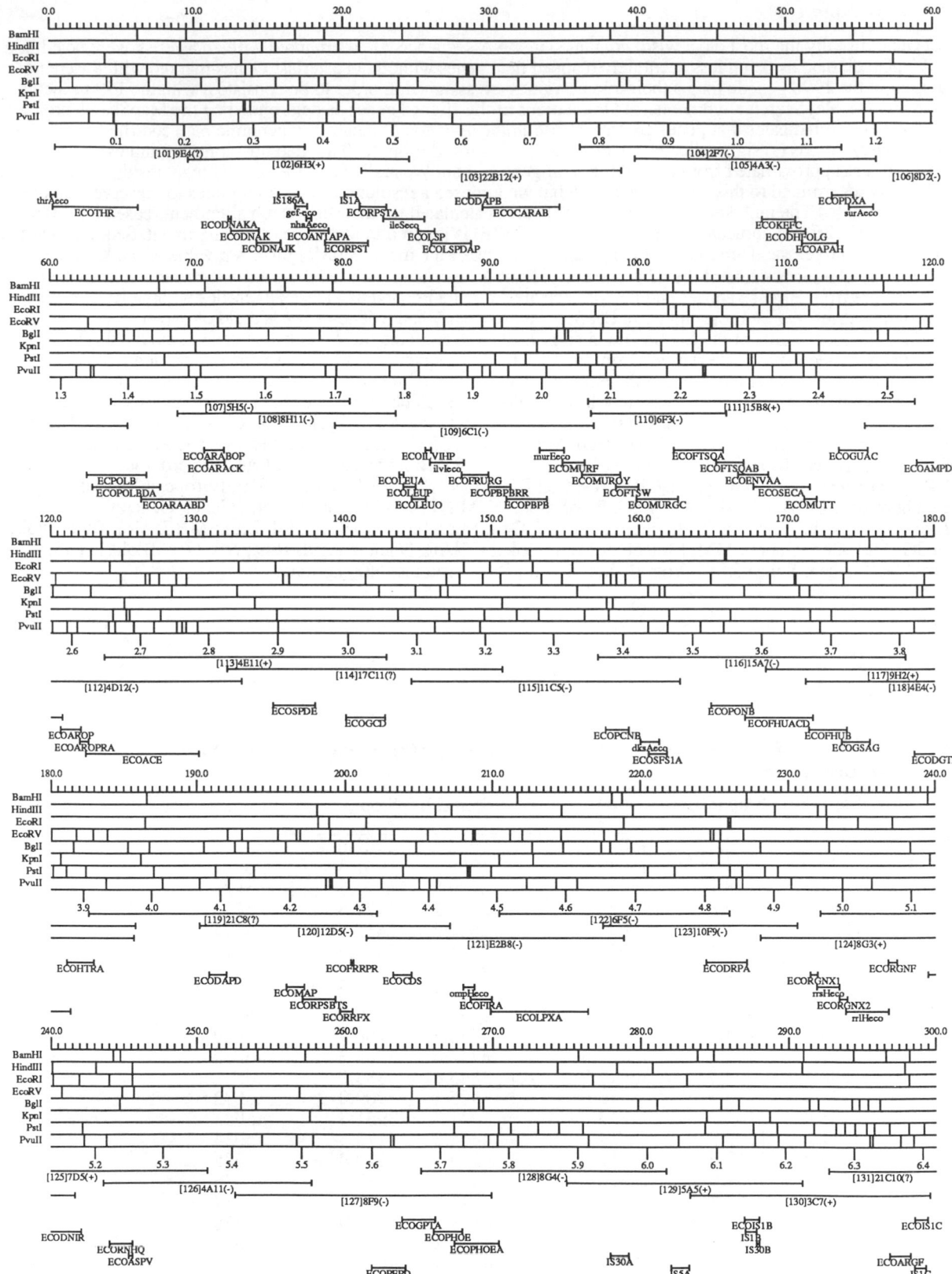

Figure 1   Revised, integrated genomic restriction map of *E. coli*.

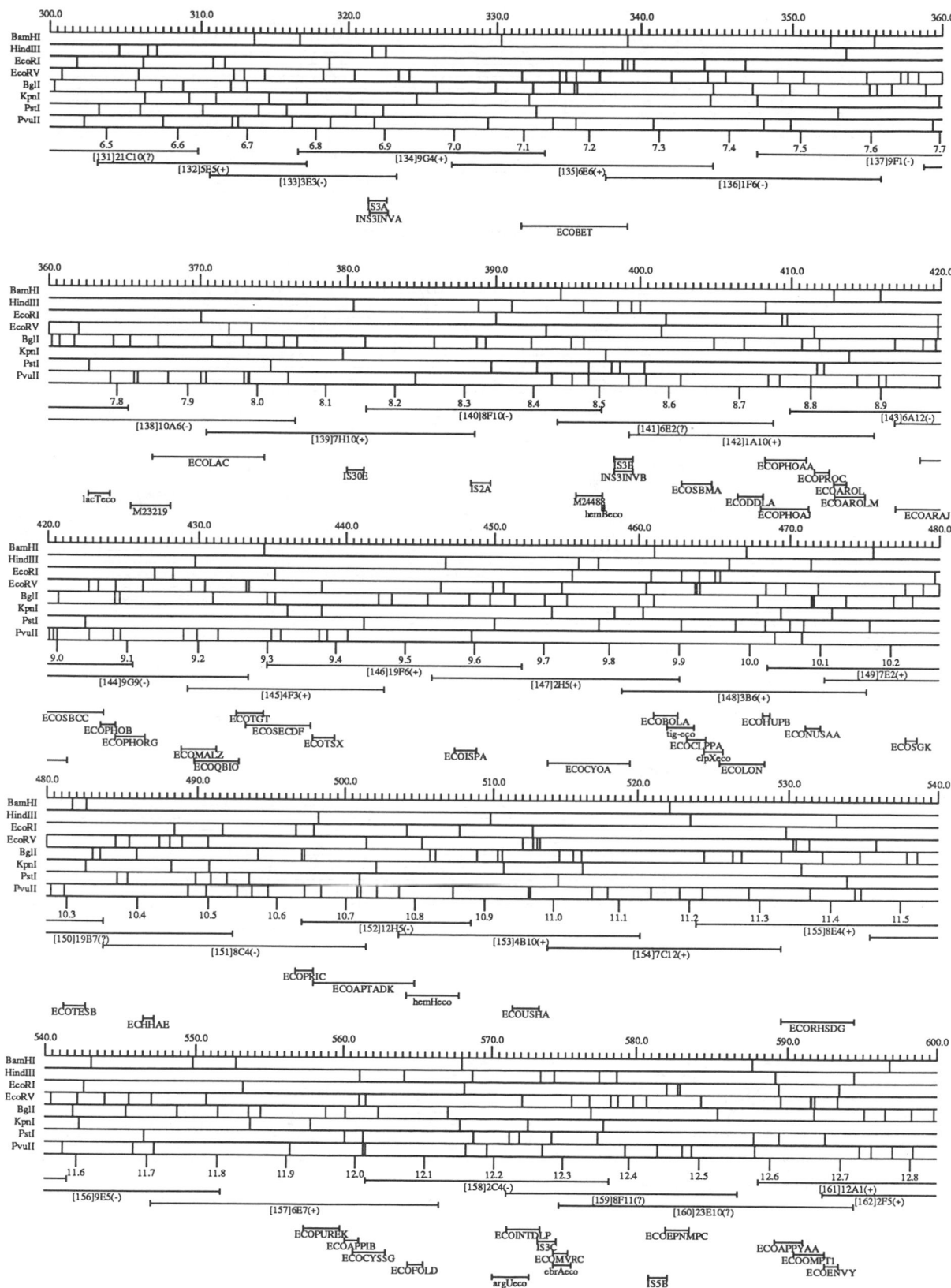

Figure 1 *Continued*

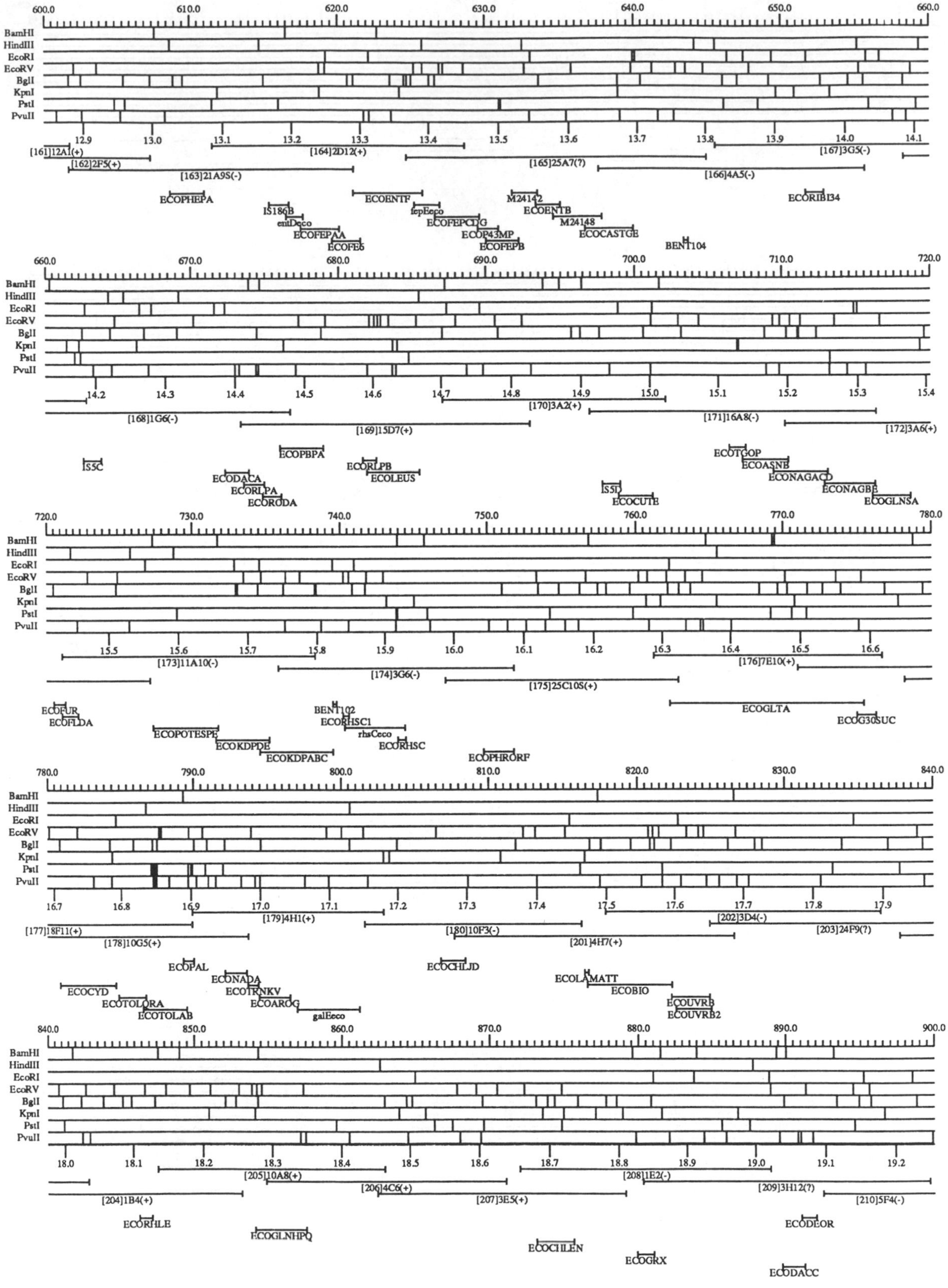

Figure 1   *Continued*

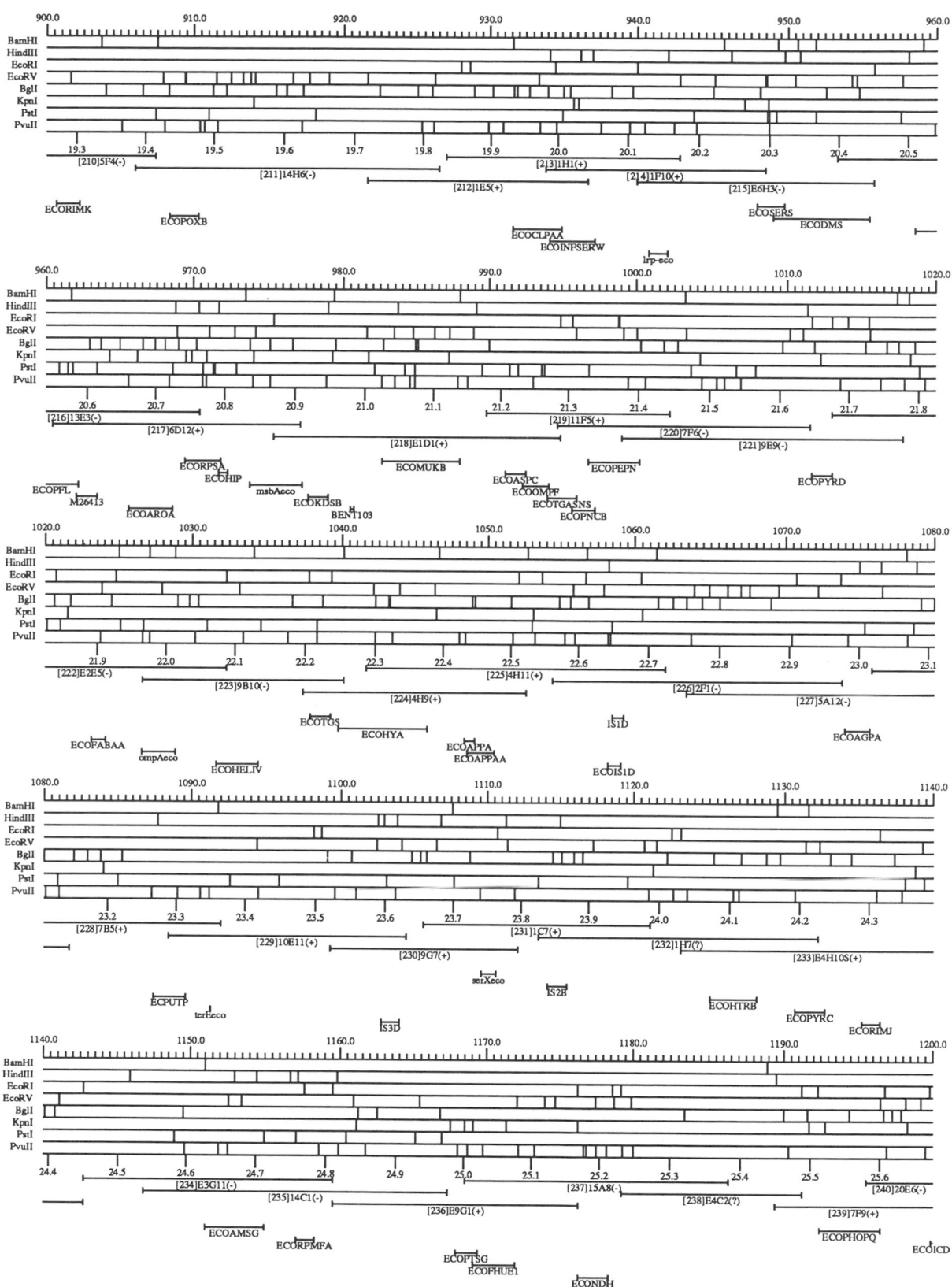

Figure 1 *Continued*

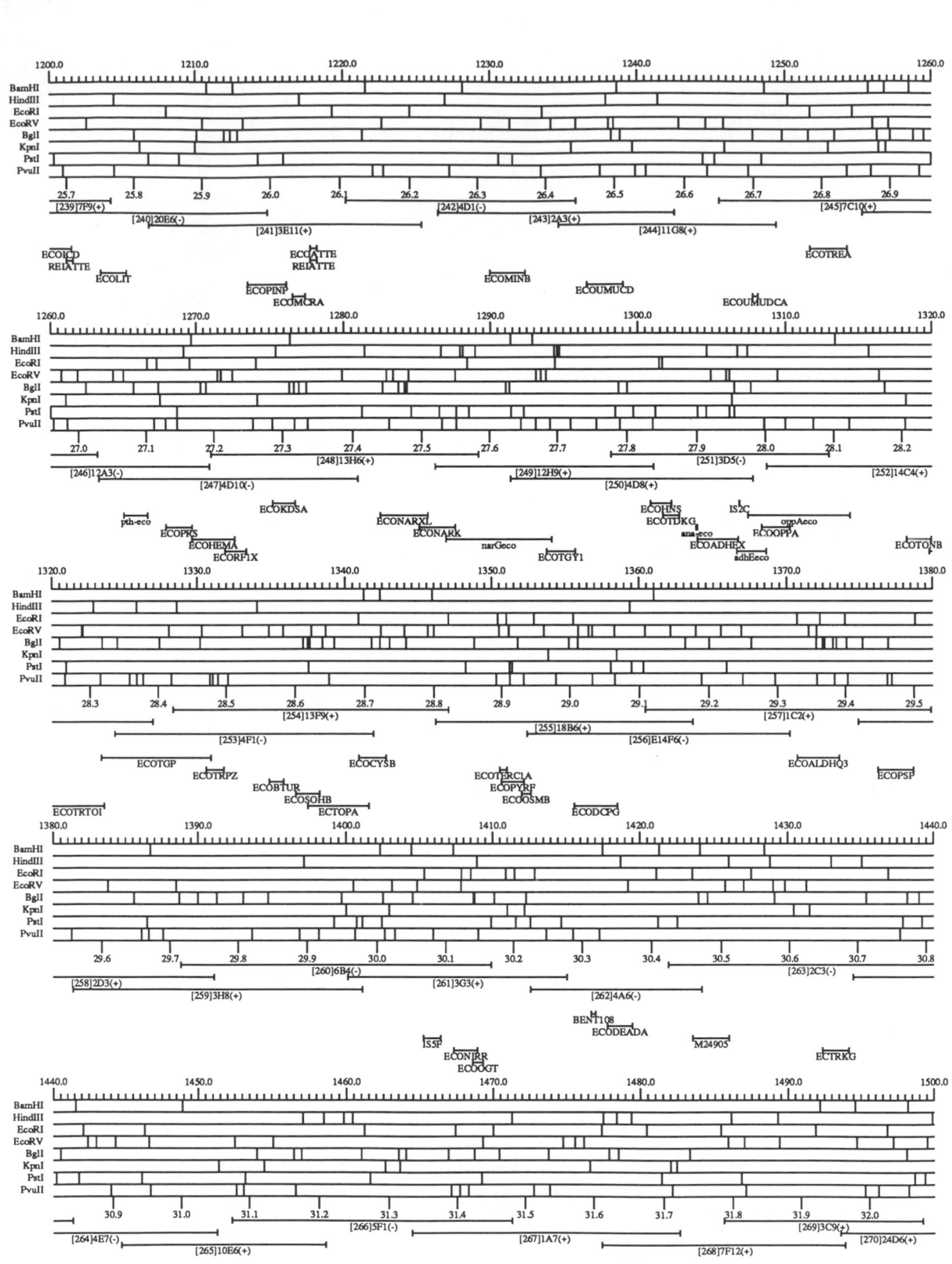

Figure 1 *Continued*

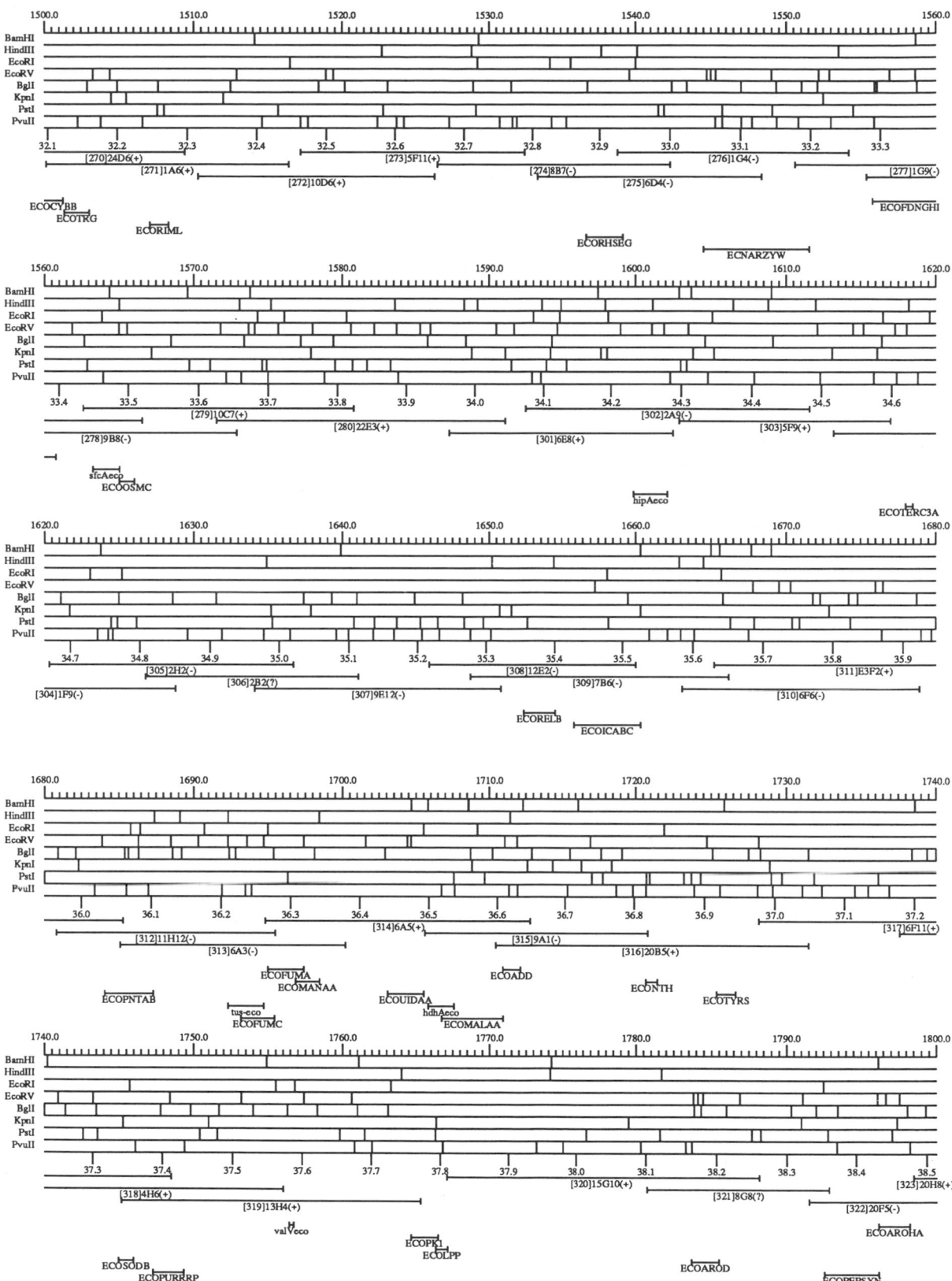

Figure 1   *Continued*

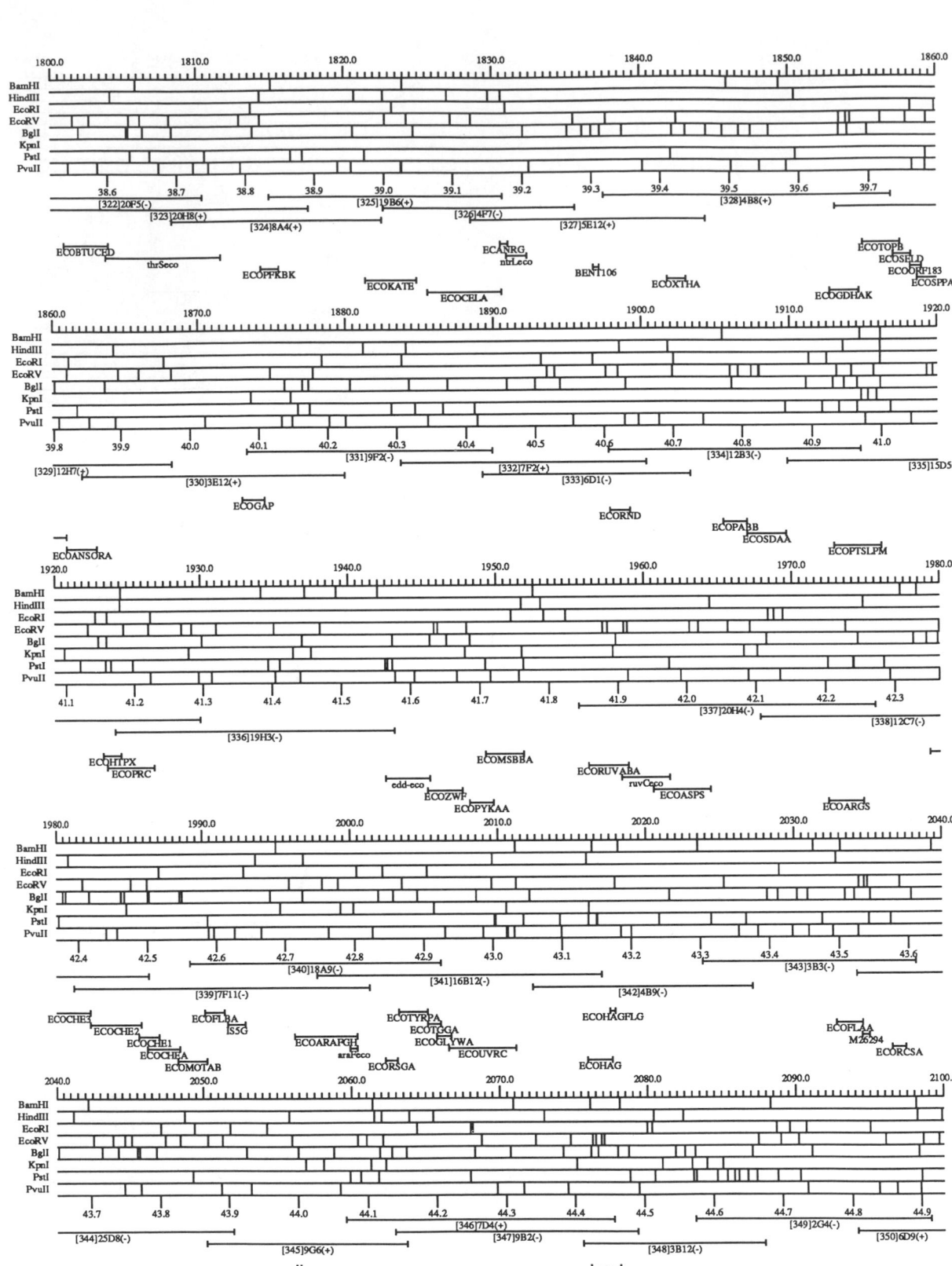

Figure 1 *Continued*

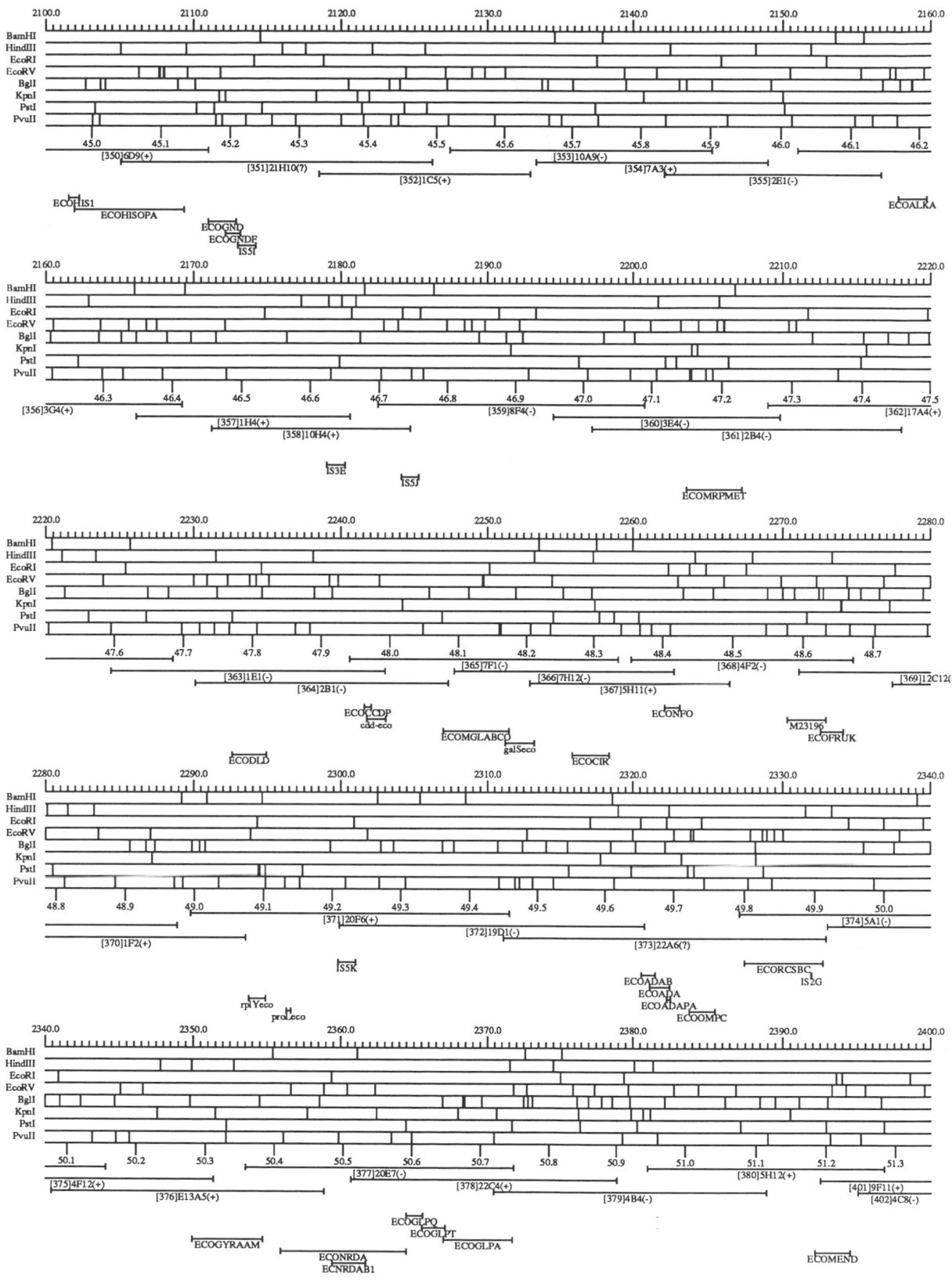

Figure 1 *Continued*

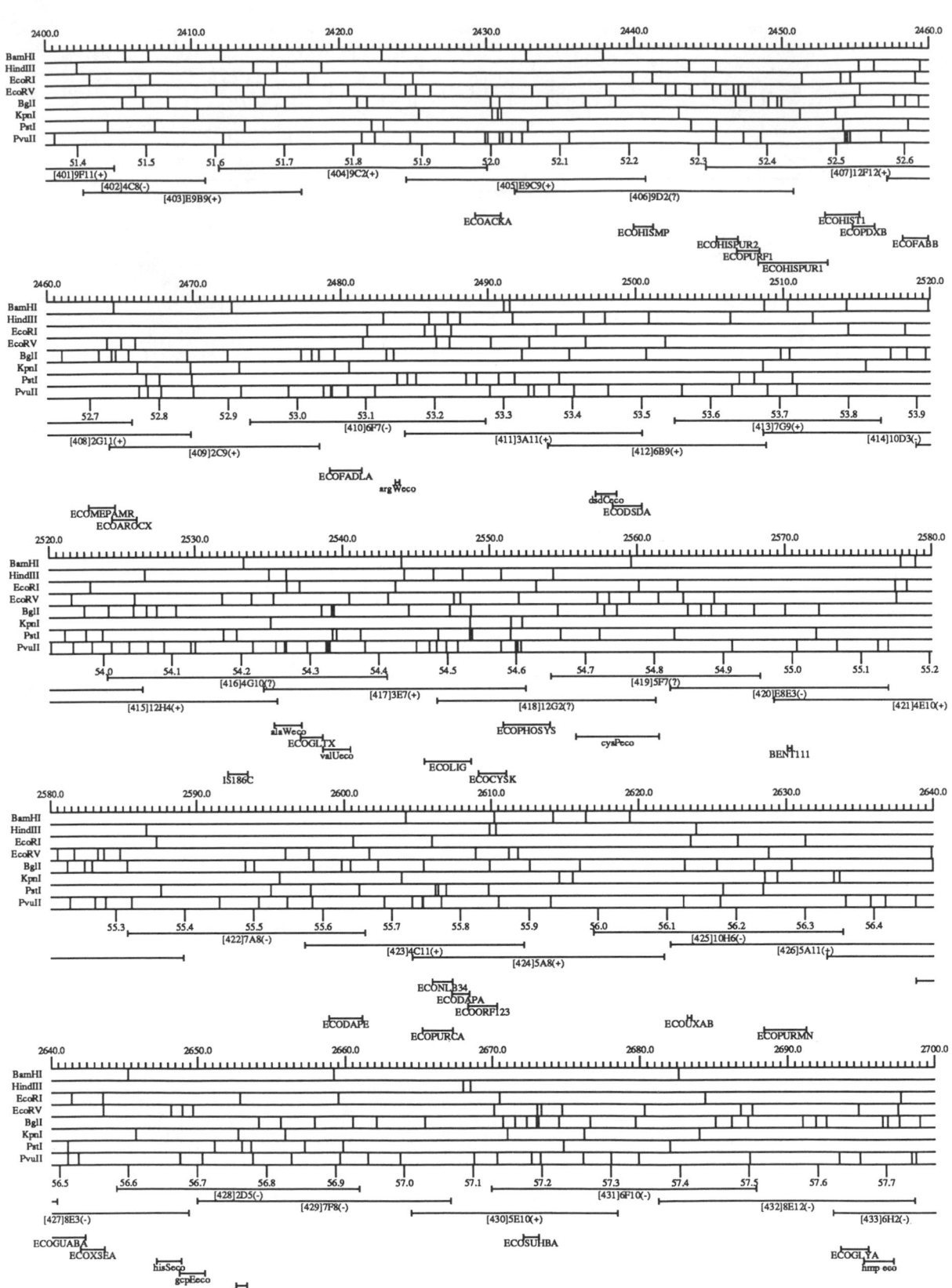

Figure 1 *Continued*

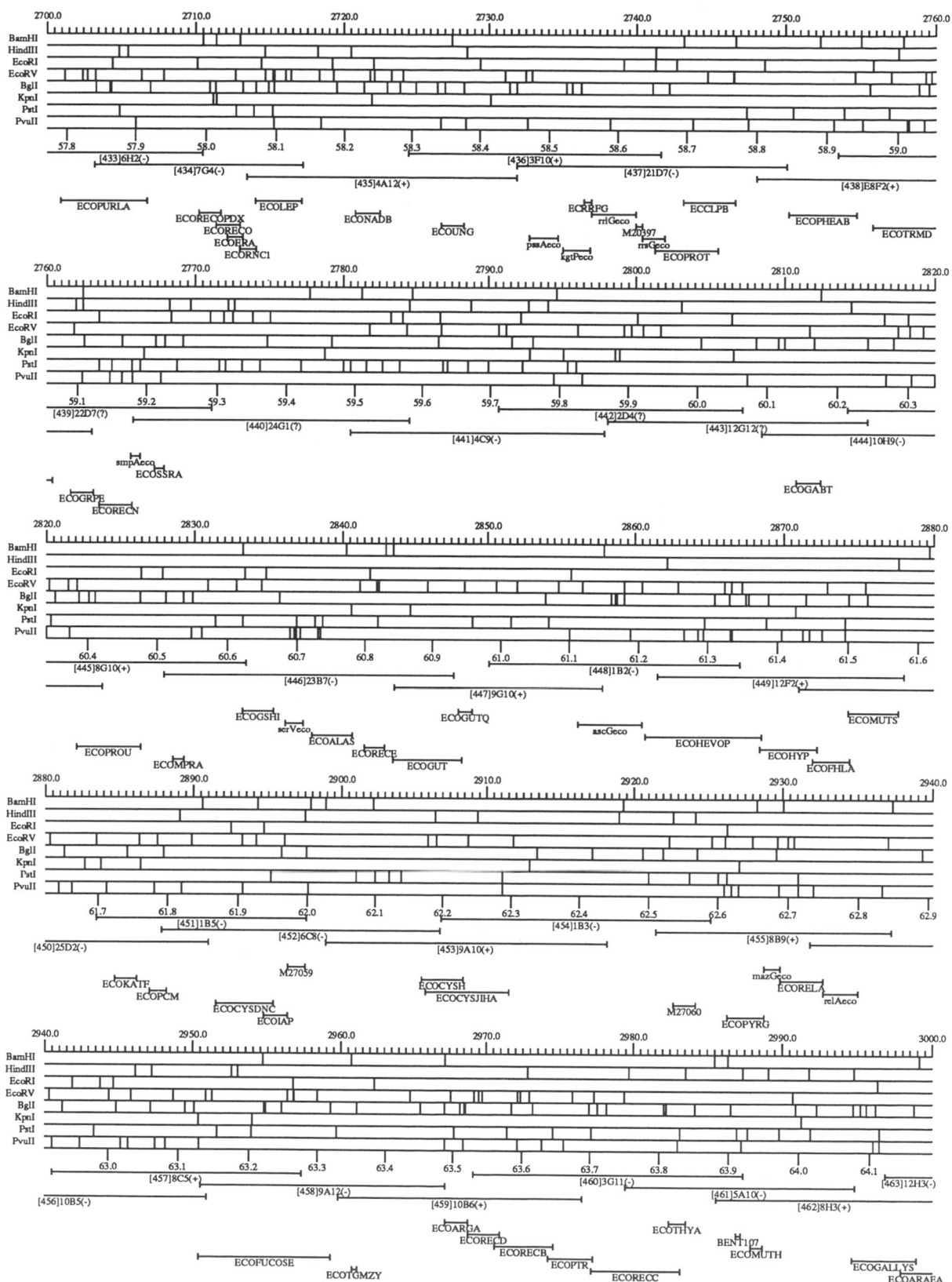

Figure 1   *Continued*

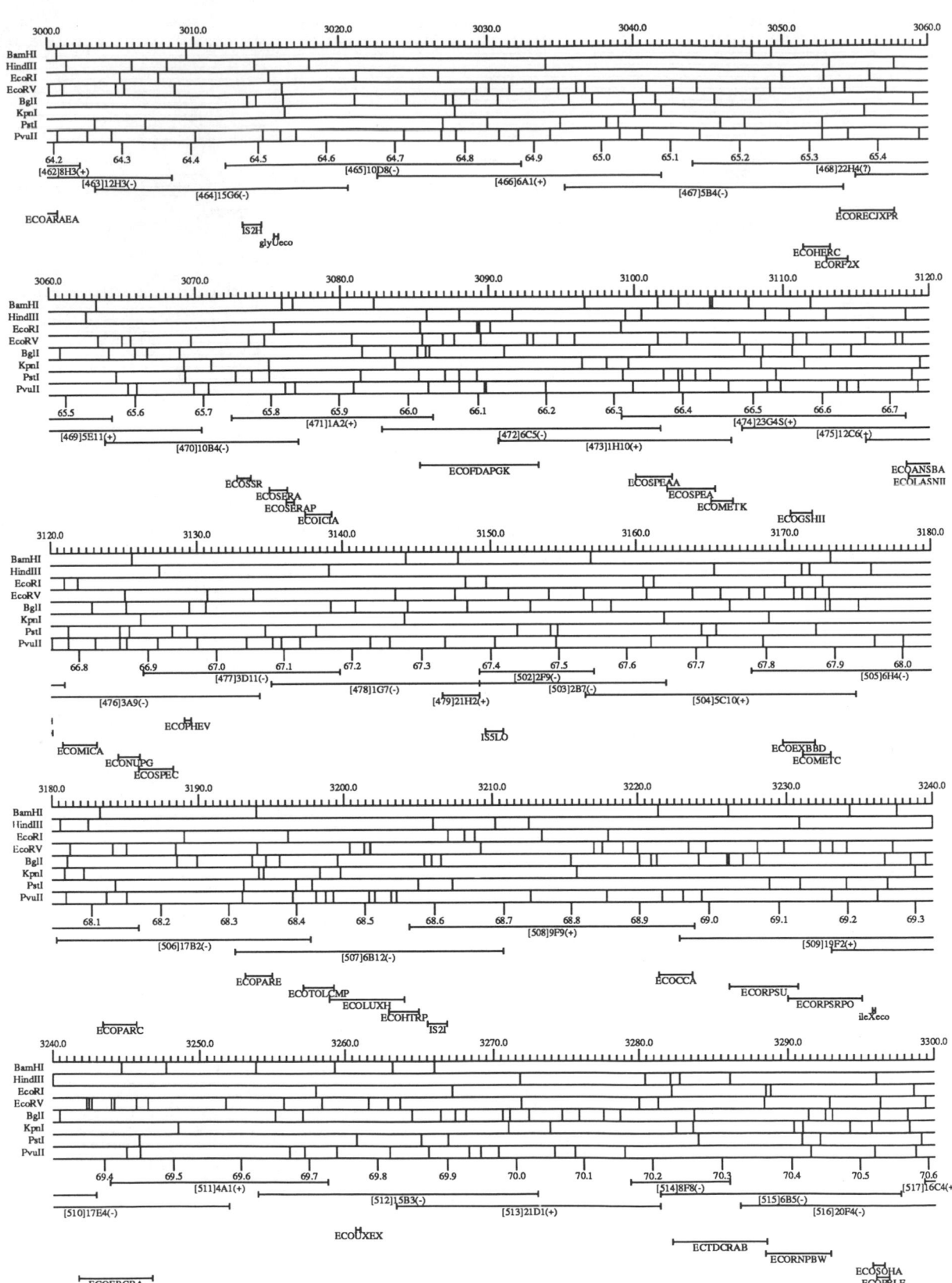

Figure 1   *Continued*

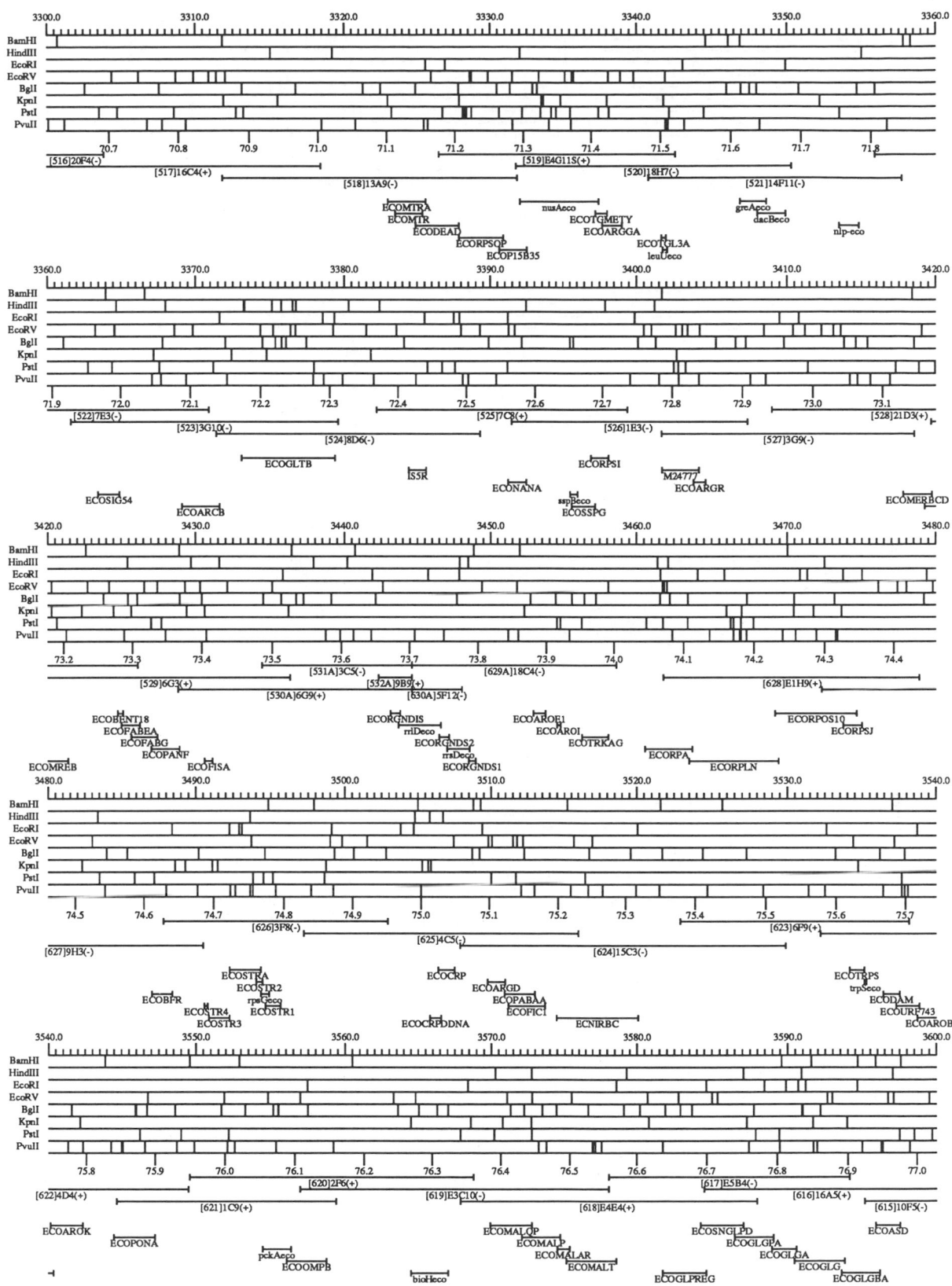

Figure 1 *Continued*

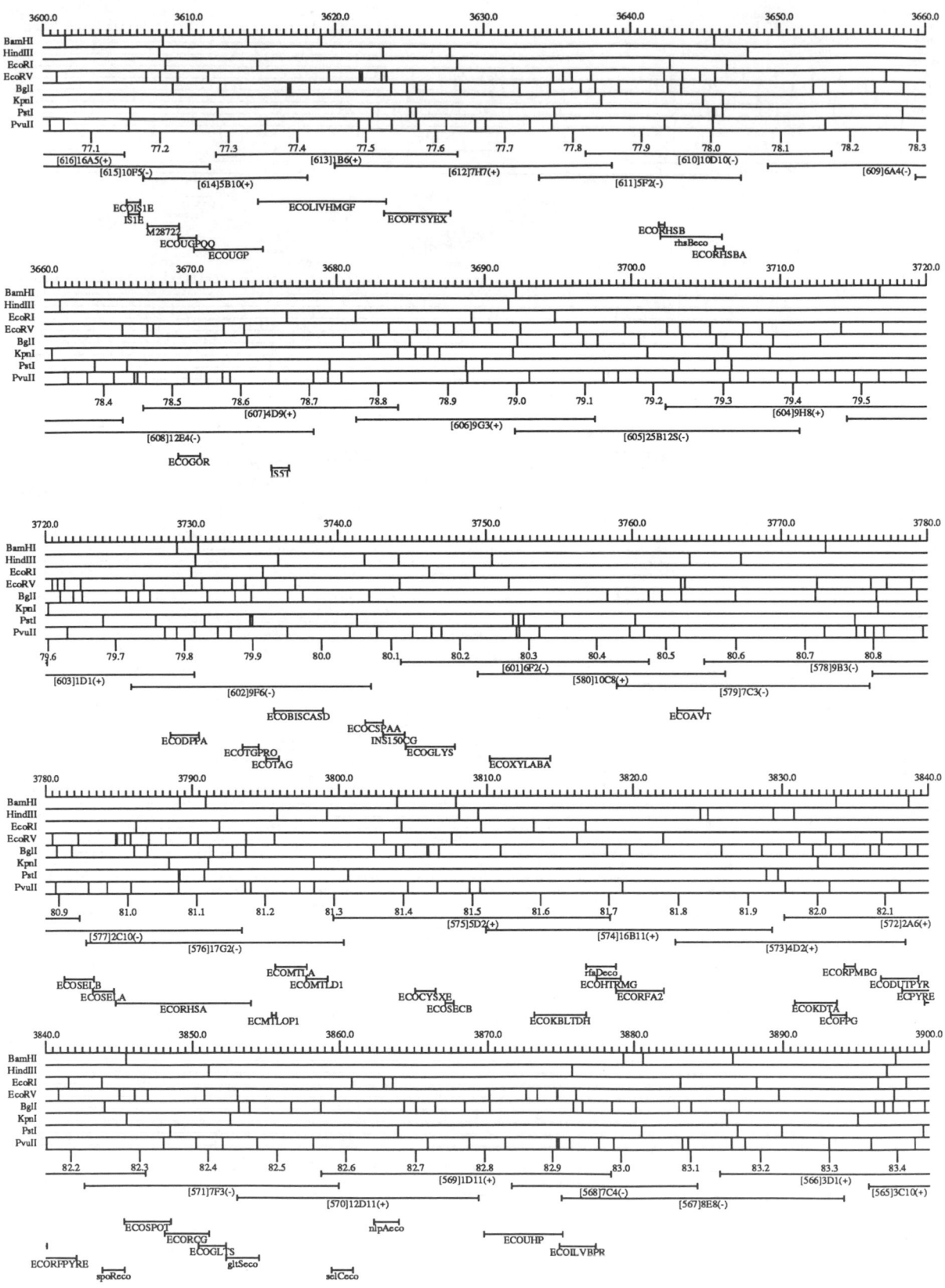

Figure 1   *Continued*

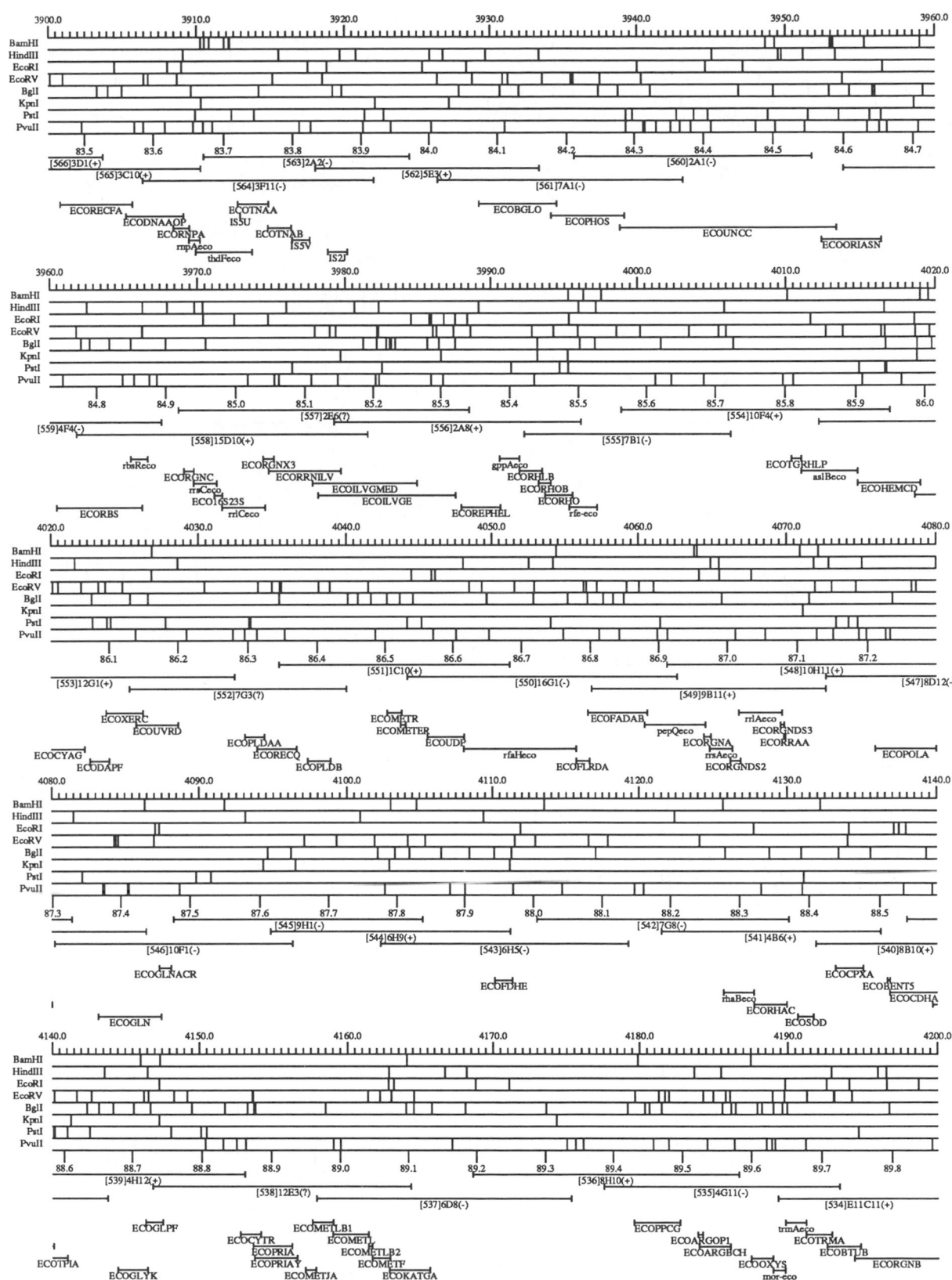

Figure 1   *Continued*

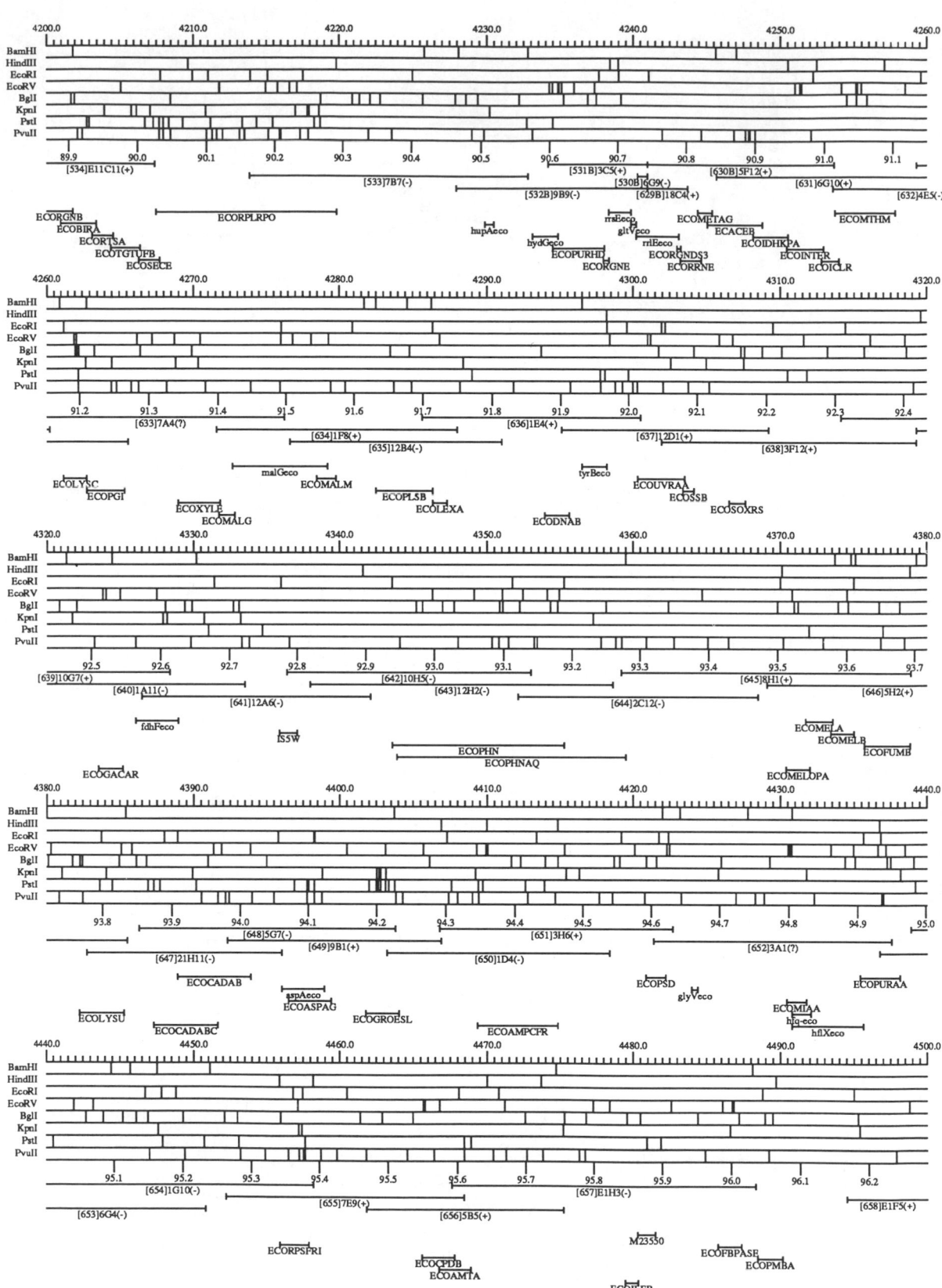

Figure 1   *Continued*

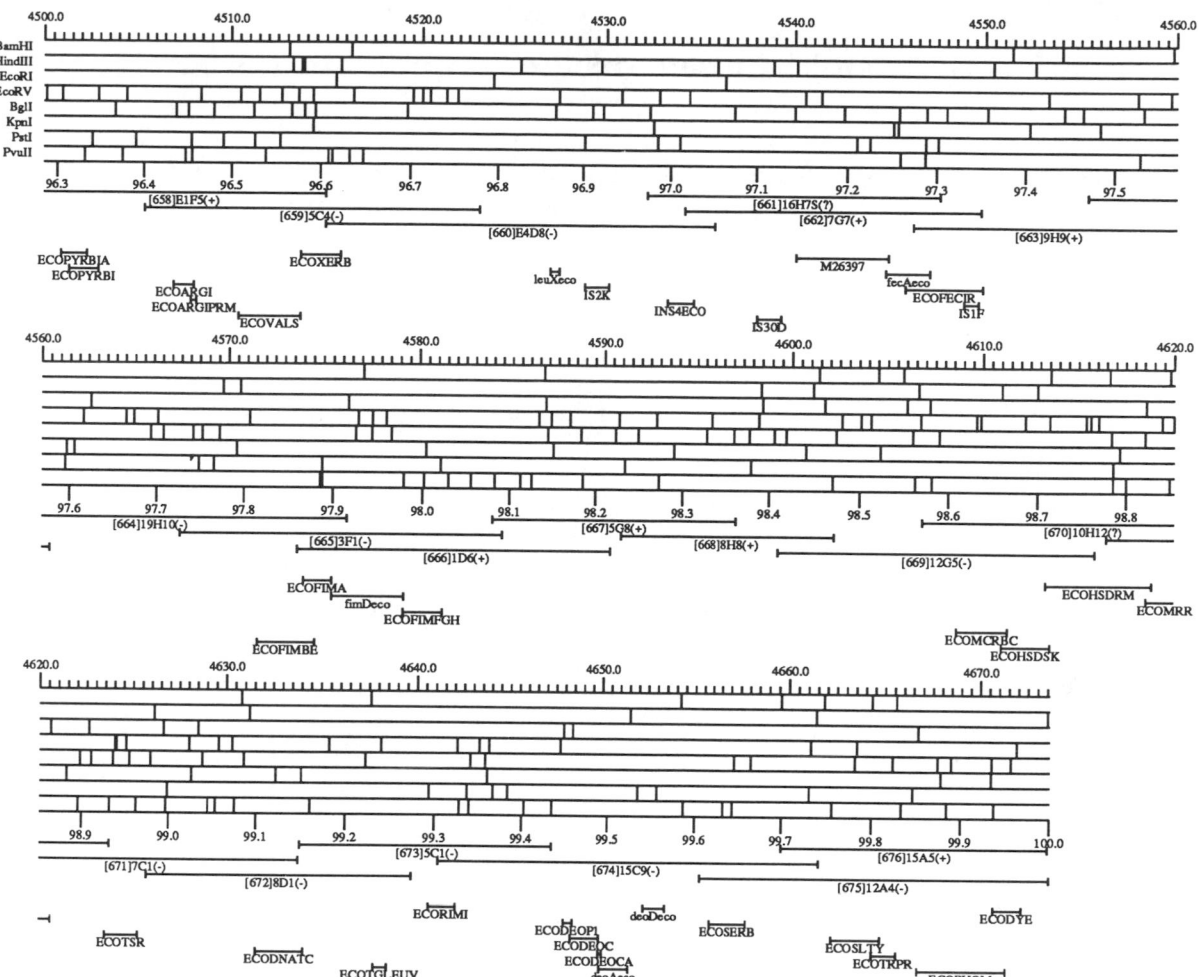

Figure 1 *Continued*

B. Location of Cosmid Clones on the Genetic Map of *E. coli*

Figure 1 is reprinted, with permission, from Tobata et al. (*J. Bacteriol.*, vol. 171, p. 1215 [1989]). The references cited in the figure caption can be found in the original paper.

Figure 1 Mutations complemented by the cloned DNA and the pLC plasmids used as the starting materials for chromosome walking are shown on the map from reference 2. The DNA regions that hybridized to rRNA, λ phage DNA (15), and cloned *terC* segment (1) are also shown and referred to as *rrn*, *rac*, and *terC*, respectively. E and S before the clone number stand for the *Eco*RI and *Sau*3AI clone, respectively.

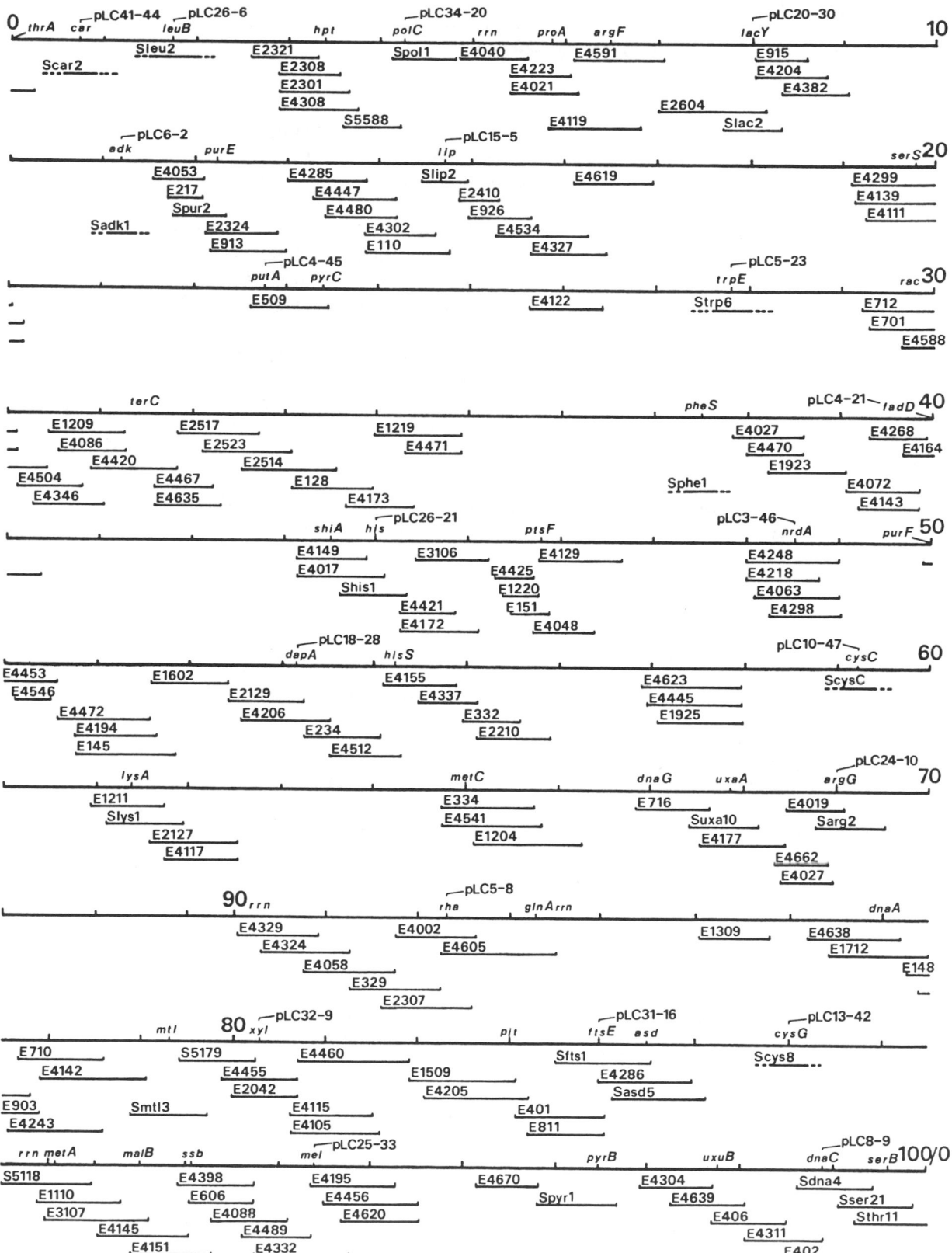

Figure 1 *See facing page for caption.*

SECTION 2

C. Restriction Maps of *E. coli* K12

The *Not*I restriction map of an *E. coli* strain was first published by Smith et al. (1987). Table 1 and Figures 1 and 2 show the *Not*I map of strain EMG2. The *Not*I map of *E. coli* K12 strain MG1655 has been published by Weinstock and co-workers (Heath et al. 1992a; see below).

The complete *Avr*II, *Not*I, *Sfi*I, and *Xba*I restriction maps of MG1655 have been published (Daniels 1990; Heath et al. 1992a,b; Perkins et al. 1992). Figure 3 presents the *Xba*I, *Not*I, and *Sfi*I restriction maps of *E. coli* strain MG1655 (Heath et al. 1992b), and Figure 4 displays a restriction analysis of several *E. coli* K12 strains (Heath et al. 1992b; see also Daniels 1990).

Table 1 and Figures 1 and 2 are reprinted, with permission, from Smith et al. (*Science*, vol. 236, pp. 1449, 1452, and 1453, respectively [1987]. © 1987 by the AAAS.) Figures 3 and 4 are reprinted, with permission, from Heath et al. (*J. Mol. Biol.* [1992b, in press]).

Table 1 Sizes of the *Not*I and *Sfi*I Fragments Seen in a Total Digest of *E. coli* EMG2

	*Not*I size (kb)		*Sfi*I size (kb)		*Not*I size (kb)		*Sfi*I size (kb)
A	1000	A	415	O	106	O	148
B	360	B	375	P	100	P	148
C	360	C	365	Q	100	Q	122
D	275	D	345	R	95	R	90
E	250	E	325	S	43	S	75
F	250	F	315	T	43	T	75
G	245	G	285	U	40	U	63
H	240	H	275	V	20	V	55
I	230	I	211			W	43
J	210	J	200		4696	X	42
K	205	K	180			Y	38
L	203	L	170			Z	35
M	191	M	158				
N	130	N	158				4711

These sizes were determined by comparing the mobility of DNA bands with that of DNA size standards composed of tandemly annealed λ DNA oligomers subjected to electrophoresis in adjacent lanes. See, for example, Figures 2 and 5 in Smith et al. (1987). The DNA sizes are accurate to about ±5 kb above 300 kb and ±2 kb below 300 kb. The apparent sizes of DNA bands run at different pulse times are generally very consistent.

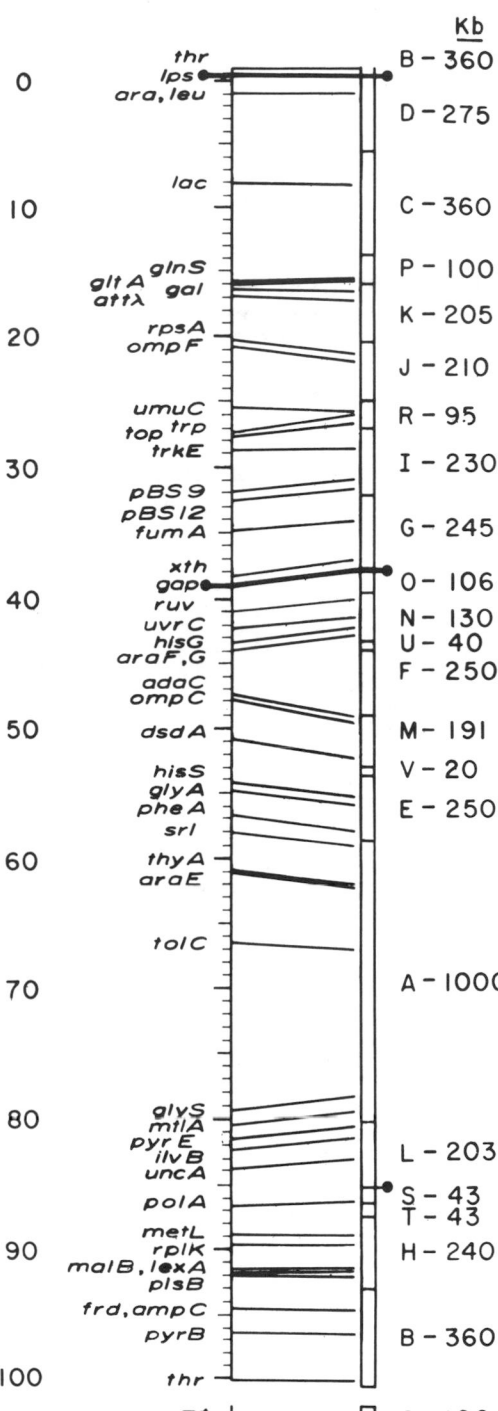

Figure 1 Some of the evidence used to construct the physical map of the *E. coli* genome. Shown are 22 *Not*I fragments drawn to scale and positioned according to their order in the genome. Also shown is an abbreviated version of the genetic map. Cloned probes used to determine the correspondence between the genetic and physical maps are indicated by the name of the gene, and linking probes are shown as bold, solid circles. The scale of the genetic map is in minutes, and the scale of the physical map is in kilobases. The sizes of the *Not*I bands are indicated. Differences in these scales are evident as nonparallel lines from mapped gene positions to the physical map. Only in the case of two of the linking probes is the exact correspondence between the maps known. In all other cases, the relative position of a probe within each restriction fragment was estimated by linear interpolation of the genetic map.

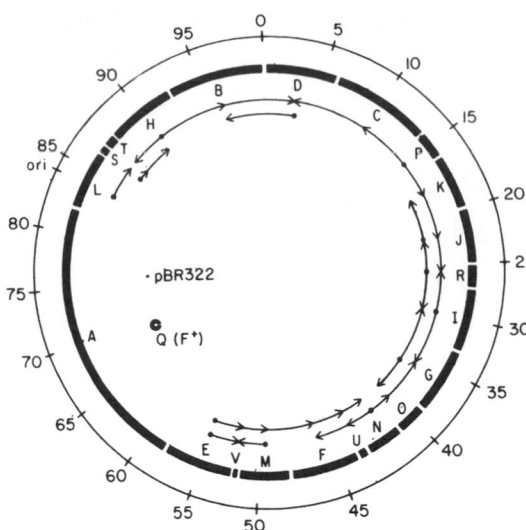

Figure 2 The physical map of *E. coli* strain EMG2, drawn to scale, with an F$^+$ plasmid (100 kb) and pBR322 (4.5 kb) shown to indicate relative sizes. The genetic map coordinates are superimposed on the physical map, with genetic distances distorted as needed to make the two maps coincident. Also shown (arrows) is a summary of some of the partial *Not*I digestion data used to complete and confirm the physical map. Each dot represents a probe, and each arrowhead indicates the size of bands seen in partial digests.

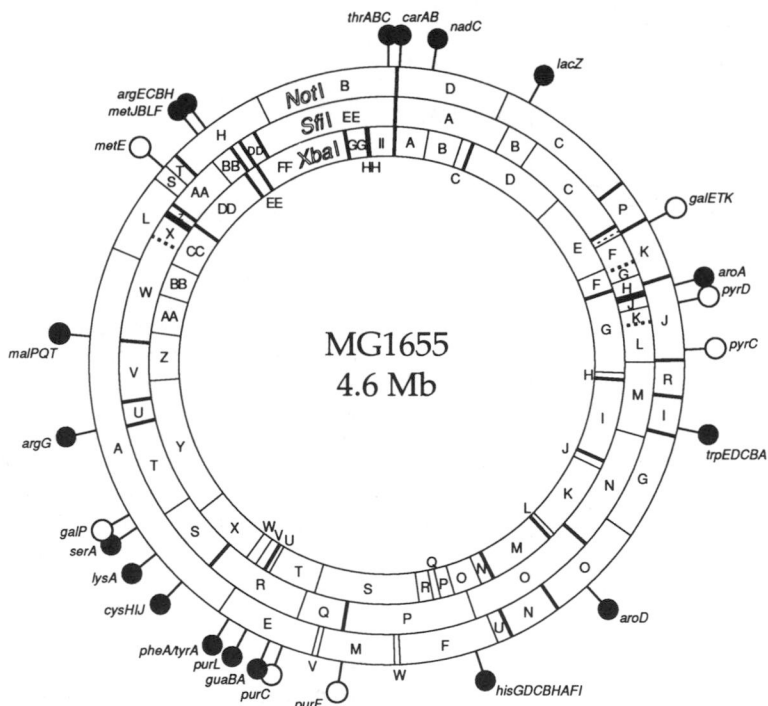

Figure 3 *Xba*I, *Not*I, and *Sfi*I restriction maps of *E. coli* strain MG1655. The map is oriented so that 0 minutes (genetic map) and 0 kb (physical map) are located at the top at locus *thrABC*. *Xba*I fragments (inner circle) are lettered consecutively from A to II beginning at site sX1 (25.6 kb) between fragments xII and xA near 0 minutes and continuing in a clockwise direction. The *Not*I map of MG1655 (outer circle) and the *Sfi*I map (middle circle) have been described previously (Heath et al. 1992a; Perkins et al. 1992). Transposon insertion locations are designated by lollipops and are labeled with the inactivated genetic locus. Filled lollipops represent insertions of Tn*10*dCamMCS, which contains *Not*I, *Sfi*I, and *Xba*I sites. Open lollipops represent insertions of Tn*10*dCamNS, which lacks a *Xba*I site. *Xba*I, *Not*I, and *Sfi*I sites present in sequenced regions of the chromosome are designated with thick lines. Sites present in unsequenced regions are designated with thin lines. The four *Sfi*I sites represented by dashed lines are only partially digested due to *dcm* methylation (Perkins et al. 1992).

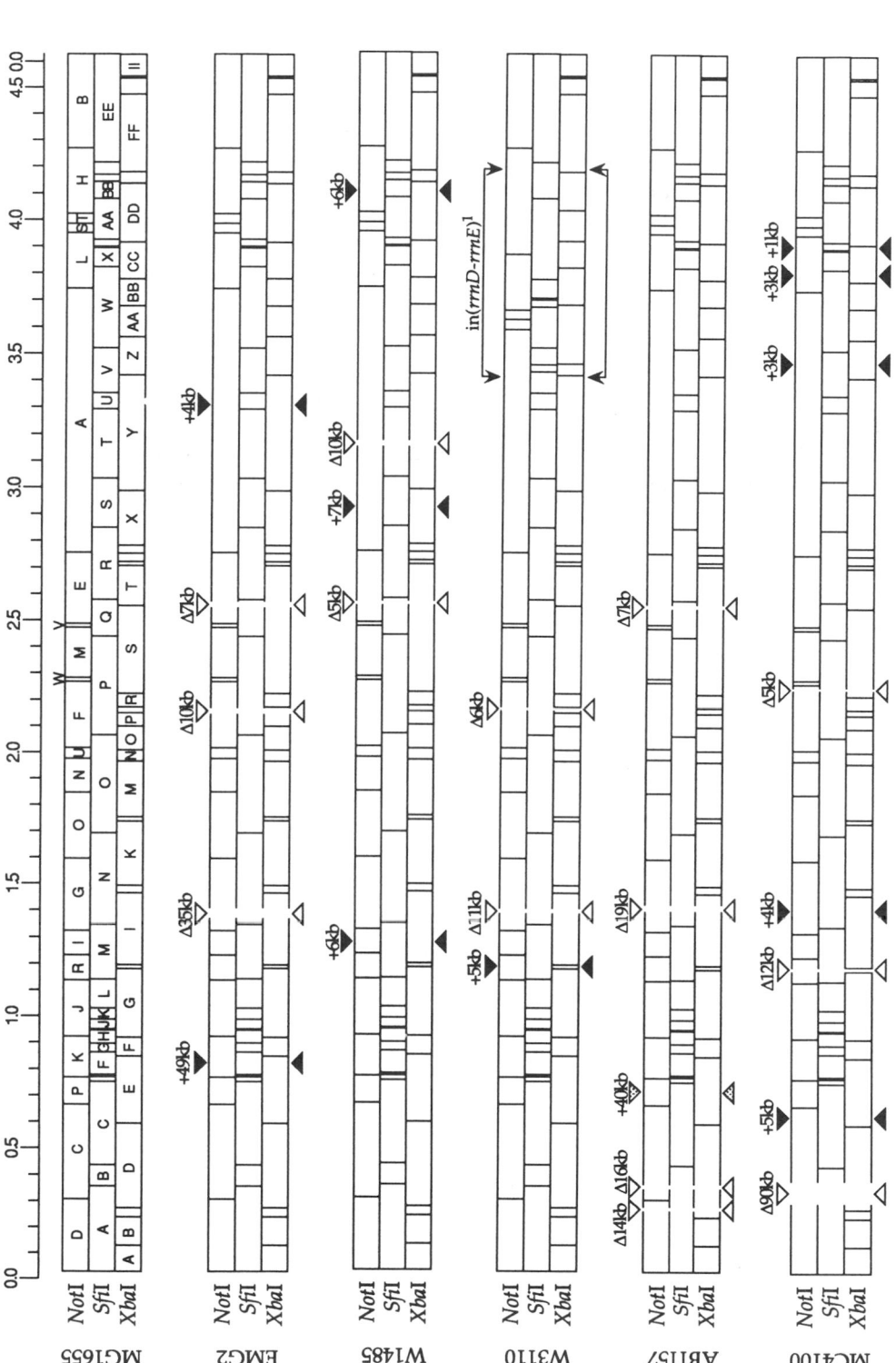

Figure 4 Restriction analysis of several *E. coli* K12 strains. Chromosomal DNA was isolated from several *E. coli* K12 strains and digested with the restriction enzymes *Not*I, *Sfi*I, and *Xba*I. Restriction maps for each strain were deduced by comparing the restriction digest patterns with that of MG1655. This figure illustrates the locations at which each strain differs from MG1655. The top of the figure is a scale numbered in megabase pairs. Each map is oriented such that 0 minutes is on the left. Filled triangles represent an increase in fragment size and open triangles a decrease in fragment size relative to MG1655. The amount of increase or decrease is indicated in kilobases above each triangle. In cases where a site has been deleted, that site is removed from the map. The dotted triangle on the map of AB1157 represents an unstable 40-kb element that is represented in approximately one half of the AB1157 chromosomes. The inversion present in strain W3110 is indicated, and the map has been adjusted accordingly.

D. *E. coli* Proteins Identified on Two-dimensional Gels and a Plasmid Map of *E. coli*

Neidhardt and co-workers have published four editions of the *E. coli* gene-protein index (Neidhardt et al. 1983; Phillips et al. 1987; VanBogelen et al. 1990; VanBogelen and Neidhardt 1991), allowing identification of protein spots on 2-D gels. This information is available as a continuously updated database (see Section 8B of this Handbook). Table 1 shows the current *E. coli* proteins identified on 2-D gels. For further information, see VanBogelen and Neidhardt (1991) and Section 8B of this Handbook. A detailed plasmid map that tells whether a particular region of interest has been cloned on plasmids in the Clarke-Carbon library (Clarke and Carbon 1976) was published by Phillips et al. (1987). Now, Nishimura et al. (1992) have correlated a subset of 518 pLC plasmids to the physical map of *E. coli* K12 (see Figure 1).

Table 1 is reprinted, with permission, from VanBogelen and Neidhardt (*Electrophoresis*, vol. 12, pp. 979–984 [1991]). Figure 1 is reprinted, with permission, from Nishimura et al. (*Microbiol. Rev.*, vol. 56, pp. 139–143 [1992]). The references and figures cited in the table, table notes, and figure caption can be found in the original papers.

Table 1   *E. coli* Proteins Identified on Two-dimensional Gels

Protein name a)	Alpha-numeric	Gene name b)	E. C. Number c)	Identification method d)	Donor or reference	Genbank Code
56Kd protein-A.L. Goldberg	G049.2			CP	A. Goldberg	
62Kd protein-J Nielson	G057.1			C	J. O. Lampen, J. Nielsen	
Acetate CoA transferase e)	C026.8	atoA	2.8.3.8	CP	F. Frerman	Ecoaspv
Acetate CoA transferase e)	G024.2	atoA	2.8.3.8	CP	F. Frerman	
Acetolactate synthase I	D057.0	ilvB	4.1.3.18	C	P.Silverman	Ecoilvbpr
Acetolactate synthase I	G010.6	ilvN	4.1.3.18	C	P.Silverman	Ecoilvbpr
Acetyl-CoA acetyltransferase e)	H038.8	atoB	2.3.1.9	CP	H. Schulz	
Acetyl-CoA acetyltransferase e)	H038.9	atoB	2.3.1.9	CP	H. Schulz	
Acetyl-CoA acyltransferase e)	H038.6	fadA	2.3.1.16	CP	H. Schulz	Ecofada
Acetyl-CoA acyltransferase e)	H038.7	fadA	2.3.1.16	CP	H. Schulz	Ecofada
Adenylate kinase	F026.0	adk	2.7.4.3	C	G. Cohen	Ecoadk
Alanine-tRNA ligase	F093.0	alaS	6.1.1.7	C	H. Weissbach	Ecoalas
Alcohol dehydrogenase	H097.3	adh	1.1.1.1		[25]	
Alkaline phosphatase	F046.6	phoA	3.1.3.1	CP	M. Schlesinger	Ecophoaa
Amidophosphoribosyltransferase	E054.1	purF	2.4.2.14	C	H. Zalkin	Ecopurf
AMP nucleosidase	G052.0	amn	3.2.2.4	CGM	V. Schramm	Ecoamn
Anthranilate phosphoribosyltranferase	H054.5	trpD	2.4.2.18	CGP	C. Yanofsky	Ecotrpx
Anthranilate synthase	E050.0	trpE	4.1.3.27	CGP	C. Yanofsky	Ecotrpx
Antigen 43	B052.0			C	P. Owen	
Antigen 47	B027.1			C	P. Owen	
Arabinose binding protein f)	F032.0	araF		CP	K. Matthews, [22]	Ecoarafgh
L-Arabinose isomerase e)	G054.1	araA	5.3.1.4	GP	R. Schleif	Ecoaraabd
AraC	H030.2	araC		CGP	N. Lee	Ecoarabop
Arginyl-tRNA ligase	E058.0	argS	6.1.1.19	CP	[3]	Ecoargs
ArgR e)	D014.3	argR		C	W. Maas	Ecoargr
Aromatic amino acid amino transferase	E036.0	tyrB	2.6.1.57	CG	S. Kiramitsu	Ecotyrb
Asparagine-tRNA ligase g)	G099.0	asnS	6.1.1.22	C	H. Weissbach	Ecotgasns
Asparate carbamoyltransferase-alpha	I017.5	pyrI	2.1.3.2	C	W. Lipscomb	Ecopyrbi
Asparate-tRNA ligase	F058.5		6.1.1.12	CP	H. Weissbach	
Aspartate aminotransferase	F039.6	aspC	2.6.1.1	CM	T. Yagi	Ecoaspc
Aspartate aminotransferase	F039.7	aspC	2.6.1.1	CM	T. Yagi	Ecoaspc
Aspartate carbamyltransferase-beta	H031.3	pyrB	2.1.3.2	C	W. Lipscomb	Ecopyrbi
ATP synthase- F1 sector, beta subunit	B046.7	atpD	3.6.1.3	C	P. Bragg	Ecouncc
ATP synthase- F1 sector, alpha subunit	G051.0	atpA	3.6.1.3	C	P. Bragg	Ecouncc
ATP-sulfurylase subunit	C058.3	cysN	2.7.7.4	C	T. Leyh	
Bacterioferritin	C014.3	bfr		CM	J. Yariv	Ecobfr
Binding Protein for leucine, isoleucine and valine	D040.7	livJ		CGP	J. Anderson	Ecolivjk1
Branched-chain amino-acid aminotransferase	F032.5	ilvE	2.6.1.42	C	E. Goldman, G. Hatfield	Ecoilve
Carbamoyl phosphate synthase-alpha subunit	G041.4	carA	6.3.5.5	CG	J. Villafranca	Ecocarab
Carbamoyl phosphate synthase-beta subunit	E133.0	carB	6.3.5.5	CG	J. Villafranca	Ecocarab
Chorismate mutase T-prephenate dehydrogenase	F037.9	tyrA	5.4.99.5	C	R. Duggleby	Ecopheab
Citrate-(si)-synthase	H047.4	gltA	4.1.3.7	GMP	H. Duckworth	Ecoglta
Citrate-(si)-synthase	H047.5	gltA	4.1.3.7	CGM	H. Duckworth	Ecoglta
ClpB i)	E072.0	clpB			[27]	
ClpB	F084.1	clpB		P	[27]	Ecoprot
ClpP	F021.5	clpP		P	[23]	Ecoclppa
Cold shock protein	B046.5			P	[26]	
Cold shock protein	G041.2			P	[26]	
Cold shock protein	G055.0			P	[26]	
CspA-cold shock protein	F010.6	cspA		P	[28]	Ecocspaa
Cyanate hydrolase e)	C013.6	cynS	3.5.5.3	C	J. Fuchs	Ecocynsa
Cyclic AMP receptor protein	I021.4	crp		CG	J. Krakow	Ecocrp
ß-cystathionase	G043.0	metC	4.4.1.8	CG	A. Martel	Ecometc
Cysteine-binding protein	C026.5			C	[22]	
Deoxy-manno-octulosonate cytidylyltransferase	C032.7	kdsB	2.7.7.38	C	W. Kohlbrenner	Ecokdsb
Dihydrofolate reductase	B020.0	folA	1.5.1.3	C	D. Smith	Ecofola
Dihydrolipoamide acetyltransferase	C070.0	aceF	2.3.1.12	CGMP	R. Perham	Ecoace
Dihydrolipoamide acetytransferase	C062.7	aceF	2.3.1.12	CGMP	R. Perham	Ecoace
Dihydrolipoamide dehydrogenase	G050.5	lpd	1.8.1.4	C	R. Perham, C. Williams	Ecoace
Dihydrolipoamide succinyltransfcrase	F050.3	sucB	2.3.1.61	CMP	R. Perham	Ecoglta
DNA-binding protein NS1	I011.5			C	A. Subramanian	
DNA-binding protein NS2	I011.3			C	A. Subramanian	
DNA-directed DNA polymerase I	F113.0	polA	2.7.7.7	C	W. Kelley	Ecopola
DNA-directed DNA polymerase III-beta	B036.1	dnaN	2.7.7.7		A. Aoyama	Ecodnaaop
DNA-directed DNA polymerase III-beta	A036.1	dnaN	2.7.7.7		A. Aoyama	Ecodnaaop
DNA-directed DNA polymerase III-delta	F032.4		2.7.7.7	C	A. Aoyama	

Table 1   *Continued*

Protein name a)	Alpha-numeric	Gene name b)	E. C. Number c)	Identifi-cation method d)	Donor or reference	Genbank Code
DNA-directed DNA polymerase III-epsilon	G027.3	*dnaQ*	2.7.7.7	C	A. Aoyama	Ecomutd
DNA-directed DNA polymerase III-gamma	H052.0	*dnaX*	2.7.7.7	C	A. Aoyama	Ecodnazx
DNA-directed DNA polymerase III-tau	H080.0	*dnaX*	2.7.7.7	C	A. Aoyama	Ecodnazx
DNA-directed RNA polymerase-alpha	B040.7	*rpoA*	2.7.7.6	CG	H. Weissbach, [14]	Ecorpa
DNA-directed RNA polymerase-beta	D157.0	*rpoB*	2.7.7.6	CG	H. Weissbach, [14]	Ecorplrpo
DNA-directed RNA polymerase-beta prime	I160.0	*rpoC*	2.7.7.6	CG	H. Weissbach, [14]	Ecorplrpo
DNA-directed RNA polymerase-sigma 32	F033.4	*rpoH*	2.7.7.6	G	[29]	Ecohtprr
DNA-directed RNA polymerase-sigma 70	B082.0	*rpoD*	2.7.7.6	G	H. Weissbach	Ecorpsrpo
DNA-topoisomerase I	I115.0	*topA*	5.99.1.2	C	N. Cozzarelli, F. Dean	Ecotopa
DNA-topoisomerase II ' subunit v	F044.6		5.99.1.3	C	A. Morrison	
DNA-topoisomerase II subunit A	D101.5	*gyrA*	5.99.1.3	C	N. Cozzarelli, R. McMacken, A.Morrison, R. Otter	Ecogyra
DNA-topoisomerase II subunit B	G094.0	*gyrB*	5.99.1.3	C	N. Cozzarelli, R. McMacken, R. Otter	
DnaA	I049.0	*dnaA*		C	J. Kaguni	Ecodnaaop
DnaB	B047.0	*dnaB*		C	R. McMacken	Ecodnab
DnaC	I030.3	*dnaC*		C	R. McMacken	Ecodnatc
DnaG--primase	G060.0	*dnaG*		C	R. McMacken	Ecorpsrpo
DnaJ	H036.5	*dnaJ*		CGP	C. Georgeopolou	Ecodnajk
DnaK (HSP-70)	B066.0	*dnaK*		G	C. Georgopoulos	Ecodnak
Ecotin	G016.3			C	A. Goldberg	
Enolase	F043.8	*eno*	4.2.1.11	C	F. Wold	
ERA protein	G031.0	*era*		CG	D. Court, C. Lemer	Ecoera
ERA protein	H032.0	*era*		CG	D. Court, C. Lemer	Ecoera
Exodeoxyribonuclease III (Endonuclease II)	G028.2	*xthA*	3.1.11.2	CG	B. Weiss	Ecoxtha
F-protein (Flavidoxin)	A019.0			C	R. Matthews	
Fatty acid oxidation complex	G073.4	*fadB*	1.1.1.35	CP	H. Schulz	
Fatty acid oxidation complex	G073.5	*fadB*	1.1.1.35	CP	H. Schulz	Ecofadab
Ferric uptake regulation repressor	G015.8	*fur*		C	J. Neilands	Ecofur
Formate acetyltransferase	G074.0	*pfl*	2.3.1.54	CP	J. Knappe	Ecopfl
Formate aetyltransferase	G070.0	*pfl*	2.3.1.54	CP	J. Knappe	Ecopfl
Fructose-bisphosphate	G038.6	*fbp*	3.1.3.11	C	F. Marcus, I. E	Ecofbpase
Galactose binding protein h)	C029.8	*mglB*		C	R. Hogg [22]	Ecomglb1
beta-Galactosidase e)	E123.0	*lacZ*	3.2.1.23	CGP	A. Fowler [14]	Ecolac
GlpR	G028.0	*glpR*		C	T. Larson	Ecoglpreg
Glucose phosphotransferase system enzyme III^GLC	B018.7	*crr*		C	S. Roseman	Ecophosys
Glucose-6-phosphate dehydrogenase	F048.8	*zwf*	1.1.1.49	CM	D. Fraenkel	Ecozwf
Glucosephosphotransferase enzyme II	H042.6	*ptsG*		C	C. Bouma	Ecoptsg
Glucosephosphotransferase enzyme II	H042.7	*ptsG*		C	C. Bouma	Ecoptsg
Glutamate decarboxylase	D046.5	*gadS*	4.1.1.15	CP	E. Boeker	
Glutamate decarboxylase	E046.5	*gadS*	4.1.1.15	CP	E. Boeker	
Glutamate dehydrogenase (NADP+)	G043.6	*gdhA*	1.4.1.4		D. Fraenkel	Ecogdhak
Glutamate dehydrogenase (NADP+)	G043.7	*gdhA*	1.4.1.4		D. Fraenkel	Ecogdhak
Glutamate synthase (NADPH) small subunit	F050.4	*gltD*	1.4.1.13	CM	W. Orme-Johnson	
Glutamate-ammonia ligase adenylylated	D050.0	*glnA*	6.3.1.2	C	R. Bender, S. Rhee	Ecoglna
Glutamate-ammonia ligase unadenylylated	D049.9	*glnA*	6.3.1.2	C	S. Rhee	Ecoglna
Glutamate-asparate-binding protein	H030.1			CG	[22]	
Glutamate-tRNA ligase	F047.8	*gltM*	6.1.1.17	C	H. Weissbach	
Glutamine-binding protein	H025.1			C	[22]	
Glutamine-binding protein	H025.0			C	[22]	
Glutamine-binding protein	H025.7			C	C. Ho	
Glutamine-tRNA ligase	G061.0	*glnS*	6.1.1.18	CP	[3]	Ecoglns
Glutaredoxin	B011.0			C	A. Holmgren	
Glutathione reductase-NAD(P)H	F045.0	*gor*	1.6.4.2	C	C. Williams	Ecogor
Glyceraldehyde-3-phosphate dehydrogenase	H034.3	*gap*	1.2.1.12	C	D. Fraenkel	Ecogap
Glyceraldehyde-3-phosphate dehydrogenase	I033.5	*gap*	1.2.1.12	C	D. Fraenkel	Ecogap
Glycerol kinase e)	E048.7	*glpK*	2.7.1.30	CP	H. Paulus	Ecoglyk
Glycerol-3-P dehydrogenase- (NAD+) aerobic e)	H047.3	*glpD*	1.1.99.5	C	J. Weiner	Ecosnglpd
Glycerol-3-phosphate diesterase	C039.0	*glpQ*		C	T. Larson	
Glycine betaine binding protein	G033.5	*proV*		C	M. Villarejo	Ecoprou
Glycine-tRNA ligase-beta subunit	E077.5	*glyS*	6.1.1.14	CGP	[3]	Ecoglys
GP2- (a glycerol-3-phosphate binding protein)	H041.0			CP	W. Boos	
GroEL	B056.5	*mopA*		CG	R. Hendrix, A. Subramanian	Ecogroels
GroES	C015.0	*mopB*		G	K. Tilly [30]	Ecogroels
GrpE	B025.3	*grpE*		GPM	[31]	Ecogrpe
H-NS	E014.0	*hns*		C	Guallerzi	Ecohns
H-NS	F014.7	*hns*		P,C	Guallerzi	Ecohns

Table 1 *Continued*

Protein name a)	Alpha-numeric	Gene name b)	E. C. Number c)	Identifi-cation method d)	Donor or reference	Genbank Code
Histidine binding protein	C028.5	hisJ		G	G. Ames	
Histidine permease	I026.2	hisP		G	G. Ames	Ecohismp
Histidine-tRNA ligase	F048.1	hisS	6.1.1.21	CP	H. Weissbach [3]	Ecohiss
Homocysteine N^5-methyltetrahydrofolate transmethylase	C137.0	metH	2.1.1.13	C	R. Matthews	Ecometh
HP33A	I034.0			C	T. Formosa, B. Alberts	
HtpE	C014.7	htpE		P	[32]	
HtpG	C062.5	htpG		P	[32]	Ecohsp
HtpH	D033.4	htpH		P	[32]	
HtpI	D048.5	htpI		P	[32]	
HtpK	F010.1	htpK		P	[32]	
HtpN	G013.5	htpN		P	[32]	
HtpO	G021.0	htpO		P	[32]	
HtpT	A029.5	htpT			This laboratory	
4-Hydroxy-2-oxoglutarate aldolase i)	F022.5		4.1.2.14	G	E. Dekker	
Hydroxydecanoyl-(acyl carrier protein)-hydrolase	H017.2	fabA	4.2.1.60	C	J. Cronan, Jr.	Ecofabaa
IHF-integration host factor j)	I010.5	himA		C	H. Nash	Ecohima
Indole-3-glycerol-phosphate synthase	F045.4	trpC	4.1.1.48	CGMP	C. Yanofsky	Ecotrpx
Indole-3-glycerol-phosphate synthase	F045.5	trpC	4.1.1.48	CGMP	C. Yanofsky	Ecotrpx
Isocitrate dehydrogenase-(NADP+)	C043.8	icd	1.1.1.42	CP	H. Reeves	Ecoicd
Isoleucine-tRNA ligase	F107.0		6.1.1.5	CG	[3]	
ß-ketoacyl-[acyl carrier protein] synthatase	F042.2	fabB	2.3.1.41	C	J. Cronan, Jr.	Ecofabb
KS protease	H083.0			C	Karen Silber	
KS protease	H083.1			C	Karen Silber	
Lac operon repressor protein	H039.0	lacI		C	K. Matthews	Ecolac
Lactamase	I035.7	ampC	3.5.2.6	CGP	B. Jaurin	Ecoampcfr
Lactate dehydrogenase	H062.0	dld	1.1.1.28	C	C. Ho	Ecodld
Leucine binding protein	B040.8	livK		CGP	D. Oxender	Ecolivjk2
Leucine regulatory repressor protein	I015.1	lrp		C	J. Calvo	Ecolrrpa
Leucine-tRNA ligase	D100.0	leuS	6.1.1.4	CG	[3]	Ecoleus
Lipoprotein k)	G011.4	lpp		CF	M. Inouye	Ecolpp
Lon (Protease La)	H094.0	lon		CG	A. Goldberg, A. Markovitz	Ecolon
Lon (Protease La)	H094.1	lon		C	A. Goldberg, A. Markovitz	Ecolon
Lysine decarboxylase e)	G075.0	cadA	4.1.1.18	CP	E. Boeker	
Lysine-tRNA ligase form I	D058.5	lysS	6.1.1.6	CG	I. Hirschfield [3]	Ecoherc
Lysine-tRNA ligase form II	D060.5	lysU	6.1.1.6	CG	I. Hirschfield	Ecolysug
Lysine-tRNA ligase form III	C058.4		6.1.1.6	CG	I. Hirschfield	
Lysine-tRNA ligase form IV	C060.5		6.1.1.6	CG	I. Hirschfield	
Maltose-binding protein	D036.0	malE		P	[22]	Ecomalb
Methionine adenosyltransferase	C044.6	metK	2.5.1.6	CG	E. Hafner	Ecometk
Methionine aminopeptidase	F029.7	map		C	A. Ben-Bassit	Ecomap
Methionine-tRNA ligase	F072.0	metG	6.1.1.10	C	H. Weissbach	Ecometg
Methionine-tRNA ligase	G072.0	metG	6.1.1.10	C	H. Weissbach	Ecometg
Methionyl-tRNA formyltransferase	F033.6		2.1.2.9	C	H. Weissbach	
Methyl-accepting chemotaxis protein I	B060.1	tsr		L	[33]	Ecotsr
Methyl-accepting chemotaxis protein I	B060.2	tsr		L	[33]	Ecotsr
Methyl-accepting chemotaxis protein I	B060.3	tsr		L	[33]	Ecotsr
Methyl-accepting chemotaxis protein I	B060.4	tsr		L	[33]	Ecotsr
Methyl-accepting chemotaxis protein I	B060.5	tsr		L	[33]	Ecotsr
Methyl-accepting chemotaxis protein I	B060.6	tsr		L	[33]	Ecotsr
Methyl-accepting chemotaxis protein I	B060.7	tsr		L	[33]	Ecotsr
Methyl-accepting chemotaxis protein I	B060.8	tsr		L	[33]	Ecotsr
Methly-accepting chemotaxis protein II	D058.1	tar		L	[33]	Ecotartap
Methly-accepting chemotaxis protein II	D058.3	tar		L	[33]	Ecotartap
Methly-accepting chemotaxis protein II	D058.2	tar		L	[33]	Ecotartap
Methly-accepting chemotaxis protein II	E058.1	tar		L	[33]	Ecotartap
Methly-accepting chemotaxis protein II	E058.2	tar		L	[33]	Ecotartap
Methly-accepting chemotaxis protein II	E058.3	tar		L	[33]	Ecotartap
Muramoyl-pentapeptide carboxypeptidase	H040.5	dacA	3.4.17.8	CZ	U. Schwarz, F.Wientjes	Ecodaca
Murein transglycosylase-soluable form	H063.0			C	U. Schwarz	
MutS	E093.0	mutS		C	P. Modrich	
NADH oxidase	B035.1			?		
NADPH: flavidoxin/ferridoxin oxidoreductase	H030.5			C	J. Knappe	
NusA	B061.0	nusA		C	J. Greenblatt via D. Friedman	Econusa
NusB	H013.8	nusB		G	D. Friedman	Econusb
O6-methylguanine-DNA methyltransferase	I039.5	ada	2.1.1.63	CP	T. Lindahl	Ecoada
OmpA k)	F024.5	ompA		C	S. Mizushima	Ecoompa

Table 1   *Continued*

Protein name a)	Alpha-numeric	Gene name b)	E. C. Number c)	Identification method d)	Donor or reference	Genbank Code
OmpA k)	F028.0	ompA		F	S. Mizushima	Ecoompa
OmpA k)	F033.0	ompA		F	S. Mizushima	Ecoompa
OmpA k)	F033.1	ompA		F	S. Mizushima	Ecoompa
OmpC	A035.5	ompC		CG	L. van Alphen, S. Mizushima	Ecoompc
OmpF	B036.0	ompF		CG	L. van Alphen, M Inouye, S. Mizushima	Ecoompf
Origin binding protein a	G055.3			C	M. Schaechter	
Ornithine carbamoyltransferase	F039.0	argI	2.1.3.3	C	W. Lipscomb	EcoargI
Ornithine decarboxylase	F080.0	speC	4.1.1.17	G	S. Boyle	Ecospec
Oxoglutarate dehydrogenase-lipoamide	G097.0	sucA	1.2.4.2	CGM	R. Perham	Ecoglta
P II-uridylyated	F012.2	glnB		CP	S. Rhee	Ecoglnb
p46	C025.4	pfs		C	S. Quirk	
Phenylalanine-tRNA ligase- alpha subunit	G036.0	pheS	6.1.1.20	CP	[3]	Ecothrinf
Phenylalanine-tRNA ligase-beta subunit	D094.0	pheT	6.1.1.20	CP	[3]	Ecothrinf
PhoE	B037.0	phoE		CP	J. Tommassen	Ecophoe
Phospho-2-keto-3-deoxygluconate aldolase i)	F022.5	eda	4.1.3.16	CM	D. Fraenkel	
Phospho-2-keto-3-deoxyheptonate aldolase	F038.1	aroF	4.1.2.15	C	A. DeLucia	Ecoarof
Phosphoenolpyruvate carboxykinase (GTP) e)	E056.0	pck	4.1.1.47	CP	W. Bridger	Ecopcka
Phosphoenolpyruvate carboxylase	F084.0	ppc	4..1.1.3	C	H. Katsuki	Ecoppcg
Phosphoenolpyruvate-protein phosphotransferase	B058.3	ptsI	2.7.3.9	C	E.B. Waygood	Ecophosys
Phosphofructokinase I	F035.8	pfkA	2.7.1.11	C	D. Fraenkel	Ecocdha
Phosphofructokinase II	E036.6	pfkB	2.7.1.11	C	D. Fraenkel	Ecopfkbk
Phosphogluconate dehydrogenase e)	C042.6	gnd	1.1.1.44	CP	D. Fraenkel	Ecognd
Phosphohistidinoprotein-hexose phosphotransferase	F007.0	ptsH	2.7.1.69	C	S. Roseman, E.B. Waygood	Ecoptsh
Phosphoribosylaminoimidazole-succinocarboxamide synthetase m)	B026.3	purC	6.3.2.6		[24]	Ecopurca
Polyhook	A039.8			C	P. Matsamura	
Polyribonucleotide nucleotidyltransferase	C078.0	pnp	2.7.7.8	CG	C. Portier	Ecorpsop
Protease III	G095.0	ptr		C	D. Zipser	Ecoptr
Protein chain elongation factor, EF-G	D084.0	fusA		CG	[14]	Ecostr2
Protein chain elongation factor, EF-P	C022.7			C	C. Ganoza	
Protein chain elongation factor, EF-Ts	C030.7	tsf		CG	[14]	Ecorpsbts
Protein chain elongation factor, EF-Ts	C031.6	tsf		CG	[14]	Ecorpsbts
Protein chain elongation factor, EF-Tu	E042.0	tufA		CG	[14]	Ecostr3
Protein chain initiation factor 2A	G117.0	infB		C	H. Weissbach	Econusa
Protein chain initiation factor 2B	F119.0	infB		C	H. Weissbach	Econusa
Protein chain initiation factor 3L	I019.5	infC		CG	D. Elhardt	Ecohima
Protein chain initiation factor 3S	I020.1	infC		CG	D. Elhardt	Ecothrinf
Protein n' Replication factor Y	H096.8	priA		C	S. Wickner via	Ecopriafy
Protein release factor 1	H021.6	prfA		C	P. Tai	Ecorrfx
Protein release factor 1	H021.7	prfA		C	P. Tai	Ecorfix
PS (phage shock) protein	D026.3			CP	J. Brisette, P. Model	
PS (phage shock) protein	E026.0			CP	J. Brisette, P. Model	
Pyruvate dehydrogenase-lipoamide	F099.0	aceE	1.2.4.1	CGM	R. Perham	Ecoace
Pyruvate kinase I	G054.7		2.7.1.40	C	M. Malcovati	
Pyruvate oxidase n)	G058.0	poxB	1.2.3.6	C	J. Cronan, Jr.	Ecopoxb
Pyruvate water dikinase	B083.0	pps	2.7.9.2	CP	W. Bridger	
R-protein	G029.7			C	R. Matthews	
RecA	C039.3	recA		CGP	J. Little [34]	Ecoreca
Rescue	D046.4			C	C. Ganoza	
Rhamnulose-1-phosphate aldolase e)	F029.4	rhaD	4.1.2.19	CMP	D. Feingold	
Rhamnulose-1-phosphate aldolase e)	G030.0	rhaD	4.1.2.19	CMP	D. Feingold	
Ribose-binding protein	H027.9	rbsB		C	[22]	Ecorbs
Ribosomal protein L7/L12 acetyltransferase	F037.8			C	H. Weissbach	
Ribosomal protein L7/L12 acetyltransferase	F038.5			C	H. Weissbach	
Ribosomal subunit protein	I010.3				[35]	
Ribosomal subunit protein	I011.6			F	[35]	
Ribosomal subunit protein	I012.3			F	[35]	
Ribosomal subunit protein	I013.3			F	[35]	
Ribosomal subunit protein	I013.7			F	[35]	
Ribosomal subunit protein	I014.2			F	[35]	
Ribosomal subunit protein	I014.3			F	[35]	
Ribosomal subunit protein	I014.5			F	[35]	
Ribosomal subunit protein	I014.7			F	[35]	
Ribosomal subunit protein L1	I026.0	rplA		CF	M. Nomura	Ecorplrpo
Ribosomal subunit protein L2 o)	I030.2	rplB		CM	M. Nomura	Ecorpos10

Table 1 *Continued*

Protein name a)	Alpha-numeric	Gene name b)	E. C. Number c)	Identifi-cation method d)	Donor or reference	Genbank Code
Ribosomal subunit protein L3	I023.0	rplC		CF	M. Nomura	Ecorpos10
Ribosomal subunit protein L4	I021.1	rplD		CF	M. Nomura	Ecorpos10
Ribosomal subunit protein L5	I016.4	rplE		CF	M. Nomura	Ecorpln
Ribosomal subunit protein L6	I017.2	rplF		CF	M. Nomura	Ecorpln
Ribosomal subunit protein L7	A013.0	rplL		CF	M. Nomura [36]	Ecorplrpoe
Ribosomal subunit protein L9	H014.0	rplI		CF	M. Nomura	Ecorpsfri
Ribosomal subunit protein L11	I014.1	rplK		CF	M. Nomura	Ecorplrpo
Ribosomal Subunit protein L12	B013.0	rplL		CF	M. Nomura [36]	Ecorplrpo
Ribosomal subunit protein L17 o)	I014.9	rplQ		C	M. Nomura	Ecorpa
Ribosomal subunit protein L21	I013.2	rplU		CF	[14]	
Ribosomal subunit protein L23	I013.0	rplW		C	M. Nomura	Ecorpos
Ribosomal subunit protein L25	I011.8	rplY		CF	M. Nomura	
Ribosomal subunit protein L27 o)	I012.8	rpmA		C	M. Nomura	
Ribosomal subunit protein L28 o)	I012.2	rpmB		C	M. Nomura	Ecorpmbq
Ribosomal subunit protein L29	I011.1	rpmC		C	M. Nomura	Ecorpos10
Ribosomal subunit protein L30	I011.4	rpmD		CF	M. Nomura	Ecorpln
Ribosomal subunit protein L32 o)	I011.7	rpmF		C	M. Nomura	Ecorpmfq
Ribosomal subunit protein S1	B065.0	rpsA		CF	M. Nomura [14]	Ecorpsa
Ribosomal subunit protein S2	H027.0	rpsB		F	[37]	Ecorpsbts
Ribosomal subunit protein S5	I014.4	rpsE		CF	M. Nomura	Ecorpln
Ribosomal subunit protein S6A	D014.7	rpsF		CF	M. Nomura [38]	Ecorpsfri
Ribosomal subunit protein S6B	C014.8	rpsF		CF	M. Nomura [38]	Ecorpsfri
Ribosomal subunit protein S6C	C015.3	rpsF		F	[38]	Ecorpsfri
Ribosomal subunit protein S8	I013.5	rpsH		CF	M. Nomura	Ecorpln
Ribosomal subunit protein S9	I014.6	rpsI		CM	M. Nomura	Ecorpsi
Ribosomal subunit protein S10	I012.1	rpsJ		CF	M. Nomura	Ecorpos10
Ribosomal subunit protein S11 o)	I013.8	rpsK		C	M. Nomura	Ecorpa
Ribosomal subunit protein S15	I012.0	rpsO		CF	M. Nomura	Ecorpsop
Ribosomal subunit protein S16	I011.9	rpsP		C	M. Nomura	Ecotrmd
Ribosomal subunit protein S17	I012.6	rpsQ		CF	M. Nomura	Ecorpos10
Ribosomal subunit protein S20 o)	I012.9	rpsT		C	M. Nomura	Ecorpsta
Ribulokinase	D055.0	araB	2.7.1.16	CP	N. Lee	Ecoaraabd
Ribulose-phosphate 4-epimerase e)	G028.1	araD	5.1.3.4	CP	N. Lee	Ecoaraabd
rRNA(m6/2A)methyltransferase	H025.2	ksgA		C	P. VanKnippenberg	Ecoksga
Serine hydroxymethyltransferase	G043.8	glyA	2.1.2.1	CG	V. Schirch	EcoglyA
Serine-tRNA ligase	E048.8	serS	6.1.1.11	C	H. Weissbach	Ecosers
Single-stranded DNA binding protein	F018.8	ssb		C	R. McMacken, K. Williams	Ecossb
SspB	A025.8	sspA		G	M. D. Williams	
Stringent starvation protein	D027.1	sspA		G	M. D. Williams	Ecossspg
Succinate dehydrogenase (fumerate reductase)-flavoprotein subunit e)	G063.2	frdA	1.3.99.1	CP	J. Weiner	Ecoampcfr
Succinate dehydrogenase (fumerate reductase)-iron-sulfur protein e)	H024.7	frdB	1.3.99.1	CP	J. Weiner	Ecoampcfr
Succinate dehyrogenase	F060.3	sdhA	1.3.99.1	C	J. Weiner	Ecoglta
Succinate-CoA ligase-(ADP-forming)-alpha subunit	H028.0		6.2.1.5	CP	W. Bridger	
Succinate-CoA ligase-(ADP-forming)-beta subunit	E039.8		6.2.1.5	CP	W. Bridger	
O-Succinylhomoserine(thiol)-lyase	G040.1	metB	4.2.99.9		A. Martel	Ecometlb1
Sulfate adenylyltransferase	H036.0	cysD	2.7.7.4	CP	T. Leyh	
Superoxide dismutase-iron protein	F021.3	sodB	1.15.1.1	C	J. Fee	Ecosodb
Superoxide dismutase-manganese protein	I021.3	sodA	1.15.1.1	CP	J. Fee, P. Reichard	Ecosoda
Tetrahydropteroyltriglutamate methylase	F088.0	metE	2.1.1.14	GP	C. Berg	
Thioredoxin	B012.0	trxA		CG	C. Lunn, M. Russel, C. Williams	Ecotrxa
Thioredoxin reductase-NADPH	D031.5	trxB	1.6.4.5	CG	M. Russel, C. Williams	Ecotrxb
Threonine dehydratase-biosynthetic	F050.1	ilvA	4.2.1.16	CM	E. Goldman, G. Hatfield	Ecoilva
Threonine-tRNA ligase	G065.0	thrS	6.1.1.3	C	[3]	Ecothrinf
TnpR (resolvase from Tn3)p)	I016.0			C	N. Cozzarelli, J. Dungan	
Transcription termination factor rho	I048.8	rho		CG	[37]	Ecorho
Trigger factor	B050.3	tig		C	W. Wickner	Ecotig
tRNA uracil-5-methyltransferase	G039.6	trmA	2.1.1.35	C	T. Ny	Ecotrma
Tryptophan synthase-alpha subunit	E025.5	trpA	4.2.1.20	CGP	E. Wilson Miles, C. Yanofsky	Ecotrpx
Tryptophan synthase-beta subunit	G040.0	trpB	4.2.1.20	CGP	E. Wilson Miles, C. Yanofsky	Ecotrpx
Tryptophan-tRNA ligase	H037.0	trpS	6.1.1.2	CG	K. Muench, C. Yanofsky	Ecotrps
Tryptophanase e)	G046.5	tnaA	4.1.99.1	CP	M. Goldberg	Ecotnaa
Tyrosine-tRNA ligase	G044.0	tyrS	6.1.1.1	C	H. Weissbach	Ecotyrs
UvrA	H124.0	uvrA		C	L. Grossman, A. Sancar	Ecouvra

Table 1   *Continued*

Protein name a)	Alpha-numeric	Gene name b)	E. C. Number c)	Identification method d)	Donor or reference	Genbank Code
UvrB	C080.0	*uvrB*		C	L. Grossman, A. Sancar	Ecouvrb
UvrC	I065.0	*uvrC*		C	L. Grossman, A. Sancar	Ecouvrca
Valine-tRNA ligase	E106.0	*valS*	6.1.1.9	CG	[3]	Ecosyntgv
W-protein	G043.2			C	C. Ganoza	
Xylose-binding protein	C032.4			P	[3]	

a) Protein names are from *Enzyme Nomenclature* [21] and Bachmann [20]. Abbreviations: kDa, Kilodalton; CoA, coenzyme A ; RNA, ribonucleic acid; tRNA transfer RNA; AMP adenosine 5'-monophospahte; DNA, deoxyribonucleic acid; NADPH, reduced nicotinamide-adenine dinucleotide phosphate; NAD(P)H, either NADH reduced nicotinamide-adenine dinucleotide) of NADPH; NAD$^+$, oxidized nicotinamide-adenine dinucleotide; NADP$^+$, oxidized nicotinamide-adenine dinucleotidephosphate; GTP, guanosine 5'-triphosphate; ADP, adenosine 5'-diphosphate.

b) Gene names are from Bachmann [20].

c) EC numbers are from *Enzyme Nomenclature* [21].

d) Letters designating the methods used to identify proteins are: C, migration with purified protein; F, migration with polpeptide or purified cellular fraction; G, genetic criterion, e.g. mutants (deletion, insertion, frameshift, nonsense, missense, regulatory), plasmid bearing strians, and in vitro synthesis of protein; L, selective labeling, e.g., methylation and phosphorylation; M, peptide map similarity; P, physiological critierion, e.g., induction and repression; Z, selective derivatization, e.g., covalent binding of penicillan.

e) Proteins repressed by growth in glucose-containing medium or induced by specific growth conditions do not appear on the autoradiograms in Figures 1 and 2.

f) Copeland et al. [22] reported arabinose-binding protein to migrate at ca. 86 x 73 (Fig 1a). The purified proteins used by our two groups were supplied by different donors; we have not yet reconciled this discrepancy.

g) The large dark spot at these coordinates is F088.0, which obscures G099.0. When cells are grown in methionine containing medium, synthesis of F088.0 is repressed and G099.0 can be clearly seen. Similarly, gels of extract of a mutant, CBK040 (metE::Tn5), that does not produce F088.0, allow the clear identification of G099.0 as asparagine-tRNA ligase.

h) Copeland et al. [22] reported galactose-binding protein to migrate at ca. 77 x 73 (Fig 1a). The small difference in spot patterns is probably due to variation in electrophoresis methods that resulted in slightly different spot patterns.

i) The unexpected finding that this spot corresponds to two different enzymes has not yet been resolved.

j) Integration host factor (IHF) subunits alpha and beta have previously been resolved only on urea-sodium dodecyl sulfate gels (H. Nash, personal communication).

k) These proteins are solubilized when the cell extract is prepared by boiling in the presence of SDS and, therefore, appear faintly, if at all, on the autoradiograms of Figs. 1 and 2.

l) Squires [23] reported this protein as F68.5 but stated that is was the same as protein E72.0 in [6].

m) Parker [24] reported this protein as C027.0 although he reports its migration at 91 x 64 as our group has. The alpha-numeric designation was originally assigned as B026.3 migrating at 91 x 64 in [14].

n) The faintness of the spot at these coordinates is misleading. Pyruvate oxidase migrates with a much more abundant polypeptide spot (intensity approximately the same as (45 x 107) located at these coordinates on other gels.

o) These proteins do not appear in the autoradiogram in Fig 2B. Some nonequilibrium gels have more proteins on the basic side of those shown in Fig. 2B, with which these ribosomal proteins comigrate. Coordinates given are approximate locations of the proteins if they were to appear on the reference gel.

p) This spot would appear only when extracts are prepared from cells carrying transposon Tn5.

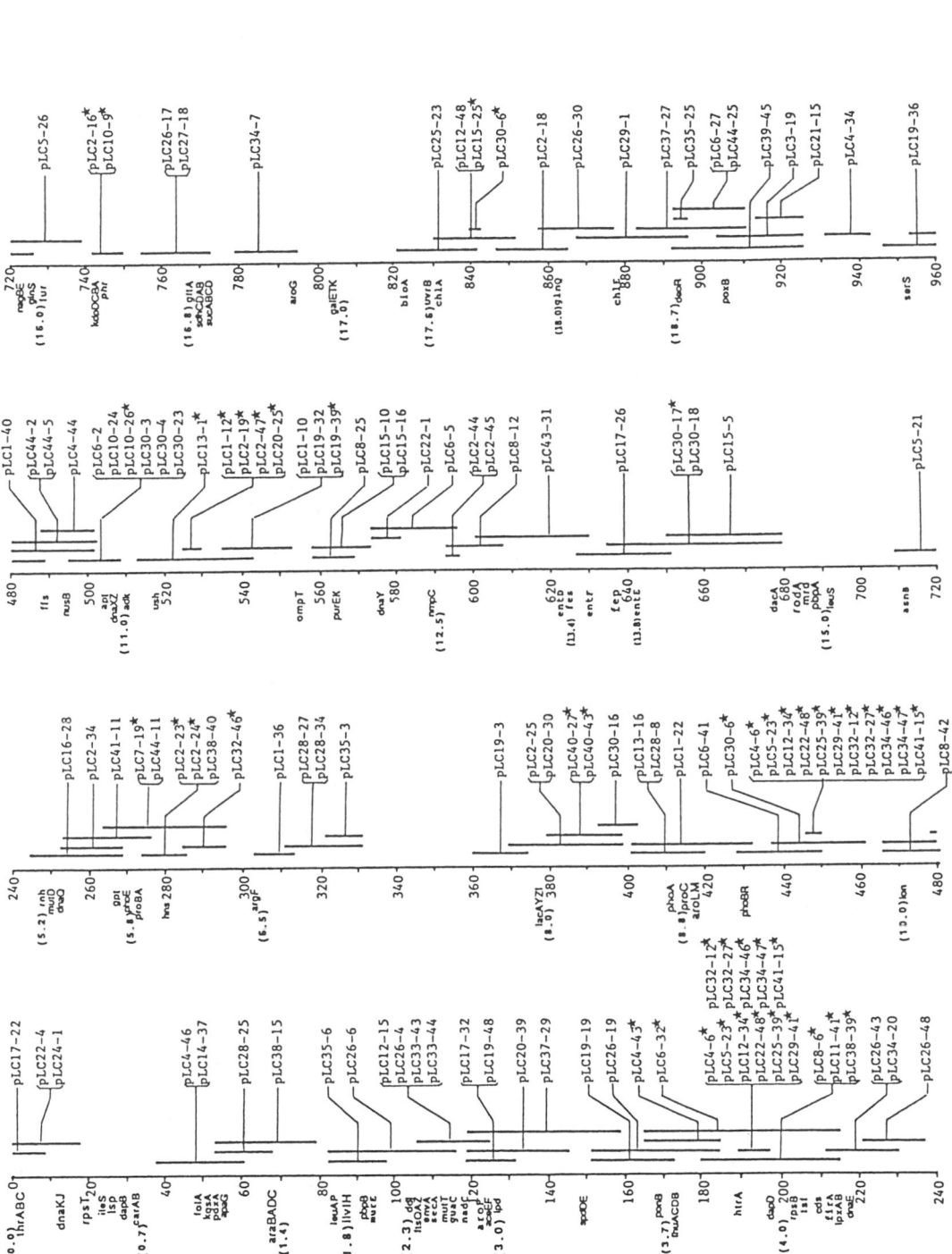

Figure 1   Location of pLC plasmids on the physical map. The vertical bars show the DNA regions carried by pLC plasmids that hybridized to the various *E. coli* ordered clones. However, the ends of each bar have not been accurately determined. The bars simply indicate the maximal length that does not overlap with the adjacent lambda clones that did not hybridize to the plasmids. The pLC plasmids that hybridized to clones from more than one location are indicated by ★. To make a map compatible with the genetic map published by Bachmann (1) and with the plasmid map of Phillips et al. (18), the revised physical map of Médigue et al. (15) was used. Médigue et al. have restored the inversion in the strain W3110 used by Kohara et al. (i.e., inversion between *rrnD* and *rrnE* (9)) to its normal orientation and have located on it the sequenced genes available in the EMBL library. Some additional genes are from a previous paper by Kohara (11).

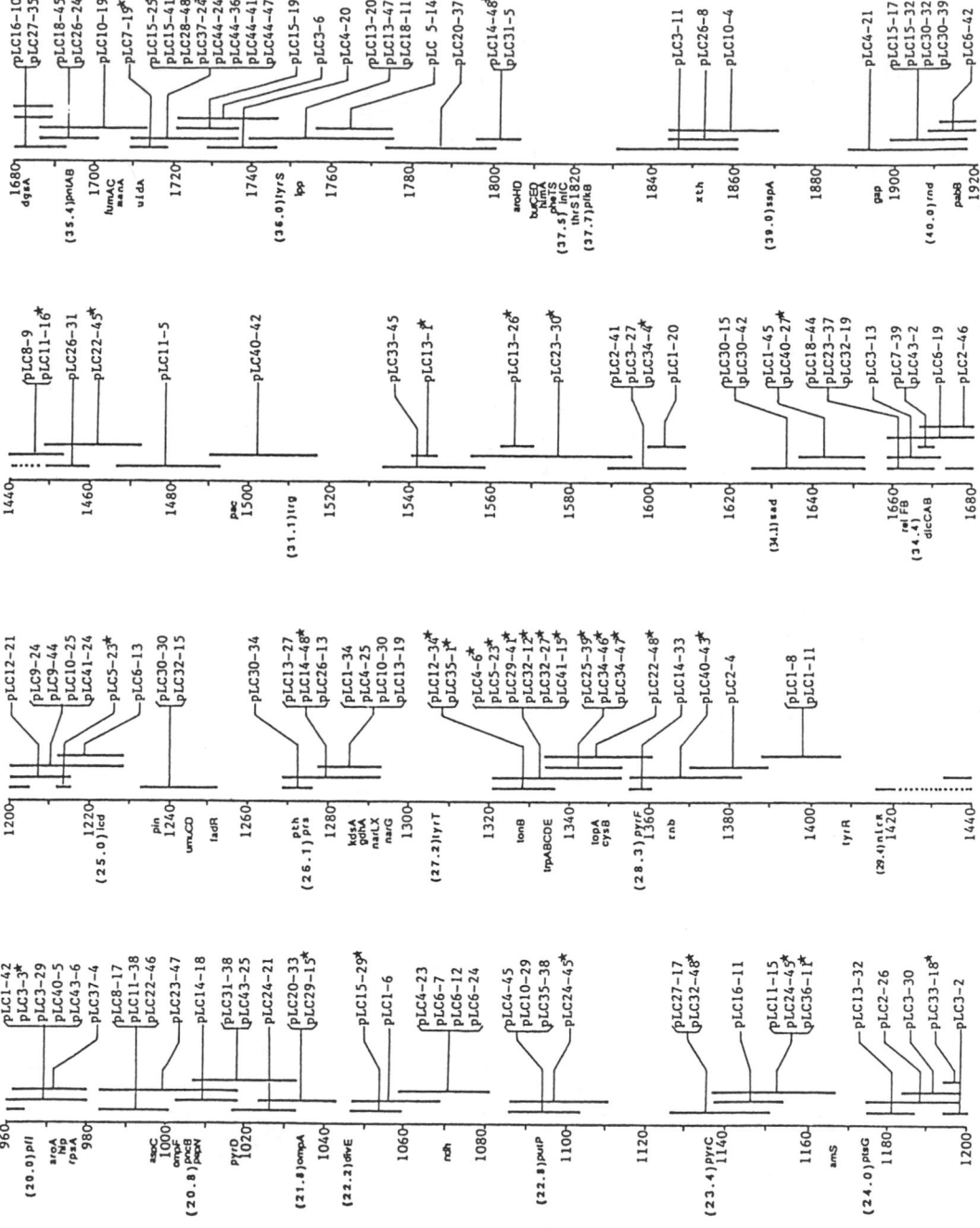

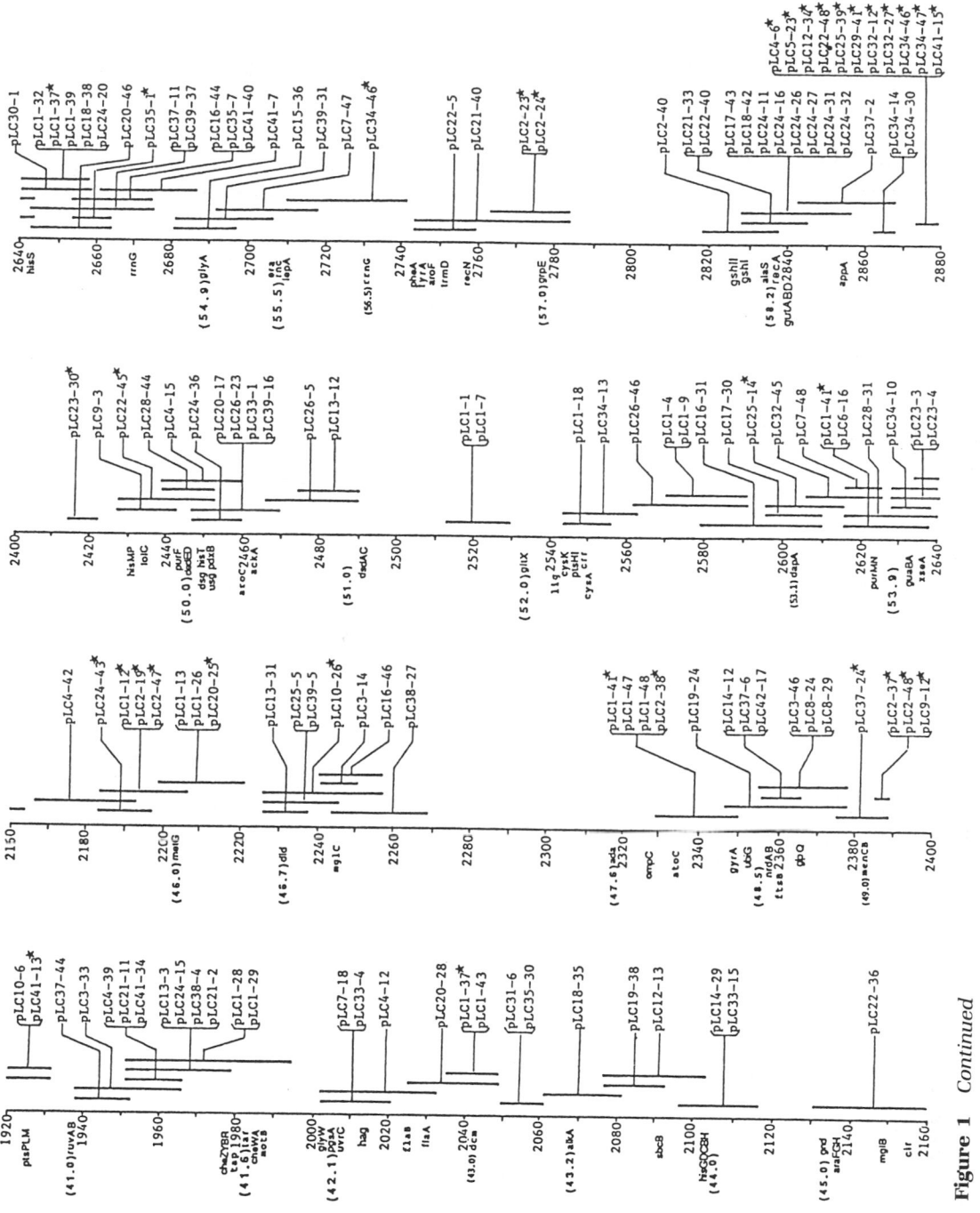

Figure 1 *Continued*

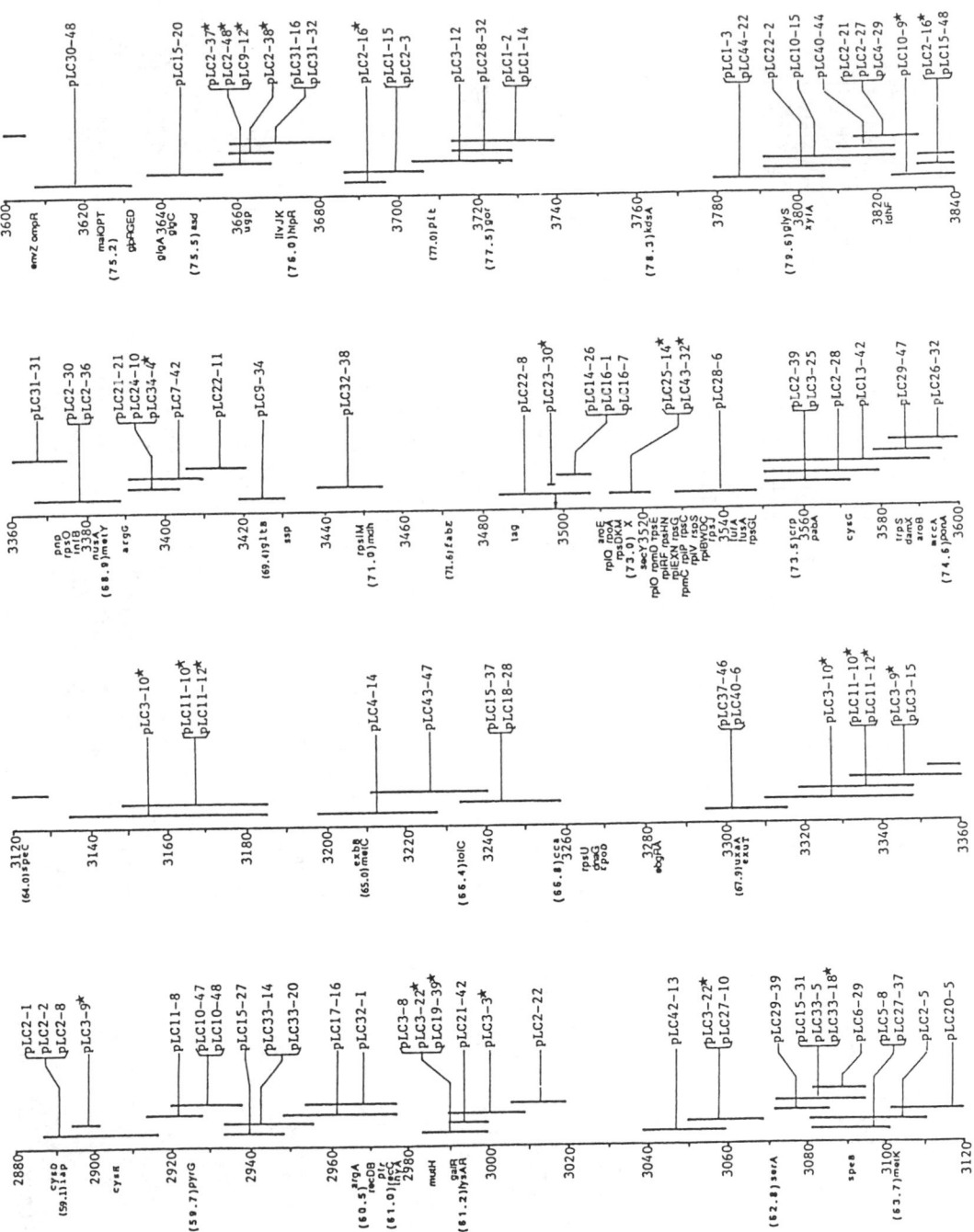

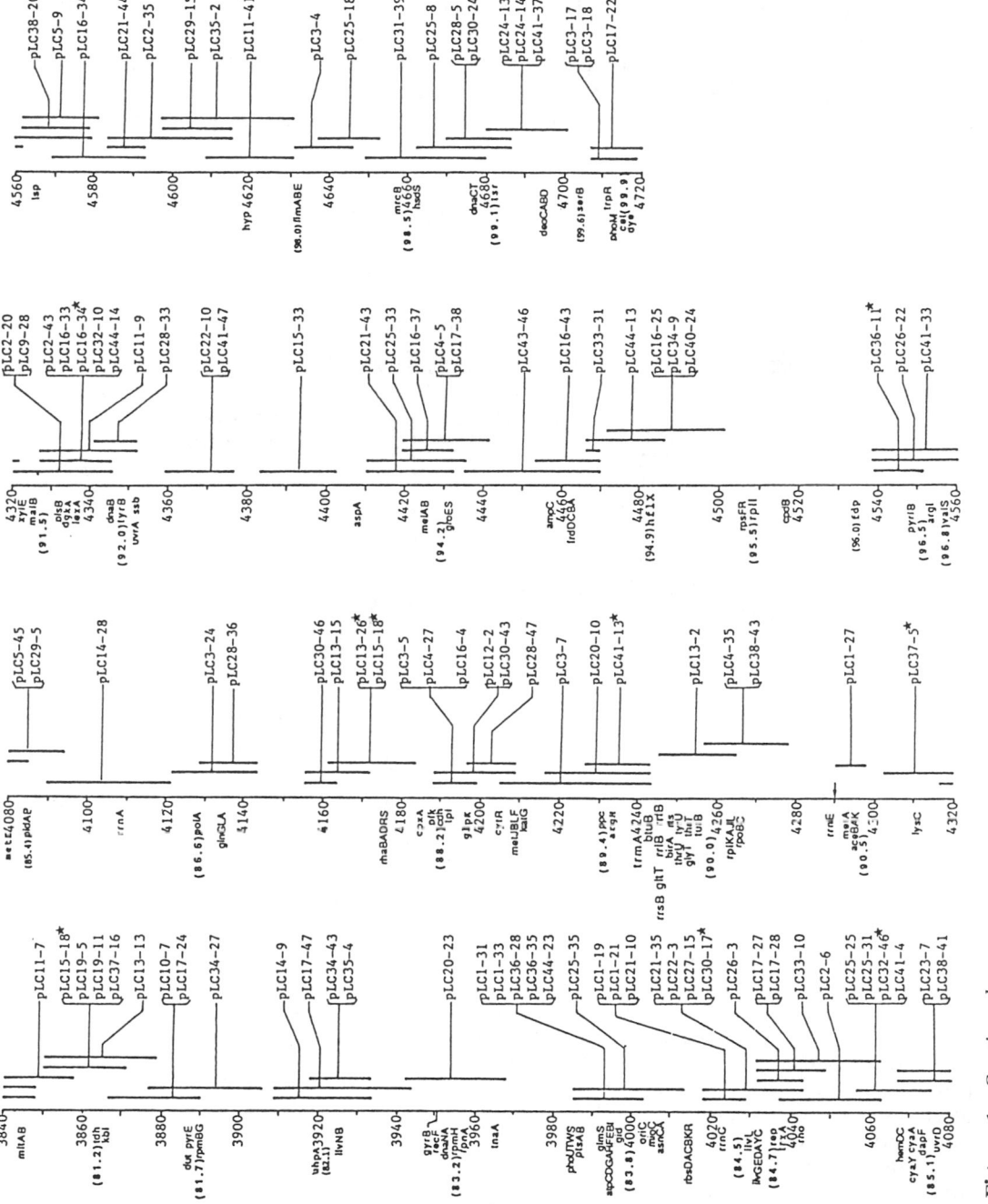

Figure 1 *Continued*

E. Alignment of Genetic Maps of *E. coli* and *S. typhimurium*

Figure 1 is reprinted, with permission, from Riley and Sanderson (*The Bacterial Chromosome* [eds. K. Drlica and M. Riley]. American Society for Microbiology, Washington, D.C., p. 90 [1990]). Figure 2 is reprinted, with permission, from Riley and Krawiec (Escherichia coli *and* Salmonella typhimurium. *Cellular and Molecular Biology* [ed. F.C. Neidhardt]. American Society for Microbiology, Washington, D.C., pp. 974–976 [1987]). The references cited in the figure captions can be found in the original papers.

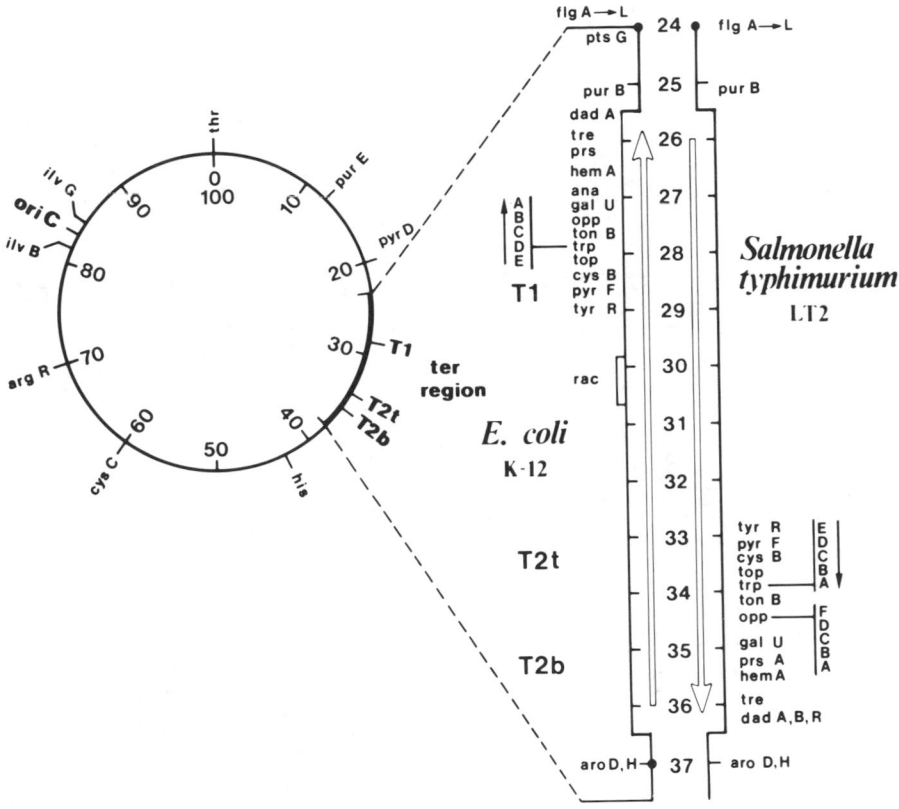

Figure 1 Inverted region in the linkage maps of *E. coli* K12 (3) and *S. typhimurium* LT2 (55). In most cases, genes are shown only if they have been mapped in both genera; other genes may be found in references 3 and 55. The positions of genes shown on the circular map at the left are the same in the two genera. Details of the structure of the *ter* (terminus) region in *E. coli* K12 (shown in the figure as T1, T2t, and T2b) are from François et al. and Kuempel et al. (in *The Bacterial Chromosome*). An inversion of the DNA has been detected (53) and estimated by Casse et al. (12) to include 10–11% of the chromosome. The region between *purB* (25 min) and *aroD,H* (37 min) is inverted; the region is shown expanded on the right. The *ter* region of *S. typhimurium* LT2 has not been mapped precisely, but T1 is expected to be at ~33 minutes on the *S. typhimurium* map, T2t is expected to be at 28 minutes, and T2b is expected to be at 29 minutes.

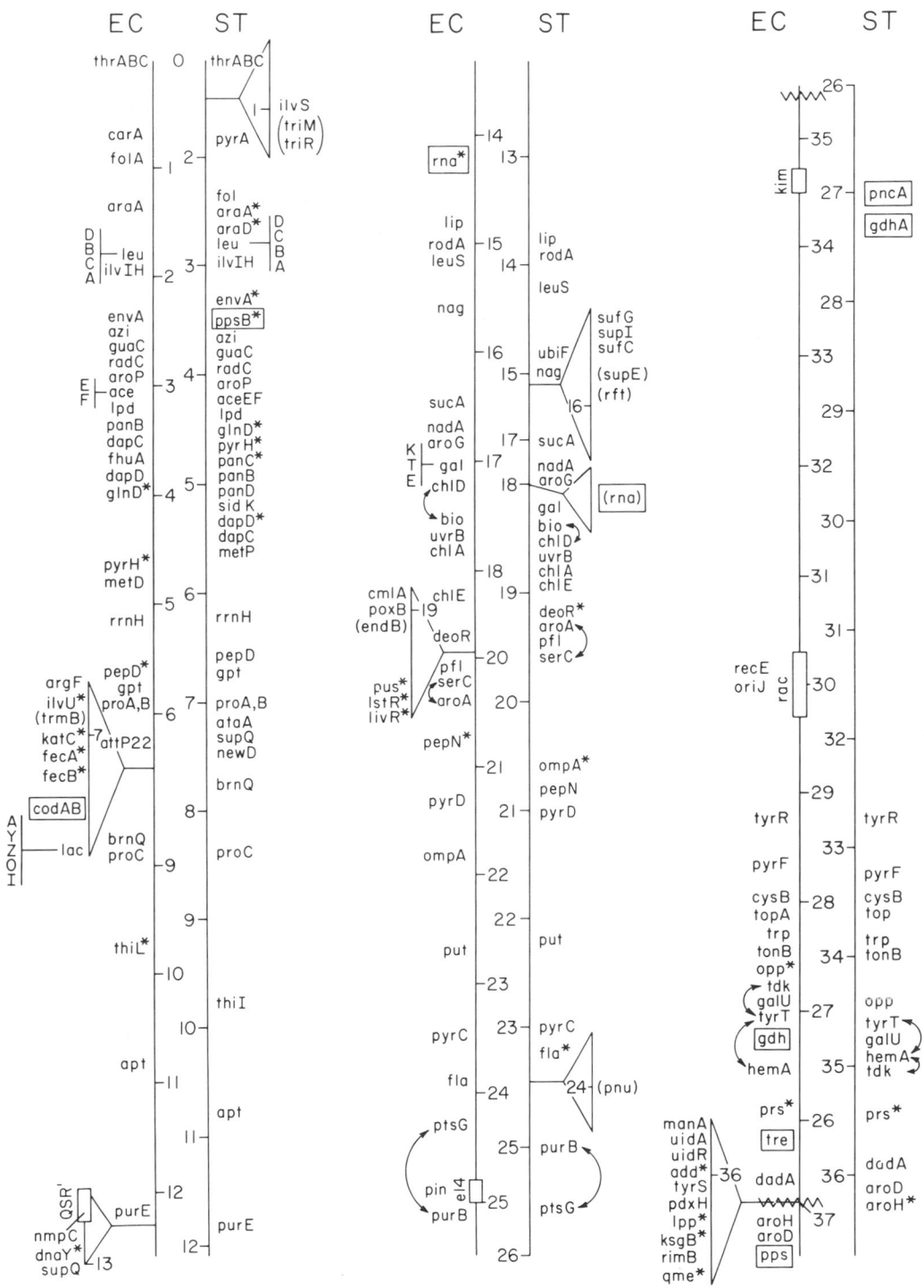

Figure 2 Alignment of genetic maps of *E. coli* (EC) and *S. typhimurium* (ST). Genes that have been mapped in both *E. coli* (5) and *S. typhimurium* (97) were aligned, starting at map position 0. Pairs of genes were not included if one of the two has been mapped only approximately (signified by parentheses in the data sources). In the map alignment, when a pair of genes was displaced by 0.6 map unit or more, adjustment was made to improve the alignment. Excess genetic distance in one of the maps is displayed as loops that balloon out from the paired regions to one side or the other. The precise position of the loops in most

(Figure 2 caption continued on the next page.)

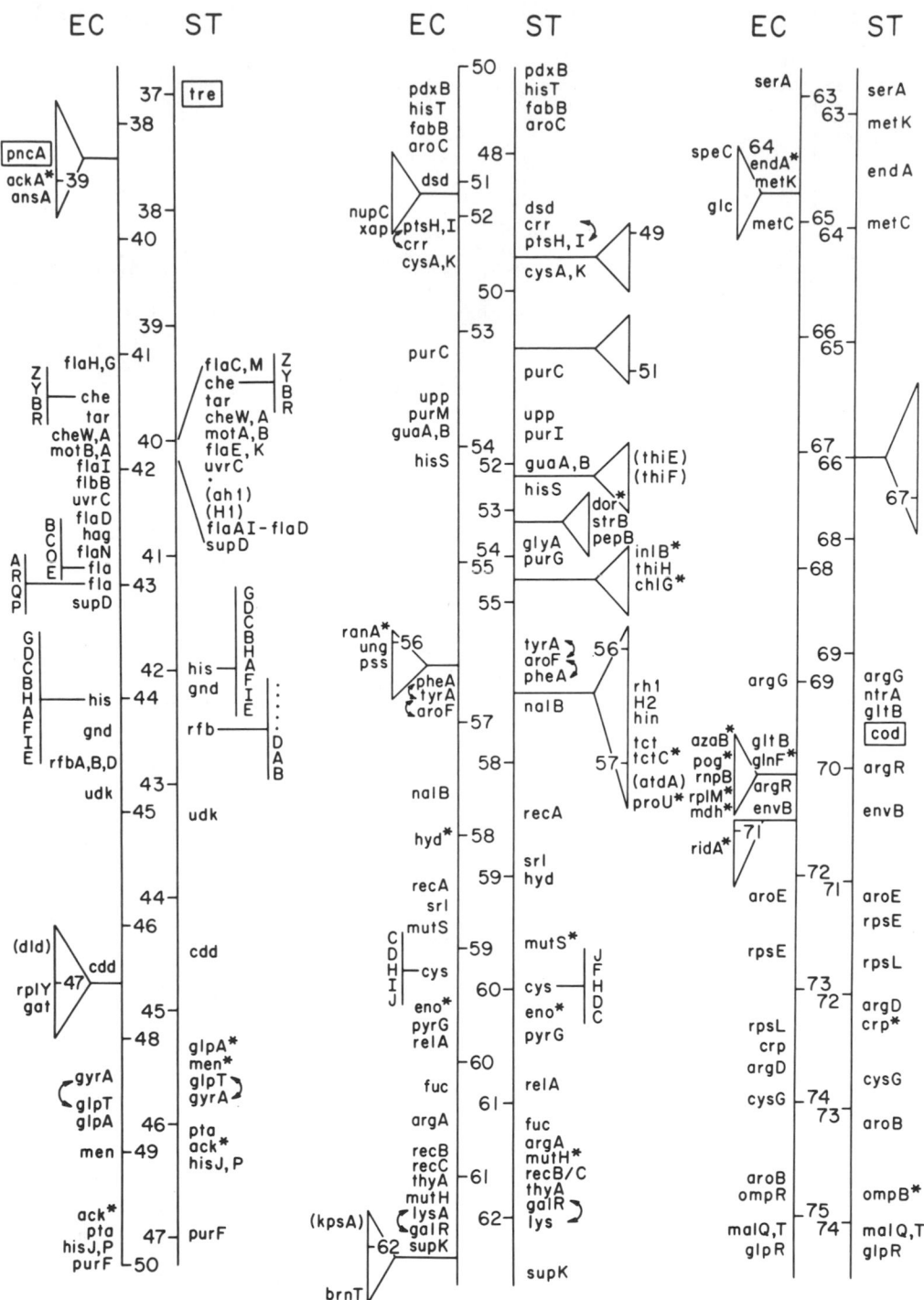

Figure 2 *Continued*

cases is not closely defined by the available genetic information. Within the constraints of available genetic data, the positions of loops were chosen so as to minimize, where possible, the numbers of known genes on the loops and the total number of loops per genome. The zigzag symbols demark the ends of the genetic region that is inverted in one map relative to the other (19). Boxed allele designations indicate genes that reside at markedly different locations in the two maps. Double-headed arrows indicate different map orders for two or more genes in *E. coli* as compared with *S. typhimurium*. (Genes that were marked with an asterisk in the data sources, indicating that they have not been precisely mapped

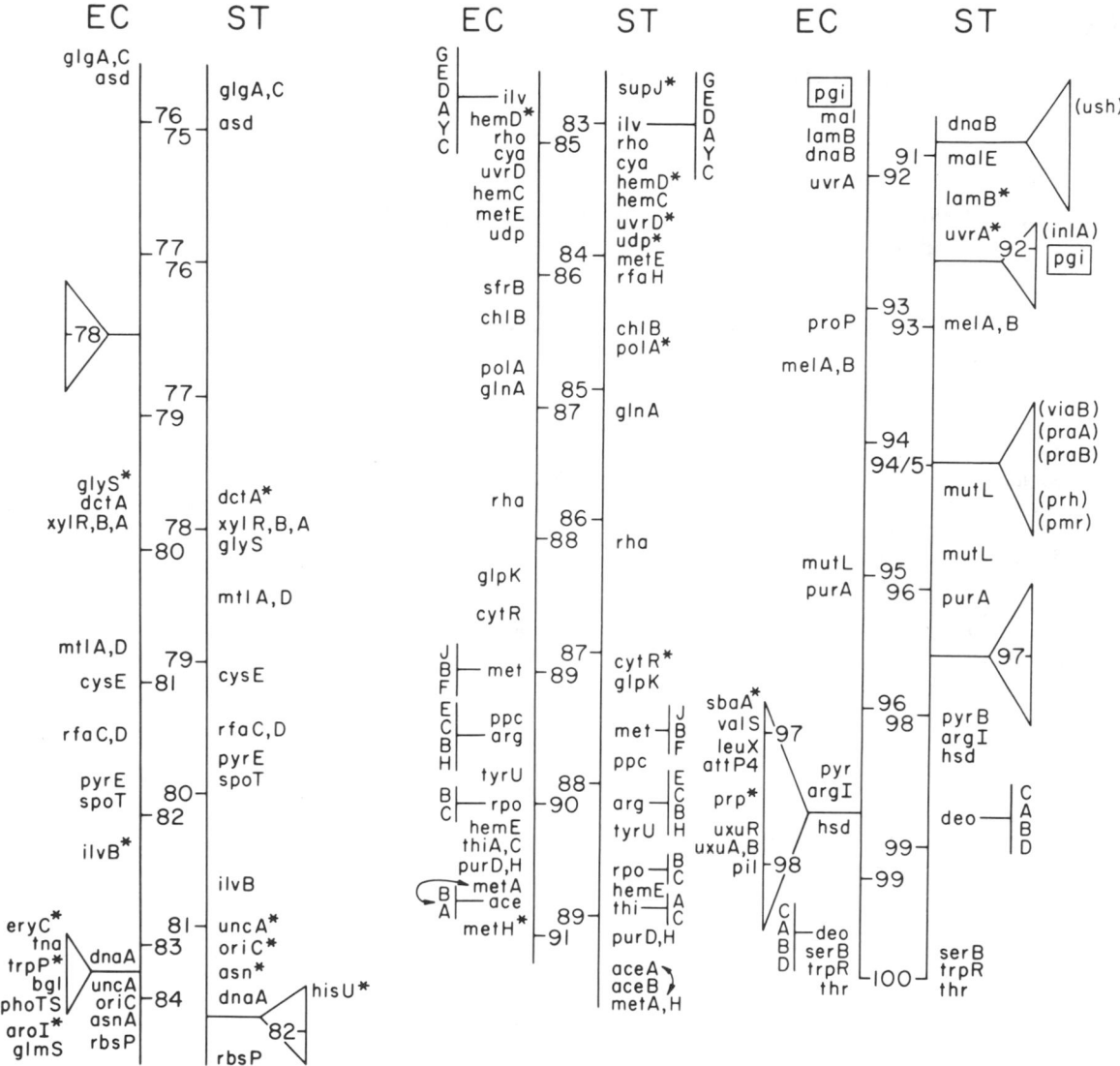

Figure 2 *Continued*

with respect to adjacent markers, we excluded from consideration in this connection.)
Rectangular boxes set into the *E. coli* map designate cryptic prophages. The genes for rRNA
(the *rrn* series) have been shown to map at analogous locations in *E. coli* and *S. typhimurium*
and are omitted from this diagram. All alleles that reside in map segments that are
designated as loops are shown in the diagram, including those genes that have been only
approximately mapped. Effort has been made to display gene locations accurately; nonetheless, for definitive information, especially in crowded sections, the original compilations of
map information should be consulted.

SECTION 2

F. Sizes of Bacterial Chromosomes

Table 1 is reprinted, with permission, from Krawiec and Riley (*Microbiol. Rev.*, vol. 54, p. 506 [1990]). The references cited in the table and table notes can be found in the original paper.

Table 1   Sizes of Bacterial Chromosomes Measured by PFGE

Organism	Size (kb)[a]	Geometry	Reference
Anabaena sp. strain PCC 7120	6,400	Circle[b]	13
Bacillus cereus	5,700	Circle[b,c]	188
Bacillus subtilis	Incomplete	Circle[b]	359
Borrelia burgdorferi B31	950	Linear	83
Brucella abortus 544[T]	2,600	ND[d]	5
Brucella melitensis 16 M[T]	2,600	ND	5
Campylobacter coli	1,714	ND	50
Campylobacter jejuni UA580	1,721	Circle[c]	50
Campylobacter laridis UA487	1,451	ND	50
Caulobacter crescentus	4,000	Circle[b]	81
Chlamydia psittaci AB7, 1H, 1B	1,450	ND	95
Chlamydia trachomatis L2	1,450	ND	95
Clostridium perfringens	3,600	Circle[b,c]	46
Enterococcus faecalis	2,600	ND	35
Escherichia coli K-12	4,700[e]	Circle[b,c]	
Haemophilus influenzae V23	1,980	Circle[b]	178
Haemophilus influenzae Rd	1,834	Circle[b]	202
Haemohilus parainfluenzae	2,340	Circle[b]	177
Lactococcus cremoris	2,600	ND	35
Lactococcus lactis	2,500	ND	35
Mycoplasma gallisepticum PG 31	1,050	ND	278
Mycoplasma genitalium	585	ND	340
Mycoplasma hypopneumoniae strain J	1,140	ND	278
Mycoplasma iowae	1,280	ND	278
Mycoplasma mobile	780	Circle[c]	17
Mycoplasma pneumoniae	785[f]	Circle[b,c]	194
Mycoplasma synoviae WVU 1853	900	ND	278
Mycoplasma mycoides			
subsp. *mycoides* Y	1,240	ND	278
subsp. *mycoides* GC 1176-2	1,330	ND	278
Myxococcus xanthus	9,454	ND	51
Porochlamydia buthi	1,550	ND	95
Porochlamydia chironomi	2,650	ND	95
Pseudomonas aeruginosa	5,900	Circle[b,c]	292
Rhodobacter sphaeroides 2.4.1			
Chromosome I	3,046	Circle[b]	343, 344
Chromosome II[g]	914	Circle[b]	343, 344
Rickettsiella grylli	2,100	ND	95
Rickettsiella melolanthae	1,720	Circle[c]	95
Staphylococcus aureus	2,860	ND	366
Staphylococcus aureus	2,748	ND	263
Streptococcus sanguis	2,300	ND	35
Streptococcus thermophilus	1,700	ND	35
Sulfolobus acidocaldarius	ca. 3,100	Circle[b]	380
Thermococcus celer	1,890	Circle[b]	250
Ureaplasma urealyticum 960[T]	900	Circle[b,c]	55

Notes to Table 1

   [a] Sizes may be the average achieved with several rare-cutting restriction endonucleases.

   [b] Geometry was determined by hybridization to electrophoretically distributed macrofragments, using as probes macrofragments produced by a different rare-cutting restriction enzyme, small linking fragments containing the rare-cutting site, fragments produced by random priming of macrofragments, specific gene probes or probes of transposons previously inserted into specific genes, or probes complementary to one end of a macrofragment hybridized to partially digested fragments.

   [c] Geometry was determined by constructing restriction maps by full or partial digestion with one or more rare-cutting enzymes in either single or double digests and determination of fragment sizes by either one- or two-dimensional gel electrophoresis.

   [d] ND, Not determined.

   [e] The sum of *Not*I fragments from *E. coli* K-12 has been estimated to be 4,550 bp (324). More refined analyses with complete sets of contigs produced by six base cutters has yielded an estimate of 4,700 kb (184, 294).

   [f] By using a complete set of *Eco*RI contigs or a complete set of (unordered) *Xho*I fragments, the size of the *M. pneumoniae* chromosome has been calculated to be 835 or 849 kb, respectively (371).

   [g] Discussed in the text (When Does a Plasmid Become a Chromosome?).

SECTION·3

Salmonella typhimurium Genetic Map

Figure 1 is reprinted from Sanderson and Douey (1990, pp. 2.4–2.5). Tables 1 and 2 are reprinted, with permission, from Sanderson and Roth (1988, pp. 488–506). Some additional data and corrections to the table as contained in Sanderson and Roth (1990) are listed at the end of the table. The references cited in the figure caption and Table 1 can be found in Sanderson and Roth (1988).

REFERENCES

Sanderson, K.E. and D. Douey. 1990. Linkage map of *Salmonella typhimurium*, edition VII. In *Genetic Maps* (ed. S.J. O'Brien), pp. 2.3–2.21. Cold Spring Harbor Laboratory Press, Cold Spring Harbor, New York.

Sanderson, K.E. and J.R. Roth. 1988. Linkage map of *Salmonella typhimurium*, edition VII. *Microbiol. Rev.* **52:** 485–532.

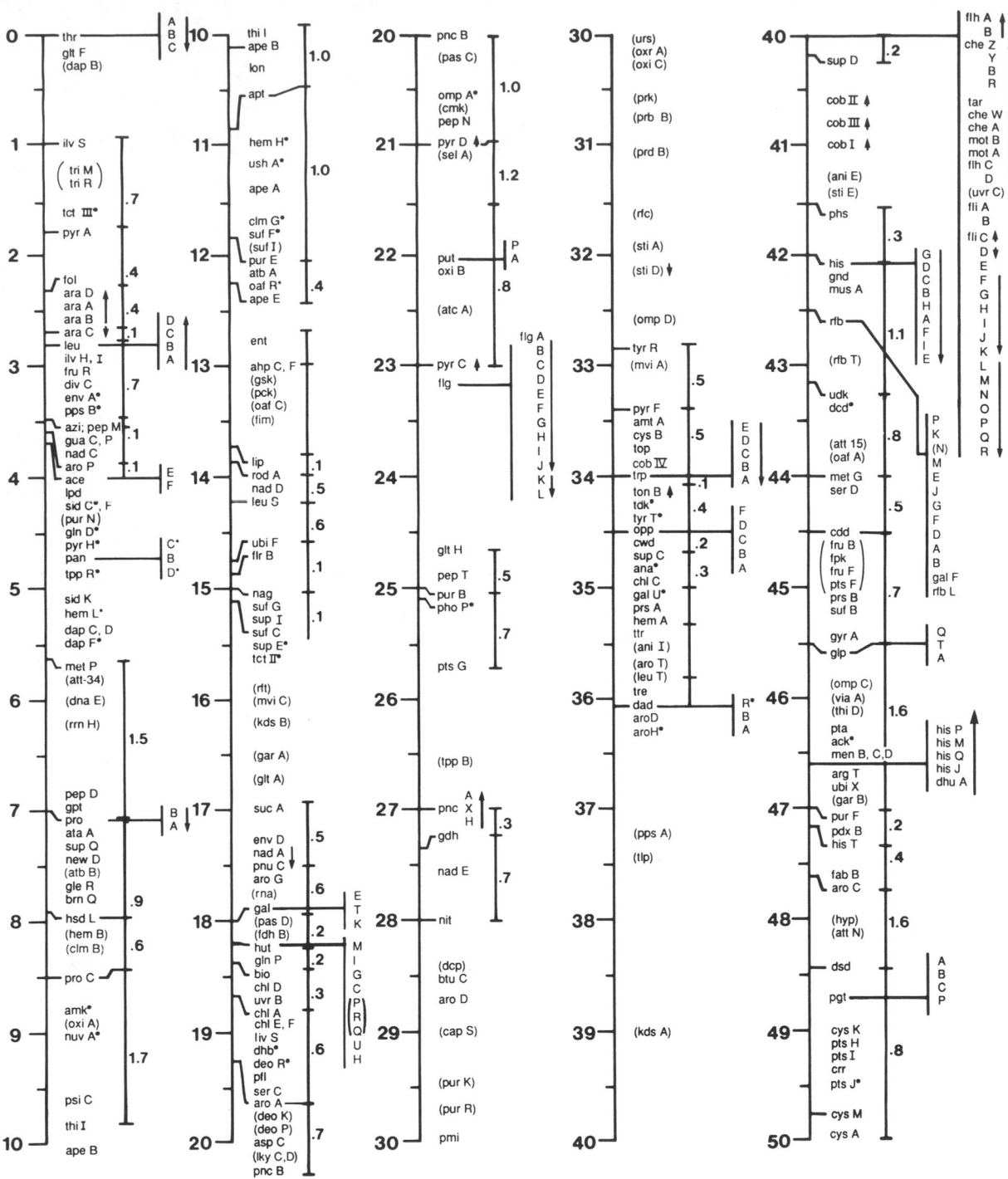

Figure 1   The circular linkage map of *S. typhimurium*, represented as ten segments. The scale of 100 minutes begins at zero for the *thr* loci, as in previous maps (418) and in the linkage maps of *E. coli* (14). The segmented line to the right of the gene symbol indicates that the genes are jointly transduced; the numbers to the right of the segmented line indicate the linear distance between genes. This linear distance was determined from the fragment of joint transduction and was calculated by assuming that the length of P22, KB1, and ES18 transducing fragments is 1 minute, whereas that of P1 is 2 minutes, and applying the formula developed by Wu (520) to convert the percentage of joint transduction to map distance. The genetic symbols are defined in Table

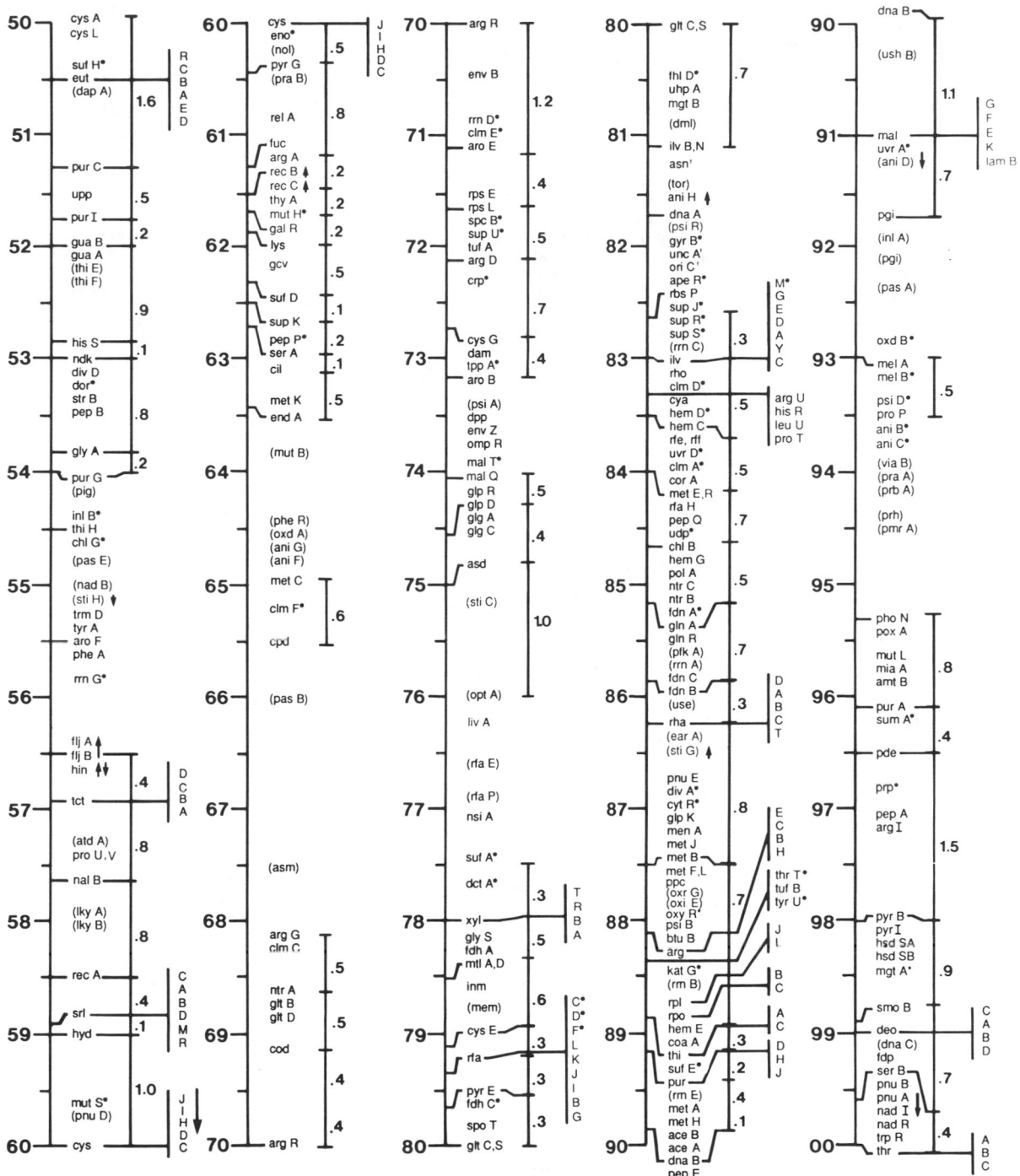

Figure 1 *Continued*

1. Parentheses around a gene symbol indicate that the location of the gene is known only approximately, usually from conjugation studies. An asterisk indicates that a gene has been mapped more precisely, usually by phage-mediated transduction, but that its position with respect to adjacent markers is not known. Arrows to the extreme right of genes and operons indicate the direction of mRNA transcription by these loci. Daggers are shown to the right of a few genes; these genes of *S. typhimurium* are carried on an F′ factor, and this plasmid was shown to complement *E. coli* K12 mutations of that gene; mutant alleles of *S. typhimurium* have not been tested directly.

Table 1   Genes of *S. typhimurium*

Genetic symbol	Mnemonic	Former or alternative symbol; enzyme deficiency or other phenotype[a]	Gene in E. coli[b]	Map (min)[c]	References[d]
aceA	Acetate	Growth on acetate or fatty acids; isocitrate lyase (EC 4.1.3.1)	*aceA*	89	418, 517
aceB	Acetate	Growth on acetate or fatty acids; malate synthase (EC 4.1.3.2)	*aceB*	89	418, 517
aceE	Acetate	Acetate requirement; pyruvate dehydrogenase (pyruvate:cytochrome b_1 oxidoreductase; EC 1.2.2.2)	*aceE*	3	412, 418
aceF	Acetate	Acetate requirement; pyruvate dehydrogenase (pyruvate lipoate oxidoreductase; EC 1.2.4.1)	*aceF*	3	412, 418
ack	Acetate kinase	Acetate kinase (ATP:acetate phosphotransferase, EC 2.7.2.1)	*ackA*	46	272, 418, 498
ahpC	Alkyl hydroperoxide	Alkyl hydroperoxide reductase, C22 subunit	NM	13	AB
ahpF	Alkyl hydroperoxide	Alkyl hydroperoxide reductase, F52a subunit	NM	13	AB
alr	Alanine racemase	Biosynthetic alanine racemase (EC 5.1.1.1)	*alr*	NM	127, 129, 145, 146
amk		AMP kinase		9	412, 418
amtA		Resistance to 40 mM 3-amino-1,2,4-triazole in the presence of histidine		33	410, 412, 418
ana		Anaerobic gas production	*ana*	35	192
aniB	Anaerobically inducible	Induced by anaerobiosis; does not reduce benzyl viologen		93	4
aniC	Anaerobically inducible	Induced by anaerobiosis; does not reduce benzyl viologen		93	4
aniD	Anaerobically inducible	Induced by anaerobiosis; does not reduce benzyl viologen		92	4
aniE	Anaerobically inducible	May be the same as *phs*		41	4
aniF	Anaerobically inducible	Induced by anaerobiosis; does not reduce benzyl viologen		63	4
aniG	Anaerobically inducible	Induced by anaerobiosis; does not reduce benzyl viologen		63	4
aniH	Anaerobically inducible	Induced by anaerobiosis; does not reduce benzyl viologen		81	4
aniI	Anaerobically inducible	Induced by anaerobiosis; does not reduce benzyl viologen		35	4
apeA	Acyl peptide esterase	Acyl amino acid esterase (hydrolyzes N-acetyl-L-phenylalanine-β-naphthyl ester)		11	412, 418
apeB	Acyl peptide esterase	Acyl amino acid esterase (hydrolyzes N-acetyl-L-phenylalanine-β-naphthyl ester)		10	412, 418
apeE	Acyl peptide esterase	Membrane-bound acyl amino acid esterase (hydrolyzes N-acetyl-L-phenylalanine-β-naphthyl ester)		12	418
apeR	Acyl peptide esterase	*apeD*; regulatory gene for *apeE*		82	418
aphA		Nonspecific acid phosphatase II		NM	418
apt		Adenine phosphoribosyltransferase		10	412, 418
araA	Arabinose	L-Arabinose isomerase (EC 5.3.1.4)	*araA*	2	285, 296, 297, 412, 418
araB	Arabinose	Ribulokinase (EC 2.7.1.16)	*araB*	2	285, 296, 410, 412, 418
araC	Arabinose	Regulatory gene for arabinose catabolic enzymes	*araC*	2	284, 285, 296, 412, 418
araD	Arabinose	L-Ribulose-phosphate 4-epimerase (EC 5.1.3.4)	*araD*	2	285, 295, 296, 418
argA	Arginine	*argB*; amino acid acetyl transferase (EC 2.3.1.1)	*argA*	61	342, 410, 412, 418, 443
argB	Arginine	*argC*; N-acetyl-γ-glutamate kinase (EC 2.7.2.8)	*argB*	88	410, 412, 418
argC	Arginine	*argH*; N-acetyl-γ-glutamyl phosphate reductase (EC 1.2.1.38)	*argC*	88	410, 412, 418
argD	Arginine	*argG*; acetylornithine aminotransferase (EC 2.6.1.11)	*argD*	72	410, 412, 418
argE	Arginine	*argA*; acetylornithine deacetylase (EC 3.5.1.16)	*argE*	88	37, 410, 412, 418
argG	Arginine	*argE*; argininosuccinate synthetase (EC 6.3.4.5)	*argG*	69	410, 412, 418
argH	Arginine	*argF*; argininosuccinate lyase (EC 4.3.2.1)	*argH*	88	410, 412, 418
argI	Arginine	Ornithine carbamoyltransferase (EC 2.1.3.3)	*argI*	98	410, 412, 418
argP	Arginine	Arginine transport	*argP*	NM	412, 418
argR	Arginine	L-Arginine regulation	*argR*	70	147, 412, 418
argS	Arginine	Arginyl-tRNA synthetase (EC 6.1.1.19)	*argS*	NM	412, 418

Table 1 *Continued*

Genetic symbol	Mnemonic	Former or alternative symbol; enzyme deficiency or other phenotype[a]	Gene in *E. coli*[b]	Map (min)[c]	References[d]
argT	Arginine	Lysine-arginine-ornithine-binding protein		46	8, 283, 418, 428, 455
argU	Arginine	*argT*; tRNA structural gene for arginine		83	41, 44
aroA	Aromatic	3-Enolpyruvylshikimate 5-phosphate synthetase	*aroA*	19	121, 196, 349, 410, 412, 418, 442, 453
aroB	Aromatic	5-Dehydroquinate synthetase	*aroB*	73	410, 412, 418
aroC	Aromatic	*aroD*; chorismate synthetase	*aroC*	47	195, 410, 412, 418
aroD	Aromatic	*aroE*; 5'-dehydroquinate dehydratase (EC 4.2.1.10)	*aroD*	36	410, 412, 418
aroE	Aromatic	*aroC*; 5-dehydroshikimate reductase	*aroE*	71	410, 412, 418
aroF	Aromatic	Tyrosine-repressible DAHP synthetase	*aroF*	55	410, 412, 418
aroG	Aromatic	Phenylalanine-repressible DAHP synthetase	*aroG*	17	410, 412, 418
aroH	Aromatic	Tryptophan-repressible DAHP synthetase	*aroH*	36	410, 412, 418
aroP	Aromatic	Aromatic amino acid transport	*aroP*	3	410, 412, 418
aroT	Aromatic	Ability to transport tryptophan, phenylalanine, tyrosine	*aroT*	35	410, 412, 418
asd		Aspartic semialdehyde dehydrogenase (EC 1.2.1.11)	*asd*	75	410, 412, 418
asm		Unable to assimilate low levels of ammonia; deficient in glutamate synthase and glutamine synthase		68	418
asn	Asparagine	Asparagine synthesis	*asn*	81	418
aspC	Aspartate	Aspartate aminotransferase (EC 2.6.1.1)	*aspC*	20	497
ataA	Attachment	*attP22 I*; attachment site for prophage P22	*attP22*	7	393, 410, 412, 418
atbA	Attachment	*attP27 I*; attachment site for prophage P27		12	410, 412, 418
atbB	Attachment	*attP27 II*; second attachment site for prophage P27		7	410, 412, 418
atcA	Attachment	*attP221*; attachment site for prophage P221		22	410, 412, 418
atdA	Attachment	*attP14*; attachment site for prophage P14 in group C *Salmonella* spp.		57	410, 412, 418
ats		Arylsulfatase		NM	418
att15	Attachment	Attachment site of phage ε[15] to chromosome in group E *Salmonella* spp.		43	412, 418
att34	Attachment	Attachment site in phage ε[34] to chromosome in group E *Salmonella* spp.		5	412, 418
attN	Attachment	Attachment site for prophage N in *S. montevideo*		48	412, 418
avtA		Alanine-valine transaminase (transaminase C)	*avtA*	NM	29, 514
azi	Azide	Resistant to 3 mM sodium azide on L-methionine	*azi*	3	410, 412, 418
bio	Biotin	Requirement	*bioA*	18	412, 418
brnQ		*ilvT*; branched-chain amino acid transport	*brnQ*	7	311, 353, 412, 418
btuB	B$_{12}$ utilization	*bfe*; transport of vitamin B$_{12}$	*btuB*	88	412, 418, H
btuC	B$_{12}$ utilization	Transport of vitamin B$_{12}$	*btuC*	28	H
capS	Capsule	Capsular polysaccharide synthesis	*capS*	29	418
cdd		Cytidine deaminase (EC 3.5.4.5)	*cdd*	44	410, 412, 418
cheA	Chemotaxis	*cheP*; chemotaxis	*cheA*	40	268, 412, 418, 456, 521
cheB	Chemotaxis	*cheX*; chemotaxis. Protein-glutamate methylesterase	*cheB*	40	39, 40, 102, 268, 385, 412, 418, 437, 438, 445, 461
cheR	Chemotaxis	Chemotaxis. Protein-glutamate methyl transferase	*cheX*	40	39, 40, 102, 268, 412, 418, 461, Y
cheS	Chemotaxis	Chemotaxis		NM	412, 418
cheW	Chemotaxis	Chemotaxis	*cheW*	40	268, 412, 418, 458
cheY	Chemotaxis	*cheQ*; chemotaxis	*cheY*	40	268, 412, 418, 445, 457
cheZ	Chemotaxis	*cheT*; chemotaxis	*cheZ*	40	385, 418, 445, 459, 460
chlA	Chlorate	Resistance; affects nitrate reductase, tetrathionate reductase, chlorate reductase, and hydrogen lyase	*chlA*	18	89, 410, 412, 418
chlB	Chlorate	Resistance; affects nitrate reductase, tetrathionate reductase, and hydrogen lyase	*chlB*	84	410, 412, 418
chlC	Chlorate	Retains sensitivity to chlorate; affects nitrate reductase	*chlC*	34	273, 410, 412, 418
chlD	Chlorate	Resistance; affects nitrate reductase, tetrathionate reductase, and hydrogen lyase	*chlD*	18	410, 412, 418
chlE	Chlorate	Resistance		18	410, 412, 418
chlF	Chlorate	Resistance		18	410, 412, 418

Table 1 *Continued*

Genetic symbol	Mnemonic	Former or alternative symbol; enzyme deficiency or other phenotype[a]	Gene in E. coli[b]	Map (min)[c]	References[d]
chlG	Chlorate	Resistance; affects nitrate reductase, tetrathionate reductase, chlorate reductase, and hydrogen lyase		20	410, 412, 418
cil	Citrate lyase	Deficiency of the essential enzyme in a temperature-sensitive mutant		63	266
clmA	Conditional-lethal mutation	Heat- or cold-sensitive mutation		84	X
clmB	Conditional-lethal mutation	Heat- or cold-sensitive mutation		8	X
clmC	Conditional-lethal mutation	Heat- or cold-sensitive mutation		69	X
clmD	Conditional-lethal mutation	Heat- or cold-sensitive mutation		83	X
clmE	Conditional-lethal mutation	Heat- or cold-sensitive mutation		71	X
clmF	Conditional-lethal mutation	Heat- or cold-sensitive mutation		64	X
clmG	Conditional-lethal mutation	Heat- or cold-sensitive mutation		11	X
cmk		Cytidylate kinase (EC 2.7.4.14)		89	412, 418
coaA	Coenzyme A	Coenzyme A synthesis; pantothenate kinase		89	412, 418
cobI	Cobalamin	Operon encoding synthesis of cobalamide (vitamin B_{12})		41	128, 141, 234, 235
cobII	Cobalamin	Operon encoding synthesis of DMB; defective in vitamin B_{12} synthesis		41	128, 234, 235
cobIII	Cobalamin	Operon encoding functions for joining cobalamide and DMB; defective in vitamin B_{12} synthesis		41	128, 234, 235
cobIV	Cobalamin	Operon including functions required for vitamin B_{12} synthesis		34	L
cod		Cytosine deaminase (EC 3.5.4.1)		69	410, 412, 418
corA	Cobalt resistance	Magnesium transport	*corA*	84	194
cpd		cAMP phosphodiesterase (EC 3.1.4.17)	*cpd*	64	45, 412, 418, 485
crp		cAMP receptor protein	*crp*	72	37, 97, 107, 110, 240, 410, 412, 418, 430, 512
crr		Factor III for sugar transport by phosphotransferase IIB′ (*ptsG*) system	*crr*	48	107, 316, 317, 318, 329, 341, 375, 377, 406, 410, 412, 418, 485
cwd	Cell wall defect	Sensitive to bile salts; mucoid		34	192
cya	cAMP	Adenylate cyclase (EC 4.6.1.1)	*cya*	83	37, 110, 240, 410, 412, 418, 485, 512
cysA	Cysteine	Sulfate-thiosulfate transport (chromate resistance)	*cysA*	49	221, 332, 341, 370, 410, 412, 418
cysB	Cysteine	Cysteine regulation; positive control of L-cystine transport	*cysB*	33	226, 227, 325, 358, 410, 412, 418
cysC	Cysteine	Adenylylsulfate kinase (EC 2.7.1.25)	*cysC*	60	332, 410, 412, 418
cysD	Cysteine	Sulfate adenylyltransferase (EC 2.7.7.4)	*cysD*	60	332, 410, 412, 418
cysE	Cysteine	Serine acetyltransferase (EC 2.3.1.30)	*cysE*	79	410, 412, 418
cysG	Cysteine	Seroheme component of sulfite reductase	*cysG*	72	234, 410, 412, 418
cysH	Cysteine	Adenylylsulfate reductase (EC 1.8.99.2)	*cysH*	60	410, 412, 418, S
cysI	Cysteine	Heme protein component of sulfite reductase	*cysI*	60	412, 418, S
cysJ	Cysteine	Flavoprotein component of sulfite reductase	*cysJ*	60	410, 412, 418, S
cysK	Cysteine	*trz*; resistance to 1,2,4-triazole; O-acetylserine sulfhydrylase A (EC 4.2.99.8)	*cysK*	49	65, 221, 332, 338, 410, 412, 418
cysL	Cysteine	Resistance to selenate		50	412, 418
cysM	Cysteine	O-Acetylserine sulfhydrylase B (EC 4.2.99.8)		49	221, 332, 338, 418
cytR		Regulatory gene for *deo* operon and *udp* and *cdd* genes	*cytR*	87	410, 412, 418
dadA		*dad*; D-histidine, D-methionine utilization; D-alanine dehydrogenase (EC 1.4.99.1)	*dadA*	36	410, 412, 418
dadB		Catabolic alanine racemase (EC 5.1.1.1)	*dadB*	36	129, 146, 418, 503
dadR		Insensitivity of *dadA* to catabolite repression		36	412, 418
dam		DNA adenine methylase	*dam*	NM	362, 394, 418
dapA	Diaminopimelate	Dihydropicolinate synthase (EC 4.2.1.52)	*dapA*	50	410, 412, 418
dapB	Diaminopimelate	Dihydropicolinate reductase	*dapB*	0	410, 412, 418

Table 1   *Continued*

Genetic symbol	Mnemonic	Former or alternative symbol; enzyme deficiency or other phenotype[a]	Gene in *E. coli*[b]	Map (min)[c]	References[d]
dapC	Diaminopimelate	Tetrahydropicolinate succinylase	*dapC*	5	410, 412, 418
dapD	Diaminopimelate	Succinyl-diaminopimelate aminotransferase	*dapD*	5	410, 412, 418
dapF	Diaminopimelate	Diaminopimelate epimerase (EC 5.1.1.7)		5	410, 412, 418
dcd		dCTP deaminase (EC 3.5.4.13)	*dcd*	43	412, 418
dcm		DNA cytosine methylation	*dcm*	NM	418
dcp		Dipeptidyl carboxypeptidase	*dcp*	28	418
dctA		Transport of dicarboxylic acids	*dctA*	77	412, 418
deoA	Deoxyribose	*tpp*; thymidine phosphorylase (EC 2.4.2.4)	*deoA*	99	410, 412, 418
deoB	Deoxyribose	*drm*; phosphopentomutase (EC 2.7.5.6)	*deoB*	99	410, 412, 418
deoC	Deoxyribose	*dra*; phosphodeoxyriboaldolase (EC 4.1.2.4)	*deoC*	99	410, 412, 418
deoD	Deoxyribose	*pnu*, *pup*; purine nucleoside phosphorylase (EC 2.4.2.1)	*deoD*	99	410, 412, 418
deoK	Deoxyribose	Deoxyribokinase		20	410, 412, 418
deoP	Deoxyribose	Deoxyribose transport		20	410, 412, 418
deoR	Deoxyribose	Constitutive for enzymes of *deoA*, *deoB*, *deoC*, and *deoD*	*deoR*	19	410, 412, 418
dhb		2,3-Dihydroxybenzoic acid requirement		19	412, 418
dhuA	D-Histidine	Utilization; increased activity of histidine-binding protein J		46	8, 281, 283, 410, 412, 418, 428, 455
divA	Division	*wrkA*; septum initiation defect		87	410, 412, 418
divC	Division	*smoA*; septum initiation defect		3	93, 410, 412, 418
divD	Division	Round cell morphology		53	11, 418
dml	D-Malate	Utilization		80	410, 412, 418
dnaA	DNA	DNA initiation	*dnaA*	81	123, 313, 412, 418, 441
dnaB	DNA	DNA synthesis	*dnaB*	89	312–314, 418, 519
dnaC	DNA	DNA synthesis initiation and cell division uncoupling	*dnaC*	99	255, 312, 313, 398, 410, 412, 418
dnaE	DNA	DNA synthesis	*dnaE*	6	312, 313, 418
dnaG	DNA	DNA biosynthesis; DNA primase	*dnaG*	NM	125, 313, 398, 418
dnaJ	DNA	DNA biosynthesis	*dnaJ*	NM	313
dnaK	DNA	DNA biosynthesis	*dnaK*	NM	313
dnaL	DNA	DNA biosynthesis	*dnaL*	NM	313
dnaN	DNA	DNA biosynthesis; DNA polymerase III, beta subunit	*dnaN*	NM	123, 313
dnaQ	DNA	DNA biosynthesis	*dnaQ*	NM	312, 313
dnaX	DNA	DNA biosynthesis	*dnaX*	NM	313
dnaY	DNA	DNA biosynthesis	*dnaY*	NM	313
dnaZ	DNA	DNA biosynthesis	*dnaZ*	NM	123, 313, 398
dor		Deletion of *r*-determinants from plasmids		53	159, 210, 399, 418
dpp		Dipeptide permease		74	AG
dsd		D-Serine sensitivity; D-serine dehydratase (EC 4.2.1.14)	*dsd*	48	410, 412, 418
dum		dUMP synthesis			412, 418
earA		Regulates expression of *aniG*		86	3
eca		Enterobacterial common-antigen synthesis		NM	412, 418
endA		Endonuclease I	*endA*	63	418
eno	Enolase	Enolase (EC 4.2.1.11)	*eno*	60	418
ent	Enterochelin	*asc*, *enb*; enterochelin (dihydroxybenzoyl-serine trimer)	*ent*	13	410, 412
envA	Envelope	Cell division defect, chain formation	*envA*	3	412, 418
envB	Envelope	*bac*; spherical cells, drug sensitivity	*envB*	70	10, 11, 410, 412, 418
envD	Envelope	Autolysis; drug sensitivity; alterations in cell morphology		17	371, 412, 418
envZ	Outer membrane protein	*ompB*, *tppB*; positive regulation of tripeptide permease and outer membrane protein	*envZ*	74	152, 294, X
eutA	Ethanolamine utilization	Required for use of ethanolamine as sole carbon or nitrogen source		50	395
eutB	Ethanolamine utilization	Ethanolamine ammonia lyase, subunit I		50	395
eutC	Ethanolamine utilization	Ethanolamine ammonia lyase, subunit II		50	395
eutD	Ethanolamine utilization	CoA-dependent acetaldehyde dehydrogenase		50	395
eutE	Ethanolamine utilization	Required for use of ethanolamine as sole carbon source		50	395
eutR	Ethanolamine utilization	Positive regulatory gene for *eut* operon		50	395, V

Table 1   *Continued*

Genetic symbol	Mnemonic	Former or alternative symbol; enzyme deficiency or other phenotype[a]	Gene in *E. coli*[b]	Map (min)[c]	References[d]
fabB	Fatty acid biosynthesis	β-Ketoacyl acyl carrier protein synthetase I (EC 2.3.1.41)		47	98, 410, 412, 418
fdhA		Formate hydrogenlyase complex; formate dehydrogenase	*fdhA*	77	410, 412, 418
fdhB		Formate hydrogenlyase complex; formate dehydrogenase		18	412, 418
fdhC		Formate hydrogenase associated with both hydrogenase and nitrate reductase		80	418
fdhF		*fhl*; formate dehydrogenase	*fdhF*	93	21
fdnA		Formate dehydrogenase associated with nitrate reductase		85	367, 418, B
fdnB		Synthesis or activation of the cytochrome associated with formate dehydrogenase		85	367, 418
fdnC		Synthesis or activation of the cytochrome associated with formate dehydrogenase		85	367
fdp		Fructose-1,6-diphosphatase (EC 3.1.3.11)	*fdp*	99	410, 412, 418
fhlD		Formate dehydrogenase 2 activity		81	21, 412, 418
fim	Pili	*pil*; fimbriae (pili)	*fim*	14	90, 130, 382, 410, 412, 418
flgA	Flagella	*flaFI*; flagellar synthesis; function unknown	*flgA*	23	222, 302, 412, 418
flgB	Flagella	*flaFII*; flagellar synthesis; function unknown	*flgB*	23	222, 302, 412, 418
flgC	Flagella	*flaFIII*; flagellar synthesis; basal-body protein	*flgC*	23	222, 302, 412, 418
flgD	Flagella	*flaFIV*; flagellar synthesis; basal-body rod modification	*flgD*	23	222, 302, 412, 418
flgE	Flagella	*flaFV*; flagellar synthesis; hook protein	*flgE*	23	2, 207, 222, 302, 354, 412, 418
flgF	Flagella	*flaFVI*; flagellar synthesis; basal-body rod protein	*flgF*	23	2, 207, 222, 302, 412, 418
flgG	Flagella	*flaFVII*; flagellar synthesis; basal-body rod protein	*flgG*	23	2, 207, 222, 302, 412, 418
flgH	Flagella	*flaFVIII*; flagellar synthesis; basal-body L-ring protein	*flgH*	23	222, 302, 412, 418
flgI	Flagella	*flaFIX*; flagellar synthesis; basal-body P-ring protein	*flgI*	23	2, 204, 222, 302, 412, 418
flgJ	Flagella	*flaFX*; flagellar synthesis; function unknown	*flgJ*	23	222, 302, 412, 418
flgK	Flagella	*flaW*; flagellar synthesis; hook-associated protein 1	*flgK*	23	198, 200–202, 205, 206, 222, 247, 302, 525
flgL	Flagella	*flaU*; flagellar synthesis; hook-associated protein 3	*flgL*	23	198, 200–202, 205, 206, 222, 247, 302, 418, 525
flhA	Flagella	*flaC*; flagellar synthesis; function unknown	*flhA*	40	222, 302, 410, 412, 418
flhB	Flagella	*flaM*; flagellar synthesis; function unknown	*flhB*	40	222, 268, 302, 410, 412, 418, 460
flhC	Flagella	*flaE*; flagellar synthesis; regulation of gene expression	*flhC*	40	222, 268, 302, 410, 412, 418
flhD	Flagella	*flaK*; flagellar synthesis; regulation of gene expression (flagellum-specific sigma factor?)	*flhD*	40	222, 268, 302, 410, 412, 418
fliA	Flagella	*flaL*; flagellar synthesis; regulation of late gene expression	*fliA*	40	222, 247, 302, 410, 412, 418
fliB	Flagella	*nml*; flagellar synthesis; N-methylation of lysine residues in flagellin		40	222, 302, 410, 412, 418
fliC	Flagella	*H1*; flagellar synthesis; phase 1 flagellin (filament structural protein)	*fliC*	40	143, 199, 205, 222, 243, 247, 302, 410, 412, 418, 507, 508
fliD	Flagella	*flaV*; flagellar synthesis; hook-associated protein 2	*fliD*	40	198, 200–202, 205, 206, 222, 246, 247, 302, 525
fliE	Flagella	*flaAI*; flagellar synthesis; function unknown	*fliE*	40	222, 302, 410, 412, 418
fliF	Flagella	*flaAII.1*; flagellar synthesis; basal-body M-ring protein	*fliF*	40	2, 105, 222, 302, 418, 523
fliG	Flagella	*flaAII.2*, *motC*, *cheV*; flagellar synthesis; motor switching and energizing	*fliG*	40	105, 222, 302, 418, 522, 523
fliH	Flagella	*flaAII.3*; flagellar synthesis; function unknown	*fliH*	40	105, 203, 222, 302, 418, 523
fliI	Flagella	*flaAIII*; flagellar synthesis; function unknown	*fliI*	40	203, 222, 302, 410, 412, 418, 523
fliJ	Flagella	*flaS*; flagellar synthesis; function unknown	*fliJ*	40	203, 222, 302, 412, 418
fliK	Flagella	*flaR*; flagellar synthesis; hook length control	*fliK*	40	203, 222, 302, 412, 418

Table 1   *Continued*

Genetic symbol	Mnemonic	Former or alternative symbol; enzyme deficiency or other phenotype[a]	Gene in *E. coli*[b]	Map (min)[c]	References[d]
fliL	Flagella	*flaQI*; flagellar synthesis; function unknown	*fliL*	40	203, 222, 302, 523
fliM	Flagella	*flaQII*, *cheC*, *cheU*; flagellar synthesis; motor switching and energizing	*fliM*	40	203, 222, 302
fliN	Flagella	*flaN*; flagellar synthesis; motor switching and energizing	*fliN*	40	222, 302, 412, 418, 522, 523
fliO	Flagella	*flaP*; flagellar synthesis; function unknown	*fliO*	40	222, 302, 410, 412, 418
fliP	Flagella	*flaB*; flagellar synthesis; function unknown	*fliP*	40	222, 302, 410, 412, 418
fliQ	Flagella	*flaD*; flagellar synthesis; function unknown	*fliQ*	40	222, 302, 410, 412, 418, 522
fliR	Flagella	*flaX*; flagellar synthesis; function unknown	*fliR*	40	222, 302
fljA	Flagella	*rh1*; flagellar synthesis; repressor of phase 1 flagellin gene (*fliC*)	None	56	143, 222, 302, 410, 412, 418, 524
fljB	Flagella	*H2*; flagellar synthesis; phase 2 flagellin (filament structural protein)	None	56	199, 222, 223, 302, 410, 412, 418, 478, 524
flrB	Fluoroleucine resistance	Leucine or isoleucine regulation or both		14	410, 412, 418
fol	Folate	Trimethoprim resistance; tetrahydrofolate dehydrogenase (folate reductase)	*folA*	2	410, 412, 418
fpk	Fructose	Fructose phosphate kinase	*fpk*	45	148
frd	Fumarate reductase	Fumurate reductase (EC 1.3.99.1)	*frd*	NM	32
fruB	Fructose	Fructose phosphotransferase enzyme IIIA		45	148
fruF	Fructose	Fructose phosphotransferase pseudo-HPr		45	148
fruR	Fructose	Regulation of the fructose regulon, regulation of gluconeogenesis; may be the same as *ppsB*	*fruR*	3	82, 148
fuc	Fucose	L-Fucose utilization	*fuc*	61	412, 418
fur	Ferrichrome	Ferrichrome uptake, regulation of iron uptake; constitutive synthesis of iron-enterochelin		NM	418
galC	Galactose	Constitutive synthesis of specific galactose permease		18	412, 418
galE	Galactose	UDP glucose 4-epimerase (EC 5.1.3.2)	*galE*	18	224, 301, 348, 349, 410, 412, 418, P
galF	Galactose	Modifier of UDP-glucose pyrophosphorylase		42	410, 412, 418
galK	Galactose	Galactokinase (EC 2.7.1.6)	*galK*	18	410, 412, 418, P
galP	Galactose	Specific galactose permease	*galP*	NM	412, 418, P
galR	Galactose	Regulation	*galR*	61	410, 412, 418, P
galT	Galactose	Galactose-1-phosphate uridylyltransferase (EC 2.7.7.10)	*galT*	18	418
galU	Galactose	Glucose-1-phosphate uridylyltransferase (EC 2.7.7.9)	*galU*	34	192, 410, 412, 418
garA	Gamma resistant	Resistant to γ and UV radiation; large cells; high RNA and protein content (may be equivalent to *rodA*)		0	418
garB	Gamma resistant	Resistant to γ and UV radiation; large cells; high RNA and protein content		0	418
gcv	Glycine cleavage	Defective in the glycine cleavage enzyme system	*gcv*	62	AA
gdh	Glutamate	Glutamate dehydrogenase (EC 1.4.1.4)	*gdh*	27	193, 327, 418
gleR		Glycyl-leucyl-resistant regulatory gene for transport of branched-chain amino acids		7	418
glgA	Glycogen	Starch (bacterial glycogen) synthase (EC 2.4.1.21)	*glgA*	74	290, 291, 412, 418
glgC	Glycogen	Glucose-1-phosphate adenylyltransferase (EC 2.7.7.27)	*glgC*	74	290, 291, 412, 418
glnA	Glutamine	Glutamine synthetase (EC 6.3.1.2)	*glnA*	85	8, 173, 231, 258, 259, 267, 315, 412, 418
glnD	Glutamine	PIIA uridyl transferase	*glnD*	5	412, 418
glnE	Glutamine	Covalent modification of glutamine synthetase; glutamine synthetase adenylyl transferase (EC 2.7.2.42)		NM	267, 418
glnH	Glutamine	Periplasmic glutamine-binding protein		NM	418
glnP	Glutamine	Glutamine transport (high-affinity system)	*glnP*	20	412, 418
glnR	Glutamine	Regulation of enzymes for glutamine metabolism		85	418
glpA	Glycerol phosphate	Glycerol-3-phosphate dehydrogenase (anaerobic) (EC 1.1.99.5)	*glpA*	45	410, 412, 418

Table 1   *Continued*

Genetic symbol	Mnemonic	Former or alternative symbol; enzyme deficiency or other phenotype[a]	Gene in *E. coli*[b]	Map (min)[c]	References[d]
glpD	Glycerol phosphate	Glycerol-3-phosphate dehydrogenase (NAD⁺) (EC 1.1.1.8)	*glpD*	74	410, 412, 418
glpK	Glycerol phosphate	Glycerol kinase (EC 2.7.1.30)	*glpK*	87	351, 410, 412, 418
glpQ	Glycerol phosphate	Glycerol-3-phosphate diesterase	*glpQ*	45	182
glpR	Glycerol phosphate	Regulatory gene for *glpD*, *glpK*, and *glpT*		74	412, 418
glpT	Glycerol phosphate	sn-Glycerol-3-phosphate transport	*glpT*	45	183, 410, 412, 418
gltA	Glutamate	Requirement	*gltA*	16	410, 412, 418
gltB	Glutamate	Glutamate synthetase (EC 2.6.1.53)	*gltB*	69	303, 418
gltC	Glutamate	Growth on glutamate as sole source of carbon		80	412, 418
gltD	Glutamate	Glutamate synthase, small subunit		69	303
gltF	Glutamate	Glutamate-specific transport system		100	5
gltH	Glutamate	Requirement	*gltH*	25	418
gltS	Glutamate	Glutamate permease	*gltS*	80	5
glyA	Glycine	Serine hydroxymethyltransferase (EC 2.1.2.1)	*glyA*	53	410, 412, 418, 488
glyS	Glycine	Glycyl-tRNA synthetase (EC 6.1.1.14)	*glyS*	78	412, 418
gnd		Phosphogluconate dehydrogenase (EC 1.1.1.43)	*gnd*	42	18, 53, 410, 412, 418
gpd		Glucosamine-6-phosphate deaminase		NM	412, 418
gpsA		sn-Glycerol-3-phosphate dehydrogenase [NAD(P)⁺] (EC 1.1.1.94)		NM	412, 418
gpt		*gxu*; guanine-hypoxanthine phosphoribosyltransferase (EC 2.4.2.8)	*gpt*	6	355, 393, 412, 418
gsk		Guanosine kinase	*gsk*	13	412, 418
guaA	Guanine	GMP synthetase (EC 6.3.4.1)	*guaA*	52	131, 410, 412, 418
guaB	Guanine	IMP dehydrogenase (EC 1.1.1.205)	*guaB*	52	131, 410, 412, 418
guaC	Guanine	GMP reductase (EC 1.6.6.8)	*guaC*	3	412, 418
guaP	Guanine	Guanine uptake	*guaC*	3	410, 412, 418
gyrA	Gyrase	*hisW*, *nalA*; resistance or sensitivity to nalidixic acid; DNA gyrase	*gyrA*	46	241, 313, 387, 412, 418, 526
gyrB	Gyrase	*hisU*, DNA gyrase	*gyrB*	81	241
hemA	Heme	5-Aminolevulinate synthase (EC 2.3.1.37)	*hemA*	34	410, 412, 418
hemB	Heme	Heme deficient	*hemB*	8	410, 412, 418
hemC	Heme	Heme deficient; urogen I synthase	*hemC*	83	412, 418
hemD	Heme	Heme deficient; uroporphyrinogen III cosynthase	*hemD*	83	412, 418
hemE	Heme	Accumulation of uroporphyrin III	*hemE*	88	412, 418
hemG	Heme	Defective in heme synthesis	*hemG*	84	J
hemH	Heme	Defective in heme synthesis	*hemH*	11	J
hemL	Heme	*popC*; defective in synthesis of aminolevulinate or heme	*popC*	5	K
hin	H inversion	*vh2*; flagellar synthesis; regulation of flagellin gene expression by site-specific inversion of DNA	None	56	57, 58, 222, 302, 418, 432, 478, 524
hisA	Histidine	N-(5′-phospho-L-ribosylformimino)-5-amino-1-(5′-phosphoribosyl)-4-imidazolecarboxamide isomerase (EC 5.3.1.16)	*hisA*	42	71, 220, 410, 412, 418
hisB	Histidine	Imidazoleglycerol-phosphate dehydratase (EC 4.2.1.19) and histidinol phosphatase (EC 3.1.3.15)		0	71, 410, 412, 418
hisC	Histidine	Histidinol-phosphate aminotransferase (EC 2.6.1.9)	*hisC*	42	390, 410, 412, 418
hisD	Histidine	Histidinol dehydrogenase (EC 1.1.1.23)	*hisD*	42	13, 59, 220, 287, 390, 410, 412, 418, 424
hisE	Histidine	Phosphoribosyl-ATP pyrophosphohydrolase	*hisE*	42	59, 71, 72, 81, 410, 412, 418
hisF	Histidine	Cyclase	*hisF*	42	71, 81, 410, 412, 418
hisG	Histidine	ATP phosphoribosyltransferase (EC 2.4.2.17)	*hisG*	42	7, 13, 20, 87, 88, 144, 391, 401, 410, 412, 418
hisH	Histidine	Amido transferase	*hisH*	42	71, 410, 412, 418
hisI	Histidine	Phosphoribosyl-AMP cyclohydrolase (EC 3.5.4.19) (may be bifunctional with *hisE*)	*hisI*	42	59, 71, 81, 410, 412, 418
hisJ	Histidine	Periplasmic histidine-binding protein J for high-affinity histidine transport system	*hisJ*	46	12, 62, 281, 283, 410, 412, 418, 535
hisM	Histidine	Histidine transport; location of protein not known		46	9, 345, 368, 418

Table 1 *Continued*

Genetic symbol	Mnemonic	Former or alternative symbol; enzyme deficiency or other phenotype[a]	Gene in *E. coli*[b]	Map (min)[c]	References[d]
hisP	Histidine	High-affinity histidine transport; P protein in the inner membrane		46	9, 12, 154, 283, 410, 412, 418
hisQ	Histidine	Histidine transport; Q, a membrane protein		46	9, 12, 283. 418
hisR	Histidine	tRNA structural gene	*hisR*	83	41, 44, 72, 410, 412, 418
hisS	Histidine	Histidyl-tRNA synthetase (EC 6.1.1.21)	*hisS*	53	410, 412, 418
hisT	Histidine	Pseudouridine modification of tRNA	*hisT*	47	340, 410, 412, 418
hpt		Hypoxanthine phosphoribosyltransferase (not EC 2.4.2.8) (see *gpt*)	*hpt*	NM	355, 412, 418
hsdL	Host specificity	*hspLT*; restriction modification system		8	410, 412, 418
hsdSA	Host specificity	*hspS*; restriction modification system	*hsd*	98	262, 410, 412, 418
hsdSB	Host specificity	Restriction modification system	*hsd*	98	61, 262, 412, 418
hutC	Histidine utilization	Utilization; repressor		18	410, 412, 418
hutG	Histidine utilization	Formiminoglutamase (EC 3.5.3.8)		18	410, 412, 418
hutH	Histidine utilization	Histidine ammonia-lyase (EC 4.3.1.3)		18	410, 412, 418
hutI	Histidine utilization	Imidazolonepropionase (EC 3.5.2.7)		18	410, 412, 418
hutM	Histidine utilization	Utilization; promoter for *hutIGC*		18	410, 412, 418
hutP	Histidine utilization	Utilization; promoter for *hutUH*		18	410, 412, 418
hutQ	Histidine utilization	Utilization; promoter for *hutUH*		18	410, 412, 418
hutR	Histidine utilization	Utilization; catabolite insensitivity of *hutUH*		18	410, 412, 418
hutU	Histidine utilization	Utilization; urocanate hydratase (EC 4.2.1.49)		18	410, 412, 418
hyd		*aniA*, *fhlB*; hydrogenase	*hyd*	59	21, 412, 418, 420
hyp	Hydrophobic peptide auxotrophy	Hydrophobic polypeptide requirement		48	418
ilvA	Isoleucine-valine	*ile*; threonine dehydratase (EC 4.2.1.16)	*ilvA*	83	116, 177, 410, 412, 418, 479
ilvB	Isoleucine-valine	Acetolactate synthetase I, large subunit (valine sensitivity) (EC 4.1.3.18)	*ilvB*	80	101, 279, 280, 410, 412, 418, 512
ilvC	Isoleucine-valine	*ilvA*; 2-acetolactate mutase (EC 5.4.99.3)	*ilvC*	83	35, 410, 412, 418
ilvD	Isoleucine-valine	*ilvB*; dihydroxyacid dehydratase (EC 4.2.1.19)	*ilvD*	83	116, 177, 410, 412, 418
ilvE	Isoleucine-valine	*ilvC*; branched-chain aminotransferase (EC 2.6.1.42)	*ilvE*	83	116, 177, 410, 412, 418
ilvG	Isoleucine-valine	Acetolactate synthase II, large subunit (feedback inhibition insensitive)	*ilvG*	83	101, 116, 177, 278, 280, 412, 418, 421
ilvH	Isoleucine-valine	Acetolactate synthase II subunit (normally inactive)	*ilvH*	3	418, 451
ilvI	Isoleucine-valine	Acetolactate synthase II subunit (normally inactive)	*ilvI*	3	418, 451
ilvM	Isoleucine-valine	Acetolactate synthase II, small subunit (feedback inhibition insensitive)		83	101, 421
ilvN	Isoleucine-valine	Acetolactate synthase I, small subunit		80	101
ilvS	Isoleucine-valine	Isoleucyl-tRNA synthetase (EC 6.1.1.5)		1	410, 412, 418
ilvY	Isoleucine	Regulation of *ilvC*	*ilvY*	83	418
inlA	Inositol	Fermentation		92	35, 410, 412, 418
inlB	Inositol	Fermentation		54	410, 412, 418
inm		Sensitivity to mutagenesis by nitrosoguanidine		79	92
katG	Catalase	*cls*; HPI and HPII catalases (EC 1.11.1.6)	*katG*	88	333, 412, 418
kdsA		Ketodeoxyoctonate synthesis		39	70, 156, 157, 172, 174, 384, 388, 412, 418, 465
kdsB		CMP ketodeoxyoctonate synthetase		16	70, 155–157, 418
lamB	Lambda	Determines a protein resembling the lambda receptor	*lamB*	91	418
leuA	Leucine	α-Isopropylmalate synthase (EC 4.1.3.12)	*leuA*	2	73, 74, 149, 150, 178, 410, 412, 418, 433
leuB	Leucine	β-Isopropylmalate dehydrogenase	*leuB*	2	410, 412, 418, 433
leuC	Leucine	α-Isopropylmalate isomerase subunit	*leuC*	2	410, 412, 418
leuD	Leucine	α-Isopropylmalate isomerase subunit	*leuD*	2	142, 396, 410, 412, 418, 464
leuS	Leucine	Leucyl-tRNA synthetase (EC 6.1.1.4)	*leuS*	14	410, 412, 418
leuT	Leucine	Leucine transport		35	410, 412, 418
leuU	Leucine	*leuT*; tRNA structural gene for leucine	*leuT*	83	41, 44
lev		Levomycetin resistance		NM	412, 418
lig	Ligase	DNA ligase	*lig*	NM	313
lip	Lipoic acid	Requirement	*lip*	13	410, 412, 418

Table 1   *Continued*

Genetic symbol	Mnemonic	Former or alternative symbol; enzyme deficiency or other phenotype[a]	Gene in E. coli[b]	Map (min)[c]	References[d]
livA	Leucine, isoleucine, valine	High-affinity branched-chain amino acid transport	*liv*	76	310, 311, 418
livS	Leucine, isoleucine, valine	Regulatory gene; high-affinity branched-chain amino acid transport	*livR*	19	337
lkyA	Leaky	Leakage of periplasmic proteins		58	412, 418
lkyB	Leaky	Leakage of periplasmic proteins		58	412, 418
lkyC	Leaky	Leakage of periplasmic proteins		20	412, 418
lkyD	Leaky	Leakage of periplasmic proteins; morphology defect		20	80, 93, 300, 412, 418
lon	Long form	*capR*; filamentous growth; radiation sensitivity; polyamine metabolism; stabilization of abnormal proteins	*lon*	9	115, 412, 418, 504
lpd		Dihydrolipoamide dehydrogenase (EC 1.8.1.4)	*lpd*	3	412, 418
lpp	Lipoprotein	Murein lipoprotein structural gene		NM	418
lys	Lysine	Requirement	*lysA*	62	95, 410, 412, 418
malE	Maltose	*malB*; maltose uptake; periplasmic maltose-binding protein	*malE*	91	410, 412, 418, A
malF	Maltose	Maltose uptake; inner membrane protein	*malF*	91	A
malG	Maltose	Maltose uptake; inner membrane protein	*malG*	91	A
malK	Maltose	Maltose uptake; inner membrane protein	*malK*	91	A
malQ	Maltose	Amylomaltase (EC 1.2.1.25)	*malQ*	74	410, 412, 418
malT	Maltose	Regulation of maltose genes	*malT*	74	418
melA	Melibiose	α-Galactosidase (EC 3.2.1.22)	*mel*	93	418
melB	Melibiose	Permease		93	418
mem	Membrane	Sugar transport and membrane protein defective		78	412, 418
menA	Menaquinone	Menaquinone deficient; defective in trimethylamine oxide reduction; grows on vitamin K$_1$	*menA*	87	270, 271, 418
menB	Menaquinone	Biosynthesis; grows on vitamins K$_1$ and K$_5$		46	89, 270, 271
menC	Menaquinone	Biosynthesis	*menC*	46	271
menD	Menaquinone	Biosynthesis	*menD*	46	271
metA	Methionine	*metI*; homoserine transsuccinylase (EC 2.3.1.46)	*metA*	89	410, 412, 418
metB	Methionine	Cystathionine γ-synthase (EC 4.2.99.9)	*metB*	87	410, 412, 418, 487, 489, 490
metC	Methionine	Cystathionine γ-lyase (EC 4.4.1.1)	*metC*	64	410, 412, 418, AA
metE	Methionine	Tetrahydropteroyltriglutamate methyltransferase (EC 2.1.1.14)	*metE*	84	372, 410, 412, 418, 431, 494
metF	Methionine	5,10-Methylenetetrahydrofolate reductase (EC 1.1.99.15)	*metF*	87	410, 412, 418
metG	Methionine	Methionyl-tRNA synthetase	*metG*	44	410, 412, 418
metH	Methionine	Vitamin B$_{12}$-dependent homocysteine-N^5-methylenetetrahydrofolate transmethylase	*metH*	89	410, 412, 418, 491, 494
metJ	Methionine	Methionine analog resistant; protein for methionine pathway regulation	*metJ*	87	410, 412, 418, 489, 490, 492
metK	Methionine	Methionine analog resistant; S-adenosylmethionine synthetase	*metK*	63	410, 412, 418
metL	Methionine	Aspartokinase II-homoserine dehydrogenase II	*metL*	87	490
metP	Methionine	High-affinity methionine transport	*metD*	5	373, 410, 412, 418
metR	Methionine	*trans*-Acting protein for expression of *metE* and *metH*		84	372, 493, 494
mglA	Methyl galactosidase	Membrane-bound protein for transport	*mglA*	NM	336, F
mglB	Methyl galactoside	Galactose-binding protein	*mglB*	NM	336, 412, 418, F
mglC	Methyl galactosidase	Membrane-bound protein for transport	*mglC*	NM	336, F
mglD	Methyl galactosidase	Repressor for *mgl* operon		NM	F
mglE	Methyl galactosidase	Transport		NM	336, F
mgtA	Magnesium transport	Magnesium transport		98	AF
mgtB	Magnesium transport	Magnesium transport		81	AF
miaA		Deficient in the nucleotide ms^{2}io^{6}A adenosine, a modified base present in some tRNAs		96	46, 60, 126
min	Minicells	Cell division	*min*	NM	412, 418
motA	Motility	Nonmotile but flagellate	*mot*	40	268, 410, 412, 418
motA	Motility	Nonmotile but flagellate	*mot*	40	268, 410, 412, 418

Table 1 *Continued*

Genetic symbol	Mnemonic	Former or alternative symbol; enzyme deficiency or other phenotype[a]	Gene in *E. coli*[b]	Map (min)[c]	References[d]
mta	*meso*-Tartaric acid	Utilization of and resistance to *meso*-tartaric acid		NM	418
mtlA	Mannitol	D-Mannitol phosphotransferase enzyme IIA	*mtlA*	78	161, 377, 405, 410, 412, 418
mtlD	Mannitol	Mannitol-1-phosphate dehydrogenase (EC 1.1.1.17)	*mtlD*	78	412, 418
musA	Mu sensitivity	Adsorption of phage Mu		42	335, 418
musB	Mu sensitivity	Adsorption of phage Mu		42	418
mutB	Mutator	Increased frequency of mutation with alkylating agents		64	418
mutG	Mutator	Increased frequency of mutation in host chromosome, not in P22		NM	412, 418
mutH	Mutator	Mutator	*mutH*	61	363, 412, 418
mutL	Mutator	Increased frequency of mutation	*mutL*	96	362, 363, 412, 418
mutS	Mutator	Increased frequency of mutation with alkylating agents	*mutS*	59	168, 362, 363, 418
mviA	Mouse virulence	Affects the growth rate of cells in mice		34	G
mviC	Mouse virulence	Affects the growth rate of cells in mice		16	G
nadA	Nicotinamide	*nicA*; requirement; quinolinic acid synthetase	*nadA*	17	19, 197, 410, 412, 418, 448, 482, 533
nadB	Nicotinamide	*nic*; L-aspartate oxidase	*nadB*	55	19, 94, 197, 412, 418, 533
nadC	Nicotinamide	Quinolinic acid PRPP phosphoribosyl transferase	*nadC*	3	19, 197, 412, 418
nadD	Nicotinamide	Requirement; NAMN adenylyl transferase		14	19, 214, 215, 418
nadE	Nicotinamide	Essential biosynthetic gene, unsupplementable; NAD synthetase		25	217, Q
nadI	Nicotinamide	Derepression of *nadA* and *nadB*		99	94, 533, AF
nadR	Nicotinamide	Controls expression of several genes for NAD synthesis		99	137, 197, O
nag	*N*-Acetylglucosamine	Nonutilization	*nag*	15	412, 418
nalB	Nalidixic acid	Resistance or sensitivity	*nalB*	57	412, 418
nap		Deficiency for nonspecific acid phosphatase I		NM	418
ndk		Nucleosidediphosphate kinase (EC 2.7.4.6)		53	412, 418
newD		Substitute gene for *leuD*		7	393, 412, 418, 464
nit	Nitrogen	Nitrogen metabolism		28	412, 418
nol	Norleucine	Norleucine resistance; possible defect in valine uptake or regulation		60	412, 418
nrdA		Ribonucleoside diphosphate reductase (EC 1.17.4.1), subunit B1	*nrdA*	NM	313
nsiA		Nicotinamide starvation inducible; NAD metabolism regulation		77	418
ntrA	Nitrogen regulation	*glnF*; repressor-activator for *glnA* expression and for other nitrogen-controlled genes	*glnF*	69	259, 315, 412, 418
ntrB	Nitrogen regulation	*glnR*; regulation of *glnA* expression and other nitrogen-controlled genes	*glnR*	85	8, 258, 259, 267, 315, 346, 347, 418
ntrC	Nitrogen regulation	*glnR*; regulation of *glnA* expression and other nitrogen-controlled genes	*glnR*	85	8, 258, 259, 315, 347, 361, 418
nuvA		Uridine thiolation factor A activity	*nuvA*	NM	261, R
oafA	O-antigen factor	*O-5*, *ofi*; lipopolysaccharide O-factor 5 (acetyl group)		43	410, 412, 418
oafC	O-antigen factor	Determines factor 1 in lipopolysaccharide of group E *Salmonella* spp.		13	410, 412, 418
oafR	O-antigen factor	Synthesis of lipopolysaccharide O antigen 12^2		12	410, 412, 418
ompA		Outer membrane protein 33K (II* of *E. coli*)	*ompA*	20	140, 412, 418
ompC		Outer membrane protein 36K (Ib of *E. coli*)	*opmC*	45	412, 418, 532
ompD		Outer membrane protein 34K		32	412, 418
ompF		Outer membrane protein 35K (Ia)	*ompF*	NM	418
ompR	Outer membrane protein	*ompB*, *tppA*; positive regulation of tripeptide permease and of outer membrane protein	*ompR*	74	152, 153, 186, 228, 293, 294, 412, 418
oppA	Oligopeptide permease	Oligopeptide-binding protein	*opp*	34	160, 186, 187, 190–192, 418
oppB	Oligopeptide permease	Oligopeptide transport system	*opp*	34	160, 186, 190, 192, 418
oppC	Oligopeptide permease	Oligopeptide transport system	*opp*	34	160, 186, 190, 192, 418
oppD	Oligopeptide permease	Oligopeptide transport system	*opp*	34	160, 186, 190, 192, 418
oppF	Oligopeptide permease	Oligopeptide transport system	*opp*	34	190, 192

Table 1   *Continued*

Genetic symbol	Mnemonic	Former or alternative symbol; enzyme deficiency or other phenotype[a]	Gene in *E. coli*[b]	Map (min)[c]	References[d]
optA	Oligopeptidase	Oligopeptidase [hydrolyzes *N*-acetyl-(L-alanyl)₄]		76	418
oriC	Origin	*poh*; origin of replication of chromosome	*oriC*	81	418
oxdA	Oxygen dependent	Gene activity controlled by *oxrA*		64	467
oxdB	Oxygen dependent	Gene activity controlled by *oxrA*		93	467
oxiA	Oxygen inducible	Induced by anaerobiosis		10	4, 230
oxiB	Oxygen inducible	Induced by anaerobiosis		22	4
oxiC	Oxygen inducible	Induced by anaerobiosis		30	4
oxiE	Oxygen inducible	Induced by anaerobiosis		88	4
oxrA	Oxygen regulation	Prevent oxygen regulation of *pepT*	*nirR*	30	467
oxrF	Oxygen regulation	Regulates expression of *aniH*		NM	3
oxrG	Oxygen regulation	Regulates expression of *aniC, I*		88	3
oxyR	Oxidative stress resistant	Positive regulator		88	85, 286, 333, 463
pabA	P-amino benzoate	Requirement; *p*-aminobenzoate synthase	*pabA*	NM	248
panB	Pantothenic acid	Ketopantohydroxymethyl transferase (EC 4.1.2.12)	*panB*	5	410, 412, 418
panC	Pantothenic acid	Pantothenate synthetase (EC 6.3.2.1)	*panC*	5	410, 412, 418
panD	Pantothenic acid	Ketopantoic acid reductase		5	412, 418, 513
panE	Pantothenic acid	Ketopantoic acid reductase		NM	418
panT	Pantothenic acid	Pantothenate transport		NM	412, 418
pasA		6-Aminonicotinic acid sensitive		92	136, 410, 412, 418
pasB		6-Aminonicotinic acid sensitive		66	136, 418
pasC		6-Aminonicotinic acid sensitive		20	136, 418
pasD		6-Aminonicotinic acid sensitive		18	136
pasE		6-Aminonicotinic acid sensitive		55	136
pck		Phosphoenolpyruvate carboxykinase (ATP) (EC 4.1.1.49)	*pck*	13	410, 412, 418
pclA	Permissive for *cly*	Permissive for lytic growth of P22 *cly*		NM	418
pclB	Permissive for *cly*	Permissive for lytic growth of P22 *cly*		NM	418
pclC	Permissive for *cly*	Permissive for lytic growth of P22 *cly*		NM	418
pde	Phosphodiesterase	2′,3′-Cyclic nucleotide 2′-phosphodiesterase		96	418
pdxB	Pyridoxine	Requirement	*pdxB*	47	410, 412, 418
pepA	Peptidase	Peptidase A (similar to aminopeptidase A of *E. coli*)		97	153, 386, 412, 418
pepB	Peptidase	Peptidase B (aminopeptidase)		53	386, 418
pepD	Peptidase	*ptdD*; Peptidase D (a dipeptidase, carnosinase)	*pepD*	6	386, 410, 412, 418
pepE	Peptidase	Peptidase E (splits Asp-X peptide bonds)		90	75, 418
pepM	Peptidase	Peptidase M; aminopeptidase that removes N-terminal methionine from proteins		3	326
pepN	Peptidase	*ptdN*; peptidase N (an aminopeptidase, naphthylamidase)	*pepN*	20	55, 386, 410, 412, 418
pepP	Peptidase	*ptdP*; peptidase P (splits X-Pro peptide bonds)		63	410, 412, 418
pepQ	Peptidase	Peptidase Q (splits X-Pro peptide bonds)		84	412, 418
pepT	Peptidase	Peptidase T (a tripeptidase)		25	418, 466, 468
pfkA		6-Phosphofructokinase (EC 2.7.1.11)	*pfkA*	85	412, 418
pfl		Pyruvate formate lyase	*pfl*	19	196, 412, 418
pgi	Phosphoglucose isomerase	*oxrC, pasA*; regulation of fermentative or biosynthetic enzymes; glucosephosphate isomerase (EC 5.3.1.9)	*pgi*	92	229, 412, 418
pgtA	Phosphoglycerate	Positive activator of phosphoglycerate transport		49	236, 407, 527, 530
pgtB	Phosphoglycerate	Protein for signal transmission for phosphoglycerate transport		49	236, 407, 527, 530
pgtC	Phosphoglycerate	Protein for signal transmission for phosphoglycerate transport		49	236, 407, 527, 530
pgtP	Phosphoglycerate	Transporter for phosphoglycerate transport		49	158, 236, 407, 527, 530
pheA	Phenylalanine	Chorismate mutase (EC 5.4.99.5)	*pheA*	55	410, 412, 418
pheR	Phenylalanine	Regulator gene for *pheA*		64	410, 412, 418
phoN	Phosphatase	Nonspecific acid phosphatase		25	412, 418
phoP	Phosphatase	Nonspecific acid phosphatase I		95	412, 418
phoS	Phosphatase	Periplasmic phosphate-binding protein		NM	23, 412, 418
phs		Hydrogen sulfide production		41	89, 412, 418
pig	Pigment	Brownish colonies		54	410, 412, 418
ply	Phage lysogeny	*pox*; control of P22 lysogeny		NM	418

Table 1 *Continued*

Genetic symbol	Mnemonic	Former or alternative symbol; enzyme deficiency or other phenotype[a]	Gene in *E. coli*[b]	Map (min)[c]	References[d]
pmi	Mannose	Mannose-6-phosphate isomerase (EC 5.3.1.8)	*manA*	30	164, 410, 412, 418
pmrA		Polymyxin resistance		94	412, 418, Z
pncA	Pyridine nucleotide cycle	Nicotinamide deamidase (EC 3.5.1.19)	*pncA*	27	193, 214, 412, 418
pncB	Pyridine nucleotide cycle	Nicotinic acid phosphoribosyltransferase (EC 2.4.2.11)		20	254, 412, 418
pncH	Pyridine nucleotide cycle	Nicotinamide used as sole nitrogen source		27	193
pncX		6-Aminonicotinamide resistant		27	193, 214
pnuA	Pyridine nucleotide uptake	*pncC*; NMN uptake deficient		99	137, 418, 449, AF
pnuB	Pyridine nucleotide uptake	Growth on lower than normal levels of NMN		99	418, 449
pnuC	Pyridine nucleotide uptake	NMN uptake deficient		17	418, 449, 482
pnuD	Pyridine nucleotide uptake	Reduced NMN uptake in *nad pncA*+ strain		60	449
pnuE	Pyridine nucleotide cycle	Failure to use exogenous NAD; periplasmic NAD pyrophosphorylase		86	366
polA	Polymerase	DNA nucleotidyltransferase (EC 2.7.7.7)		0	167, 410, 412, 418
poxA	Pyruvate oxidase	Hypersensitivity to antimicrobial agents; lower levels of pyruvate oxidase and acetolactate synthase deficiency in α-ketobutyrate metabolism	*poxA*	95	499
ppc		Phosphoenolpyruvate carboxylase (EC 4.1.1.31)	*ppc*	87	404, 410, 412, 418
ppsA		Phosphoenolpyruvate synthase	*ppsA*	37	148
ppsB		Deficiency in phosphoenolpyruvate synthase; may be identical to *fruR*	*ppsA*	3	412, 418
praA		Phage P221 receptor function		94	412, 418
praB		Phase P221 receptor function		60	412, 418
prbA		Phage ES18 receptor function		92	412, 418
prbB		Phase ES18 receptor function		30	412, 418
prdB		Phage PH51 receptor function		31	412, 418
prh		Phage HK009 receptor function		94	412, 418
prk		Phage HK068 receptor function		30	412, 418
proA	Proline	Glutamate to glutamic-γ-semialdehyde	*proA*	7	304, 393, 410, 412, 418
proB	Proline	Glutamate to glutamic-γ-semialdehyde	*proB*	7	304, 410, 412, 418
proC	Proline	Pyrroline-5-carboxylate reductase (EC 1.5.1.2)	*proC*	8	51, 410, 412, 418
proP	Proline	Proline permease II; betaine and proline; low affinity	*proP*	93	67, 68, 120, 242, 412, 418
proT	Proline	tRNA structural gene for proline	*proT*	83	41, 44
proU	Proline	Proline/glycine betaine permease (high-affinity betaine uptake)	*proU*	57	12, 66, 118, 120, 185, 242, 418, 473
proV	Proline	Periplasmic betaine-binding protein	*proV*	57	188
prp	Propionate	Propionate metabolism	*prp*	97	M
prsA		Phosphoribosylpyrophosphate synthetase	*prs*	35	48, 237, 418
prsB		Phosphoribosylpyrophosphate synthetase		44	418
psiA		Phosphate starvation inducible		74	138
psiB		Phosphate starvation inducible		88	138
psiC		Phosphate starvation inducible		10	138
psiD		Phosphate starvation inducible		93	138
psiR		Regulates *psiC* activity		82	138
psuA		Suppressor of polarity		NM	412, 418
pta	Phosphotransacetylase	Acetyl-CoA:orthophosphate acetyltransferase (EC 2.3.1.8)		46	272, 418, 498
ptsF	Phosphotransferase system	*fruA*; fructose phosphotransferase enzyme IIA	*ptsF*	NM	148, 377, 405, 412, 418
ptsG	Phosphotransferase system	*glu, gpt*; glucose phosphotransferase enzyme IIB′-factor III (*crr*) system (methyl-β-D-glucoside)	*ptsG*	25	47, 161, 377, 405, 412, 418, 462
ptsH	Phosphotransferase system	*carB*; phosphohistidine protein-hexose phosphotransferase (EC 2.7.1.69	*ptsH*	49	25, 65, 162, 189, 330, 332, 377, 378, 405, 410, 412, 418, 505, 510
ptsI	Phosphotransferase system	*carA*; enzyme I of the phosphotransferase system	*ptsI*	49	65, 161, 265, 330, 332, 377, 405, 410, 412, 418, 509, 511

Table 1   *Continued*

Genetic symbol	Mnemonic	Former or alternative symbol; enzyme deficiency or other phenotype[a]	Gene in *E. coli*[b]	Map (min)[c]	References[d]
ptsJ	Phosphotransferase system	Enzyme I* of the phosphotransferase system, not expressed in wild type	*ptsJ*	49	83
ptsM	Phosphotransferase system	*manA*; mannose-glucose phosphotransferase enzyme IIA (2-deoxyglucose)	*ptsM*	NM	377, 405, 412, 418, 462
purA	Purine	Adenylosuccinate synthetase (EC 6.3.4.4)	*purA*	96	410, 412, 418
purB	Purine	Adenylosuccinate lyase (EC 4.3.2.2)	*purB*	25	410, 412, 418
purC	Purine	Phosphoribosylaminoimidazole-succinocarboxamide synthetase (EC 6.3.2.6)	*purC*	51	410, 412, 418
purD	Purine	Phosphoribosylglycinamide synthetase (EC 6.3.1.13)	*purD*	89	121, 410, 412, 418
purE	Purine	Phosphoribosylaminoimidazole carboxylase (EC 4.1.1.21)	*purE*	11	355, 410, 412, 418
purF	Purine	Amidophosphoribosyltransferase (EC 2.4.2.14)	*purF*	47	114, 410, 412, 418
purG	Purine	Phosphoribosylglycinamidine synthetase (EC 6.3.5.3)	*purL*	54	410, 412, 418
purH	Purine	Phosphoribosylaminoimidazolecarboxamide formyltransferase (EC 2.1.2.3)	*purH*	89	121, 410, 412, 418
purI	Purine	Phosphoribosylaminoimidazole synthetase (EC 6.3.3.1)	*purM*	51	410, 412, 418
purJ	Purine	IMP cyclohydrolase (EC 3.5.4.10)		89	410, 412, 418
purN	Purine	Cryptic *purF* analog; synthesis of phosphoribosylamine		4	I, AD
purR	Purine	Constitutive high expression of *pur* genes		30	I
putA	Proline	*putB*; utilization; bifunctional enzyme; proline oxidase and pyrroline-5-carboxylate dehydrogenase	*putA*	22	12, 100, 171, 307, 410, 412, 418, T
putP	Proline	Utilization; major L-proline permease		22	12, 68, 108, 120, 307, 410, 412, 418, T
pyrA	Pyrimidine	*argD*, *ars*; arginine + uracil requirement; carbamoyl-phosphate synthase (glutamine) (EC 6.3.5.5)	*car*	1	342, 410, 412, 418
pyrB	Pyrimidine	Aspartate carbamoyltransferase (EC 2.1.3.2)	*pyrB*	98	133, 233, 322, 323, 410, 412, 418
pyrC	Pyrimidine	Dihydro-orotase (EC 3.5.2.3)	*pyrC*	23	250, 343, 410, 412, 418, 486
pyrD	Pyrimidine	Dihydro-orotate oxidase (EC 1.3.3.1)	*pyrD*	20	250, 410, 412, 418, 486
pyrE	Pyrimidine	Orotate phosphoribosyltransferase (EC 2.4.2.10)	*pyrE*	79	233, 250, 344, 410, 412, 418
pyrF	Pyrimidine	Orotidine-5′-phosphate decarboxylase (EC 4.1.1.23)	*pyrF*	33	410, 412, 418, 481
pyrG	Pyrimidine	CTP synthetase	*pyrG*	60	410, 412, 418
pyrH	Pyrimidine	UMP kinase	*pyrH*	5	232, 249, 342, 410, 412, 418
pyrI	Pyrimidine	Regulatory polypeptide for aspartate transcarbamylase (EC 2.1.3.2), regulatory subunit	*pyrI*	98	133, 323
rbsP	Ribose	Ribose-binding protein	*rbsP*	82	412, 418
recA		Recombination deficient; degrades DNA	*recA*	58	17, 117, 159, 165, 216, 357, 359, 410, 412, 418, 425
recB		Recombination deficient; exonuclease V	*recB*	61	117, 216, 412, 418, 443
recC		Recombination deficient; exonuclease V	*recC*	61	117, 216, 412, 418, 443
relA	RNA relaxed	*RC*; regulation of RNA synthesis	*relA*	61	224, 261, 412, 418
rfaB	Rough	UDP-D-galactose:lipopolysaccharide α-1,6-D-galactosyl transferase		79	245, 418
rfaC	Rough	Lipopolysaccharide core defect; proximal heptose deficient	*rfa*	79	49, 54, 356, 412, 418
rfaD	Rough	D-Glycero-D-manno-heptose epimerase	*rfaD*	79	412, 418, 518
rfaE	Rough	Lipopolysaccharide core defect; proximal heptose deficient		76	49, 339, 383, 410, 412, 418, 480
rfaF	Rough	Lipopolysaccharide core defect; distal heptose deficient		79	410, 412, 418
rfaG	Rough	Lipopolysaccharide core defect; glucose I transferase		79	50, 245, 410, 412, 418

Table 1 *Continued*

Genetic symbol	Mnemonic	Former or alternative symbol; enzyme deficiency or other phenotype[a]	Gene in E. coli[b]	Map (min)[c]	References[d]
rfaH	Rough	Deficient in lipopolysaccharide core synthesis and in F-factor expression; transcription control factor	sfrB	84	99, 410, 412, 418
rfaI	Rough	Lipopolysaccharide core defect; galactose I deficient		78	244, 245, 418
rfaJ	Rough	Lipopolysaccharide core defect; glucose II transferase		79	244, 245, 410, 412, 418
rfaK	Rough	Lipopolysaccharide core defect; N-acetylglucosamine transferase deficient		79	410, 412, 418
rfaL	Rough	Lipopolysaccharide core defect; O-translocase		79	410, 412, 418
rfaP	Rough	Lipopolysaccharide core defect; phosphorylation of heptose		77	410, 412, 418
rfbA	Rough	TDP-glucose pyrophosphorylase		42	52, 53, 292, 389, 410, 412, 418, 501
rfbB	Rough	TDP-glucose oxidoreductase		42	52, 53, 410, 412, 418
rfbD	Rough	TDP-rhamnose synthetase		42	52, 53, 410, 412, 418
rfbE	Rough	Lipopolysaccharide side chain defect; CDP paratose synthesis in S. typhi		42	53, D
rfbF	Rough	Glucose-1-phosphate cytidylyltransferase (EC 2.7.7.33)		42	52, 53, 410, 412, 418
rfbG	Rough	CDP-glucose oxidoreductase		42	52, 53, 410, 412, 418
rfbH	Rough	CDP-abequose synthetase		42	52, 53, 410, 412, 418
rfbJ	Rough	CDP-4-keto-3,6-D-glucose dehydrogenase		42	53, D
rfbK	Rough	Lipopolysaccharide side chain defect	rfb	42	52, 53, 410, 412, 418
rfbL	Rough	Phosphomannomutase B		42	52, 53, 410, 412, 418
rfbM	Rough	Mannose-1-phosphate guanylyltransferase (EC 2.7.7.22)		42	52, 53, 410, 412, 418
rfbN	Rough	Galactose-diphosphoglycosyl carrier lipid synthetase		42	52, 53, 412, 418
rfbT	Rough	O-Translocase		42	52, 410, 412, 418
rfc	Rough	rouC; O-repeat unit not polymerized		31	369, 410, 412, 418
rfe	Rough	Defect in synthesis of enterobacterial common antigen, the T1 antigen, and O-side chains of Salmonella groups L and C1	rfe	83	292, 410, 412, 418
rff	Rough	Block in synthesis of enterobacterial common antigen	rff	84	292, 319, 412, 418
rft	Rough	"Transient" T1 forms		15	410, 412, 418
rfu	Rough	"Transient" T1 forms		NM	418
rhaA		L-Rhamnose isomerase (EC 5.3.1.14)	rhaA	86	6, 410, 412, 418
rhaB		L-Rhamnulokinase (EC 2.7.1.5)	rhaB	86	6, 410, 412, 418
rhaC		Regulation	rhaC	86	6, 410, 412, 418
rhaD		L-Rhamnulose-1-phosphate aldolase (EC 4.1.2.19)	rhaD	86	6, 410, 412, 418
rhaT		L-Rhamnose transport		86	6, 410, 412, 418
rho		psu; polarity suppressor; transcription terminator factor Rho	rho	83	418
rna		rnsA; RNase I	rna	17	412, 418
rnc		RNase III	rnc	NM	412, 418
rnpB	RNase	RNase P, RNA component	rnpB	NM	15
rodA	Rod	Round cell morphology; mecillinam resistant	rodA	13	11, 418
rplJ	Ribosomal protein, large	Ribosomal protein subunit	rplJ	88	475, 483
rplL	Ribosomal protein, large	Ribosomal protein subunit	rplL	88	475, 483
rpoB	RNA polymerase	rif; RNA polymerase, β subunit (EC 2.7.7.6)	rpoB	89	233, 342, 410, 412, 418, 474, 475, 483, 484
rpoC	RNA polymerase	RNA polymerase, β′ subunit (EC 2.7.7.6)	rpoC	89	233, 412, 418, 475, 483, 484
rpoD	RNA polymerase	RNA polymerase, σ subunit	rpoD	NM	125, 180, 418
rpsE	Ribosomal protein, small	spcA; 30S ribosomal subunit protein S5	rpsE	71	410, 412, 418
rpsL	Ribosomal protein, small	strA; 30S ribosomal subunit protein S12	rpsL	71	410, 412, 418
rpsU	Ribosomal protein, small	30S ribosomal subunit protein S21	rpsU	NM	125

Table 1   *Continued*

Genetic symbol	Mnemonic	Former or alternative symbol; enzyme deficiency or other phenotype[a]	Gene in E. coli[b]	Map (min)[c]	References[d]
rrnA	rRNA	rRNA operon	rrnA	86	288, 289, 418
rrnB	rRNA	rRNA operon	rrnB	89	288, 289, 418, 446
rrnC	rRNA	rRNA operon	rrnC	82	288, 289, 418
rrnD	rRNA	rRNA operon	rrnD	73	288, 418
rrnE	rRNA	rRNA operon	rrnE	89	288, 289, 418
rrnG	rRNA	rRNA operon	rrnG	55	288
rrnH	rRNA	rRNA operon	rrnH	6	288, 418
selA	Selenium	Selenium incorporation		21	260
serA	Serine	Phosphoglycerate dehydrogenase (EC 1.1.1.95)	serA	62	410, 412, 418
serB	Serine	Phosphoserine phosphatase (EC 3.1.3.3)	serB	19	137, 410, 412, 418
serC	Serine	Requirement	serC	19	196, 418
serD	Serine	Requirement for pyridoxine plus L-serine or glycine		44	412, 418
sidC		Siderochrome utilization; ferrichrome transport; albomycin resistance		4	412, 418
sidF		Siderochrome utilization; ferrichrome transport; albomycin resistance		4	412, 418
sidK		Siderochrome utilization; albomycin resistance; receptor of phage ES18 in *S. typhimurium* and of T5 in *S. paratyphi* B	tonA	5	412, 418
smoB		Smooth colony morphology in histidine-constitutive mutants		99	410, 412, 418
spcB	Spectinomycin	Resistance (nonribosomal)		72	410, 412, 418
spoT	Spot	Guanosine 5′-diphosphate, 3′-diphosphate pyrophosphatase	spoT	79	400, 418
srlA	Sorbitol	*gut*; D-glucitol-specific enzyme II of the phosphotransferase system	srlA	59	419, W
srlB	Sorbitol	*gut*; D-glucitol-specific enzyme III of the phosphotransferase system		59	419, W
srlC	Sorbitol	*gut*; Regulatory gene	srlC	59	410, 412, 418, 419, W
srlD	Sorbitol	*gut*; sorbitol-6-phosphate dehydrogenase (EC 1.1.1.140)	srlD	59	W
srlM	Sorbitol	*gut*; DNA-binding protein which activates transcription of *srl*		59	W
srlR	Sorbitol	*gut*; regulatory gene	srlR	59	W
ssb	Single-strand binding	Single-strand DNA-binding protein	ssb	NM	313
stiA	Starvation inducible	*sinA*; starvation for carbon source or other requirements causes induction; repressed in *relA*		32	135, 450
stiB	Starvation inducible	Starvation for carbon source or other requirements causes induction		NM	450
stiC	Starvation inducible	Starvation for carbon source or other requirements causes induction		75	450
stiD	Starvation inducible	Starvation for carbon source or other requirements causes induction		32	450
stiE	Starvation inducible	Starvation for carbon source or other requirements causes induction		41	450
stiF	Starvation inducible	Starvation for carbon source or other requirements causes induction		NM	450
stiG	Starvation inducible	Starvation for carbon source or other requirements causes induction		86	450
stiH	Starvation inducible	Starvation for carbon source or other requirements causes induction		55	450
strB	Streptomycin	Low-level resistance plus auxotrophy; nonribosomal		53	410, 412, 418
strC		Streptomycin resistance, not *strA* or *strB*		NM	412, 418
stx	*Salmonella* toxin	Enterotoxin		84, U	
sucA	Succinate	*lys, suc*; succinate requirement; α-ketoglutarate dehydrogenase (decarboxylase component)	sucA	17	410, 412, 418
sufA		Frameshift suppressor affecting proline tRNA and correcting +1 frame shifts at runs of C in the mRNA		77	256, 263, 410, 412, 418
sufB		Frameshift suppressor affecting proline tRNA and correcting +1 frame shifts at runs of C in the mRNA		45	256, 263, 410, 412, 418

Table 1   *Continued*

Genetic symbol	Mnemonic	Former or alternative symbol; enzyme deficiency or other phenotype[a]	Gene in *E. coli*[b]	Map (min)[c]	References[d]
sufC		Recessive suppressor of +1 frameshift mutations at runs of C in the mRNA		15	410, 412, 418
sufD		Frameshift suppressor affecting glycine tRNA and correcting +1 frameshift mutations at runs of G in the mRNA		62	256, 410, 412, 418
sufE		Frameshift suppressor correcting +1 frameshift mutations at runs of G in the mRNA		89	256, 410, 412, 418
sufF		Recessive frameshift suppressor correcting +1 frameshift mutations at runs of G in the mRNA		11	410, 412, 418
sufG		Frameshift suppressor correcting +1 frameshift mutations at runs of A in the mRNA		15	256, 412, 418
sufH		Frameshift suppressor		50	256, 412, 418
sufI		Frameshift suppressor		11	256, 412, 418
sulB	Suppressor	Suppressor of *lon*	*sulB*	NM	96
sumA		Suppressor of missense		96	412, 418
supC	Suppressor	Ochre suppressor	*supC*	34	410, 412, 418
supD	Suppressor	Amber suppressor; serine insertion	*supD*	40	42, 218, 410, 412, 418
supE	Suppressor	*supY*; amber suppressor; glutamine insertion	*supE*	15	27, 42, 218, 410, 412, 418
supF	Suppressor	See *tyrT*			42, 218, 412, 418
supG	Suppressor	Ochre suppressor; lysine insertion	*supG*	NM	412, 418
supI	Suppressor	Nonsense suppressor induced by ICR-191 and allelic to *sufG*		15	412, 418
supJ	Suppressor	*supH*; amber suppressor; leucine insertion	*supJ*	82	42, 412, 418
supK	Suppressor	*supT*; recessive UGA suppressor; also corrects some frameshift mutations		62	410, 412, 418
supM	Suppressor	See *tyrU*			410, 412, 418
supQ	Suppressor	Suppressor of nonsense and deletion mutations of *leuD*		7	410, 412, 418, 464
supR	Suppressor	Amber suppressor; haploid lethal		82	410, 412, 418
supS	Suppressor	UGA suppressor; haploid lethal		82	410, 412, 418
supU	Suppressor	Suppressor of UGA muations; may be due to alteration of ribosome structure		72	418
tar	Taxis-associated receptor	Chemotaxis transduction polypeptide; aspartate receptor	*tar*	40	134, 331, 402, 418
tctA	Tricarboxylate transport	Membrane protein		57	418, 515, 516
tctB	Tricarboxylate transport	Membrane protein		57	515, 516
tctC	Tricarboxylate transport	Tricarboxylate-binding protein		57	412, 418, 447, 476, 515, 516
tctD	Tricarboxylate transport	Regulatory protein		57	515, 516
tctII	Tricarboxylate transport	Transport		15	AE
tctIII	Tricarboxylate transport	Transport		1	AE
tdk		Thymidine kinase (EC 2.7.1.21)	*tdk*	34	410, 412, 418
thiA	Thiamine	*thiG*; thiamine or thiazole moiety	*thiA*	89	410, 412, 418
thiC	Thiamine	*thiA*; thiamine or pyrimidine moiety	*thiC*	89	410, 412, 418
thiD	Thiamine	Thiamine requirement		46	410, 412, 418
thiE	Thiamine	Thiazole type		52	410, 412, 418
thiF	Thiamine	Thiazole type		52	410, 412, 418
thiH	Thiamine	*thiB*; thiamine requirement		54	412, 418
thiI	Thiamine	*thiC*; thiazole type		10	412, 418
thrA	Threonine	*thrC, thrD*; aspartokinase (EC 2.7.2.4) and homoserine dehydrogenase I (EC 1.1.1.3)	*thrA*	0	410, 412, 418
thrB	Threonine	*thrA*; and homoserine kinase (EC 2.7.1.39)	*thrB*	0	410, 412, 418
thrC	Threonine	*thrB*; and homoserine synthase (EC 4.2.99.2)	*thrC*	0	410, 412, 418
thrT	Threonine	*sufJ*; threonine tRNA		88	43, 256, 257
thyA	Thymine	Requirement	*thyA*	61	412, 418, 443
tip	Taxis-involved protein	Methyl-accepting chemotaxis protein (aspartate receptor)	*tap*	NM	403
tkt		Transketolase (EC 2.2.1.1)	*tkt*	NM	412, 418
tlp		Loss of protease II		37	412, 418

Table 1   *Continued*

Genetic symbol	Mnemonic	Former or alternative symbol; enzyme deficiency or other phenotype[a]	Gene in *E. coli*[b]	Map (min)[c]	References[d]
tlr		Thiolutin resistance; P22 development at high temperature		NM	418
tonB		*chr*; regulates levels of some outer membrane proteins; resistance to ES18; determines a salmonellocin; affects iron transport	*tonB*	34	111, 192, 410, 412, 418
top	DNA topoisomerase I	*supX*; Topoisomerase	*topA*	33	185, 241, 357, 359, 360, 379–381, 387, 410, 412, 418, 477, 526
tor		Trimethylamine oxide reductase		80	269, 270, 418
tppA	Tripeptide permease	Resistance to alafosfalin, regulator of *tppB*		74	153, 186, 228
tppB	Tripeptide permease	Resistance to alafosfalin; tripeptide permease		27	153, 186, 228, 229
tppR	Tripeptide permease	Regulator of tripeptide permease		3	229
traT	Transfer	Membrane protein cross-reacts immunologically with TraT protein of F; restores permeability mutants to normal		pSLT	469–472, 495, AC
tre	Trehalose	Utilization	*tre*	37	376, 377, 410, 412, 418
triM		Tricarboxylic acid metabolism; see *tctIII*		1	412, 418
triR		Tricarballylic acid transport; see *tctIII*		1	412, 418
trmD		Likely to be defective in the tRNA (guanine-1-)-methyltransferase (EC 2.1.1.31)	*trmD*	55	E
trpA	Tryptophan	*trpC*; tryptophan synthetase, component alpha (EC 4.2.1.20)	*trpA*	34	1, 320, 410, 412, 418, 452, 531
trpB	Tryptophan	*trpD*; tryptophan synthetase, component beta (EC 4.2.1.20)	*trpB*	34	410, 412, 418
trpC	Tryptophan	*trpE*; *N*-(5-phosphoribosyl) anthranilate isomerase and indole-3-glycerol phosphate synthase (EC 4.1.1.48)	*trpC*	34	209, 410, 412, 418
trpD	Tryptophan	*trpB*; anthranilate phosphoribosyltransferase (EC 2.4.2.18)	*trpD*	34	209, 410, 412, 418
trpE	Tryptophan	*trpA*; anthranilate synthase (EC 4.1.3.27)	*trpE*	34	103, 139, 410, 412, 418
trpR	Tryptophan	Resistance to 5-methyltryptophan; derepression of tryptophan enzymes	*trpR*	99	410, 412, 418
tsr		Chemotaxis receptor; serine specificity		NM	
ttr		Tetrathionate reductase		35	410, 412, 418
tufA		Protein chain elongation factor EF-Tu	*tufA*	71	211–213, 418
tufB		Protein chain elongation factor EF-Tu	*tufB*	88	211–213, 418
tyn		Tyramine oxidase		NM	418
tyrA	Tyrosine	Requirement	*tyrA*	55	410, 412, 418
tyrR	Tyrosine	Regulator gene for *aroF* and *tyrA*	*tyrR*	32	410, 412, 418
tyrT	Tyrosine	*supC*; ochre suppressor; tyrosine tRNA1	*tyrT*	34	42, 46, 126, 412, 418
tyrU	Tyrosine	*supM*; ochre suppressor; tyrosine tRNA2	*tyrU*	88	412, 418
ubiF	Ubiquinone	*cad*; deficient in ubiquinone synthesis; accumulates 2-octaprenyl-3-methyl-6-methoxy-1,4-benzoquinone	*ubiF*	14	418
ubiX	Ubiquinone	Growth stimulation by *p*-hydroxybenzoic acid; polyprenyl *p*-hydrobenzoate carboxylase		46	412, 418
udk		Uridine kinase (EC 2.7.1.48)	*udk*	43	410, 412, 418
udp		Uridine phosphorylase (EC 2.4.2.3)	*udp*	84	410, 412, 418
uhpA		Utilization of hexose phosphate		80	412, 418
uhpT		Hexosephosphate transport	*uhpT*		412, 418
umuC		Induction of mutations by UV; sensitivity to UV	*umuC*		184, 418, 440
uncA	Uncoupling	Membrane-bound (Mg^{2+}, Ca^{2+})ATPase	*unc*	81	418, 454
upp		Uracil phosphoribosyltransferase (EC 2.4.2.9)	*upp*	51	410, 412, 418
urs		Uracil catabolism defect		30	513
use	Uracil sensitivity	Altered expression of genes *pyrA*, *pyrC*, *pyrD*, and *argI*		84	418
ushA	UDP sugar hydrolase	UDP-sugar hydrolase (5′-nucleotidase) (silent gene in *Salmonella* spp.)	*ushA*	11	62, 63, 412, 418, C
ushB	UDP sugar hydrolase	UDP-sugar hydrolase (membrane associated)		90	62, 63, 412, 418, C
usp	Ureidosuccinate	Permeability to ureidosuccinate (i.e., carbamyl asparatate)		NM	410, 412, 418
uvrA	UV	Repair of UV damage to DNA; UV endonuclease, component B	*uvrA*	91	410, 412, 418
uvrB	UV	Repair of UV damage to DNA; UV endonuclease component B	*uvrB*	18	410, 412, 418

Table 1 *Continued*

Genetic symbol	Mnemonic	Former or alternative symbol; enzyme deficiency or other phenotype[a]	Gene in *E. coli*[b]	Map (min)[c]	References[d]
uvrC	UV	Repair of UV damage to DNA	*uvrC*	40	412, 418
uvrD	UV	Repair of UV damage to DNA; increased sensitivity to mutagenesis by alkylating agents	*uvrD*	84	363–365, 412, 418
valS		Valyl-tRNA synthetase (EC 6.1.1.9)	*valS*		412, 418
viaA		*ViA*; Vi antigen		46	410, 412, 418
viaB		*ViB*; Vi antigen (in *S. typhosa*)		94	410, 412, 418
xylA	D-Xylose	Xylose isomerase (EC 5.3.1.5)	*xyl*	78	151, 412, 418
xylB	D-Xylose	Xylulokinase (EC 2.7.1.17)		78	151, 412, 418
xylR	D-Xylose	Regulation		78	151, 412, 418
xylT	D-Xylose	Transport		78	151, 412, 418

[a] Abbreviations: AMP, adenosine monophosphate; ATP, adenosine triphosphate; ATPase, adenosine triphosphatase; cAMP, cyclic AMP; CDP, cytidine diphosphate; CMP, cytidine monophosphate; CoA, coenzyme A; CTP, cytidine triphosphate; DAHP, 3-deoxy-D-arabinoheptulsonic acid 7-phosphate; dCTP, deoxycytidine triphosphate; dUMP deoxyuridine monophosphate; DMB, dimethylbenzimidazole; GMP, guanosine monophosphate; HP, hydrogen peroxide; IMP, inosine monophosphate; NAD(P), nicotinamide adenine dinucleotide (phosphate); PRPP, phosphoribosyl pyrophosphate; NAMN, nicotinic acid mononucleotide; NMN, nicotinamide mononucleotide; RNase, ribonuclease; TDP, thymidine diphosphate; UDP, uridine diphosphate; UMP, uridine monophosphate; UV, ultraviolet.

[b] The homologous gene in *E. coli* is described by Bachmann (14).

[c] Map positions in minutes are shown in Fig. 1, from 9 to 99 min. NM indicates that the gene is not mapped. The symbol pSLT indicates that the gene is on the plasmid of LT2.

[d] The numbers refer to references in Literature Cited. References 410, 412, and 418 refer to earlier editions of the linkage map in which other references to the indicated gene are given. Sanderson and Hurley (413) list all major references up to 1983 in a single source. There are many papers in the summary of the cellular and molecular biology of *E. coli* and *S. typhimurium* (346) which have important information on the genes of *S. typhimurium*. Letters A through AG refer to personal communications from the following sources: A, G. F.-L. Ames; B, E. Barrett; C, I. Beacham, D, P. R. Reeves; E, G. Björk; F, W. Boos; G, J. R. Curtiss and W. H. Benjamin, Jr.; H, T. Doak and J. R. Roth; I, D. Downs and J. R. Roth; J, T. Elliott, J. Delling, and J. R. Roth; K, T. Elliott and J. R. Roth; L, J. Escalante-Semerena and J. R. Roth; M, A. Fernandez-Briera and A. Garrido-Pertierra; N, P. Fields, E. Groisman, and F. Heffron; O, J. W. Foster; P, H.-S. Houng, D. J. Kopecko, and L. S. Baron; Q, K. T. Hughes, B. M. Olivera, and J. R. Roth; R, G. Kramer and B. N. Ames; S, N. D. Kredich; T, S. Maloy; U, J. W. Peterson; V, D. Roof and J. R. Roth; W, M. Saier; X, M. Schmid; Y, S. A. Simms and J. Stock; Z, J. K. Spitznagel; AA, G. V. Stauffer; AB, G. Storz and B. N. Ames; AC, S. Sukupolvi; AD, G.-M. Tang and J. R. Roth; AE, K. Widenhorn, J. Somers, and W. W. Kay; AF, N. Zhu and J. R. Roth; AG, C. Higgins.

Additional Data and Corrections to Table 1

The following are updated mapping data:

aniD = 91 min	*dml* = 81 min	*hemA* = 35 min	*prsB* = 45 min
aniF = 65 min	*dum* = NM	*hisB* = 42 min	*pyrD* = 21 min
aniG = 65 min	*fdhA* = 78 min	*lon* = 10 min	*rfaI* = 79 min
apt = 11 min	*fdhF* = NM	*metC* = 65 min	*rft* = 16 min
argG = 68 min	*fdnB* = 86 min	*musB* = NM	*rrnD* = 71 min
argI = 97 min	*fdnC* = 86 min	*mviA* = 33 min	*serB* = 99 min
chlC = 35 min	*galC* = NM	*nadE* = 27 min	*tor* = 81 min
chlG = 55 min	*galU* = 35 min	*oriC* = 82 min	*tppA* = 73 min
clmF = 65 min	*garA* = 16 min	*oxiA* = 9 min	*tre* = 36 min
cmk = 21 min	*garB* = 47 min	*phoN* = 95 min	*uhpT* = NM
cpd = 65 min	*glyA* = 54 min	*phoP* = 25 min	*umuC* = NM
crr = 49 min	*gpt* = 7 min	*olA* = 85 min	*use* = 86 min
cysA = 50 min	*gyrB* = 82 min	*orbA* = 94 min	*valS* = NM

The following are corrections or additions to mnemonic designations:

envZ = Envelope; *fim* = Fimbriae; *omp* = Outer membrane protein; *nuE* = Pyridine nucleotide uptake; *putA* and *putP* = Proline utilization; *sufA–sufI* = Suppressor frameshift

The following are corrections or additions to former or alternative symbols:

fliE = *cheV, flaAI, motC*; *fliH* = *cheV, flaAII.3, motC*; *fliL* = *cheC, cheU, flaQI*; *fliQ* = *flaD, flaQ*

Miscellaneous corrections:

For *fdhC*, column 3 should read "Formate dehydrogenase." The last entry on page 3.12 should read "*motB*" not *motA*. For *strC*, column 3 should read "not *rpsL* or *strB*." For *thrC*, column 3 should read "threonine" synthase.

Table 2   Alternative Gene Symbols[a]

Former or alternative symbol	Current symbol	Former or alternative symbol	Current symbol
aniA	hyd	gut	srlD
apeD	apeR	gut	srlA
argA	argE	gut	srlB
argB	argA	gut	srlM
argC	argB	gut	srlR
argD	pyrA	gxu	gpt
argE	argG	H1	fliC
argF	argH	H2	fljB
argG	argD	hisU	gyrB
argH	argC	hisW	gyrA
argT	argU	hspLT	hsdL
aroC	aroE	hspS	hsdSA
aroD	aroC	ile	ilvA
aroE	aroD	ilvA	ilvC
ars	pyrA	ilvB	ilvD
asc	ent	ilvC	ilvE
attP14	atdA	ilvT	brnQ
attP22 I	ataA	leuT	leuU
attP22I	atcA	lys	sucA
attP27 I	atbA	malB	malE
attP27II	atbB	manA	ptsM
bac	envB	metI	metA
bfe	btuB	nalA	gyrA
cad	ubiF	nic	nadB
capR	lon	nicA	nadA
carA	ptsI	nml	fliB
carB	ptsH	O-5	oafA
cheP	cheA	ofi	oafA
cheQ	cheY	ompB	envZ
cheT	cheZ	ompB	ompR
cheX	cheB	oxrC	pgi
chr	tonB	pasA	pgi
cls	katG	pil	fim
dad	dadA	pncC	pnuA
dra	deoC	pnu	deoD
drm	deoB	poh	oriC
enb	ent	popC	hemL
fhl	fdhF	pox	ply
fhlB	hyd	psu	rho
flaAI	fliE	ptdD	pepD
flaAII.1	fliF	ptdN	pepN
flaAII.2	fliG	ptdP	pepP
flaAII.3	fliH	pup	deoD
flaAIII	fliI	putB	putA
flaB	fliP	rhl	fljA
flaC	flhA	rif	rpoB
flaD	fliQ	rnsA	rna
flaE	flhC	rouC	rfc
flaFI	flgA	sinA	stiA
flaFII	flgB	smoA	divC
flaFIII	flgC	spcA	rpsE
flaFIV	flgD	strA	rpsL
flaFIX	flgI	suc	sucA
flaFV	flgE	sufJ	thrT
flaFVI	flgF	supC	tyrT
flaFVII	flgG	supH	supJ
flaFVIII	flgH	supM	tyrU
flaFX	flgJ	supT	supK
flaH	flgH	supX	top
flaK	flhD	supY	supE
flaL	fliA	thiA	thiC
flaM	flhB	thiB	thiH
flaN	fliN	thiC	thiI
flaP	fliO	thiG	thiA
flaQI	fliL	thrA	thrB
flaQII	fliM	thrB	thrC
flaR	fliK	thrC	thrA
flaS	fliJ	thrD	thrA
flaU	flgL	tpp	deoA
flaV	fliD	tppA	ompR
flaW	flgK	tppB	envZ
flaX	flgF	trpA	trpE
flaX	fliR	trpB	trpD
fruA	ptsF	trpC	trpA
galE	galF	trpD	trpB
glnF	ntrA	trpE	trpC
glnR	ntrB	trz	cysK
glnR	ntrC	vh2	hin
glu	ptsG	ViA	viaA
gpt	ptsG	ViB	viaB
gut	srlC	wrkA	divA

[a] The alternative symbols have been used in past publications. It is recommended that their use be abandoned and that the current symbols, listed and described in Table 1 and in the references referred to there, be used in the future.

SECTION·4

Bacillus subtilis Genetic Map

Figure 1 and Tables 1 and 2 are from Ziegler (1990, pp. 2.28–2.46). The references cited in Table 1 can be found in the original publication. Figure 2 is reprinted, with permission, from Piggot (1989, p. 2).

REFERENCES

Piggot, P.J. 1989. Revised genetic map of *Bacillus subtilis* 168. In *Regulation of Prokaryotic Development* (ed. R. Smith et al.), pp. 1–41. American Society for Microbiology, Washington, D.C.

Ziegler, D.R. 1990. *Bacillus subtilis* 168. In *Genetic Maps*, Fifth Edition (ed. S.J. O'Brien), pp. 2.28–2.53. Cold Spring Harbor Laboratory Press, Cold Spring Harbor, New York.

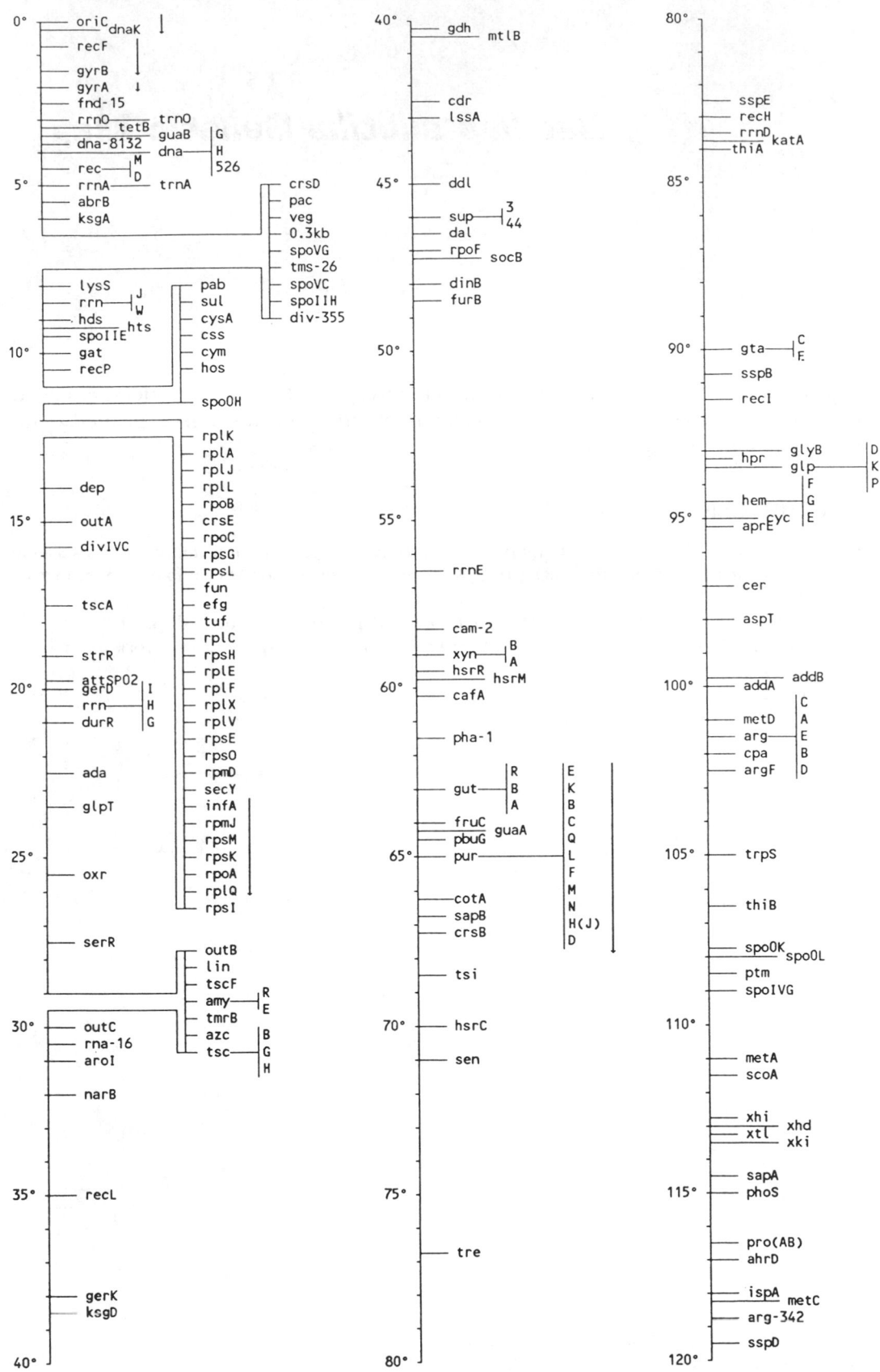

Figure 1 Genetic map of *B. subtilis*.

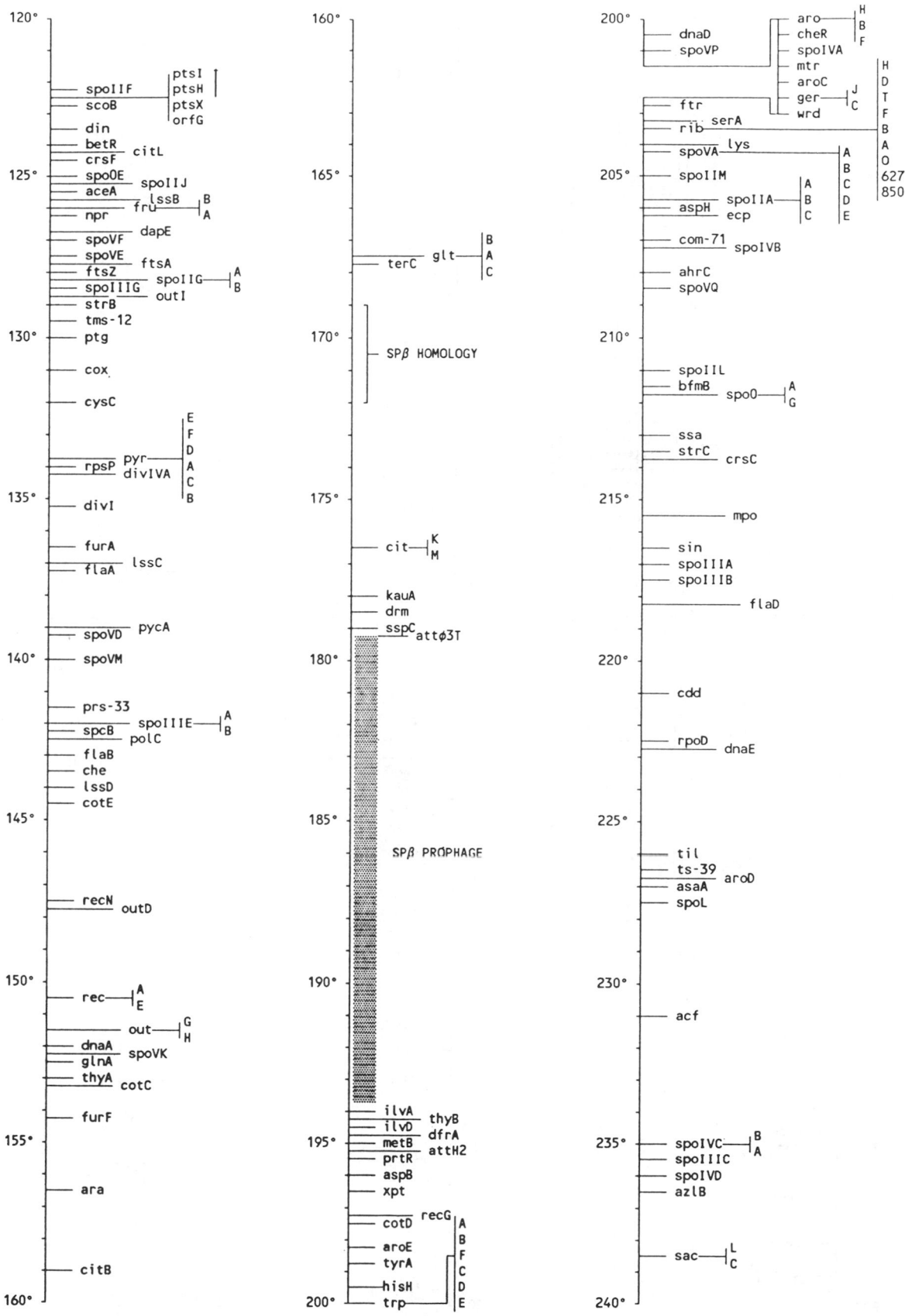

Figure 1 *Continued*

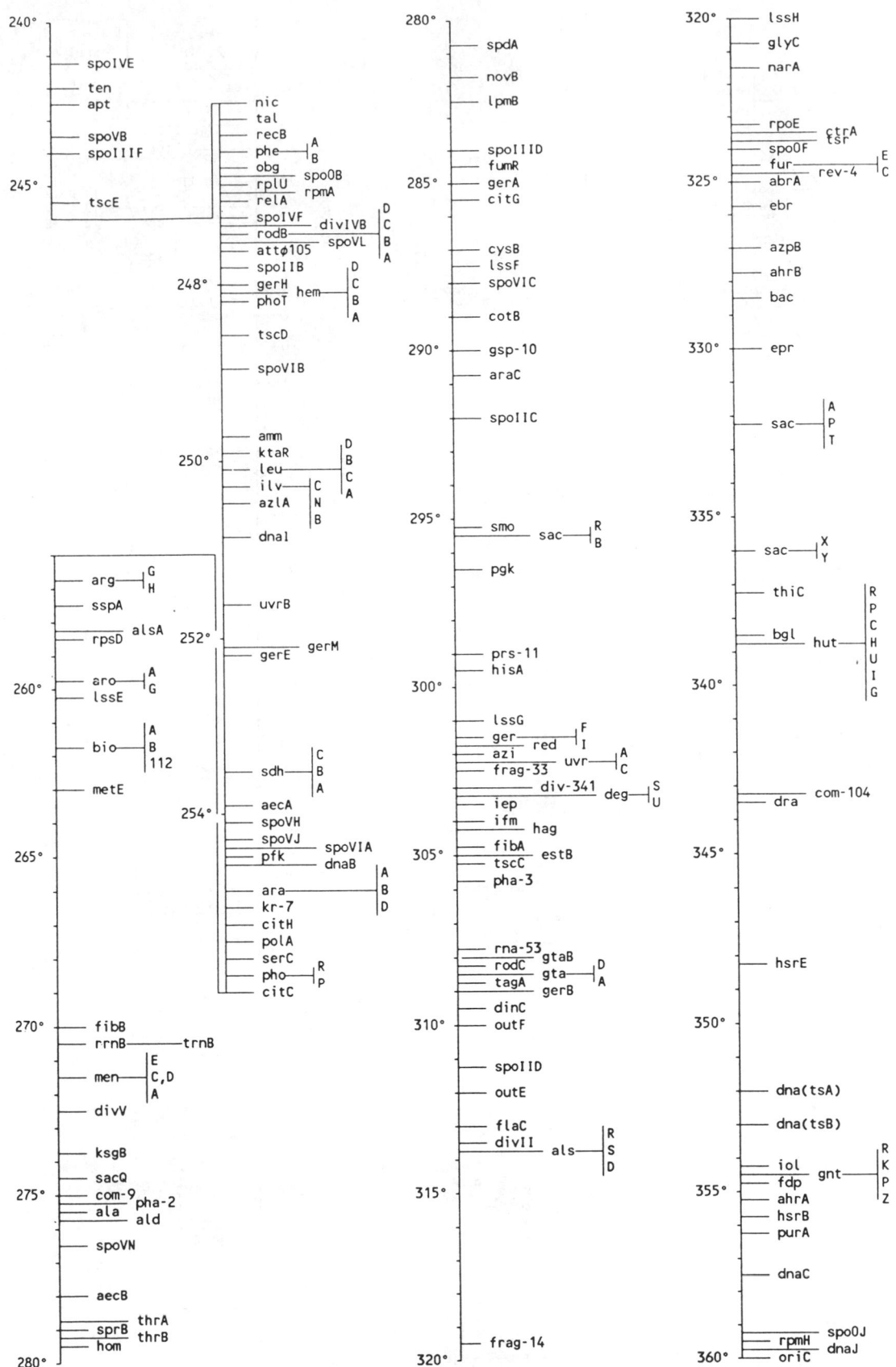

Figure 1   *Continued*

Table 1 Genes of *B. subtilis*

GENE	MNEMONIC	POSITION[a]	DESCRIPTION	REFERENCE[b]
abn	antibiotic	U	affects production of surfactin and antibiotics	A
abrA	antibiotic resistance	325 B	partial suppressor of spo0 mutant phenotypes	1
abrB	antibiotic resistance	6 B	encodes 10.7 kDal DNA binding protein; "ambiactive regulator" that repressess spo0E, aprE and stimulates hpr; binds upstream from genes it represses; mutants suppress some spo0A phenotypes but not asporogeny; probably allelic to absA, absB, cpsX, tolA	1-4 A[d] B[de]
absC	antibiotic sensitivity	U	partial suppressor of spo0A; sensitivity to antibiotics and and ϕ2, resistance to ϕ15, deficiency in protease secretion	5-6
absD	antibiotic sensitivity	U	partial suppressor of spo0A; sensitivity to antibiotics and and ϕ2, resistance to ϕ15, deficiency in protease secretion	5-6
aceA	acetate	126 C	pyruvate decarboxylase; subunit E3 of pyruvate dehydrogenase/branched 2-oxo 6,7 dehydrogenase complex; identical to bfmA, aceB	7-8
acf	acriflavine	231 B	acriflavine resistance	9
ada	adaptive	23 C	hypersensitivity to methylating agents	10
addA	ATP-dependent DNase	100 C	ATP-dependent DNase; mutation add-5 (formerly recE5) leads to sensitivity to DNA-damaging agents	11 12 13[d]
addB	ATP-dependent DNase	100 C	ATP-dependent DNase; sensitivity to DNA-damaging agents	12 13[d]
aecA	aminoethyl cysteine	254 C	aminoethyl cysteine resistance (700 μg/ml); regulation of aspartokinase II	14
aecB	aminoethyl cysteine	278 B	aminoethyl cysteine resistance (700 μg/ml); probable structural gene for aspartokinase II, the α and β subunits of which are encoded by in-phase, overlapping genes	14 15[de]
afo	antifolate	U	antifolate resistance (100 μg/ml methotrexate)	16
ahrA	arginine hydroxamate resistance	355 C	arginine hydroxamate resistance; allele aut-1 is resistant resistant only in the presence of ornithine	17
ahrB	arginine hydroxamate resistance	328 B	arginine hydroxamate resistance	17
ahrC	arginine hydroxamate	208 C	arginine hydroxamate resistance; probably repressor of argAF	18 19[c]
ahrD	arginine hydroxamate resistance	117 B	arginine hydroxamate resistance in the presence of citrulline	18
ala	alanine	276 C	alanine requirement	14
ald	alanine dehydrogenase	276 A	L-alanine dehydrogenase; no growth on L-alanine as carbon	20
alsA	acetolactate synthase	258 B	positive regulator of alsRS operon	21
alsR	acetolactate synthase	314 C	constitutive α-acetolactate synthase synthesis	c[d] 21
alsS	acetolactate synthase	314 C	structural gene for α-acetolactate synthase	c[d]
amm	ammonia	250 D	glutamate requirement; marker may be spurious	22
amyE	amylase	29 B	amylase structural gene; allele (+M) Marburg-type amylase, (+N) Natto-type, (H) hyperactivity; formerly called amyA	23[de] 24 25[c] 26[de] 27[de] 28
amyR	amylase	29 B	control of amylase synthesis, overproduction; alleles 1 and 21 are from B. subtilis 168; allele 2 is from B. subtilis natto; gra mutations abolishing glucose repression of amylase synthesis, lie in an operator-like sequence; locus formerly called amyH	23[de] 26[de] 29[de]
aprE	alkaline protease	95 B	subtilisin (alkaline protease); also called sprE	30[de] 31[de]
apt	adenine phosphoribosyl-transferase	243 C	adenine phosphoribosyl transferase; resistance to 2-fluoroadenine	32
ara	arabinose	157 C	arabinose utilization; linkage group III, mistakenly called called araC in previous editions of the map	33 D
araA	arabinose	255 C	L-arabinose isomerase; part of linkage group II, mistakenly called araB in previous editions of the map	33 D
araB	arabinose	255 C	L-ribulokinase	33 D
araC	arabinose	291 C	putative regulatory gene for araDBA; linkage group I, mistakenly called araA in previous editions of the map; may be allelic to constituive mutations	33 34 D
araD	arabinose	255 C	L-ribulose-5-phosphate-4-epimerase	33 D
argA	arginine	102 B	amino-acid acetyltransferase; arginine requirement; originally called argO	35[d] 36[d] 37 38
argB	arginine	102 B	acetylglutamate kinase; arginine requirement; originally called argO	35[d] 36[d] 37 38

Table 1   *Continued*

GENE	MNEMONIC	POSITION[a]	DESCRIPTION	REFERENCE[b]
argC	arginine	102 B	N-acetyl- -glutamyl-phosphate reductase; arginine requirement; originally called argO	35[d] 36[d] 37[d] 38[d]
argD	arginine	102 B	acetylornithine aminotransferase; arginine requirement; originally called argO	35[d] 36[d] 37[d] 38[d]
argE	arginine	102 B	acetylornithine deacetylase; arginine requirement; originally called argO	35[d] 36[d] 37[d] 38[d]
argF	arginine	102 B	ornithine carbamoyltransferase; arginine or citrulline requirement; originally called argC	35[d] 36[d] 37 38[d]
argG	arginine	257 B	arginine, ornithine or citrulline requirement; originally called argA	36-38
argH	arginine	257 B	arginine, ornithine or citrulline requirement; originally called argA	36-38
arg-342	arginine	119 C	arginine, ornithine, citrulline, or proline requirement	40
aroA	aromatic	260 B	3-deoxy-D-arabinoheptalosonic 7-phosphate synthase; requires shikimic acid (or phenylalanine, tyrosine and tryptophan)	41-42
aroB	aromatic	202 B	dehydroquinate synthase; requires shikimic acid (or phenylalanine, tyrosine and tryptophan)	41-43[d]
aroC	aromatic	202 B	dehydroquinate dehydratase; requires shikimic acid or (phenylalanine, tyrosine, and tryptophan)	41[d]42[d] 44[d]45[d]
aroD	aromatic	227 A	shikimate dehydrogenase; requires shikimic acid	41 42 46
aroE	aromatic	198 B	3-enolpyruvyl shikimate-5-phosphate synthase; requires phenylalanine, tyrosine and tryptophan	41-43
aroF	aromatic	202 B	chorismate synthase; requires phenylalanine, tyrosine and tryptophan	41 42
aroG	aromatic	260 A	chorismate mutase; isozyme 3 requires phenylalanine, tyrosine and tryptophan	41 42
aroH	aromatic	202 B	chorismate mutase; isozyme 1,2 (from B. subtilis W23) requires phenylalanine, tyrosine and tryptophan	41 42
aroI	aromatic	31 A	shikimate kinase; requires phenylalanine, tyrosine and tryptophan	23 26 42 47[d]
aroJ	aromatic	U	requires tyrosine and phenylalanine	42 48
asaA	arsenate	227 B	arsenate sensitive (1 mM); asaA2 is a large deletion	49 50
aspB	aspartate	196 B	aspartate aminotransferase	40 51
aspH	aspartate	206 C	constitutive aspartase	52
aspT	aspartate	98 B	deficient in high affinity aspartate transport	51
attɸ3T	attachment ɸ3T	179 C	integration site for ɸ3T; probably identical to attSPβ	53 54 C
attɸ105	attachment ɸ105	248 B	integration site for ɸ105	55 56
attH2	attachment H2	196 C	integration site for phage H2; prophage bears thymidilate synthase gene, converts Thy⁻ hosts to Thy⁺, and produces specialized transducing phages for metB, ilvD, and ilvA	57
attSP02	attachment SP02	20 B	integration site for SP02; prophage responsible for the RecC phenotype	58-60
azc	azetidin carboxylic acid	29 C	resistance to L-azetidin-2 carboxylic acid (500 µg/ml)	61
azi	azide	302 C	resistance to sodium azide	62
azlA	azaleucine	250 B	4-azaleucine resistance (40 µg/ml); may be promoter of ilvBC-leu operon	63 64[d] 65
azlB	azaleucine	237 C	4-azaleucine resistance (40 µg/ml)	63
azpB	azopyrimidine	327 B	resistance to azopyrimidines (50 µg/ml HPUra)	66
bac	bacilysin	329 C	production of bacilysin	67
betR	betacin	124 C	tolerance to betacin	68
bfmB	branched fatty acids	212 C	requires branched chain fatty acid or valine or isoleucine formerly called bfa	7
bgl	beta-glucanase	334 B	endo-β-1,3-1,4-glucanase; encodes 55 kDal protein	69 70[de]
bioA	biotin	262 B	7-KAP:DAP aminotransferase	71
bioB	biotin	262 B	biotin synthetase	71
bio-112	biotin	262 B	early defect in biotin synthesis	71
but	bromouracil tolerance	U	5-bromouracil tolerance	72
cafA	caffeine	60 C	caffeine resistance	73
cam-2	chloramphenicol	58 C	chloramphenicol resistance, probably altered permeability	74
car	carbohydrate	U	no growth on trehalose, maltose, or sucrose	75
cdd	cytidine deaminase	221 C	encodes 14 kDal monomer of deoxycytidine-cytidine deaminase; promoter previously designated P43	76[de] 77[de] 78[de] 79[de]
cdr	Cd⁺⁺ resistance	43 B	reduced cadmium uptake	80
cer	cerulenin	97 C	resistance to cerulenin	E
che	chemotaxis	144 C	chemotaxis mutants; 20 complementation groups known	81[d] 82
cheF	chemotaxis	144 C	encodes 20 kDal chemotaxis protein; part of che cluster	83[de]
cheR	chemotaxis	202 B	chemotaxis	81
citB	citric acid cycle	159 A	aconitate hydratase; require glutamate for growth	82 84[d] 85[e] 86[d]
citC	citric acid cycle	256 B	isocitrate dehydrogenase; autolytic, requires glutamate	83
citG	citric acid cycle	286 B	fumarate hydratase; encodes 50 kDal protein	83 87[d] 88[d] 89[e]
citH	citric acid cycle	255 B	malate dehydrogenase; also called mdh	83 90

Table 1   *Continued*

GENE	MNEMONIC	POSITION[a]	DESCRIPTION	REFERENCE[b]
citK	citric acid cycle	177 B	encodes 2-ketoglutarate dehydrogenase, subunit E1	82[c] 91[d] 92 93[c] 94[d]
citL	citric acid cycle	124 C	lipoamide dehydrogenase; subunit E3 of both the 2-keto-glutarate dehydrogenase, pyruvate dehydrogenase/branched 2-oxo acid dehydrogenase complexes	82 95
citM	citric acid cycle	176 A	encodes 417-amino acid protein, 2-ketoglutarate dehydrogen-ase subunit E2; citM1 was originally termed citK1	92[d] 93[c] 94[d]
com-9	competence	275 C	reduced DNA binding and transformation; normal transduction and protoplast transformation	96
com-30	competence	U	reduced DNA binding and transformation; normal transduction and protoplast transformation	96
com-71	competence	207 C	reduced DNA binding and transformation; normal transduction and protoplast transformation	96
com-104	competence	343 C	reduced DNA binding and transformation; normal transduction and protoplast transformation	96
comI	competence	247 D	reduced competence for transformation	97
comII	competence	300 D	reduced competence for transformation	97
comIII	competence	280 D	reduced competence for transformation	97
comIV	competence	225 D	reduced competence for transformation	97
comV	competence	210 D	reduced competence for transformation	97
comVI	competence	212 D	reduced competence for transformation	97
comVII	competence	229 D	reduced competence for transformation	97
cotA	coat	66 C	encodes 63 kDal spore coat protein; responsible for spore pigment but not required for its resistance properties; allelic to pig	98[de] 99
cotB	coat	289 C	encodes 59 kDal spore coat protein; not required for spore resistance properties	98[de]
cotC	coat	153 C	encodes 12 kDal spore coat protein; not required for spore resistance properties	98[de]
cotD	coat	198 C	encodes 11 kDal spore coat protein; not required for spore resistance properties; slow germination of spores	98[de]
cotE	coat	145 C	encodes 24 kDal spore coat protein; deletion mutant has pleiotropic loss of outer spore coat proteins, defective germination, lysozyme sensitivity	100[d]
cox	cytochrome	131 C	mutants lack cytochromes a and a$_3$, fail to grow on succinate	101
cpa	carbamoyl phoshate arginine	102 B	carbamoyl phosphate synthetase, arginine repressed; presumably allelic to urs	35[d] 36[d] 102
crk	cytidine kinase sporulation	U	cytidine kinase	76 77
crsB	catabolite resistant sporulation	67 C	catabolite resistant sporulation	103
crsC	catabolite resistant sporulation	214 C	catabolite resistant sporulation	103
crsD	catabolite resistant sporulation	7 C	catabolite resistant sporulation; smooth colony morphology on TBAB	103
crsE	catabolite resistant sporulation	12 B	catabolite resistant sporulation; antibiotic resistance during sprulation; probably allelic to rpoC	103 104
crsF	catabolite resistant sporulation	125 C	catabolite resistant sporulation	103
css	cysteine	11 C	cysteine sensitivity	105
ctc		U	cloned gene transcribed during stationary phase by σ^B (σ^{37})	106[de]
ctrA	cytidine	323 A	encodes 59.7 kDa CTP synthetase; allele -1 causes requirement for cytidine in the absence of NH_4^+	107 108[de]
cyc	cycloserine	95 C	cycloserine resistance	62
cym	cysteine-methionine	11 C	requirement of cysteine or methionine	105 109
cysA	cysteine	11 A	serine transacetylase; cysteine requirement	47[d] 107
cysB	cysteine	287 B	cysteine requirement	22
cysC	cysteine	132 C	cysteine, methionine, sulfite, or sulfide requirement	40 110
dacA	D-alanine carboxypeptidase	U	penicillin binding protein 5; major D-alanine carboxypep-tidase activity	111[de]
dal	D-alanine	47 A	encodes 43 kDa polypeptide, the major D-alanine racemase	47[d] 95 109 112[de]
dapE	diaminopimelate	127 C	N-acetyl-L-diaminopimilate deacylase dCMP dehydrogenase	113
dcd	dCMP	U	dCMP dehydrogenase	114
dck	deoxyditidine kinase	U	deoxycytidine kinase	76 77
ddd	deoxycytidine diphosphate deaminase	U	deoxycytidine diphosphate deaminase	76 77
ddl	D-alanine ligase	45 C	D-alanine ligase	113
degS	degradative	303 C	encodes 45 kDa polypeptide, probably sensor of two-component system regulating synthesis of degradative enzymes (pro-teases, α-amylase, β-glucanase); Hy alleles increase enzymes and decrease flagella and transformation competence, and alter sporulation regulation	24 115 116 117[d] 118[de] 119[de]

Table 1   *Continued*

GENE	MNEMONIC	POSITION[a]	DESCRIPTION	REFERENCE[b]
degU	degradative	303 C	encodes 26 kDa polypeptide, probably transducer of two-component system regulating synthesis of degregative enzymes (proteases, α-amylase, β-glucanase); Hy alleles increase enzymes, decrease flagella and transformation competence, and alter sporulation regulation	22 115[d] 116[de] 117[d] 118[de] 119[de]
den	DNA entry nuclease	30 D	encodes 17 kDa DNA entry nuclease; cotransduced with aroI; deletion mutant has 5% wild type transformation frequency; overlaps ORF encoding 18 kDa protein that may regulate Den	120[d] 121[de]
dep	dependence	14 C	confers kasugamycin or spectinomycin dependence on some resistant strains	122
dfrA	dihydrofolate reductase	195 C	dihydrofolate reductase, trimethoprim resistance; encodes 168 amino acid polypeptide; lies in operon with and overlaps thyB; also called tmp	123 124[de]
din	DNase inhibitor	124 C	excretion of inhibitor of sporulation exonuclease; mutants are phenotypically Spo+	125
dinA	damage inducible	302 D	induced by DNA damaging agents; may be allelic to uvrA	126
dinB	damage inducible	48 C	induced by DNA damaging agents	126
dinC	damage inducible	309 C	induced by DNA damaging agents	126
divI	division	135 C	temperature-sensitive cell division; formerly called divD	127
divII	division	314 C	temperature-sensitive cell division; formerly called divC	127
divIVA	division	134 C	minicell production	128
divIVB	division	247 B	minicell production	128
divIVC	division	16 C	minicell production; formerly called divA	129
divV	division	273 D	temperature-sensitive cell division; formerly called divB	129
div-341	division	303 B	filamentous growth at 45°C; oligosporogenous (stage 0) but partially catabolite resistant sporulation at 37°C	129
div-355	division	8 C	septum initiation mutant	130
dnaA	DNA	152 C	DNA synthesis; may encode ribonucleotide reductase	131-134[d]
dnaB	DNA	255 C	DNA synthesis; initiation of chromosome replication; 472 amino acid protein; may be part of a four gene operon	131-133 134[d] 135[e]
dnaC	DNA	357 B	DNA synthesis	131 133 134[d] 136
dnaD	DNA	201 C	DNA synthesis; initiation of chromosome replication	131 136
dnaE	DNA	223 B	DNA synthesis; encodes 68 kDal protein homologous with E. coli DNA primase	131 137[de] 138[e] 139
dnaG	DNA	4 C	DNA synthesis	131 133 134[d]
dnaH	DNA	4 B	DNA synthesis	131 133
dnaI	DNA	251 C	DNA synthesis	131 133
dnaJ	DNA	360 B	DNA synthesis; encodes protein of 446 amino acids; homologous to E. coli dnaA	140[de]
dnaK	DNA	0 B	DNA synthesis; encodes protein of 378 amino acids; homologous to E. coli dnaN	140[de]
dna(ts)A	DNA	352 C	DNA synthesis	133 141
dna(ts)B	DNA	353 C	DNA synthesis	133 141
dna-526	DNA	4 C	DNA elongation; cell filamentation	130
dna-8132	DNA	4 B	DNA synthesis; initiation of chromosome replication	62 142
dns	deoxyribonucleoside	U	requirement for dAdo or dGuo plus dCyd, suppressed by cdd, ddd, or cdd, thyB	76
dra	deoxyriboaldolase	344 C	deoxyriboaldolase; unable to use deoxyribonucleotides	143
drm	deoxyribomutase	179 C	phosphodeoxyribomutase; unable to use ribonucleotides or deoxyribonucleotides	143
durR	duramycin	21 C	duramycin resistance; membranes profoundly altered in lipid composition	144
ebr	ethidium bromide	326 B	ethidium bromide resistance (10 μg/ml)	145
ecp	ethoxycarbonyl pyrimidine	206 C	resistance to 2-amino-5-ethoxycarbonyl pyrimidine-4(3H)-one (300-800 μg/ml)	146
efg	elongation factor G	12 B	elongation factor G; fusidic acid resistance	147-149 150[c] 151[de]
epr	extracellular protease	330 C	minor extracellular protease, 645 amino acids; some homology to aprE, isp; not required for growth or sporulation	151[de]
estA	esterase	U	esterase A defect	152
estB	esterase	305 C	esterase B defect	152
ethA	ethionine	U	resistance to ethionine; increased DNA methylation, relaxed RNA synthesis	153
fdp	fructose biphosphate	355 B	deficient in D-fructose-1,6-bisphosphate 1-phoshohydrolase; fails to grow on gluconeogenic carbon sources in the presence of the bypass mutation bfd	154 155
fibA	fiber	305 C	helical macrofiber production, transient division suppression	156
fibB	fiber	270 C	helical macrofiber production, persistent division suppression	156
fil	filamentous	135 D	filamentous, flagellaless, and decreased autolysin activity; may be allelic to flaA	157
flaA	flagella	137 C	flagellaless	158 159

Table 1 *Continued*

GENE	MNEMONIC	POSITION[a]	DESCRIPTION	REFERENCE[b]	
flaB	flagella	143 C	flagellaless	158	159
flaC	flagella	313 C	flagellaless	158	159
flaD	flagella	218 C	flagellaless; lyt mutants are autolysin deficient; rgn mutants hyperproduce proteases and regenerate rods from protoplasts more efficiently than do wild type cells; in high copy number converts smooth to rough colony morphology	157 159 160 161[d]	
fnd-15	fluoroindol	3 C	fluoroindol resistance (50 μg/ml)	162	
frag-14	fragment	320 C	2.05 kb fragment; contains post-exponential promoter	163[d]	
frag-33	fragment	303 C	3.32 kb fragment; contains post-exponential promoter	163[d]	
fruA	fructose	126 C	fructose transport	164	165
fruB	fructose	126 C	fructose 1-phosphate kinase	164	165
fruC	fructose	64 C	fructokinase	164	165
ftsA	filamentation	131 C	encodes 48.1 kDa protein homologous to E. coli ftsA	166[de]	
ftsZ	filamentation	131 C	encodes 40.3 kDa protein homologous to E. coli ftsZ; lethal to E. coli (causes filamentation) in high copy number	166[de]	
ftr	fluorotryptophan	203 C	fluorotryptophan resistance; may be allelic to fnd-7, fnd-8	162	
fumR	fumarase	285 C	regulation of fumarase	62	
fun		12 C	streptomycin resistance (100 μg/ml); higher resistance in conjunction with rspL1 (475 μg/ml) and strR	167	
furA	fluorouracil	137 B	5-fluorouracil resistance (1 μg/ml)	168	
furB	flourouracil	49 D	5-flourouracil resistance (40 μg/ml) in the presence of 40 μg/ml uracil	169	
furC	flourouracil	325 B	5-flourouracil resistance (40 μg/ml) in the presence of 40 μg/ml uracil	1	
furE	flourouracil	325 B	5-flourouracil resistance (40 μg/ml) in the presence of 40 μg/ml uracil	1	
furF	flourouracil	154 C	5-flourouracil resistance (40 μg/ml) in the presence of 40 μg/ml uracil	170	
gap	glyceraldehyde phosphate	U	encodes glyceraldehyde-3-phosphate, 334 amino-acid protein	171[de]	
gat	glutamate anthranilate	10 B	glutamine binding protein; common subunit to anthranilate synthetase and p-amino-benzoate synthetase	2 123 172 173	
gcaA	glucosamine	U	L-glutamine-D-fructose-6-phosphate aminotransferase	174	
gdh	glucose dehydrogenase	40 B	encodes 28 kDal protein, glucose dehydrogenase; second gene of operon transcribed by forespore-specific polymerase, EσG	175[d]-177 178[e]	
gerAI	germination	285 B	germination defective; encodes 54 kDal protein, may be membrane protein	179[d] 180[e] 181-182	
gerAII	germination	285 B	germination defective; encodes 41 kDal protein, may be membrane protein	181-184[de]	
gerAIII	germination	285 B	germination defective; encodes 42 kDal protein with apparent signal peptide	181-184[de]	
gerB	germination	309 C	germination defective	185	186
gerC	germination	202 B	germination defective	185	186
gerD	germination	20 C	germination defective	185-187	
gerE	germination	252 B	germination defective; encodes 8.5 kDal protein; may be DNA binding protein	185 186 188[d] 189[d] 190[e] 191	
gerF	germination	302 C	germination defective	185	186
gerH	germination	248 C	germination defective	192	
gerI	germination	302 C	germination defective	192	
gerJ	germination	202 B	germination defective, allelic to tzm; spores are heat sensitive, have defective cortex	44[d] 185 193 194	
gerK	germination	38 C	germination defective	195	
gerM	germination	252 B	germination defective; oligosporogenous, many of the spores blocked at stage II	196	
glnA	glutamine	152 B	encodes glutamine synthetase, 444 amino-acid protein; second gene in operon; mutations in first gene lead to constitutive expression of operon, mutations in glnA lead to glutamine auxotrophy and relief of catabolite repression	197-199 200[d] 201[c] 202[e]	
glpD	glycerol phosphate	94 B	glycerol-3-phosphate dehydrogenase	203 204[d]	
glpK	glycerol phosphate	94 B	glycerol kinase	203 204[d]	
glpP	glycerol phosphate	94 B	positive regulation of glpD and glpK	203 205	
glpT	glycerol phosphate	24 B	fosfomycin resistance; glycerol phosphate transport	206	
gltA	glutamate	168 B	glutamate synthase	40 170 207[d]	
gltB	glutamate	168 B	glutamate synthase	207[d] 208	
gltC	glutamate	168 C	trans-acting, positive transcription factor for gltAB operon	F	
glyA	glycine	201 D	glycine requirement	209	
glyB	glycine	93 A	glycine requirement	47[d] 210	
glyC	glycine	321 B	glycine requirement	211	
gntK	gluconate	355 B	gluconate kinase; encodes 57 kDal protein	154 155 212[cd] 213[e] 214[e]	
gntP	gluconate	355 B	gluconate permease; encodes 47 kDal protein	144 145 212[cd] 213[e] 214[e]	

Table 1   *Continued*

GENE	MNEMONIC	POSITION[a]	DESCRIPTION	REFERENCE[b]
gntR	gluconate	355 B	encodes 28 kDal protein; possibly repressor of gnt operon	214[e]
gntZ	gluconate	355 B	encodes 52 kDal protein of unknown function	214[e]
gsp-10		290 C	outgrowth defective, may belong to gerA locus	192 193
gtaA	glucoteichoic acid	308 B	glucosylation of teichoic acid; lacks UDP-glucose-poly-(glycerol phosphate)-glucosyltransferase	216-218 219[d]
gtaB	glucoteichoic acid	308 B	glucosylation of teichoic acid; lacks UDP-glucose pyro-phosphorylase; smooth colony morphology	216-218 219[d]
gtaC	glucoteichoic acid	90 B	glucosylation of teichoic acid; lacks phosphoglucomutase; smooth colony morphology	216-218
gtaD	glucoteichoic acid	308 B	glucosylation of teichoic acid	218 219[d]
gtaE	glucoteichoic acid	90 B	glucosylation of teichoic acid; smooth colony morphology	218
guaA	guanine	64 C	GMP synthetase; formerly guaB	62 220
guaB	guanine	4 A	IMP dehydrogenase; formerly guaA	1-2 220
gutA	glucitol	63 C	D-glucitol permease	221 222
gutB	glucitol	63 C	D-glucitol dehydrogenase	221 222
gutR	glucitol	63 C	regulation of gutA, gutB; constitutive mutation allows for growth on xylitol	221 222
gyrA	gyrase	2 B	DNA gyrase, subunit A; encodes protein of 821 amino acids; expresses naladixic acid resistance	140[de] 223[d] 224
gyrB	gyrase	2 B	DNA gyrase, subunit B; encodes protein of 638 amino acids; expresses novobiocin resistance	140[de] 223[d]
hag	H antigen	304 C	encodes flagellin, 304 amino-acid protein with no leader; amplification of promoter causes loss of motitlity, filamentation, failure to synthesize flagellin; transcribed by RNAP with σ^b; allele 1 from strain 168, allele 2 from W23, allele 3 is straight filament	158[de] 225[de] 226[de]
hds	host DNA suppressor	9 C	extragenic suppressor of several DNA mutations; resistance to arylaxopyrimidines	227
hemA	heme biosynthesis	248 C	δ-aminolevulinic acid synthetase; requires heme	228 229
hemB	heme biosynthesis	248 C	δ-aminolevulinic acid dehydrase	229-231
hemC	heme biosynthesis	248 C	prophobilinogen deaminase	229-231
hemD	heme biosynthesis	248 C	uroporphyrinogen III cosynthase	229-231
hemE	heme biosynthesis	95 C	uroporphyrinogen decarboxylase	231 232
hemF	heme biosynthesis	95 C	coproporphyrinogen oxidase	231 232
hemG	heme biosynthesis	95 C	ferrochelatase	231 232
hisA	histidine	300 A	histidine requirement	105 211 233[d]
his cluster		300 D	histidine synthetic genes, except for hisH, are probably all linked to hisA	234
hisH	histidine	200 B	histidine phosphate aminotransferase, tyrosine-phenyl-alanine aminotransferase; originally called hisB	48 235[e]
hom	homoserine	280 B	homoserine dehydrogenase	236
hos	heterogeneity of sporulation	11 C	in strain 168, production of dark brown pigment	237
hpr	hyperproduction	93 B	hyperproduction of alkaline and neutral proteases; encodes 23.8 kDal protein, probably a negative regulator of protease production and at least one sporulation gene; scoC4 and catA7 are alleles	238-240[de]
hsrB	restriction	356 C	restriction enzyme Bsu 1247I	241
hsrC	restriction	70 C	restriction enzyme Bsu 1247II	241
hsrE	restriction	348 C	restriction enzyme Bsu 1231I	241
hsrM	host specificity	60 C	methylation component of Bsu restriction/modification; encodes 50 kDal protein	242[de]
hsrR	host specificity	60 C	endonuclease component of Bsu restriction/modification; encodes 66 kDal protein	241 242[de] 243
hts	H₂S	9 C	overproduction of hydrogen sulfide; probably regulation of cysteine desulfhydrylase	105
hutC	histidine utilization	339 B	constitutive synthesis of histidine-degrading enzymes	244 245
hutG	histidine utilization	339 B	formiminoglutamic acid (FGA)-hydrolase	244-246
hutH	histidine utilization	339 B	structural gene for histidase; encodes 55.7 kDal protein	244[c] 245 246[de]
hutI	histidine utilization	339 B	imidazolone propionate hydrolase	245
hutP	histidine utilization	339 B	positive regulator required for hut operon expression; encodes 16.6 kDal protein	244 247[de]
hutR	histidine utilization	339 B	insensitivity of hut operon to catabolite reporession	244
hutU	histidine utilization	339 B	urocanase	245 246
iep	inhibitor of exoprotease	304 C	bifunctional regulator of exoprotease production; N-terminal domain inhibits production, C-terminal domain augments it	248[de]
ifm	flagella	304 C	increased flagella and motility	158
ilvA	isoleucine-valine	194 B	threonine dehydratase; requires only isoleucine; vas allele causes valine sensitivity	95 249 250
ilvB	isoleucine-valine	250 B	condensing enzyme	64[d] 65[d] 250 251
ilvC	isoleucine-valine	250 B	β-hydroxy-α-ketoacid reductoisomerase	64[d] 65[d] 250 251

Table 1 *Continued*

GENE	MNEMONIC	POSITION[a]	DESCRIPTION	REFERENCE[b]
ilvD	isoleucine-valine	195 B	dihydroxyacid dehydratase	252 [c][de]
ilvN	isoleucine-valine	250 B	subunit of condensing enzyme of branched fatty acid biosynthesis	c [de]
infA	initiation factor	13 B	encodes a 72-amino acid polypeptide, initiation factor IF-1	253[de]
iol	inositol	354 B	unable to grow on inositol	154 155
ispA	intracellular serine protease	118 C	encodes major intracellular serine protease; not required for growth or sporulation	254[de] 255[de]
iur	inhibition uracil	U	inhibition by uracil, relation to pyrA unknown	G
katA	catalase	84 C	catalase 1; synthesized only in vegetative cells	256
kauA	keto-acid uptake	178 B	branched chain α-keto acid transport	257
kdpA	potassium dependence	U	requires 0.25 M K⁺ for growth	C
kr-7	kasugamycin	255 C	kasugamycin resistance	114
ksgA	kasugamycin	6 B	high-level kasugamycin resistance	258
ksgB	kasugamycin	274 B	kasugamycin resistance permeability; cross resistance to gentamycin, kanamycin	258
ksgC	kasugamycin	U	fumarase defective, kasugamycin resistance	62
ksgD	kasugamycin	39 C	kasugamycin resistance	62
ktaR	ketothiaisoleucine	250 C	resistance to ketothiaisoleucine; possibly isoleucyl-tRNA synthetase	259
leuA	leucine	250 A	α-isopropylmalate synthase	64[c] 65[d] 260[d] 261[d] 262[d] 263[d] 264[c]
leuB	leucine	250 B	isopropylmalate isomerase; leucine requirement	64[c] 65[d] 251[c] 262[d]
leuC	leucine	250 B	β-isopropylmalate dehydrogenase	64[d] 65[d] 241[c] 252 262[d]
leuD	leucine	250 B	probably isopropylmalate isomerase	64[d] 65[d] 265
lin	lincomycin	29 B	lincomycin resistance	266 267
liv	leucine isoleucine valine	250 B	auxotrophy for leucine, isoleucine, and valine; generated by Tn917 insertions in ilvBNC	268
lpmB	lipiarmycin	283 C	lipiarmycin resistance	269
lssA	lysis	43 C	peptidoglycan synthesis; temperature-sensitive lysis	270
lssB	lysis	126 C	peptidoglycan synthesis; temperature-sensitive lysis; may be allelic to dapE	270
lssC	lysis	137 D	peptidoglycan synthesis; temperature-sensitive lysis	270
lssD	lysis	144 D	peptidoglycan synthesis; temperature-sensitive lysis	270
lssE	lysis	260 C	temperature-sensitive lysis	270
lssF	lysis	288 C	peptidoglycan synthesis; temperature-sensitive lysis	270
lssG	lysis	301 C	temperature-sensitive lysis; cell wall anomalies	270
lssH	lysis	320 C	temperature-sensitive lysis	270
lys	lysine	204 A	diaminopimelate decarboxylase; lysine requirement	52 64[d] 208 221[d] 271[d]
lysS	lysine	8 B	lysyl-tRNA synthase	272
mal	maltose	256 D	maltose-inducible α-glucosidases; slow growth on maltose	H
menB	menaquinone	272 B	napthoate syntase, menaquinone deficiency	261[d] 273 274
menC,D	menaquinone	272 B	multiple aminoglycoside resistance; menaquinone deficiency	273 274[d]
menE	menaquinone	272 B	OSB-coenzyme A synthetase; menaquinone deficiency	273 274[d]
metA	methionine	111 B	requirement for methionine, cystathionine, or homocysteine	103
metB	methionine	195 B	requirement for methionine or homocysteine	103 260[d]
metC	methionine	118 A	requirement for methionine	103 260[d] 47[d]
metD	methionine	101 B	requirement for methionine	216
metE	methionine	263 B	SAM synthetase deficiency	153 275
mit	mitomycin	U	resistance to mitomycin C (0.25 µg/ml)	276 277
mpo	membrane proteins	216 C	overproduction of membrane proteins MP32 and MP18; temperature-sensitive sporulation	278
mth	methionine threonine	279 B	auxotrophy for threonine and methionine or homoserine; generated by Tn917 insertions in thr-hom region	268
mtlA	mannitol	41 D	deficiency in mannitol transport	62
mtlB	mannitol	41 B	mannitol-1-phosphate dehydrogenase	107
mtr	5-methyltryptophan	202 B	resistance to 5-methyl-tryptophan (1 mg/ml); derepression of the tryptophan biosynthetic pathway	162 279 280
mut-1207	mutator	U	strong SOS-independent mutator phenotype at elevated temperatures; probably editing defect	281
nea	neamine	12 D	neamine resistance	266 267
narA	nitrate	322 B	inability to use nitrate as nitrogen source	21 107
narB	nitrate	32 B	inability to use nitrate as nitrogen source	21 107
neo	neomycin	U	neomycin resistance (0.5 mg/ml)	266 267
nic	nicotinic acid	246 B	nicotinic acid requirement	168
nonA	non-permissive	U	permissive for SP10 and NR2 infection in hsrM1 background	282
novB	novobiocin	282 B	resistance to novobiocin (2 µg/ml)	224

Table 1   *Continued*

GENE	MNEMONIC	POSITION[a]	DESCRIPTION	REFERENCE[b]
nprE	neutral protease	126 C	encodes 521 amino acid protein, pre-pro-neutral protease; nprR mutations are in regulatory region	283 1[de]
obg	spo0B-associated GTP-binding protein	246 B	encodes 47.7 kDal GTP-binding protein essential for cell viability; part of spo0B operon	284[de]
ole	oleandomycin	12 D	oleandomycin resistance	266[de] 267
oriC	origin	0 B	origin of replication; clusters of nine and four DnaA-Box sequences flank the "dnaA"-like gene termed here dnaJ; in trans, cluster of four competes with chromosomal replication	140[de] 202 225 285 286 287
outA	outgrowth	15 C	temperature-sensitive outgrowth; impaired division	192 215
outB	outgrowth	29 C	temperature-sensitive outgrowth; no resumption of RNA synthesis; encodes 30 kDal protein	192 288[d] 289 290[e]
outC	outgrowth	30 C	temperature-sensitive outgrowth; no resumption of DNA synthesis	192 287
outD	outgrowth	148 C	temperature-sensitive outgrowth; incomplete resumption of protein synthesis	192 215
outE	outgrowth	312 D	temperature-sensitive outgrowth; imcomplete resumption of protein synthesis	192 287
outF	outgrowth	310 C	temperature-sensitive outgrowth; no resumption of RNA synthesis	192 287
outG	outgrowth	152 B	expressed primarily during outgrowth	291[d]
outH	outgrowth	152 B	expressed primarily during outgrowth	291[d]
outI	outgrowth	129 C	expressed primarily during outgrowth	291[d]
oxr	oxolinic acid	26 B	oxolinic acid resistance; reduced permeability	292
pab	p-aminobenzoic acid	11 B	p-aminobenzoic synthase, subunit A; p-aminobenzoic acid requirement	173
pac	pactamycin	7 B	resistance to pactamycin (5 μg/ml)	267[d]
pbuG		65 B	hypoxanthine-guanine transport system; resistance to 8-azaguanine	220[d] 293 32
pdp		U	pyrimidine nucleotide phosphorylase	62
pfk	phosphofructokinase	254 C	phosphofructokinase	167
pgk	phosphoglycerol kinase	297 B	3-phosphoglycerol kinase; germination defective	185 192
pha-1	phage	62 C	resistance to phage SPO1	107
pha-2	phage	275 C	resistance to phage SPP1	294
pha-3	phage	306 C	resistance to most group III bacteriophages	295
pheA	phenylalanine	246 A	encodes 31.9 kDal prephrenate dehydrastase; insertion mutants are phenylalanine auxotrophs; may be part of spo0B operon	41 261[d] 284[e] 296 297[d]
pheB	phenylalanine	246 B	encodes 16.7 kDal protein presumed to be chorismate mutase; insertion mutants are phenylalanine auxotrophs; may be part of spo0B operon	284[de]
pheS	phenylalanine	U	encodes 42 kDal α-subunit of phenylalanyl-tRNA synthetase	298[de]
pheT	phenylalanine	U	encodes 97 kDal β-subunit of phenylalanyl-tRNA synthetase	298[de]
phoP	phosphatase	256 B	positive regulation of alkaline phosphatase and alkaline phosphodiesterase; 27.7 kDal product highly homologous to E. coli phoB	299[c]-301[de]
phoR	phosphatase	256 B	regulation of alkaline phosphatase; encodes 579 amino acid peptide homologous to E. coli phoR; probably sensory component in conjunction with phoP	299[c] 301[d] 302 303[de]
phoS	phosphatase	115 B	constitutive alkaline phosphatase; probably negative regulator	304 305
phoT	phosphatase	248 C	constitutive alkaline phosphatase	193
pla	plaque	U	improves plaque forming ability of SPβ in Thy$^+$ background and H2 in Thy$^+$ background	57
polA	polymerase	255 C	DNA polymerase I	90 134[d] 306 307
polC	polymerase	143 C	DNA polymerase III; a fragment homologous to dnaZX, which encodes E. coli DNA polymerase III, has been cloned and sequenced	90 131 305-311[d] 312[de]
pro(AB)	proline	117 B	proline requirement	305 313
pro(H)	protease	U	high production of neutral and alkaline proteases	314
pro(L)	protease	U	high production of neutral and alkaline proteases	314
prs-11	protein secretion	299 C	decreased secretion of exoproteins; oligosporogenic	315
prs-33	protein secretion	142 C	decreased secretion of exoproteins; asporogneous, rough colony, decreased motility	315
prtR	protease regulation	196 C	enhancement of exoenzyme production in high copy number; not essential for growth or exoenzyme production	316[de]
ptg	peptidoglycan	130 C	defect in peptidoglycan synthesis	113
ptm	pyrithymine	109 C	pyrithymine resistance	62
ptsH	phosphotransferase system	123 C	encodes phosphoenolpyruvate phosphotransferase system phosphocarrier protein (Hpr), an 88-amino acid protein; mutation at ser-46 causes shortened log phase of growth on pts sugars; an overlapping, opposite orientation ORF exists	317[d] 318[e] 319[e]
ptsI	phosphotransferase system	123 C	encodes phosphenolpyruvate phosphotransferase enzyme I; mutant shows no growth on pts sugars	317[d] 165 320 319[e] 321

Table 1 *Continued*

GENE	MNEMONIC	POSITION[a]	DESCRIPTION	REFERENCE[b]
ptsX	phosphotransferase system	123 C	encodes 15.6 kDal protein that complements E. coli crr mutants; preceeded by an ORF homologous to E. coli ptsG	317[d] 319[e]
pupA		U	adenosine phosphorylase	34 322
pupI		U	inosine phosphorylase	322
purA	purine	356 A	adenyl succinate synthetase	222 248[d] 323 324 47[d]
purB	purine	65 B	adenylosuccinate lyase; formerly purE; encodes 49.5 kDal protein	325-327[de] 220[d] 328[de]
purC	purine	65 B	SAICAR synthetase; encodes 27.4 kDal protein	220[d] 328[de]
purD	purine	65 B	GAR synthetase; encodes 45.3 kDal protein	220[d] 328[de]
purE	purine	65 A	AIR carboxylase I; purE1 was formerly purB33; other purE alleles were formerly denoted purC or purD; encodes 17.2 kDal protein; first gene in purEKBCQLFMNH(J)D operon	220[d] 260[d] 324 325 47[d] 327[de] 329
purF	purine	65 B	PRPP amidotransferase; purF6 was formerly denoted purB6 encodes 51.6 kDal protein	220[d] 327[de]
purH(J)	purine	65 B	AICAR formyltransferase; encodes 45.3 kDal protein	220[d] 327[de]
purK	purine	65 B	AIR carboxylase II; encodes 42.2 kDal protein	327[de]
purL	purine	65 B	FGAM synthetase II; encodes 80.3 kDal protein	220[d] 327[de]
purM	purine	65 B	AIR synthetase; requirement for adenine and thiamine; formerly ath; encodes 37.1 kDal protein	167 220[d] 327[de]
purN	purine	65 B	GAR formyltransferase; encodes 21.8 kDal protein	327[de]
purQ	purine	65 B	FGAM synthetase I; encodes 24.8 kDal protein	327[de]
pycA	pyruvate carboxylase	139 C	pyruvate carboxylase; temperature-sensitive requirement for aspartate	328
pyrA	pyrimidine	134 B	carbamyl phosphate synthetase (pyrimidine repressible); encodes 110 kDal protein	330 331[d]
pyrB	pyrimidine	134 B	aspartate transcarbamylase; encodes 305 amino acid protein	330 331[d] 330 331[de]
pyrC	pyrimidine	134 B	dihydroorotase; encodes 46 kDal protein	280[d] 331[d]
pyrD	pyrimidine	134 A	dihydroorotate dehydrogenase; encodes 34 kDal protein	260[d] 330 331 332[de]
pyrE	pyrimidine	134 B	orotate:PRPP pyrophoribosyltransferase; encodes 26 kDal protein	330 333[d] 332[de]
pyrF	pyrimidine	134 B	orotidylate decarboxylase; encodes 27 kDal protein	330 331[de] 333[d]
pyrG	pyrimidine	U	CTP synthetase	77
pyrR	pyrimidine	U	resistance to pyrimidine	C
recA	recombination	151 B	transformation defective, transduction proficient, repair defective; may lie in same operon as recE	168 334 K
recB	recombination	246 B	transformation, transduction, and repair defective	168 334
recD	recombination	5 B	transformation, transduction, and repair defective	334
recE	recombination	151 C	transformation, transduction, and repair defective; may lie in same operon as recA	47[d] 334 335[d] K
recF	recombination	1 B	encodes protein of 323 amino acids; named by homology with E. coli gene	134 140[de] 334 336[d]
recG	recombination	197 C	transformation proficient, transduction and repair defective	134[d] 334
recH	recombination	83 B	ATP-dependent nuclease; recombination and repair defective	185 337
recI	recombination	92 C	transformation defective, repair defective	133 334
recL	recombination	35 D	recombination defective, repair defective	133 334
recM	recombination	5 C	recombination defective, repair defective	133
recN	recombination	148 C	recombination and repair defective, possibly allelic to recE	62 133
recP	recombination	11 C	mitomicin C sensitivity; moderate transformation impairment	12 338
red	red	302 C	red or red-purple colony	339
relA	relaxed	247 C	relaxed RNA synthesis	340 341
relG	relaxed	3 D	relaxed RNA synthesis	342
rev-4	revertant	325 C	restoration of Spo phenotype to antibiotic resistant, Spo strains; catabolite resistant sporulation; encodes protein not required for growth or sporulation	104 343 108[de]
ribA	riboflavin	203 B	riboflavin requirement	344
ribB	riboflavin	203 B	riboflavin synthetase; riboflavin requirement	344
ribC	riboflavin	203 B	prototroph, accumulates riboflavin; lumichrome, lumiflavin resistance	344-346
ribD	riboflavin	203 B	riboflavin requirement	344 346[c]
ribF	riboflavin	203 B	constitutive production of MERL	346
ribH	riboflavin	203 B	riboflavin requirement; accumulates ribulose-substituted pteridines	346[c]
ribO	riboflavin	203 B	prototroph; accumulates riboflavin	344
ribT	riboflavin	203 B	riboflavin requirement; accumulates ribityl-substituted pteridines	346[c]

Table 1   *Continued*

GENE	MNEMONIC	POSITION[a]	DESCRIPTION	REFERENCE[b]
rib-627	riboflavin	203 B	riboflavin requirement; probably encodes one of first two enzymes of riboflavin synthesis	347
rib-850	riboflavin	203 B	riboflavin requirement; probably encodes one of first two enzymes of riboflavin synthesis	347
rna-16	RNA	31 C	temperature-sensitive RNA synthesis; delayed spore outgrowth at permissive temperature	348
rna-53	RNA	308 C	temperature-sensitive RNA synthesis	339 348
rodB	rod	247 B	cell wall defective; salt dependence	349 350
rodC	rod	309 C	salt-dependent morphology; encodes 89 kDal protein; may be a tag gene (alternate nomenclature of tagB proposed); essential for viability	217 231[de] 350[d] 351[de] 219
rplA	ribosomal protein	12 B	protein BL1 (= E. coli L1); chloramphenicol resistance II	352 353
rplC	ribosomal protein	12 B	protein BL7 (= E. coli L3); possible micrococcin resistance	352
rplE	ribosomal protein	12 C	protein BL6 (= E. coli L5)	352
rplF	ribosomal protein	12 C	protein BL8 (= E. coli L6)	352
rplJ	ribosomal protein	12 B	protein BL5 (= E. coli L10)	352 353
rplK	ribosomal protein	12 B	protein BL11 (= E. coli L11); bryamycin (thiostrepton) resistance; some mutants have relaxed phenotype	333 353 354
rplL	ribosomal protein	12 B	protein BL9 (= E. coli L12)	353 355 356
rplO	ribosomal protein	12 D	protein BL15; chloramphenicol resistance III	356[de]
rplQ	ribosomal protein	13 B	encodes 120-amino acid polypeptide, ribosomal protein L17	252[de]
rplU	ribosomal protein	247 C	protein L20	357
rplV	ribosomal protein	12 C	protein BL22; erythromycin resistance	356 358
rplX	ribosomal protein	12 C	protein BL23 (= E. coli L24)	352
rpmA	ribosomal protein	247 B	ribosomal protein homologous with E. coli L27	357 359[e]
rpmD	ribosomal protein	13 B	protein BL27 (= E. coli L30); kasugamycin resistance	352[K][d] 360
rpmH	ribosomal protein	360 B	encodes protein of 44 amino acids homologous with E. coli ribosomal protein L34	140[de]
rpmJ	ribosomal protein	13 B	encodes 37-amino acid polypeptide, ribosomal protein B	253[de]
rpoA	RNA polymerase	13 B	encodes 314-amino acid polypeptide, RNA polymerase subunit α; alpha operon has gene order infA-rpmJ-rpsM-rpsK-rpoA-rplQ	361[de] 253[de]
rpoB	RNA polymerase	12 B	RNA polymerase subunit β; rifampin resistance	362[d] 363 364
rpoC	RNA polymerase	12 B	RNA polymerase subunit β'; lipiarmycin and streptolydigin resistant; crsE may be allelic	103 365[d] 363 364 366
rpoD	RNA polymerase	223 B	RNA polymerase major σ factor, 43 kDal; in operon with dnaE and "P23"; crsA alleles suppress some spo0 mutations and allow sporulation in the presence of glucose and other sporulation inhibitors	103 139[de] 367[d] 368
rpoE	RNA polymerase	323 B	RNA polymerase subunit δ; deletion mutant has normal growth, sporulation, and SP01 growth	369[de]
rpsD	ribosomal protein	259 C	protein S4; restores fun, strR mutants to Spo⁺ phenotype	370
rpsE	ribosomal protein	13 B	protein S5; spectinomycin resistance	352 371 K[d] 372
rpsF	ribosomal protein	1 D	protein S6	373
rpsG	ribosomal protein	12 B	protein S7	148 352
rpsH	ribosomal protein	12 C	protein S8	352
rpsI	ribosomal protein	12 C	protein S9	352 360
rpsJ	ribosomal protein	12 D	protein S10; tetracycline resistance	374
rpsK	ribosomal protein	13 B	protein S11, 131 amino acids	352 360 361[de]
rpsL	ribosomal protein	12 B	protein S12; streptomycin resistance	352 361[de]
rpsM	ribosomal protein	13 C	protein S13, 121 amino acids	K[d]
rpsO	ribosomal protein	13 B	protein S15	373
rpsP	ribosomal protein	134 B	protein S16	374
rpsT	ribosomal protein	12 D	protein S20	356
-	ribosomal protein cluster	12 C	proteins BL4, BL14, BL16, BL17, BL25, S3, S17, S19; resistance to bryamycin, kanamycin, neamycin, oleandomycin, streptomycin	
rrnA	ribosomal RNA	5 C	ribosomal RNA gene set	286[de] 375 376
rrnB	ribosomal RNA	271 C	ribosomal RNA gene set	377[de] 378 379[de] 380[e]
rrnD	ribosomal RNA	84 B	ribosomal RNA gene set	379
rrnE	ribosomal RNA	57 C	ribosomal RNA gene set	379
rrnG	ribosomal RNA	21 B	ribosomal RNA gene set	378 382 383
rrnH	ribosomal RNA	21 B	ribosomal RNA gene set	378 382 383
rrnI	ribosomal RNA	21 B	ribosomal RNA gene set	378 382 383
rrnJ	ribosomal RNA	9 C	ribosomal RNA gene set	382 383

Table 1 *Continued*

GENE	MNEMONIC	POSITION[a]	DESCRIPTION	REFERENCE[b]
rrnO	ribosomal RNA	3 B	ribosomal RNA gene set	285[d] 286[de] 384 385
rrnW	ribosomal RNA	9 C	ribosomal RNA gene set	381 383
sacA	sucrose	332 A	sucrase	129[c] 386[c]
sacB	sucrose	296 B	levansucrase; encodes 53 kDal precursor with 29 residue signal peptide; sacR is regulatory site	115 387 388 389[d] 390[e] L
sacC	sucrose	239 C	encodes 75.9 kDal precursor of exocelluar levanase	
sacL	sucrose	239 C	allow detection of exocellular levanase	391 392[e]
sacP	sucrose	332 B	encodes enzyme II of the sucrose phosphotransferase system, a 48.9 kDal polypeptide homologous to enteric bacterial enzyme II genes	391 M[de]
sacQ	sucrose	275 B	hypersecretion of exoenzymes; encodes protein of 46 amino acids; probably allelic to amyB, pap	387[de] 393 394
sacR	sucrose	296 B	constitutive sucrase and/or levansucrase	115 387 389[d]
sacT	sucrose	332 B	constitutive sucrase and/or levansucrase	115 395
sacV	sucrose	U	regulation of levansucrase expression	N
sacX	sucrose	336 B	encodes repressor regulating sacB; previously sacS	115 395
sacY	sucrose	336 B	encodes antiterminator regulating sacB; previously sacS	396[d] N
sapA	suppressor alkaline phosphatase	115 B	allows synthesis of alkaline phosphatase in spo⁻ mutants; possibly allelic to scoA, phoS	305
sapB	suppressor alkaline phosphatase	67 C	allows synthesis of alkaline phosphatase in spo⁻ mutants	305
scoA	sporulation control	112 B	delayed sporulation; elevated proteolytic activity; may be allelic to sapA	236
scoB	sporulation control	123 C	delayed sporulation; elevated proteolytic activity	237
"scRNA"		143 D	gene encoding RNA moiety homolgous to E. coli 4.5S RNA, adjacent to a gene encoding a polymerase III-like protein	312[de] 397[e]
sdhA	succinate dehydrogenase	254 B	flavoprotein subunit of succinate dehydrogenase; encodes 65 kDal protein	171[d] 398[ce] 399[e]
sdhB	succinate dehydrogenase	254 B	iron protein subunit of succinate dehydrogenase; encodes 28 kDal protein	171[d] 398[e]
sdhC	succinate dehydrogenase	254 B	encodes cytochrome b₅₅₈, 23 kDal protein	171[d] 400[d] 401[d]
secY	secretion	13 B	homologous to E. coli gene encoding component of secretory system	253[de] K[de]
sen		71 C	encodes highly basic polypeptide of 65 amino acids; in high copy number stimulates expression of exoenzymes; not required for growth or sporulation; overproduction deleterious to growth and sporulation	O[de]
serA	serine	203 B	requirement for serine or glycine; may be allelic to glyA	44[d] 277 268
serC	serine	256 C	requirement for serine	268
serR	serine	28 C	resistance to serine	C
sigB	RNA polymerase	47 B	RNA polymerase minor σ factor σᴮ (σ³⁷); 30 kDal; not required for growth or sporulation; adjacent to three open reading frames, orfV (12 kDal) and orfW (18 kDal) of unkown function, and socB (orfX); sig terminology supercedes rpoF	402[de] 403[de] K[de]
sigD	sigma factor	U	encodes a 254-amino-acid polypeptide, σ (σ²⁸); homologous to E. coli fibB; null mutant fails to synthesize flagellin and grows in filaments	404[de] 250[de]
sin	sporulation inhibition	217 C	inhibits sporulation and exoenzyme production in high copy number; encodes 111 amino acids; deletion mutant is Spo⁺, overproduces exoenzymes, grows in chains	405[de]
smo	smooth	295 B	smooth-rough colony morphology; long chains of small, oval cells	349 389[d]
socB	suppression of ctc	47 C	alteration in utilization of ctc promoter; in operon with sigB; null mutations cause small colony phenotype; may regulate sigB; originally called orfX	403[de] 406 407[de]
--	"SPβ Homology"	170 B	region of homology with SPβ prophage	408[d]
spcB	spectinomycin	142 B	resistance to spectinomycin (100 μg/ml)	301 409
spcD	spectinomycin	U	spectinomycin dependence	410
spdA	sporulation	281 C	continual sporulation during exponential growth; altered in pyruvate carboxylase specific activity	411
spg	sporangiomycin	U	sporangiomycin resistance; 50S ribosome alteration	412
spo0A	sporulation	212 B	sporulation, stage 0; encodes 30 kDal protein homologous to Spo0F, E. coli OmpR, SfrA; allelic to spo0C; rvt and crs alleles restore sporulation to spo0B, spo0E mutants and to alcohol-treated cells; abrC alleles are weak intragenic suppressors	1 40[c] 103 260 40[d] 413 414 415[d] 416 417[e] 418[e] 419
spo0B	sporulation	247 B	sporulation, stage 0; encodes 23 kDal protein; allelic to spo0D	40 260[d] 359[e] 414 420[d] 421[d] 422[e] 284[e]

Table 1   *Continued*

GENE	MNEMONIC	POSITION[a]	DESCRIPTION	REFERENCE[b]
spoOD	sporulation	247 B	stage 0 sporulation; probably allelic to spoOB	423[d]
spoOE	sporulation	125 C	stage 0 sporulation	424
spoOF	sporulation	324 B	stage 0 sporulation; encodes 14 kDal protein homologous to SpoOA and E. coli OmpR, SfrA; protein inhibits sporulation when gene is present in 4 or more copies; monocistronic operon	233[d] 414 424 425[d] 426[cde] 427[e] 428 122[de]
spoOG	sporulation	212 B	stage 0 sporulation; probably allelic to spoOA	414 423[d]
spoOH	sporulation	11 B	stage 0 sporulation; encodes 26.1 kDal σ factor, σ^H (σ^{30}); soc alleles affect transcription of the ctc promoter; also called sigH	414 429 406 430[d] 431[e]
spoOJ	sporulation	359 C	stage 0 sporulation; two or more complementation groups; allele -87 disengages sequential induction of operons from morphological development; allele -93 originally spoCM1	2 413[d] 414 423 432 P[d]
spoOK	sporulation	108 C	stage 0 sporulation	414 P[d]
spoOL	sporulation	108 C	stage 0 sporulation	192
spoIIAA	sporulation	206 B	stage II sporulation; encodes 13.1 Mdal protein	414 433[d] 434[c] 435 436[c] 437[e]
spoIIAB	sporulation	206 B	stage II sporulation; encodes 16.3 Mdal protein	414 433[d] 434[c] 435 436[d] 437[e]
spoIIAC	sporulation	206 B	stage II sporulation; encodes σ^F, 22.2 Mdal sporulation σ factor; alternate nomenclature sigF	414 433[d] 434[c] 435 436[c] 437[e]
spoIIB	sporulation	248 B	stage II sporulation	414
spoIIC	sporulation	292 C	stage II sporulation; only known mutant presumed lost	414
spoIID	sporulation	311 C	stage II sporulation; encodes 343 amino acid protein likely to be a DNA binding protein	190 414 438[d] 439[de]
spoIIE	sporulation	9 B	stage II sporulation; two or more complementation groups; single transcription unit 2.5 kb in size	414 423[d] 440[de]
spoIIF	sporulation	122 C	stage II sporulation	411 Q[d]
spoIIGA	sporulation	128 C	stage II sporulation; required for processing of the σ^E (σ^{29}) precursor, P31	441[d] 442 443
spoIIGB	sporulation	128 C	stage II sporulation; encodes 27.7 kDal precursor to σ^E, P31; also denoted sigE	357[d] 423[d] 442[de] 443
spoIIH	sporulation	7 B	stage II sporulation	414
spoIIJ	sporulation	125 C	stage II sporulation; Tn917 mutant oligosporogenous	444
spoIIL	sporulation	211 C	stage II sporulation	444
spoIIM	sporulation	205 C	stage II sporulation; homologous to E. coli ftsA	444 Q[de]
spoIIIA	sporulation	217 C	stage III sporulation; two or more complementation groups	414 423[d]
spoIIIB	sporulation	218 C	stage III sporulation; status uncertain--major lesion in some mutants actually lies in spoIIIE	259[d] 414
spoIIIC	sporulation	236 C	stage III sporulation; comprised of 138 codons which are joined to spoIVCB by site-specific recombination in the mother cell chromosome during sporulation; encodes the carboxy terminus of sigma factor σ^K	414 445[de] 446[de] 447
spoIIID	sporulation	284 C	stage III sporulation; probably only one gene	414 P[de]
spoIIIEA	sporulation	142 C	stage III sporulation; encodes 518 amino acid polypeptide	414 423[d] 446[de]
spoIIIEB	sporulation	142 C	stage III sporulation; encodes 252 amino acid polypeptide	414 423[d] 448[de]
spoIIIF	sporulation	244 B	stage III sporulation	294 423[d]
spoIIIG	sporulation	129 C	stage III sporulation; encodes σ^G, a 30.1 kDa sporulation σ-factor that reads ssp genes; initially cotranscribed with spoIIG, after 4 hr of sporulation transcribed independently	449[de] 450[de] 451
spoIVA	sporulation	202 B	stage IV sporulation	44[d] 414 P[d]
spoIVB	sporulation	207 C	stage IV sporulation	414
spoIVCA	sporulation	235 C	stage IV sporulation; may encode recomginase which catalyzes site-specific recombination between spoIIIC and spoIVCB	414 423[d] 447 452 453 456[d]
spoIVCB	sporulation	235 C	stage IV sporulation; joined to spoIIIC by site-specific recombination in the mother cell chromosome during sporulation; encodes amino terminus of sigma factor σ^K	414 423[d] 453 454[d] 446[e]
spoIVD	sporulation	236 C	stage IV sporulation	414
spoIVE	sporulation	241 C	stage IV sporulation	414
spoIVF	sporulation	247 B	stage IV sporulation	296 416 423[d]
spoIVG	sporulation	109 C	stage IV sporulation	414
spoVAA	sporulation	204 B	stage V sporulation; encodes 23 kDal protein	455 456[de]
spoVAB	sporulation	204 B	stage V sporulation; encodes 15 kDal protein	455 456[de]
spoVAC	sporulation	204 B	stage V sporulation; encodes 16 kDal protein	455 456[de]
spoVAD	sporulation	204 B	stage V sporulation; encodes 36 kDal protein	455 456[de]
spoVAE	sporulation	204 B	stage V sporulation; encodes 34 kDal protein	455 456[de]

Table 1 *Continued*

GENE	MNEMONIC	POSITION[a]	DESCRIPTION	REFERENCE[b]
spoVB	sporulation	243 C	stage V sporulation	296[d] 414 423[d]
spoVC	sporulation	7 C	stage V sporulation	414 457
spoVD	sporulation	139 C	stage V sporulation	414 457 P[d]
spoVE	sporulation	128 C	stage V sporulation; encodes basic, 32 kDal protein; probably a membrane protein	414 458[d] 459[e] 460
spoVF	sporulation	127 C	stage V sporulation; requires dipicolonic acid for heat-resistant spores	414 461
spoVG	sporulation	7 B	stage V sporulation; expressed in spore compartment; may be polycistronic operon	462[d] 463 464
spoVH	sporulation	254 B	stage V sporulation; defective DPA; lysozyme-resistant heat-sensitive spores; status uncertain--major lesion in original mutant actually lies in spoVA	250 423[d] P
spoVJ	sporulation	254 B	stage V sporulation; lysozyme-resistant, heat-sensitive spores	226 423[d]
spoVK	sporulation	152 B	stage V sporulation; dark pigmentation, phase gray spores, partially heat resistant	444
spoVL	sporulation	248 C	stage V sporulation	444
spoVM	sporulation	140 C	stage V sporulation; DPA$^+$; phase bright, partially heat-resistant spores	444
spoVN	sporulation	277 C	stage V sporulation; DPA$^-$; partially heat-resistant spores	444
spoVP	sporulation	201 C	stage V sporulation; DPA$^+$	444
spoVQ	sporulation	209 C	stage V sporulation; DPA$^+$; phase bright, heat-sensitive spores	444
spoVIA	sporulation	254 C	stage VI sporulation; coat lacks 36kDa peptide; lysozyme-sensitive spores, delayed germination	465
spoVIB	sporulation	249 C	state VI sporulation; coat has reduced amounts of 36 kDal peptide, delayed germination	466
spoVIC	sporulation	288 C	sporulation, stage VI; slow sporulation and germination; disorganized spore coat	423[d] 467
spoL1	sporulation	228 B	'decadent' sporulation	468
sprA	suppressor	U	derepression of monoserine kinase, homoserine dehydrogenase and the minor threonine dehydratase	469
sprB	suppressor	279 B	suppression of isoleucine auxotrophy; sensitivity to methionine	469 470
srf	surfactin	U	abolishes production of surfactin	A
srh	suppressor recH	125 D	restores recH mutants to near wild-type transformation efficiency	471
srm	spectinomycin resistance modifier	12 D	growth and sporulation of rpsE mutants	372
ssa		213 C	alcohol-resistant sporulation; may be an allele of spoOA	472 473
ssp-1	spore specific	U	endonuclease excising spore photoproducts	474
sspA	small spore protein	258 C	encodes SASP-α, 69 amino acid protein; required for spore UV light resistance	475 476[de] 477
sspB	small spore protein	91 C	encodes SASP-β, 67 amino acid protein	475 476[e]
sspC	small spore protein	179 C	encodes 72 amino acid spore protein	475--478[de]
sspD	small spore protein	120 C	encodes 64 amino acid spore protein	475 476[de]
sspE	small spore protein	83 C	encodes SASP- , 84 amino acid protein	479[de]
stb	stability	U	reported to enhance stability of some plasmids	480
strB	streptomycin	129 C	streptomycin resistance (200 µg/ml)	481
strC	streptomycin	214 C	streptomycin resistance (1 mg/ml); requires glucose	481
strD	streptomycin	12 C	streptomycin dependence (0.5 mg/ml); probably allelic to rpsL	482
strR	streptomycin	19 C	streptomycin dependence; lethality suppressed by fun or rpr-21	167
suf	suppressor of frameshift	U	suppresses leuA169	C
suh	suppressor of histidine	U	may be leaky hisH mutation	R
sul	sulfonilamide	11 B	sulfanilamide resistance	58[d] 173 483[d]
sup-3	suppressor	46 C	nonsense suppressor	484
sup-44	suppressor	46 C	nonsense suppressor	485
tab		U	converts phenotype of spoIIIA mutant to a stage IV-V mutant	S
tagA	teichoic acid	309 B	decreased synthesis of teichoic acid; essential for growth	212 219[d]
tal	thienylalanine	246 C	resistance to β-thienylalanine	486
ten	transfection enhancement	242 C	transfection enhancement for phage SP82 DNA	T
tem-1	temperature-sensitive	U	temperature-sensitive protein and RNA synthesis; cotransduces with purE	G
terC	terminus	168 B	terminus of replication; fork immobilized 3-5 kb from gltA in an AT-rich region very close to one of two inverted repeats; a linked ORF encoding a 122 amino acid protein may participate in fork termination	487-490[d] 491-493[d] 494 495[e] 496

Table 1   *Continued*

GENE	MNEMONIC	POSITION[a]	DESCRIPTION	REFERENCE[b]
tetB	tetracycline	12 C	resistance to tetracycline (20 μg/ml); reduced uptake; may correspond to a tet determinant naturally occurring in the chromosome of strain 168-derived cells, which encodes a protein 80% homologous to that of plasmid pNS1981 (=pBC16); may also correspond to a region which can mutate to high tetracycline resistance by chromosomal amplification and which encodes a 72 kDal protein	497 498[d] 499[e] 449 500[d]
thiA	thiamine	84 B	thiamine requirement	501
thiB	thiamine	107 B	thiamine requirement	40 502
thiC	thiamine	337 B	thiamine requirement	242
thrA	threonine	279 B	homoserine kinase	236 265[d] 47[d]
thrB	threonine	279 B	threonine synthetase with minor dehydrogenase activity	236
thyA	thymidine	153 B	thymidylate synthetase A; aminopterin sensitive, 46°-resistant	503
thyB	thymidine	194 A	thymidylate synthetase B; 46°-sensitive; encodes 267 amino acid polypeptide; lies in operon with and overlaps dfrA	124[de] 503
til	tilerone	226 C	tilerone resistance	62
tmrA	tunicamycin resistance	29 B	not a gene, but sequence causing amplification of tmrB	504-506[de]
tmrB	tunicamycin resistance	29 B	encodes 97 or 99 amino acid polypeptide which causes tunicamycin resistance when amplified to high copy number	26[de] 504 506
tms-12	temperature-sensitive	130 C	temperature-sensitive septum initiation; allelic to ts-12	130[d] 507
tms-26	temperature-sensitive	7 B	temperature-sensitive growth; rigidity	244[d] 508 509
tmsA	temperature-sensitive	U	temperature-sensitive growth	510
tmsB	temperature-sensitive	U	temperature-sensitive growth	510
tolB	tolerance	U	tolerance to φ29 in spo0 strains	5-6
tre	trehalose	77 A	inability to utilize trehalose as carbon source	107
trnA	tRNA	5 B	tRNA genes within rrnA: Ile, Ala	511
trnB	tRNA	271 C	formerly trrnE; 21 genes: Val, Thr, Lys, Leu, Gly, Leu, Arg, Pro, Ala, Met, Ile, Ser, fMet, Asp, Phe, His, Gly, Ile or Met, Asn, Ser, Glu	379[de] 511 512
trnI	tRNA	21 B	formerly trrnB or trnH; 6 genes: Asn, Thr, Gly, Arg, Pro, Ala	512 513
trnO	tRNA	3 B	tRNA genes within rrnO: Ile, Ala	511
trnY	tRNA	U	tRNA genes: Lys, Glu, Asp, Phe	511
trpA	tryptophan	200 B	tryptophan synthase; encodes 29 kDal protein	235[e] 514[c]
trpB	tryptophan	200 B	tryptophan synthase; encodes 44 kDal protein	235[e] 514 515[d]
trpC	tryptophan	200 B	indoleglycerol phosphate synthase; encodes 28 kDal protein	44[d] 235[e] 514 515[c] 47[d]
trpD	tryptophan	200 B	anthranilate phosphoribosyl transferase; encodes 36 kDal protein	235[e] 514 515 516[de]
trpE	tryptophan	200 B	anthranilate synthase; encodes 58 kDal protein	235[e] 470 514[de] 515[c] 516[de] 517
trpF	tryptophan	200 B	N-(5'-Phosphoribosyl)-isomerase; encodes 24 kDal protein	235[e] 514[c] 516[de]
trpS	tryptophan	105 B	encodes 330 amino-acid protein, tryptophanyl-tRNA synthase; may mutate to resistance to 5-fluorotryptophan (50 μg/ml)	518 519[de]
trrnA	tRNA, rRNA	U	2-3 tRNA genes, probably flanked by rRNA gene sets	513[d]
trrnC	tRNA, rRNA	U	8-12 tRNA genes, probably flanked by rRNA gene sets	513
trrnD	tRNA, rRNA	U	16 tRNA genes, linked to an rRNA set containing the minor 5S rRNA species gene	513[de]
ts-2	temperature-sensitive	U	temperature-sensitive growth	U
ts-355	temperature-sensitive	U	temperature-sensitive growth	V
ts-39	temperature-sensitive	227 C	temperature-sensitive synthesis of phosphatidyl-ethanolamine originally called ts39-2	520
tscA	temperature-sensitive	18 C	temperature-sensitive growth; delayed stop of DNA, protein synthesis at 45°C	521
tscB	temperature-sensitive	30 C	temperature-sensitive growth; delayed stop of DNA, protein synthesis at 45°C	286[d] 521
tscC	temperature-sensitive	305 C	temperature-sensitive growth; protein, DNA synthesis at 45°C	521
tscD	temperature-sensitive	249 C	temperature-sensitive growth; normal protein and DNA synthesis at 45°C	521
tscE	temperature-sensitive	245 C	temperature-sensitive growth; normal protein and DNA synthesis at 45°C	521
tscF	temperature-sensitive	29 C	temperature-sensitive growth; normal DNA synthesis, delayed stop of protein synthesis at 45°C	521
tscG	temperature-sensitive	30 C	temperature-sensitive growth; normal DNA and protein synthesis at 45°C	288[d] 521
tscH	temperature-sensitive	30 C	temperature-sensitive growth; normal DNA and protein synthesis at 45°C	288[d] 521
tsi	temperature-sensitive	69 C	temperature-sensitive induction of PBSX	522-524

Table 1 *Continued*

GENE	MNEMONIC	POSITION[a]	DESCRIPTION	REFERENCE[b]
tsr	temperature-sensitive RNA synthesis	324 B	encodes 22.8 kDa protein required for RNA synthesis; in operon with ORF encoding 20 kDa protein not required for growth or sporulation	122[de]
tuf	Tu-factor	12 B	elongation factor Tu; kirromycin resistance (200 µg/ml)	525
tyrA	tyrosine	199 B	prephenate dehydrogenase; may mutate to auxotrophy (tyr), to D-tyrosine resistance (D-tyrR), to aminotyrosine resistance (amt), or to histidine sensitivity (inh)	22[d] 42 44 45[d] 526 527[d]
upp		U	uracil phosphoribosyl transferase	528
urg	uracil glycosidase	U	N-glycosidase	529 530
urc	uracil cysteine	133 B	auxotrophy for uracil and cysteine or methionine; generated by Tn917 insertion causing a deletion in the cysC-pyr region	268
uvrA	UV repair	302 B	excision repair defective; no initial incision; originally called hcr	474[c] J
uvrB	UV repair	252 C	excision of UV-induced pyrimidine dimers in DNA	474 J
uvrC	UV repair	302 B	excision repair defective; gaps not closed; formerly uvrA1	J
veg	vegetative	7 B	transcribed actively during growth and sporulation	531
virM	virginamycin M	U	resistance to virginamycin M (50 µg/ml)	532 533
virS	virginamycin S	U	resistance to virginamycin S (13 µg/ml)	532 533
wrd	weird	202 C	slow growth on PGYE medium, normal on MA medium	187
xglA	X-gal	U	structural gene for endogenous β-galactosidase	P
xglR	X-gal	U	regulatory gene for endogenous β-galactosidase; mutations cause overproduction of "X-gal-ase," interfering with studies involving regulation of lacZ-gene fusions	P
xhd	PBSX head	113 B	induced PBSX lacks heads	534
xhi	PBSX heat inducibility	113 B	heat inducible PBSX	502
xin	PBSX induction	113 B	induction defective PBSX	534
xki	PBSX kill	113 B	fails to kill B. subtilis W23	279
xpt	xanthine phosphoribosyl-transferase	197 C	xanthine phosphoribosyltransferase; resistance to 8-azaxanthine	32
xtl	PBSX tail	113 B	induced PBSX lacks tails	534
xylA	xylose	U	xylose isomerase	535[d] 536[e]
xylB	xylose	U	xlylulokinase	536[d]
xynA	xylan	59 C	β-xylanase; encodes 22 kDal protein	537[d] 538 539[de]
xynB	xylan	59 C	β-xylosidase	538 540[d]
0.3 kb		7 B	expressed in forespore compartment beginning at t_4 of sporulation; encodes 61 amino acid polypeptide	64 531[de] 541

[a]The letters A-D following each position refer to the degree of precision with which the map position is known. "A" indicates that the gene is a landmark in its region of the chromosome. "B" designates a gene that has been ordered with respect to nearby loci, whereas "C" indicates that the gene, while reliably demonstrated to reside at the given map position, has not been ordered with respect to its neighbors. "D" indicates that published mapping data are incomplete or contradictory and that the map position given is somewhat speculative; not all genes identified with a "D" are included on the map diagram. "U" indicates unmapped genes.

[b]Unpublished references:
[A] Zuber P, Nakano NM, Khosravifar R, Robertson JB, Abstr 10th Intl Spores Conference
[B] Parego M, Hoch JA, Abstr 10th Intl Spores Conference
[C] Zahler S
[D] Paveia PhD Thesis, University of Lisbon, Lisbon, Portugal, 1987
[E] Zeigler D
[F] Bohannon DE, Sonenshein AL, Abstr 10th Intl Spores Conference
[G] Siegel E
[H] Gonzalez DS, Chambliss GH (1986) Abstr Amer Soc Microbiol
[I] Yang M, Henner D, Abstr 9th Int Spores Conference
[J] Yasbin R
[L] Stewart G
[K] Price CW
[L] Martin I, DeBarbouille M, Klier A, Rapoport G, Abstr 10th Intl Spores Conference
[M] Fouet A, Arnaud M, Klier A, Rapoport G, Abstr 10th Intl Spores Conference
[N] Arnaud M, DeBourbouille M, Fouet A, Klier A, Kunst F, Martin T, Msadek T, Rapoport G, Dedonder R, Abstr 10th Intl Spores Conference
[O] Wang L-F, Wong SL, Doi RH, Abstr 10th Intl Spores Conference
[P] Errington J
[Q] Leighton T, Louie P, Lee A, Malak I, Tucker J, Fluss L, Skogen B, Straigier P, Abstr 10th Intl Spores Conference
[R] Nester EW
[S] Cutting S, Abstr 10th Intl Spores Conference
[T] Green D
[U] Takahashi I
[V] Mendelson N

[c]reference containing fine-structure genetic map of a locus.
[d]reference describing molecular cloning of a gene.
[e]reference giving DNA sequence data (sometimes partial) for a gene.

Table 2   Alternate Gene Nomenclature

Alternate	Preferred	Alternate	Preferred	Alternate	Preferred
abrC	spo0A	ery	rplV	sacU	degSU
absA	abrB	fus	efg	sas	spoIIA
absB	abrB	gerG	pgk	scoC	hpr
aceB	aceA	gra	amyR	sigA	rpoA
amt	tyrA	gsp	out	sigE	spoIIGB
amyA	amyE	hcr	uvrA	sigF	spoIIAC
amyB	sacQ	hsdR	hsrR	sigG	spoIIIG
amyH	amyR	inh	tyrA	sigH	spo0H
aspA	pycA	kir	tuf	sof	spo0A
ath	purM	lpm	rpoC	sorR	gutR
aut-1	ahrA	lyt	flaD	spcA	gutR
azpA	polC	mdh	citH	spo0C	spo0A
bfa	bfmB	mic	rplC	spoCM	spo0J
bfmA	aceA	mut	polC	sprE	aprE
bsr	hsrR	nalA	gyrA	std	rpoC
cafB	gyrA	nonB	hsrM	strA	rpsL
catA	scoC	novA	gyrB	tagB	rodC
citF	sdh	odhA	citK	tmp	dfrA
cmlII	rplA	odhB	citM	tolA	abrB
cmlIII	rplO	orfX	socB	trnH	trnI
cmlIV	rplL	pap	degSU	trpX	gat
cpsX	abrB	pig	cotA	trrnB	trnH
crsA	rpoD	pyrX	pyrE	trrnE	trnB
cytR	pyrR	recC	attSPO2	ts-8132	dna-8132
divA	divIVC	recO	recB	tsA	dna(tsA)
divB	divV	relC	rplK	tsB	dna(tsB)
divC	divII	rfm	rpoB	tsp	rplK
divD	divI	rgn	flaD	tzm	gerJ
dnaF	polC	rif	rpoB	urs	cpu
dnaP	polC	rou	rpoF	vas	ilvA
drp	rpoB	sacS	sigB	0.4 kb	spoVG
dtyrR	tyrA		sacX and sacY		

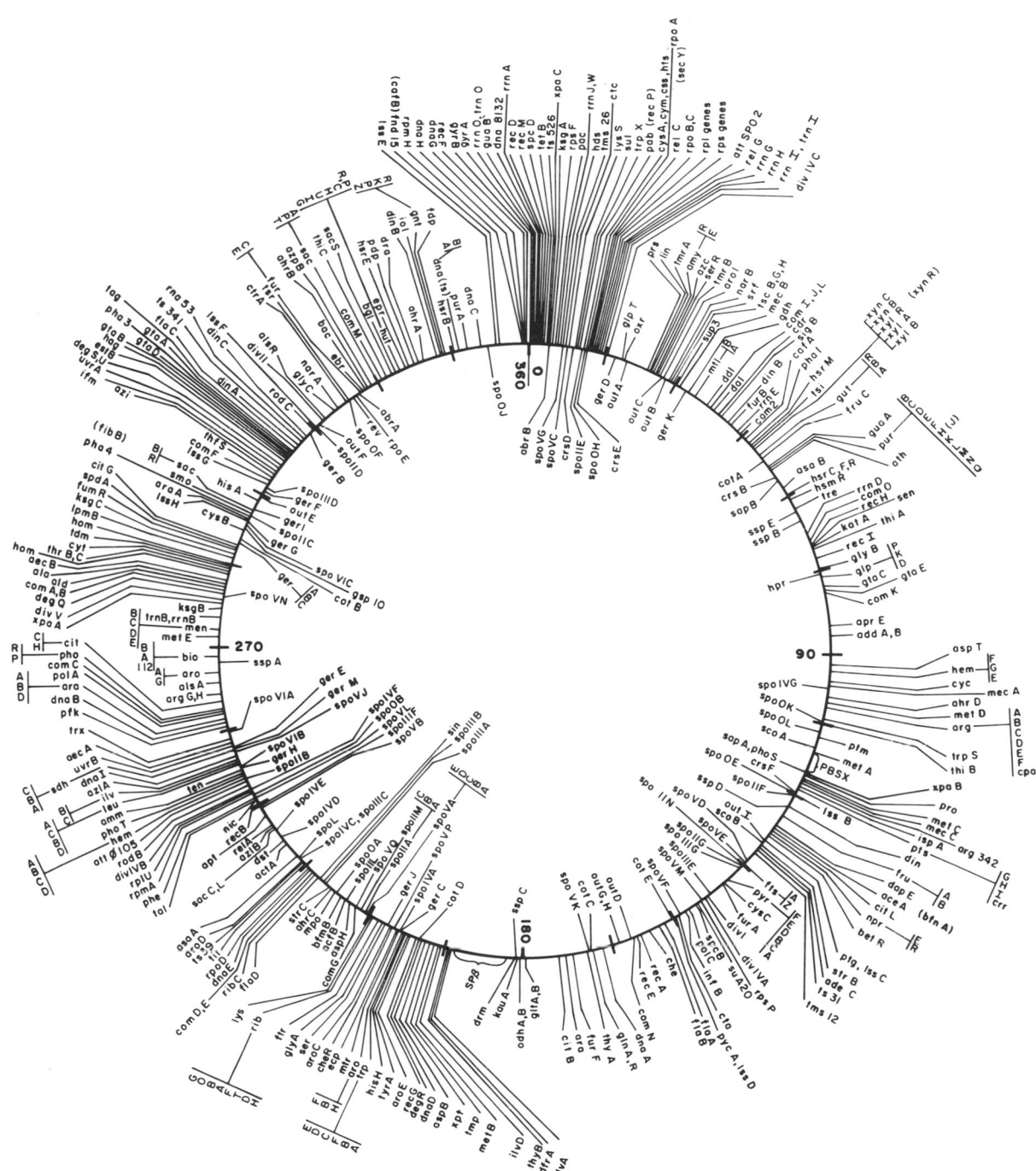

Figure 2   Genetic map of *B. subtilis* 168. For clarity, many ribosomal loci in the region 10° to 15° are not shown individually.

SECTION·5

F Factor and F Derivatives

A. Genetic and physical map of the F plasmid and table of F plasmid genes

B. Hfr points of origins in *E. coli* and useful Hfr strains

C. Hfr strains of *S. typhimurium* and *S. abony*

D. Transduction mapping functions

E. Genetic map locations of some commonly used F' factors

REFERENCES

Holloway, B. and K.B. Low. 1987. F-prime and R-prime factors. In Escherichia coli *and* Salmonella typhimurium. *Cellular and Molecular Biology* (ed. F.C. Neidhardt), pp. 1145–1153. American Society for Microbiology, Washington, D.C.

Kemper, J. 1974. Gene order and co-transduction in the *leu-ara-fol-pyrA* region of the *Salmonella typhimurium* linkage map. *J. Bacteriol.* **117:** 94–99.

Low, K.B. 1987a. Hfr strains of *Escherichia coli* K-12. In Escherichia coli *and* Salmonella typhimurium. *Cellular and Molecular Biology* (ed. F.C. Neidhardt), pp. 1134–1137. American Society for Microbiology, Washington, D.C.

Low, K.B. 1987b. Mapping techniques and chromosome size. In Escherichia coli *and* Salmonella typhimurium. *Cellular and Molecular Biology* (ed. F.C. Neidhardt), pp. 1184–1189. American Society for Microbiology, Washington, D.C.

Sanderson, K.E. and P.R. MacLachlan. 1987. F-mediated conjugation, F$^+$ strains, and Hfr strains of *Salmonella typhimurium* and *Salmonella abony*. In Escherichia coli *and* Salmonella typhimurium. *Cellular and Molecular Biology* (ed. F.C. Neidhardt), pp. 1138–1144. American Society for Microbiology, Washington, D.C.

Willetts, N. and R. Skurray. 1987. F factor and conjugation. In Escherichia coli *and* Salmonella typhimurium. *Cellular and Molecular Biology* (ed. F.C. Neidhardt), pp. 1110–1133. American Society for Microbiology, Washington, D.C.

Wu, T.T. 1966. A model for three-point analysis of random general transduction. *Genetics* **54:** 405–410.

SECTION 5

A. Genetic and Physical Map of the F Plasmid and Table of F Plasmid Genes

Figure 1 and Table 1 are reprinted, with permission, from Willetts and Skurray (Escherichia coli *and* Salmonella typhimurium. *Cellular and Molecular Biology* [ed. F.C. Neidhardt]. American Society for Microbiology, Washington, D.C., pp. 1111 and 1112–1113, respectively [1987]). The references cited in the table can be found in the original paper.

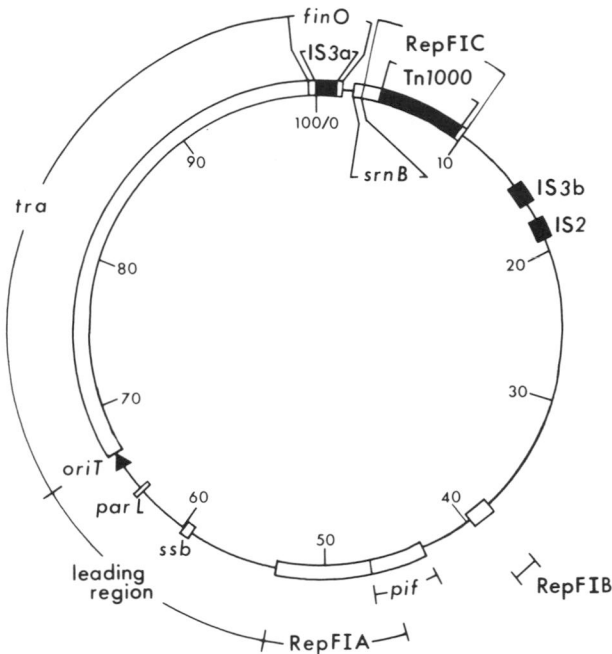

Figure 1 Genetic and physical map of the F plasmid. The coordinate 100/0F has been assigned to the end of IS3a (see Physical and Genetic Structure of F in Willetts and Skurray [1987]), and coordinates for previously mapped genes and sites have been progressively adjusted in a counterclockwise manner. All symbols are described in Willetts and Skurray (1987) or in Table 1 below and Table 2 in Willetts and Skurray (1987), as are the coordinates of genes, regions, and sequences and the references from which this map was compiled. Detailed maps of the RepFIA replication region and the transfer (*tra*) region are provided in Figures 2 and 3, respectively, in Willetts and Skurray (1987). The arrow at the origin of transfer, *oriT*, indicates the direction of F plasmid conjugational DNA transfer, with the leading region entering the recipient first and the *tra* region last. It should be noted that *finO* occupies coordinates 99.5–100F and 1.3–1.6F, which are represented as open blocks on either side of IS3a. Likewise, the RepFIC replication region (3.0–4.2F and 9.9–10.0F) is represented as open blocks on either side of Tn*1000*.

5.2

Table 1   F Plasmid Genes

Symbol	Derivation	Map position (kb)	Alternative symbols, function(s)	Reference(s)
ccdA	Coupled cell division	47.3–47.9	*incH*, *letA*, *lynA*; suppression of *ccdB*	25, 33, 197, 199, 226
ccdB	Coupled cell division	47.3–47.9	*letD*, *lynA*; inhibition of host cell division; activation of SOS functions	25, 33, 40, 138, 144, 197, 199, 226
chr	Chromosomal replication	53.7–57.1	Mutation affects Hfr chromosomal replication	51, 52
ecp	Enhanced Cop phenotype	48.4–49.15	Increased copy number of *copA*(Ts) mutations at low temperature	120
fcr	F cointegrate resolution	47.1–47.3	Site-specific resolution of plasmid cointegrates formed at *rfsF*; coincident with *oriV*	148, 223, 224
finO	Fertility inhibition	99.5–100/0 and 1.3–1.6	FinOP repression of transfer, in combination with *finP*; IS3a (100/0–1.3) is inserted within *finO* on F	90, 322; Cheah and Skurray, in press
finP	Fertility inhibition	67.6	FinOP repression of transfer, in combination with *finO* product; may be an RNA molecule	90, 218, 322
fisO	Fertility inhibition site	67.6	*traO*, *o_J*; site of action of FinOP system; presumably the operator site for *traJ*	90, 218, 322
fisQ	Fertility inhibition site	77.0–81.9	Site of action of FinQ fertility inhibition system specified by R62 and R820a	96, 100, 321, 322
incB	Incompatibility	49.3–49.46	Incompatibility (RepFIA) associated with four 19-bp iterons; coincides with *oriS*; binding sites for RepFIA protein E	219, 220, 253, 293; Matsubara and Murotsu, personal communication
incC	Incompatibility	50.3–50.55	*aos*, *copB*, *pom*; incompatibility (RepFIA) associated with five 19-bp iterons; binding sites for RepFIA protein E	219, 220, 261, 289, 293; Matsubara and Murotsu, personal communication
incE	Incompatibility	38.0–38.45	Incompatibility associated with RepFIB (F secondary replicon)	30, 97, 252
letB	Lethal	51.6–53.0	Defective growth of mutant plasmid-carrying cells; located in region encoding ParFIA functions	197, 199
oriS	Origin of replication	49.4	*oriII*, *ori-2*, *oriE*, *oriV2*; origin of RepFIA replication; unidirectional	29, 80, 85, 179, 220, 290
oriT	Origin of transfer	66.75	Site of initiation of conjugal DNA transfer; postulated site of action of *traYI* "nickase"; site-specific recombination with chromosome	82, 83, 111, 112, 124, 263, 286, 307
oriV	Origin of replication	46.9	*oriI*, *ori-1*, *oriV1*; origin of RepFIA replication; bidirectional; see also *fcr*, *rfsF* for other *oriV*-mediated functions	78, 164, 290
parA	Partitioning	50.8–51.9	*sopA*; required for ParFIA partitioning	20, 22, 115, 148, 225
parB	Partitioning	51.9–52.9	*sopB*, *incG*; required for ParFIA partitioning	20, 22, 115, 148, 225, 244
parC	Partitioning	52.9–53.3	*incD*, *sopC*; *cis*-acting site required for ParFIA partitioning; 11 43-bp iterons; binds *parB* product	20, 99, 115, 148, 225, 244
parL	Partitioning	63.65–63.85	*cis*-acting site within leading region; stabilizes p15A-derived pACYC184 plasmid vector	248; S. Loh, A. Ray, and R. Skurray, unpublished data
pifA	Phage inhibition by F	43.7–45.9	Inhibition of development of T7 and other female-specific phages	59, 60, 201, 254
pifB	Phage inhibition by F	43.3–43.7	As for *pifA*	254
pifO	Phage inhibition by F	46.8–47.2	*cis*-acting site required for autoregulation of *repC*; presumably lies within *repC* promoter	164, 202
prtA	Protection against damaging agents	57.1–64.7	Integrated F reduces damage by UV irradiation and monofunctional alkylating agents; may be equivalent to *rsf*	51, 52
repC	Replication	45.9–46.9	*pifC*; encodes RepFIA protein C; essential for replication from *oriV*; repressor of *pifA* and *pifB*	79, 164, 281
repE	Replication	49.46–50.3	*repA*, *copA*; encodes RepFIA protein E; essential for replication from *oriS* and *oriV*; autoregulatory	148, 164, 176, 177, 253, 272, 300; Matsubara and Murotsu, personal communication
resD	Resolution	47.9–48.7	Encodes RepFIA protein D; required for resolution of cointegrates at *fcr*	Lane, personal communication; Malamy, personal communication

Table 1 *Continued*

Symbol	Derivation	Map position (kb)	Alternative symbols, function(s)	Reference(s)
rfsF	Replicon fusion site of F	47.0–47.2	Site-specific, *recA*-independent formation of cointegrates between two *oriV*-carrying plasmids; *oriV*-mediated mobilization; potentially involved in *oriV*-mediated integration of F and mini-F into chromosome	146, 164, 167, 217, 223, 224
rriA	Rifampin-resistant initiation	49.16	Primosome assembly site; potentially involved in RNA priming of DNA replication from *oriS*	129
rriB	Rifampin-resistant initiation	47.1	Possible primosome assembly site; potentially involved in RNA priming of DNA replication from *oriV*; seen only as sequences homologous with *rriA*	164
rsf	Recombination stimulating factor	57.1–64.7	Mutation affects yield of recombinants after Hfr chromosomal transfer	51, 52
srnB	Stable RNA negative	2.6–3.0	Degradation of stable RNA at 42°C in presence of rifampin or *rpoC* initiation; complements lytic function of λ *S* gene; potentially acts to promote membrane-induced changes leading to release of RNase I from periplasm to cytoplasm	133, 229-231; Ohnishi, personal communication; Bergquist, personal communication
ssb	Single-strand binding	59.2–59.8	*ssf*; single-stranded DNA-binding protein (SSB-F)	49, 107, 153
traA	Transfer	68.8–69.2	Structural gene for F prepropilin	95, 132, 162, 206, 214, 322
traB	Transfer	70.7–72.2	F pilus biosynthesis/assembly	162, 322
traC	Transfer	74.6–77.0	F pilus biosynthesis/assembly	162, 322
traD	Transfer	90.9–92.1	Conjugal DNA metabolism; F-specific RNA phage penetration	241, 322, 323
traE	Transfer	69.5–70.0	F pilus biosynthesis/assembly	95, 162, 322
traF	Transfer	81.9–82.6	F pilus biosynthesis/assembly	162, 322
traG	Transfer	86.3–89.0	Bifunctional; N terminus, F pilus biosynthesis/assembly, possibly *N*-acetylation of pilin; C terminus, mating pair stabilization	162, 322
traH	Transfer	85.1–86.2	F pilus biosynthesis/assembly	162, 322
traI	Transfer	93.9–98.8	DNA helicase I; conjugal DNA metabolism; component of *oriT*-specific nickase (?); see *traZ*	3, 322, 323
traJ	Transfer	67.5–68.25	Positive regulatory gene; product required for transcription of *traYZ* operon and possibly *traM* operon	64, 92, 96, 218, 285, 322
traK	Transfer	70.0–70.7	F pilus biosynthesis/assembly	162, 322
traL	Transfer	69.2–69.5	F pilus biosynthesis/assembly	95, 322
traM	Transfer	67.0–67.45	Initiation conjugal DNA transfer (?)	285, 322, 323
traN	Transfer	78.6–80.4	Stabilization of mating pairs	322; Ippen-Ihler, personal communication
traP	Transfer	72.2–72.9	Defined as encoding 23.5-kDa protein; function unknown	322
traQ	Transfer	83.5–83.9	F pilus biosynthesis/assembly; cleavage of prepropilin	162, 214; Ippen-Ihler, personal communication
traS	Transfer	89.2–89.7	Surface exclusion; reduction of recipient ability; inhibits DNA transfer	322
traT	Transfer	89.9–90.65	Surface exclusion; reduction of recipient ability; inhibits mating pair formation; serum resistance	204, 207, 211, 322
traU	Transfer	78.0–78.6	F pilus biosynthesis/assembly	322; Ippen-Ihler, personal communication
traV	Transfer	73.2–73.8	F pilus biosynthesis/assembly	322; Ippen-Ihler, personal communication
traW	Transfer	(74.6–77.0)	F pilus biosynthesis/assembly; may overlap or be identical to *traC*	322
traY	Transfer	68.25–68.8	Conjugal DNA metabolism; postulated component of *oriT*-specific nickase	92, 218, 322, 323
traZ	Transfer	96.6–98.8	Conjugal DNA metabolism; postulated component of *oriT*-specific nickase; may be a portion of *traI*	322, 323; Minkley, personal communication

B. Hfr Points of Origins in *E. coli* and Useful Hfr Strains

Figure 1 and Table 1 are reprinted, with permission, from Low (Escherichia coli *and* Salmonella typhimurium. *Cellular and Molecular Biology* [ed. F.C. Neidhardt]. American Society for Microbiology, Washington, D.C., pp. 1134 and 1135, respectively [1987a]. The references cited in the table and table notes can be found in the original paper. The chapters cited can be found in the original volume.

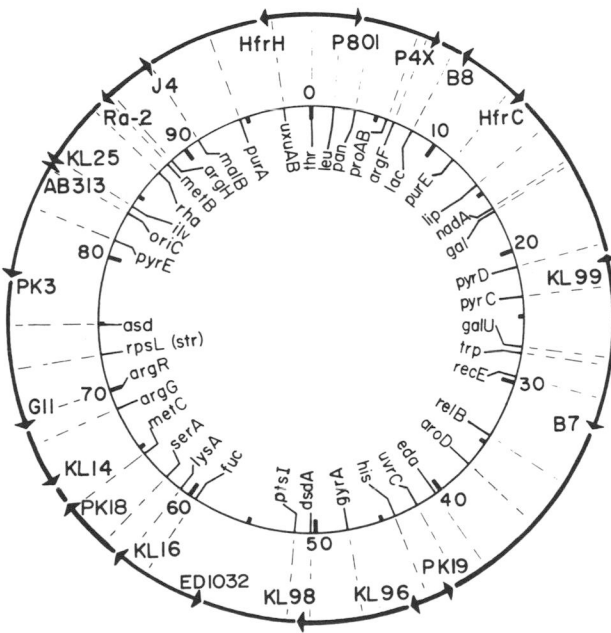

Figure 1 Map positions of integrated sex factors (F, F$_{ts}$-*lac*, or ColV) for some Hfr strains. Commonly used derivatives of these Hfr strains are listed in Table 1. Each arrowhead indicates the position and orientation of integration. The sequence of markers transferred from a given strain begins behind the arrowhead; e.g., HfrH transfers genes in the order *uxuAB*, *thr*, *leu*, etc.

Table 1   Useful *E. coli* K12 Hfr Strains

Hfr strain	Ancestral Hfr strain	PO no.[a]	Earliest marker known	Latest marker known	Site of Tn10[b]	Comment(s)	Reference(s)
3000[c] (=HfrH *thi*⁻λ⁻)	HfrH	1	*uxuBA*	*valS*			24, 33; chapter 72
NK6051[d]	HfrH	1	*uxuBA*	*valS*	*purE*		48
P801[c]	P801	120	*leu*	*pan*			B. J. Bachmann, personal communication from F. Jacob
BW6165[d]	P801	120	*leu*	*pan*	*argE*		48
BW113[c]	P4X	3	*argF*	*lac*			17, 37
BW6156[d]	P4X	3	*argF*	*lac*	*zje*		48
B8[c]	B8	118	*tsx*	*lac*			7
BW6160[d]	B8	118	*tsx*	*lac*	*zdh*		48
KL226[c]	HfrC	2A	*gsk*	*fep*			29, 37, 42
BW7261[d]	HfrC	2A	*gsk*	*fep*	*leu*		48
KL99[c]	KL99	42	*pyrC*	*pyrD*			37
BW7623[d]	KL99	42	*pyrC*	*pyrD*	*zed*		48
KL208[c]	B7	43	*rac*	*trg*		F deleted from 33–43 kb	7, 37, 47
BW7620[d]	B7	43	*rac*	*trg*	*purE*	F deleted from 33–43 kb	48
PK191[c]	PK19	66	*supD*	*cheC*		Fertility factor is ColV	30, 37
BW5660[d]	PK19	66	*supD*	*cheC*	*srlC*	Fertility factor is ColV	48
KL96[c]	KL96	44	*his*	*purF*			35, 37
BW7622[d]	KL96	44	*his*	*purF*	*trpB*		48
KL983[c]	KL98	53	*dsdA*	*supN*			37
BW5659[d]	KL98	53	*dsdA*	*supN*	*zdh*		48
ED1032	ED1032	201	*thy*	*his*		Transposition of F42-114; transfers *tra* genes early	8
KL16[c]	KL16	45	*lysA*	*serA*			35, 37
BW6163[d]	KL16	45	*lysA*	*serA*	*zed*		48
PK18	PK18	132	*metC*	*argG*		Fertility factor is ColV	23, 30
KL14[c]	KL14	68	*cca*	*tolC*			20, 37
BW6159[d]	KL14	68	*cca*	*tolC*	*ilv*		48
G11	G11	124	*str*	*argG*			38
KL800[c]	PK3	131	*xyl*	*malA*		Fertility factor is ColV	30; K. B. Low, unpublished data
BW6175[d]	PK3	131	*xyl*	*malA*	*argE*	Fertility factor is ColV	48
KL228[c]	AB313	13	*rbs*	*ilvE*			37
BW6169[d]	AB313	13	*rbs*	*ilvE*	*argA*		48
KL25[c]	KL25	46	*ilvE*	*pyrE*		Very unstable Hfr	35, 37
Ra-2[c]	Ra-2	48	*metB*	*rha*		Very unstable Hfr	33, 34, 37
BW6164[d]	Ra-2	48	*metB*	*rha*	*thr*	Very unstable Hfr	48
KL209[c]	J4 (=P10)	18	*argE*	*purA*		F factor in *malB*	37, 45
BW6166[d]	J4 (=P10)	18	*argE*	*purA*	*zhf*	F factor in *malB*	48

[a] Point of origin (PO) numbers were arbitrarily assigned to distinguish independent F factor insertion events in the formation of Hfr strains. These numbers do not relate to map position (see Fig. 1).

[b] Sites of Tn10 insertion in these strains lie within the range of 10 to 30 min from the points of origin. Some of these strains also carry a deletion of the *lac* region, as well as certain other markers. The "zxy" system for naming the sites of unknown transposon insertions is given in reference 10. The second letter (x) indicates the appropriate 10-min interval of the map (a = 0 to 10, b = 10 to 20, etc.), and the third letter (y) indicates the particular minute within that interval (a = 0 to 1, b = 1 to 2, etc.). All other gene symbols are as given on the *E. coli* K-12 map (chapter 53).

[c] Strain is included in the Hfr kit available from the *E. coli* Genetic Stock Center, c/o B. J. Bachmann, Department of Biology, Osborn Laboratory, Yale University, P.O. Box 6666, New Haven, CT 06511.

[d] Strain is included in the Hfr::Tn10 kit from the address given in footnote c.

C. Hfr Strains of *S. typhimurium* and *S. abony*

Figure 1 is reprinted, with permission, from Sanderson and MacLachlan (Escherichia coli *and* Salmonella typhimurium. *Cellular and Molecular Biology* [ed. F.C. Neidhardt]. American Society for Microbiology, Washington, D.C., p. 1143 [1987]. The references cited in the figure caption can be found in the original paper, and the chapter cited can be found in the original volume.

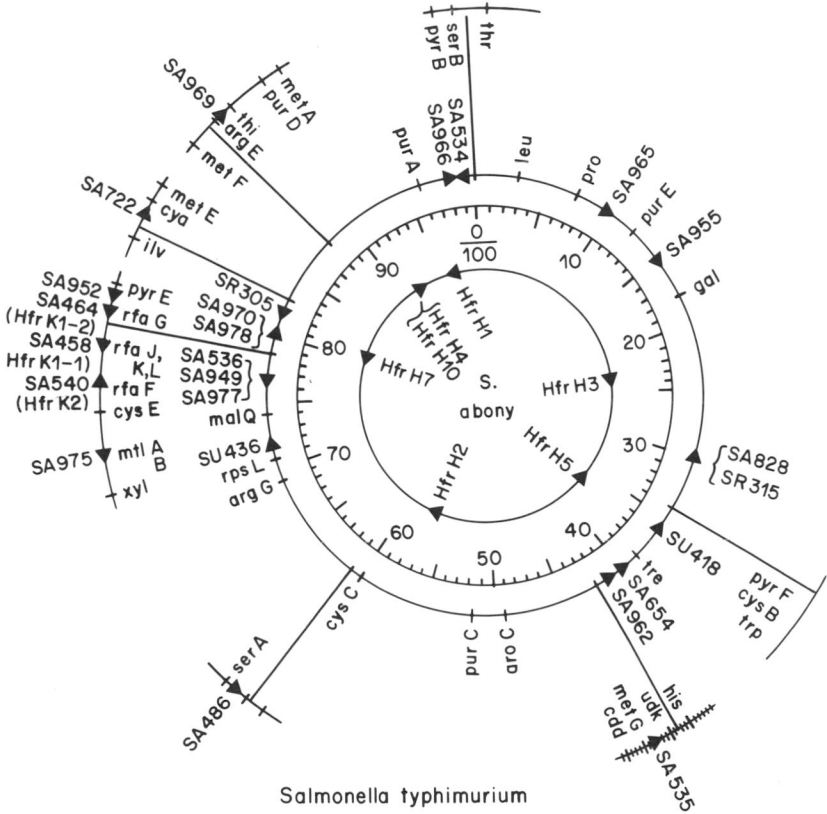

Salmonella typhimurium

Figure 1 Hfr strains of *S. typhimurium* (shown outside the circle) and of *S. abony* (shown inside the circle). The numbers in the circle represent time of entry for Hfr strains. The point of origin and orientation of transfer of each strain are indicated by the arrows. For *S. typhimurium*, the strain number (e.g., SR305) is given, and Table 1 in Sanderson and MacLachlan (1987) gives the Hfr designation (HfrA) and the genotype of each strain. In those cases in which F is inserted into a known transduction linkage group, the Hfr is displayed on an arc outside the main circle; the genes shown on the cross-hatched arc are a P1-mediated transduction linkage group, but all other linkage groups are for P22 phage. Where more than one Hfr strain number is shown in a gene interval (e.g., SA536, SA949, and SA977, shown within a bracket), these represent independent isolates for which the points of origins are not proven different, though differences may exist. Not all strains with points of origins in the *rfa* genes (HfrK1 and HfrK2, at 79 min) are shown. For more details, see Table 1 in Sanderson and MacLachlan (1987). Discussions of experiments revealing points of origins and other properties of Hfr strains are provided by Sanderson et al. (28), and a few corrections in the original data are provided in the footnotes of Table 1 in Sanderson and MacLachlan (1987). Only those genes whose positions are known with respect to the points of origins of the Hfr strains are shown here; the complete linkage map is shown elsewhere (29; Chapter 54).

SECTION 5

D. Transduction Mapping Functions

Figure 1 is reprinted, with permission, from Low (Escherichia coli *and* Salmonella typhimurium. *Cellular and Molecular Biology* [ed. F.C. Neidhardt]. American Society for Microbiology, Washington, D.C., p. 1188 [1987b]). The references cited in the figure caption can be found here in the reference list on page 2.1.

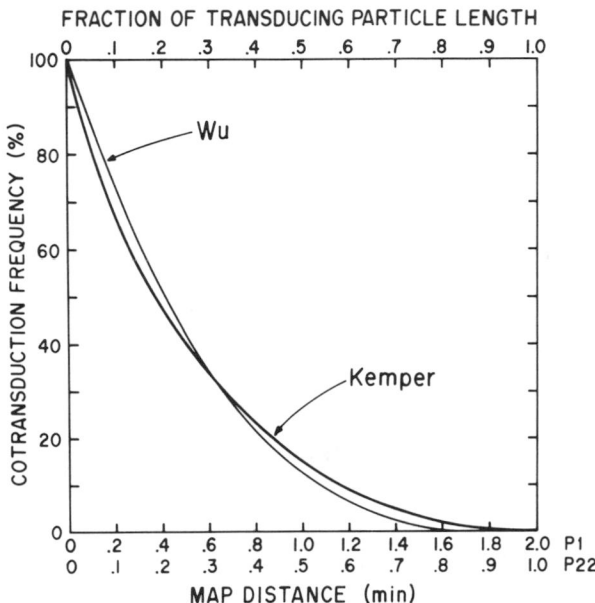

Figure 1 Transduction mapping functions used in recent editions of the maps of *E. coli* K12 (Wu 1966) or *S. typhimurium* (Kemper 1974). In future editions of the *S. typhimurium* map, the function of Wu will be used (K. Sanderson, pers. comm.).

E. Genetic Map Locations of Some Commonly Used F′ Factors

Figure 1 is reprinted, with permission, from Holloway and Low (Escherichia coli *and* Salmonella typhimurium. *Cellular and Molecular Biology* [ed. F.C. Neidhardt]. American Society for Microbiology, Washington, D.C., p. 1149 [1987]).

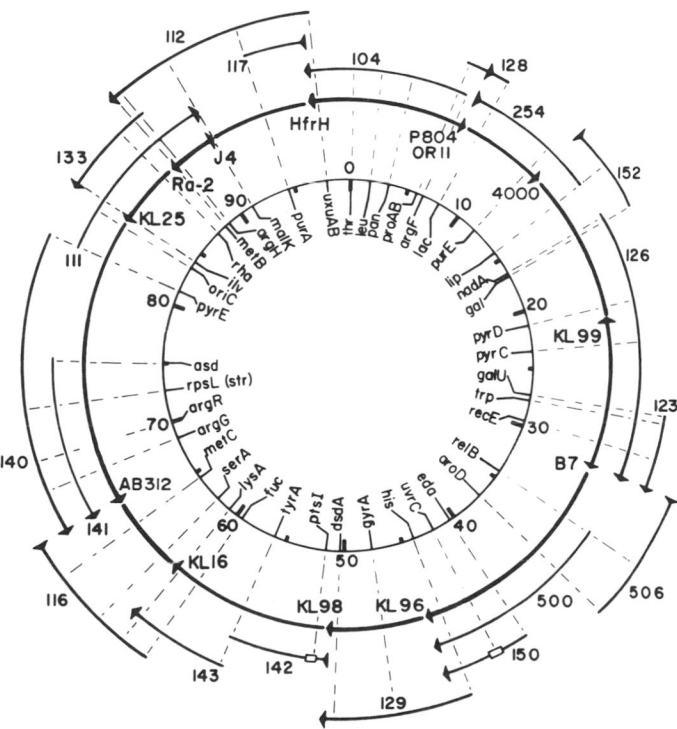

Figure 1 These and certain other individual F′s, or the entire "kit" shown, are available from the *E. coli* Genetic Stock Center (see Section 6B of this Handbook).

SECTION·6

Bacterial Strains and How to Obtain Them

A. General sources
1. *E. coli* Genetic Stock Center
2. American Type Culture Collection
3. National Institute of Genetics (Japan)
4. *Salmonella* Genetic Stock Center
5. *Bacillus* Genetic Stock Center

B. Strain sets
1. CSH collection
2. Singer/Gross collection of antibiotic-resistant strains
3. Singer/Gross Hfr collection
4. Wanner collection of Hfr strains
5. *E. coli* Genetic Stock Center Hfr collection
6. *E. coli* Genetic Stock Center F′ kit

C. Additional useful strains for recombinant DNA experiments

REFERENCES

Bachmann, B. 1983. Linkage map of *Escherichia coli* K-12, edition 7. *Microbiol. Rev.* **47:** 180–230.

Baker, T.A., M.M. Howe, and C.A. Gross. 1983. Mu dX, a derivative of Mu d1 (*lac* Ap^r) which makes stable *lacZ* fusions at high temperatures. *J. Bacteriol.* **156:** 970–974.

Brickman, E. and J.R. Beckwith. 1975. Analysis of the regulation of *Escherichia coli* alkaline phosphatase synthesis using deletions and φ80 transducing phages. *J. Mol. Biol.* **96:** 307–316.

Calos, M.P. and J.H. Miller. 1981. Genetic and sequence analysis of frameshift mutations induced by ICR-191. *J. Mol. Biol.* **153:** 39–66.

Cupples, C.G. and J.H. Miller 1988. Effects of amino acid substitutions at the active site in *Escherichia coli* β-galactosidase. *Genetics* **120:** 637–644.

Cupples, C.G. and J.H. Miller. 1989. A set of *lacZ* mutations in *Escherichia coli* that allow rapid detection of each of the six base substitutions. *Proc. Natl. Acad. Sci.* **86:** 5345–5349.

Groisman, E.A. and M.J. Casadaban. 1986. Mini-Mu bacteriophage with plasmid replicons for *in vivo* cloning and *lac* gene fusing. *J. Bacteriol.* **168:** 357–364.

Heffron, F., P. Bedinger, J.J. Champoux, and S. Falkow. 1977. Deletions affecting the transposition of an antibiotic resistance gene. *Proc. Natl. Acad. Sci.* **74:** 702–706.

Hershfield, V., H.W. Boyer, C. Yanofsky, M.A. Lovett, and D.R. Helinski. 1974. Plasmid ColE1 as a molecular vehicle for cloning and amplification of DNA. *Proc. Natl. Acad. Sci.* **71:** 3455–3459.

Kleina, L.G., J.-M. Masson, J. Normanly, J. Abelson, and J.H. Miller. 1990. Construction of *Escherichia coli* amber suppressor tRNA genes. II. Synthesis of additional tRNA genes and improvement of suppressor efficiency. *J. Mol. Biol.* **213:** 705–717.

Kushner, S.R. 1987. Useful host strains and techniques for recombinant DNA experiments. In Escherichia coli *and* Salmonella typhimurium. *Cellular and Molecular Biology* (ed. F.C. Neidhardt), pp. 1225–1230. American Society for Microbiology, Washington, D.C.

Miller, J.H. 1972. *Experiments in Molecular Genetics.* Cold Spring Harbor Laboratory, Cold Spring Harbor, New York.

Miller, J.H. 1985. Mutagenic specificity of ultraviolet light. *J. Mol. Biol.* **182:** 45–68.

Miller, J.H., D. Ganem, P. Lu, and A. Schmitz. 1977. Genetic studies of the *lac* repressor. I. Correlation of mutational sites with specific amino acid residues: Construction of a colinear gene-protein map. *J. Mol. Biol.* **109:** 275–301.

Miller, J.H., M.P. Calos, D. Galas, M. Hofer, D.E. Buchel, and B. Müller-Hill. 1980. Genetic analysis of transpositions in the *lac* region of *Escherichia coli. J. Mol. Biol.* **144:** 1–18.

Müller-Hill, B. and J. Kania. 1974. *lac* repressor can be fused to β-galactosidase. *Nature* **249:** 561–563.

Sambrook, J., E.F. Fritsch, and T. Maniatis. 1989. *Molecular Cloning. A Laboratory Manual,* Second Edition. Cold Spring Harbor Laboratory Press, Cold Spring Harbor, New York.

Singer, M., T.A. Baker, G. Schnitzler, S.M. Deischel, M. Goel, W. Dove, K.J. Jaacks, A.D. Grossman, J.W. Erickson, and C.A. Gross. 1989. A collection of strains containing genetically linked alternating antibiotic resistance elements for genetic mapping of *Escherichia coli. Microbiol. Rev.* **53:** 1–24.

Wang, B., L. Liu, E.A. Groisman, M.J. Casadaban, and C.M. Berg. 1987. High-frequency generalized transduction by mini-Mu plasmid phage. *Genetics* **116:** 201–206.

Wanner, B.L. 1986. Novel regulatory mutants of the phosphate regulon in *Escherichia coli* K-12. *J. Mol. Biol.* **191:** 39–58.

Way, J.C., M.A. Davis, D. Morisato, D.E. Roberts, and N. Kleckner. 1984. New Tn*10* derivatives for transposon mutagenesis and for construction of *lacZ* operon fusions by transposition. *Gene* **32:** 369–379.

SECTION 6

A. General Sources

1. E. coli *Genetic Stock Center (CGSC)*

There is no charge for strains. Requests can be for a specific strain, a strain with a mutation in a known locus, or a strain with a transposon integrated near a given locus. The stock center will also deposit strains. Strains can be obtained by writing to:

> Dr. Barbara Bachmann
> *E. coli* Genetic Stock Center
> Department of Biology, 255 OML
> Yale University, P.O. Box 6666
> New Haven, CT 06511-7444

2. *American Type Culture Collection (ATCC)*

A wide variety of bacterial strains, plasmids, and phage vectors are stored and sent out by the ATCC, which also publishes a "Bacteria" catalog and separate catalogs for each collection ("Animal and Plant Viruses," "Cells," "Fungi," etc.) every four years, together with yearly updates. The 17th edition of the "Catalog of Bacteria and Bacteriophages" was published in 1989. In general, their fees are $45 per strain for U.S. and Canadian non-profit institutions and $70 for other U.S. and foreign institutions. Information can be obtained from:

> American Type Culture Collection
> 12301 Parklawn Drive
> Rockville, MD 20852
> Customer service: 301-881-2600
> Phone orders: 1-800-638-6597
> FAX orders: 301-231-5826

3. *National Institute of Genetics (Japan)*

This institute maintains an extensive stock of *E. coli* strains. Their catalog can be obtained by writing to:

> National Institute of Genetics
> Genetic Stocks Research Center
> Mishima, 411
> Japan

4. Salmonella *Genetic Stock Center*

The center sends out strains on request to qualified investigators. Requests can be made to:

> Dr. Kenneth E. Sanderson
> *Salmonella* Genetic Stock Center
> Department of Biology
> University of Calgary
> Calgary, Alberta
> Canada T2N 1N4

5. Bacillus *Genetic Stock Center*

The center puts out a catalog describing all of the strains in their collection, including several mapping kits, as well as additional information useful for *Bacillus* geneticists. There is no charge for non-profit institutions and a $50 charge per item for industrial users. Requests can be made to:

> *Bacillus* Genetic Stock Center
> c/o Dr. Daniel R. Ziegler
> The Ohio State University
> Department of Biochemistry
> 484 West 12th Avenue
> Columbus, OH 43210

B. Strain Sets

1. *CSH collection*

The 44 strains used with the Laboratory Manual portion of this volume can be obtained as a kit from Cold Spring Harbor Laboratory Press. The strains are prepared in the author's laboratory. The properties of these strains are listed below in Table 1.

2. *Singer/Gross collection of antibiotic-resistant strains*

Singer, Gross, and their co-workers have prepared a set of *E. coli* strains with Tn*10* and Tn*10kan* markers inserted at approximately every minute on the chromosome, as shown in Figure 1, which is reprinted, with permission, from Singer et al. (*Microbiol. Rev.*, vol. 53, pp. 20–21 [1989]). Any of the strains carrying these markers, and in some cases the entire set, can be obtained from:

> Dr. Mitchell Singer
> Department of Bacteriology
> University of Wisconsin-Madison
> Madison, WI 53706

3. *Singer/Gross Hfr collection*

Singer, Gross, and their co-workers have also compiled a collection of Hfr strains carrying either Tn*10* or Tn*10kan* markers at different positions on the *E. coli* chromosome. These strains are listed below in Table 2, which is excerpted from Singer et al. (*Microbiol. Rev.*, vol. 53, p. 3 [1989]). This collection can be obtained by writing to Dr. Mitchell Singer at the address above.

4. *Wanner collection of Hfr strains*

The *E. coli* Genetic Stock Center (CGSC) distributes a set of 17 Hfr strains (compiled by Wanner [1986]) that contain Tn*10* markers at different positions on the chromosome. Most of these strains are depicted below in Figure 2. Strains can be requested from:

> *E. coli* Genetic Stock Center
> Department of Biology, 255 OML
> Yale University, P.O. Box 6666
> New Haven, CT 06511-7444

5. E. coli *Genetic Stock Center Hfr collection*

In addition to the Wanner collection of Hfr strains mentioned above, the *E. coli* Genetic Stock Center also sends out a kit of Hfr strains without integrated Tn*10* elements. The Hfr strains included in this kit are indicated by a superscript "c" in Table 1 in Section 5B of this Handbook. These strains can be obtained by writing to the *E. coli* Genetic Stock Center at the address above.

6. E. coli *Genetic Stock Center F' kit*

A set of strains carrying different F' factors, indicated in Figure 1 in Section 5E of this Handbook, has been prepared by K.B. Low and can be obtained by writing to the *E. coli* Genetic Stock Center at the address above.

Table 1   Properties of CSH Strains

Number	Mating type	Genotype	Comments[a]
CSH100	F'$lacproA^+,B^+$ ($lacI^Q$ $lacPL8$)	$ara\ \Delta(gpt\text{-}lac)5$	Derived from CSH142; see Miller et al. (1977).
CSH101–106	F'$lacproA^+,B^+$ (carry $I^-\ Z^-$ mutations in the lac region)	$ara\ \Delta(gpt\text{-}lac)5$	Correspond to CC101–106 (Cupples and Miller 1989). Derived from CSH142. See also Figure 4.54 in the Laboratory Manual portion of this volume.
CSH107	F'$lacproA^+,B^+$ (carries a +1 frame-shift mutation in the I portion of a $lacI$-Z fusion)	$ara\ \Delta(gpt\text{-}lac)5$	F' derived from strain 7156-14 (Müller-Hill and Kania 1974); see Miller (1985). Strain derived from CSH142. The F'$lacpro$ episome carries a +1 frameshift mutation ($lacI50$) (Calos and Miller 1981).
CSH108	F'$lacproA^+,B^+$ (carries $I^-\ Z^-$ mutations in the lac region)	$ara\ \Delta(gpt\text{-}lac)5$ $gyrA\ argE_{am}\ rpoB$	Equivalent to strain XAC-1, carrying amber mutation at position 17 of $lacZ$ (Kleina et al. 1990).
CSH109	F$^-$	$ara\ \Delta(gpt\text{-}lac)5$ $rpsL$	Corresponds to S90C (Miller et al. 1977), an Str^r derivative of CSH142 by P1vir transduction from X5097 of J. Beckwith.
CSH110	F$^-$	$ara\ \Delta(gpt\text{-}lac)5$ $supE\ gyrA$ $argE_{am}\ metB$ $rpoB$	Derived from spontaneous Nal^r ($gyrA$) derivative of P90 (see CSH142) by transduction from E210R of M. van Montagu (Hfr P4X $metB\ argE_{am}\ rpoB$) to produce $metB\ argE_{am}\ rpoB$ derivative. Transduced to Arg$^+$ from X7026 ($supE$) of J. Beckwith and then heat-cured of the prophage.
CSH111	F$^-$	$ara\ \Delta(gpt\text{-}lac)5$ $supF\ gyrA$ $argE_{am}\ metB$ $rpoB$	Same as CSH110, except P1 donor was X7150 ($supF$) of J. Beckwith.
CSH112	F$^-$	$ara\ \Delta(gpt\text{-}lac)5\ supB$ $gyrA\ argE_{am}\ metB$ $rpoB$	Same as CSH111, except P1 donor was CA165 ($supB$) of J. Beckwith.

CSH113	F⁻	*ara* Δ(*gpt-lac*)5 *supC gyrA argE*ₐₘ *metB rpoB*	Same as CSH111, except P1 donor was X7151 (*supC*) of J. Beckwith.
CSH114	F⁻	*ara* Δ(*gpt-lac*)5 *rpsL mutT*	Derived from CSH109 by first transducing with P1*vir* from EE273 of E. Eisenstadt (P90C *mutT leu*::Tn*10* from ES1438 of E. Siegel), selecting for Tet^r, and then with P1*vir* from CSH142 for Leu⁺.
CSH115	F⁻	*ara* Δ(*gpt-lac*)5 *rpsL mutS*:: mini-Tn*10*	Derived by insertion of mini-Tn*10* from λ1098 (Way et al. 1984) in CSH142 and transfer by P1*vir* to CSH109 (T. Phuong and J.H. Miller, unpubl.).
CSH116	F⁻	*ara mutD5 zae-502*:: Tn*10* Δ(*gpt-lac*)5 *rpsL*	Derived by P1*vir* transduction of CSH109 to Tet^r from PF1044 (*mutD5 zae-502*::Tn*10*) of P. Foster.
CSH117	F⁻	*ara* Δ(*gpt-lac*)5 *rpsL mutY*::mini-Tn*10*	Derived by insertion of mini-Tn*10* from λ1098 (Way et al. 1984) in CSH109 (C. Cruz and J.H. Miller, unpubl.).
CSH118	F⁺ *kan* (carries determinants for Kan^r)	*ara* Δ(*gpt-lac*)5	The F⁺ *kan* is RSF2001 (Heffron et al. 1977) transferred from C600 into CSH142. It is a derivative of F into which a 3.2-kb kanamycin resistance sequence has been transposed from the R plasmid pML21 (Hershfield et al. 1974).
CSH119	HfrC	*car-96*::Tn*10* Δ(*gpt-lac*)5 *cysG303 relA? spoT1? metB1*	Corresponds to EG333 (CGSC 6211) from K.B. Low.
CSH120	Hfr P4X	*relA1 spoT1 metB1 zje-2005*::Tn*10*	Corresponds to CGSC 6754.
CSH121	HfrH	*ara ilvJ? zae-502*::Tn*10* Δ(*gpt-lac*)5	Derived from CA70921, an *ara val^r* derivative of CA7033 (J. Beckwith), by P1*vir* transduction of the *zae-502*::Tn*10* mutation from strain 18436 of M. Singer (see Singer et al. 1989). The designation of *ilvJ* as causing the Val^r phenotype is uncertain.

Table 1 *Continued*

Number	Mating type	Genotype	Comments
CSH122	Hfr KL16	zed-977::Tn10 relA1 spoT1 thi-1	Corresponds to CGSC 6758.
CSH123	Hfr KL14	relA1 spoT1 thi-1 thiA::Tn10	Corresponds to CAG12208 (Singer et al. 1989), derived from CGSC 4294.
CSH124	Hfr PK191	Δ(gpt-lac)5 nupG-511::Tn10	Derived by P1 transduction of PK191 (CGSC 4316) of the nupG-511::Tn10 mutation from CGSC 6568.
CSH125	F⁻	ara leu lacY purE gal supE trp his argG malA rpsL xyl mtl ilv metA	From L. Caro.
CSH126	F⁻	ara Δ(gpt-lac)5 Δ(recA-srl)306 srl-301::Tn10-84	Derived by P1vir transduction to Tetʳ of CSH142 from JC10289 of A.J. Clark. The Δ306 deletes part of the Tn10 which had inserted into srlR.
CSH127			P1vir lysate of CSH142.
CSH128	F⁻	ara Δ(gpt-lac)5 (P1clr100)	A derivative of CSH142 lysogenic for P1clr100.
CSH129	F⁻	ara Δ(gpt-lac)5/pCClac5	CSH142 carrying a pMB9 derivative containing the lacPOZ region (Cupples and Miller 1988).
CSH130	F⁻	ara Δ(gpt-lac)5 gyrA rpoB	Derived by P1vir transduction of Nalʳ (gyrA) derivative of P90 (see CSH142) from E210R (see CSH110) to yield argEₐₘ rpoB derivative. Heat-cured of prophage and then transduced to Arg⁺ from CSH142.
CSH131	Hfr KL14	relA1 spoT1 thiA::Tn10kan	Corresponds to CAG12201 of Singer et al. (1989).
CSH132	Hfr KL16	zed::Tn10kan relA1 spoT1 thi-1	Corresponds to CAG12200 of Singer et al. (1989).
CSH133	F⁻	pBCO::Mud5005/ araD169 araB::Mucts Δ(lac)ₓ₉₄ galU galR rpsL	Corresponds to MC1040-2/pEG5005 (Groisman and Casadaban 1986).

CSH134	F−	$\Delta(argR\text{-}lac)_{U169}$ trp $\Delta(brnQ, phoA$ $proC$ $phoB$ $phoR)_{24}$ Mucts62	Corresponds to XPh43 Mucts (Wang et al. 1987; see also Brickman and Beckwith 1975).
CSH135	F′ pro^+ $lacZ8305$ ::Mucts62	MudX $\Delta(gpt\text{-}lac)5$ his met tyr $gyrA$ $rpsL$	Corresponds to CAG5050 (Baker et al. 1983).
CSH136	lysate of λ1098		λ1098 from Way et al. (1984).
CSH137	lysate of λ1105		λ1105 from Way et al. (1984).
CSH138	F′ $lacproA^+,B^+$ ($lacI^Q$ $lacPL8$)	ara $ilvJ?$ $\Delta(gpt\text{-}lac)5$ Tn5	Corresponds to MPC4 (Miller et al. 1980). The designation of $ilvJ$ as causing the Val^r phenotype is uncertain.
CSH139	F′ $lacproA^+,B^+$ ($lacI^Q$ $lacPL8$)	ara $ilvJ?$ $\Delta(gpl\text{-}lac)5$ gal his trp lys $rpsL$ Tn10	Corresponds to MPC5 (Miller et al. 1980). The designation of $ilvJ$ as causing the Val^r phenotype is uncertain.
CSH140	F′ lac^+ $proA^+,B^+$ (carries I^- mutation in the lac region)	ara $\Delta(gpt\text{-}lac)5$	Derived from CSH142. F′ carries $I378$, an ICR 191-induced frameshift (Calos and Miller 1981).
CSH141	F′ lac^+ $proA^+,B^+$	$\Delta(gpt\text{-}lac)5$ $supE$ $rpsE$	Corresponds to CSH23 (Miller 1972) and E5014 from J. Beckwith.
CSH142	F−	ara $\Delta(gpt\text{-}lac)5$	Corresponds to P90C. Derived from G90 of W. Gilbert [$\Delta(lac)RV$ (λCI857t68h80dlac)], crossed with ara derivative of CA7033 of J. Beckwith to yield P90, an ara $\Delta(gpt\text{-}lac)5$ derivative. Strain then heat-cured of prophage to yield P90C.
CSH143	F′ $lacproA^+,B^+$ (carries I^-Z^- mutations in the lac region)	ara $\Delta(gpt\text{-}lac)5$ $gyrA$	Spontaneous Nal^r ($gyrA$) derivative of CSH142, carrying F′ from CSH108, which carries amber mutation at position 17 of $lacZ$.

[a]See pp. 6.1–6.2 for references.

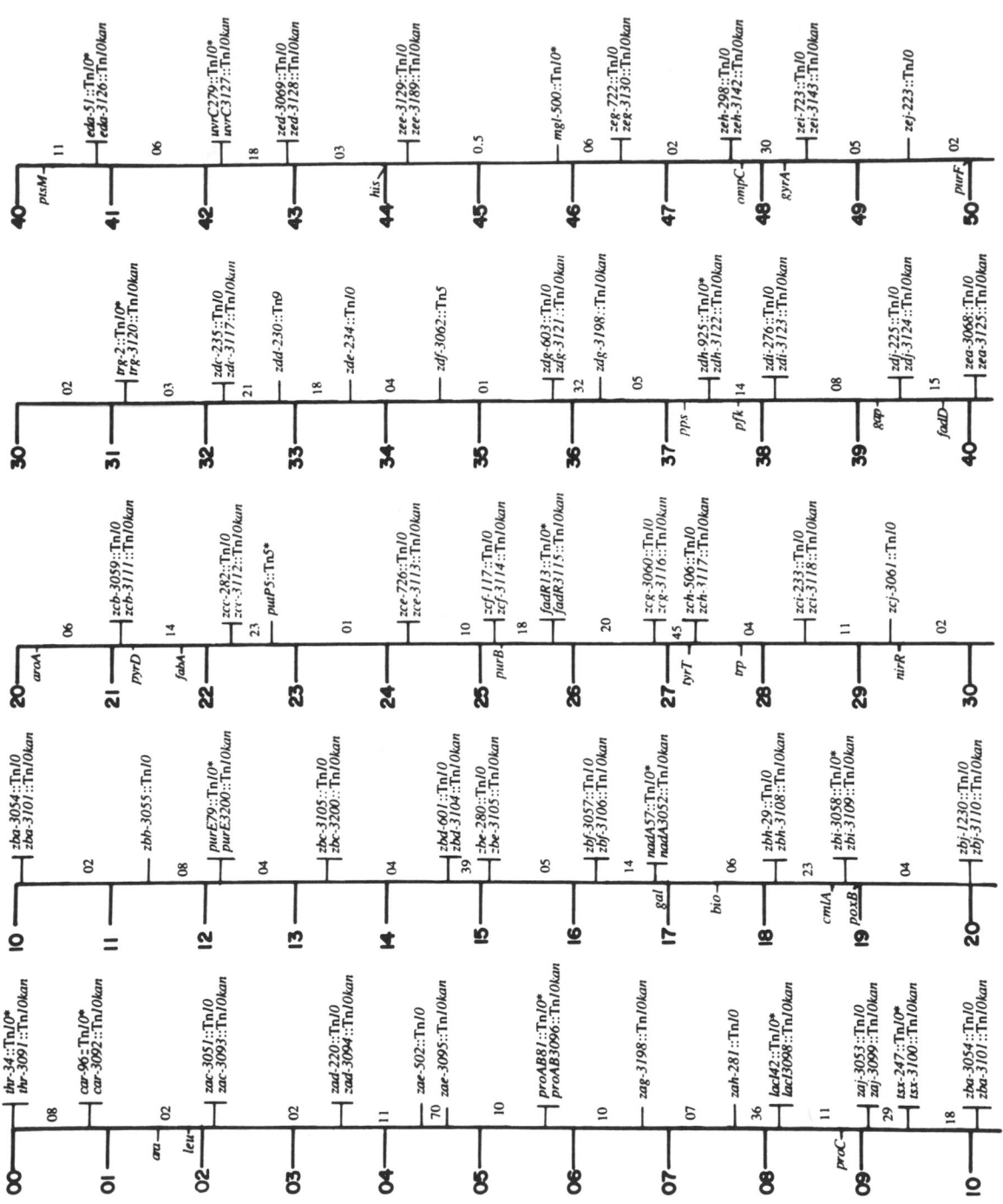

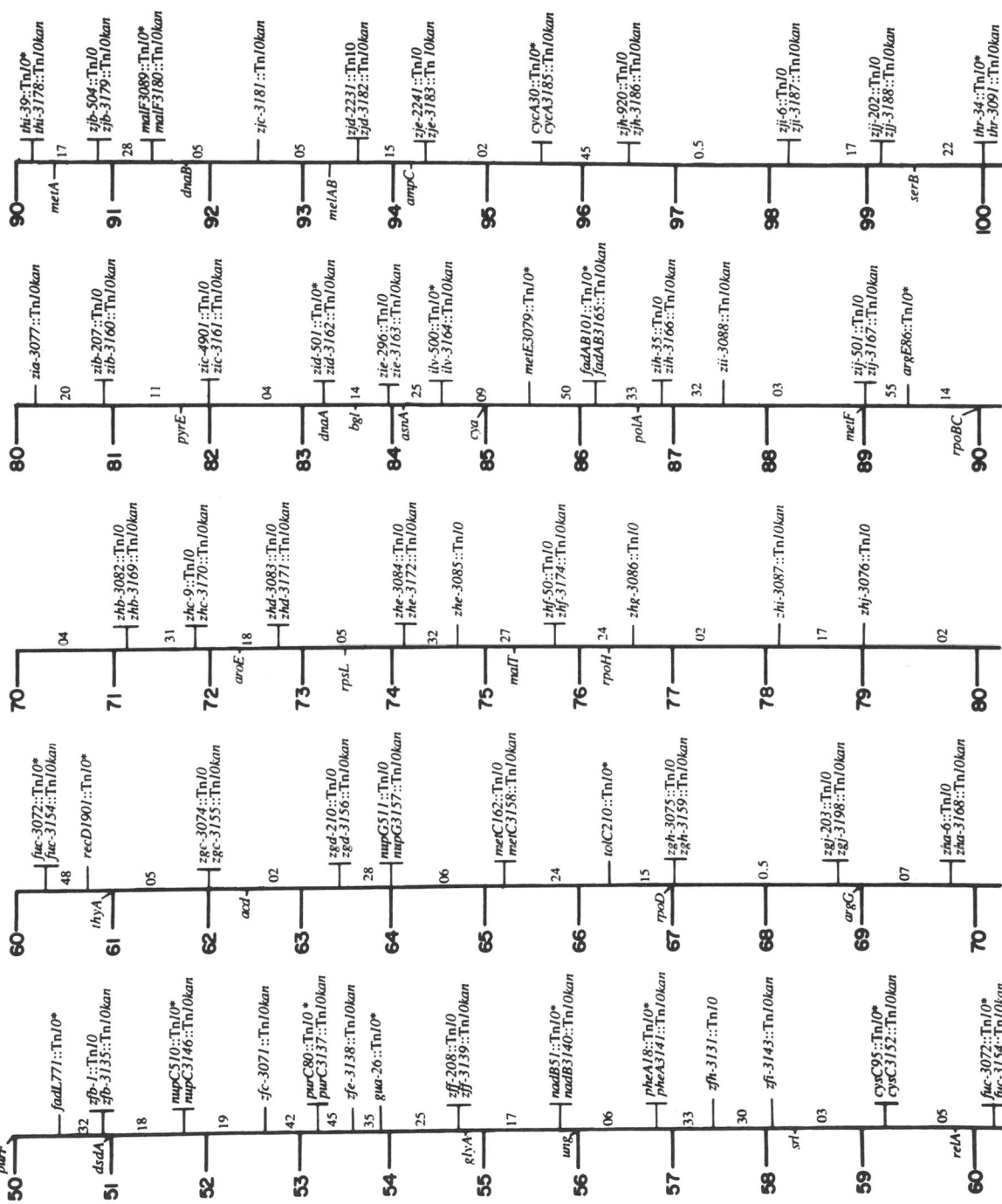

Figure 1 Insertion map. Asterisks indicate insertions used to align the insertion map to the standard *E. coli* genetic map of Bachmann (1983). Each insertion was placed onto the standard genetic map as described in Singer et al. (1989). Numbers between insertions represent the P1 cotransduction frequency indicated as percentage of cotransduction (see Table 2 in Singer et al. 1989). Minute designations are to the left of the bar.

Table 2   Hfr Strains

Line	Strain	Genotype	Source or construction
1	HfrH	PO1, *thi-1, relA1, spoT1, supQ80*	B. Bachmann
2	CAG12206	HfrH *nadA3052*::Tn*10kan*	P1(CAG12147) × HrfH
3	CAG5051	HfrH *nadA57*::Tn*10*	P1(CAG18341) × HfrH
4	KL227	PO3, *metB1, relA1*	B. Bachmann
5	CAG12204	KL227 *btuB3192*::Tn*10kan*	P1(CAG8408) × KL227
6	CAG5052	KL227 *btuB3191*::Tn*10*	P1(CAG3032) × KL227
7	KL208	PO43, *relA1*	B. Bachmann
8	CAG12203	KL208 *zbc-3105*::Tn*10kan*	P1(CAG12046) × KL208
9	CAG5053	KL208 *zbc-280*::Tn*10*	P1(CAG12123) × KL208
10	KL96	PO44, *relA1, thi-1*	B. Bachmann
11	CAG12202	KL96 *trpB3193*::Tn*10kan*	P1(CAG18579) × KL96
12	CAG5054	KL96 *trpB83*::Tn*10*	P1(CAG18458) × KL96
13	KL16	PO45, *relA1, thi-1*	B. Bachmann
14	CAG12200	KL16 *zed-3120*::Tn*10kan*	P1(CAG18451) × KL16
15	CAG5055	KL16 *zed-3069*::Tn*10*	P1(CAG18563) × KL16
16	KL228	PO13, *thi-1, leu-6, gal-6 lacY1 or lacZ4 supE44*	B. Bachmann
17	CAG12205	KL228 *zgh-3159*::Tn*10kan*	P1(CAG18164) × KL228
18	CAG8209	KL228 *zgh-3075*::Tn*10*	P1(CAG18574) × KL228
19	KL14	PO68, *leu, relA1*	B. Bachmann
20	CAG12201	KL14 *thi-3178*::Tn*10kan*	P1(CAG18500) × KL14
21	CAG8160	KL14 *thi-39*::Tn*10*	P1(CAG18616) × KL14

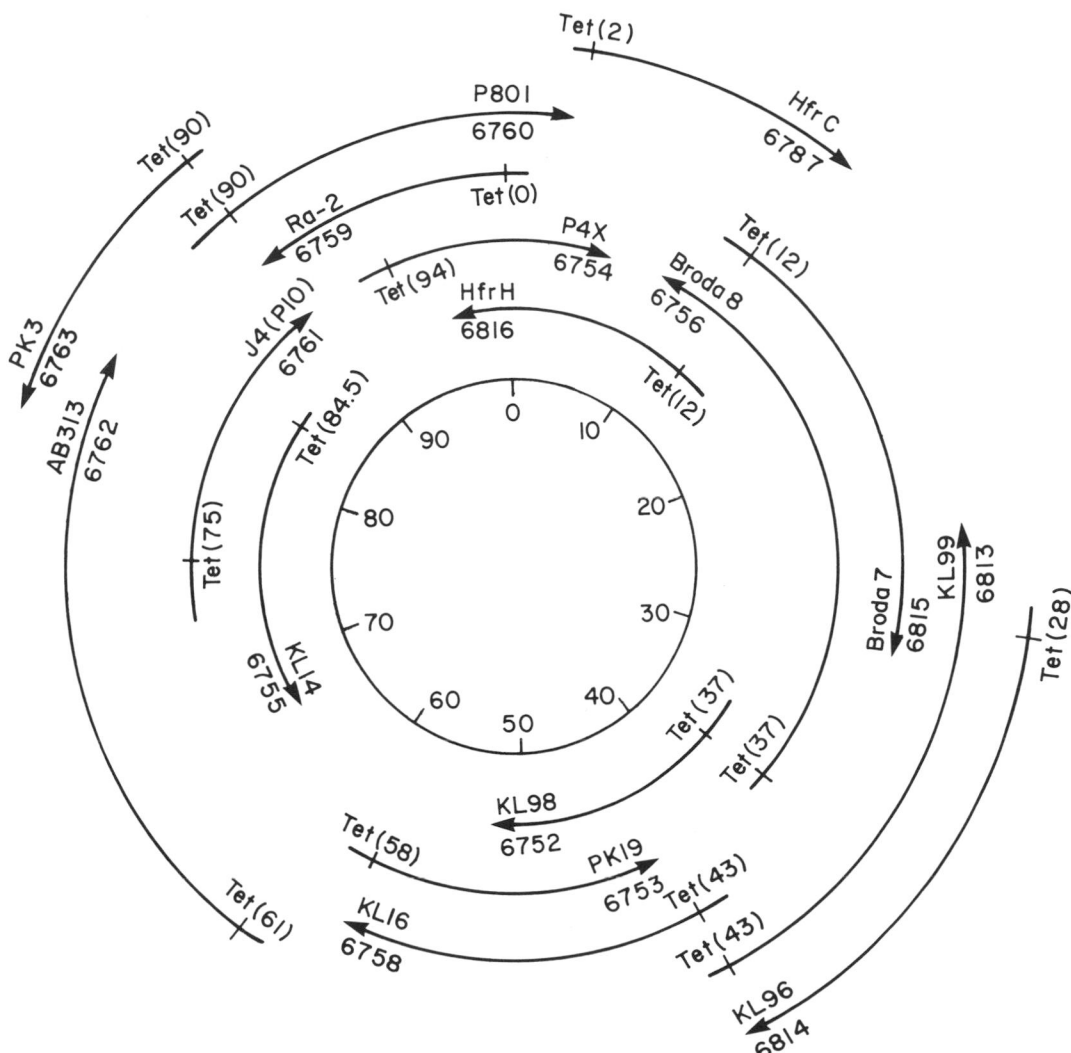

Figure 2 Hfr's from the CGSC collection. The Hfr point of origin is given above the line and the CGSC strain designation is given below the line.

SECTION 6

C. Additional Useful Strains for Recombinant DNA Experiments

Table 1 lists some important genes in *E. coli* host strains useful for recombinant DNA experiments, and Tables 2 and 3 list some useful strains for recombinant DNA experiments. Tables 1 and 2 are reprinted, with permission, from Kushner (*Escherichia coli and* Salmonella typhimurium. *Cellular and Molecular Biology* [ed. F.C. Neidhardt]. American Society for Microbiology, Washington, D.C., pp. 1226 and 1227, respectively [1987]). Table 3 is reprinted from Sambrook et al. (*Molecular Cloning. A Laboratory Manual*, Second Edition. Cold Spring Harbor Laboratory Press, Cold Spring Harbor, New York, Book 3, pp. A.9–A.13 [1989]). The references cited in Tables 2 and 3 can be found in the original publications.

Table 1   Important Genes in *E. coli* Host Strains

Gene	Product	Properties associated with mutant alleles	
		Advantages	Disadvantages
dam	Structural gene for DNA adenosine methylase	Deficient strains do not methylate GATC sequences. Permits use of such restriction enzymes as *Bcl* I and *Cla*I.	DNA isolated from mutants contains extra single-strand breaks
dcm	Structural gene for DNA cytosine methylase	Deficient strains do not methylate CCAGG or CCTGG sequences. Permits use of such restriction enzymes as *Eco*RII and *Stu*I.	None
galK	Structural gene for galactokinase	Deficiency permits use of galactose-linked expression vector systems.	None
hsdM	Modification subunit of *Eco*K restriction enzyme	No modification of heterologous DNA sequences.	Cloned DNA is not modified. Will be degraded when transformed into a *hsdR*⁺ host.
hsdR	Catalytic subunit of *Eco*K restriction enzyme	Deficient strains allow heterologous DNA sequences to be modified by the *hsdM*-encoded methylase. Cloned fragments are not degraded.	None
htpR	Structural gene for sigma factor controlling heat shock genes	Deficiency results in increased stability of a variety of proteins.	None yet reported
lacZ	Structural gene for β-galactosidase	Deficiency permits use of vectors employing X-Gal color selection.	None
lon	Structural gene for Lon protease	Deficiency results in increased stability of certain fusion proteins.	Cells are mucoid. Some plasmids are unstable in *lon* background. Mucoid phenotype can be suppressed by *gal* deletions.
mutD	Structural gene for DNA polymerase III subunit	Deficiency results in increased spontaneous mutation frequency.	Only useful for mutagenesis work.
pnp	Structural gene for polynucleotide phosphorylase	Deficiency results in increased stability of certain eucaryotic mRNAs. Increased expression of eucaryotic proteins.	Overproduction of certain eucaryotic proteins results in plasmid instability.
recA	Synaptase, specific protease	Deficiency in homologous genetic recombination stabilizes certain cloned DNA sequences. Prevents recombination between homologous sequences on chromosome and cloning vehicle.	Strains grow more slowly and transform at lower efficiencies.
recB, *recC*	Structural genes for exonuclease V	Deficient strains in the presence of *sbcB* mutations stabilize certain cloned palindromic sequences.	ColE1 plasmids are unstable in *recB sbcB* double mutants. Strains grow more slowly and are difficult to transform.
sbcB	Structural gene for exonuclease I	Deficient strains in the presence of *recB* and *recC* mutations stabilize certain cloned palindromic sequences.	ColE1 plasmids are unstable in *recB sbcB* double mutants.

Table 2   Useful Strains for Recombinant DNA Experiments[a]

Strain no.	Important genotype	Important characteristics	References
DH1	*hsdR17 endA1 recA1 gyrA96*	Restriction- and recombination-deficient, high-transforming strain.	23
ES1578	*mutD5*	Useful host of mutagenesis of cloned DNA fragments; *mutD* is cotransducible with *proA*.	13
HB101	*hsdS20 recA13*	Restriction-, modification-, and recombination-deficient host used in many early recombinant DNA experiments.	6
JC7623	*recB21 recC22 sbcB15*	Palindromic sequences appear more stable in this genetic background.	29
JC7689	*his⁺ sbcB15*	Useful for transferring *sbcB15* by cotransduction with *his*.	29
JC10241	*srlA::Tn10 recA⁺*	*recA* is closely linked to *srlA* by cotransduction.	12
JM103	*hsdR4 Δ(lac-pro)/F′ traD36 proAB lacI^q M15*	Excellent host for M13 cloning.	36
K165	*htpR165*	*htpR* can be cotransduced with *malT*.	10
LE392	*hsdR514 supE44 supF58*	A useful host for cosmid and bacteriophage cloning.	L. Enquist, personal communication
NK5992	*argA::Tn10*	*recB* and *recC* are closely linked by P1 transduction.	N. Kleckner, personal communication
SK107	*pnp-7 rna-19*	Increased expression of cloned eucaryotic genes.	24
SK484	Tn*10*	Tn*10* is closely linked to *proA* and *mutD*.	S. R. Kushner, unpublished data
SK1967	*serB22 zjj-202::Tn10*	*serB22* is 80% linked to Tn*10* insertion.	28
SK2251	*zgi::Tn10*	Contains Tn*10* closely linked to *his* and *sbcB*.	27
SK2881	*hsdR4 recA1 endA sbcB15*	Restriction- and recombination-deficient, high-transforming strain.	42
SG20332	*lon::Tn10 Δ16Δ17*	Nontransposable Tn*10* insertion in *lon*.	S. Gottesman, personal communication

[a] Representative strains carrying mutations of interest or the means of moving them between different genetic backgrounds.

Table 3 Bacterial Strains

Strain	Genotype	Remarks
71/18	*supE thi* Δ(*lac-proAB*) F′ [*proAB⁺ lacIq lacZ*ΔM15]	A strain used for growth of phagemids. It makes high levels of *lac* repressor and is used for inducible expression of genes that are under the control of the *lac* promoter. This strain can be used for detection of recombinants expressing β-galactosidase fusion proteins (Messing et al. 1977; Dente et al. 1983; Rüther and Müller-Hill 1983).
BB4	*supF58 supE44 hsdR514 galK2 galT22 trpR55 metB1 tonA* Δ*lac*U169 F′ [*proAB⁺ lacIq lacZ*ΔM15 Tn*10*(*tetr*)]	A *recA⁺* strain used for growth of λZAP and other λ bacteriophages. The F′ in this strain carries *lacZ*ΔM15, which permits α-complementation with the amino terminus of β-galactosidase encoded in λZAP. The F′ allows superinfection with an M13 helper bacteriophage, a step required for converting a recombinant λZAP to a pBluescript plasmid (Bullock et al. 1987).
BHB2688	(N205 *recA* [λ*imm*434 *c*Its *b2 red E*am *Sam*/λ])	A bacteriophage λ lysogen used to prepare packaging extracts (Hohn and Murray 1977; Hohn 1979).
BHB2690	(N205 *recA* [λ*imm*434 *c*Its *b2 red D*am *Sam*/λ])	A bacteriophage λ lysogen used to prepare packaging extracts (Hohn and Murray 1977; Hohn 1979).
BL21(DE3)	*hsdS gal* (λ*c*Its857 *ind*1 *Sam*7 *nin*5 *lac* UV5-T7 *gene 1*)	A strain employed for high-level expression of genes cloned into expression vectors containing bacteriophage T7 promoter. Bacteriophage T7 RNA polymerase is carried on the bacteriophage λ DE3, which is integrated into the chromosome of BL21 (Studier and Moffatt 1986).
BNN102 (C600*hflA*)	*supE44 hsdR thi-1 thr-1 leuB6 lacY1 tonA21 hflA*150[*chr::*Tn*10*(*tet*)]	An *hflA* strain used to select λgt10 recombinants. The high frequency lysogeny mutation suppresses plaque formation by *c*I⁺ bacteriophages but allows plaque formation by recombinant *c*I⁻ bacteriophages (Young and Davis 1983a).
C-1a	A wild-type strain.	A clone of *E. coli* strain C wild type maintained on minimal medium for several years. *E. coli* C is F⁻ and lacks host restriction and modification activity. It is a nonsuppressing host strain used in complementation tests with amber mutants of bacteriophage λ (Bertani and Weigle 1953; Borck et al. 1976).
C600 (BNN93)	*supE44 hsdR thi-1 thr-1 leuB6 lacY1 tonA21*	A suppressing strain often used for making lysates (Appleyard 1954) and for propagation of λgt10 (Young and Davis 1983a).
CES200	*sbcB15 recB21 recC22 hsdR*	A strain used for growth of Spi⁻ bacteriophages (Nader et al. 1985).
CES201	*recA sbcB15 recB21 recC22 hsdR*	A recombination-deficient strain used for growth of Spi⁻ bacteriophages (Wyman and Wertman 1987).
CJ236	*dut1 ung1 thi-1 relA1*/pCJ105(*cam*r F′)	A *dut⁻ ung⁻* strain used to prepare uracil-containing DNA for site-directed mutagenesis experiments (Kunkel et al. 1987). pCJ105 carries an F′ and *cam*r; growth of CJ236 in the presence of chloramphenicol selects for retention of the F′.
CSH18	*supE thi* Δ(*lac-pro*) F′ [*proAB⁺ lacZ⁻*]	A suppressing strain used to screen recombinants made in bacteriophage λ vectors carrying a *lacZ* gene in the stuffer fragments. These vectors give rise to blue plaques in the presence of the chromogenic substrate X-gal; recombinants in which the stuffer fragment has been replaced by foreign DNA give rise to white plaques (Miller 1972; Williams and Blattner 1979).
DH1	*supE44 hsdR17 recA1 endA1 gyrA96 thi-1 relA1*	A recombination-deficient suppressing strain used for plating and growth of plasmids and cosmids (Low 1968; Meselson and Yuan 1968; Hanahan 1983).

Table 3 *Continued*

Strain	Genotype	Remarks
DH5	*supE*44 *hsdR*17 *recA*1 *endA*1 *gyrA*96 *thi*-1 *relA*1	A recombination-deficient suppressing strain used for plating and growth of plasmids and cosmids (Low 1968; Meselson and Yuan 1968; Hanahan 1983). This strain has a higher transformation efficiency than DH1.
DH5α	*supE*44 Δ*lac*U169 (φ80 *lacZ*ΔM15) *hsdR*17 *recA*1 *endA*1 *gyrA*96 *thi*-1 *relA*1	A recombination-deficient suppressing strain used for plating and growth of plasmids and cosmids. The φ80 *lacZ*ΔM15 permits α-complementation with the amino terminus of β-galactosidase encoded in pUC vectors (Hanahan 1983; Bethesda Research Laboratories 1986).
DP50*supF*	*supE*44 *supF*58 *hsdS*3(r⁻_B m⁻_B) *dapD*8 *lacY*1 *glnV*44 Δ(*gal-uvrB*)47 *tyrT*58 *gyrA*29 *tonA*53 Δ(*thyA*57)	A strain used for isolation and propagation of bacteriophage λ recombinants (Leder et al. 1977; B. Bachmann, pers. comm.).
ED8654	*supE supF hsdR metB lacY gal trpR*	A suppressing strain commonly used to propagate bacteriophage λ vectors and their recombinants (Borck et al. 1976; Murray et al. 1977).
ED8767	*supE*44 *supF*58 *hsdS*3(r⁻_B m⁻_B) *recA*56 *galK*2 *galT*22 *metB*1	A recombination-deficient suppressing strain used for propagation of bacteriophage λ vectors (Murray et al. 1977).
HB101	*supE*44 *hsdS*20(r⁻_B m⁻_B) *recA*13 *ara*-14 *proA*2 *lacY*1 *galK*2 *rpsL*20 *xyl*-5 *mtl*-1	A suppressing strain commonly used for large-scale production of plasmids. It is an *E. coli* K12 × *E. coli* B hybrid that is highly transformable (Boyer and Roulland-Dussoix 1969; Bolivar and Backman 1979).
HMS174	*recA*1 *hsdR* rif^r	A recombination-deficient nonsuppressing strain used for high-level expression of genes cloned into expression vectors containing bacteriophage T7 promoter. Bacteriophage T7 RNA polymerase is provided by infection with a bacteriophage λ that carries bacteriophage T7 gene *1* (Campbell et al. 1978; Studier and Moffatt 1986).
JM101[a]	*supE thi* Δ(*lac-proAB*) F′ [*traD*36 *proAB*⁺ *lacI*^q *lacZ*ΔM15]	A strain that will support growth of vectors carrying amber mutations (Messing 1979).
JM105	*supE endA sbcB*15 *hsdR*4 *rpsL thi* Δ(*lac-proAB*) F′ [*traD*36 *proAB*⁺ *lacI*^q *lacZ*ΔM15]	A strain that will support growth of vectors carrying amber mutations and will modify but not restrict transfected DNA (Yanisch-Perron et al. 1985).
JM107[b]	*supE*44 *endA*1 *hsdR*17 *gyrA*96 *relA*1 *thi* Δ(*lac-proAB*) F′ [*traD*36 *proAB*⁺ *lacI*^q *lacZ*ΔM15]	A strain that will support growth of vectors carrying amber mutations and will modify but not restrict transfected DNA (Yanisch-Perron et al. 1985).
JM109[b,c]	*recA*1 *supE*44 *endA*1 *hsdR*17 *gyrA*96 *relA*1 *thi* Δ(*lac-proAB*) F′ [*traD*36 *proAB*⁺ *lacI*^q *lacZ*ΔM15]	A recombination-deficient strain that will support growth of vectors carrying amber mutations and will modify but not restrict transfected DNA (Yanisch-Perron et al. 1985).
JM110	*dam dcm supE*44 *hsdR*17 *thi leu rpsL lacY galK galT ara tonA thr tsx* Δ(*lac-proAB*) F′ [*traD*36 *proAB*⁺ *lacI*^q *lacZ*ΔM15]	A strain that will not modify *Bcl*I sites and will support growth of vectors carrying amber mutations (Yanisch-Perron et al. 1985).
K802	*supE hsdR gal metB*	A suppressing strain used to propagate bacteriophage λ vectors and their recombinants (Wood 1966).
KK2186	*supE sbcB*15 *hsdR*4 *rpsL thi* Δ(*lac-proAB*) F′ [*traD*36 *proAB*⁺ *lacI*^q *lacZ*ΔM15]	A strain that will support growth of vectors carrying amber mutations and will modify but not restrict transfected DNA (Zagursky and Berman 1984).

Strain	Genotype	Description
LE392	supE44 supF58 hsdR514 galK2 galT22 metB1 trpR55 lacY1	A suppressing strain commonly used to propagate bacteriophage λ vectors and their recombinants. LE392 is a derivative of ED8654 (Borck et al. 1976; Murray et al. 1977).
LG90	Δ(lac-proAB)	A strain in which lacZ is deleted that is used for detection of recombinants expressing β-galactosidase fusion proteins (Guarente and Ptashne 1981).
M5219	lacZ trpA rpsL (λbio252 cIts857 ΔH1)	A strain used for regulated expression of genes cloned downstream from the bacteriophage λ p_L promoter. It contains a defective λ prophage that encodes the bacteriophage λcIts857 repressor and N protein, which is an antagonist of transcription termination (Remaut et al. 1981; Shimatake and Rosenberg 1981).
MBM7014.5	hsdR2 mcrB1 zij202::Tn10(tet^r) araD139 araCU25am ΔlacU169	An mcrB strain used for λORF8 primary libraries. Libraries are made with DNA treated with methylases to protect HindIII and BamHI sites. M.AluI methylase is used to protect HindIII sites since M.HindIII methylase is not available commercially. This strain is defective in the restriction system that recognizes AluI-methylated DNA sites (Raleigh and Wilson 1986).
MC1061	hsdR mcrB araD139 Δ(araABC-leu)7679 ΔlacX74 galU galK rpsL thi	An mcrB strain used for λORF8 primary libraries as described for the strain MBM7014.5 (Meissner et al. 1987).
MM294	supE44 hsdR endA1 pro thi	A suppressing strain used for large-scale production of plasmids. It is highly transformable (Meselson and Yuan 1968).
MV1184[d]	ara Δ(lac-proAB) rpsL thi (φ80 lacZΔM15) Δ(srl-recA)306::Tn10(tet^r) F' [traD36 proAB$^+$ lacI^q lacZΔM15]	A recombination-deficient strain used to propagate phagemids pUC118/pUC119 and to obtain single-stranded copies of phagemids (Vieira and Messing 1987).
MV1193	Δ(lac-proAB) rpsL thi endA spcB15 hsdR4 Δ(srl-recA)306::Tn10(tet^r) F' [traD36 proAB$^+$ lacI^q lacZΔM15]	A recombination-deficient strain used to propagate phagemids pUC118/pUC119 and to obtain single-stranded copies of phagemids (Zoller and Smith 1987).
MZ-1	galKΔ8attLΔBamN$_7$N$_{53}$cIts857ΔH1 his ilv bio N$^+$	A temperature-sensitive lysogenic strain used as a host for plasmids containing the bacteriophage λ p_L promoter (Nagai and Thøgersen 1984).
NM531	supE supF hsdR trpR lacY recA13 metB gal	A recombination-deficient suppressing strain used for propagation of bacteriophage λ vectors (Arber et al. 1983).
NM538	supF hsdR trpR lacY	A strain used for assay and propagation of bacteriophage λ (Frischauf et al. 1983).
NM539	supF hsdR lacY (P2cox)	A strain used for selection of Spi$^-$ bacteriophages. NM539 is a derivative of NM538 (Frischauf et al. 1983).
Q358	supE hsdR φ80^r	A supE host used for growth of bacteriophage λ vectors (Karn et al. 1980).
Q359	supE hsdR φ80^r P2	A supE host used to select Spi$^-$ recombinants (Karn et al. 1980).
R594	galK2 galT22 rpsL179 lac$^-$	A nonsuppressing strain used as a nonpermissive host for vectors containing amber or ochre mutations (Campbell 1965).
RB791	W3110 lacI^{q}L8	A strain that makes high levels of lac repressor and is used for inducible expression of genes under the control of the lac and tac promoters (Brent and Ptashne 1981).
RR1	supE44 hsdS20(r_B^- m_B^-) ara-14 proA2 lacY1 galK2 rpsL20 xyl-5 mtl-1	A recA$^+$ derivative of HB101 that can be transformed with high efficiency (Bolivar et al. 1977; Peacock et al. 1981; B. Bachmann, pers. comm.).
SMR10	E. coli C (λcos2 ΔB xis1 red3 gamam210 cIts857 nin5 Sam7/λ)	A bacteriophage λ lysogen used to prepare packaging extracts (Rosenberg 1985).

Table 3 *Continued*

Strain	Genotype	Remarks
TAP90	supE44 supF58 hsdR pro leuB thi-1 rpsL lacY1 tonA1 recD1903::mini-tet	A host strain used for production of high-titer bacteriophage λ lysates. This restriction-deficient supE supF strain has a mini-tet insertion in recD, which improves growth of Spi⁻ λ bacteriophages (Patterson and Dean 1987).
TG1	supE hsdΔ5 thi Δ(lac-proAB) F′[traD36 proAB⁺ lacIq lacZΔM15]	An EcoK⁻ derivative of JM101 that neither modifies nor restricts transfected DNA. It will support growth of vectors carrying amber mutations (Gibson 1984).
TG2	supE hsdΔ5 thi Δ(lac-proAB) Δ(srl-recA)306::Tn10(tetr) F′[traD36 proAB⁺ lacIq lacZΔM15]	A recombination-deficient derivative of TG1 (M. Biggin, pers. comm.).
XL1-Blue	supE44 hsdR17 recA1 endA1 gyrA46 thi relA1 lac⁻ F′[proAB⁺ lacIq lacZΔM15 Tn10(tetr)]	A recombination-deficient strain that will support the growth of vectors carrying some amber mutations, but not those with the Sam100 mutation (e.g., λZAP). Transfected DNA is modified but not restricted. XL1-Blue is used to propagate λZAPII recombinants, which are unstable in BB4. The F′ in this strain allows blue/white screening on X-gal and permits bacteriophage M13 superinfection (Bullock et al. 1987).
XS101	recA1 hsdR rpoB331 F′[kan]	A recombination-deficient strain that modifies but does not restrict transfected DNA. It carries an episome conferring resistance to kanamycin and is used for growth of phagemids (Levinson et al. 1984).
XS127	gyrA thi rpoB331 Δ(lac-proAB) argE F′[traD36 proAB⁺ lacIq lacZΔM15]	A strain used for growth of phagemids (Levinson et al. 1984).
Y1089	araD139 ΔlacU169 proA⁺ Δlon rpsL hflA150[chr::Tn10(tetr)] pMC9	A strain used for protein production from λgt11 and λgt18–23 recombinants. Expression of the foreign protein is controlled by the high levels of lac repressor made by pMC9, which carries lacIq Y1089 is deficient in the lon protease, which may allow increased stability of the foreign proteins. Lysogens are formed at a high frequency in this strain (Young and Davis 1983b).
Y1090hsdR	supF hsdR araD139 Δlon ΔlacU169 rpsL trpC22::Tn10(tetr) pMC9	A strain used for immunological screening of expression libraries and propagation of λgt11 and λgt18–23 (Young and Davis 1983b; Jendrisak et al. 1987). Expression of the foreign protein is controlled by the high levels of lac repressor made by pMC9, which carries lacIq. Detection of proteins toxic to E. coli can be achieved by adding IPTG several hours after initiation of plaque formation. Some proteins are unstable in E. coli. Y1090hsdR is deficient in the lon protease, which may allow increased stability of antigens for antibody screening. The supF marker suppresses Sam100 to allow cell lysis (Young and Davis 1983b).
YK537	supE44 hsdR hsdM recA1 phoA8 leuB6 thi lacY rpsL20 galK2 ara-14 xyl-5 mtl-1	A recombination-deficient suppressing strain used for regulated expression of genes cloned downstream from the phoA promoter (Oka et al. 1985).

[a]Strain JM103 (Messing et al. 1981) is a restrictionless derivative of JM101 that has been used to propagate bacteriophage M13 recombinants. However, some cultivars of JM103 have lost the hsdR4 mutation (Felton 1983) and are lysogenic for bacteriophage P1 (which codes for its own restriction/modification system). JM103 is therefore no longer recommended as a host for bacteriophage M13 vectors. Strain KK2186 (Zagursky and Berman 1984) is genetically identical to JM103 except that it is nonlysogenic for bacteriophage P1.

[b]Strains JM106 and JM108 are identical to JM107 and JM109, respectively, except that they do not carry an F′ episome. These strains will not support the growth of bacteriophage M13 but may be used to propagate plasmids. However, JM106 and JM108 do not carry the lacI[q] marker (normally present on the F′ episome) and are therefore unable effectively to suppress the synthesis of potentially toxic products encoded by foreign DNA sequences cloned into plasmids carrying the lacZ promoter.

[c]Strains JM108 and JM109 are defective for synthesis of bacterial cell walls and form mucoid colonies on minimal media. This does not affect their ability to support the growth of bacteriophage M13.

[d]The original strain of MV1184, constructed by M. Volkert (pers. comm.), did not carry an F′ episome. However, the strain of MV1184 distributed by the Messing laboratory clearly carries an F′ episome. It is therefore advisable to check strains of MV1184 on their arrival in the laboratory for their ability to support the growth of male-specific bacteriophages.

SECTION·7

Clone Banks and Libraries

1. Kohara et al. λ library of entire E. coli chromosome

The library described by Kohara et al. (1987) consists of a set of ordered *E. coli* restriction fragments. The cloned DNA is on a nonlysogenic λ vector. A gene-mapping membrane containing a representative set of 476 Kohara clones is now commercially available through Takara Biochemical Inc. The membrane is prepared by immobilizing the phage clones on a sheet of positively charged nylon membrane. The membrane is suited for hybridization against ^{32}P-labeled DNA probes and has been used to map cloned *E. coli* genes. Individual samples from this set can be obtained from Drs. Donna Daniels and Fred Blattner (see address below). This library is derived from strain W3110, which has since been found to carry a 50-kb insertion (see, e.g., Section 2A of this Handbook). An entire set of 476 phage is also available, but arrangements must be made to go to Wisconsin to transfer each phage stock; write to Dr. Donna Daniels at the following address:

> Drs. Donna Daniels and Fred Blattner
> Department of Genetics
> University of Wisconsin-Madison
> Madison, WI 53705

2. Blattner library of entire E. coli chromosome

Blattner and co-workers (unpubl.) have constructed an ordered library of the entire *E. coli* chromosome using various Charon phages. Individual members, as well as the whole collection, will be available from Dr. Donna Daniels (at the address above) under the same conditions that apply for the Kohara collection.

3. Clarke-Carbon bank of hybrid plasmids

This bank was prepared by using sheared *E. coli* DNA (average size 14 kb) and cloning them into a ColE1 vector with AT tailing (Clarke and Carbon 1975, 1976). Individual clones can be obtained by writing to:

> *E. coli* Genetic Stock Center
> Department of Biology, 255 OML
> Yale University, P.O. Box 6666
> New Haven, CT 06511-7444

Correlations of genes in this library with known proteins have been compiled by Phillips et al. (1987), VanBogelen et al. (1990), and VanBogelen and Neidhardt (1991) (see Sections 2D and 8B of this Handbook). Nishimura et al. (1992) have correlated a subset of 518 pLC plasmids to the physical map of *E. coli* K12 (see Section 2D of this Handbook).

REFERENCES

Clarke, L. and J. Carbon. 1975. Biochemical construction and selection of hybrid plasmids containing specific segments of the *Escherichia coli* genome. *Proc. Natl. Acad. Sci.* **72:** 4361–4365.

Clarke, L. and J. Carbon. 1976. A colony bank containing synthetic ColE1 hybrid plasmids representative of the entire *E. coli* genome. *Cell* **9:** 91–99.

Kohara, Y., K. Akiyama, and K. Isono. 1987. The physical map of the whole *E. coli* chromosome: Application of a new strategy for rapid analysis and sorting of a large genomic library. *Cell* **50:** 495–508.

Nishimura, A., K. Okiyama, Y. Kohara, and K. Horiuchi. 1992. Correlation of a subset of the pLC plasmids to the physical map of *Escherichia coli* K-12. *Microbiol. Rev.* **56:** 137–151.

Phillips, T.A., V. Vaughn, P.L. Bloch, and F.C. Neidhardt. 1987. Gene-protein index of *Escherichia coli* K-12, edition 2. In Escherichia coli *and* Salmonella typhimurium. *Cellular and Molecular Biology* (ed. F.C. Neidhardt), pp. 919–966. American Society for Microbiology, Washington, D.C.

VanBogelen, R.A. and F.C. Neidhardt. 1991. The gene-protein database of *Escherichia coli*: Edition 4. *Electrophoresis* **12:** 955–994.

VanBogelen, R.A., M.E. Hutton, and F.C. Neidhardt. 1990. Gene-protein database of *Escherichia coli* K-12: Edition 3. *Electrophoresis* **11:** 1131–1166.

SECTION·8

Databases

A. DNA sequence banks

B. Gene-protein database of *E. coli*

C. Bibliographic database

REFERENCES

Benton, D. 1990. Recent changes in the GenBank On-line Service. *Nucleic Acids Res.* **18:** 1517–1520.

Bairoch, A. 1991. PROSITE: A dictionary of sites and patterns in proteins. *Nucleic Acids Res.* **19:** 2241–2245.

Bairoch, A. and B. Boeckmann. 1991. The SWISS-PROT protein sequence data bank. *Nucleic Acids Res.* **19:** 2247–2249.

Burks, C., M. Cassidy, M.J. Cinkosky, K.E. Cumella, P. Gilna, J.E.-D. Hayden, G.M. Keen, T.A. Kelley, M. Kelly, D. Kristofferson, and J. Ryals. 1991. GenBank. *Nucleic Acids Res.* **19:** 221–225.

Kröger, M., R. Wahl, P. Rice. 1991. Compilation of DNA sequences of *Escherichia coli* (update 1991). *Nucleic Acids Res.* **19:** 2023–2043.

Stöhr, P.J. and G.N. Cameron. 1991. The EMBL data library. *Nucleic Acids Res.* **19:** 2227–2230.

VanBogelen, R.A. and F.C. Neidhardt. 1991. The gene-protein database of *Escherichia coli*: Edition 4. *Electrophoresis* **12:** 955–994.

VanBogelen, R.A., M.E. Hutton, and F.C. Neidhardt. 1990. Gene-protein database of *Escherichia coli* K-12: Edition 3. *Electrophoresis* **11:** 1131–1166.

SECTION 8

A. DNA Sequence Banks

1. *GenBank*

The Genetic Sequence Data Bank provides access to all published DNA and RNA sequences, as well as much unpublished sequence data. For a recent review of all aspects of GenBank, see Burks et al. (1991). Written questions about access to the data should be addressed to:

GenBank
IntelliGenetics, Inc.
700 East El Camino Real
Mountain View, CA 94040
(e-mail: genbank@genbank.bio.net; FAX 415-962-7302; telephone 415-962-7364)

Questions about submitting new data should be addressed to:

GenBank
Theoretical Biology and Biophysics Group
T-10, MS K710
Los Alamos National Laboratory
Los Alamos, NM 87545
(e-mail: genbank@life.lanl.gov; FAX 505-665-3493; telephone 505-665-2177)

The GenBank On-line Service (see also Benton 1990) gives access to GenBank, EMBL, and GenPept (protein translation of GenBank) data either through direct login or by e-mail (Burks et al. 1991). For information on e-mail sequence retrieval from GenBank, EMBL, GenPept, and SWISS-PROT, send a message with the word "HELP" to retrieve@genbank.bio.net. For FASTA similarity searches send a HELP message to search@genbank.bio.net.

2. *The EMBL Data Library*

The European Molecular Biology Laboratory Data Library maintains and distributes a nucleotide sequence database and also distributes a protein sequence database (SWISS-PROT; Bairoch and Boeckmann 1991) as well as other databases (see Stöhr and Cameron 1991). These include the PROSITE pattern database (Bairoch 1991), the ENZYME database of EC nomenclature (Blairoch 1990 as cited in Stöhr and Cameron 1991), and the ECD-*E. coli* map database (Kröger et al. 1991). The EMBL Library can be contacted as follows:

Network:

datasubs@embl.bitnet (for data submission); datalib@embl.bitnet (for questions requiring a personal response); NETSERV@EMBL-Heidelberg.DE (Internet) (for a general Help message)

Postal Address:

Data Submissions
EMBL Data Library
Postfach 10.2209
6900 Heidelberg, Germany
(FAX [49]-6221-387519 or 387306; telephone [49]-6221-387258; telex 461613 [embl d])

3. *General*

For a more complete description of the various databases and their access by electronic mail or FTP, contact Dr. Peter Markiewicz for a hypercard stack (Macintosh) at: JMILLER%BIOVX1.SPAN@STAR.STANFORD.EDU

SECTION 8

B. Gene-Protein Database of *E. coli*

The gene-protein database of *E. coli*, compiled by Neidhardt and co-workers, has as its core an index that links each of the protein spots from a 2-D polyacrylamide gel to the gene that encodes the protein (see also Sections 2D and 7 of this Handbook). Additional information about each protein and its gene is generated from 2-D gel analysis or collated from the literature to form the database. Earlier editions of the database have provided periodic updates of information (i.e., VanBogelen et al. 1990). The current edition (VanBogelen and Neidhardt 1991) not only does this, but also introduces a new reference gel image produced by a new electrophoresis system. The new gel format is more compatible with computer-assisted image analysis. The new edition continues the use of the former reference gel images but adds a reference image of an equilibrium gel of *E. coli* strain W3110 produced by the new standardization gel system.

For additional information, contact:

Dr. Frederick C. Neidhardt
Department of Microbiology and Immunology
University of Michigan
Ann Arbor, MI 48109-0620
(FAX 313-764-3562; telephone 313-763-1209)

C. Bibliographic Database

Medline is the National Library of Medicine's (NLM) bibliographic database covering medicine, nursing, dentistry, veterinary medicine, and the preclinical sciences dating back to 1966.

Information about Medline may be obtained from the MEDLARS service desk. Hours: Monday–Friday 8:30 AM–5:00 PM ET

MEDLARS Management Section
National Library of Medicine
8600 Rockville Pike
Bethesda, MD 20894
(Telephone 800-638-8480 or in Maryland 301-496-6193)

SECTION·9

Phage

A. Phage λ maps

B. Induction

C. Replication

D. *E. coli* hosts

E. General properties

F. Specialized transducing phages

G. Vectors

REFERENCES

Arber, W., L. Enquist, B. Hohn, N.E. Murray, and K. Murray. 1983. Experimental methods for use with lambda. In *Lambda II* (ed. R.W. Hendrix et al.), pp. 433–466. Cold Spring Harbor Laboratory, Cold Spring Harbor, New York.

Daniels, D.L., J.L. Schroeder, W. Szybalski, F. Sanger, and F.R. Blattner. 1983a. A molecular map of coliphage lambda. In *Lambda II* (ed. R.W. Hendrix et al.), pp. 469–517. Cold Spring Harbor Laboratory, Cold Spring Harbor, New York.

Daniels, D.L., J.L. Schroeder, W. Szybalski, F. Sanger, and F.R. Blattner. 1990. A molecular map of coliphage lambda. In *Genetic Maps*, Fifth Edition, Book 1 "Viruses" (ed. S.J. O'Brien), pp. 1.3–1.23. Cold Spring Harbor Laboratory Press, Cold Spring Harbor, New York.

Daniels, D.L., J.L. Schroeder, W. Szybalski, F. Sanger, A.R. Coulson, G.F. Hong, D.F. Hill, G.B. Petersen, and F.R. Blattner. 1983b. Complete annotated lambda sequence. In *Lambda II* (ed. R.W. Hendrix et al.), pp. 519–676. Cold Spring Harbor Laboratory, Cold Spring Harbor, New York.

Dunn, I.S. and F.R. Blattner. 1987. Charons 36–40: Multi-enzyme, high-capacity, recombination-deficient replacement vectors with polylinkers and polystuffers. *Nucleic Acids Res.* **15:** 2677–2698

Furth, M.E. and S.H. Wickner. 1983. Lambda DNA replication. In *Lambda II* (ed. R.W. Hendrix et al.), pp. 145–173. Cold Spring Harbor Laboratory, Cold Spring Harbor, New York.

Hershey, A.D. and W. Dove. 1971. Introduction to lambda. In *The Bacteriophage Lambda* (ed. A.D. Hershey), pp. 3–11. Cold Spring Harbor Laboratory, Cold Spring Harbor, New York.

Murray, N.E. 1983. Phage lambda and molecular cloning. In *Lambda II* (ed. R.W. Hendrix et al.), pp. 395–432. Cold Spring Harbor Laboratory, Cold Spring Harbor, New York.

Murray, N.E. 1991. Special uses of lambda phage for molecular cloning. *Methods Enzymol.* **204:** 280–301.

Roberts, J.W. and R. Devoret. 1983. Lysogenic induction. In *Lambda II* (ed. R.W. Hendrix et al.), pp. 123–144. Cold Spring Harbor Laboratory, Cold Spring Harbor, New York.

Sambrook, J., E.F. Fritsch, and T. Maniatis. 1989. *Molecular Cloning. A Laboratory Manual*, Second Edition. Cold Spring Harbor Laboratory Press, Cold Spring Harbor, New York.

Sanger, F., A.R. Coulson, G.F. Hong, D.F. Hill, and G.B. Petersen. 1982. Nucleotide sequence of bacteriophage lambda DNA. *J. Mol. Biol.* **162:** 729–773.

Young, R.A. and R.W. Davis. 1983. Efficient isolation of genes by using antibody probes. *Proc. Natl. Acad. Sci.* **80:** 1194–1198.

Weisberg, R.A. 1987. Specialized transduction. In Escherichia coli *and* Salmonella typhimurium. *Cellular and Molecular Biology* (ed. F.C. Neidhardt), pp. 1169–1176. American Society for Microbiology, Washington, D.C.

SECTION 9

A. Phage λ Maps

The genetic and restriction maps and the DNA sequence of phage λ are available in several forms. The map of bacteriophage λ DNA in Figure 1 is reprinted from Daniels et al. (*Lambda II* [ed. R.W. Hendrix et al.]. Cold Spring Harbor Laboratory, Cold Spring Harbor, New York, p. 473 [1983a]). All of the sites on the λ map are presented in tabular form in Daniels et al. (1983a, 1990). These tables also list the DNA sequence coordinates and restriction sites, together with the appropriate references. The complete sequence of phage λ, originally eludicated by Sanger et al. (1982), is presented in annotated form in Daniels et al. (1983b).

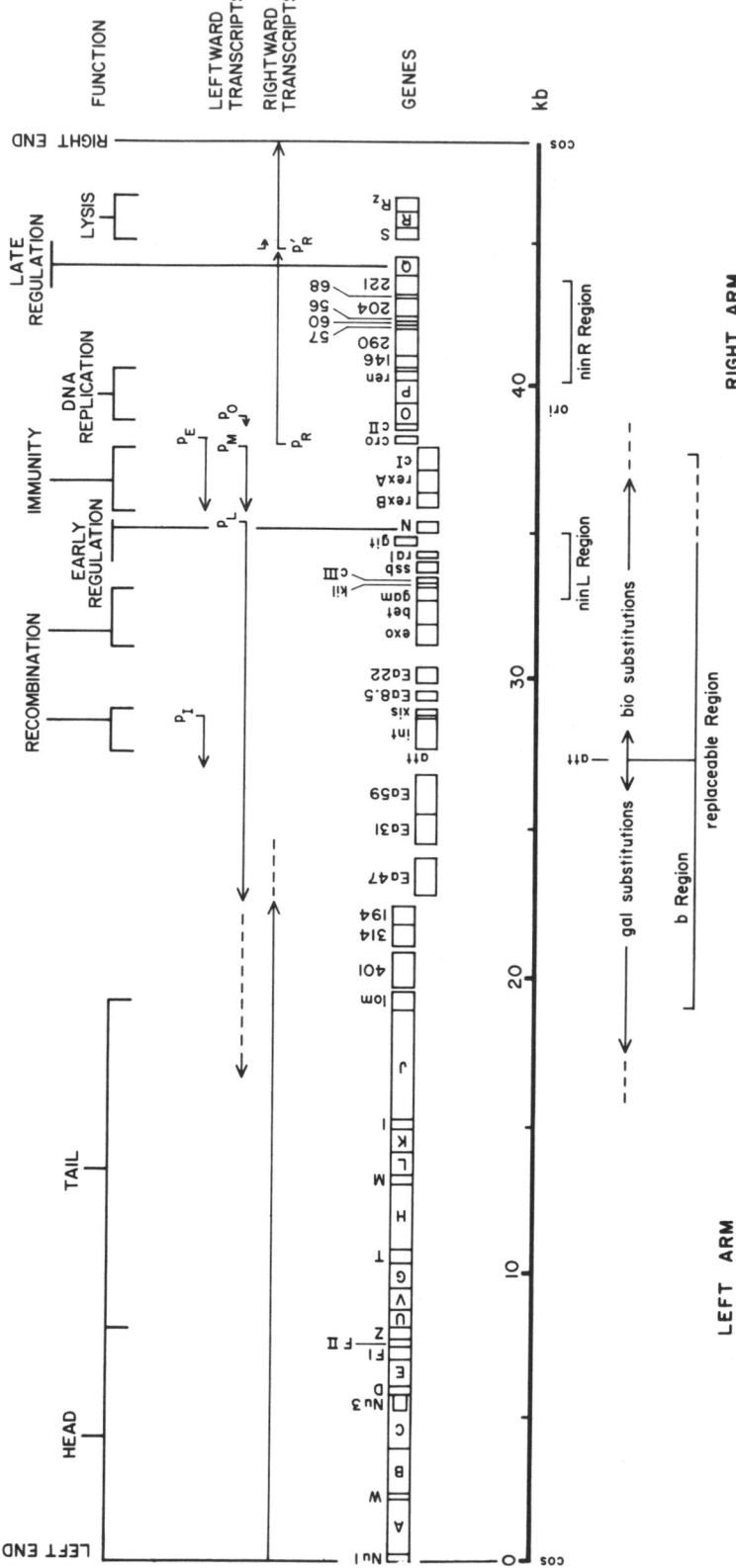

Figure 1 Map of phage λ DNA. A scale drawing of the molecular map presented in Table 1 in Daniels et al. (1983a) is presented above the kilobase scale, beginning and ending at the cohesive-end site (cos). Gene clusters with related functions are indicated above the brackets, and the regulatory genes N (involved in early regulation) and Q (involved in late regulation) are indicated by vertical lines. Known promoters are denoted by p with a subscript to indicate their unique points of origin. (p_I) int protein promoter; (p_E) establishment promoter for cI; (p_M) maintenance promoter for cI; (p_L) major leftward promoter; (p_R) major rightward promoter; (p_O) oop promoter; (p'_R) late promoter. (→) Extent and direction of transcription; (---) readthrough. Map positions of genes as determined by analysis of open reading frames (ORFs) in the DNA sequence are indicated. Known genes (identified either genetically or functionally by mutation analysis or functionally by SDS gel electrophoresis of protein product, or both) are indicated with letter names, whereas ORFs, presumed from the sequence to code for protein products but with no previously known gene assignments, are given numbers corresponding to the coding capacity of the ORF. Major areas of substitution and deletion mutations are indicated below the scale. (att) Attachment site; (ori) origin of replication.

SECTION 9

B. Induction

Tables 1 and 2 are reprinted from Roberts and Devoret (*Lambda II* [ed. R.W. Hendrix et al.]. Cold Spring Harbor Laboratory, Cold Spring Harbor, New York, pp. 124 and 126, respectively [1983]). The references cited in the table notes can be found in the original paper.

Table 1   Inducibility of Some Prophages in Three Enterobacteriaceae

Enterobacteriaceae	Inducible prophages	Noninducible prophages
Escherichia coli	λ, 434, 21, 80, 82, 170, 186[a], 424[b]	18, Mu, 62, 299
Salmonella typhimurium	P22, L	
Shigella dysenteriae	Pl[c]	P2, P4

[a]Fully inducible by UV light, prophage 186 does not display zygotic induction, because the conjugating female is temporarily refractory to phage infection. The delay enables repression to be established before the cell recovers its sensitivity to phage 186 infection (Woods and Egan 1981).

[b]Prophage 424 has properties similar to phage 186 (P. Morand and R. Devoret, unpubl.).

[c]Data from Melechen and Skaar (1962).

Table 2   Examples of Agents or Treatments That Induce Prophage λ

Agents that damage DNA
  physical carcinogens: UV light, X-rays, α-rays
  activated chemical carcinogens: benzapyrene, 7,12-dimethylbenzanthracene, aflatoxins
  antitumor drugs: bleomycin, daunorubicin, mitomycin C, neocarzinostatin
Agents or treatments that disrupt DNA replication
  antifolates: aminopterin, trimethoprim
  antigyrase drugs: nalidixic acid, oxolinic acid
  inhibitors of thymidylate synthetase synthesis: 5-fluorouracil
  thymine deprivation of *thy*⁻ mutants
  growth of mutants in nonpermissive conditions
Introduction of replicons into the lysogen
  UV-damaged plasmids and phages: F sex factor, colicinogenic factor I, R factors, phageP1
  intact plasmids: *lynR* mutants of mini F

For general references, see Heinemann (1971), Borek and Ryan (1973), Moreau and Devoret (1977), Bradner (1978), and Anderson et al. (1980).

SECTION 9

C. Replication

Table 1 is reprinted from Furth and Wickner (*Lambda II* [ed. R.W. Hendrix et al.]. Cold Spring Harbor Laboratory, Cold Spring Harbor, New York, p. 147 [1983]).

Table 1 Proteins Required for λ DNA Replication

Gene	Essential for replication of		Biochemical function
	E. coli	λ	
E. coli initiation proteins			
dnaA	yes	no	unknown
dnaC	yes	no	interacts physically and functionally with *dnaB* protein; involved in prepriming reaction
dnaB	yes	yes[a]	ribonucleoside triphosphatase activity stimulated by single-stranded DNA; involved in prepriming reaction
rpoB	yes	yes[a]	RNA polymerase β-subunit; transcriptional activation of replication
E. coli elongation proteins			
dnaC	yes	no	see above
dnaB	yes	yes[a]	see above
dnaG	yes	yes[a]	primase, synthesizes ribonucleotides, deoxyribonucleotides, and mixed oligonucleotides (primers for DNA synthesis)
ssb	yes	n.t.[a,b]	binds to single-stranded DNA
dnaE	yes	yes	α-component of DNA polymerase III
dnaZ	yes	yes	γ-component of DNA polymerase III elongation complex
dnaN	yes	n.t.[b]	β-component of DNA polymerase III elongation complex
gyrA, gyrB	yes	n.t.[a]	subunits of DNA gyrase, catalyzes ATP-dependen negative supercoiling of relaxed closed circular DNA
lig	yes	yes	DNA ligase, covalently closes nicks in double-stranded DNA
polAex	yes	n.t.	5'-3' exonuclease of DNA polymerase I
dnaJ	yes	yes	unknown
dnaK	yes	yes	DNA-independent ATPase
dnaQ	yes	n.t.	ε-component of DNA polymerase III elongation complex
dnaX	yes	yes	δ-component of DNA polymerase III holoenzyme
dnaY	yes	yes	unknown
E. coli dispensable proteins			
grpD	no	yes	unknown
grpE	no	yes	unknown
λ replication proteins			
O	no	yes[a]	origin-specific DNA-binding protein
P	no	yes[a]	interacts with *dnaB* protein

[a]Shown to be required in vitro.
[b]n.t. indicates not tested.

9.6

D. *E. coli* Hosts

Table 1 is reprinted from Arber et al. (*Lambda II* [ed. R.W. Hendrix et al.]. Cold Spring Harbor Laboratory, Cold Spring Harbor, New York, pp. 434–435 [1983]). The references cited in the table and table notes can be found in the original paper.

Table 1   Some *E. coli* Strains Commonly Used as Hosts for Phage λ

Strain	Partial genotype	References	Main features and use
C600[a]	*supE tonA lacY*	Appleyard (1954)	permissive indicator; host for lysates
5K[a]	*supE tonA lacY hsdR*	Hubacek and Glover (1970)	r^-m^+ permissive indicator; good transfection host
R594[a]	*sup⁰*	Campbell (1965)	nonpermissive host for λ*am* phage
W3101[a]	*sup⁰*	Bachmann (1972)	nonpermissive host for λ*am* phage
Y-mel[a]	*supF*	see Goldberg and Howe (1969)	indicator for λ*Sam7*
N1323(λ)[b]	*lop8* (λ*cI857Sam7*)	see Gottesman et al. (1973) for *lop8*	source of λ DNA
WA803[a]	*supE hsdS*	Wood (1966)	r^-m^-; indicator and transfection host
ED8654[a]	*supE supF hsdR trpR lacY*	Borck et al. (1976)	r^-m^+; indicator and transfection host
NM531[b]	*supE supF hsdR trpR lacY recA13*	N.E. Murray (pers. comm.)	as ED8654 but Rec⁻
ED8767[b]	*supE supF hsdS lacY recA56*	Murray et al. (1977)	r^-m^- Rec⁻ indicator; good for cosmids
NM538[b,c]	*supF hsdR trpR lacY*	A.-M. Frischauf et al. (in prep.)	r^-m^+ Su⁺ host; good for *Sam7* and *Aam Bam* phages
NM539[b,c]	*supF hsdR lacY* (P2cox3)	A.-M. Frischauf et al. (in prep.)	derivative of NM538; selection of λ Spi⁻ for better growth of *cI*⁺ phages
SM32[g]	*lon∇galE sulA strA*	Mizusawa and Ward (1982)	
NM514[b]	*hsdR lyc7* (*lyc7* is an *hfl* allele)	for *lyc7*, see Lecocq and Gathoye (1973)	r^-m^+; selection of λ*imm*⁴³⁴ *cI*⁻ (see Plaque Morphology Mutants)
JC8679[a,b,c]	*recB21recC22sbcA23*	Gillen et al. (1981)	host for *red*⁻ and Spi⁻ Chi⁰ phage
NM519[a,b,c]	*recB21recC22sbcA23hsdR*	N.E. Murray (pers. comm.)	r^-m^+; for recovery of Spi⁻ Chi⁰ phage
GC507[h]	*supE supF hsdS trpR met recA13/pgam ∇1 bla*	G.F. Crouse (pers. comm.)	Rec⁻ host for Spi⁻ phage (see Recombination-deficient Mutants; Fec⁻ and Spi⁻ Phenotypes
K388[e]	*rpoB* (Snu1)	Baumann and Friedman (1976)	scoring *nin5* and Chi
LE289[f]	(λ[*int-FII*]) *galT supF*	L. Enquist and R. Weisberg (1976 and pers. comm.)	scoring *xis* or *int*
WA2127[a]	*ptsM supE hsdS tonA lac*	Elliott and Arber (1978)	enriching for phage with longer genomes
LE30[f]	*mutD5*	Enquist and Weisberg (1977)	mutator host
BHB2688[d]	*recA* (λ*imm*⁴³⁴*cIts b2red3Eam4Sam7*)/λ	Hohn (1979)	in vitro packaging extracts
BHB2690[d]	*recA* (λ*imm*⁴³⁴*cIts b2red3Dam15Sam7*)/λ		

[a] From B. Bachmann (Yale University School of Medicine, New Haven, Connecticut.
[b] From N.E. Murray (University of Edinburgh, Scotland).
[c] Note that λ*Aam32* phages and, hence, Charon 4A, grow poorly on *recBC supE* hosts and λ*Aam32 Bam1 sbam*⁰ phages grow poorly on *supE* hosts.
[d] From B. Hohn (Friedrich Miescher Institut, Basel, Switzerland).
[e] From D. Friedman (University of Michigan School of Medicine).
[f] From L. Enquist (Molecular Genetics, Inc., Minnetonka, Minnesota).
[g] From D.F. Ward (Uniform Services University of the Health Sciences, Bethesda, Maryland).
[h] From G.F. Crouse (National Cancer Institut, Frederich, Maryland).

SECTION 9

E. General Properties

Table 1 is reprinted from Hershey and Dove (*The Bacteriophage Lambda* [ed. A.D. Hershey]. Cold Spring Harbor Laboratory, Cold Spring Harbor, New York, p. 8 [1971]). The references and chapter cited in the table notes can be found in the original publication.

Table 1   Properties of Phage λ and Some Relatives

Phage	Immunity	Host range	Tail antigen	Prophage location[a]	N gene product	DNA ends
λ	unique[b]	—	—	*gal–bio*[c]	—	—
21	unique[b]	like λ[b]	like λ[b]	near *trp*[b]	unlike λ[d]	like λ[e]
φ80	unique[f]	like T1[f]	unlike λ[f]	*tdk–trp*[c,g]	unlike λ[h]	like λ[i]
φ81	unique[j]	like T1[j]	?	*gal–bio*[j]	?	like λ[i]
82	unique[b]	like 434[b]	like λ[b]	*gal–bio*[c]	?	?
424	unique[b]	like 434[k]	like λ[b]	near *his*[b]	?	like λ[e]
434	unique[b]	unlike λ[b]	like λ[b]	*gal–bio*[c]	like λ[d]	like λ[e]

[a] Where two bacterial loci are mentioned, the prophage is inserted between them. For more detail, see Taylor (1970) and Chapter 2.
[b] Jacob and Wollman (1961)
[c] Taylor (1970).
[d] Thomas (1970).
[e] Baldwin et al. (1966).
[f] Matsushiro (1963).
[g] Igarashi et al. (1967).
[h] Szpirer and Brachet (1970).
[i] Yamagishi et al. (1965).
[j] Takeda et al. (1970).
[k] J. S. Parkinson (personal communication).

SECTION 9

F. Specialized Transducing Phages

Table 1 is reprinted, with permission, from Weisberg (Escherichia coli *and* Salmonella typhimurium. *Cellular and Molecular Biology* [ed. F.C. Neidhardt]. American Society for Microbiology, Washington, D.C., p. 1171 [1987]). The references cited in the table and table notes can be found in the original paper.

Table 1   *E. coli* Specialized Transducing Phages[a]

Map position	Gene	Reference(s)	Map position	Gene	Reference(s)
00–01	deo-car	21, 22, 46, 49, 51, 54, 87, 146, 152, 176, 187, 191	52	lig	14, 23, 56
			53	dapE-purC	141
01–02	ara-envA	16, 32, 41, 58, 81, 92, 101, 112, 115, 116, 139, 156	54	upp-hisS	40, 155, 180
			55	glyA	170
04	glnD-polC	12, 45, 111, 152, 158	57	grpE-pheA	27, 60, 99, 149, 195
05	dnaQ	119	58	recA-srl	15, 124, 125
06	proB	156	59	cysDJ	39, 78, 156
07	argF	95	60	pyrG-relA	47
08	lac	10, 80, 140	60	fuc	163
09	phoA-proC	18, 150	61	galR-recC	14, 61, 72, 142, 156
09	nusB-tsx	52, 173	61	argA	34
10	lon	129; M. Maurizi, P. Trisler, and S. Gottesman, manuscript in preparation	65	metC	156
			67	rpsU-rpoD	64
			68	exuR-uxa	121, 145
11	dnaZ	182	69	nusA	110, 129; S. Adhya, J. Levin, and R. Haber, personal communication
12	purE	156			
15	lip-leuS	167			
16–18	glnS-uvrB	1, 11, 36, 55, 79, 93, 131, 137, 144, 166, 188, 190, 192, 193	71	rpsI-rplM	82
			72–73	aroE-rpsL	83, 85, 152
20	serC-hip	96, 99	74	crp-cysG	31; S. Adhya, personal communication
21	aspC-asnS	136			
22	divE	174	75	ompR-envZ	147, 175
24	fla-rne	103, 143	75	malT-malQ	73
26	fadR	33	75	glpRD	153
27–28	tyrT-cysB	3, 14, 120, 122	81	rpm-spoT	2, 78, 99
29	nirR	154	83–84	gyrB-asnA	14, 67, 68, 91, 126, 127, 132, 134, 165, 181, 194
29	tyrR	28			
30	recE-sbcA	42, 53, 57, 90, 186	84	rbs	113
35–36	fumA-manA	65	84–85	rrnC-rho	9, 62, 66, 86, 88, 130, 152, 156, 157, 179
38	infC-himA	70, 71, 128, 168, 169			
41	ruv	159	87	polA-glnA	94
41–43	fla-hag; resA	59, 102, 105, 106, 162	88–89	cytR-metF	63, 80, 84, 109
42	uvrC	6	90–91	argE-rpoC	77, 97, 98, 107, 123, 189
43	serU	14, 171	92	malG-lamB	118
44	his-gnd	7, 14, 138	92	uvrA	6
47–48	ompC-resBC	151; J. Brill and S. Gottesman, personal communication	94	mop-frdA	29, 35, 50, 69, 177
			95	leuX	192, 193
48–49	gyrA-glpT	108	97	valS-pyrE	95
50	purF	164	98	fimD	43
51	dsdA	13	99	dnaC	74, 75
52	crr	19, 20	99	hsd	14, 148

[a] Map coordinates are, for the most part, taken from the review of Bachmann (8). A pair of genes separated by a hyphen (for example, *deo-car*) indicates that the entire segment bounded by and including the genes in question has been incorporated into one or more transducing phage lines.

SECTION 9

G. Vectors

Phage λ vectors have been employed extensively in recombinant DNA research. For a complete description of these vectors, see Murray (1983, 1991) and Sambrook et al. (1989). Tables 1 and 2 list some of the more widely used λ-based vectors, and Figures 1 and 2 and Tables 3 and 4 depict two specific vectors: λgt11 and Charon 40. λgt11 carries a unique *Eco*RI restriction site in *lacZ*, which permits immunological screening for a fusion protein resulting from the in-frame insertion of a DNA fragment; an amber mutation in the λ *S* gene permits accumulation of high amounts of the fusion protein. An amber suppressing host (*supF*) is required for propagation of the phage. Table 3 and Figure 1 describe λgt11.

Charon 40 employs a polystuffer which is cleaved with *Nae*I, facilitating insertion of up to 24 kb at any of a set of restriction sites. Table 4 and Figure 2 (for a two-copy polystuffer) describe this vector (see also Dunn and Blattner 1987; Sambrook et al. 1989).

Tables 1 and 2 are reprinted, with permission, from Murray (*Methods Enzymol.* vol. 204, pp. 292 and 297, respectively [1991]), and Tables 3 and 4 and Figures 1 and 2 are reprinted from Sambrook et al. (*Molecular Cloning. A Laboratory Manual*, Second Edition. Cold Spring Harbor Laboratory Press, Cold Spring Harbor, New York, pp. 2.43, 2.31, 2.42, and 2.30, respectively [1989]). The references cited in Tables 1 and 2 can be found in the original paper.

Table 1   Insertion Vectors[a]

Vector	Space	Cloning sites	Genotype of recombinants	Biological features	Refs.
λNM641	10.7	*Eco*RI	*att⁻ int⁻ red⁻ imm⁴³⁴ cI⁻ nin5 χ°*	Selection on Hfl⁻; *red⁻ gam⁺*	33, 34, 35
λgt10	6.7	*Eco*RI	*att⁻ imm⁴³⁴ cI⁻ χ°*	Selection on Hfl⁻	7, 36
λNM1149	10.3	*Eco*RI, *Hin*dIII	*att⁻ int⁻ imm⁴³⁴ cI⁻ χ°*	Selection on Hfl⁻	37
λNM1150	11.3	*Eco*RI, *Hin*dIII	*att⁻ int⁻ red⁻ imm⁴³⁴ cI⁻ χ°*	Selection on Hfl⁻; *red⁻ gam⁺*	37
λecc and λjac	~10	Use adapters	*att⁻ int⁻ red⁻ gam⁻cI857 nin5 χ⁺*	Spi selection; T3 and T7 promoters	38
Charon 21A	8.1	*Eco*RI, *Hin*dIII, *Xho*I	*Wam43 Eam1100 att⁻ int⁻ imm⁸⁰ nin5*	—	7
λgt11	6.3	*Eco*RI	*cI857 nin5 Sam100*	LacZ screen; fusion poly-peptides	7, 36
λgt22–23	7.3	*Not*I, *Xba*I, *Sac*I, *Sal*I, *Eco*RI	*red⁻ gam⁻ cI857 nin5 Sam100 χ⁺*	LacZ screen; fusion poly-peptides, multiple cloning sites	7
λorf8	7.2	*Eco*RI, *Bam*HI, *Hin*dIII	*intam imm²¹ nin5*	LacZ screen; fusion poly-peptides	7
λZAP	10.1	*Sac*I, *Not*I, *Xba*I, *Spe*I, *Eco*RI, *Xho*I	*cI857 nin5 Sam100*	LacZ screen; fusion poly-peptides; pBluescript SK (−); T3 and T7 promoters	7
Charon BS	7.9	*Eco*RI, *Hin*dIII	*att⁻ int⁻ imm⁸⁰ nin5*	LacZ screen; fusion poly-peptides; pBluescript SK (+ and −); T3 and T7 promoters	39
λNM1151	10.8	*Hin*dIII, *Eco*RI, *Bam*HI	(*int⁻* for *Bam*HI) *imm²¹ts nin5 χ°*	Integration proficient	37
λNM1151ABS	10.8	*Hin*dIII, *Eco*RI, *Bam*HI (derivative with *Not*I in place of *Bam*HI is also available)	*Aam32 Bam1 Sam7* derivative of λNM1151	Used in jumping libraries	40

[a] Vectors listed include those mentioned in the text and a few with special properties.

Table 2   Replacement Vectors

Name	Space[a] (kb)	Cloning sites in polylinkers	Biological features[b]	Ref.
EMBL3	9–21	*Sal*I, *Bam*HI, *Eco*RI (flanking polylinkers inverted)[d]	Recombinants *red⁻ gam⁻* ChiD, needs Gam in trans in *recA⁻* host	7
EMBL4	9–21	*Eco*RI, *Bam*HI, *Sal*I (inverted)[d]	As for EMBL3	7
λ2001	9–21	*Xba*I, *Sac*I, *Xho*I, *Bam*HI, *Hind*III, *Eco*RI (inverted)[d]	As for EMBL3	7
EMBL301	9–21	*Sal*I, *Bgl*II (*Not*I, *Xma*III), *Nae*I (*Sfi*I, *Bgl*I/*Xho*I), *Bam*HI, *Eco*RI (inverted)[d]	As for EMBL3	51
λDASH	9–21	*Xba*I, *Sac*I, *Xho*I, *Bam*HI, *Hind*III, *Eco*RI (inverted)[d]	λ2001 with T3 and T7 promoters	7
Charon 40	9–21	*Eco*RI, *Sac*I, *Kpn*I, *Sma*I, *Xba*I, *Sal*I, *Hind*III, *Not*I, *Xma*III, *Avr*II, *Spe*I, *Xho*I, *Apa*I, *Bam*HI, *Sfi*I, *Nae*I (inverted)[d]	Central fragment made up of *Nae*I repeats, recombinants *red⁻ gam⁺* and will grow in any *recA⁻* host	7
EMBL3*cos–Not*	9–21	*Sal*I, *Bam*HI, *Eco*RI—central fragment—*Eco*RI, *Bam*HI, *Not*I, *Sal*I	λ *cos* adjacent to left-hand polylinker	20
EMBL3*cos*W	9–21	*Sal*I, *Sfi*I, *Xho*I, *Bam*HI, *Xma*III, *Eco*RI—central fragment—*Eco*RI, *Bam*HI, *Xho*I (*Not*I, *Xma*III), *Sfi*I, *Sal*I	As for EMBL3 *cos Not*, but with SP6 and T7 promoters and transcription terminators	c

[a] There is no significant difference in the cloning capacity of any of these vectors; the figure of 21 kb is a cautious one based on an upper limit of 50 kb for a λ genome. Larger genomes (~52 kb) have been recovered, but they may be less stable than those of normal size.

[b] With the exception of Charon 40, the above vectors can be traced back to λ1059 where the χ sequence was identified as Chi3 (i.e., ChiD rather than ChiC as indicated in many references).

[c] Paul Whittaker, personal communication, 1990.

[d] Inverted indicates that the orientation of the right-hand polylinker is inverted relative to the left.

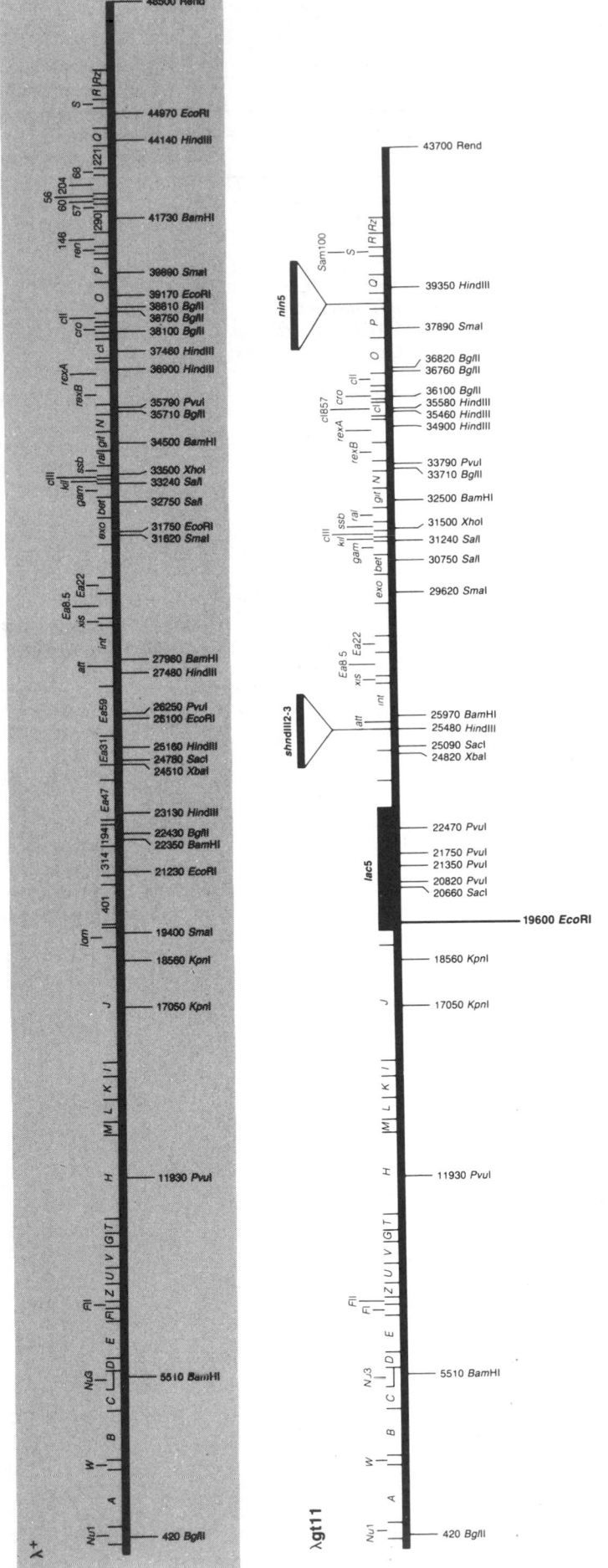

9.14

Figure 1 Map of λgt11.

Table 3   λgt11

GENOTYPE

λ*lac*5 Δ*shn*dIIIλ2-3 *sr*Iλ3° *c*I*ts*857 *sr*Iλ4° *nin*5 *sr*Iλ5° Sam100

GENETIC CHARACTERISTICS

Amber mutations	*S*am100
Suppressor required	*supF*
Deletions	*shn*dIIIλ2-3, *nin*5
Substitutions	*lac*5
Insertions	no
red/gam status of vector	*red$^+$ gam$^+$*
red/gam status of recombinants	*red$^+$ gam$^+$*
chi site present in recombinants	no
Propagates on *recA$^-$* hosts	yes
Spi selection	no
Recommended hosts	Y1090*hsdR*

CLONING SITES

Site	Insertion/ substitution	Size (kb)			Recombinants
		left arm	right arm	insert	
*Eco*RI	Insertion	19.5	24.2	0–7.2	Lac$^-$

REFERENCE

Young and Davis 1983

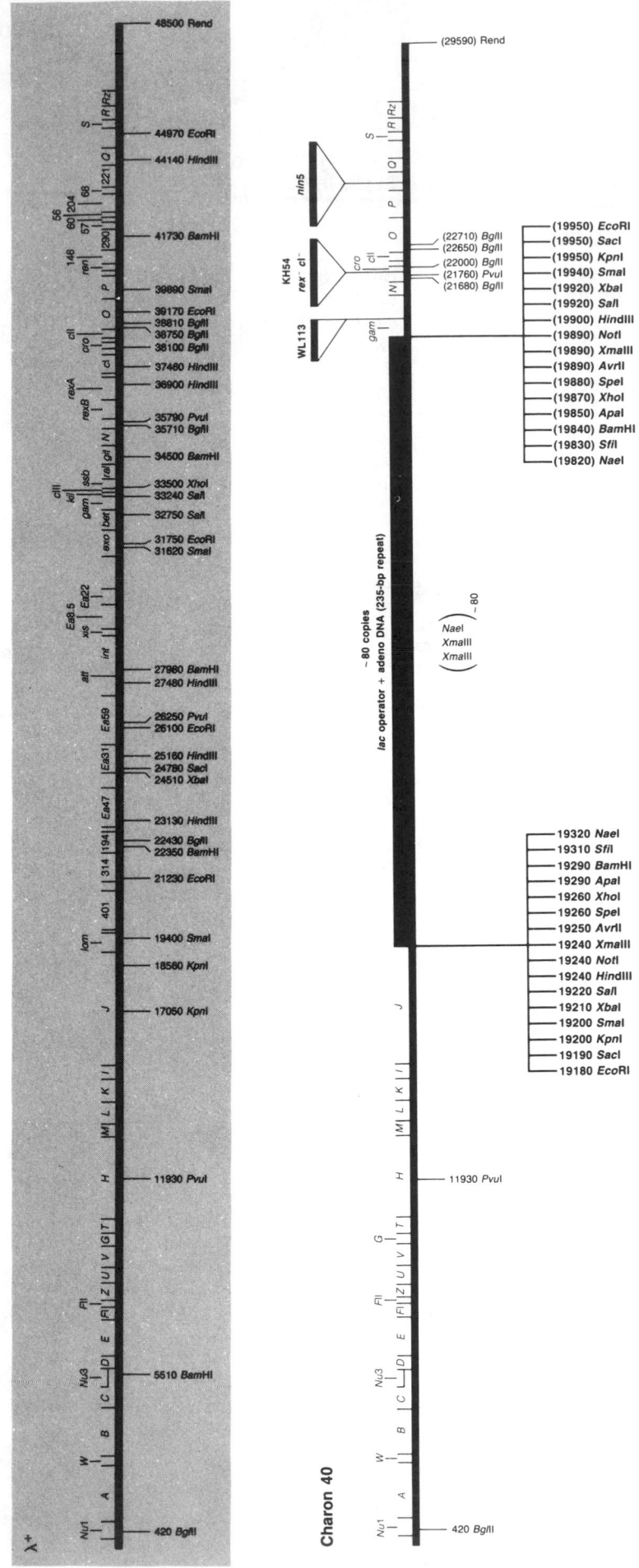

9.16

Figure 2 Map of Charon 40.

Table 4   Charon 40

GENOTYPE

λ*sbh*Iλ1° *sap*Iλ1° *sk*Iλ1° *sk*Iλ2° *ssm*Iλ1 ⟨polycloning site–polystuffer–polycloning site⟩ *ssl*Iλ2 ΔWL113 KH54 *ssm*Iλ3° *nin*5 *shn*dIIIλ6° *sr*Iλ5°

GENETIC CHARACTERISTICS

Amber mutations	no
Suppressor required	no
Deletions	WL113, KH54, *nin*5
Substitutions	polystuffer (*lac* operator + adenovirus DNA fragment)
	~80×
Insertions	no
red/gam status of vector	*red*⁻ *gam*⁺
red/gam status of recombinants	*red*⁻ *gam*⁺
chi site present in recombinants	no
Propagates on *recA*⁻ hosts	yes
Spi selection	no
Recommended hosts	LE392

CLONING SITES

Site[a]	Insertion/ substitution	Size (kb)			Recombinants
		left arm	right arm	insert	
*Eco*RI, *Apa*I *Xba*I, *Sfi*I *Sac*I, *Avr*II *Sal*I, *Spe*I *Hin*dIII, *Xho*I *Bam*HI, *Nae*I *Kpn*I, *Not*I *Sma*I, *Xma*III	Substitution	19.2	9.6	9.2–24.2	Most plaques are recombinants. Stuffer fragments are efficiently removed.

[a]The sites are not necessarily used as the pairs listed. One or two sites can be used; each site could be employed with any other site.

REFERENCE

Dunn and Blattner 1987

Some Widely Used Plasmids

A. pBR322

B. pACYC177

C. pACYC184

D. pUC19/18

E. Replicons carried by currently used plasmid vectors

The restriction maps and tables of restriction sites in *A–D* are reprinted, with permission, from *New England BioLabs 1990–1991 Catalog* (pp. 110–111 [pBR322], 106–107 [pACYC177], 108–109 [pACYC184], and 112–113 [pUC19]). Figure 1 in *D* and Table 1 in *E* are reprinted from Sambrook et al. (1989, pp. 1.13 and 1.4, respectively).

REFERENCES

New England BioLabs 1990–1991 Catalog. 1990. New England BioLabs, Inc., Beverly, Massachusetts.

Sambrook, J., E.F. Fritsch, and T. Maniatis. 1989. *Molecular Cloning. A Laboratory Manual*, Second Edition. Cold Spring Harbor Laboratory Press, Cold Spring Harbor, New York.

A. pBR322

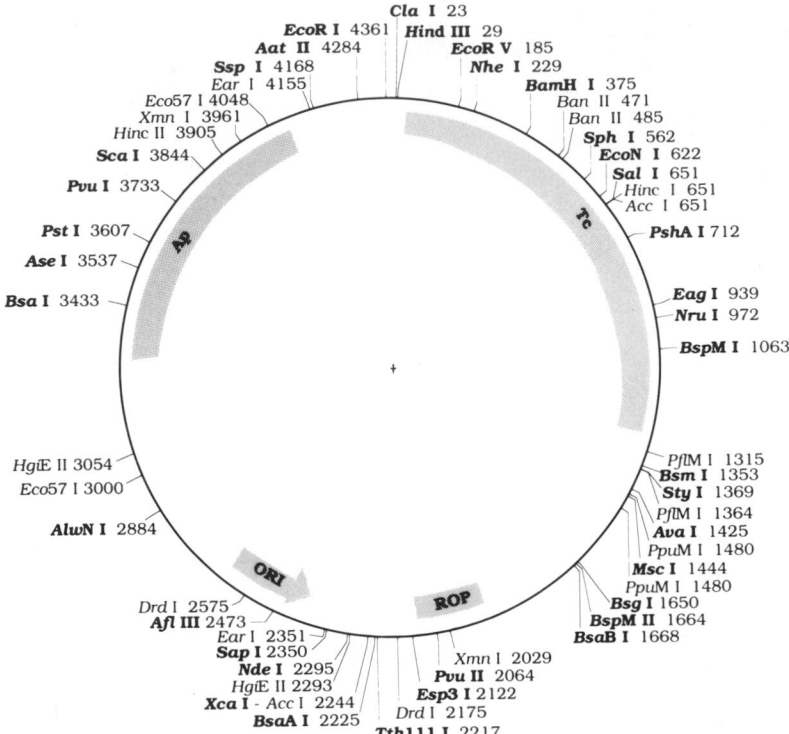

Cla I 23
EcoR I 4361   Hind III 29
Aat II 4284   EcoR V 185
Ssp I 4168   Nhe I 229
Ear I 4155   BamH I 375
Eco57 I 4048   Ban II 471
Xmn I 3961   Ban II 485
Hinc II 3905   Sph I 562
Sca I 3844   EcoN I 622
Pvu I 3733   Sal I 651
Pst I 3607   Hinc I 651
Ase I 3537   Acc I 651
Bsa I 3433   PshA I 712
Eag I 939
Nru I 972
BspM I 1063
HgiE II 3054   PflM I 1315
Eco57 I 3000   Bsm I 1353
AlwN I 2884   Sty I 1369
PflM I 1364
Ava I 1425
PpuM I 1480
Msc I 1444
PpuM I 1480
Bsg I 1650
BspM II 1664
BsaB I 1668
Drd I 2575
Afl III 2473   Xmn I 2029
Ear I 2351   Pvu II 2064
Sap I 2350   Esp3 I 2122
Nde I 2295   Drd I 2175
HgiE II 2293   Tth111 I 2217
Xca I - Acc I 2244
BsaA I 2225

Tc
Ap
ORI
ROP

pBR322 is an *E. coli* plasmid cloning vector. Recent sequencing data from Watson (confirmed at New England Biolabs) has shown its length to be 4361 base pairs not 4363 base pairs as previously reported. pBR322 was constructed in vitro using the tetracycline resistance gene (Tc) from pSC101, the origin of DNA replication start (ORI) and *rop* gene from the ColE1 derivative pMB1, and the ampicillin resistance gene (Ap) from transposon Tn3. Numbering of the sequence begins within the unique *Eco*RI site: the first T in the sequence ...GAATTC... is designated as nucleotide number 1. Numbering then continues around the molecule in the direction of Tc to Ap.

The map shows the restriction sites of those enzymes that cut the molecule once or twice; the unique sites are shown in **bold** type. The table lists the sites of those enzymes that cut a moderate number of times. The coordinates refer to the position of the 5′ base in each recognition sequence. The map also shows the relative positions of the antibiotic resistance genes, *rop* (mediates the activity of RNase I), and the origin of replication. The exact positions are: tetracycline resistance (Tc) 86–1268; β-lactamase (Ap) 3296–4084; ROP 1918–2105; origin of DNA replication start (ORI) 2535.

References
1. Bolivar, F. et al. 1977. *Gene* **2:** 95–113.
2. Sutcliffe, J.G. 1978. *Cold Spring Harbor Symp. Quant. Biol.* **43:** 77–90.
3. Sutcliffe, J.G. 1978. *Proc. Natl. Acad. Sci.* **75:** 3737–3741.
4. Peden, K.W.C. 1983. *Gene* **22:** 277–280.
5. Backman, K. and and H.W. Boyer. 1983. *Gene* **26:** 197–203.
6. Lathe, R., M.P. Kieny, S. Skory, and J.P. Lecocoq. 1984. *DNA* **3:** 173–182.
7. Heusterspreute, M. and J. Davison. 1984. *DNA* **3:** 259–264.
8. *GenBank* 1987. 50.0 VB0001.
9. Watson, N. 1988. *Gene* **70:** 399–403.

pBR322 DNA: Location of Restriction Sites

Enzyme	#	Locations				
Aat II	1	4284				
Afl III	1	2473				
AlwN I	1	2884				
Ase I	1	3537				
Ava I	1	1425				
BamH I	1	375				
Bsa I	1	3433				
BsaA I	1	2225				
BsaB I	1	1668				
Bsg I	1	1650				
Bsm I	1	1353				
BspM I	1	1063				
BspE I	1	1664				
Cla I	1	23				
Eag I	1	939				
EcoN I	1	622				
EcoR I	1	4359				
EcoR V	1	185				
Esp3 I	1	2122				
Hind III	1	29				
Msc I	1	1444				
Nde I	1	2295				
Nhe I	1	229				
Nru I	1	972				
PshA I	1	712				
Pst I	1	3607				
Pvu I	1	3733				
Pvu II	1	2064				
Sal I	1	651				
Sap I	1	2350				
Sca I	1	3844				
Sph I	1	562				
Ssp I	1	4168				
Sty I	1	1369				
Tth111 I	1	2217				
Xca I	1	2244				
Acc I	2	651	2244			
Ban II	2	471	485			
Drd I	2	2162	2575			
Dsa I	2	528	1447			
Ear I	2	2351	4155			
Eco57 I	2	3000	4048			
HgiE II	2	2293	3054			
Hinc II	2	651	3905			
PflM I	2	1315	1364			
PpuM I	2	1438	1480			
Xmn I	2	2029	3961			
ApaL I	3	2289	2787	4033		
Bbs I	3	737	1600	4351		
BceF I	3	594	1151	2959		
Bcg I	3	708	2063	3882		
Bgl I	3	929	1163	3480		
BsmA I	3	2122	3433	4198		
Dra I	3	3230	3249	3941		
Rsa I	3	164	2280	3845		
BspH I	4	489	3193	4201	4306	
Eci I	4	1393	2545	2691	3519	
Eco47 III	4	232	494	775	1727	
EcoO109 I	4	523	1438	1480	4341	
Fin I	4	538	888	1084	1761	
Fsp I	4	260	1356	1454	3586	
Gsu I	4	811	1401	1981	3451	
Mly I	4	632	2373	2844	3361	
Mme I	4	197	284	2663	2847	
Nae I	4	401	769	929	1283	
Nar I/ Kas I	4	413	434	548	1205	
Nsp I	4	562	1816	2108	2473	
Ple I	4	632	2373	2844	3361	
Sfe I	4	138	2738	2929	3607	
Gdi II	5	295	399	531	939	3754
Mae I	5	230	1489	2968	3221	3556
Tth111 II	5	7	1920	3047	3080	3086
Aha II	6	413	434	548	1205	3902
		4284				
BstN I	6	130	1059	1442	2500	2621
		2634				
Eae I	6	295	399	531	939	1444
		3754				
NspB II	6	1139	2064	2183	2813	3058
		3999				
Taq II	6	654	2385	3724	3883	4036
		4079				
Tfi I	6	852	1006	1304	1525	2029
		2448				
Cfr10 I	7	160	401	410	769	929
		1283	3446			

Enzyme	#	Locations				
Mcr I	7	286	653	939	2386	2810
		3733	3882			
Taq I	7	24	339	652	1127	1268
		2573	4017			
Ava II	8	799	887	1136	1439	1481
		1760	3504	3726		
BsaJ I	8	115	129	528	534	1167
		1369	1447	2633		
BstY I	8	375	1667	3114	3125	3211
		3223	3991	4008		
Dde I	8	1581	1743	2283	2748	3157
		3323	3863	4289		
HgiA I	8	276	587	1174	1465	2289
		2787	3948	4033		
TspE I	8	59	252	1320	1334	3234
		3540	3795	4360		
Ban I	9	76	119	413	434	548
		766	1205	1289	3314	
Tsp45 I	9	125	213	881	1148	1915
		2128	2223	3623	3834	
Bsp1286 I	10	276	471	485	587	1174
		1465	2289	2787	3948	4033
Fau I	10	181	703	890	1037	1232
		1497	1763	1949	2185	2195
Hinf I	10	632	852	1006	1304	1525
		2029	2373	2448	2844	3361
Mae II	10	901	957	1546	1570	1800
		2226	3176	3592	3965	4285
Nci I	10	170	534	1258	1484	1812
		2118	2153	2852	3548	3899
Hae II	11	232	413	434	494	548
		775	1205	1644	1727	2347
		2717				
Hga I	11	390	649	944	976	1240
		1390	2002	2179	2575	3153
		3903				
Mbo II	11	464	738	1009	1601	2352
		3123	3214	3969	4047	4156
		4352				
Alw I	12	375	376	1097	1667	3040
		3114	3126	3211	3224	3688
		3991	4009			
Fok I	12	112	133	987	1032	1681
		1770	1848	2007	2148	3346
		3527	3814			
Hph I	12	126	408	453	1307	1528
		2083	2092	3217	3444	3840
		4066	4081			
Mse I	15	34	56	1720	1940	1972
		2254	3179	3231	3236	3250
		3303	3538	3577	3942	4314
Sau96 I	15	172	524	799	887	1136
		1260	1439	1481	1760	1947
		3408	3487	3504	3726	4342
ScrF I	16	130	170	534	1059	1258
		1442	1484	1812	2118	2153
		2500	2621	2634	2852	3548
		3899				
Alu I	17	15	30	686	1089	1997
		2054	2065	2114	2133	2414
		2640	2730	2776	3033	3554
		3654	3717			
Mae III	17	125	213	881	1148	1808
		1831	1915	2128	2223	2830
		2893	3009	3292	3623	3681
		3834	4022			
Bsr I	19	210	305	610	680	884
		1133	1510	1781	1805	2220
		2248	2876	2889	3006	3412
		3530	3573	3837	4012	

There are no restriction sites in pBR322 DNA
for the following enzymes:

Afl II	Age I	Apa I	Avr II	Bcl I	Bgl II
BssH II	BstB I	BstE II	BstX I	Bsu36 I	Dra III
Drd II	Esp I	Fse I	Hpa I	Kpn I	Mfe I
Mlu I	Nco I	Not I	Nsi I	Pac I	Pml I
Rsr II	Sac I	Sac II	Sfi I	Sma I/Xma I	
SnaB I	Spe I	Spl I	Stu I	Xba I	Xcm I
Xho I/PaeR7 I					

B. pACYC177

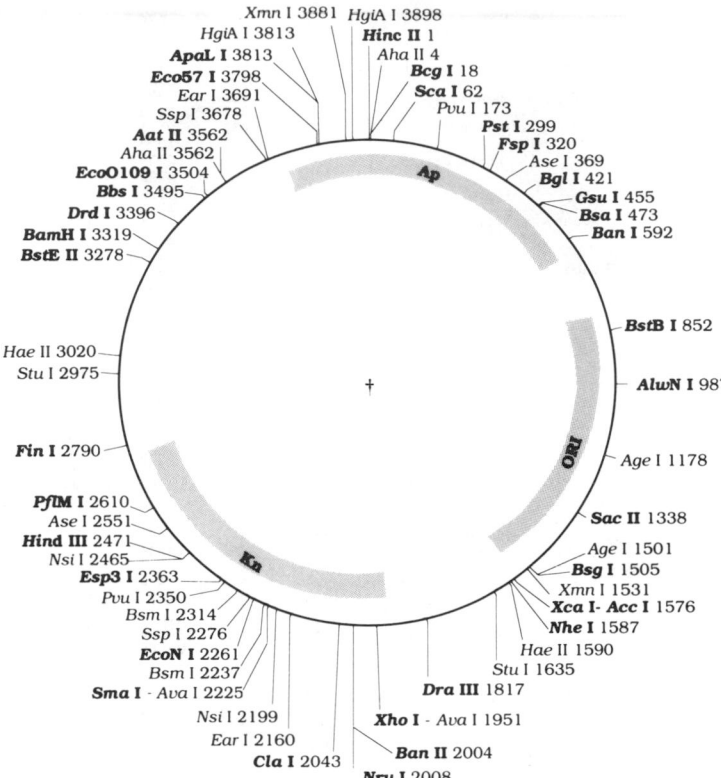

Xmn I 3881 HgiA I 3898
HgiA I 3813 Hinc II 1
ApaL I 3813 Aha II 4
Eco57 I 3798 Bcg I 18
Ear I 3691 Sca I 62
Ssp I 3678 Pvu I 173
Aat II 3562 Pst I 299
Aha II 3562 Fsp I 320
EcoO109 I 3504 Ase I 369
Bbs I 3495 Bgl I 421
Drd I 3396 Gsu I 455
BamH I 3319 Bsa I 473
BstE II 3278 Ban I 592

Hae II 3020 BstB I 852
Stu I 2975 AluN I 987

Fin I 2790 Age I 1178

PflM I 2610 Sac II 1338
Ase I 2551 Age I 1501
Hind III 2471 Bsg I 1505
Nsi I 2465 Xmn I 1531
Esp3 I 2363 Xca I- Acc I 1576
Pvu I 2350 Nhe I 1587
Bsm I 2314 Hae II 1590
Ssp I 2276 Stu I 1635
EcoN I 2261
Bsm I 2237 Dra III 1817
Sma I - Ava I 2225
Nsi I 2199 Xho I - Ava I 1951
Ear I 2160 Ban II 2004
Cla I 2043 Nru I 2008

pACYC177 is a low-copy-number *E. coli* plasmid cloning vector 3940 base pairs in length. It carries the origin of replication from plasmid p15A, enabling it to co-exist with vectors like pBR322 and pUC19 that carry the ColE1 origin. This feature of compatibility makes it useful for cloning experiments that require the presence of more than one recombinant plasmid per cell. pACYC177 carries the kanamycin resistance gene (Kn) from Tn*903* and the β-lactamase gene (Ap) from Tn3.

The map shows the locations of sites for enzymes that cleave the molecule once or twice; the unique sites are shown in **bold** type. The table lists the sites of those enzymes that cut a moderate number of times. The coordinates refer to the position of the 5′ base in each recognition sequence. Nucleotide number 1 of pACYC177 is the first G of the unique *Hinc*II site GTTGAC. The map also shows the relative positions of the antibiotic resistance genes and the origin of replication. The exact positions are: aminoglycoside 3′-phosphotransferase (Kn) 1920–2732; β-lactamase (Ap) 612–3696; origin of DNA replication start (ORI) 1326–1328.

The plasmid can be maintained in *E. coli* cells in media containing ampicillin (50–100 μg/ml) and kanamycin (35–50 μg/ml). To obtain large amounts of plasmid DNA, it is necessary to amplify using chloramphenicol due to the low copy number of pACYC177.

References
1. Chang, A.C.Y. and S.N. Cohen. 1978. *J. Bacteriol.* **134:** 1141–1156.
2. Rose, R.E. 1988. *Nucleic Acids Res.* **16:** 356.
3. *GenBank* VB0109 (GenBank sequence is 3941 base pairs).

pACYC177 DNA: Location of Restriction Sites

Enzyme	#	Locations				
Aat II	1	3562				
Acc I	1	1576				
*Alw*N I	1	987				
ApaL I	1	3813				
*Bam*H I	1	3319				
Ban I	1	592				
Ban II	1	2004				
Bbs I	1	3495				
Bcg I	1	18				
Bgl I	1	421				
Bsa I	1	473				
Bsg I	1	1505				
*Bst*B I	1	852				
*Bst*E II	1	3278				
Cla I	1	2043				
Dra III	1	1817				
Drd I	1	3396				
*Eco*57 I	1	3798				
*Eco*N I	1	2261				
*Eco*O109 I	1	3504				
*Esp*3 I	1	2363				
Fau I	1	3469				
Fin I	1	2790				
Fsp I	1	320				
Gsu I	1	455				
Hinc II	1	1				
Hind III	1	2471				
Nhe I	1	1587				
Nru I	1	2008				
Nsp I	1	1013				
*Pfl*M I	1	2610				
Pst I	1	299				
Sac II	1	1338				
Sca I	1	62				
Sfe I	1	299				
Sma I/	1	2225				
Xma I						
Xca I	1	1576				
Xho I/	1	1951				
*Pae*R7 I						
Age I	2	1178	1501			
Aha II	2	4	3562			
Ase I	2	369	2551			
Ava I	2	1951	2225			
Bsm I	2	2237	2314			
Ear I	2	2160	3691			
*Eco*47 III	2	1590	3020			
Hae II	2	1590	3020			
*Hgi*A I	2	3813	3898			
Nsi I	2	2199	2465			
Pvu I	2	173	2350			
Ssp I	2	2276	3678			
Stu I	2	1635	2975			
Xmn I	2	1531	3881			
*Bce*F I	3	1329	2656	2793		
Dra I	3	657	676	3905		
Eci I	3	387	1311	1392		
*Tth*111 II	3	1298	2173	2326		
Ava II	4	181	403	1705	2906	
*Bsp*1286 I	4	2004	3428	3813	3898	
*Bsp*H I	4	713	1871	3540	3645	
Eae I	4	152	1157	3164	3377	
Gdi II	4	152	1157	3164	3377	
Mae I	4	352	687	1588	3324	
*Nsp*B II	4	1061	1166	1338	3847	
Rsa I	4	63	2187	3250	3388	
*Bsm*A I	5	474	1147	1825	2364	3649
*Cfr*10 I	5	460	945	1178	1501	2307
Hga I	5	4	754	1282	1496	3413
*Tsp*45 I	5	73	284	912	2501	3279
*Bst*Y I	6	683	695	2602	3319	3838
		3855				
Mcr I	6	24	173	1064	1433	2350
		3155				
Mly I	6	546	1031	1447	1625	2577
		2986				
Nci I	6	8	359	1023	1120	2225
		2226				
Ple I	6	546	1031	1447	1625	2577
		2986				
Taq II	6	23	182	1434	2591	3767
		3810				
Bbv I	7	124	301	490	992	995
		1060	1507			
Dsa I	7	1338	1676	1737	2873	2934
		3391	3475			
Mae II	7	316	732	1818	1943	3403
		3563	3883			

Enzyme	#	Locations				
Mme I	7	1027	1066	1970	2164	2570
		2579	3187			
Tfi I	7	1641	2264	2320	2492	2583
		2970	3367			
*Bst*N I	8	1224	1236	1371	1382	1633
		2242	2599	2978		
*Sau*96 I	9	181	403	420	499	1342
		1705	2906	3341	3505	
*Bst*U I	10	478	802	1155	1339	1959
		2009	2354	3168	3593	3925
Fok I	10	93	380	561	1653	1978
		2611	2958	3070	3112	3284
Nla IV	10	246	457	498	592	1217
		1314	1672	2938	3319	3596
Taq I	10	853	1428	1798	1952	2044
		2318	2721	2814	3154	3831
Mwo I	11	301	421	998	1159	1408
		1637	2059	2091	2305	2968
		3098				
Alu I	12	191	254	354	867	1100
		1690	2472	2922	3044	3151
		3358	3457			
*Bsa*J I	12	885	1224	1338	1676	1737
		2224	2225	2626	2873	2934
		3391	3475			
Alw I	13	219	683	696	1776	2221
		2602	2835	3185	3244	3319
		3320	3838	3856		
Hae III	13	153	420	500	1158	1342
		1636	1956	2439	2976	3165
		3342	3378	3506		
Hinf I	13	546	1031	1447	1625	1641
		2264	2320	2492	2577	2583
		2970	2986	3367		
Bsr I	14	70	334	377	495	893
		986	999	1573	1708	2139
		2763	2903	3432	3835	
Dde I	14	44	584	750	864	1127
		1594	1786	2370	2825	3017
		3108	3211	3459	3558	
Mae III	14	73	226	284	615	890
		912	1045	1842	2114	2206
		2501	2758	3279	3825	
Scr F I	14	8	359	1023	1120	1224
		1236	1371	1382	1633	2225
		2226	2242	2599	2978	
*Tsp*E I	14	113	368	674	981	1967
		2151	2333	2550	2700	2742
		2750	3373	3446	3488	
Mbo II	15	693	778	1367	1378	1610
		1761	2161	2252	2850	3001
		3146	3495	3691	3800	3878
Hha I/	16	321	414	801	907	949
*Hin*P I		1058	1208	1591	2066	2288
		2305	2375	3021	3329	3592
		3924				
Hpa II/	16	8	250	360	427	461
Msp I		946	1023	1049	1121	1179
		1502	2226	2308	2489	3073
		3314				
Mse I	16	331	370	605	658	672
		677	729	766	942	1360
		1964	2343	2552	2748	3534
		3906				
Nla III	16	99	135	213	223	714
		952	1014	1528	1872	1975
		2087	2202	2507	3182	3541
		3646				

There are no restriction sites in pACYC177 DNA
for the following enzymes:

Afl II	*Afl* III	*Apa* I	*Avr* II	*Bcl* I	*Bgl* II
*Bsa*A I	*Bsa*B I	*Bsp*M I	*Bsp*E I	*Bss*H II	*Bst*X I
*Bsu*36 I	*Drd* II	*Eag* I	*Eco*R I	*Eco*R V	*Esp* I
Fse I	*Hgi*E II	*Hpa* I	*Kpn* I	*Mfe* I	*Mlu* I
Msc I	*Nae* I	*Nar* I/*Kas* I		*Nco* I	*Nde* I
Not I	*Pac* I	*Pml* I	*Ppu*M I	*Psh*A I	*Pvu* II
Rsr II	*Sac* I	*Sal* I	*Sap* I	*Sfi* I	*Sna*B I
Spe I	*Sph* I	*Spl* I	*Sty* I	*Tth*111 I	*Xba* I
Xcm I					

C. pACYC184

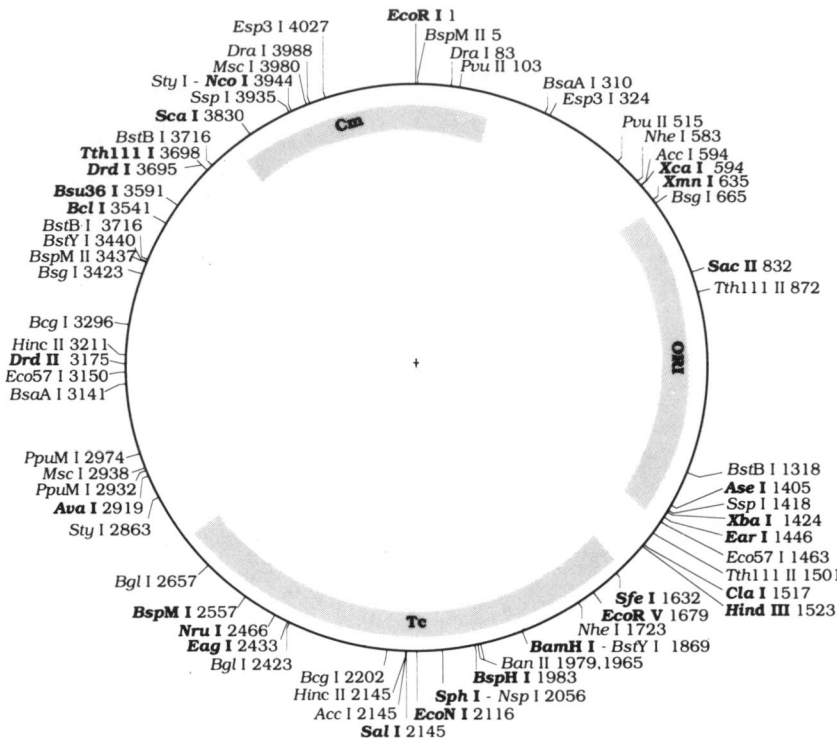

pACYC184 is a small, medium-copy-number *E. coli* plasmid cloning vector 4244 base pairs in length. It carries the origin of replication from plasmid p15A, which enables it to co-exist with vectors like pBR322 and pUC19 that carry the ColE1 origin. This compatibility makes it useful for cloning experiments that require the presence of more than one recombinant plasmid per cell.

pACYC184 carries the chloramphenicol resistance gene (Cm) from Tn*9* and the tetracycline resistance gene (Tc) from pSC101. The map shows the location of sites for enzymes that cleave the molecule once or twice; unique sites are shown in **bold** type. The table lists the sites of those enzymes that cut a moderate number of times. The coordinates refer to the position of the 5′ base in each recognition sequence. Nucleotide number 1 of pACYC184 is the first G of the unique *Eco*RI site GAATTC. The map also shows the relative positions of the antibiotic resistance genes and the origin of replication. The exact positions are: tetracycline resistance (Tc) 1580–2770; chloramphenicol resistance (Cm) 219–3804; origin of DNA replication start (ORI) 845–847.

The plasmid can be maintained in *E. coli* cells in media containing tetracycline (25–50 μg/ml) and chloramphenicol (35–50 μg/ml). To obtain large amounts of plasmid DNA, it is necessary to amplify using spectinomycin (50 μg/ml) due to the low copy number of pACYC184 and the presence of the chloramphenicol resistance gene.

References

1. Chang, A.C.Y. and S.N. Cohen. 1978. *J. Bacteriol.* **134:** 1141–1156.
2. Rose, R.E. 1988. *Nucleic Acids Res.* **16:** 355.

pACYC184 DNA: Location of Restriction Sites

Enzyme	#	Locations				
Ase I	1	1405				
Ava I	1	2919				
Bam H I	1	1869				
Bcl I	1	3541				
Bsa B I	1	3441				
Bsp H I	1	1983				
Bsp M I	1	2557				
Bsu36 I	1	3591				
Cla I	1	1517				
Drd I	1	3695				
Drd II	1	3175				
Eag I	1	2433				
Ear I	1	1446				
EcoN I	1	2116				
EcoR I	1	1				
EcoR V	1	1679				
Hind III	1	1523				
Nco I	1	3944				
Nru I	1	2466				
Sac II	1	832				
Sal I	1	2145				
Sca I	1	3830				
Sfe I	1	1632				
Sph I	1	2056				
Tth111 I	1	3698				
Xba I	1	1424				
Xca I	1	594				
Xmn I	1	635				
Acc I	2	594	2145			
Ban II	2	1965	1979			
Bcg I	2	2202	3296			
Bgl I	2	2423	2657			
Bsa A I	2	310	3141			
Bsg I	2	665	3423			
BspE I	2	5	3437			
BstB I	2	1318	3716			
BstY I	2	1869	3440			
Dra I	2	83	3988			
Eco57 I	2	1463	3150			
Esp3 I	2	324	4027			
Hinc II	2	2145	3211			
Msc I	2	2938	3980			
Nhe I	2	583	1723			
Nsp I	2	1157	2056			
PpuM I	2	2932	2974			
PshA I	2	2206	3300			
Pvu II	2	103	515			
Ssp I	2	1418	3935			
Sty I	2	2863	3944			
Tth111 II	2	872	1501			
AlwN I	3	532	1180	3518		
Bbs I	3	2231	3094	3325		
Bsm I	3	14	2847	3851		
BsmA I	3	325	1024	4027		
Eci I	3	778	859	2887		
Fsp I	3	1754	2850	2948		
Rsa I	3	125	1658	3831		
Taq II	3	736	2148	4011		
Age I	4	669	992	3181	3641	
Aha II	4	1907	1928	2042	2699	
Alw I	4	1869	1870	2591	3440	
Eco O109 I	4	2017	2932	2974	3387	
Gsu I	4	493	2305	2895	4144	
Mae I	4	584	1425	1724	2983	
Mly I	4	724	1140	2126	3209	
Nar I/	4	1907	1928	2042	2699	
Kas I						
PflM I	4	337	2809	2858	4014	
Ple I	4	724	1140	2126	3209	
Ava II	5	2293	2381	2630	2933	2975
Eco47 III	5	580	1726	1988	2269	3500
HgiA I	5	490	1770	2081	2668	2959
Mme I	5	298	1104	1143	1691	1778
Nae I	5	1895	2263	2423	2777	3357
Tfi I	5	2346	2500	2798	3019	3896
BceF I	6	92	556	842	2088	2645
		3880				
Dsa I	6	832	2022	2941	3383	3565
		3944				
Gdi II	6	1013	1789	1893	2025	2433
		3380				
Mcr I	6	737	1106	1780	2147	2433
		3697				
Fin I	7	371	520	2032	2382	2578
		3135	3374			
Fok I	7	8	1606	1627	2481	2526
		3454	3733			
Tsp45 I	7	410	1259	1619	1707	2375
		2642	3615			

Enzyme	#	Locations				
Bsp1286 I	8	490	1770	1965	1979	2081
		2668	2959	3778		
Eae I	8	1013	1789	1893	2025	2433
		2938	3380	3980		
NspB II	8	103	515	832	1004	1109
		2633	3262	3903		
Dde I	9	229	1044	1307	3075	3205
		3242	3578	3592	4025	
Hga I	9	437	675	889	1884	2143
		2438	2470	2734	2884	
Hinf I	9	724	1140	2126	2346	2500
		2798	3019	3209	3896	
Mae II	9	311	323	2395	2451	3040
		3064	3142	3985	4160	
Ban I	10	1570	1613	1907	1928	2042
		2260	2699	2783	3354	3779
Fau I	10	25	1675	2197	2384	2531
		2726	2991	3159	3291	3810
Nci I	10	266	358	1051	1148	1664
		2028	2752	2978	3378	3700
Taq I	10	246	744	1319	1518	1833
		2146	2621	2762	3705	3717
Hae II	11	580	1726	1907	1928	1988
		2042	2269	2699	3363	3417
		3500				
Mbo II	11	404	793	804	1393	1447
		1958	2232	2503	3095	3326
		3963				
Sau96 I	11	346	829	1666	2018	2293
		2381	2630	2754	2933	2975
		3388				
BstN I	12	349	377	789	800	935
		947	1624	2553	2936	3766
		4014	4070			
Cfr10 I	12	669	992	1225	1654	1895
		1904	2263	2423	2777	3181
		3357	3641			
Mse I	12	84	812	1230	1406	1528
		1550	3493	3674	3774	3795
		3847	3989			
Alu I	13	104	223	232	516	533
		1072	1305	1509	1524	2180
		2583	3274	4219		
Mae III	13	410	553	1126	1259	1281
		1619	1707	2375	2642	3246
		3615	4083	4188		
TspE I	13	2	1191	1439	1493	1553
		1746	2814	2828	3554	3708
		3720	3802	3841		
Bbv I	15	449	664	1111	1176	1179
		1720	2109	2267	2900	2924
		3053	3056	3146	3361	3458
Bsr I	15	200	495	598	1172	1185
		1278	1704	1799	2104	2174
		2378	2627	3004	3586	3999
Mbo I/	15	275	318	402	1385	1396
Sau3A I		1843	1870	1961	2320	2592
		2623	2638	2955	3441	3542
Hph I	16	291	443	1258	1620	1902
		1947	2801	3022	3184	3948
		4002	4010	4078	4196	4214
		4222				
SfaN I	16	168	446	796	875	1628
		1698	1741	1887	1899	2152
		2520	2527	2915	3446	3455
		3908				
BsaJ I	18	348	349	832	946	1285
		1609	1623	2022	2028	2661
		2863	2941	3377	3383	3565
		3944	4013	4014		
BstU I	18	833	1017	1370	1840	2196
		2311	2440	2467	2472	2533
		2599	2728	2738	2883	2909
		3031	3128	3290		

There are no restriction sites in pACYC184 DNA
for the following enzymes:

Aat II	Afl II	Afl III	Apa I	Apa L I	Avr II
Bgl II	Bsa I	BssH II	Bst E II	Bst X I	Dra III
Esp I	Fse I	HgiE II	Hpa I	Kpn I	Mfe I
Mlu I	Nde I	Not I	Nsi I	Pac I	Pml I
Pst I	Pvu I	Rsr II	Sac I	Sap I	Sfi I
Sma I/Xma I		SnaB I	Spe I	Spl I	Stu I
Xcm I	Xho I/PaeR7 I				

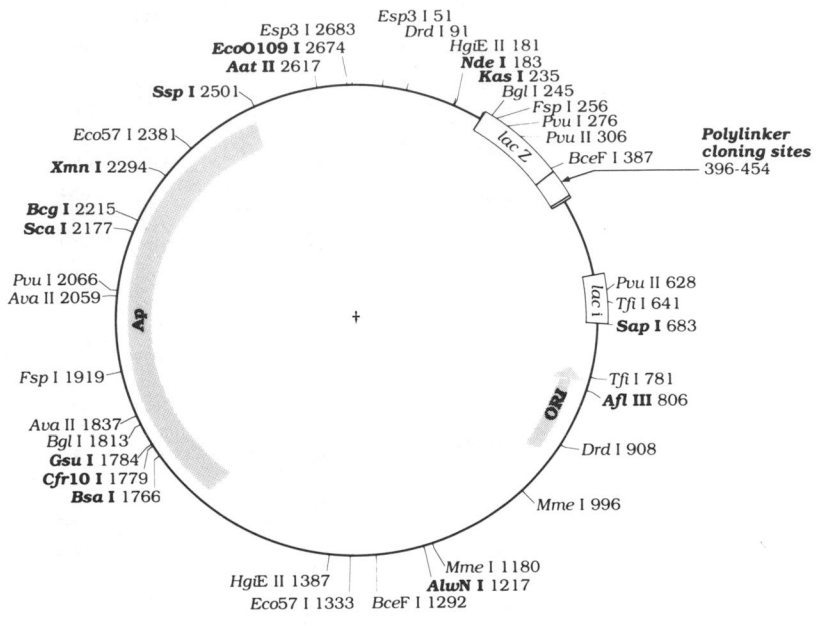

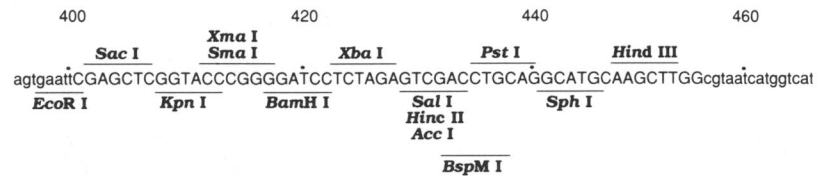

LacZ' ◄──Ala Leu Ser Ala His Arg Cys Thr Ser Glu Leu Pro Asp Gly Pro Val Ser Ser Asn Thr Ile Met Thr (Met)

pUC19 is a small, high-copy-number *E. coli* plasmid cloning vector 2686 base pairs in length. pUC19 is part of a series of related plasmids constructed by Messing and co-workers that contains portions of pBR322 and M13mp19. It can be amplified with chloramphenicol and carries a 54-bp multiple cloning site polylinker. Nucleotide numbering starts at the first T in the sequence ...TCGCGCGTTT... (this was originally an *Eco*RI/*Pvu*II site but it was changed during construction) and proceeds clockwise around the molecule in the direction *lac* to Ap.

The map shows the restriction sites of those enzymes that cut the molecule once or twice; the unique sites are shown in **bold** type. The table lists the sites of those enzymes that cut a moderate number of times. The polylinker is shown below the map. The coordinates refer to the position of the 5' base in each recognition sequence. The map also shows the relative positions of the coding sequences and the origin of DNA replication. The exact positions are: β-lactamase (Ap, mature form) 1629–2417; *lacZ* and polylinker 238–469; ColE1 origin of DNA replication start (ORI) 867.

References

1. Yanisch-Perron, C., J. Vieira, and J. Messing. 1985. *Gene* **33:** 103–119.
2. *GenBank* VB0026 (Vecbase:pUC19c).

pUC19 DNA: Location of Restriction Sites

Enzyme	#	Locations				
Aat II	1	2617				
Acc I	1	429				
Afl III	1	806				
AluN I	1	1217				
Ava I	1	412				
BamH I	1	417				
Ban II	1	402				
Bcg I	1	2215				
Bsa I	1	1766				
BspM I	1	433				
Cfr10 I	1	1779				
EcoO109 I	1	2674				
EcoR I	1	396				
Gsu I	1	1784				
Hinc II	1	429				
Hind III	1	447				
Kpn I	1	408				
Nar I/ Kas I	1	235				
Nde I	1	183				
Pst I	1	435				
Sac I	1	402				
Sal I	1	429				
Sap I	1	683				
Sca I	1	2177				
Sma I/ Xma I	1	412				
Sph I	1	441				
Ssp I	1	2501				
Xba I	1	423				
Xmn I	1	2294				
Ava II	2	1837	2059			
BceF I	2	387	1292			
Bgl I	2	245	1813			
Drd I	2	91	908			
Eco57 I	2	1333	2381			
Esp3 I	2	51	2683			
Fsp I	2	256	1919			
HgiE II	2	181	1387			
Mme I	2	996	1180			
Pvu I	2	276	2066			
Pvu II	2	306	628			
Tfi I	2	641	781			
Aha II	3	235	2235	2617		
ApaL I	3	177	1120	2366		
Ase I	3	576	635	1870		
BspH I	3	1526	2534	2639		
Dra I	3	1563	1582	2274		
Eae I	3	388	645	2087		
Ear I	3	290	684	2488		
Eci I	3	878	1024	1852		
Gdi II	3	388	645	2087		
Hae II	3	235	680	1050		
Nsp I	3	37	441	806		
Rsa I	3	168	409	2178		
Tth111 II	3	1380	1413	1419		
Ban I	4	235	408	550	1647	
BsmA I	4	51	1766	2531	2684	
Hga I	4	108	908	1486	2236	
Mae I	4	424	1301	1554	1889	
Mly I	4	427	706	1177	1694	
Ple I	4	427	706	1177	1694	
Sfe I	4	435	1071	1262	1940	
Taq I	4	400	430	906	2350	
Tsp45 I	4	57	368	1956	2167	
BsaJ I	5	354	412	413	545	966
Bsp1286 I	5	177	402	1120	2281	2366
BstN I	5	354	545	833	954	967
Fau I	5	114	124	284	598	655
Fok I	5	77	321	1679	1860	2147
HgiA I	5	177	402	1120	2281	2366
Mae II	5	374	1509	1925	2298	2618
Mcr I	5	276	719	1143	2066	2215
Taq II	5	718	2057	2216	2369	2412
Dde I	6	171	1081	1490	1656	2196
		2622				
Hinf I	6	427	641	706	781	1177
		1694				
NspB II	6	112	306	628	1146	1391
		2332				
Sau96 I	6	286	1741	1820	1837	2059
		2675				
BstY I	7	417	1447	1458	1544	1556
		2324	2341			
Hph I	7	12	21	1550	1777	2173
		2399	2414			
Mbo II	7	291	685	1456	1547	2302
		2380	2489			

Enzyme	#	Locations				
Nci I	7	47	82	412	413	1185
		1881	2232			
TspE I	7	397	487	504	579	1567
		1873	2128			
SfaN I	8	78	153	208	229	894
		1946	2156	2386		
Alw I	10	417	418	1373	1447	1459
		1544	1557	2021	2324	2342
BstU I	10	2	4	107	652	654
		852	1433	1763	2256	2588
Bsr I	11	365	391	606	1209	1222
		1339	1745	1863	1906	2170
		2345				
Hae III	11	287	389	646	820	831
		849	1283	1741	1821	2088
		2675				
Mae III	11	57	348	368	1163	1226
		1342	1625	1956	2014	2167
		2355				
Nla III	11	38	442	461	807	1527
		2018	2028	2106	2142	2535
		2640				
Nla IV	11	235	408	417	550	836
		875	1647	1741	1782	1993
		2583				
Bbv I	12	41	254	327	630	711
		729	1148	1213	1216	1422
		1750	2116			
ScrF I	12	47	82	354	412	413
		545	833	954	967	1185
		1881	2232			
Hpa II/ Msp I	13	48	82	413	524	1013
		1160	1186	1376	1780	1814
		1881	1991	2233		
Mnl I	13	29	289	421	662	695
		921	978	1245	1645	1726
		1856	2062	2673		
Mse I	13	142	339	577	636	1512
		1564	1569	1583	1636	1871
		1910	2275	2647		
Mwo I	13	179	229	236	245	275
		445	544	588	672	739
		853	1425	1813		
Mbo I/ Sau3A I	15	277	418	1373	1448	1459
		1467	1545	1557	1662	2003
		2021	2067	2325	2342	2378
Alu I	16	43	62	307	403	448
		470	565	629	747	973
		1063	1109	1366	1887	1987
		2050				
Hha I/ HinP I	17	3	106	236	257	588
		653	681	714	984	1051
		1151	1325	1434	1827	1920
		2257	2589			
Fnu4H I	19	41	150	254	327	630
		711	729	732	850	1005
		1148	1213	1216	1422	1750
		2089	2116	2211	2440	
CviJ I	45	43	62	95	135	139
		253	287	307	389	403
		448	470	522	543	565
		629	646	728	747	820
		831	849	875	973	1063
		1109	1114	1139	1218	1283
		1294	1337	1366	1729	1741
		1782	1808	1812	1821	1887
		1977	1987	2050	2088	2675

There are no restriction sites in pUC19 DNA for the following enzymes:

Afl II	Age I	Apa I	Avr II	Bbs I	Bcl I
Bgl II	BsaA I	BsaB I	Bsg I	Bsm I	BspE I
BssH II	BstB I	BstE II	BstX I	Bsu36 I	Cla I
Dra III	Drd II	Dsa I	Eag I	Eco47 III	EcoN I
EcoR V	Esp I	Fin I	Fse I	Hpa I	Mfe I
Mlu I	Msc I	Nae I	Nco I	Nhe I	Not I
Nru I	Nsi I	Pac I	PflM I	Pml I	PpuM I
PshA I	Rsr II	Sac II	Sfi I	SnaB I	Spe I
Spl I	Stu I	Sty I	Tth111 I	Xca I	Xcm I
Xho I/PaeR7 I					

Polycloning Sites
pUC18

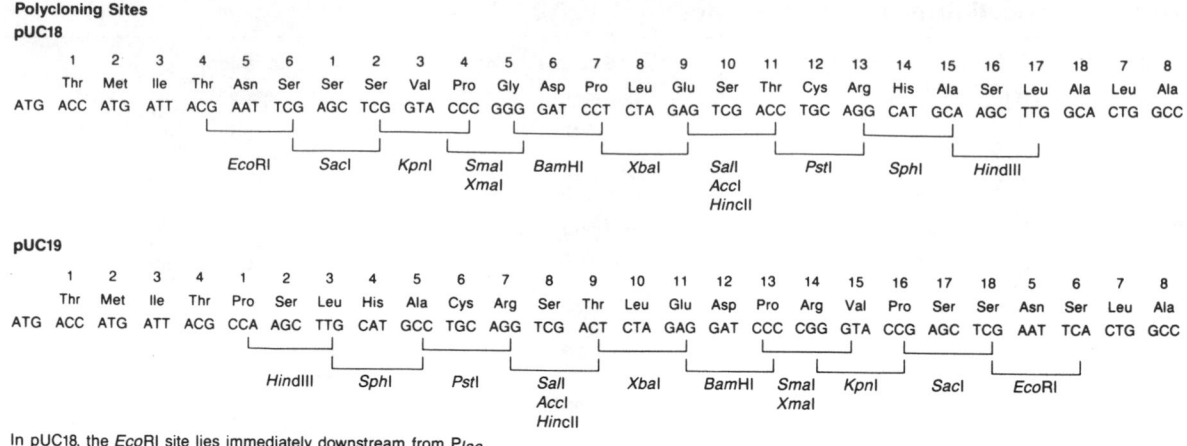

In pUC18, the *Eco*RI site lies immediately downstream from P*lac*.
In pUC19, the *Hind*III site lies immediately downstream from P*lac*.

Figure 1   pUC18 is identical to pUC19 except that the polycloning site is in the opposite direction.

E. Replicons Carried by Currently Used Plasmid Vectors

Table 1   Replicons Carried by Currently Used Plasmid Vectors

Plasmid	Replicon	Copy number
pBR322 and its derivatives	pMB1	15–20
pUC vectors	pMB1	500–700
pACYC and its derivatives	p15A	10–12
pSC101 and its derivatives	pSC101	~5
ColE1	ColE1	15–20

A. M13mp18 and the M13mp series of vectors

B. Bluescript M13+, M13−

The restriction maps of M13mp18 and the M13mp series of vectors and the table of restriction sites in *A* are reprinted, with permission, from *New England BioLabs 1990–1991 Catalog* (pp. 103–105). The map of Bluescript M13− in *B* is reprinted from Sambrook et al. (1989, p. 1.20).

REFERENCES

New England BioLabs 1990–1991 Catalog. 1990. New England BioLabs, Inc., Beverly, Massachusetts.
Sambrook, J., E.F. Fritsch, and T. Maniatis. 1989. *Molecular Cloning. A Laboratory Manual,* Second Edition. Cold Spring Harbor Laboratory Press, Cold Spring Harbor, New York.

SECTION 11

A. M13mp18 and the M13mp Series of Vectors

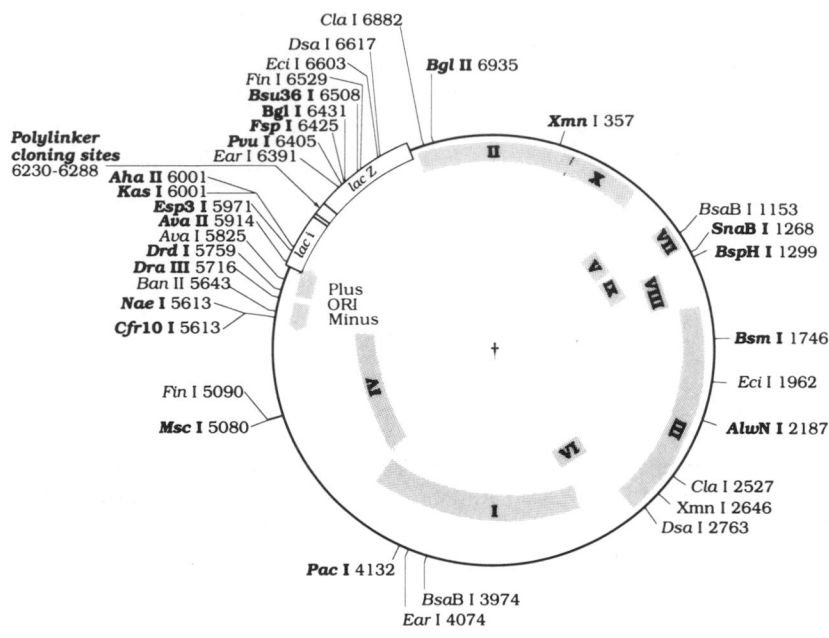

1

(Met) Thr Met Ile Thr Asn Ser Ser Ser Val Pro Gly Asp Pro Leu Glu Ser Thr Cys Arg His Ala Ser Leu Ala → LacZ'

```
        6230              6250              6270              6290
                        Xma I
              Sac I     Sma I      Xba I         Pst I        Hind III
atgaccatgattacgaattCGAGCTCGGTACCCGGGGATCCTCTAGAGTCGACCTGCAGGCATGCAAGCTTGGcact
             EcoR I    Kpn I    BamH I        Sal I          Sph I
                                              Hinc II
                                              Acc I
```

M13 is a filamentous, male-specific *E. coli* bacteriophage. The DNA molecule is a single-stranded circle 6407 bases in length. Double-stranded forms arise as intermediates during DNA replication. The M13mp series of vectors are derivatives of M13 constructed by Messing and co-workers to facilitate the cloning and sequencing of DNA fragments. M13mp18 and M13mp19 are the most recent vectors. They are 7250 bases in length and differ only in the orientation of the 54-base polylinker that they carry. The polylinker includes 10 discrete hexanucleotide recognition sites for 13 different enzymes. Altogether, it will accept fragments generated by over 50 enzymes. Nucleotide numbering begins at the first A in the sequence ...GTTAAT... (originally a *Hpa*I site that was changed during construction) and continues clockwise around the viral strand in the 5' to 3' direction.

The map shows the restriction sites of enzymes that cut the molecule once or twice; the unique sites are shown in **bold** type. The table on page 11.4 lists the sites of those enzymes that cut a moderate number of times. The coordinates refer to the position of the 5' base in each recognition sequence. The polylinker is shown below the map; a list of the polylinkers in other M13mp vectors can be found in the M13 sequencing section of the *New England BioLabs 1990–1991 Catalog*. The map also shows the positions of gene products (I–X) and the origins of plus- and minus-strand replication. All coding sequences are transcribed clockwise; they are shown arranged in concentric rings according to their reading frames.

References

1. Messing, J. et al. 1977. *Proc. Natl. Acad. Sci.* **74:** 3652–3646.
2. Gronenborn, B. and J. Messing. 1978. *Nature* **272:** 375–377.
3. van Wezenbeck, P. and J.G.G. Schoenmakers. 1979. *Nucleic Acids Res.* **6:** 2799–2818.
4. van Wezenbeck, P.M.G.F., T.J.M. Hulsebos, and J.G.G. Schoenmakers. 1980. *Gene* **11:** 129–148.
5. Messing, J., R. Crea, and P.H. Seeburg. 1981. *Nucleic Acids Res.* **9:** 309–321.
6. Messing, J. 1981. In *Third Cleveland Symposium on Macromolecules: Recombinant DNA*, pp. 143–153. Elsevier, Amsterdam.
7. Messing, J. 1983. In *Methods in Enzymology*, vol. 101 (part C): *Recombinant DNA* (ed. R. Wu, et al.), pp. 20–78. Academic Press. New York.
8. Yanisch-Perron, C., J. Vieira, and J. Messing. 1985. *Gene* **33:** 103–119.
9. Roberts, R.J. 1987. *Nucleic Acids Res.* (Suppl.) **15:** r189–r217.

M13mp18: Location of Restriction Sites

Enzyme	#	Locations				
Acc I	1	6264				
Aha II	1	6001				
AlwN I	1	2187				
Ava II	1	5914				
BamH I	1	6252				
Bgl I	1	6431				
Bgl II	1	6935				
Bsm I	1	1746				
BspH I	1	1299				
Bsu36 I	1	6508				
Cfr10 I	1	5613				
Dra III	1	5716				
Drd I	1	5759				
EcoR I	1	6231				
Esp3 I	1	5971				
Fsp I	1	6425				
HgiE II	1	6468				
Hinc II	1	6264				
Hind III	1	6282				
Kpn I	1	6243				
Msc I	1	5080				
Nae I	1	5613				
Nar I/ Kas I	1	6001				
Pac I	1	4132				
Pst I	1	6270				
Pvu I	1	6405				
Sac I	1	6237				
Sal I	1	6264				
Sma I/ Xma I	1	6247				
SnaB I	1	1268				
Sph I	1	6276				
Xba I	1	6258				
Ava I	2	5825	6247			
Ban II	2	5643	6237			
BsaB I	2	1149	3974			
Cla I	2	2527	6882			
Dsa I	2	2763	6617			
Ear I	2	4074	6391			
Eci I	2	1962	6603			
Eco47 III	2	2710	3039			
Fin I	2	5090	6529			
Gdi II	2	6036	6293			
Gsu I	2	6493	6899			
Xmn I	2	357	2646			
Afl III	3	195	3618	3717		
BspM I	3	1113	2258	6268		
BstY I	3	2220	6252	6935		
Eae I	3	5080	6036	6293		
HgiA I	3	4743	5465	6237		
Nde I	3	2723	3803	6846		
Pvu II	3	5960	6053	6375		
Alw I	4	1382	2221	6252	6253	
Fok I	4	239	3547	6361	7244	
Mcr I	4	1422	3879	6405	6521	
Mme I	4	300	5441	5762	6627	
Nci I	4	1924	6247	6248	6838	
NspB II	4	1631	5960	6053	6375	
Sau96 I	4	5724	5914	5938	6396	
Taq II	4	1423	2445	3279	5704	
BsaA I	5	1268	4446	5037	5462	5716
BsmA I	5	2013	2189	4036	5972	7017
Bsp1286 I	5	2088	4743	5465	5643	6237
Dra I	5	189	472	4622	6784	7074
Mae I	5	3827	5565	6259	6861	6979
Ase I	6	4131	4135	4238	4628	6046
		6105				
Hae II	6	2710	3039	5559	5567	6001
		6446				
Nsp I	6	195	3531	3618	3717	6276
		6856				
Ssp I	6	499	2660	5023	5213	6767
		6788				
Ban I	7	1249	5677	6001	6131	6243
		6465	6477			
BceF I	7	1334	1952	3440	4719	5675
		6295	6606			
BstN I	7	1014	1966	5941	5998	6137
		6328	6455			
Drd II	7	320	2301	2391	4471	4948
		5711	5868			
Hga I	7	526	2164	2479	3237	4083
		5158	6687			
Mbo I/ Sau3A I	7	1382	1714	2221	6253	6406
		6502	6936			
SfaN I	7	25	388	1354	3979	4850
		6546	6559			

Enzyme	#	Locations				
Sfe I	7	205	1329	1751	2470	5495
		6270	6673			
Tth111 II	7	3395	3446	4257	4506	4671
		5878	6893			
Mly I	8	2011	2845	4072	5329	5766
		5788	6262	6904		
Ple I	8	2011	2845	4072	5329	5766
		5788	6262	6904		
Tsp45 I	8	1377	1775	2543	2621	2744
		5111	5541	6314		
BsaJ I	9	2763	2894	5997	6136	6247
		6248	6327	6617	6838	
Bbv I	10	931	1367	2521	3132	4871
		5536	5923	6052	6355	6428
Fau I	10	1385	1984	3324	5430	5517
		5571	6027	6084	6398	7131
ScrF I	11	1014	1924	1966	5941	5998
		6137	6247	6248	6328	6455
		6838				
Mbo II	12	781	977	2218	3912	4075
		4271	4937	5255	5587	6391
		6504	6805			
Taq I	13	336	975	1127	1508	1949
		2528	3455	3694	4665	5683
		6235	6265	6883		
Hae III	15	1396	2245	2554	5081	5239
		5345	5414	5725	5939	6037
		6294	6396	6513	6682	7023
Nla III	16	149	196	1107	1300	1799
		2034	2157	2855	3532	3619
		3718	5179	6222	6277	6857
		6965				
Fnu4H I	17	931	1367	1394	2285	2288
		2312	2357	2521	3132	4871
		5500	5514	5536	5923	6052
		6355	6428			
Bsr I	18	534	797	811	1771	2130
		2171	2578	2859	3214	3688
		3781	5102	5114	5802	6076
		6291	6317	6539		
BstU I	18	43	347	1119	1176	2466
		3355	3409	3599	3952	4313
		4994	5489	5513	5533	6029
		6031	6689	6752		
Hpa II/ Msp I	18	314	965	1095	1924	2378
		2396	2552	3370	3842	4018
		5614	6159	6248	6463	6481
		6838	6961	7021		
Hph I	18	1376	1503	1774	1909	2398
		2542	2581	2620	2626	2635
		4847	4923	5117	5706	5947
		5980	7005	7031		
Nla IV	18	1062	1249	1541	1551	1803
		2051	2374	2392	5644	5656
		5677	5867	6001	6131	6243
		6252	6465	6477		
Rsa I	19	173	280	1022	1165	1769
		1796	1889	1905	1970	2133
		3467	3668	4190	4380	5384
		5486	6244	6843	7165	
Tfi I	19	136	216	490	511	723
		2497	3258	3418	3742	3838
		4117	4349	5120	5375	5438
		6041	6624	6885	7041	
Mwo I	20	1325	1388	3218	3598	5503
		5505	5547	5574	5604	5910
		5968	6004	6088	6132	6280
		6401	6431	6440	6466	7125

There are no restriction sites in M13mp18 DNA for the following enzymes:

Aat II	Afl II	Age I	Apa I	ApaL I	Avr II
Bbs I	Bcg I	Bcl I	Bsa I	Bsg I	BspE I
BssH II	BstB I	BstE II	BstX I	Eag I	Eco57 I
EcoN I	EcoO109 I	EcoR V	Esp I	Fse I	Hpa I
Mfe I	Mlu I	Nco I	Nhe I	Not I	Nru I
Nsi I	PflM I	Pml I	PpuM I	PshA I	Rsr II
Sac II	Sap I	Sca I	Sfi I	Spe I	Spl I
Stu I	Sty I	Tth111 I	Xca I	Xcm I	
Xho I/PaeR7 I					

M13mp Series

6171   6191   6211   6231   6251   6271   6291
                                            1   10   20

mp1

Mer Thr Met Ile Thr Asp Ser Leu Ala Val Val Gln Arg Arg Asp Trp Glu Asn Pro Gly ──→ Lac Z

5' . . . TCGTATGTTGTGTGGAATTGTGAGCGGATAACAATTTCACACAGGAAACAGCTATGACCATGATTACGGATTCACTGGCCGTCGT TTTACAACGTCGTGACTGGGAAAACCCTGGCGTTACCCAACTTAAT GGCG

mp2

                                                            EcoRI

5' . . . TCGTATGTTGTGTGGAATTGTGAGCGGATAACAATTTCACACAGGAAACAGCTATGACCATGATTACGGAATTCACTGGCCGTCGTTTTACAACGTCGTGACTGGGAAAACCCTGGCGTTACCCAACTTAAT GGCG

mp7

                                            EcoRI   Sall Accl HincII   Sall   EcoRI
                                                         BamHI   PstI   BamH I

5' . . . TCGTATGTTGTGTGGAATTGTGAGCGGATAACAATTTCACACAGGAAACAGCTATGACCATGATTACGGAATTCCCCGGATCCGTCGACCTGCAGGTCGACGGATCCGGGGAATTCACTGGCCGTCGTTTTACAACGTCGTGACTGGGAAAACCCT GGCG

mp8

                                        Smal Xmal   Sall Accl HincII   Hind III
                                            EcoRI   BamHI   PstI

5' . . . TCGTATGTTGTGTGGAATTGTGAGCGGATAACAATTTCACACAGGAAACAGCTATGACCATGATTACGAATTCCCGGGGATCGTCGACCTGCAGCCAAGCTTGGCACTGGCCGTCGTTTTACAACGTCGTGACTGGGAAAACCCT GGCG

mp9

                                            HindIII   Sall Accl HincII   Smal Xmal
                                                        PstI   BamHI   EcoRI

5' . . . TCGTATGTTGTGTGGAATTGTGAGCGGATAACAATTTCACACAGGAAACAGCTATGACCATGATTACGCCAAGCTTGGCTGCAGGTCGACGGATCCCCGGGAATTCACTGGCCGTCGTTTTACAACGTCGTGACTGGGAAAACCCT GGCG

mp10

                                        EcoRI   Smal Xmal   Xbal   PstI
                                        SacI   BamHI   Sall Accl HincII   Hind III

5' . . . TCGTATGTTGTGTGGAATTGTGAGCGGATAACAATTTCACACAGGAAACAGCTATGACCATGATTACGAATTCGAGCTCGCCCGGGGATCCTCTAGAGTCGACCTGCAGCCCAAGCTTGGCACTGGCCGTCGTTTTACAACGTCGTGACTGGGAAAACCCT GGCG

mp11

                                            HindIII   Sall Accl HincII   BamHI   SacI   EcoRI
                                                        PstI   Xbal   Smal Xmal

5' . . . TCGTATGTTGTGTGGAATTGTGAGCGGATAACAATTTCACACAGGAAACAGCTATGACCATGATTACGCCAAGCTTGGGCTGCAGGTCGACTCTAGAGGATCCCCGGGCGAGCTCGAATTCACTGGCCGTCGTTTTACAACGTCGTGACTGGGAAAACCCT GGCG

mp18

24mer Reverse Sequencing Primer (−48) #1233   16mer Reverse Sequencing Primer (−21) #1201 →   PstI   Xbal   Smal Xmal   17mer Sequencing Primer (−20) #1211   17mer Sequencing Primer (−40) #1212

AGCGGATAACAATTTCACACAGGA   AACAGCTATGACCATG   EcoRI   KpnI   BamHI   Sall Accl HincII   SphI   TGACCGGCAGCAAAATG   CAGCACTGACCCTTTTG

5' . . . TCGTATGTTGTGTGGAATTGTGAGCGGATAACAATTTCACACAGGAAACAGCTATGACCATGATTACGAATTCGAGCTCGGTACCCGGGGATCCTCTAGAGTCGACCTGCAGGCATGCAAGCTTGGCACTGGCCGTCGTTTTACAACGTCGTGACTGGGAAAACCCT GGCG

←CAACACACCTTAACAC   SacI   Smal Xmal   Xbal   PstI   HindIII   TGCAGCACTGACCCT

Hybridization Probe-Primer #1202   TTCGAACCGTGACCGGCAGC   Pentadecamer (15mer) Sequencing Primer #1200

                                    T
                                T   ←RD 29 Primer #1221   AGCACTGACCCTTTTGGGACCGC
                                T   24mer Sequencing Primer (−47) #1224
                            TTTTTT

mp19

24mer Reverse Sequencing Primer (−48) #1233   16mer Reverse Sequencing Primer (−21) #1201 →   HindIII   PstI   Xbal   Smal Xmal   SacI   17mer Sequencing Primer (−20) #1211   17mer Sequencing Primer (−40) #1212

AGCGGATAACAATTTCACACAGGA   AACAGCTATGACCATG   TGACCGGCAGCAAAATG   CAGCACTGACCCTTTTG

5' . . . TCGTATGTTGTGTGGAATTGTGAGCGGATAACAATTTCACACAGGAAACAGCTATGACCATGATTACGCCAAGCTTGCATGCCTGCAGGTCGACTCTAGAGGATCCCGGGTACCGAGCTCGAATTCACTGGCCGTCGTTTTACAACGTCGTGACTGGGAAAACCCT GGCG

←CAACACACCTTAACAC   SphI   Sall Accl HincII   BamHI   KpnI   EcoRI   TGCAGCACTGACCCT

Hybridization Probe-Primer #1202   C   CTTAAGTGACCGGCAGC   Pentadecamer (15mer) Sequencing Primer #1200

                                C/   RD 20 Primer #1220   AGCACTGACCCTTTTGGGACCGC
                            C   24mer Sequencing Primer (−47) #1224

laci' lacZ'

B. Bluescript M13+, M13−

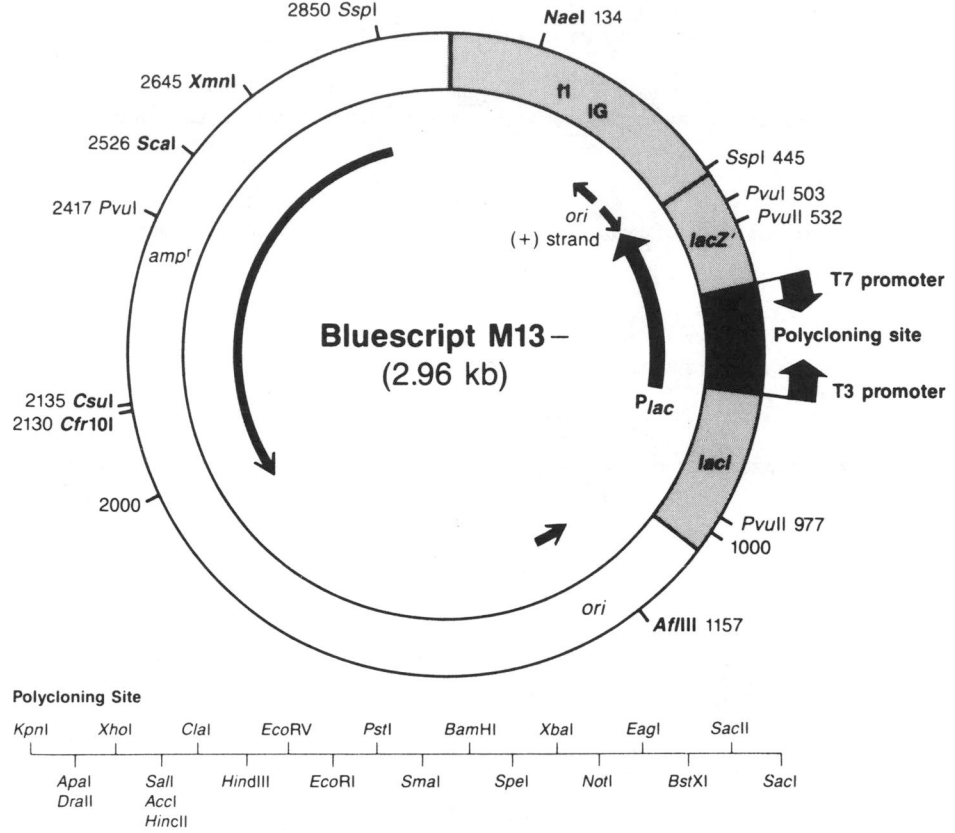

Polycloning Site

KpnI	XhoI	ClaI	EcoRV	PstI	BamHI	XbaI	EagI	SacII
ApaI	SalI	HindIII	EcoRI	SmaI	SpeI	NotI	BstXI	SacI
DraII	AccI							
	HincII							

In Bluescript SK (M13−), the SacI site lies immediately downstream from the bacteriophage T3 promoter and the KpnI site lies immediately downstream from the bacteriophage T7 promoter. In Bluescript KS (M13−), the polycloning site is in the opposite orientation.

Bluescript M13+ and M13− are vectors that contain the bacteriophage M13 origin of DNA replication inserted in opposite orientations into a vector derived from a pUC plasmid that contains phage T3 and T7 promoters. These vectors can thus be used to generate single-stranded DNA in vivo, or RNA in vitro, that is complementary to either of the two strands of foreign DNA inserted into the polycloning site.

References
1. Short et al. 1988. *Nucleic Acids Res.* **16:** 7583–7600.
2. Stratagene Cloning Systems. Product literature for M13−.

SECTION·12

Useful Expression Vectors

A. Factors affecting overall product yield of a plasmid-directed expression system in *E. coli*

B. Bacterial strains with mutations in major protease systems

C. Gene fusion systems used to facilitate protein purification

D. Chemical and enzymatic agents that have been used to cleave fusion proteins site-specifically

E. Some representative expression vectors
1. pKK177-3
2. pET-3a
3. pMON 5743
4. pKC30

F. Some improved fusion vectors

Expressing cloned genes in different organisms is now an advanced science with a rapidly expanding literature of new methodologies. The reader is referred to two recent treatises on this subject: (1) the laboratory manual by Sambrook and co-workers (Sambrook et al. 1989), which details many of the most widely used techniques; and (2) the volume "Gene Expression Technology" (Goeddel 1990), which contains a series of up-to-date reviews on many different aspects of the expression of cloned genes. The latter volume is divided into sections on expression in *Escherichia coli*, *Bacillus subtilis*, yeast, and mammalian cells. Also, the inaugural issue of *Current Opinions in Biotechnology* (1990) is devoted to reviews on this subject. Here, some tables and some figures of selected expression vectors are reproduced from Goeddel (1990) and Sambrook et al. (1989).

REFERENCES

Amann, E. and J. Brosius. 1985. "ATG vectors" for regulated high-level expression of cloned genes in *Escherichia coli*. *Gene* **40:** 183–190.

Balbas, P. and F. Bolivar. 1990. Design and construction of expression plasmid vectors in *Escherichia coli*. *Methods Enzymol.* **185:** 14–37.

Current Opinions in Biotechnology. 1990. **1:** 1–130.

de Lorenzo, V., M. Herrero, U. Jakubzik, and K.N. Timmis. 1990. Mini-Tn5 transposon derivatives for insertion mutagenesis, promoter probing, and chromosomal insertion of cloned DNA in gram-negative eubacteria. *J. Bacteriol.* **172:** 6568–6572.

Goeddel, D.V., ed. 1990. Gene Expression Technology. *Methods in Enzymology*, vol. 185. Academic Press, San Diego.

Gottesman, S. 1990. Minimizing proteolysis in *Escherichia coli:* Genetic solutions. *Methods Enzymol.* **85:** 119–129.

Horii, T., T. Ogawa, and H. Ogawa. 1980. Organization of the *recA* gene of *Escherichia coli. Proc. Natl. Acad. Sci.* **77:** 313–317.

Mead, D.A., E. Szczesna Skorupa, and B. Kemper. 1985. Single stranded DNA SP6 promoter plasmids for engineering mutant RNAs and proteins: Synthesis of a "stretched" pre-parathyroid hormone. *Nucleic Acids Res.* **13:** 1103–1118.

Olins, P.O., and S.H. Rangwala. 1990. Vector for enhanced translation of foreign genes in *Escherichia coli. Methods Enzymol.* **185:** 115–129.

Olins, P.O., C.S. Devine, S.H. Rangwala, and K.S. Kavka. 1988. The T7 phage gene *10* leader RNA, a ribosome-binding site that dramatically enhances the expression of foreign genes in *Escherichia coli. Gene* **73:** 227–235.

Sambrook, J., E.F. Fritsch, and T. Maniatis. 1989. *Molecular Cloning. A Laboratory Manual*, Second Edition. Cold Spring Harbor Laboratory Press, Cold Spring Harbor, New York.

Shimatake, H. and M. Rosenberg. 1981. Purified lambda regulatory protein *c*II positively activates promoters for lysogenic development. *Nature* **292:** 128–132.

Simons, R.W., F. Housman, and N. Kleckner. 1987. Improved single and multicopy *lac*-based cloning vectors for protein and operon fusions. *Gene* **53:** 85–96.

Soberon, X., L. Covarrubias, and F. Bolivar. 1980. Construction and characterization of new cloning vehicles. IV. Deletion derivatives of pBR322 and pBR325. *Gene* **9:** 287–305.

Studier, F.W. and B.A. Moffatt. 1986. Use of bacteriophage T7 RNA polymerase to direct selective high-level expression of cloned genes. *J. Mol. Biol.* **189:** 113–130.

Sussman, J.K., C. Masada-Pepe, E.L. Simons, and R.W. Simons. 1990. Vectors for constructing *kan* gene fusions: Direct selection of mutations affecting IS*10* gene expression. *Gene* **90:** 135–140.

Uhlén, M. and T. Moks. 1990. Gene fusions for purpose of expression: An introduction. *Methods Enzymol.* **185:** 129–144.

SECTION 12

A. Factors Affecting Overall Product Yield of a Plasmid-directed Expression System in *E. coli*

Table 1 is reprinted, with permission, from Balbas and Bolivar (*Methods Enzymol.*, vol. 185, p. 17 [1990]). The references cited in the table note can be found in the original paper.

Table 1 Factors Affecting Overall Product Yield of a Plasmid-directed Expression System in *E. coli*[a]

Factors influencing final protein concentration	Molecular aspects of plasmid vector that may be manipulated to increase product yield	Strain elements used in conjunction with expression vector
Plasmid-related elements		
Plasmid/gene copy number	Replicon	Elements for copy-number regulation
	Genes for selection	Selection-sensitive genotype
	Genes for retention	Elements for cell suicide
	Genes for partition	Not identified
	Instability-inducing DNA sequences	RecA⁻ genotype
mRNA-related elements		
Initiation rate of mRNA synthesis	Promoter and regulatory sequences	trans-Acting regulatory proteins (repressors and activators)
Elongation and termination of mRNA synthesis	Transcriptional terminators and attenuators	trans-Acting termination factors (rho, NusA, NusB)
	Antiterminators	trans-Acting antitermination factors (λN and λQ, NusA)
mRNA degradation rate	RNase recognition sequences	RNase⁻ or RNase *ts* genotype
	REP sequences	Not identified
	Other sequences	Not identified
Protein-related elements		
Translation initiation rate	RBS and AUG sequences	Ribosomes with altered specificity
	Distance and sequence between RBS and translation initiation	Ribosomes with altered specificity
	Specialized ribosomes	Not identified
	Other sequences	Not identified
Elongation and termination of protein synthesis	Gene codon usage	tRNA availability
	Termination codon usage	Availability of termination factors
Protein degradation rates	Fusion proteins	Altered proteolytic activities
	Secretion proteins	Not identified

[a] Under the assumption that the substrates (nucleotides and amino acids), catalytic components (RNA polymerase, DNA polymerase, tRNAs, and cofactors), and energy requirements are not rate limiting. Compiled from text Refs. 14–86.

B. Bacterial Strains with Mutations in Major Protease Systems

Table 1 is reprinted, with permission, from Gottesman (*Methods Enzymol.*, vol. 185, p. 122 [1990]).

Table 1 Bacterial Strains with Mutations in Major Protease Systems

Strain	Relevant genotype	Parental strain	Comments
lon mutant hosts			
SG1117	Δ*gal* Δ*lac lon*-146::ΔTn10 *leu sup*⁺*rec*⁺	HB101[a]	Tetracycline resistant; good transformation recipient
SG12036	Δ*gal* Δ*lon*-510 *sulA*	C600	UV and complex medium resistant; good transformation recipient
SG12041	Δ*gal* Δ*lon*-510 *sulA recA*	C600	Nonfilamenting; Rec⁻
SG1611	F′iᴽ *lacZ* ΔM15/Δ(*lac-pro*) Δ*gal* Δ*lon*-510 *supE*	JM101	Can be used for M13 or pUC vectors
SG13646	Δ*gal* ΔBAM cI857 ΔH1 *lon*-146::ΔTn10 *pro his ilv*	N5340[b]	Useful for plasmids with p$_L$ promoters
lon htpR mutant hosts			
SG21163	Δ*lon*-510 *supFts htpR*am165	MC4100[c]	Temperature sensitive; will not grow at 39°
lon clp mutant hosts			
SG12044	Δ*gal* Δ*lon*-510 *sulA clpA*319::Δ*kan*	C600	Kanamycin resistant
SG21165	Δ*gal* Δ*lac* Δ*lon*-510 Δ*clpA*500	MC4100	
lon clp htpR mutant host			
SG21173	Δ*lon*-510*supFts hptR*am165	MC4100	Kanamycin resistant; temperature sensitive
	Δ*clpA*319::Δ*kan* Δ*lac*		

[a]SA2692, the immediate parent of SG1117, was derived by S. Adhya from HB101.
[b]S. Gottesman, M. Gottesman, J. E. Shaw, and M. L. Pearson, *Cell* **24**, 225 (1981).
[c]M. J. Casadaban and S. N. Cohen, *Proc. Natl. Acad. Sci. U.S.A.* **76**, 4530 (1979).

C. Gene Fusion Systems Used to Facilitate Protein Purification

Table 1 is reprinted, with permission, from Uhlén and Moks (*Methods Enzymol.*, vol. 185, p. 138 [1990]). The references cited in the table can be found in the original paper.

Table 1 Gene Fusion Systems Used to Facilitate Protein Purification[a]

Gene product	Origin	Molecular Weight ($\times 10^3$)	Sec.	Ligand	Refs.
β-Galactosidase	*Escherichia coli*	116	−	TPEG, APTG	7, 8, 39
Protein A	*Staphylococcus aureus*	31	+	IgG	11, 20, 40
CAT	*Escherichia coli*	24	+	Chloramphenicol	9
Poly(Arg)	Synthetic	1–3	−	Ion-exchange	15
Streptavidin	*Streptomyces*	13	+	Biotin	10
Poly(Glu)	Synthetic	1–2	−	Ion-exchange	41
Z	Synthetic	7	+	IgG	22, 23, 34
PhoS	*Escherichia coli*	36	+	Hydroxylapatite	42
Cysteine	Synthetic	<1	+	Thiol	43
Protein G	Streptococci	28	+	Albumin	44
MBP	*Escherichia coli*	40	+	Starch	45
GST	*Escherichia coli*	26	−	Glutathione	46
Flag peptide	Synthetic	2–5	+	Specific IgG	47
Poly(His)	Synthetic	1–7	+	Zn^{2+}, Cu^{2+}	48

[a] A few references for each gene fusion are listed. The molecular weight of the most common fusion part is indicated, as well as if the secretion (Sec.) of the fusion proteins has been demonstrated (+). CAT, Chloramphenicol acetyltransferase; Z, IgG-binding fragment based on staphylococcal protein A; PhoS, phosphate-binding protein; MBP maltose-binding protein; GST, glutathione *S*-transferase; TPEG, (*p*-aminophenyl-β-D-thiogalactosidase); APTG, *p*-aminophenyl-β-D-thiogalactoside.

D. Chemical and Enzymatic Agents That Have Been Used to Cleave Fusion Proteins Site-specifically

Table 1 is reprinted, with permission, from Uhlén and Moks (*Method Enzymol.*, vol. 185, p. 141 [1990]). The references cited in the table can be found in the original paper.

Table 1 Chemical and Enzymatic Agents That Have Been Used to Cleave Fusion Proteins Site-specifically[a]

Cleavage method	Recognition sequence	Refs.
Chemical		
Cyanogen bromide	-Met▼-	2, 3, 35
Formic acid	-Asp▼Pro-	17, 40, 49
Hydroxylamine	-Asn▼Gly-	22, 34
Enzymatic		
Collagenase	-Pro-Val▼Gly-Pro-	8, 39
Enterokinase	-Asp-Asp-Asp-Lys▼-	47
Factor Xa	-Ile-Glu-Gly-Arg▼-	13, 46
Thrombin	-Gly-Pro-Arg▼-	46, 50
Trypsin	-Arg▼-	51
Clostripain	-Arg▼-	9
Ala⁶⁴-subtilisin	-Gly-Ala-His-Arg▼-	43

[a] A few references for each cleavage method are listed. The recognition sequences used in the references are shown and the cleavage site is indicated with an arrowhead.

SECTION 12

E. Some Representative Expression Vectors

Figures 1, 2, and 4, which depict the plasmids pKK177-3, pET-3a, and pKC30, are reprinted from Sambrook et al. (*Molecular Cloning. A Laboratory Manual*, Second Edition. Cold Spring Harbor Laboratory Press, Cold Spring Harbor, New York, pp. 17.12, 17.14, and 17.10, respectively [1989]).

Figure 3, which depicts the plasmid pMON 5743, is reprinted, with permission, from Olins and Rangwala (*Methods Enzymol.*, vol. 185, p. 116 [1990]). This plasmid includes the inducible *recA* promoter, the *g10*-L RBS (the phage T7 gene-*10* leader which is highly efficient for translating many different foreign genes in *E. coli*), and an origin of single-stranded DNA replication.

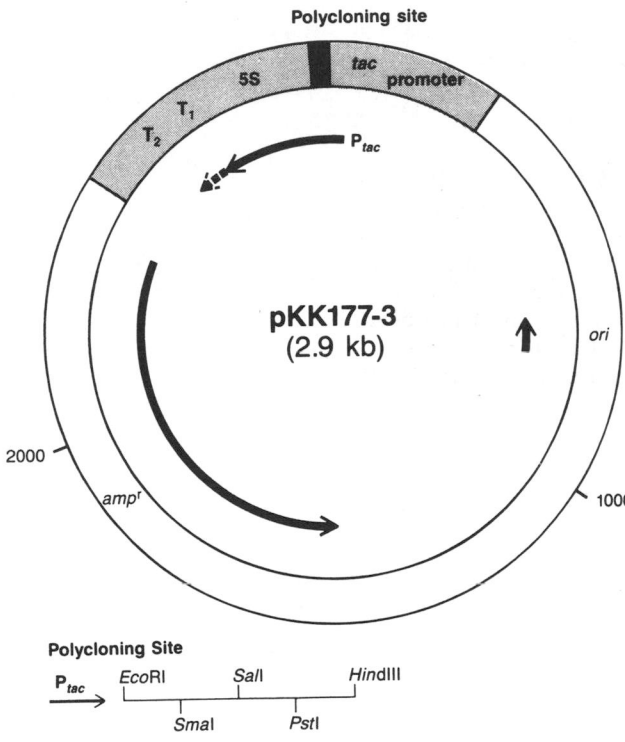

Figure 1 pKK177-3 is a *tac* vector containing multiple sites downstream from the *tac* promoter into which a gene can be cloned. Downstream from these sites is *rrn*B, which contains an *E. coli* 5S gene and the T$_1$ and T$_2$ terminators (Amann and Brosius 1985).

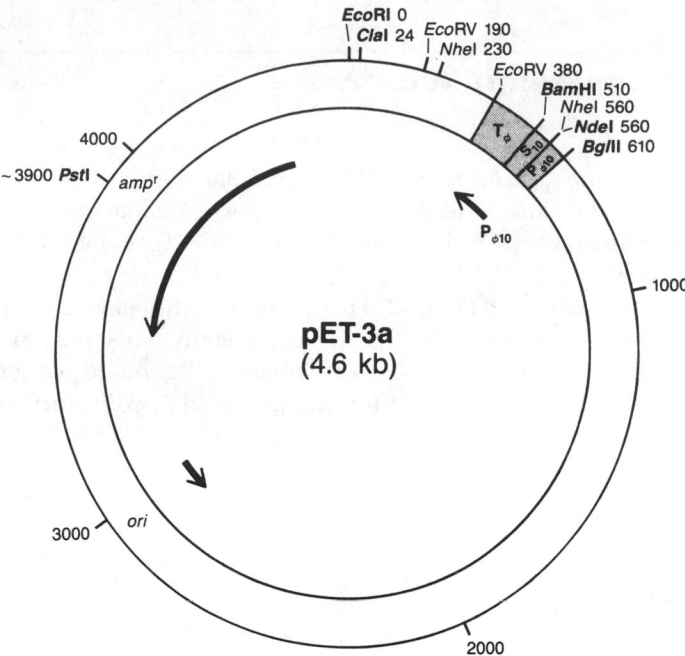

Figure 2   pET-3 carries the phage T7 ϕ10 promoter ($P_{\phi 10}$) and ϕ terminator (T_ϕ). The terminator may make the transcripts more resistant to exonucleolytic degradation (Studier and Moffatt 1986). pET-3a is a derivative of pET-3 into which the translation start (S_{10}) of phage T7 ϕ10 (the major capsid protein of phage T7) with a *Bam*HI site at codon 11 has been inserted. The *Nde*I site (CATATG) is located at the translation start site and can be used to construct a plasmid that directs the expression of native proteins.

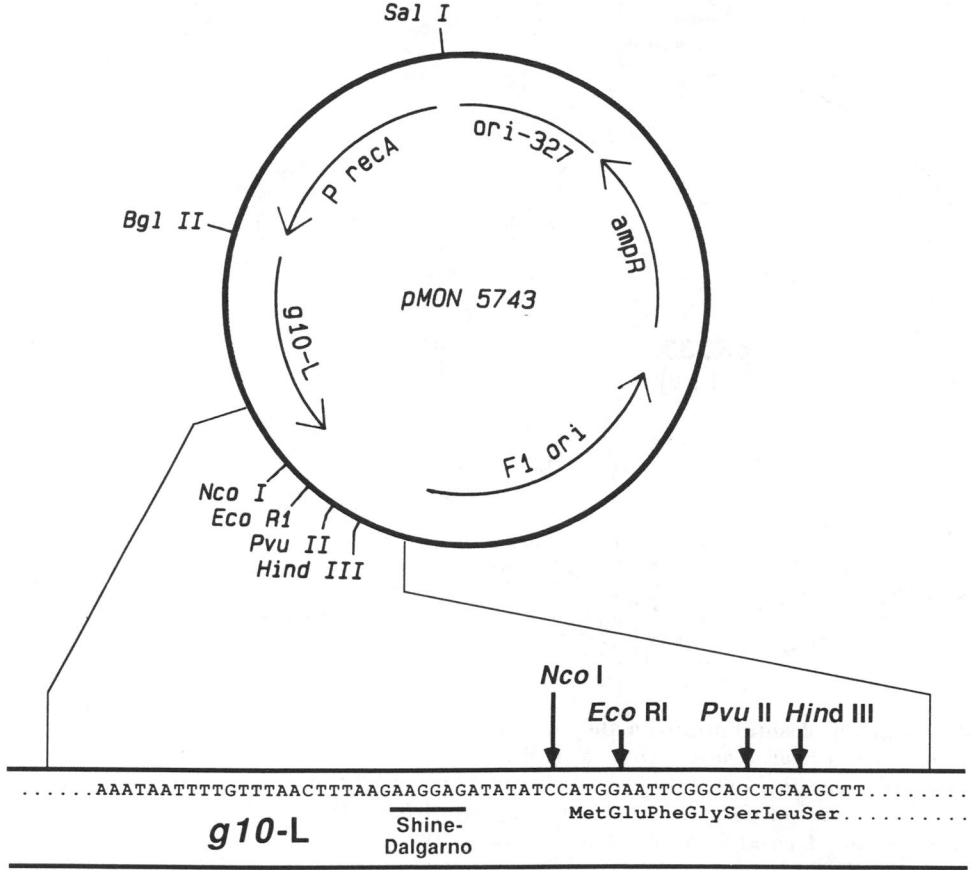

Figure 3   Physical map of plasmid pMON 5743. The *ampR* and *ori-327* regions are derived from the *Sal*I to *Eco*RI fragment of pBR327 (Soberon et al. 1980). The *recA* promoter (Horii et al. 1980), *g10*-L (Olins et al. 1988), and multi-linker segments were constructed as synthetic DNAs, and the phage F1 origin of single-stranded DNA replication is derived from the 512-bp *Rsa*I fragment of the phage (Mead et al. 1985). All restriction sites shown are unique. The map is not to scale.

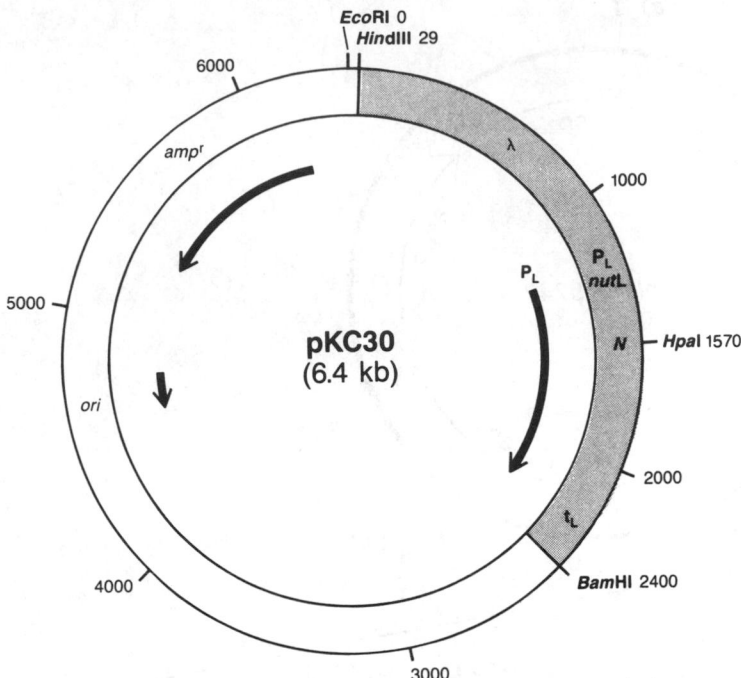

Figure 4 pKC30, a plasmid approximately 6.4 kb in length, carries the p_L promoter of phage λ and a *Hpa*I recognition site located 321 nucleotides downstream from the p_L transcriptional start site. The plasmid is a derivative of pBR322 and contains a *Hin*dIII-*Bam*HI fragment (gray box) derived from phage λ inserted between the *Hin*dIII and *Bam*HI sites within the tetracycline resistance gene (*tet*^{r}). The insertion contains the promoter signal, p_L, a site recognized by the *N* gene product (*nut*L), the *N* gene itself, and the strong *rho*-dependent transcription-termination signal t_L. The *Hpa*I recognition site lies within the coding region of the *N* gene. Sequences inserted into the *Hpa*I site can be regulated by introducing the recombinant plasmid into a temperature-sensitive phage λ lysogen (*c*I*ts*857). The cells are grown to mid-log phase at 30°C and then shifted to 40°C to inactivate the *c*I gene product and to turn on the p_L promoter. This vector has been used to express the phage λ *c*II protein at a level such that the protein comprises 4% of the total protein of the cell (Shimatake and Rosenberg 1981).

F. Some Improved Fusion Vectors

Fusions and vectors that generate fusions are described in Unit 6 of the Laboratory Manual portion of this volume (see also Sections 13 and 14 of this Handbook). Some additional vectors useful for fusions to *lac* or *kan* have been described by Simons, Kleckner, and their co-workers (Simons et al. 1987; Sussman et al. 1990). The advantage of these vectors is that the resulting fusions have low background levels of *lac* or *kan* enzymes and are useful for selecting regulatory mutants with increased expression of different genes. These vectors are depicted below.

Table 1 and Figure 1 are reprinted, with permission, from Simons et al. (*Gene*, vol. 53, pp. 90 and 88–89, respectively [1987]). Figure 2 and Tables 2 and 3 are reprinted, with permission, from Sussman et al. (*Gene*, vol. 90, pp. 136 and 138 [1990]). Figure 3 is reprinted, with permission, from de Lorenzo et al. (*J. Bacteriol.*, vol. 172, p. 6570 [1990]). The references, tables, and figures cited in the table notes and figure captions can be found in the respective original papers.

Table 1   List of Plasmid and Phage Vectors

(A) Ap^R plasmid vectors for constructing multicopy fusions

Plasmid	Fusion type	Order of cloning sites and structure of flanking regions[a]	βGal activity[b]	Size (kb)	Maximum insert size (kb) compatible with transfer to ind^+ phage vectors[c]
pRS415	operon	bla-Tl_4-EcoRI-SmaI-BamHI-$lacZ^+$	50	10.8	4.5
pRS414	protein	bla-Tl_4-EcoRI-SmaI-BamHI-$lacZ^{\cdot}$	<0.01	10.7	4.6
pRS528	operon	bla-Tl_4-BamHI-SmaI-EcoRI-$lacZ^+$	50	10.8	4.5
pRS591	protein	bla-Tl_4-BamHI-SmaI-EcoRI-$lacZ^{\cdot}$	<0.01	10.7	4.6
pRS551	operon	bla-kan-Tl_4-EcoRI-SmaI-BamHI-$lacZ^+$	50	12.5	2.8
pRS552	protein	bla-kan-Tl_4-EcoRI-SmaI-BamHI-$lacZ^{\cdot}$	<0.01	12.4	2.9
pRS550	operon	bla-kan-Tl_4-BamHI-SmaI-EcoRI-$lacZ^+$	50	12.5	2.8
pRS577	protein	bla-kan-Tl_4-BamHI-SmaI-EcoRI-$lacZ^{\cdot}$	<0.01	12.4	2.9

(B) Phage vectors for transferring fusions to single copy

Phage	Structure of recombination region[d]	Genotype of immunity region	Plaque color[e]	Lysogenic phenotypes		
				Color[e]	βGal activity[b]	Size (% wt λ)
λRS45	bla'-$lacZ_{sc}$	imm21 ind^+	white	white	<0.01	89
λRS74	bla'-$plac$UV5-$lacZ^+$	imm21 ind^+	dark blue	dark blue	600[f]	95
λRS88	bla'-$lacZ_{sc}$	imm434 ind^-	white	white	<0.01	98
λRS91	bla'-$plac$UV5-$lacZ^+$	imm434 ind^-	dark blue	dark blue	600[f]	104

(C) Ap^R Km^R plasmid vector for recovering fusions from single copy

Plasmid	Structure of recombination region[d]	*lac* genes	Transformant phenotypes		
			Color[e]	βGal activity[b]	Size (kb)
pRS308	$bla^+$$lacZ_{sc}$	$lacZ_{sc}Y^+A^+$	white	<0.01	8.0

[a] The *Sma*I site is not unique in the Km^R plasmids; *lac'*, the truncated *lacZ* segment of Casadaban et al. (1980).
[b] Activities determined as described in Table II.
[c] Bacteriophage λ is limited in the amount of additional sequence that it can accommodate: Phage genomes larger than about 51 kb (105% of wt λ) are not efficiently packaged into virions (Weil et al., 1972; Murray, 1983). The values shown are estimates of the maximum sizes of fragments that can be cloned into the respective plasmid vectors, and subsequently transferred to the $cI$$ind^+$ phage vectors, without exceeding this limit. Similar estimates indicate that the $cI$$ind^-$ phage vectors will accept approx. 4.0 kb less insert than will the ind^+ vectors. See also Fig. 2.
[d] *bla'*; the 5'-terminal half of *bla*; *lacZ*$_{sc}$, the distal one third of *lacZ* (see Figs. 1 and 2).
[e] In the presence of XGal.
[f] Induced with 1 mM IPTG.

A. OPERON FUSION VECTORS:

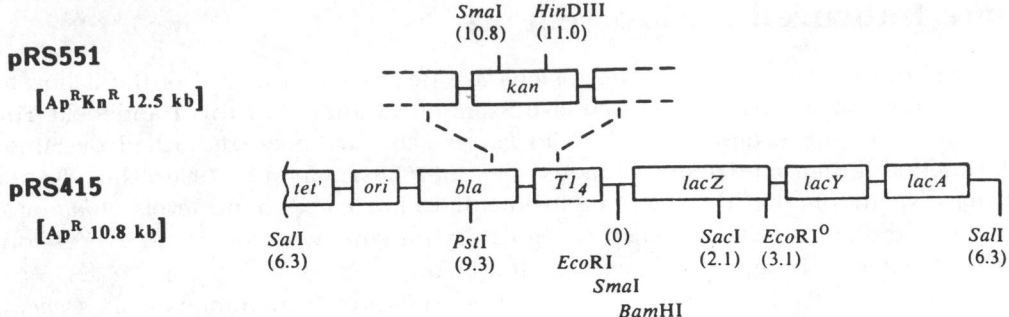

B. PROTEIN FUSION VECTORS:

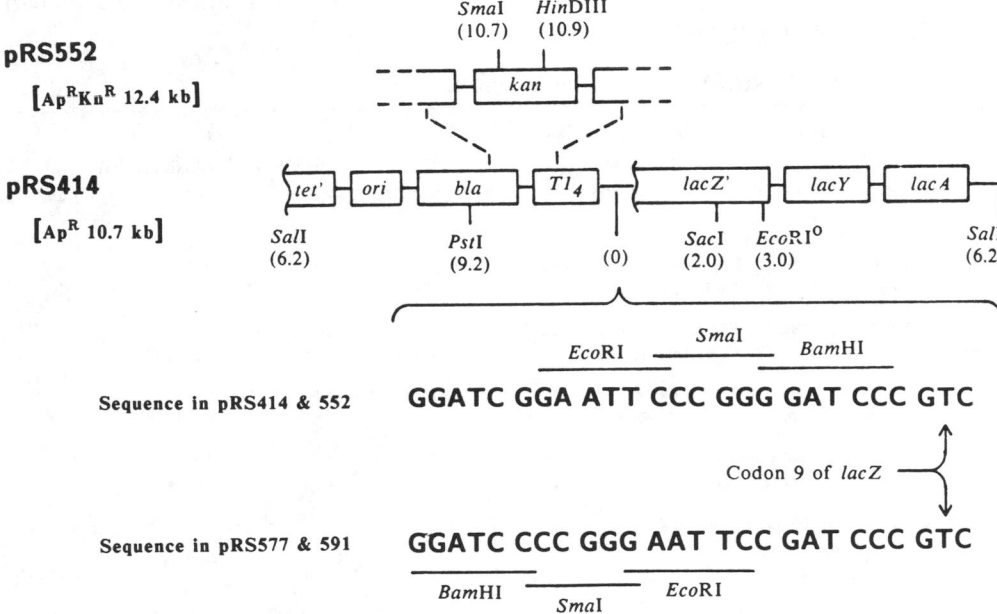

C. VECTOR FOR RECOVERING SINGLE COPY FUSIONS:

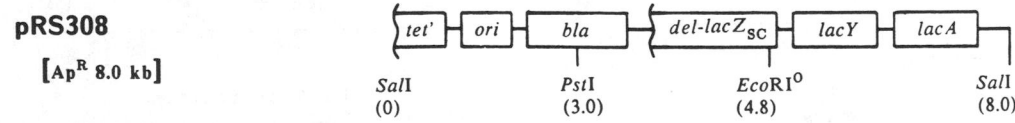

Figure 1 The structures of various multi-copy fusion vectors are shown (not drawn to scale), and a rudimentary restriction map (in kb) of each is provided. (*A*) Operon fusion vectors. pRS415 was derived from pNK678 (Simons et al. 1983) in several steps, which can be summarized as two basic modifications. pNK678 contains a portion of the W205 *trp-lac* fusion (Mitchell et al. 1975; Casadaban et al. 1980). In the first modification, utilizing an *Rsa*I site in the *trp* sequence (at bp 6655 of Yanofsky et al. 1981), all *trp* sequence upstream of bp 6657 was deleted and replaced by an *Eco*RI-*Sma*I-*Bam*HI polylinker. This deletion left 75 bp of *trp* sequence, including the distal 28 bp of the *trpA* gene, the *trpT* sequence, and the W205 fusion junction. The sequence of the W205 junction has been determined (R. Zagursky, unpubl.). It joins bp 6731 of *trp* (Yanofsky et al. 1981) to bp 105 (see Fig. 5 of Dickson et al. 1975) of the *lac* operon leader. This modification reduced background *lacZ* gene expression

by 50%. Presumably, the deleted *trpA* sequence contains a weak, uncharacterized promoter. In the second modification, four tandem copies ($T1_4$) of a 169-bp *Bst*NI fragment (bp 6610–6778 of Brosius et al. 1981) bearing the strong transcriptional terminator, *T1*, of the *E. coli rrnB* ribosomal RNA operon (Brosius et al. 1981) were inserted (in their native orientation with respect to normal transcription of the *lac* genes) at the *Eco*RI site of the polylinker, such that a unique *Eco*RI site was restored at the polylinker. This modification reduced *lacZ* expression to 3% of pNK678 (compare pRS415 and pNK678, Table II), presumably by termination of transcription originating in the plasmid backbone. The remaining low level of expression is of unknown origin, but it is not sufficient for growth on minimal lactose (see Table II). pRS551 was constructed by inserting a 1696-bp *Pvu*II fragment containing the *kan* gene and promoter from Tn*903* (bp 700–2395 of Oka et al. 1981) at the junction between *bla* and $T1_4$ of pRS415. Note that *kan* contains a *Sma*I site. (*B*) Protein fusion vectors. pRS414 was constructed from pRS415 and pMC1403. pMC1403 was originally derived by Casadaban et al. (1980) for the construction of protein fusions to the *lacZ* gene. It was engineered such that several unique cloning sites were inserted into the eighth codon of the *lacZ* gene. Simons et al. (1987) merely recombined *Eco*RI-*Sal*I fragments from pMC1403 and pRS415 so that in pRS414, $T1_4$ would be upstream of these cloning sites. pRS552 was derived in an analogous fashion from pRS551 and pMC1403. pRS528, 550, 591, and 577 (not shown) are identical to pRS415, 551, 414, and 552, respectively, except that the cloning sites are in reverse order. The nucleotide sequence of the polylinker and adjacent portion of the *lacZ'* gene of the protein fusion vectors is shown for use in the design of protein fusion constructions. (*C*) Vector for recovering single-copy fusions. pRS308 carries a deletion of sequence between the unique *Eco*RI and *Sac*I sites in pNK678 (essentially a deletion of the *T1* terminators and proximal two thirds of the *lacZ* gene of pRS415). It retains the *bla* gene and that portion of the *lac* operon distal to the *Sac*I site. The complete sequence of these vectors can be inferred from published sources taking into account the manipulations that were predicted to occur during construction. pRS551 (12460 bp, numbered from the *Eco*RI site in the polylinker): bp 1–18 = GAATTCCCGGGGATCCGG; bp 19–93 = bp 6657–6731 of *trp* (Yanofsky et al. 1981); bp 94–113 = bp 105–124 of the *lac* operon leader (Fig. 5 in Dickson et al. 1975); bp 114–3191 = bp 1–3078 of *lacZ* (Kalnins et al. 1983); bp 3192–4490 = bp 61–1359 of the *lacY* region (Buchel et al. 1980); bp 4491–6298 = bp 4419–6226 of the *lacA* region (Hediger et al. 1985); bp 6299–10011 = bp 651–4363 of pBR322 (Sutcliffe 1979; Peden 1983); bp 10012–10019 = TTGATCCG; bp 10020–11715 = bp 700–2395 of Tn*903* (Oka et al. 1981); bp 11716–11728 = CGGATCAATTCCC; bp 11729–12452 = four tandem copies of the sequence AATTCC + bp 6610–6778 of the *rrnB*1 *T1* terminator (Brosius et al. 1981) + GGAATT; bp 12453–12460 = GGGGATCG. pRS552 (12340 bp) = pRS551 deleted for bp 17–136. pRS415 (10752 bp) = pRS551 deleted for bp 10014–11721. pRS414 (10632 bp) = pRS415 deleted for bp 17–136. pRS550 (12460 bp), pRS528 (10752 bp), pRS577 (12340 bp), and pRS591 (10632 bp) have the sequence CCCGGGAATTCC substituted for, respectively, bp 12460–11 of pRS551, bp 10752–11 of pRS415, bp 12340–11 of pRS552, and bp 10632–11 of pRS414. pRS308 (7952 bp) = pRS415 in which bp 8306–2065 are replaced by the sequence AATT. Simons et al. (1987) have not confirmed the sequences of the junctions created during any of the constructions, except for those in the polylinker regions. A complete description of all plasmid constructions, along with their inferred sequences and restriction maps, can be obtained from R.W.S. *Eco*RI°, *Eco*RI site removed by an uncharacterized point mutation (Bertrand et al. 1984); *tet'*, the distal half of *tet*; *lacZ'*, the *lacZ* segment truncated by Casadaban et al. (1980); *lacZ*~sc~, a deletion removing the proximal two thirds of *lacZ*.

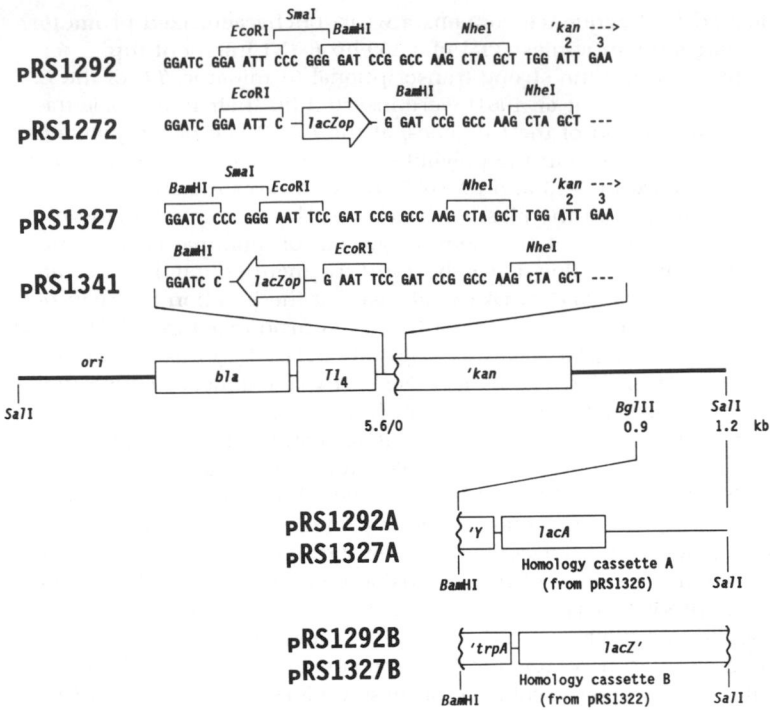

Figure 2 Structures and partial restriction maps of multi-copy 'kan vectors are shown. Plasmids pRS1292 and pRS1327 contain four unique cloning sites upstream of 'kan (the second and third codons of native kan are indicated). pRS1272 and pRS1341 are pRS1292 and pRS1327 with lacZopUV5 (large arrow). Other derivatives contain "homology cassettes" for transferring fusions to phage vectors (see Fig. 2). pRS1292 was constructed from pKM109-9 (Reiss et al. 1984), which contains the kan gene with its first codon replaced by a BamHI site (and unique downstream SmaI and SalI sites). Sussman et al. (1990) converted the SmaI site to BglII (pKM109-9 S/B) and combined the 1.2-kb BamHI-SalI 'kan fragment with the 4.5-kb SalI-BamHI backbone from pRS415. pRS1327 is pKM109-9 S/B with the BamHI site converted to EcoRI and the 1.2-kb EcoRI-SalI 'kan fragment combined with the 4.5-kb SalI-EcoRI plasmid backbone from pRS528. The A and B derivatives were constructed by inserting "homology cassettes" A or B (from pRS1326 or pRS1322) between the BglII and SalI sites. pRS1322 contains the BamHI-SalI lac fragment from pMC1633 (M. Casadaban, pers. comm.) + the SalI-BamHI backbone of pRS591, with subsequent replacement of sequence between EcoRI in lacZ and a downstream SalI site with a SalI linker. pRS1326 was constructed by replacing lac sequence between the BamHI and the distal ClaI site in pRS415 with a BamHI site. pRS415, pRS528, and pRS591 have been described (Simons et al. 1987). Plasmid sequences, which can be inferred from published sources, and details of constructions will be provided with all requests for vectors.

Table 2   Multi-copy Amp^r Plasmids for Constructing 'kan Protein Fusions

Plasmid[a]	Cloning sites[a]	Homology cassette[b]	Plasmid size (kb)	Max. insert for transfer to λ (kb)[c]
pRS1292	R1-Sm-Bm-Nh	none	7.1	na
pRS1327	Bm-Sm-R1-Nh	none	7.1	na
pRS1272	R1-*lacZop*UV5-Bm-Nh	none	7.2	na
pRS1341	Bm-*lacZop*UV5-R1-Nh	none	7.2	na
pRS1292A	R1-Sm-Bm-Nh	A	9.9	7.1
pRS1292B	R1-Sm-Bm-Nh	B	11.0	2.9
pRS1327A	Bm-Sm-R1-Nh	A	9.9	7.1
pRS1327B	Bm-Sm-R1-Nh	B	11.0	2.9

[a] See Fig. 1. R1, *Eco*RI; Sm, *Sma*I; Bm, *Bam*HI; Nh, *Nhe*I.

[b] See Figs. 1, 2 and sections **a, b**.

[c] Values are estimates of the maximum size of fragments that can be cloned in the plasmid vectors and subsequently crossed to λRS45 or λRS74 (Simons et al., 1987). na, not applicable.

Table 3   Properties of 'kan Vectors with and without Inserts and Comparison with Fusions to the 'lacZ Gene

IS10 genotype[a]	Amount of tnp gene (codons)[e]	tnp'-'kan fusions		tnp'-'lacZ fusions	
		Element[b]	Km^R (μg Km/ml)[c]	Element[b]	βGal (units) [Lac$^{+/-}$][d]
1 (no insert)	—	pRS1292	≈1	pRS414	<0.01 [−]
2 (*lacZop*UV5)	—	pRS1272	≈2	pRS476	20 [−]
3 R5	75	pRS967	6	pRS556	40 [±]
4 mci30 R5	75	pRS1087	70	pRS1325	565 [+]
5 HH104 R5	75	pRS1088	260	pRS993	1560 [+]
6 mci30 HH104	75	pRS1323	>350	pCJ291	14525 [+]
7 HH104	25	pRS1362	170	pRS1301	1450 [+]
8 HH104	75	pRS1328	130	pRS709	1235 [+]
9 HH104	166	pRS1353	80	pRS1352	420 [+]
10 (no insert)	—	λRS434	1	λRS194	<0.01 [−]
11 HH104 R5	75	λRS414	15	λRS382	?0 [−]

[a] R5 abolishes MCI; mci30 and HH104 increase pIN strength (Case et al., 1988; 1989).

[b] See Figs. 1 and 3 and Simons et al. (1987). λRS434, λRS414, λRS194 and λRS382 were obtained by crossing λRS45 with pRS1292B, pRS1356 (= pRS1344 + cassette B), pRS552 and pRS993.

[c] Km^R estimated as the μg Km/ml resulting in ≈50% EOP in *E. coli* strain DR459 [ΔlacX74 galOP308 rpsL trpR Δ(tonB-trpA)905]; in each case, twice the 50% EOP concentration reduced EOP > 100-fold.

[d] βGal assayed in DR459 according to Simons et al. (1987); symbols −, ± and +, no, slow, and normal growth on minimal lactose plates, respectively.

[e] N-terminal codons.

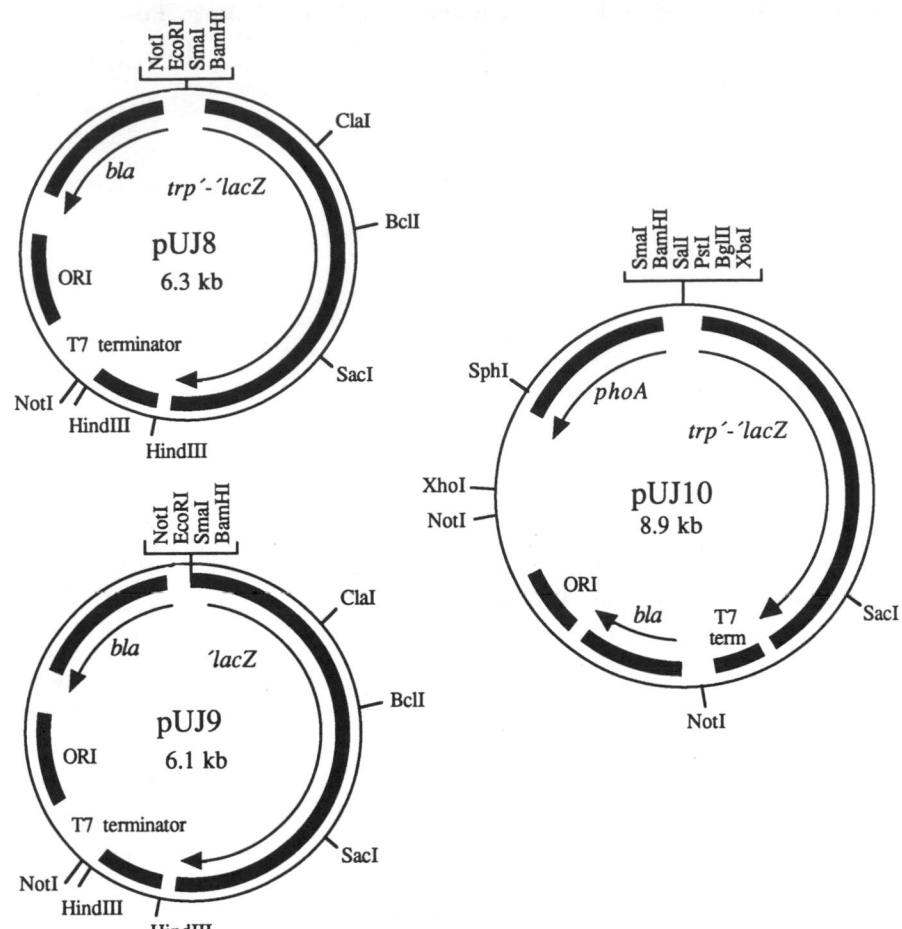

Figure 3 Auxiliary plasmids for insertion of gene fusions into the chromosomes of target bacteria. pUJ8 allows the generation in vitro of operon fusions with *lacZ* (Type I) in a fashion identical to that of other plasmids like pRZ5605 (18), whereas pUJ9 is used for Type-II fusion construction identical to that of pMLB1034 (29). Both plasmids, however, allow the excision of the fusions obtained as a *Not*I fragment, cloning of the fusions at the *Not*I site of the mini-Tn5 elements (see Section 13D of this Handbook), and its eventual chromosomal integration through transposition to the chromosome. The same is true for pUJ10, which allows the construction in vitro of divergent Type-I fusions to *lacZ/phoA* reporters and their insertion in the chromosome as *Not*I fragments. The *Not*I fragment of pUJ8 (devoid of inserts) containing the *lacZ* gene is 4.1 kb, that of pUJ9 is 3.9 kb, and the corresponding fragment in pUJ10 with both *phoA* and *lacZ* is 6.7 kb. The other *Not*I fragment, containing in all cases an origin of replication and an ampicillin resistance gene derived from pUC18, is 2.2 kb. Strong phage T7 transcriptional terminators (T) have been added in all cases downstream from the *lacZ* sequence at a former *Dra*I site.

SECTION · 13

Transposable Elements

A. Useful transposable elements and structures of representative transposable elements

B. Useful Tn*10* derivatives

C. Selected restriction sites in Tn*10*

D. Mini-Tn*5* transposon derivatives

E. Copy numbers of IS elements in the chromosomes of *E. coli* and other bacteria

F. Locations of some IS elements on the *E. coli* K12 genetic map

REFERENCES

Berg, D.E. and M. Howe, eds. 1989. *Mobile DNA*. American Society for Microbiology, Washington, D.C.

Berg, C.M., D.E. Berg, and E.A. Groisman. 1989. Transposable elements and the genetic engineering of bacteria. In *Mobile DNA* (ed. D.E. Berg and M. Howe), pp. 879–925. American Society for Microbiology, Washington, D.C.

de Lorenzo, V., M. Herrero, U. Jakubzik, and K.N. Timmis. 1990. Mini-Tn*5* transposon derivatives for insertion mutagenesis, promoter probing, and chromosomal insertion of cloned DNA in gram-negative eubacteria. *J. Bacteriol.* **172:** 6568–6572.

Deonier, R.C. 1987. Locations of native insertion sequence elements. In Escherichia coli *and* Salmonella typhimurium. *Cellular and Molecular Biology* (ed. F.C. Neidhardt), pp. 982–989. American Society for Microbiology, Washington, D.C.

Herrero, M., V. de Lorenzo, and K.N. Timmis. 1990. Transposon vectors containing non-antibiotic resistance selection markers for cloning and stable chromosomal insertion of foreign genes in gram-negative bacteria. *J. Bacteriol.* **172:** 6557–6567.

Kleckner, N., J. Bender, and S. Gottesman. 1991. Uses of transposons with emphasis on Tn*10*. *Methods Enzymol.* **204:** 139–180.

Way, J.C., M.A. Davis, D. Morisato, D.E. Roberts, and N. Kleckner. 1984. New Tn*10* derivatives for transposon mutagenesis and for construction of *lacZ* operon fusions by transposition. *Gene* **32:** 369–379.

SECTION 13

A. Useful Transposable Elements and Structures of Representative Transposable Elements

Table 1 and Figure 1 are reprinted, with permission, from Berg et al. (*Mobile DNA* [ed. D.E. Berg and M. Howe]. American Society for Microbiology, Washington, D.C., pp. 881–889 [1989]). The references cited in the table, table notes, and figure caption can be found in the original paper.

Table 1 Useful Transposable Elements

Designation	Marker(s)[a]	Size (kb) and structure	Comments[b] (references)
Gram-negative elements			
Tn3[c]	*amp*	5; no IS elements	Duplication: 5 bp. IR: 38 bp. Transposition: Replicative; high frequency to plasmids, low to chromosome; AT-rich regions preferred. Stimulates adjacent deletion formation. Polar in one orientation, stimulates distal transcription in other orientation. Depicted in Fig. 1. (157, 169; Sherratt, this volume)
Tn3-Ap,Km	*amp, kan*	11.7	Derivative of Tn3 with replication region and *kan* gene of plasmid R1. (148)
Tn3-*trp*	*amp, trpB,C,D*		Derivative of Tn3 with *trpB,C,D* genes. (148)
Tn3-HoHo1 (Tn3-*lac*)	*amp, 'lac*	14.25	Derivative of Tn3. *lac* operon lacks a functional promoter and Shine-Dalgarno sequences, so both type I and type II fusions are obtained. Transposition defective. (367)
m-Tn3-*lac*	*amp, 'lac*	4.6	Derivative of Tn3. Forms type II protein fusions. Resolution via P1 *loxP*; no *res* site. Because of transposition immunity, cannot be used to mutagenize pBR322-derived plasmids. Also derivatives with genes selectable in *S. cerevisiae*. (342)
Tn2301 (Tn1-*oriT*)	*amp*		Derivative of Tn1 (closely related to Tn3) with cloned F-factor conjugation genes. Makes F'-like plasmid insertions and Hfr-like chromosomal insertions. (192)
Tn951	*lacZY*	16.6	Duplication: 5 bp. IR: 40 bp, first 38 bp identical to those of Tn3. Transposition defective, complemented by Tn3 (but not Tn2501) transposase. Tn2501 ("minor" transposon) and IS1 are present in Tn951. Resolution of Tn951 cointegrates by Tn2501 resolvase. (86)
Tn2501	None	6.3	Duplication: 5 bp. IR: 38 bp. Minor transposon of Tn951. In Tn3 family, IRs most closely related to Tn21. Tn2501 promotes its own transposition, but not that of Tn951. (263)
Tn2505 (Tn951-kan)	*kan*	8.2	Derived from Tn951 by replacement of 10.5-kb *Bam*HI fragment containing IS1, *lac*, and right IR of Tn2501 with 2.1-kb *kan* fragment from Tn5. Transposition defect is complementable by Tn3 transposase. Is the minor transposon of Tn2515 (below) in pMBLG2. Can be used to generate deletions by recombination with a copy of Tn2506 (below) transposed to the same plasmid. (262)
Tn2515	*kan, tet*		Derived from a plasmid containing Tn2505 by insertion of a 4.8-kb *Eco*RI fragment containing the Tn3 transposase and IR. Forms a transposition-proficient major transposon containing the minor transposon Tn2505. Distinguishable from Tn2505 by ability to transpose further and by Tet[r]. (262)
Tn2506	*cam*	7.3	Derived in several steps from Tn2505. Contains the *cam* gene from pSA. Minor transposon of Tn2516. (262)
Tn2516	*cam, tet*		Major transposon of Tn2506. Analogous to Tn2515. (262)
Tn501	*mer*	8.2	Duplication: 5 bp. IR: 38. In Tn3 family. Transposase, resolvase, and resistance genes are mercury inducible. Depicted in Fig. 1. (169, 336, 337)
Tn1721	*tet*	11.2	Duplication: 5 bp. IR: 38 bp. In Tn3 family, related to Tn501, identical to Tn1771. Contains minor transposon, Tn1722. *Eco*RI sites 15 bp from end are useful for mapping and for deleting most of transposon. *tet* gene is amplified by homologous recombination involving long flanking direct repeats and is selected by growth with high levels of tetracycline. Depicted in Fig. 1. (75, 337, 387)
Tn1722	None	5.6	Minor transposon of Tn1721 (see Fig. 1). (337, 387)
Tn4431	*tet, 'lux*	15	Derivative of Tn1721 with promoterless *lux* operon of *Vibrio fischeri*. Contains *lux* in a 7.5-kb *Bam*HI fragment inserted into *Bam*HI site of Tn1721. (349)
Tn1721-PATE	*tet, pate*	22	Derivative of Tn1721 with polygalacturonate *trans*-eliminase I and II genes from *Klebsiella oxytoca*. (395)

Table 1 *Continued*

Designation	Marker(s)[a]	Size (kb) and structure	Comments[b] (references)
Tn1725	*cam*	8.9	Derivative of Tn1722 with 3.3-kb *Hind*III *cam* fragment of plasmid Sa inserted in *Hind*III site. (387)
Tn1731-*bam*	*tet*	7.6	Derivative of Tn1722 with 2.4-kb *Eco*RI fragment containing *tet* of Tn1721 from pJOE452 replacing 0.2-kb *Hpa*I fragment of Tn1722 and with a *Bam*HI site introduced 57 bp from the left end. (387, 388; R. Schmitt, personal communication)
Tn1732	*kan*	6.7	Derivative of Tn1722 with 1.45-kb *Mlu-Pst*I *kan* fragment of Tn2680 replacing 0.4-kb *Mlu-Pst*I fragment of Tn1722. (387)
Tn1733	*kan, ble, str*	8.9	Derivative of Tn1722 with 3.3-kb *Hind*III *kan ble str* fragment of Tn5 into *Hind*III site. (387)
Tn1735Cm	*cam, p_{tac}*	9.6	Derivative of Tn1731-*bam* with 1.75-kb *Bam*HI-*Hind*III *p_{tac}-lacI^q* fragment from plasmid pFD115 replacing 3-kb *Bam*HI-*Hind*III fragment. High-level, outward transcription inducible by IPTG. Contains 3.2-kb *Hind*III *cam* fragment of plasmid SA inserted in *Hind*III site. (388)
Tn1735Km	*kan, p_{tac}*	9.8	As in Tn1735Cm, but with 3.3-kb *Hind*III *kan ble str* fragment of Tn5 inserted in *Hind*III site. (388)
Tn1735Sm	*str, p_{tac}*	8.4	As in Tn1735Cm, but with 2.0-kb *Hind*III-*aodA* fragment of pHP45 inserted in *Hind*III site. (388)
Tn1736Tc	*tet, 'cam*	8.4	Derivative of Tn1731-*bam* with 0.8-kb promotorless *cam* gene of Tn9 from pCM4 inserted in *Bam*HI site. Forms type I fusions. (388; Schmitt, personal communication)
Tn1736Km	*kan, 'cam*	6.8	Derivative of Tn1731-*bam* with 1.45-kb *Bam*HI-*Hind*III *kan* fragment of pNeo replacing 3.0-kb *Bam*HI-*Hind*III fragment and with promoterless *cam* gene as in Tn1736Tc. (388; Schmitt, personal communication)
Tn1737Km	*kan, 'lac*	9.8	Derivative of Tn1731-*bam* as in Tn1736Km but with 3.8-kb promoterless *lac* fragment from pFD127 in *Bam*HI site. Forms type I fusions. (388; Schmitt, personal communication)
Tn1737Cm	*cam, 'lac*	11.1	Derivative of Tn1731-*bam* in several steps with *lac* as in Tn1737Km and *cam* from plasmid SA. Forms type I fusions. (388; Schmitt, personal communication)
Tn1737Sm	*str, 'lac*	10.6	Derivative of Tn1731-*bam* in several steps with *lac* as in Tn1737Km and *str* from pHP45. Forms type I fusions. (388; Schmitt, personal communication)
Gamma-delta (Tn1000)	Unknown	6.0	Duplication: 5 bp. IR: 35 bp. Related to Tn3. Present on F factor. Transposition: high frequency to plasmid, quite random, but strong preference for AT-rich sequences. Cointegrate formation results in mobilization of chromosome or nonconjugative plasmid. (157, 162, 241; R. Reed, personal communication)
Tn5	*kan, ble, str^d*	5.7; 1.5-kb IS50 elements inverted	Duplication: 9 kb. IR: 19 bp. Transposition: Conservative; high frequency; quite random but some hot spots found. Transposition to multicopy plasmid is selected by high neomycin resistance. Sometimes stimulates distal transcription in either orientation. *kan* gene encodes *npt-II*. Streptomycin resistance is cryptic in *E. coli*. Depicted in Fig. 1. (34, 36, 41, 43, 256, 297; Berg, this volume)
Tn5-*lac*	*kan, 'lac*	12	Type I operon fusions. *lac* replacing most of IS50_L. (218)
Tn5ORF*lac*	*kan, 'lac*		Type II protein fusions. *lac* replacing most of IS50_L. (215)
Tn*phoA* (Tn5*phoA*)	*kan, 'phoA*	7.7	Type II protein fusions with *phoA*. *phoA* replacing most of IS50L. Fusions used to detect membrane or exported proteins. Depicted in Fig. 1. (253)
Tn5seq1	*kan*	3.2	Has outward-facing subterminal T7 and SP6 and transposase gene of IS50R). Useful for DNA sequencing and for in vivo and in vitro T7- or SP6-directed transcription. Depicted in Fig. 1. (275)

Table 1 *Continued*

Designation	Marker(s)[a]	Size (kb) and structure	Comments[b] (references)
Tn5seq2	*supF*	0.26	Has one O and one I end of IS*50* and unique subterminal sequences for use as primer binding sites for DNA sequencing. Transposase gene in *cis* on donor plasmid. Used for mutagenesis of λ cloning vectors. Insertions selected by suppression of λ or chromosomal nonsense mutation. Depicted in Fig. 1. (S. Phadnis, H. Huang, and D. E. Berg, in preparation)
Tn5tac1	*kan, p_tac*	4.6	Has O and I ends and transposase gene of IS*50*R and *lacI*q and outward-facing *p_tac*. High-level transcription is inducible by IPTG. Depicted in Fig. 1 (78)
Tn5-VB32	*tet, 'kan*		Type I operon fusions with *kan*. (22)
Tn5-131, -132	*tet*	5.7	Tn*10 tet* gene in place of Tn*5* resistance genes. Tn*5*-131 is transposition deficient; Tn*5*-132 is transposition proficient. (37, 324)
Tn5-133, -134	*kan, tet*	8.4	Contains Tn*10 tet* gene. Tn*5*-133 is transposition deficient; Tn*5*-134 is transposition proficient. (37, 324)
Tn5-Tc	*kan, tet*		Contains *Sau*3A fragment of RP4 in *Bam*HI site. (356)
Tn5-CM	*cam*	3.9	Contains 0.8-kb *Bam*HI promoterless *cam* cartridge from Tn9 inserted downstream of the *kan* promoter, in place of the Tn*5* central region. (334)
Tn5-GM	*gen*	7.5	Contains 4.4-kb *Hind*III fragment from pOX38-GM in place of Tn*5* central region. (334)
Tn5-TC2	*tet*	5.1	Contains 2.1-kb *Eco*RI-*Pvu*II *tet* fragment from pBR322 in place of Tn*5* central *Bcl*I segment. (334)
Tn5-TP	*tmp*	5.2	Contains 2.1-kb *Bam*HI *tmp* fragment from R388 in place of Tn*5* central region. (334)
Tn5-751	*kan, tmp*	9.0	Contains 3.3-kb *Bam*HI *tmp* fragment from R751. (310)
Tn5-SM	*str*	5.2	Contains 2.1-kb *Hinc*II fragment with promoterless *str* gene from RSF1010 inserted downstream of *kan* promoter, in place of Tn*5* central region. (334)
Tn5-AP	*amp*	5.0	Contains 2.2-kb *Bam*HI fragment from a pSC101::Tn3 plasmid in place of Tn*5* central region. (334)
Tn5-oriT	*kan*	6.5	Contains transfer origin of plasmid RK2 inserted as a 760-bp fragment into the *Bam*HI site of Tn*5*. Chromosomal insertions are Hfr-like in presence of helper RK2. Depicted in Fig. 1. (412)
Tn5-mob	*kan*		Contains transfer origin of plasmid RP4 inserted in *Bam*HI site of Tn*5*. Chromosomal insertions are Hfr-like, and plasmid insertions are F'-like in the presence of helper RP4. (355)
Tn5-V	*kan*		Contains replication origin of plasmid pSC101. Useful for in vitro cloning of target genes from other species into *E. coli*, although the active replication origin may limit its usefulness for insertion mutagenesis in the enterics. (137)
Tn5-PV	*kan*		Contains transfer and replication origins of plasmid RK2 and T-DNA borders of *Agrobacterium tumefaciens* Ti plasmid. Mediates transfer of target plasmids from *Agrobacterium* sp. to plant cells. Can be complemented for autonomous replication in *trans*. (213)
Tn5-luxAB	*kan*		Derivative of Tn*5*-PV with promoterless *luxAB* genes replacing most of IS*50*L. Type I fusion probe. (Koncz-Kalman and Schell, personal communication)
Tn5tox	*kan, tox*	10.5	Has 4.6-kb *Bam*HI *Bacillus thuringiensis* delta endotoxin gene fragment in place of Tn*5* central region. Used for introducing antilepidopteran toxin gene into root-colonizing pseudomonads. Also a 7.3-kb derivative in which the *kan* gene and part of IS*50*R are deleted has been constructed. Transposition defective. (285, 286)
Tn5-233	*gen/kan, str/spc*	6.6	Central *Bgl*II fragment of Tn*5* replaced with 3.5-kb *Bam*HI-*Bgl*II *gen/kan-str/spc* fragment of plasmid Sa. Unlike Tn*5*, does not confer resistance to neomycin. Does not suppress transposition of a newly introduced Tn*5*. (106)

Table 1   *Continued*

Designation	Marker(s)[a]	Size (kb) and structure	Comments[b] (references)
Tn5-GmSpSm	*gen, spc, str*	6.8	Central *Bgl*II fragment of Tn5 replaced with 3.7-kb *Bam*HI fragment of Tn*1696*. (176)
Tn5-235	*kan, 'lacZY*	10	*lacZY* inserted into *Bam*HI site in *str*; *lac* transcribed as part of the *kan* operon. (106)
Tn5-410	*trpE*	8	Derivative of Tn5 with *E. coli trpE* gene. (261)
Tn5.7	*tmp, str, spc*	7.6	Central *Bgl*II fragment of Tn5 replaced by 4.6-kb *Sau*3A fragment of Tn7. (423)
Tn5 *Bgl*⁺	*kan, bgl*	11.3	Derivative of Tn5 with 5.6-kb *Bam*HI insert containing *bgl*⁺ (cellobiose utilization) gene from *Klebsiella oxytoca*. (395)
Tn5 *Amy*⁺	*kan, amy*	8.7	Derivative of Tn5 with 3.0-kb fragment containing *amy*⁺ (amylase) gene from *Klebsiella oxytoca*. (395)
Tn7	*tmp, str, spc*	14	Duplication: 5 bp. IR: About 30 bp, but needs about 105 bp from left end and 75 bp from right end for transposition. Transposes to a single site in the *E. coli* chromosome with a very high frequency and to secondary sites with a lower frequency. Transposes to plasmids. Useful for introducing DNA fragments to a single defined chromosomal site in *E. coli*. Depicted in Fig. 1. (238; Craig, this volume)
Tn7-*lac*	*lacZYA*	11.2	Transposition defective. Also 6.7- to 8.6-kb derivatives with *lac* under the control of different promoters, with different restriction sites or with part of *lacA* deleted, or both. (16; G. F. Barry, personal communication)
Tn9	*cam*	2.6; 0.8-kb IS*1* elements as direct repeats	Duplication: 9 (rarely 8) bp. Transposition to many sites, but significant hot spots. Stimulates adjacent deletion formation. One IS element and *cam* can be lost by crossover between IS elements. Usefulness in *E. coli* is compromised by presence of multiple IS*1* elements in genome. Depicted in Fig. 1. (187)
Tn*10*	*tet*	9.3; 1.3-kb IS*10* elements inverted	Duplication: 9 bp. IR: 23. Transposition: To many sites, but hot spots found. Stimulates adjacent deletion and inversion formation. Stimulates low-level distal transcription. Depicted in Fig. 1. (97, 204; Kleckner, this volume)
Tn*10*-kan	*kan*	8.2	Derivative of Tn*10* with Tn5 *kan* gene. Called Kan^r-Tn*10*. (402)
Tn*10*-amp	*amp*	7.1	Inverse Tn*10* transposon with pBR322 *amp* gene in place of *tet*. Transposition reduced. Called *amp* hopper. (130, 402)
Tn*10*-cam	*cam*	9.0	Inverse Tn*10* transposon with Tn9 *cam* gene in place of *tet*. Transposition reduced. Called *cam* hopper or TnHACIO. (130, 402)
Tn*10*HH	*tet*	7.7	Like Tn*10*, but IS*10*L partially deleted; high hopper mutation. Called Tn*10* del4HH104. Also a 9.4-kb *tet kan* derivative called high hopper *kan-tet*. (402)
Mini-Tn*10*	*tet*	4.0	Transposase gene outside element. Therefore, stable after transposition. Called *p_tac* mini-*tet* (depicted in Fig. 1). Also a 3.5-kb derivative called *del16del17 tet*, a 2.8-kb *kan* derivative called *p_tac* mini-*kan*, and a 5.4-kb *kan* derivative called *del16del17kan*^r. Can be complemented in *trans* by strongly expressed transposase gene. (402)
Mini-Tn*10*-LK	*kan, 'lac*	4.9	Contains ends of Tn*10*, promoterless *lacZ* gene (forms type II protein fusions), and *kan*. (185)
Mini-Tn*10*-LUK	*kan, 'lac, URA3*	6.1	Derivative of mini-Tn*10*-LK with *S. cerevisiae URA3* gene inserted at unique *Bam*HI site in mini-Tn*10*-LK. (185)
Tn*10*-trplac	*tet, 'lac*	11	Type I operon fusions. Called *trp-lac* fusion hopper. *trp-lac* replacing most of IS*10*L. (402)
m-Tn*10*/URA3/supF	*URA3, supF*	2.1	Contains *S. cerevisiae URA3* gene and *E. coli supF*. Also a 4.2-kb *URA3 tet* and a 3.2-kb *TRP1 kan* derivative. (364)
Mu	*imm*^Mu	37.5	Duplication: 5 bp. IR: 2 bp. Bacteriophage. Transposition: High frequency to chromosome, not recovered in high-copy-number plasmid; quite random, but some hot spots and regional specificity are found. Depicted in Fig. 1. (182, 371, 382; Pato, this volume)

Table 1 *Continued*

Designation	Marker(s)[a]	Size (kb) and structure	Comments[b] (references)
MupAp1	*amp*	37.5	Derivative of Mu generated by Tn*3* transposition to Mu and spontaneous deletion of nonessential Mu and Tn*3* segments. Plaque forming. G loop locked into (+) orientation. (232)
Mu dI1	*amp*, '*lac*	37	Derivative of Mu with promoterless *lac* genes near the S (right) end of Mu. Type I operon fusions. Useful for studying operon expression. Carries 1.2-kb IS*121* element. Originally called Mu d1. Depicted in Fig. 1. Also derivatives with conditional or complementable transposition defects. (13, 66, 183, 287)
Mu dII301	*amp*, '*lac*	35.6	Derivative of Mu with *lac* genes deleted for promoter and translational start sequences. Type II protein fusions. (65)
Mu dI770-1	*amp*, *kan*, '*lac*		Derivative of Mu dI1 with *kan* gene. Type I fusions. (67)
Mu dII770-301	*amp*, *kan*, '*lac*	35.6	Derivative of Mu dII301 with *kan* gene. Type II fusions. Equivalent to Mu dI770-1. (67)
Mu18-1	*amp*		Mini-Mu derivative of Mu-*amp*, probably generated by inverse transposition of Tn9. Retains one IS*1* element which generates further deletions. *amp* inserted into Mu18. Complementable transposition defect. Also a MuA⁺ transposition-proficient derivative. (311, 381)
Mini-Mu-*kan*	*kan*	16	Mini-Mu derivative of Mu with Tn*903 kan* gene. Complementable transposition defect. (70)
Mini-Mu-*amp*	*amp*	6	Mini-Mu derivative of Mu with Tn3-pBR322 hybrid *amp* gene. Also a 10-kb derivative with Tn*3 amp* gene. Complementable transposition defect. (70)
Mini-Mu-*lac*	*lac*	17	Mini-Mu derivative of Mu with functional *lac* cluster. Also a 22-kb derivative with G region that has broader host range. Complementable transposition defect. (70)
Mu dI1678	*amp*, '*lac*		Mini-Mu derivative of Mu dI7701-1 with *kan* and adjacent segment deleted. Type I fusions. (67)
Mu dII1678	*amp*, '*lac*	22.4	Mini-Mu derivative of Mu dII7701-301 with *kan* and adjacent segment deleted. Type II fusions. Equivalent to Mu dI1678. (67)
Mu dI1681	*kan*, '*lac*		Mini-Mu derivative of Mu dI7701-1 with *amp* and adjacent segment deleted. Type I fusions. (67)
Mu dII1681	*kan*, '*lac*	14.2	Mini-Mu derivative of Mu dI7701-301 with *amp* and adjacent segment deleted. Type II fusions. Equivalent to Mu dI1681. (67)
Mu dI1734	*kan*, '*lac*		Mini-Mu derivative of Mu dI1681 with a Mu A and B genes deleted. Type I fusion. Complementable transposition defect. Also called Mu dJ. (67)
Mu dII1734	*kan*, '*lac*	9.7	Mini-Mu derivative of Mu dII1681 with Mu A and B genes deleted. Type II fusion. Complementable transposition defect. Also called Mu dK. Equivalent to Mu dI1734. (67)
Mu d4041	*kan*	7.8	Mini-Mu derivative of Mu dII7701-301 with *amp* and *lac* genes deleted (67). Also a derivative with a cloned mammalian thymidine kinase gene for selection in mammalian cells. (190)
Mu dE	*kan*, *erm*		Derivative of Mu d4041 with *Bam*HI *erm* fragment of pTS19E inserted in *Bam*HI site. *erm* expressed in gram-positive organisms. (225)
Mu dIIPR3	*cam*, '*kan*	4.2	Mini-Mu derivative of Mu with promoterless Tn*903 kan* gene and Tn9 *cam* gene. Forms type II fusions with *kan*. Useful for gene fusions in eucaryotes in which *lac* fusions cannot be used. (307)
Mu dIIPR13	*cam*, '*lac*	8.0	Mini-Mu with promoterless *lacZ* gene and Tn9 *cam* gene. Forms type II fusions. (308)
Mu dPR40	*gen*, *cam*, '*lac*	16	Derivative of Mu dIIPR13 with *gen* gene. (308)
Mu dIIZZ1	*cam*, '*lac*, *LEU2*		Derivative of Mu dIIPR13 with *S. cerevisiae LEU2* gene and 2μm plasmid origin of replication for selection in *S. cerevisiae*. Yeast origin lies in a removable cassette. (92)

Table 1   *Continued*

Designation	Marker(s)[a]	Size (kb) and structure	Comments[b] (references)
Mu dIIPR46	*gen, cam, 'lac*	17	Derivative of Mu dIIPR40 with *oriT* from pSUP5011. (308)
Mu dIIPR48	*gen, cam, 'lac*	21.7	Derivative of Mu dIIPR46 with *oriRiHRI* from pLJbB11. (308)
Mu d(*lacZ npt-II*)	*'lac, 'kan*		Mini-Mu derivative of Mu with promoterless *lac* and *kan* genes in tandem. Forms type II protein fusions with *lac* and type I operon fusions with *kan* in the same transcript. (231)
Mini-Mu-*tet*	*tet, 'lac*		Mini-Mu derivative of Mu dI1681 with Tn*10 tet* replacing *kan.* (19)
Mini-Mu-*lux*	*kan, 'lux*	15.5	Mini-Mu with *V. fischeri luxAB* (luceriferase) genes. Type I operon fusions; emits light when expressed. (122)
Mini-Mu-*lux*Tc[r]	*tet, 'lux*	18.5	*kan* gene of mini-Mu-*lux* replaced by *tet* gene. (122)
Mini-Mu-*cat*	*cam*		Contains synthetic MuL-end (204 bp) and MuR-end (115 bp) bracketing *cam* gene from pTAPI. Mu A and B genes in *cis* on same plasmid. (291)
Mu d13-1	*kan, 'lac*	10.3	Mini-Mu with transfer origin of RK2. Type I operon fusions. Mu A and B genes deleted. Complementable transposition defect. Useful for both plasmid and chromosomal insertions. (Groisman and Heffron, unpublished data)
Mu d5345	*cam, 'lac*	10.5	Contains *oriT* and *lac* segment for transcriptional fusions. Mu A and B genes deleted. Complementable transposition defect. Useful for both plasmid and chromosomal insertions. (Groisman and Heffron, unpublished data)
m-Mu-*tac*	*kan, p_{tac}*	7	Derivative of Mu dII1681 with *lac* replaced by 270-bp *Bam*HI p_{tac} fragment from pKK223-3. (155)
Mu d-5-3	*kan, p_{T7}*		Contains T7 promoter for regulated expression of both cloned and chromosomal genes. Mu A and B genes deleted. Complementable transposition defect. (Groisman and Heffron, unpublished data)
Mu dII79 (*cos*-mini-Mu)	*amp*		Derivative of Mu dII1681 containing cloned cosmid pHC79. Useful for in vivo cloning and packaging into λ heads. (154)
Mu dII4042	*cam, 'lac*	16.7; plasmid	Derivative of Mu dII1681 with origin of replication from multicopy plasmid P15A. Type II protein fusions. Depicted in Fig. 1. Useful for cloning host genes in vivo and for generalized transduction (Fig. 6). (161, 397)
Mu dII5085	*cam, 'lac*	13.4; plasmid	Derivative of Mu dII4042 with Mu A and B genes deleted. Complementable transposition defect. (159)
Mu d5005	*kan*	7.9; plasmid	Derivative of Mu dII4042 with ColE1-type origin of replication from multicopy plasmid pMB1. *cam*, *ori*-P15A, and *lac* deleted. Useful for cloning large (up to 25 to 30 kb) segments of host genome in vivo. Compatible with P15A-derived plasmids. (159)
Mu dI5086	*kan, 'lac*	14.9; plasmid	Derivative of Mu d5005 with promoterless *lac* operon. Type I operon fusions. (159)
Mu dI5155	*kan, 'lac*	15.6; plasmid	Derivative of Mu dI5086 with origin of transfer of broad-host-range plasmid RK2. Mini-Mu plasmid transferred conjugally in presence of helper RK2. (159)
Mu dI5166	*cam, lac*	15.8; plasmid	Derivative of Mu dI5155 with *cam* in place of *kan.* (159)
Mu dII5117	*spc/str, kan, lac*	21.7; plasmid	Derivative of Mu dII1678 with origin of replication and resistance genes of the low-copy-number plasmid SA. Useful for in vivo cloning of genes that are harmful on high-copy-number plasmids. Compatible with P15A- and pMB1-derived plasmids. (159)
Mu d5294	*kan, p_{T7}*	7.7; plasmid	Derivative of Mu d5005 with outward-facing T7 promoter. Useful for high-level in vitro and in vivo T7-directed transcription of cloned fragment. (Groisman et al., in preparation)
Mu d-P22	*cam*	36.4	Derivative of temperate *Salmonella typhimurium* phage P22 with ends of Mu. Transposes like Mu. Upon induction, DNA replicates in situ in one direction from P22 *pac* site and sequential headfuls are packaged in P22 particles. Two forms,

Table 1 *Continued*

Designation	Marker(s)[a]	Size (kb) and structure	Comments[b] (references)
			called P and Q, differ in orientation of P22 and consequently in direction of packaging. Useful for amplifying adjacent host DNA, for DNA sequencing, and for transduction. Depicted in Fig. 1 and Fig. 7. (415)
D3112	*imm*[D3112]	35	Mu-like phage, useful for transposon mutagenesis of *Pseudomonas aeruginosa*. Transposes poorly in *E. coli*. (111, 219)
Mini-D171	*tet*	5.0	Derivative of D3112. Contains *tet* gene from pSC101. Useful for in vivo cloning and generalized transduction. (Darzins and Casadaban, in preparation)
Mini-D165	*kan*	5.1	Derivative of D3112. Contains *kan* gene from Tn5. (Darzins and Casadaban, in preparation)
Mini-D214	*tet*	10.9; plasmid	Derivative of D3112. Contains *tet*, *oriT*, *oriV*, and *trfA** from RK2. Useful for in vivo cloning and generalized transduction. (Darzins and Casadaban, in preparation)
Mini-D948	*amp*	11.4; plasmid	Derivative of D3112. Contains *oriT*, *oriV*, and *trfA** from RK2 and the ColE1-type replication origin of pMB1. Useful for in vivo cloning and generalized transduction. (Darzins and Casadaban, in preparation)
Mini-D366	*gen*, *kan*	12	Derivative of D3112. Contains *gen* and *kan* genes, replication origin of plasmid Sa, and transfer origin of plasmid RK2. Useful for in vivo cloning and generalized transduction. (Darzins and Casadaban, in preparation)
Mini-D385.5	*amp*	10.2	Derivative of D3112. Contains *amp* gene and replication origin of *Pseudomonas* sp. plasmid pVSI. Useful for in vivo cloning and generalized transduction. (Darzins and Casadaban, in preparation)
Lambda (λ)	*imm*[λ]	49	Transposition due to λ-specific integration. Does not generate duplication of target DNA sequences. Depicted in Fig. 1. Insertions into secondary sites found only when *attB* is deleted. (171, 351; Thompson and Landy, this volume)
λ p*lac*Mu1	*imm*[λ], '*lac*		Derivative of Mu dII301 with lambda genes replacing *amp* and most of the Mu genome. Transposes like Mu and packages like λ. Type II protein fusions. Useful for insertion mutagenesis and in vivo cloning. Depicted in Fig. 1. (55)
λ p*lac*Mu3	*imm*[21]		Derivative of λ p*lac*Mu1 with *imm*[21]. (55)
λ p*lac*Mu50	*imm*[λ], '*lac*		Derivative of Mu dI1 with λ genes replacing *amp* and most of the Mu genome. Like λ p*lac*Mu1, but type I operon fusions. (54)
λ p*lac*Mu51	*imm*[21], '*lac*		Derivative of λ p*lac*Mu50 with *imm*[21]. (54)
λ p*lac*Mu9	*kan*, '*lac*		Derivative of λ p*lac*Mu1 with *kan* gene. Type II protein fusions. (54)
λ p*lac*Mu15	*kan*, '*lac*		Derivative of λ p*lac*Mu9 with Mu *Aam*1093 substituted for truncated (and partially functional) Mu *A* allele. (E. Bremer, T. J. Silhavy, and G. M. Weinstock, *Gene*, in press)
λ p*lac*Mu53	*kan*, '*lac*		Derivative of λ p*lac*Mu51 with *kan* gene. Type I operon fusions. (54)
λ p*lac*Mu55	*kan*, '*lac*		Derivative of λ p*lac*53 with Mu *Aam*1093 substituted for truncated Mu *A* allele. (Bremer et al., in press)
Gram-positive elements			
Tn916	*tet*	16.4; no IS elements	Duplication: None. IR: 26 bp. Conjugative transposon (transfer 10^{-8} to 10^{-5} per donor). Related to Tn918, Tn919, Tn925, and Tn1545. Inserts into many sites. Transposition can be zygotically induced. Transposition is associated with excision from donor. Circular transposition intermediate transposes. In *E. coli*, precise excision occurs at high frequency and transposition at a much lower frequency. (82–84, 142, 341, 343)

Table 1 *Continued*

Designation	Marker(s)[a]	Size (kb) and structure	Comments[b] (references)
Tn1545	tet, erm, kan	25.3	Conjugative transposon related to Tn916. More stable in *E. coli* than Tn916. Can transpose in *E. coli*. (60, 87)
Tn917	erm	5.3; no IS elements	Duplication: 5 bp. IR: 38 bp. In Tn3 family. Related to Tn551, Tn3871, Tn4430, Tn4451, and Tn4452. Transposition induced by erythromycin. Transposes to chromosome or plasmid. Can transpose in *E. coli*. (347, 379; Murphy, this volume)
Tn917-lac	erm, 'lac	8.3	Forms type I fusions. (294, 417)
Tn917-cam	erm, 'cam	6.5	Contains promoterless *cat-86* gene from *Bacillus pumilus*. Forms type I fusions. (417)
Tn917-lac-cam	erm, 'lac, 'cam	9.6	Contains promoterless *lac* and *cat-86* genes in tandem. Forms type I fusions. (417)
Tn551	erm	5.3	Duplication: 5 bp. IR: 40 bp. Related to Tn917; member of Tn3 family. Transposes preferentially to the chromosome, with low insertion specificity. (196, 243; Murphy, this volume)
Tn554	erm, spc	6.7	Duplication: None. IR: None. Transposes by site-specific recombination preferentially to one site (*att554*). No excision detected. Has three transposase genes. (217, 272, 299; Murphy, this volume)

[a] Resistance determinants: *amp*, ampicillin; *ble*, bleomycin; *cam*, chloramphenicol or fusidic acid; *erm*, erythromycin; *gen*, gentamicin; *kan*, kanamycin or neomycin; *mer*, mercury; *spc*, spectinomycin; *str*, streptomycin; *tet*, tetracycline; *tmp*, trimethoprim; *tra*, F-factor transfer genes; *trp*, *E. coli trp* operon; *imm*, phage immunity.

[b] Duplication, Number of base pairs of target DNA duplicated during transposition. Lambda, unlike the other elements described here, inserts by a site-specific mechanism, without duplicating host sequences. The designations of Casadaban and Chou (65) for fusion elements are adopted here. I designates operon fusion elements in which the promoterless reporter gene has a translation start site; II designates gene (protein) fusion elements in which the reporter gene has neither a transcriptional nor a translational start site.

[c] Tn3 is the most commonly used member of a number of identical or nearly identical ampicillin resistance transposons (e.g., Tn1, Tn2, Tn801, Tn901, and Tn1701) (Sherratt, this volume).

[d] The *str* gene is cryptic in *E. coli* but can be activated by mutation (257).

Figure 1 Structures of representative transposable elements (not to scale). The vertical filled bars are sites of action of transposition proteins. The hatched bars are the sites of action of resolution proteins (tnpR). Open boxes represent genes involved in transposition and resolutions: Mu, *A* and *B*; λ, *int* and *xis*; Tn5, *tnp/inh*; Tn10, *tnp*; Tn9, *A* and *B*; Tn7, *A–E*; and Tn3 family (including γδ, Tn501, Tn1721, and Tn917), *tnpA* and *tnpR*. IS50L of Tn5 contains an ochre allele of the *tnp/inh* gene, which is expressed only in ochre-suppressing strains. IS10L of Tn10 contains numerous sequence changes that render its *tnp* gene nonfunctional. Curved arrows indicate the site of action of the repressor of transposition (the Tn501 repressor may also act at a second site close to *tnpR*). The resistance determinants are: *amp*, ampicillin; *cam*, chloramphenicol; *kan*, kanamycin; *ble*, bleomycin; *str*, streptomycin; *tet*, tetracycline; *dhfr*, dihydrofolate reductase (trimethoprim resistance); *mer*, mercury; *erm*, erythromycin. Truncated *lacZ* and *phoA* reporter genes and also the truncated P22 *sieA* and Tn1721 *tnpA* genes are indicated by a ' (prime) at the 5' end. P_{T7}, P_{SP6}, and P_{tac} designate outward-facing promoters from phages T7 and SP6 and a *trp-lac (tac)* hybrid promoter, respectively. Other symbols: *rep*, plasmid replication origin; *cI*, *R*, *A*, and *J*, λ genes; *cos*, cohesive end of λ; *erf*, essential recombination function of P22; *pac*, packaging initiation site of P22; O and I, outside and inside IS ends, respectively, of compound transposon; *inh*, inhibitor of Tn5 transposition; *URA3*, *S. cerevisiae* uracil biosynthetic gene. A minor transposon in Tn917 is suggested by the DNA sequence (347). The transposase genes and end sequences of most of the elements are not related. The transposase genes and end sequences of members of the Tn3 family, although related, have diverged sufficiently that they do not interact (although the Tn3 and γδ resolvases will compensate for each other). For detailed descriptions of these and other elements, see the references listed in Table 1, recent books about λ (171) and Mu (371), and the following chapters in Berg and Howe (1989): Mu (Pato); Tn5 (Berg); Tn10 (Kleckner); Tn9 (Galas and Chandler); Tn7 (Craig); Tn3, γδ, Tn501, and Tn1721 (Sherratt); Tn917 (Murphy); λ (Thompson and Landy).

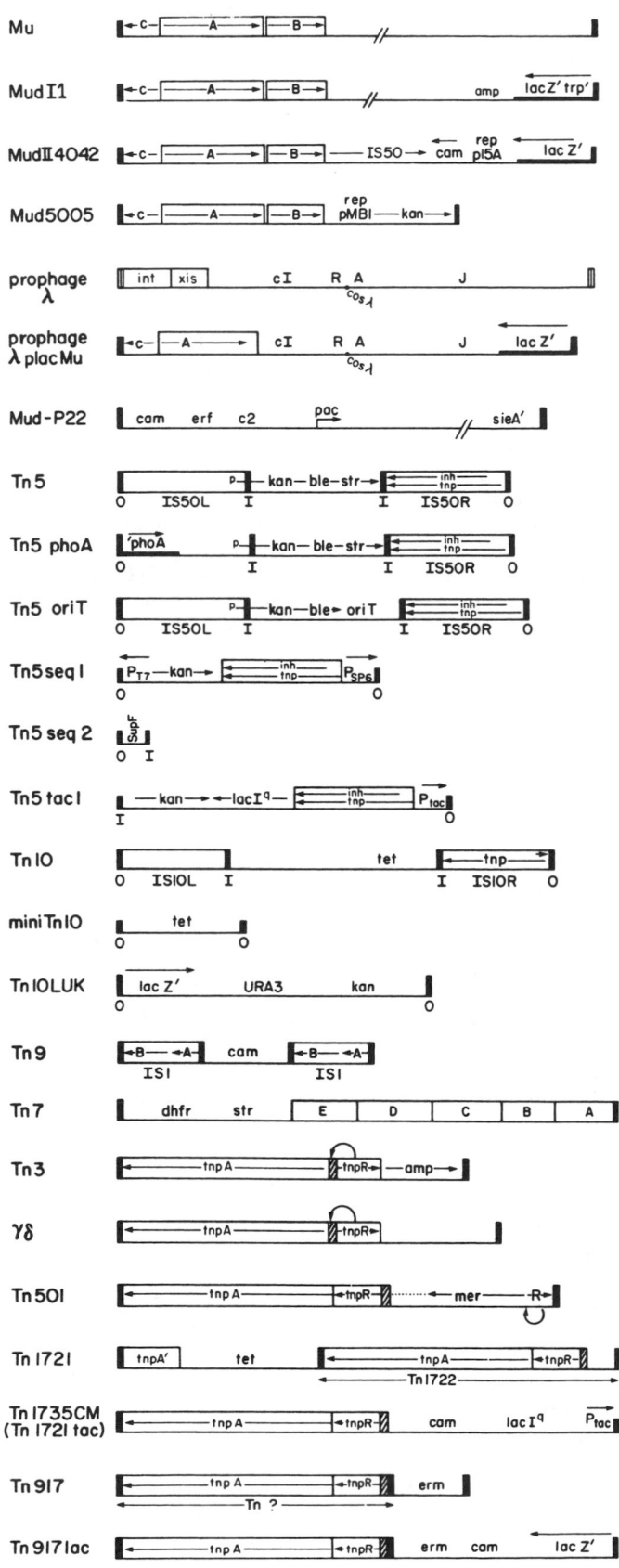

Figure 1 (*See facing page for caption.*)

B. Useful Tn*10* Derivatives

Figure 1 is reprinted, with permission, from Kleckner et al. (*Methods Enzymol.*, vol. 204, pp. 152–155 [1991]). The figures cited in the figure caption can be found in the original paper.

Figure 1 The structure of each transposon-containing restriction fragment (or for derivatives 101, 112, and 113 the transposon itself) is drawn to scale. The backbones into which these restriction fragments (or transposons) are inserted to create each transposon vehicle are described in Fig. 3. In *lacI*⁺ or *lacI*^Q strains, *Ptac* or *Plac-UV5* promoters on these constructions can be fully induced by the addition to the medium of IPTG (isopropyl-β-D-thiogalactopyranoside) to a final concentration of 1 mM. Derivative 101 (wild-type Tn*10*) constructions have been described previously (Way et al., *Gene 32:* 369 [1984]). The open reading frame for transposase protein is bp 108–1313 of IS*10* Right (TRN10IS1R.BACTERIA, GenBank J01829). Derivative 109 (*Ptac*–wild-type transposase) was constructed by cleaving Tn*10* in IS*10* Right with *BclI* (bp 66, Fig. 4a) and ligating a *PvuII* to *EcoRI* fragment containing the *Ptac* promoter (Amann et al., *Gene 25:* 167 [1983]) to this filled-in site so that the transposase gene is under *Ptac* control. The derivative extends to the *EcoRI* site at bp 3140 of Tn*10* (Fig. 4a). Derivative 109 provides a backbone for derivatives 110 and 111. Derivative 102 (*Ptac*–ATS transposase) is identical to derivative 109 with two exceptions. First, the transposase gene in derivative 102 carries two ATS mutations (*ats1*, a G to A transition at bp 508 of IS*10* Right, and *ats2*, a G to A transition at bp 853 of IS*10* Right). Second, the sequence between an *XhoII* site at bp 1319 and a *BglII* site at bp 1942 of Tn*10* has been deleted from derivative 109 and an *XbaI* linker inserted at this deletion junction. Derivative 102 provides a backbone for derivatives 103–108. Mini-Tn*10* derivatives 103–108, 110, and 111 are each bounded by identical inverted repeats of the outermost 70 bp of IS*10* Right (generated by cleaving IS*10* Right with *BclI* and converting the *BclI* site to a *BamHI* site). The 70-bp transposon end in these derivatives is embedded in 40 bp of λ *cI* gene sequence terminating in a *HindIII* site. Thus, each complete transposon is carried on a *HindIII* fragment which is inserted into the *HindIII* site (bp 2272, Fig. 4a) of derivative 102 (*Ptac*–ATS transposase) or of derivative 109 (*Ptac*–wild-type transposase) to put the mini-Tn*10* in *cis* to a transposase source. Derivative 103 (mini-Tn*10kan*/*Ptac*–ATS transposase) carries a *BamHI* Kan^r fragment from Tn*903* (Fig. 4b, TRN903.BACTERIA, GenBank J01839, bp 697 to 2392, *PvuII* fragment converted to a *BamHI* fragment with linkers) oriented in the backbone so that the *kan* gene promoter is transcribing in the same direction as the *Ptac* promoter. Derivative 104 (mini-Tn*10tet*/*Ptac*–ATS transposase) carries a *BamHI* Tet^r fragment from Tn*10* (Fig. 4a, bp 1942–4717, TRN10TETR.BACTERIA, GenBank J01830, bp 3402 to 627, *BglII* fragment converted to a *BamHI* fragment with linkers) oriented in the backbone so that the *tetR* gene promoter is transcribing in the same direction and the *tetA* gene promoter is transcribing in the opposite direction as the *Ptac* promoter. Derivative 105 (mini-Tn*10cam*/*Ptac*–ATS transposase) carries a *BamHI* Tn9-derived Cam^r fragment from pACYC184 (P18XCYC18.SYN, GenBank X06403, bp 3500 to 580, *HaeII* fragment converted to a *BamHI* fragment with linkers) oriented in the backbone so that the *cam* gene promoter is transcribing in the same direction as the *Ptac* promoter. Derivative 106 (mini-Tn*10kan* URA3/*Ptac*–ATS transposase) is identical to derivative 103, and derivative 107 (mini-Tn*10cam* URA3/*Ptac*–ATS transposase) is identical to derivative 105 except that in each case the element also carries an *S. cerevisiae* *BglII* to *BamHI* URA3 gene fragment from YEP24 (YEP24.VEC, GenBank VB0067, *EcoRI* site at bp 2241 converted to *BglII* to the *BamHI* site at bp 3784). The URA3 gene is inserted upstream of the *kan* gene (for derivative 107) or the *cam* gene (for derivative 108) and is oriented so that it is transcribed in the opposite direction from the *kan* or *cam* gene. Derivative 108 (mini-Tn*10kan* *Plac*/*Ptac*–ATS transposase) is identical to derivative 103 except that it also carries a *Plac-UV5* *BamHI* fragment downstream from the *kan* gene oriented so that the promoter is transcribing in the same direction as the *kan* gene promoter out across the transposon end. The sequence of the transposon end through the promoter fragment is shown in Fig. 5e. Derivative 110 (mini-Tn*10* *supF*/*Ptac*–wild-type transposase) carries a 248-bp *XhoII* *supF*

Caption is continued after the figure.

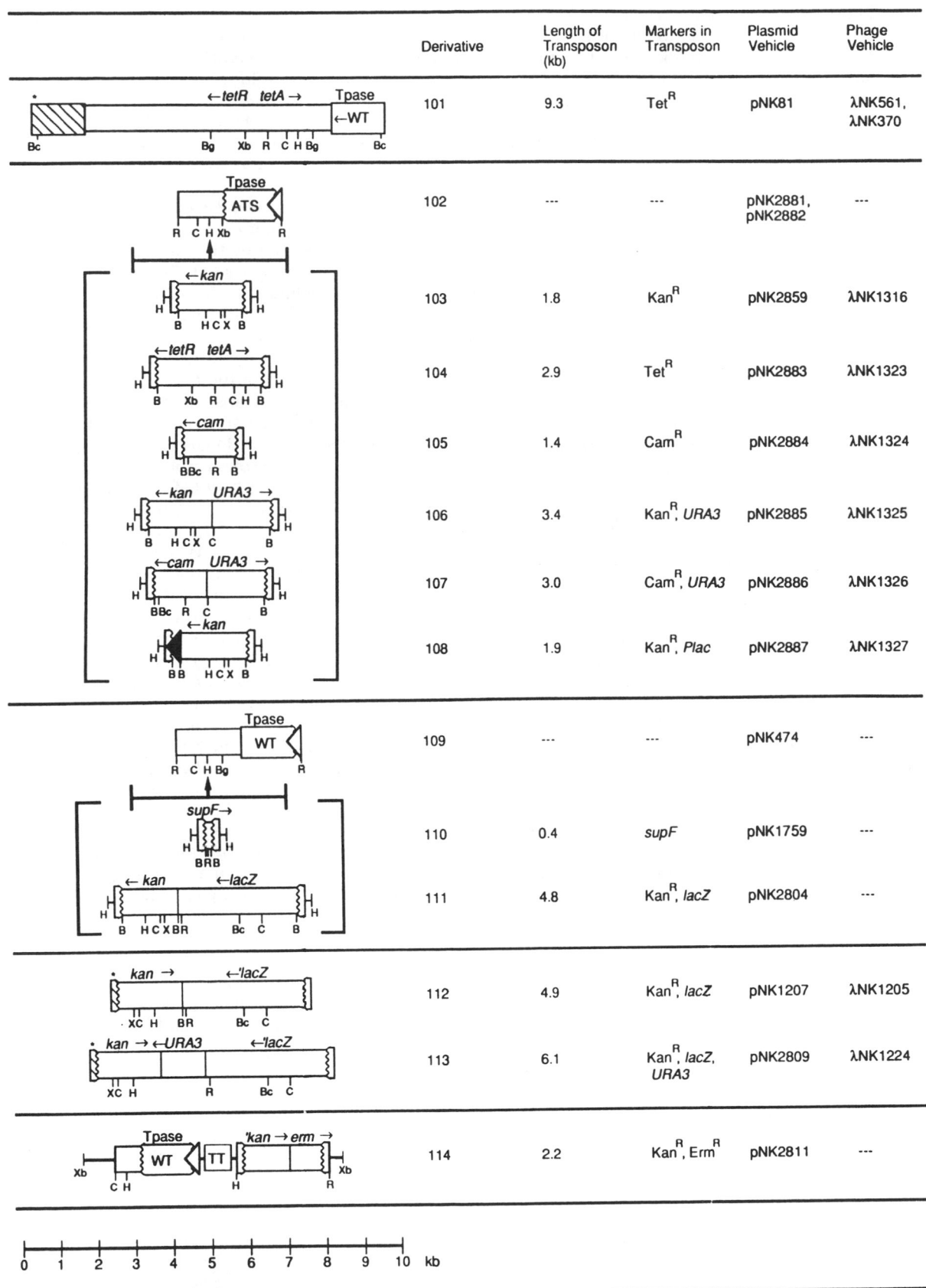

	Derivative	Length of Transposon (kb)	Markers in Transposon	Plasmid Vehicle	Phage Vehicle
	101	9.3	Tet^R	pNK81	λNK561, λNK370
	102	---	---	pNK2881, pNK2882	---
	103	1.8	Kan^R	pNK2859	λNK1316
	104	2.9	Tet^R	pNK2883	λNK1323
	105	1.4	Cam^R	pNK2884	λNK1324
	106	3.4	Kan^R, URA3	pNK2885	λNK1325
	107	3.0	Cam^R, URA3	pNK2886	λNK1326
	108	1.9	Kan^R, Plac	pNK2887	λNK1327
	109	---	---	pNK474	---
	110	0.4	supF	pNK1759	---
	111	4.8	Kan^R, lacZ	pNK2804	---
	112	4.9	Kan^R, lacZ	pNK1207	λNK1205
	113	6.1	Kan^R, lacZ, URA3	pNK2809	λNK1224
	114	2.2	Kan^R, Erm^R	pNK2811	---

Figure 1 (*See facing page for caption.*)

Figure 1   *Continued*

fragment from bp 845 to 208 of PiAN7 (PIAN7.VEC, GenBank VB0066). Ligation of this *Xho*II fragment between the *Bam*HI sites of the transposon ends restores the *Bam*HI sites. The *supF* gene is oriented in the backbone so that its promoter transcribes in the opposite direction from the *Ptac* promoter. Derivative 111 (mini-Tn*10 lacZ kan*/*Ptac*–wild-type transposase) carries a promoterless *lacZ Bam*HI fragment and a Kan^r *Bam*HI fragment. The promoterless *lacZ* fragment consists of leader sequences from the *trpA* gene followed by the ribosome binding site and coding sequence of the *lacZ* gene (Simons et al., *Gene 53:* 85 [1987]). The sequence from the transposon end through the leader region into the *lacZ* gene is shown in Fig. 5d. The fragment extends to the end of the *lacZ* gene (converted to a *Bam*HI site at bp 4373 of ECOLAC.BACTERIA, GenBank J01636). The *lacZ* fragment is oriented in the backbone so that it would be transcribed in the same direction as read by the *Ptac* promoter. Downstream from this promoterless *lacZ Bam*HI fragment is a 1.5-kb Tn*903*-derived Kan^r *Bam*HI fragment from pUC4K (Pharmacia LKB Biotechnology Inc.) oriented so that the *kan* gene promoter is transcribing in the same direction as the *Ptac* promoter. (A version of derivative 111 that is marked only with the promoterless *lacZ* fragment is also available as pNK2803.) Derivative 112 (Tn*10*–LK) carries the '*lacZ* fragment (ECOLAC.BACTERIA, GenBank J01636, bp 1309 to 4373, missing the first eight codons of the *lacZ* gene) oriented so that the gene is fused to the IS*10* Right-derived end of the transposon. The sequence across the transposon end into the '*lacZ* gene is shown in Fig. 5b. The same Kan^r '*Bam*HI fragment that marks derivative 103 is carried within the element downstream from the '*lacZ* gene but oriented so that it is transcribed in the opposite direction from the '*lacZ* gene. The second end of this element consists of the outermost 70 bp of IS*10* Left. Derivative 113 (Tn*10*–LUK) is identical to derivative 112 except that an *S. cerevisiae* URA3 *Bgl*II fragment (YSCODCD.PL, GenBank K02206, bp 1 to 1170) has been inserted at the *Bam*HI site between '*lacZ* and the Kan^r markers oriented to be transcribed in the same direction as the '*lacZ* gene. Derivative 114 carries a 'Kan^r fragment and an Erm^r fragment between the outermost 70 bp of IS*10* Right. The 'Kan^r fragment consists of a leader sequence and the Tn*5* neomycin resistance gene starting at the second codon (TRN5NEO.BACTERIA, GenBank J01834, bp 154) and extending through a *Sal*I site 1130 bp downstream (Rothstein et al., *Cell 19:* 795 [1980]). The sequence of the transposon end through the start of the '*kan* gene is shown in Fig. 5c. A 1-kb selectable erythromycin resistance (Erm^r) fragment (Martin et al., *Plasmid 18:* 250 [1987]; Josson et al., *Plasmid 21:* 9 [1989]) has been inserted into the *Sma*I site 970 bp away from the start of the '*kan* gene oriented so that it is transcribed in the same direction as the '*kan* gene. The entire transposon has been cloned on a *Hind*III to *Eco*RI fragment into *Salmonella hisG* and *hisD* sequences (indicated by a heavy line). Cloned upstream of the '*kan* end of the transposon are four tandem repeats of a 180-bp transcriptional termination sequence from the *rrnB* operon (Brosius et al., *J. Mol. Biol. 148:* 107 [1981]) that prevent expression of the '*kan* gene by nonspecific transcription from the vector (represented by a box containing TT). Transposase is provided to the transposon from a *Ptac*–wild-type transposase fusion (analogous to derivative 109) missing the innermost end of IS*10* Right from a *Xho*II site at bp 1319 to a *Bgl*II site at bp 1942 of Tn*10* and extending to the *Cla*I site at bp 2591 of Tn*10* (Fig. 4a). This *Ptac*–transposase fusion is inserted immediately upstream of the transcriptional terminators and is oriented so that it is transcribed in the opposite direction from the '*kan* gene. Beyond the transposase-proximal end of the construction is another 740 bp of *Salmonella his* DNA terminating in an *Xba*I site. Beyond the mini-Tn*10*-proximal end of the construction is another 250 bp of *his* DNA terminating in an *Xba*I site. B, *Bam*HI; Bc, *Bcl*I; Bg, *Bgl*II; C, *Cla*I; H, *Hind*III; R, *Eco*RI; X, *Xho*I; Xb, *Xba*I. Open triangle, *Ptac*; filled triangle, *Plac*–UV5.

C. Selected Restriction Sites in Tn*10*

Figure 1 is reprinted, with permission, from Way et al. (*Gene*, vol. 32, p. 374 [1984]). The references cited in the figure caption can be found in the original paper.

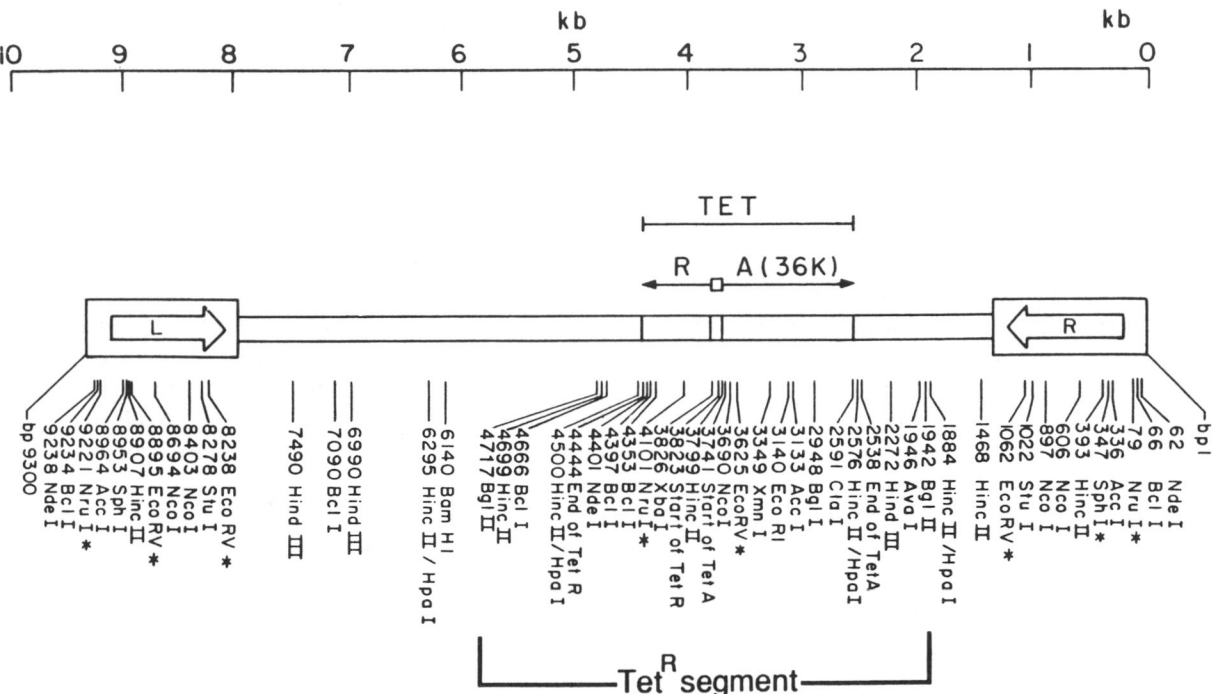

Figure 1 Tn*10* includes IS*10* Right, IS*10* Left, and the tetracycline resistance determinant, which is composed of two divergently transcribed genes: *tetR*, the repressor, and *tetA*, the structural gene for a 36-kD (K) protein. The positions and orientations of these two genes are indicated by the two correspondingly marked divergent arrows. Information compiled from Kleckner et al. (1978), Halling et al. (1982), Jorgensen and Reznikoff (1979), Jorgensen et al. (1979b), Moyed and Bertrand (1983), Wray et al. (1981), N. Kleckner (unpubl.), Hillen and Schollmeier (1983), K. Bertrand (pers. comm.), and Schollmeier and Hillen (1984). Complete DNA sequences are available for all of IS*10* Right and portions of IS*10* Left (Halling et al. 1982) and for the region between the *Hind*III site at bp 2230 and the *Hinc*II/*Hpa*I site at bp 4464 (K. Bertrand, pers. comm., and references above). Numerical positions have been assigned to sites outside of this region for convenience, but these are clearly approximations. Numbers begin with IS*10* Right. Numbers were then assigned to sites within the sequenced *tet* region by designating the *Hind*III site as bp 2300 on the map and calculating the positions of sites in that region accordingly from the DNA sequence of K. Bertrand (pers. comm.). Numbers in IS*10* Left were assigned by assuming the length of the element to be exactly 9300 bp, also clearly an approximation, and assuming that IS*10* Left contains exactly the same number of bp as IS*10* Right. Asterisks indicate that Tn*10* contains *Eco*RV and/or *Nru*I sites in addition to those indicated.

D. Mini-Tn5 Transposon Derivatives

Timmis and co-workers have described a collection of Tn5-derived mini-transposons that simplifies the generation of insertion mutants, in vivo fusions with reporter genes, and the introduction of foreign DNA fragments into the chromosomes of a variety of gram-negative bacteria, including the enteric bacteria and typical soil bacteria such as *Pseudomonas* species. The mini-transposons, pictured in Figure 1, consist of genes specifying resistance to kanamycin, chloramphenicol, streptomycin-spectinomycin, and tetracycline as selection markers and a unique *Not*I cloning site flanked by 19-bp terminal repeat sequences of Tn5. Further derivatives, depicted in Figure 2, also contain *lacZ*, *phoA*, *luxAB*, or *xylE* genes devoid of their native promoters located next to the terminal repeats in an orientation that affords the generation of gene-operon fusions. The transposons are located on a R6K-based suicide delivery plasmid, the pUT plasmid (Herrero et al. 1990), which is depicted in Figure 3. The plasmid provides the IS50$_R$ transposase *tnp* gene in *cis* but external to the mobile element and whose conjugal transfer to recipients is mediated by RP4 mobilization functions in the donor (de Lorenzo et al. 1990). Auxiliary plasmids for insertion of gene fusions into the chromosomes of target bacteria (Herrero et al. 1990) are depicted in Section 12F of this Handbook. Timmis and co-workers have also described transposon vectors containing nonantibiotic resistance selection markers for cloning and stable chromosomal insertion of foreign genes in gram-negative bacteria.

Figures 1 and 2 are reprinted, with permission, from de Lorenzo et al. (*J. Bacteriol.*, vol. 172, p. 6569 [1990]). Figure 3 is reprinted, with permission, from Herrero et al. (*J. Bacteriol.*, vol. 172, p. 6564 [1990]). The references and figures cited in the figure captions can be found in the respective original papers.

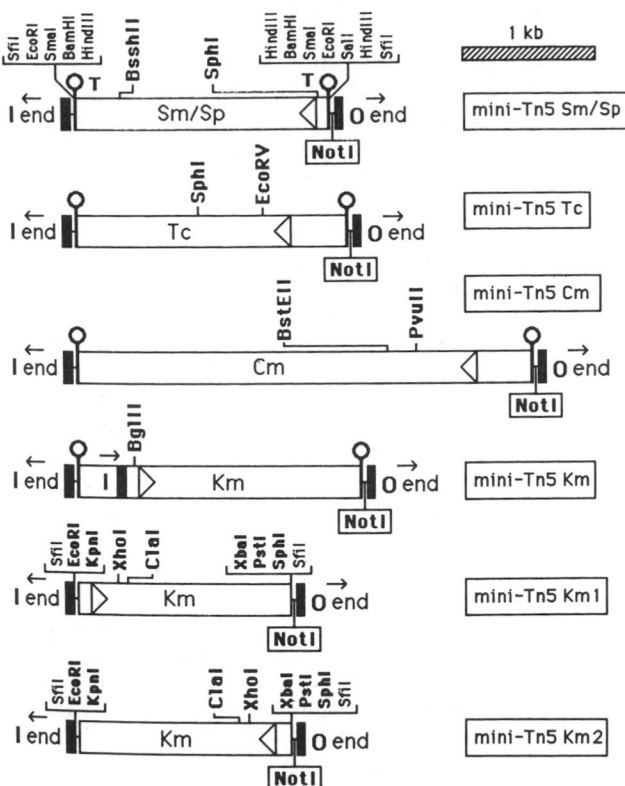

Figure 1 Structure of mini-Tn5 elements. Transposons were constructed in vitro by standard recombinant DNA techniques (19). The determinants for streptomycin-spectinomycin (Sp/Sm), tetracycline (Tc), chloramphenicol (Cm), and kanamycin (Km) resistances were obtained as *Eco*RI fragments from the plasmids bearing them as interposons (8). These fragments were subsequently cloned into the single *Eco*RI site of p18Sfi (11), excised as an *Sfi*I fragment, and inserted between the Tn5 19-bp termini in pUT (11) so that the mobile unit is present in all cases as an *Xba*I-*Eco*RI (partial) portion of the delivery plasmid. The resulting elements were named mini-Tn5Sm/Sp, mini-Tn5Tc, mini-Tn5Cm, and mini-Tn5Tm, respectively. Mini-Tn5Km has one more I end at the left extreme of the kanamycin resistance gene, which itself originates from Tn5 (8). These four transposons carry strong transcriptional terminators (labeled with a T and a circle) flanking the resistance gene, as well as all the restriction sites indicated in mini-Tn5Sm/Sp at their termini. Two further mini-Tn5Km transposons were constructed that contained the Kam^r determinant of Tn903, which was excised as a 1.7-kb *Bam*HI fragment from the mini-Tn10Km of phage λ1105 (33) and inserted into the *Bam*HI site of pUC18Sfi (11) in both orientations before introduction into the pUT plasmid. The resulting elements, mini-Tn5Km1 and mini-Tn5Km2, do not carry terminators flanking the resistance gene. Unique sites within the transposons are indicated by boldface type, but note that some of them might not be unique in the delivery plasmid pUT (11). All elements shown can be used for insertional mutagenesis or as transposon vectors for the cloning of DNA fragments flanked by *Not*I sites (readily isolated by first cloning DNA fragments into the pUC18 derivatives p18Not and pUC18Not; see reference 11 and the text for explanations).

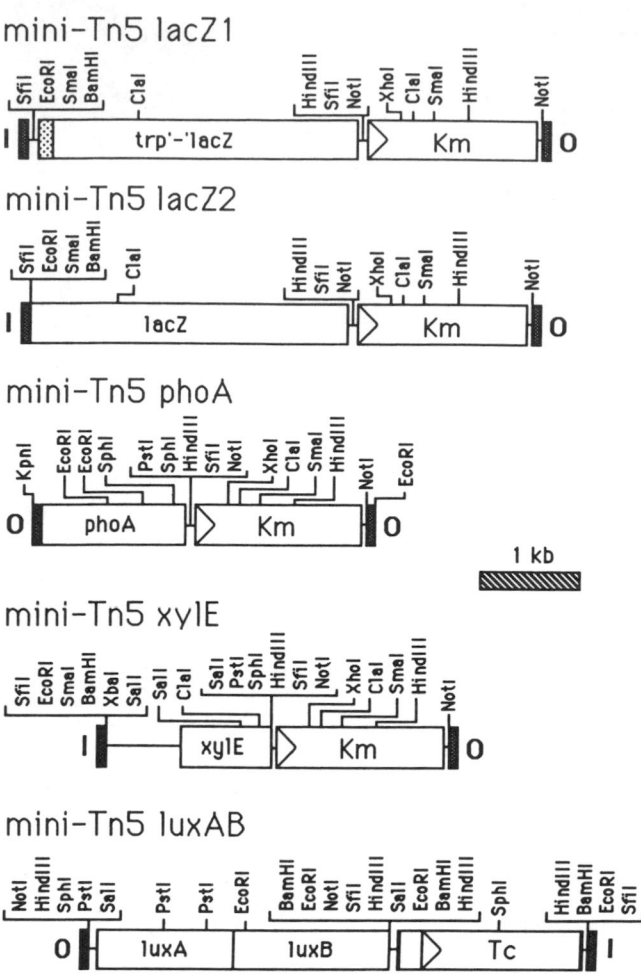

Figure 2 Organization of promoter-probe mini-transposons. Different reporter genes were cloned in the proper orientation at permissive sites next to one of the terminal ends of the mini-Tn5 elements described in de Lorenzo et al. (1990). Since the termini have neither transcriptional terminators nor stop codons in several of the possible frames, they allow generation of Type-I or Type-II gene fusions (depending on the element used) with target genes. The figure summarizes the structures of the elements. Mini-Tn5lacZ1 was made by cloning a promoterless *trp'-'lacZ* fusion with pRZ5605 (18) into p18Sfi as a 3.3-kb *Eco*RI-*Hind*III (the latter being formerly a *Dra*I site) fragment, with subsequent excision as an *Sfi*I restriction fragment and insertion at the *Sfi*I site of pUT/Km (11). The same strategy was used to create mini-Tn5lacZ2, but in this case the *lacZ* structural genes devoid of either transcriptional or translational signals are those of pMLB1034 (29). Construction of mini-Tn5phoA involved the cloning of a 1.5-kb *Xba*I-*Bst*EII fragment of pPHO7 (4, 10), containing the structural *phoA* gene, into *Xba*I-*Hinc*II-digested pUC18Sfi and subsequent transfer of an *Xba*I-*Sfi*I fragment to the corresponding sites of pUT/Km. This reconstructs the 5′ terminus of the original *phoA* transposon (10, 22) and leaves the mobile element flanked by two Tn5 O ends. The *xylE* gene present in mini-Tn5xylE was introduced as a partial 1.7-kb *Sal*I fragment of the TOL plasmid of *P. putida* (kindly provided by A. Wasserfallen) into p18Sfi and further cloned into the corresponding site of pUT/Km (11). Finally, mini-Tn5luxAB was made by reconstructing the promoterless *luxAB* unit from pFIT001 and pPALE001 (17) as a 3.2-kb *Sal*I-*Bam*HI fragment, which was cloned into the corresponding sites of pUC18Not and further cloned at the *Not*I site of the mini-Tn5Tc element. All mini-transposons are present as partial *Xba*I-*Eco*RI fragments in the delivery plasmid pUT/Km (11), except mini-Tn5luxAB, which is inverted with respect to the rest. Deduced restriction sites present in the mobile elements are indicated.

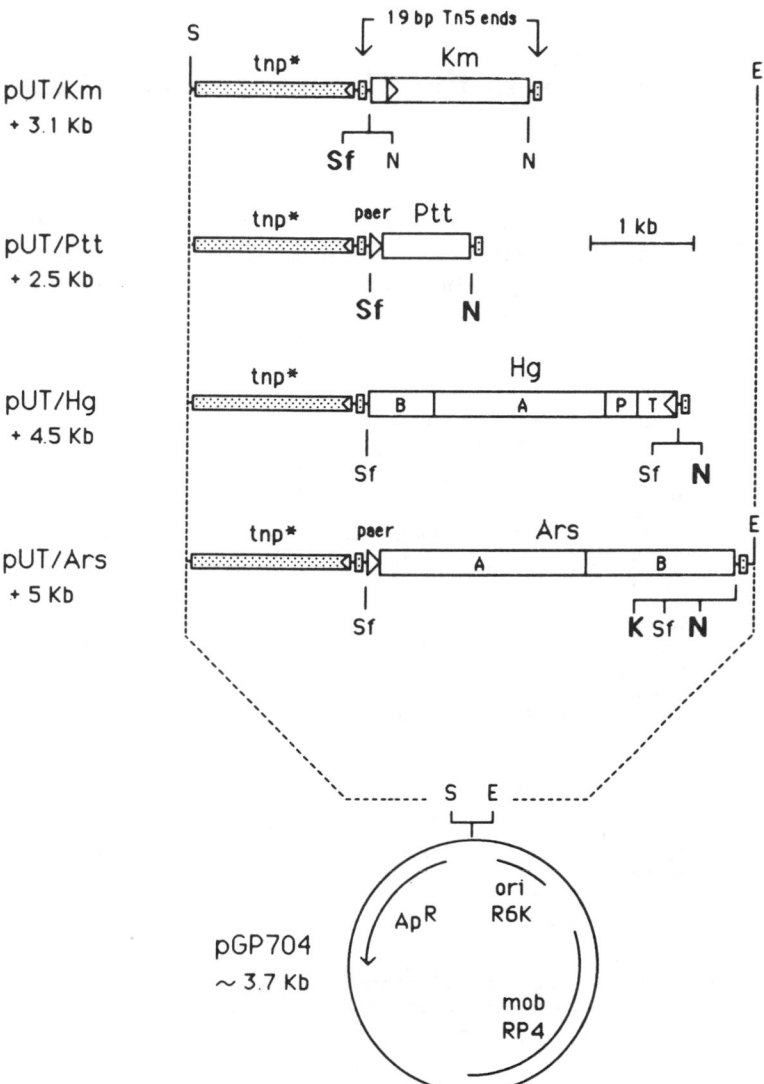

Figure 3 Tn5-based insertion delivery plasmids. The drawing displays the Tn5-based counterparts of the vectors shown in Figure 3 in Herrero et al. (1990). The common portion of the constructions corresponding to the delivery plasmid pGP704 (see Figure 2, Herrero et al. 1990) is shown at the bottom. The elements of the transposition system include the Tn5 19-bp terminal ends and an IS50$_R$ *tnp* gene devoid of *Not*I sites (*tnp**) oriented divergently to the I end for optimal transposition efficiency (11). The resistances provided by each of the elements are also indicated with their names to the left of the figure. Plasmid pUT/Km is the control mini-transposon donor with a standard antibiotic resistance (kanamycin). Important restriction sites are indicated: N, *Not*I; E, *Eco*RI; Sf, *Sfi*I; S, *Sal*I; K, *Kpn*I. Unique sites for the insertion of foreign DNA fragments are indicated by boldface type.

SECTION 13

E. Copy Numbers of IS Elements in the Chromosomes of *E. coli* and Other Bacteria

Table 1 is reprinted, with permission, from Deonier (Escherichia coli *and* Salmonella typhimurium. *Cellular and Molecular Biology* [ed. F.C. Neidhardt]. American Society for Microbiology, Washington, D.C., p. 983 [1987]). The references cited in the table and table notes can be found in the original paper.

Table 1   Copy Numbers of IS Elements in the Chromosomes of *E. coli* and Other Bacteria[a]

| Element | *E. coli* K-12 strains[b] | | | | *E. coli* B | *E. coli* C | *S. typhimurium* | *Serratia marcescens* | *Citrobacter freundii* | *Enterobacter aerogenes* |
	Wild type	C600	W1485	Other K-12						
IS*1*	6–7 (40); 8 (58)	8 (40)	6 (58); 6–7 (40); 7 (66)	6, 10, 7, 8 (58); 4 (7); 7 (65); 9 (66); 5–6 (15)	14–19 (40)	3 (58); 3 (40)	0 (58); + (57)	0[c](7); 2[c] (58); 0 (57)	0 (58); 0 (57)	0 (58); 0 (57)
IS*2*	6 (40)	7 (40)	7–8 (40); 4 (66)	7 (7); 12 (40); 6 (66)	1 (40); >1 (50)	0 (40); 1[d] (50)	0[e] (50)	0 (7); 0 (50)	0 (50)	0 (50)
IS*3*	5 (40)	6 (25)	5 (25, 40)	6 (7)	4 (25, 40)	5[f] (40)		0 (7)		
IS*4*				1, 1, 1, 1, 1, 2, (45)	>1 (50)	0 (50)	0[e] (50)	0 (50)	0 (50)	0 (50)
IS*5*		10–11 (70, 71)	11 (78)	10–12 (64); 9–10 (56); 8–11 (32)	0 (71)	~10 (71)	0[g] (71)		3 (71); + (57)	
IS*30*		2 (14); 3[h] (59); 4[h] (79)	3[h] (79)	3, 2, 6 (14); 3[h] (59)		2 (14)		0 (14)	0 (14)	0 (14)
IS*200*				0, 0 (50)	0 (50)	0 (50)	6[e] (49, 50)	0 (50)	0 (50)	0 (50)

[a] Parentheses indicate reference.
[b] *E. coli* K-12 strains for which IS*2* or IS*3* copy numbers were determined are F⁻.
[c] In one case (7) IS*1* was detected on a plasmid. In the other, presence of detected copies on a plasmid was not ruled out.
[d] Hybridization not extensive.
[e] Data for *S. typhimurium* LT-2.
[f] Result of one determination.
[g] Data for *S. typhimurium* STL-2.
[h] Data for IS*30.121*.

F. Locations of Some IS Elements on the *E. coli* K12 Genetic Map

Figure 1 is reprinted, with permission, from Deonier (Escherichia coli *and* Salmonella typhimurium. *Cellular and Molecular Biology* [ed. F.C. Neidhardt]. American Society for Microbiology, Washington, D.C., p. 985 [1987]).

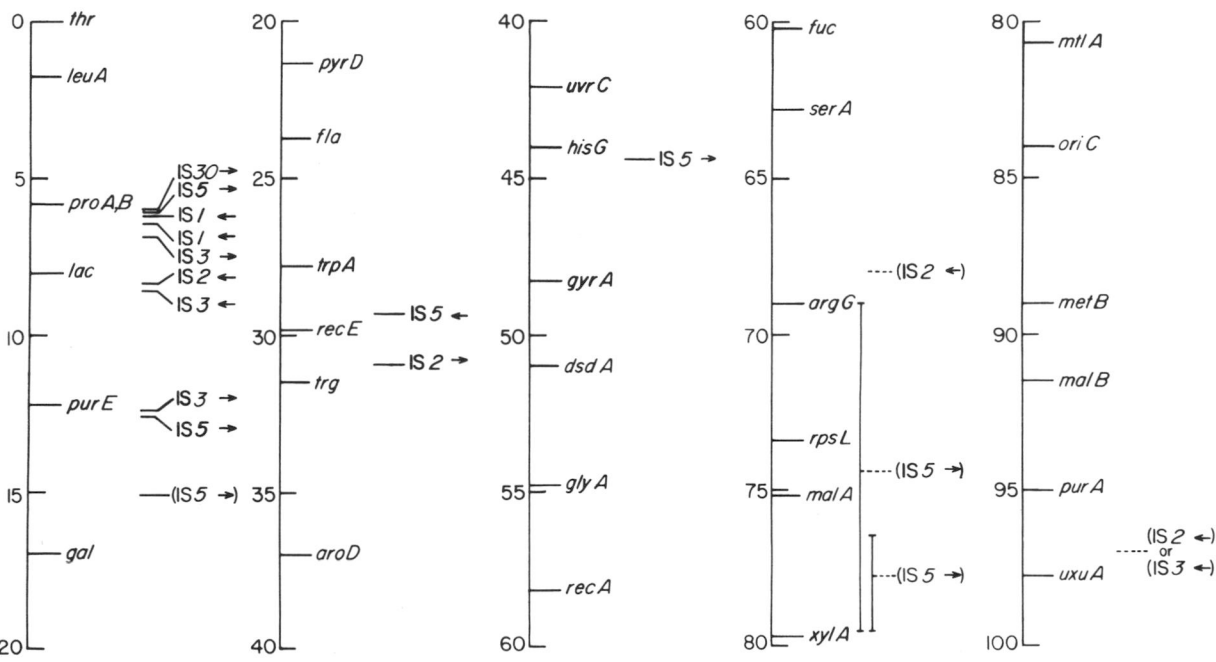

Figure 1 The orientation of each element is denoted by an arrow using the convention described in Deonier (1987). The positions of accurately mapped elements are indicated by solid connector lines. Broken lines (with or without brackets) indicate the positions of elements that have been located with less precision. Elements whose presence has not been confirmed in independently propagated *E. coli* K12 sublines are indicated with parentheses. The section on "Mapping Notes" in Deonier (1987) should be consulted for references to mapping data and accuracy of location for any element depicted in this figure.

Phage Mu

A. Genetic and physical maps of phage Mu

B. Physical maps of useful Mu and mini-Mu derivatives
1. Physical maps of the most widely used plaque-forming Mu derivatives that carry Amp^r or Kan^r
2. Physical maps of the Mu derivatives MudI(Ap, *lac*) and MudII(Ap, *lac*) that allow the generation of gene (operon) fusions and protein fusions
3. Physical maps of the most widely used mini-Mu phages
4. Physical maps of the three most widely used RP4::mini-Mu

C. Phage Mu as a genetic tool
1. Mini-Mu and mini-Mu replicons and mini-Mu and Mu fusion elements
2. p*lac*Mu1
3. Broad-host-range plasmids with mini-Mu or mini-D108 insertion

REFERENCES

Faelen, M. 1987. Useful Mu and mini-Mu derivatives. In *Phage Mu* (ed. N. Symonds et al.), pp. 309–316. Cold Spring Harbor Laboratory, Cold Spring Harbor, New York.

Howe, M.M. 1987. Genetic and physical maps. In *Phage Mu* (ed. N. Symonds et al.), pp. 271–273. Cold Spring Harbor Laboratory, Cold Spring Harbor, New York.

Silhavy, T.J. and J.R. Beckwith. 1985. Uses of *lac* fusions for the study of biological problems. *Microbiol. Rev.* **49:** 398–418.

van Gijsegem, F., A. Toussaint, and M. Casadaban. 1987. Mu as a genetic tool. In *Phage Mu* (ed. N. Symonds et al.), pp. 215–250. Cold Spring Harbor Laboratory, Cold Spring Harbor, New York.

A. Genetic and Physical Maps of Phage Mu

Figure 1 is reprinted from Howe (*Phage Mu* [ed. N. Symonds et al.]. Cold Spring Harbor Laboratory, Cold Spring Harbor, New York, p. 272 [1987]).

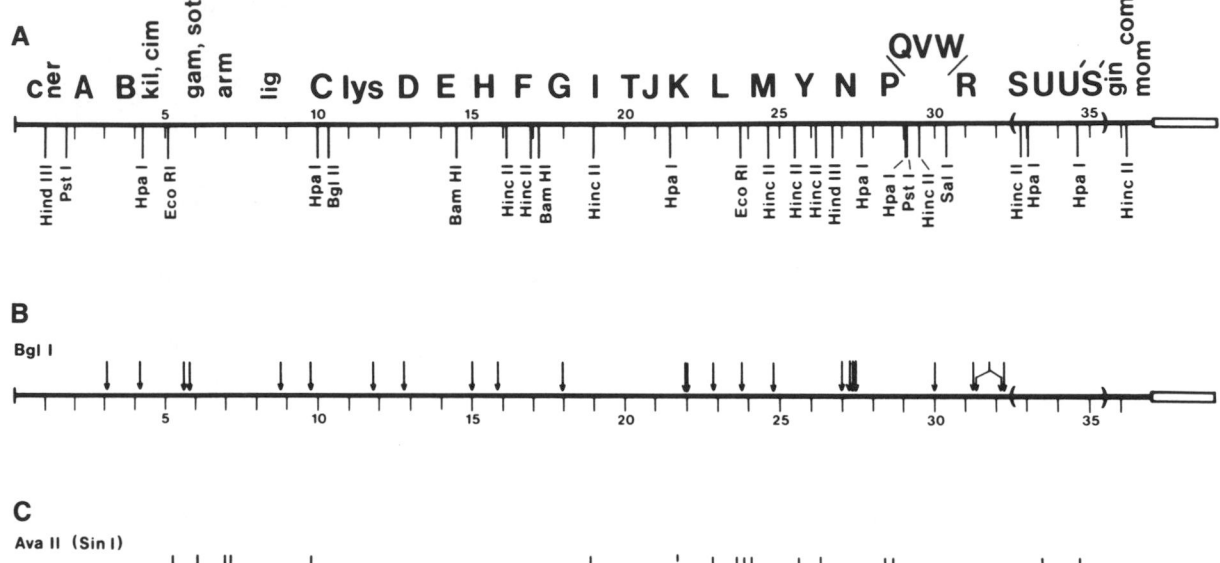

Figure 1 Bold lines represent Mu DNA; the vertical line at the left and the open box at the right end represent attached host sequences present in mature Mu DNA. Vertical tick marks denote 1-kb intervals. Parentheses indicate the invertible G segment, which is shown in the G(+) orientation. In *B*, the connected arrows at 31.3 and 32.1 kb represent two possible locations of a *Bgl*I restriction site that was not defined by the restriction analysis. Since the DNA sequence does not show a *Bgl*I site at 32.1 kb, this site is most likely located at 31.3 kb. The dashed arrow in *C* represents an *Ava*II (*Sin*I) site that is at least partially modified in *E. coli* K12.

SECTION 14

B. Physical Maps of Useful Mu and Mini-Mu Derivatives

Figures 1–4 are reprinted from Faelen (*Phage Mu* [ed. N. Symonds et al.]. Cold Spring Harbor Laboratory, Cold Spring Harbor, New York, pp. 309–313 and 316 [1987]). The references cited in the figure captions can be found in the original paper.

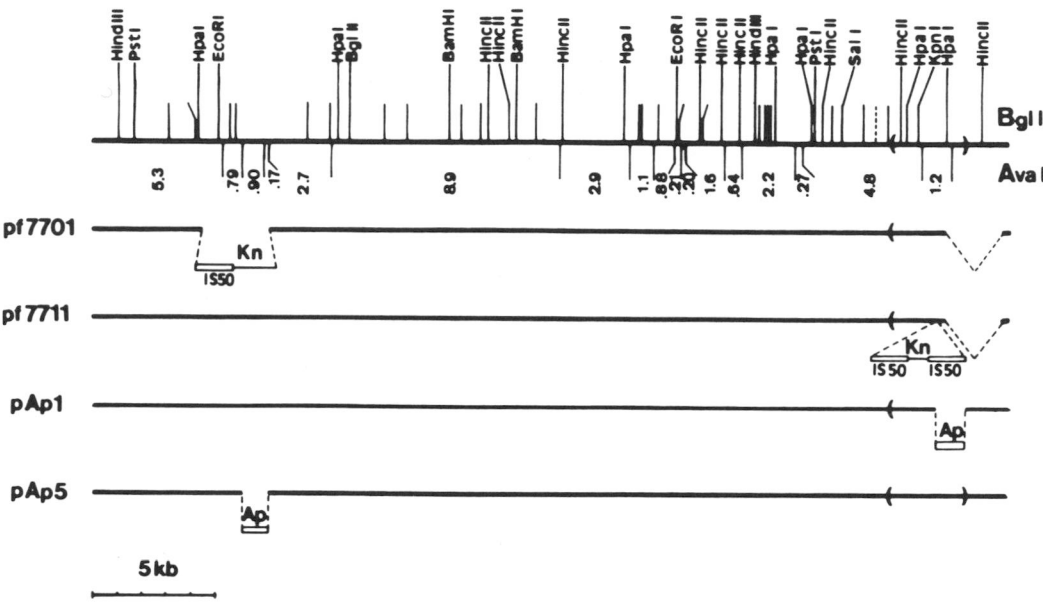

Figure 1 Physical maps of the most widely used plaque-forming Mu derivatives that carry Amp[r] or Kan[r].

Mupf7701 (also called Mu Kn7701). This map was determined by C.J. Thompson and M. Howe (unpubl.). Phage Mupf7701 carries (1) the 445-3 deletion (Chow et al. 1977b), which removes part of the G region, including genes *U'* and *S'*v, and part of the β region, which makes it *mom⁻*, and (2) a substitution of 2.8 kb of Mu DNA in the SE region, located 4.4–7.2 kb from the *c* end, by a nontransposable segment of Tn5 conferring kanamycin resistance. Mupf7701 with a normal G-β region and Mupf7701 with the P1 host range have been constructed (L. Desmet and A. Toussaint, unpubl.).

Mupf7711 (also called Mu Kn7711). This map was determined by C.J. Thompson and M. Howe (unpubl.). Phage Mupf7711 also carries the 445-3 deletion (see above) and has a deleted but transposable Tn5 inserted in the remaining part of the *U'* gene.

MupAp1 and MupAp5. This map was determined by Leach and Symonds (1979). Both phages carry a substitution of ~1.1 kb of Mu DNA by a segment of ~1.1 kb from Tn3, which still confers ampicillin resistance but is defective for transposition. The substitution in pAp1 is located to the right of the G region (in the + orientation), removing the right G inverted repeat. In pAp5, the substitution removes Mu DNA extending from ~6.1 kb to ~7.2 kb from the *c* end (A. Résibois, unpubl.).

All of the phages described in this figure carry the *cts*62 mutation. A Mu restriction map is shown at the top of the figure.

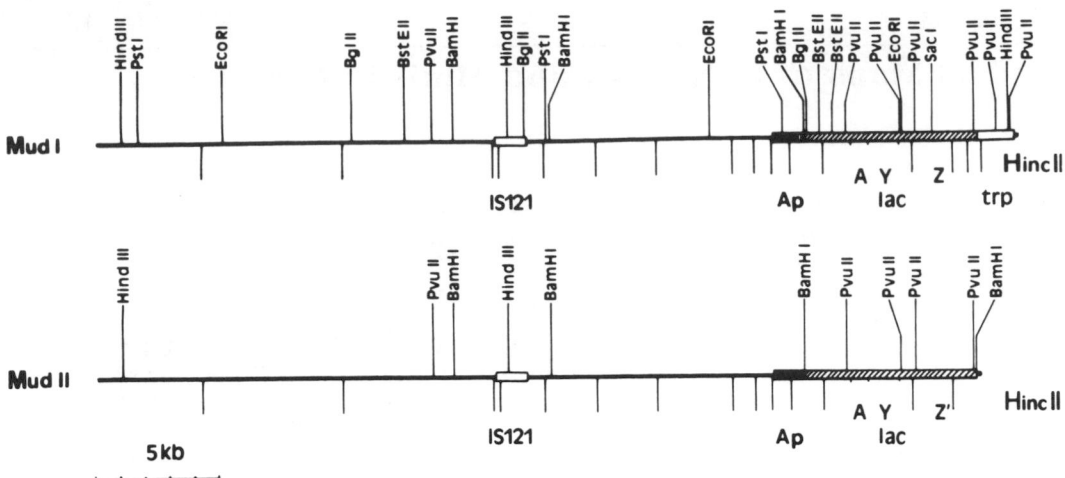

Figure 2 Physical maps of the Mu derivatives MudI(Ap,*lac*) and MudII(Ap,*lac*) that allow the generation of gene (operon) fusions and protein fusions. This map was determined by O'Connor and Malamy (1983). Both phages carry the *cts*62 mutation and an IS*121* insertion in the *H* gene, at ~16 kb from the *c* end.

MudI(Ap,*lac*) carries (from right to left in the figure) ~200 bp from the end of the β end, ~1200 bp from the *E. coli trp* operon, the *E. coli lac* operon deleted from its promoter, and a nontransposable segment of ~1.4 kb from Tn3 conferring ampicillin resistance. Due to the absence of some of the late genes, lysates of either Mud(Ap,*lac*) can only be prepared in the presence of a helper Mu.

MudII(Ap,*lac*) is the same as MudI(Ap,*lac*) except that it has only 116 bp from the β end and no *trp* fragment and its *lac* operon is deleted not only from its promoter, but also from the first eight amino acids of the *lacZ* gene.

For more information about these phages, see van Gijsegem et al. (1987).

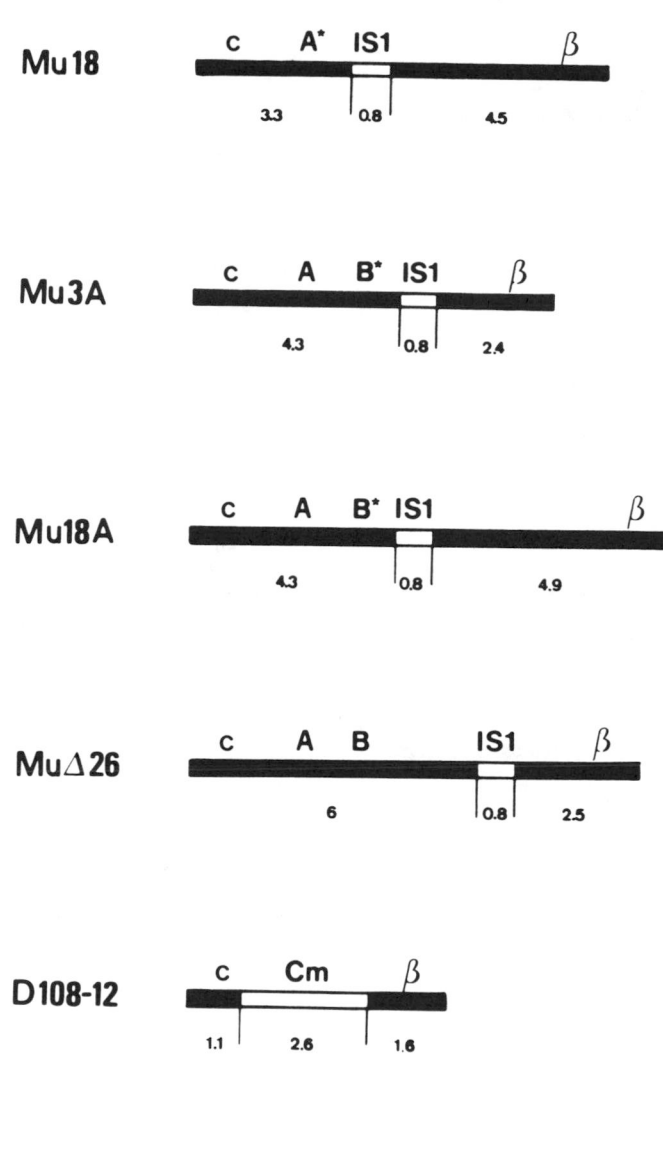

Figure 3 Physical maps of the most widely used mini-Mu phages. All of the mini-Mu phages carry the *cts*62 mutation, and the two mini-D108 phages carry the D108 *cts*10 mutation. The characteristics of these mini-Mu phages are summarized in Table 1 in Section 14C of this Handbook. For further information, refer to van Gijsegem et al. (1987). Symbols used in the maps represent *Eco*RI (E), *Hind*III (H), *Pst*I (P), *Bam*HI (B), *Sal*I (S), *Bgl*II (G), *Xho*I (X), and *Sma*I (M).

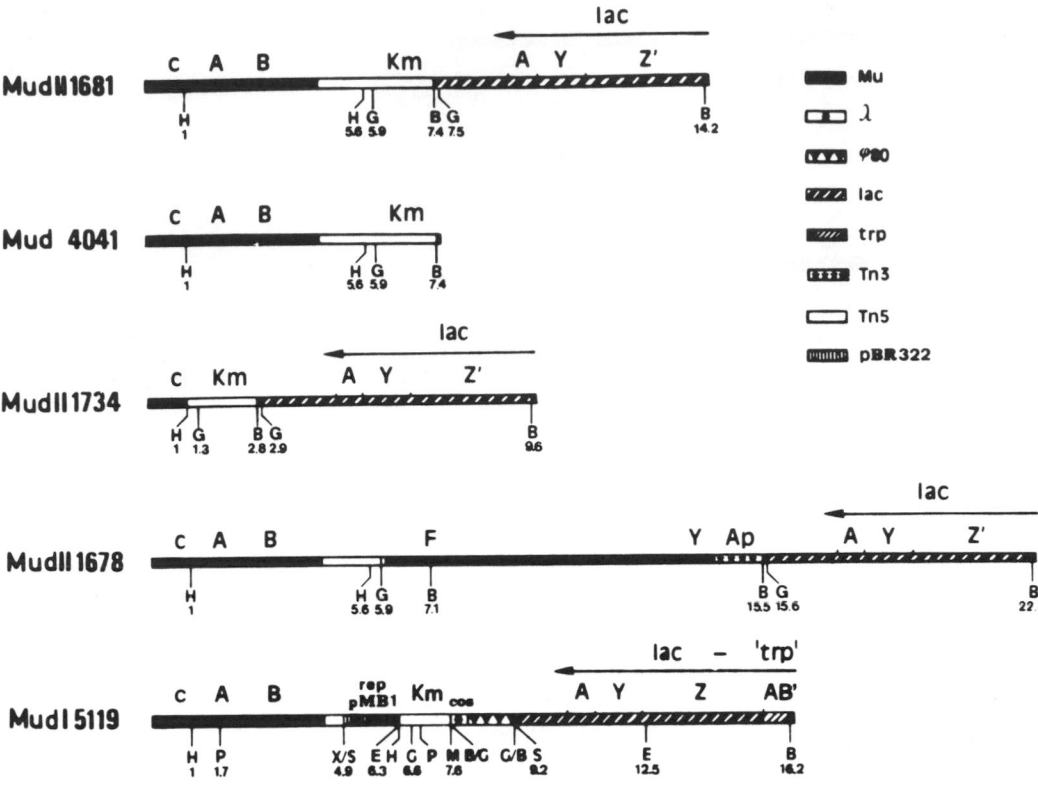

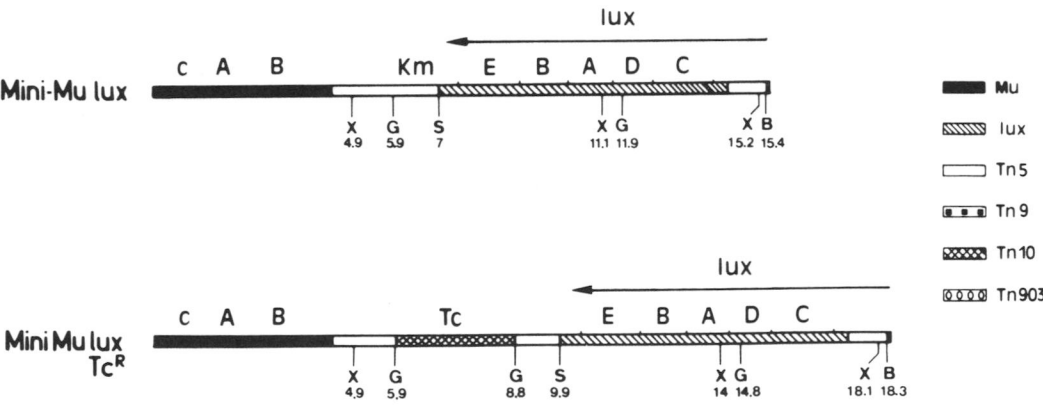

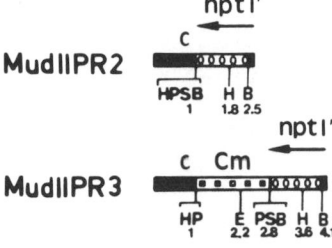

Figure 3 *Continued*

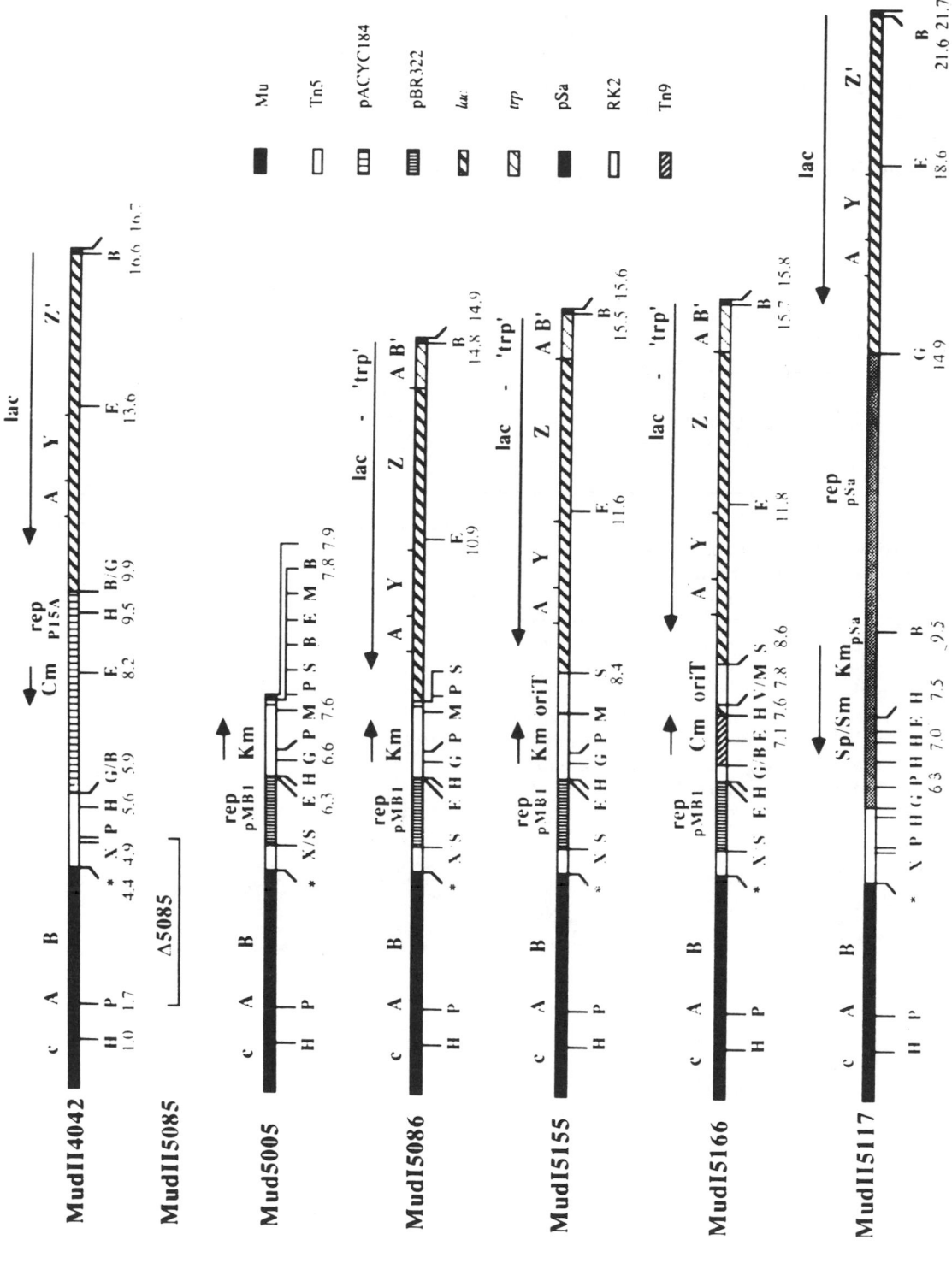

Figure 3 *Continued*

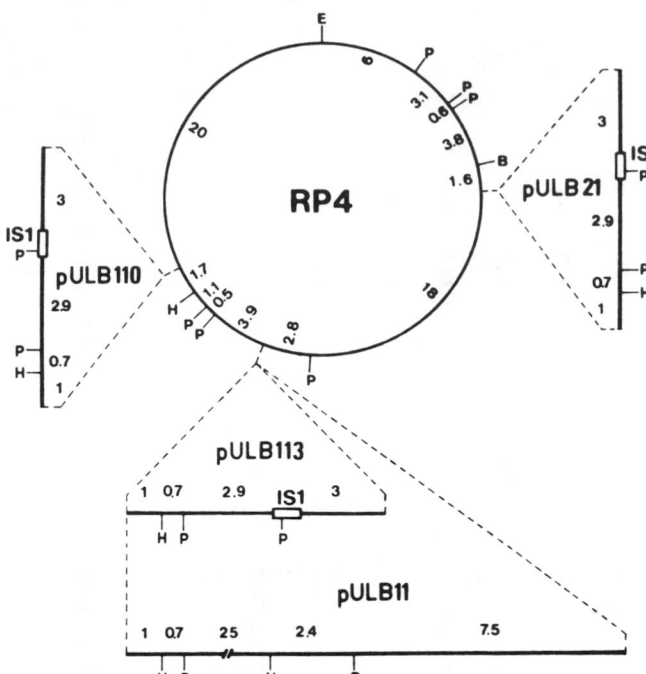

Figure 4   Physical maps of the three most widely used RP4::mini-Mu. These maps were determined by van Gijsegem and Toussaint (1982). pULB11 carries a Mu*cts*62 insertion whose *c* end is at the left of the figure. pULB113 was derived from pULB11 by homogeneitization between the Mu*cts*62 and Mu3A (see Figure 3). pULB21 carries a Mu3A insertion that inactivates the RP4 gene, conferring tetracycline resistance, whereas the Mu3A insertion in pULB110 inactivates the RP4 gene, conferring kanamycin resistance. Symbols used in the figure represent *Eco*RI (E), *Hin*dIII (H), *Pst*I (P), and *Bgl*II (B).

C. Phage Mu as a Genetic Tool

Tables 1 and 2 are reprinted from van Gijsegem et al. (*Phage Mu* [ed. N. Symonds et al.]. Cold Spring Harbor Laboratory, Cold Spring Harbor, New York, pp. 218–219 and 235, respectively [1987]). The references and figure cited in the tables and table notes can be found in the original paper.

Figure 1 is reprinted, with permission, from Silhavy and Beckwith (*Microbiol. Rev.*, vol. 49, p. 403 [1985]). Additional Mu derivatives can be found in Unit 6 of the Laboratory Manual portion of this volume.

Table 1   Mini-Mu and Mini-Mu Replicons and Mini-Mu and Mu Fusion Elements

Phage	Mu genes expressed	Selectable gene	Replicon	Other features	Fusion segment	Length (in kb)	Reference[a]
Mini-Mu							
Mu3A	cts, ner, A	—	—	—	—	7.5	1
Mu18A	cts, ner, A	—	—	—	—	10	1
Mu18	cts, ner	—	—	—	—	8.6	1
Mu18A-1	cts, ner, A	Ap	—	—	—	10	1
Mu18-1	cts, ner	Ap	—	—	—	8.6	1
Mu18A-2	cts, ner, A	Cm	—	—	—	not known	1
Mud4041	cts, ner, A, B	Kn	—	—	—	7.8	8
MuαTK[b]	cts, ner, A, B	Kn, TK	—	—	—	10	15
Mini-D108							
D108-12	cts, ner	Cm	—	—	—	5.2	1
D108-1	cts, ner, A	Cm	—	—	—	7.6	1
Gene fusion elements							
Mu							
Mud1	cts, ner, A-Y	Ap	—	H::IS121	lacZYA	37	2
Mud1-8	cts, ner, Aam, B-Y	Ap	—	H::IS121	lacZYA	37	3
MuXdIAplac	cts, ner, A+B::Tn9, C-Y	Ap, Cm	—	H::IS121	lacZYA	39.5	4
MudIIAplac301	cts, ner, A-Y	Ap	—	H::IS121	lac'ZYA	35.6	5
λplacMudI	cts, ner, A	Ap	λ	plaque-forming λ	lacZYA	~49	6
λplacMudII	cts, ner, A	Ap	λ	forming λ	lac'ZYA	~49	7

Mini-Mu

	genes	resistance			reporter	size	ref
MudI1681	cts, ner, A, B	Kn	—	—	lacZYA	15.8	8
MudII1681	cts, ner, A, B	Kn	—	—	lac'ZYA	14.3	8
MudI1734[c]	cts	Kn	—	—	lacZYA	11.2	8
MudII1734[c]	cts	Kn	—	—	lac'ZYA	9.7	8
MudI1678	cts, ner, A, B	Ap	—	—	lac'ZYA	22.4	8
mini-Mulux	cts, ner, A, B	Kn	—	—	lux	15.5	9
mini-Mulux (Tc[R])	cts, ner, A, B	Tc	—	—	lux	18.5	9
MudIIPR13	cts	Cm	—	—	lac'ZYA	9.2	16
MudIIPR3	cts	Cm	—	—	nptI'	4.2	10
MudIIPR2	cts	—	—	—	nptI'	2.6	10

Mini-Mu replicons

	genes	resistance	replicon		reporter	size	ref
MudII4042	cts, ner, A, B	Cm	p15A	—	lac'ZYA	16.7	11
Mud5005	cts, ner, A, B	Kn	pMB1	—	—	7.9	12
MudII5085	cts, ner	Cm	p15A	—	lac'ZYA	13.4	12
Mud5086	cts, ner, A, B	Kn	pMB1	—	lacZYA	14.9	12
MudI5117	cts, ner, A, B	Sm/Sp, Kn	pSa	—	lac'ZYA	21.7	12
Mud5155	cts, ner, A, B	Kn	pMB1	RK2, oriT	lacZYA	15.6	12
MudII5166	cts, ner, A, B	Cm	pMB1	RK2, oriT	lacZYA	15.8	12
Mud5060	cts, ner, A, B	Kn	pMB1	Tn3Res	—	8.3	13
Mud5260	cts, ner, A, B	Kn	pMB1	P1loxP	—	7.8	13
MudI5119	cts, ner, A, B	Kn	pMB1	λcos	lacZYA	16.3	14

All of the elements carry the *cts62* repressor mutation. A "prime" indicates that a particular gene is truncated from a few amino acids at its amino terminus. Physical maps of the mini-Mu prophages can be found in the Appendix.

[a]References: [1]Résibois et al. (1981); [2]Casadaban and Cohen (1979); [3]Hughes and Roth (1984); [4]Baker et al. (1983); [5]Casadaban and Chou (1984); [6]Bremer et al. (1984); [7]Bremer et al. 1985; [8]Castilho et al. (1984); [9]Engebrecht et al. (1985); [10]Ratet and Richaud (1986); [11]Groisman et al. (1984); [12]Groisman and Casadaban (1986a); [13]E.A. Groisman and M. Casadaban (unpubl.); [14]Groisman and Casadaban (1986b); [15]Jenkins et al. (1985); [16]P. Ratet and F. Richaud (unpubl.).

[b]MuαTK carries a mammalian thymidine kinase gene inserted at the *Bam*HI site in Mud4041. The thymidine kinase gene can be used for selection in mammalian cells.

[c]Mud1734 and its derivatives are deleted from the promoter for repressor synthesis.

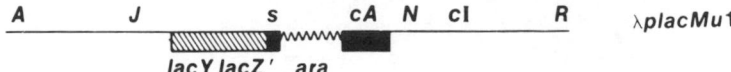

Figure 1 Lac⁻ λ-Mu*lac* hybrid phage, λp*lac*Mu1. (Zigzag line) Bacterial DNA; (closed rectangles) Mu sequences; (thin line) λ DNA; (hatched rectangle) sequences *lacY lacZ'*. The prime symbol denotes the fact that a particular gene is not completely present or is interrupted by other DNA sequences. The right Mu attachment site is designated s, and the left Mu attachment site is designated *cA*.

Table 2 Broad-host-range Plasmids with Mini-Mu or Mini-D108 Insertion

	Markers	
Plasmid	on the plasmid	in the Mini-Mu
pGMI228, RP4(ΔKn)	Tc, Ap	
PRK24, RP4 with a *trp* insertion in the *Hin*dIII site in the Kn^R gene	Tc, Ap, *trp*	
pUZ8	Kn, Tc, Hg	
pULB113, RP4::Mu3A	Kn, Tc, Ap	
pULB11, RP4::Mu*cts*62*mom*3452	Kn, Tc, Ap	
pULB18, pUZ8::Mu18A-1	Kn, Tc, Hg	Ap
pULB21, RP4Tc::Mu3A	Kn, Ap	
pULB106 R751::D108-1	Tp	Cm
pULB107, pRM5::D108-1[a]	Kn, Tc, Ap	Cm
pULB108, pRM5::Mu3A, ΔTc	Kn, Ap	
pULB110, RP4Kn::Mu3A	Tc, Ap	
pULB19, pUZ8::Mu*c*⁺	Kn, Tc, Hg	
pULB20, R300B::Mu18A-2	Cm, Su, Sm	
pULB47, pGMI228::Mu*cts*62Kn7711	Ap, Tc	Kn
pULB9, RP4::SuA1P2[b]		
pRE62, R388::Mu*cts*62pAp1	Tp	Ap
pRK24M4, pRK24::Mu*cts*62	Ap, Tc, *trp*	
pEG5150, pRK24::MudII5117	Ap, Tc, *trp*	Kn, Sp/Sm
pEG5152, pRK24::MudI5086	Ap, Tc, *trp*	Kn
pRK241, pRK24::MudI1681	Ap, Tc, *trp*	Kn

Only pULB113, pULB21, and pULB110 have been physically characterized (see Fig. 5 and Appendix 3). pULB113 mobilizes host DNA about 10 times more efficiently than all the other plasmids (F. van Gijsegem et al., unpubl.). The parental plasmids RP4, RP1, R751, and R300B (IncQ) were described by Bukhari et al. (1977), pUZ8 by Hedges and Matthew (1979), and pRM5 by Robinson et al. (1980). RP4::Mu*cts*62 plasmids were described by Murooka et al. (1981), Bialy et al. (1980), and van Gijsegem and Toussaint (1982). pREG2, pRK24M4::Mu*cts*62, pEG5150, and pEG5052 were described by Groisman and Casadaban (1986a). pRM241 was isolated by H. Shuman (unpubl.).
[a]pRM5 is RP1, with temperature-sensitive replication.
[b]These plasmids contain a β-*sup*-β' structure; the suppressor is a mutant *sup*^F that incorporates glutamine.

SECTION · 15

Other Phages

REFERENCES

Bertani, L.E. and E.W. Six. 1988. The P2-like phages and their parasite P4. In *The Bacteriophages* (ed. R. Calendar), vol. 2, pp. 73–143. Plenum Press, New York.

Calendar, R., ed. 1988. *The Bacteriophages*, 2 volumes. Plenum Press, New York.

Casjens, S., M. Hayden, E. Jackson, and R. Deans. 1983. Additional restriction endonuclease cleavage sites on the bacteriophage P22 genome. *J. Virol.* **45:** 864–867.

Dunn, J.J. and F.W. Studier. 1983. The complete nucleotide sequence of bacteriophage T7 DNA and the locations of T7 genetic elements. *J. Mol. Biol.* **166:** 477–535.

Hayashi, M., A. Aoyama, D.L. Richardson, Jr., and M.N. Hayashi. 1988. Biology of the bacteriophage φX174. In *The Bacteriophages* (ed. R. Calendar), vol. 2, pp. 1–71. Plenum Press, New York.

Kutter, E., B. Guttman, G. Mosig, and W. Rüger. 1990. Genomic map of bacteriophage T4. In

Genetic Maps, Fifth Edition (ed. S.J. O'Brien), pp. 1.24–1.51. Cold Spring Harbor Laboratory Press, Cold Spring Harbor, New York.

Mathews, C.K., E.M. Kutter, G. Mosig, and P. Berget, eds. 1983. *Bacteriophage T4*. American Society for Microbiology, Washington, D.C.

Mosig, G. and F. Eiserling. 1988. Phage T4 structure and metabolism. In *The Bacteriophages* (ed. R. Calendar), vol. 2, pp. 521–606. Plenum Press, New York.

Poteete, A.R. 1988. Bacteriophage P22. In *The Bacteriophages* (ed. R. Calendar), vol. 2, pp. 647–682. Plenum Press, New York.

Susskind, M.M. and D. Botstein. 1978. Molecular genetics of bacteriophage P22. *Microbiol. Rev.* **42:** 385–413.

Studier, F.W. and J.J. Dunn. 1983. Organization and expression of bacteriophage T7 DNA. *Cold Spring Harbor Symp. Quant. Biol.* **47:** 999–1007.

van Duin, J. 1988. Single-stranded RNA bacteriophages. In *The Bacteriophages* (ed. R. Calendar), vol. 1, pp. 117–167. Plenum Press, New York.

Wood, W.B. and H.R. Revel. 1976. The genome of bacteriophage T4. *Bact. Rev.* **40:** 847–868.

Yarmolinsky, M.B. 1990. Bacteriophage P1. In *Genetic Maps*, Fifth Edition (ed. S.J. O'Brien), pp. 1.52–1.62. Cold Spring Harbor Laboratory Press, Cold Spring Harbor, New York.

Yarmolinsky, M.B. and N. Sternberg. 1988. Bacteriophage P1. In *The Bacteriophages* (ed. R. Calendar), vol. 1, pp. 291–438.

Youderian, P. and M.M. Susskind. 1980a. Identification of the products of bacteriophage P22 genes, including a new late gene. *Virology* **107:** 258–269.

Youderian, P. and M.M. Susskind. 1980b. Bacteriophage P22 proteins specified by the region between genes 9 and *erf*. *Virology* **107:** 270–282.

SECTION 15

A. P1

Table 1 is reprinted, with permission, from Yarmolinsky and Sternberg (*The Bacteriophages* [ed. R. Calendar], vol. 1. Plenum Press, New York, p. 311 [1988]). Figure 1 and Table 2 are reprinted from Yarmolinsky (*Genetic Maps*, Fifth Edition [ed. S.J. O'Brien]. Cold Spring Harbor Laboratory Press, Cold Spring Harbor, New York, pp. 1.52–1.58 [1990]). The references cited in the tables, table notes, and figure caption can be found in the respective original publications.

Table 1 Host Range of P1[a]

Bacterium	P1 DNA injection	P1 Phage production	References[b]
Escherichia coli K12, C, B	+	+	2
Shigella dysenteriae	+	+	1
Shigella flexneri	+		6
Salmonella typhimurium	(−) +	+	(4),5,9,11
Salmonella typhi and abony	+		4
Klebsiella aerogenes	+	+	8,11
Klebsiella pneumoniae	+	+	7,8,11
Citrobacter freundii	+	+	8,11
Serratia marcescens	+	(−) +	(3),11
Enterobacter aerogenes	+	+	8,11
Enterobacter liquefaciens and cloacae	+	+	8
Erwinia carotovora	+	+	11
Erwinia amylovora	+	+	8
Yersinia pestis and pseudotuberculosis	+	−	7
Proteus mirabilis, vulgaris and inconstans	+	+	11
Pseudomonas putida	−		8
Pseudomonas aeruginosa	+	(−) +	(3),11
Pseudomonas amyloderamosa	(−) +	+	(8),11
Flavobacterium sp. M64	+	−	11
Agrobacterium tumefaciens	+	−	11
Acetobacter suboxydans	−		11
Alcaligenes faecalis	+	−	11
Myxococcus xanthus	+	−	10

[a]Symbols + and − indicate that P1 injection or phage production was detected and not detected, respectively, in either wild type members of the bacterial species or in mutants of that species. The assay for injection was generally based on detection of antibiotic-resistant colonies following exposure of the bacteria to P1Cm or P1Km. The efficiency of acquisition of antibiotic resistance by the tested strains relative to *E. coli* was as low as 10^{-8} (in *Myxococcus xanthus*). In the experiments of reference 11, phage production was based on the detection of particles capable of plaque formation on *E. coli* following exposure of bacteria having acquired kanamycin resistance from P1Kmc1.100 to an "inducing" heat pulse and a further incubation to allow phage maturation. The phage yield from P1 lysogens of the tested *Serratia*, *Proteus*, and *Pseudomonas* species was no more than about 10^{-4} the yield obtained with *E. coli*.

[b]References: (1) Bertani, 1951; (2) Bertani, 1958; (3) Amati, 1962; (4) Kondo and Mitsuhashi, 1966; (5) Okada and Watanabe, 1968; (6) Godard *et al.*, 1971; (7) Lawton and Molnar, 1972; (8) Goldberg *et al.*, 1974; (9) Ornellas and Stocker, 1974; (10) Kaiser and Dworkin, 1975; (11) Murooka and Harada, 1979. See also Tominaga and Enomoto, 1986.

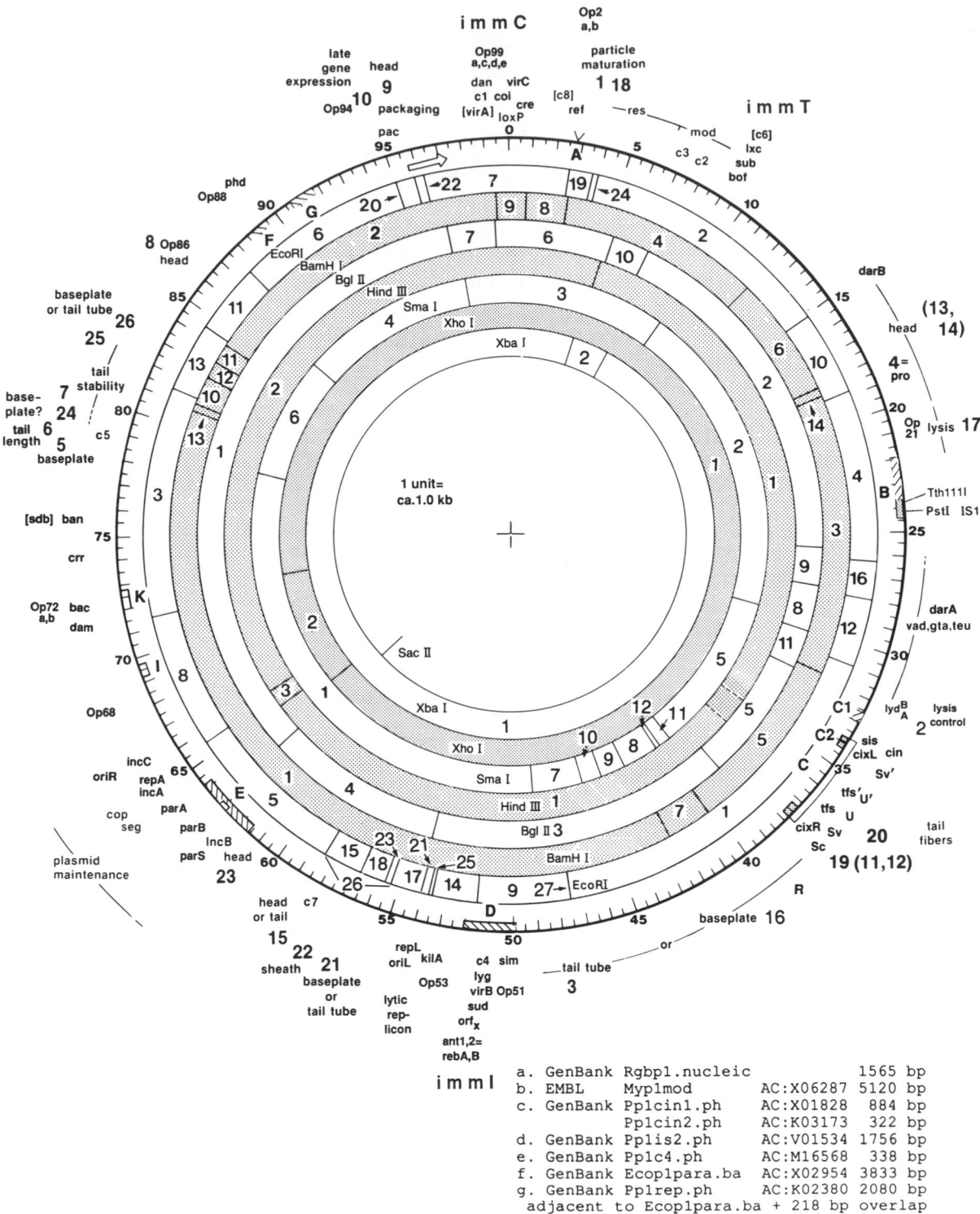

a. GenBank   Rgbp1.nucleic   1565 bp
b. EMBL   Myp1mod   AC:X06287 5120 bp
c. GenBank   Pp1cin1.ph   AC:X01828 884 bp
             Pp1cin2.ph   AC:K03173 322 bp
d. GenBank   Pp1is2.ph   AC:V01534 1756 bp
e. GenBank   Pp1c4.ph   AC:M16568 338 bp
f. GenBank   Ecop1para.ba   AC:X02954 3833 bp
g. GenBank   Pp1rep.ph   AC:K02380 2080 bp
             adjacent to Ecop1para.ba + 218 bp overlap

Figure 1   *(See facing page for caption.)*

Figure 1 Genetic and physical map of P1 (as of July 1989). Genetic symbols lie outside the circles. If bracketed, they refer to genes identified only in P1s close relative P7 (99). Several cistron designations are no longer in use, but their associated allele numbers have been reassigned in many cases (128). Cistron numbers 11 through 16 of ref. 91 were reassigned (85), as was 17 (127). Allele numbers of several conditional mutations can be found in ref. 139. Map coordinates have undergone a minor change. At a time when *lox*P (now coordinate 0/100) was not mapped precisely relative to physical markers, the unique *Pst*I cleavage site was assigned position 20 so as to roughly align the physical and genetic maps (138). Linkage cluster boundaries (127, 128) are omitted here, as are coordinates of deletion prophages, some of which appeared in previous versions of the map (127, 138). Each *c*1 protein-binding site or operator (Op) is assigned the integral number portion of its map position. In any one interval, Op sites are lettered in alphabetical and clockwise order. Approximate positions of nonhomologies between P1 and P7 (18, 57, 67, 79, 144) are indicated by **bold** letters and striped bars: insertions relative to P1 (A, the Amp^r transposon Tn*902*; F) and substitutions (B, C1, D, G, E); the invertible segment of P1 (and P7) is designated C. Stippled bars flanking C represent inverted repeats within one of which a deletion relative to the P1 sequence is marked as a line (C2). IS*1* (part of B) is also stippled. Regions of partial nonhomology between P1 and P7, within E and at positions I and K (77, 79), are indicated by open bars. Restriction maps of P7 and P1/P7 hybrids can be found in refs. 17 and 56. A linear denaturation map of P1 appears in ref. 78 and is reproduced in circular form in a comprehensive review of P1 biology (142).

The cleavage sites of *Eco*RI, *Bam*HI, *Bgl*II, *Hin*dIII, and the unique sites of *Sac*II and of *Pst*I (in the IS*1*) were mapped by ref. 8, *Eco*RI sites that generate fragments 26 and 27 by ref. 45, *Xho*I by ref. 61, *Sma*I and *Xba*I by ref. 76, and an additional unique site within IS*1*, for *Tth*111I, by ref. 53. No site was found for *Sal*I (8).

An overlapping library of P1 DNA in a λ vector has been constructed and ordered with respect to the restriction map (83). The DNA sequences of selected regions of P1 that have been determined are: from a region including *lox*P through *cre* (119[a]); *ref* (72) and its leader sequence (136); *res* and *mod* (52[b]); *bof* (44, 90); an 85-bp region that includes Op21 (20); IS*1* (82); part of the *dar* operon, i.e., parts of *Bgl*II-9, -8, and -11 (47); *cin* and its substrates (48[c], 58); from *cix*L through the invertible C segment to the *Bgl*II:5–3 junction (67, 89); from the *Bgl*II:5-3 junction to the *Bam*HI:5-7 junction (106[d], revised in ref. 35); the *c*4 gene (11[e]); from 176 bp upstream of *c*4 (including Op51) into the first gene beyond Op53 (45); from the middle of Op51 through a potential *c*1-binding site beyond *repL* (40) within which lies the *repL* operon, sequenced independently (113); the entire *par* (2[f]) and *rep* (3[g]) regions; an 85-bp region that includes Op68 (20); *dam* (v.22); a 62-bp Op72a,b sequence (43, 126); a 59-bp Op86 sequence (126); an 85-bp region that includes Op88 (20); from the unique *Kpn*I site within *Eco*RI-6 (at about map position 90) through the *Eco*RI-20 (71), including a site identified as Op94 (70); *Eco*RI-20, -22, and a 1.5-kb proximal and 0.3-kb distal region of *Eco*RI-7 (110) of which 237 bp around *pac* (within *Eco*RI-20) is published (115); and from the *Bgl*II:6-7 junction of *lox*P (30), within which lies a part of *c*1, sequenced entirely by ref. 12. Footnotes are listed below the map.

Table 2   Genes of P1

Gene, orientation[1]	Probable function, references
loxP	Locus of cross(x)over in plasmid, cre substrate (50,108)
virC	Class of vir mutations that are presumed to confer virulence by overproduction of Coi (42,101,142)
cre (+)	Cyclization recombinase; cointegrate resolvase (7,49,103,116,117,119)
Op2a,b	Control of ref expression (30,136)
[c8]	Establishment of lysogeny (98)
ref (+)	Recombination enhancement function (72,135,136)
1 (−)	Particle maturation (62,80,91,109,131)
18	Lytic growth (91)
res (−) (=hsdR)	Restriction component of host specificity DNA system (34,41,52,62)
mod (−) (=hsdMS)	Modification and site recognition component of hsd (34,41,52,62)
c2,c3	Intracistronic complementation groups within mod; killing of host by c2 and c3 mutants leads to plaque clarity (52,84a,88,91,92)
[c6][2] (−)	Maintenance of lysogeny (98)
bof[2] (−)	Regulatory function that acts cooperatively with c1, detected as ban on function, one of several effects (44,90,109,124,135,136)
sub[2] (−)	Suppressor of bof-1 (a bof allele?) (124)
lxc[2] (−)	Mutants constitutively express P1 function that complements lexC (=ssbA) defect, one of several pleiotropic effects (66,135)
immT	Tertiary immunity region: the gene(s) of footnote 2 (142)
darB	Defense (only in cis) against a subset of type I restriction enzymes (e.g. EcoB, EcoK) (64)
13,14	Head (85,131)
pro (=4)	Protein processing, required for head morphogenesis and maturation of the DarA precursor protein (80,91,109,120,127,128,131)
17	Lysis (endolysin?) (109,127,128,129,131)
Op21	Unknown (20)
IS1	Facilitates cointegration (26,59,60)

1. Listing is clockwise from loxP. Names of sites and gene clusters are indented. Transcription: (+) = clockwise, (−) = counterclockwise.

2. bof, sub, lxc, and possibly c6 may be the same gene. The direction of transcription for each is assumed to be that of bof.

Table 2 *Continued*

Gene, orientation	Probable function, references
darA[1] (−)	Defense (only in *cis*) against a subset of type I restriction enzymes (e.g. EcoA) (64,120)
vad[1] (−)	Viral architecture determination (defect increases proportion of small head particles) (55,64)
teu[2] (−)	Transduction enhancement by UV (possibly by binding to DNA and offering protection against nuclease) (64,142)
gta[1] (−)	Generalized transduction affected (frequency decreased) (64)
lydB[1] (−)	lydA,B prevent lysis delay (as does[?] λ gene S) (35,53,55,64)
lydA[1] (−)	lydA is first gene of dar operon (35,53,55,64)
2	Lysis delay; relation to lyd unclear (80,91,128,129,131)
cin (+)	C-segment inversion (and genome fusions) providing for variation in tail fibers (19,38,39,48,63,68,69)
sis	Sequence for inversion stimulation, enhances cix recombination in cis by binding to FIS (37,38,51)
cixL	C-segment inversion cross(x)-over site (left) (57,63)

Gene, orientation	Probable function, references
S_v' (−)	
U'(tfs')(−)	
U (tfs,20)(−)	
Sv (−)	Tail fibers[3], v=variable (54,58,67,80,105,125,128,131)
cixR	C-segment inversion cross(x)over site (right) (57,63)
Sc (−)	c=constant
S (19) (−)	S = S_c+Sv or Sv' (54,80,125,128,131)
11,12	Tail fibers[3] (85,131)
R (−)	Tail fiber structure or assembly. Positive regulator of translation of S (35,106)

1. The dar operon, under late promoter control (36), determines the presence of four internal proteins (120,130) which are dispensable for plaque formation (55). Relationships among members of the dar operon (darA, vad, teu, gta, and lyd) are unclear, but gta is genetically distinct from lyd (53). Unmapped mutations of (132) might be in gta or lyd.
2. Sus50 of (137) is possibly in teu (v.142).
3. Gene symbols S_c, S_v, S'_v, U, U', and tfs and tfs' (tail fiber specificity [67]) are derived from names of homologs in Mu (58). Possible synonyms of 11 and 12 are undetermined.

Table 2 *Continued*

Gene, orientation	Probable function, references
16 (-)?	Baseplate or tail tube (128,131). Distal end of gene that is upstream of R is postulated in (35) to be 16 of (85)
3	Baseplate or tail tube (80,91,109,128,131)
sim	Unknown, confers superimmunity in high copy number (27)
Op51	immI control by immC (12,30,45,126,142)
c4 (+)	Immunity-specific prevention of ant/reb expression, hence control of plaque clarity (11,45,80,91,92,99,109)
lyg	Establishment and/or maintenance of lysogeny. lyg is possibly a class of c4 alleles that fail to confer plaque clarity (94)
virB	Class of vir mutations (e.g. vir^s [73] in c4 [45]) that confer virulence by constitutive Ant synthesis (42,91,96,109,134,142)
sud	Mutation suppresses dan-1, causes a weak constitutive expression of ant and is probably a weak virB mutation (23)
orf$_x$	Unknown, apparently required for expression of ant (45)
ant/reb (+) (ant1, ant2= rebA, rebB)[1]	Antagonism of c1 repression (133,142)/c1-repressor bypass (102, 134); alternatives reflect differences in interpretation of function. The single orf that encodes both gene products (40,42, 45) is downstream of orf$_x$. Included within the ant/reb orf are the sites of reb-22 and presumably other ant/reb mutations, although ant-16 is within orf$_x$ and ant-17 is upstream of c4 (45)
immI	Cluster of immunity functions listed above (responsible for immunity difference between P1 and P7) (16,40,45,112,133,142)
Op53	Control of repL operon (30,40,45,113,119,126)
kilA (+)	Unknown; first gene of repL operon, partially homologous to ant, expression can kill host (40,113)
repL (+)	Lytic replication; acts to initiate replication at oriL (21,40, 113); minor contributor to plasmid replication (140)
oriL	Origin of lytic replication. (Originally L referred to location "left" of EcoRI-5) (21,40,111,113)
21	Baseplate or tail tube (109,128,131)
22	Sheath (109,128,131)
15	Head or tail (85,131)

1. Unmapped mutations pla or p of (121,122) that, like ant mutations (143), permit P1 to lysogenize E. coli lon mutants, are possibly ant alleles.

Table 2 *Continued*

Gene, orientation	Probable function, references
$\underline{c}$7	Unclear, mutation affects plaque $\underline{c}$larity and plasmid maintenance (80,97,98)
$\underline{23}$	Head (109,128,131)
incB	Incompatibility determinant associated with intact parS ($\underline{v}$.infra)
parS	Centromere analog required in $\underline{cis}$ for partioning. The intact site, but not all functional versions of it, bind IHF cooperatively with ParB (2,4,5,25,32,33,75)
parB (−)	⎫ Plasmid partitioning (2,5,142), cooperatively autoregulated (31)
parA (−)	⎬ ($\underline{v}$. also parS)
seg[1]	Prevention of $\underline{seg}$regation of plasmid-free cells (65,107,109,118)
cop[2]	Control of plasmid $\underline{cop}$y number (9,100,118)
incA	Set of nine 19-bp iterons downstream of repA which bind RepA and control plasmid replication, conferring group Y $\underline{inc}$ompatibility when it $\underline{trans}$ (1,3,5,13,14,84,118)
repA (−)	Plasmid $\underline{rep}$lication; acts in $\underline{trans}$ to initiate replication from $\underline{oriR}$ and autoregulate $\underline{repA}$ transcription (1,3,5,14)
incC[3]	Set of five 19-bp iterons upstream of $\underline{repA}$ which bind RepA and participate in $\underline{oriR}$ function and autoregulation of $\underline{repA}$. Can confer group Y $\underline{inc}$ompatibility in $\underline{trans}$ (3,15)
oriR	Origin of ($\underline{cis}$-acting region required for) plasmid replication of repressed P1 prophage. (Originally R referred to location "$\underline{right}$" of EcoRI-5) (1,3,6,15,111)
Op68	Control of $\underline{dam}$ operon (20)
dam (+)	Methylation of adenine residues of GATC sequences (analog of $\underline{E. coli}$ gene); implicated in control of late gene expression, packaging (20,22,142)
Op72a,b	Control of $\underline{ban}$ operon (43,46,74,112)

1. $\underline{seg}$ includes $\underline{par}$ mutants (e.g. $\underline{seg}$-101 = $\underline{par}$-101 [4] and $\underline{repA}$ mutants (e.g. $\underline{seg}$-103 = $\underline{repA}$103 [6]).
2. $\underline{cop}$ includes point mutations in $\underline{repA}$ and (in principle) deletions in $\underline{incA}$. Phenotype designations $\underline{seg}$ and $\underline{cop}$ should be replaced by genotypically more precise symbols, once the genotype is established.
3. $\underline{incC}$ is part of $\underline{oriR}$ which, in turn, is part of the basic replicon or $\underline{rep}$ region, hence the designation $\underline{rep}$-11 for an $\underline{incC}$ mutation (6).

Table 2   *Continued*

Gene, orientation	Probable function, references
bac	Cis-acting ban-control element; certain bac mutations render ban expression constitutive (24,43,46,81,126)
crr	P1crr prophage overproduces Ban, conferring cryoresistance on an unsuppressed dnaB266(am) P1bac-1 crr strain (29,46,86,123)
ban (+)[sdb]	DnaB analog (24,46,81,86,109,123) [suppressor of dnaB (104)]
5	Baseplate (91,109,128,131)
6	Tail length determination (91,109,128,131)
c5	Control of plaque clarity; establishment or maintenance of lysogeny (65,93,96,118)
24	Baseplate or tail stability (109,127,128,131)
7 / 25	Tail stability (91,109,128,131)
26	Baseplate or tail tube (109,128,131)
8	Head (91,109,128,131)
Op86	Unknown (30,126,142)
Op88	Unknown (20)
phd	Prevention of host death; antidote to product of unmapped gene(s) (doc) responsible for death on curing of P1 (141)
Op94	Control of 10 (70)
10 (+)	Late gene expression (36,42,80,91,109,110,128,129,131,142)
pac	Clustered sites from which clockwise packaging of DNA commences (8,114,115)
9 (+)	DNA cutting at pac (80,91,109,110,114,128,131,142)
c1 (−)	Repressor of lytic functions (10,28,87,92,96,98,109)
dan	dan-1 suppresses the thermolability of c1.100 repressor and is closely linked to c1.100 (23)
[virA]	Class of vir mutations, of which the only (presumed) member is vir-12, that confer virulence by a dominant c1 defect (98,99)
Op99a,c, d,e	Autoregulation of c1 (10,12,30,142). Highly degenerate site Op99b is omitted here as it fails to bind c1 repressor (110)
coi (−)	c one inactivator (96,142)
immC	Cluster of immunity genes flanking lox-cre of which c1 is the critical element (95,96,112,133,142)

B. P22

Figure 1 and Table 1 are reprinted, with permission, from Poteete (*The Bacteriophages* [ed. R. Calendar], vol. 2. Plenum Press, New York, pp. 650 and 657, respectively [1988]).

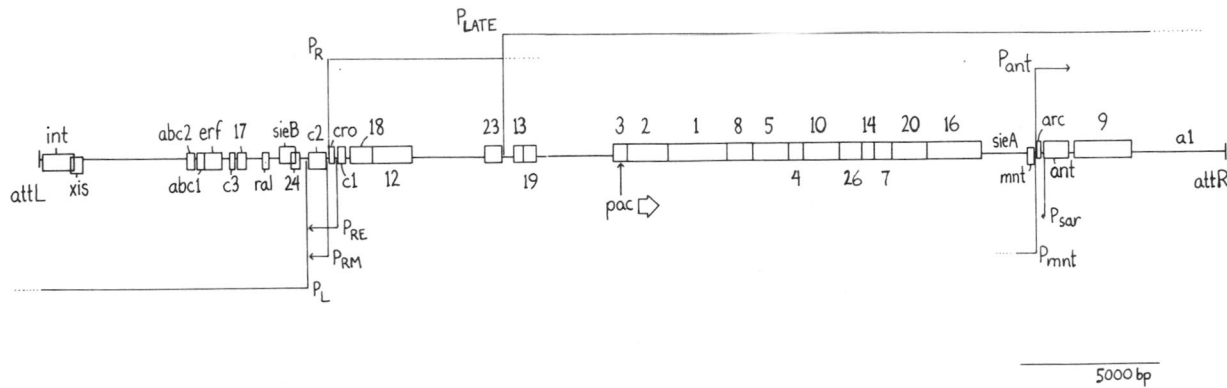

Figure 1 Genetic map of P22. All of the genes and sites indicated, except genes *4* through *sieA* and *a1*, are located in blocks of determined sequence (references as given in Poteete [1988]; S. Casjens, pers. comm.). Blocks of sequence are spaced according to the restriction map of Casjens et al. (1983). The map order of genes *4* through *16* is known from genetic experiments; their lengths were calculated from the molecular weights of their protein products (see Table 1 below; 1 kD protein = 27.3 bp); it was assumed that they are closely spaced, without significant overlaps (see Susskind and Botstein 1978; Youderian and Susskind 1980b). The start points and extents of transcripts are shown, labeled with the names of the relevant promoters. In the cases of P_{sar} and P_{ant}, the ends of the transcripts are known; P_{RM} and P_{RE} transcripts are assumed to terminate between P_L and *c2*; other 3' termini are unknown, as indicated by the dotted lines. The DNA replication origin (not shown) is thought to be located in gene *18* (see Poteete 1988).

Table 1   Genes and Proteins of P22

| Gene | Function | λ Analog | Subunit mol. wt. (kD) | |
			Sequence[a]	SDS gel[b]
int	Integration, excision	*int*	44.3	42
xis	Excision	*xis*	12.8	ND[c]
abc2	Modulation of RecBCD	*gam*	11.6	ND
abc1	activity		10.9	ND
erf	Homologous recombination	*bet*	22.9	27.5
c3	Establishment of lysogeny	*cIII*	5.7	ND
17	Escape from Fels-2 exclusion		12.2	18.5
ral	Modulation of host restriction	*ral*	7.4	ND
*sie*B	Superinfection exclusion	*sie*B	22.3	ND
24	Early transcriptional control	*N*	10.6	ND
c2	Maintenance of lysogeny	*cI*	24.0	29
cro	Early transcriptional control	*cro*	6.8	ND
c1	Establishment of lysogeny	*cII*	10.1	ND
18	DNA replication	*O*	30.6	31
12		*P*	50.1	46.5
23	Late-transcriptional control	*Q*	22.3	24
13	Lysis	*S*	11.5	10.5
19		*R + RZ*	16.0	15
3		*Nu1*	18.9	17
2	DNA encapsulation	*A*	(59)[d]	67
1		*B*	(80)	100
8	Prohead assembly; scaffolding	*Nu3*	(33.4)	39
5	Capsomere	*E + D*	ND	45.5
4			ND	19
10	Head stabilization	*W + FII*	ND	45.5
26			ND	28.5
14	Unknown step in head assembly		ND	15.5
7			ND	22
20	DNA injection		ND	43.5
16			ND	69
*sie*A	Superinfection exclusion		ND	ND
mnt	Maintenance of lysogeny	None	9.7	ND
arc	Early-transcriptional control	None	6.2	ND
ant	Antirepressor	None	34.7	33
9	Baseplate or tail	*ZUVGTHMLKIJ*	71.8	76
a1	Antigen conversion		ND	ND

[a]Molecular weights calculated from gene sequences; references as indicated in the text.
[b]Molecular weights estimated from electrophoretic mobilities in SDS polyacrylamide gels, as reported by Youderian and Susskind (1980a).
[c]Not determined.
[d]Figures in parentheses are estimates derived from incomplete DNA sequence determinations (S. Casjens, personal communication).

C. P2, 186, and P4

Figure 1 and Table 1 are reprinted, with permission, from Bertani and Six (*The Bacteriophages* [ed. R. Calendar], vol. 2. Plenum Press, New York, pp. 83 and 84–85, respectively [1988]). The references cited in the table can be found in the original publication.

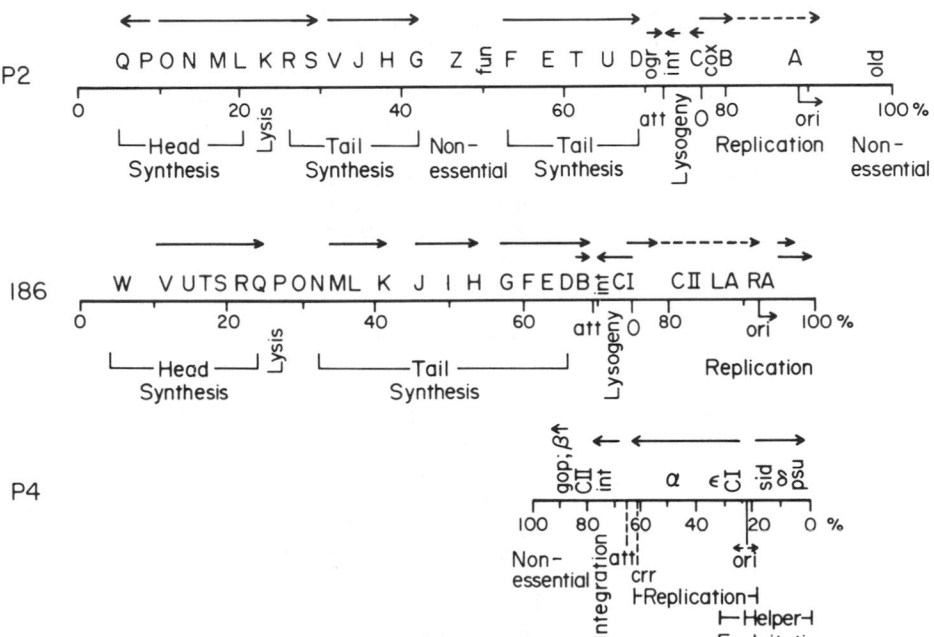

Figure 1 Genetic maps of P2, 186, and P4. The continuous lines represent double-stranded DNA molecules broken at the *cos* sites. The three maps are in scale with each other and aligned so that the left ends have a homologous single-stranded terminus (see Appendix B in Calendar 1988). The genes and their directions of transcription, when known, lie above the line. Gene products are listed in Table 1 below. Sites necessary for the initiation of replication (*ori* and *crr*), repressor binding regions (*O*), and phage attachment sites (*att*), as well as the general functions of the genes, are given below the line. In the cases of P2 and 186, the genes have been oriented on the DNA molecule using specific physical sites (see Appendix A in Calendar 1988) measured from the left ends of the molecules. Phage P4, on the other hand, has been partially sequenced starting from the right end, so the distances are given in percentage from that end instead.

Table 1   Genes and Gene Products of P2, 186, and P4

Gene	Product mol. wt. (kD)	Function	Reference
Phage P2			
old	29.5	λ exclusion	Lindahl *et al.*, 1970; Gibbs *et al.*, 1983
A	86	Replication	Lindahl, 1974; Haggård-Ljungquist, personal communication
B	18	Replication	Lindahl, 1974
cox	10	Excision	Lindahl and Sunshine, 1972
C	11	Repression	Ljungquist *et al.*, 1984
int	35	Integration	Lindahl, 1969; Ljungquist and Bertani, 1983
ogr	8	Late-gene control	Sunshine and Sauer, 1975; Birkeland and Lindqvist, 1986; Christie *et al.*, 1986
D	47	Tail	Ljungquist and Bertani, 1983
U	28	Tail	Ljungquist and Bertani, 1983
T	90	Tail fibers?	Lengyel *et al.*, 1974; Gibbs *et al.*, 1983
E	—	Stabilization of *T* protein	Lengyel *et al.*, 1974
FI	46	Sheath	Lengyel *et al.*, 1974
FII	20	Tube	Lengyel *et al.*, 1974; Gibbs *et al.*, 1983
fun	—	FUDR sensitivity	Bertani, 1964
Z	—	*fun* control	Bertani, 1976
G	—	Tail	Lengyel *et al.*, 1974
H	71	Collar	Lengyel *et al.*, 1974
J	—	Tail	Lengyel *et al.*, 1974
V	23	Tail	Lengyel *et al.*, 1974; Gibbs *et al.*, 1983
S	—	Tail length	Lengyel *et al.*, 1974
R	—	Tail length	Lengyel *et al.*, 1974
K	—	Lysis	Lindahl, 1971
L	—	DNase-resistant heads	Lengyel *et al.*, 1973; Pruss and Calendar, 1978
M	28	Packaging	Lengyel *et al.*, 1973; Bowden and Modrich, 1985
N	44	Major capsid protein	Lengyel *et al.*, 1973
O	30	Cleavage of *N* gene product	Lengyel *et al.*, 1973; Gibbs *et al.*, 1983
P	65	Packaging	Lengyel *et al.*, 1973; Bowden and Modrich, 1985
Q	32	Packaging	Lengyel *et al.*, 1973; Pruss and Calendar, 1978; Barrett, as quoted in Bowden and Modrich, 1981; Christie and Calendar, 1985

Table 1 *Continued*

Gene	Product mol. wt. (kD)	Function	Reference
Phage 186			
RA	72	DNA synthesis	Hocking and Egan, 1982b; Sivaprasad, 1984
LA	38	DNA synthesis	Hocking and Egan, 1982b; Sivaprasad, 1984
CII	—	Lysogeny	Hocking and Egan, 1982c
CI	21	Repressor	Woods and Egan, 1974; Kalionis *et al.*, 1986a
int	37	Integration	Bradley *et al.*, 1975; Kalionis *et al.*, 1986a
B	8	Late-gene control	Hocking and Egan, 1982b; Kalionis *et al.*, 1986b
N–D	—	Tail synthesis	Hocking and Egan, 1982a
P	—	Lysis	Hocking and Egan, 1982b
W–Q	—	Head synthesis	Hocking and Egan, 1982a
Phage P4			
psu	21	Polarity suppression (protein 4)	Sauer *et al.*, 1981; Dale *et al.*, 1986
δ	19	Transactivation	Souza *et al.*, 1977; Lin, 1984
sid	27	Head size determination (protein 3)	Shore *et al.*, 1978; Lin, 1984
CI	15	P4 repression	Calendar *et al.*, 1981; Lin, 1984
ε	11	Derepression of helper	Geisselsoder *et al.*, 1981; Lin, 1984
Orf106	12	DNA replication	Flensburg and Calendar, 1987, and personal communication
α	85	DNA primase (protein 1)	Gibbs *et al.*, 1973; Barrett *et al.*, 1976, 1983; Flensburg and Calendar, 1987
int	51	Integration	Pierson and Kahn, 1984; Pierson and Kahn, in preparation
CII	31	Host survival[a]	Calendar *et al.*, 1981; Dehò, 1983; Dehò and Zangrossi, personal communication
β	—	Host survival[b]	Gibbs *et al.*, 1973; Dehò, 1983
gop	—	Blocks growth in *E. coli* *pin* mutants[b]	Ghisotti *et al.*, 1983; Dehò, 1983; Dehò and Zangrossi, personal communication

[a]Possibly a repressor of *gop*.
[b]β and *gop* may be identical.

D. T4

The genome of phage T4 contains about 166,000 base pairs (Wood and Revel 1976). The mature DNA molecules are linear, but they are cut with a 3% terminal redundancy from concatenated precursors and are thus circularly permuted, producing a circular genetic map (Figure 1). About 85% of the genome has been sequenced, and the genetic and restriction maps have been adjusted accordingly from earlier versions. Genetic map distances agree with physical distances in most regions but are quite different in some areas around genes *43*, *denV*, *uvsW*, *34*, and *35*.

Table 1 gives data for all of the characterized T4 genes in standard map order, with zero, by convention, being the border between genes *rIIA* and *rIIB*. Genes that are essential on standard laboratory strains are designated by numbers; those whose relative positions are uncertain are in brackets. Much of the data is summarized in Wood and Revel (1976) and Mathews et al. (1983); only more recent references are listed in the table. Molecular weights of protein products were determined approximately from gel data or precisely from the DNA sequence (underlined).

Several factors must be considered in using this map. The precise numbers will continue to shift as sequences are completed and regions joined; thus, in referring to specific coordinates, it is crucial to refer to the date of the map being used. Although the sites for *Bgl*II, *Cla*I, *Eco*RI, *Kpn*I, *Sal*I, and *Xho*I were mapped in several laboratories, many others have been directly mapped only relative to sites for a few other enzymes in unsequenced regions, and occasional inversions are seen, as well as discrepancies (indicated by *) which may be due to strain differences, sequencing errors, restriction-site masking, or some other problem. Also, closely spaced sites may be mistaken for a single site. Furthermore, most restriction enzymes will not cut normal T4 DNA, which contains hydroxymethylcytosine instead of cytosine, so most mapping work has used cytosine-containing T4dC strains; most of these contain a deletion in the *denB* region, at about 165 kb on the map. *Eco*RV, *Nde*I, and *Sph*I also cleave glucosylated HMC-T4 DNA.

Figure 1 and Table 1 are reprinted from Kutter et al. (*Genetic Maps*, Fifth Edition [ed. S.J. O'Brien]. Cold Spring Harbor Laboratory Press, Cold Spring Harbor, New York, pp. 1.24–1.25 and 1.34–1.42, respectively [1990]). The references cited in the table can be found in the original publication.

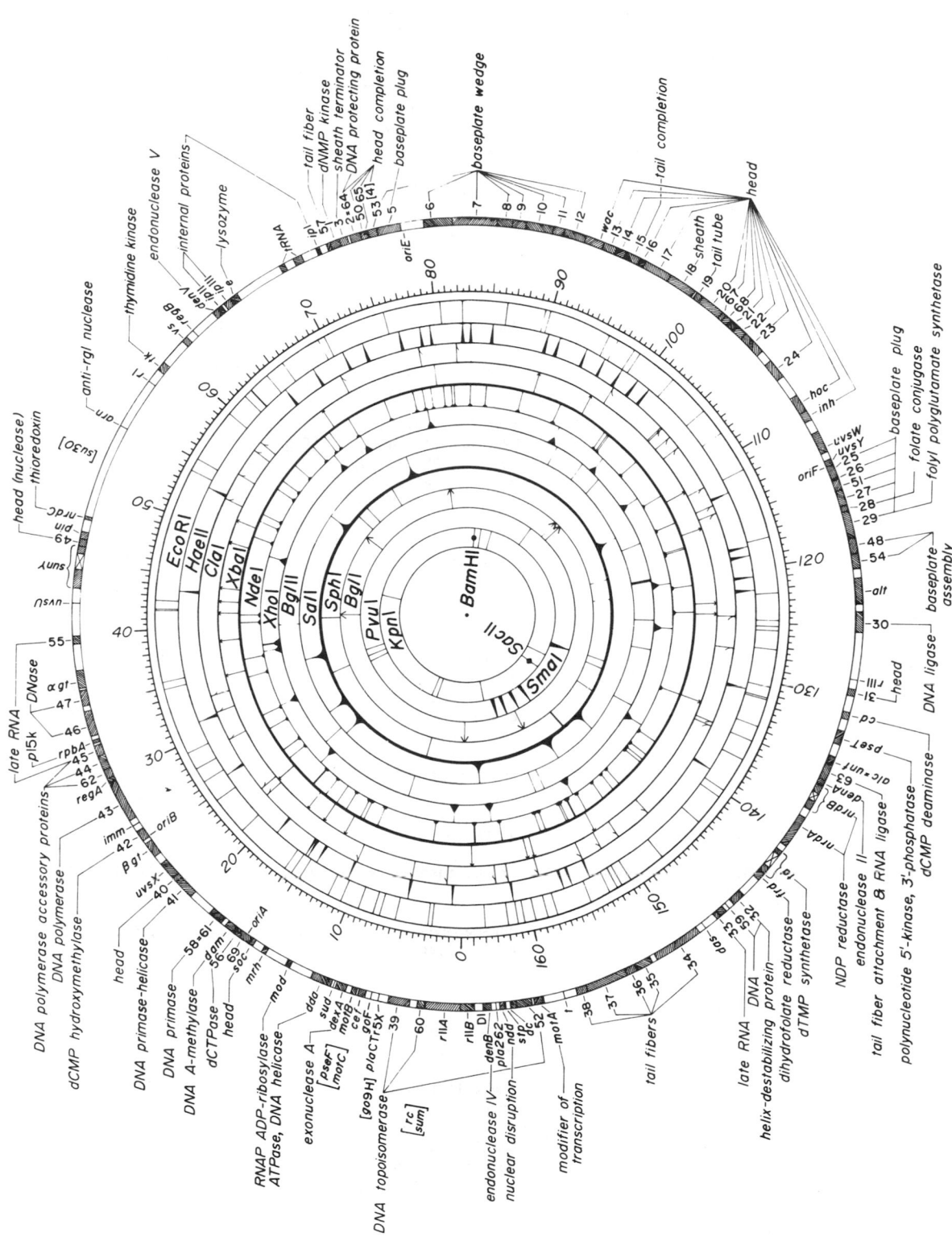

Figure 1 Genetic map of T4

Table 1   Genes of T4

Gene	Mutant Phenotype	Restrictive Host	Function and Comments	Mol Wt (x 1000)	References
rIIA	Rapid lysis; suppresses lig⁻ and some 32⁻; nonessential	rex⁺ lysogens, tabR	Membrane protein; affects membrane ATPase of E. coli	72, 74, 83, 95, **82.8**	24, 118, 181a, 26a
[m = sum]	Suppresses lig⁻; nonessential				118
[rc]	Acriflavine resistance				118
60	DNA delay	S/6, 25°	Membrane protein, DNA topoisomerase subunit	16, 18, **18.6**; gene contains untranslated sequence	71, 72
39	DNA delay	S/6, 25°	Membrane protein, DNA topoisomerase subunit, DNA-dependent ATPase	63 → 22 **58.5**	70, 71, 73, 118
plaCTr5x	Nonessential	CTr5x			118
[goF = go9H = comCα ≈ motC]	Nonessential		Mutations overcome block in HDF (rho) hosts		118, 155, 190
cef = mb = M1 ≈ motC	Nonessential	roc⁻, CT439	Modifier of suppressor tRNAs and of species 1, 2 and 3 RNA	18	118, 150, 155
[del(39-56)₄ ≈ motB]	Nonessential		Modifier of transcription?	12	118, 155
[pseF]	Nonessential		5'-phosphatase		118
dexA ≈ sud	Nonessential; suppresses 32⁻	optA	Exonuclease A	**26.0**	41, 111, 118
dda	Nonessential		DNA-dependent ATPase, DNA helicase	56	41, 58, 75, 88, 111, 118
[mod]	Nonessential		Adenylribosylation of RNA polymerase		118
mrh	Nonessential		Inhibits growth on rpoH mutants	**13.4**	37
soc	Nonessential		Small outer capsid protein	**9**	9, 112, 113, 118
{oriA in 69}					118, 134
69, 69* (overlaps 56)	Does not use oriA	25°		**44**, **26**; has untranslated sequence	17, 112, 118, 133, 134

Table 1 *Continued*

Gene	Mutant Phenotype	Restrictive Host	Function and Comments	Mol wt (x 1000)	References
56 (overlaps 69)	DNA negative		dCTPase, dUTPase, dCDPase, dUDPase	20	40, 112, 133
dam	Nonessential		DNA adenine methylase	30	112, 118, 173, 174
58 = 61	DNA delay	S/6, 25°	Primase subunit	40, 39.5	11, 14, 20, 118, 136
41	DNA arrest; single-stranded DNA, UV-sensitive		GTPase, dGTPase-, ATPase, dATPase; helicase-primase subunit	57, 59, 63, 66, 53.8	14, 65, 66, 67, 136, 163, 176
40 = sp = rIV	Polyheads; suppresses e⁻		genetic exclus.; helps head vertex assembly; lysis	14, 18, 13.3	38, 65, 67, 118, 126, 139, 140
uvsX = fdsA	UV-sensitive, recombination deficient; suppresses 49⁻		RecA-like recombination protein; DNA-dependent ATPase	40, 43.8	25, 38, 51, 65, 88, 118, 223, 224, 225
X.1	Nonessential		DNA-binding protein	25	35
βgt	No β-glucosylation of HMC-DNA		β-glucosyltransferase	40.6	118, 198
42	DNA negative		dCMP-Hydroxymethylase	28.4	101, 102, 110, 118, 197
imm	Immunity to superinfection exclusion		Plasma membrane; inhibits injec. of superinf. DNA	9.3	102, 110, 118, 139
43	DNA negative		DNA polymerase	112, 103.6	118, 187
dsd (in 43)	DNA delay	optA	DNA polymerase		26
regA			Translational regulation of (early) protein synthesis	14.6	1, 118, 129, 200
62	DNA negative		DNA-polymerase accessory protein; ssDNA-dependent ATPase, dATPase	21.4	118
44				35.9	118
45	DNA negative, no late mRNA		Accessory protein of DNA- and RNA-polymerases	24.7	118, 189
rpbA	Nonessential		RNA polymerase binding, "15K", protein	11.4	68, 216

Table 1 *Continued*

Gene	Mutant Phenotype	Restrictive Host	Function and Comments	Mol wt (x 1000)	References
46	DNA arrest; recombination deficient; reduced host DNA degradation		Recombination nuclease	63.5	49, 118
47				39.1	
αgt	No α-glucosyla-tion of HMC-DNA		α-Glucosyl transferase	46.6	49, 118, 198
[gor]	Suppresses RNA polymerase defect				118
55	No late RNA synthesis		RNA polymerase σ factor for late T4 promoters	21.5	34, 49, 118
sunY			Unknown	67.9; contains intron	47, 179, 197a, 199
I-TevII			In sunY intron; intron mobility endonuclease	30.4	7b, 156b
49,	Partially filled heads, highly branched DNA		Endonuclease VII; cleaves recombi-nation junctions	18.1	6, 86, 118, 197a, 199
49*				12	
pin	Nonessential		Inhibitor of E. coli protease	18.8	180, 183, 197a, 199
nrdC	Nonessential		Thioredoxin	10.0	103, 118, 197a, 199
arn	DNA degradation under certain conditions		Antirestriction nuclease (vs. E. coli rglB = mcrB)		27, 118
[Su30]	Nonessential; suppresses lig⁻				118
rI	Nonessential; rapid lysis				118
tk	Nonessential		Thymidine kinase	28, 21.6	118, 128, 207
vs	Nonessential		Modifies valyl-tRNA synthetase	13.1	116, 118, 128, 142, 143, 207
regB	Nonessential; folate analogue resistant		Regulation of gene expression; site-specific ribonuclease	18.0	118, 168a, 203, 207
[stI]	Nonessential "star"				118
[stIII]	Nonessential; suppresses e⁻ and t⁻; star				118
denV	UV-sensitive		Endonuclease V, N-glycosidase	16.1	118, 157, 162, 205, 206, 207
ipII	Nonessential		Internal protein	11.1 → 9.9	118, 205, 207

Table 1 *Continued*

Gene	Mutant Phenotype	Restrictive Host	Function and Comments	Mol wt (x 1000)	References
ipIII	Semi-essential		Internal protein	$\underline{21.7} \rightarrow \underline{20.4}$	118, 207
e	No lysis		Endolysin	$\underline{18.7}$	118, 124, 145
goF3	Grow on HDF hosts				118, 124
Stable species 1 (= C) RNA			unknown		10, 118, 150
Stable species 2 (= D) RNA			unknown		10, 118, 150
tRNAs	Nonessential (nonsense suppressors)	CT439	tRNA precursors		10, 56, 59, 118, 120
arg	$\underline{psu_4}$, $\underline{op}$				
ile					
thr					
ser	$\underline{psu_a}^+$, $\underline{psu_b}^+$, $\underline{psu_1}^+$				
pro					
gly					
leu	$\underline{psu_3}^+$				
gln	$\underline{psu}^+$, $\underline{psu_2}SB$				
ipI		CT596	Internal protein	$\underline{10} \rightarrow 8.5$	10, 118
57B	Poor tail fiber assembly, bypassed in some host mutants		Morphogenetic catalyst of long and short tail fiber assembly	$\underline{16}$	10, 118
57A			DNA-binding protein	$\underline{7}$	10, 36, 118
1	DNA negative		dHMP-kinase	22, 25, $\underline{27.3}$	10, 109, 118
3	Unstable tails		Tail tube, proximal tip	23.3, $\underline{19.6}$	10, 87a, 109, 118
2 = 64	Inactive filled heads, noninfectious particles		Head completion; terminal DNA-protecting protein	25, 27, $\underline{31.6}$	87a, 118, 109,
4 = 50 = [65]	Inactive filled heads, noninfectious particles		Head completion; function unknown	$\underline{17.6}$	87a, 105, 118
65	Encoded on complementary strand of 4		DNA-binding protein	$\underline{7.8}$	105, 169
53	Defective tails		Baseplate, (wedge)	23, $\underline{22.9}$	87a, 118, 132

Table 1 *Continued*

Gene	Mutant Phenotype	Restrictive Host	Function and Comments	Mol wt (x 1000)	References
5	Defective tails		Baseplate hub, baseplate lysozyme	63.0 → 44,	118, 132, 135
dbpB	Encoded on complementary strand of 5		DNA-binding protein	6.1	105
{oriE}					105, 106, 118
6	Defective tails, permit fiberless plating		Baseplate, 1/6 arm	74.4, 78, 86	118, 177
7	Defective tails, permit fiberless plating		Baseplate, 1/6 arm	127, 140, 119.2	117, 118, 177
8	Defective tails		Baseplate, 1/6 arm	39, 46, 38.0	99, 117, 118, 177
9	Defective tails fiberless particles		Baseplate, long tail fiber attachment	30, 34, 31.0	99, 118, 153
10	Defective tails		Baseplate, 1/6 arm	88, 90	118, 149, 153
11	Defective tails		Baseplate, 1/6 arm	24, 25, 26, 23.7	118, 149, 153
12	Defective tails		Baseplate, short tail fibers	54, 56, 57, 56.2	118, 178
wac	Nonessential		Whisker antigen	52, 51.9	118, 153
13			Head completion	33, 34.7	118, 178a
14			Head completion	30, 29.6	118, 178a
15	Defective tails		Proximal tail sheath stabilizer	32, 35, 31.4	118, 178a
16	DNA packaging defective; empty heads		Head filling packasome component	18.3	118, 151, 160
17 = q	DNA packaging defective; suppresses some 32⁻ or 20⁻ mutations; quinacrine resistant		Head filling packasome	69, 70.0	118, 151, 160
18	Defective tails		Tail sheath monomer	70, 80, 71.4	3, 4, 118
19	Defective tails		Tail tube monomer	20, 21, 18.5	3, 4, 118
20	Polyheads		Head plug protein (connector to neck)	63, 65, 67, 61.0	118, 161, 117a

Table 1   *Continued*

Gene	Mutant Phenotype	Restrictive Host	Function and Comments	Mol wt (x 1000)	References
67 = pip	Head defect		Core protein, precursor to internal peptides	9.1	83, 84a, 118, 199a
68	Isometric heads		Prohead core protein	17	82, 83, 84, 199a
21	Faulty heads		Head assembly core; maturation protease	23.2 → 18.4 cleaved to small peptides	81, 118
22	Faulty heads		Head assembly core (later degraded)	29.8	83, 118, 147
23	No or faulty heads		Major head subunit; cleaved to a packaging-related DNA-dependent ATPase-endonuclease	56.0 → 48.7 → 43	118, 147, 220a
gol (in 23)	grow on lit hosts (CTr5x)		Disrupt general translation and gene-23 transcription		118, 186a
24 = os	Faulty heads, osmotic shock resistance		Vertex head subunit	48.4 → 46	118, 221
hoc = eph	Nonessential		Minor capsid protein	39.1, 40	20, 77, 118, 195
inh			Inhibitor of gene-21 protease	35	77, 118, 195
dar = uvsW	Suppresses 59⁻, 46⁻; UV-sensitive				25, 28, 77, 107, 118, 220
uvsY = fdsB	UV-sensitive; DNA synthesis reduced, recombination deficient; suppresses 49⁻		Recombination protein	15.8	25, 28, 30, 55, 61, 87, 118, 194. 220, 225
(oriF)					56, 93, 94, 95, 118, 222
25	Tail defects		Baseplate, 1/6 arm; lysozyme	15	54, 55, 61, 107, 118, 135, 225
26	Tail defects		Baseplate central hub	23.8 → 42	13, 18, 89, 118, 132
51	Tail defects		Baseplate central hub protease	29.3 → 16.5	89, 118, 172, 193
27	Tail defects, permit fiberless plating		Baseplate central hub	47, 49	13, 118, 172

Table 1 *Continued*

Gene	Mutant Phenotype	Restrictive Host	Function and Comments	Mol wt (x 1000)	References
28	Tail defects		Baseplate, distal surface of central hub, gamma glutamyl hydrolase	24, 25, <u>20.1</u>	13, 118
29	Tail defects		Baseplate, tail "bulge", folyl-poly-glutamyl-synthetase	77, <u>64.5</u>	31, 74, 118
48	Tail defects		Baseplate, tail tube fibre, length "measure"?	37, 44, <u>39.7</u>	31, 32, 74, 118
54	Tail defects		Baseplate, tail tube; polymerization initiator?	36, <u>35.0</u>	31, 32, 74, 118
<u>alt</u>	Nonessential		Adenylribosyl-ation of RNA polymerase, (packaged with DNA)	<u>76.8</u> → <u>75.9</u>	63a, 118, 169
30 = <u>lig</u>	DNA arrest, hyper-rec		DNA ligase	<u>55.2</u>	5, 118
<u>r</u>III	Nonessential, rapid lysis				118
31	Capsid protein lumps; suppresses mutations in host-defective (e.g., <u>groE</u>) <u>E. coli</u> genes		Organizes head protein assembly and DNA topoiso-merase complex	<u>12.0</u>	29, 84b, 118, 137, 138, 181
<u>cd</u>	Nonessential		dCMP deaminase	21	118, 114
<u>pseT</u>	Nonessential	CTr5x (<u>lit</u>⁻)	Deoxyribonu-cleotide 3' phosphatase, 5' polynucleotide kinase	30, <u>34</u>	76, 79, 80, 118, 127
<u>alc</u> = <u>unf</u>	Allows transcription of cytosine-containing T4 DNA	<u>E. coli</u> (pR386)	Abolishes transcript elongation on dC-DNA; unfolding of host DNA; DNA and RNA polymerase-binding protein	<u>19.1</u>	62, 63, 98, 118, 159, 185a
63	Poor tail fiber attachment		RNA ligase; helps tail fiber attachment	<u>43.5</u>	15, 79, 80, 118, 159
<u>denA</u>	Nonessential; defective in host DNA degradation		Endonuclease II	<u>15.8</u>	118, 159

Table 1 *Continued*

Gene	Mutant Phenotype	Restrictive Host	Function and Comments	Mol wt (x 1000)	References
nrdB	Nonessential		Ribonucleotide reductase B subunit (split gene)	35, 40, 45.3	46, 47, 118, 182
nrdA	Nonessential		Ribonucleotide reductase A subunit	80, 85	118, 119, 201
td	Nonessential		Thymidylate synthase; baseplate hub component (split gene)	32, 33.0	21, 22, 33, 47, 118, 156
I-TevI	Nonessential		In td intron; intron mobility endonuclease	28.1	7a, 156b, 214
frd			Dihydrofolate reductase; baseplate wedge component	21.6	156
32	DNA arrest; recombination and excision repair deficient; UV-sensitive	tab-32	ssDNA-binding (= helix-destabilizing) protein	33.5	118, 131, 42
59	DNA arrest		positions gp41 on ssDNA	26.0	59, 118
33	No late RNA synthesis		RNA-polymerase-binding-protein	12.8	59, 118
dbpA			dsDNA-binding protein	10.4	59, 60
das-suα	Suppresses 46⁻, uvsX⁻		das, sur, and/or dbpA may be the same gene		59, 118, 169
[sur]	Suppresses 46⁻, 47⁻, uvsX⁻				209, 210
34	Fiberless particles		Proximal tail fiber subunit (A antigen)	145	45, 59, 118
{oriG in 34}					93, 94, 95, 118
35	Fiberless particles		Hinge tail fiber subunit	39, 40	45, 118, 130b
36	Fiberless particles		Small distal tail fiber subunit	24.3	118, 130b
37	Fiberless particles, host range		Large distal tail fiber subunit	112.8	118, 130b, 165
38	Fiberless particles		Assembly catalyst of gp37	26, 27, 28	118, 164, 165b, 130b

Table 1   *Continued*

Gene	Mutant Phenotype	Restrictive Host	Function and Comments	Mol wt (x 1000)	References
t = stII	Lysis defective; suppresses 63⁻ and rII mutations			_25.2_	118, 130a, 165a
motA = sip	Suppresses rII⁻ in K(\) hosts		Regulates middle gene expression, activates middle promoters	24, _23.6_	118, 202, 203, 204
[rV]	Temperature-dependent rapid lysis				118
52	DNA delay	S/6, 25°	Membrane protein, DNA topoisomerase subunit	52, _50.6_	69, 118, 168
ac	acriflavine resistant			_5.0_	118, 15
[ama] [rs]	Nonessential, acriflavine resistant			_5.4_	118, 15
stp	suppresses pseT mutations			_3.2_	118, 15
ndd = D2b	Nuclear disruption defective	CT447		15, _11.1_	118, 184, 185, 15
pla262	Nonessential	CT262			118, 15
denB	Nonessential; allows production of T4 with dC in DNA		Endonuclease IV	22	118
D1	Nonessential				118
rIIB	Nonessential, rapid lysis; suppresses 30⁻ and some 32⁻ mutations	rex⁺ \ lysogens, tabR	Membrane protein; affects membrane ATPase of E. coli	33, 41, _35.5_	24, 69, 118

Note: Genes originally given different names and now known to be identical are signified by =; those that are probably identical are signified by ≈. Genes within the introns of other genes are denoted by I-. Genes 69 and 49 also encode smaller proteins, denoted 69* and 49* that begin within the gene and end at the same site as the longer sequence. Origins of replication, ori, are listed in braces.

E. T7

Figures 1 and 2 are reprinted from Studier and Dunn (*Cold Spring Harbor Symp. Quant. Biol.*, vol. 47, pp. 1001 and 1002, respectively [1983]).

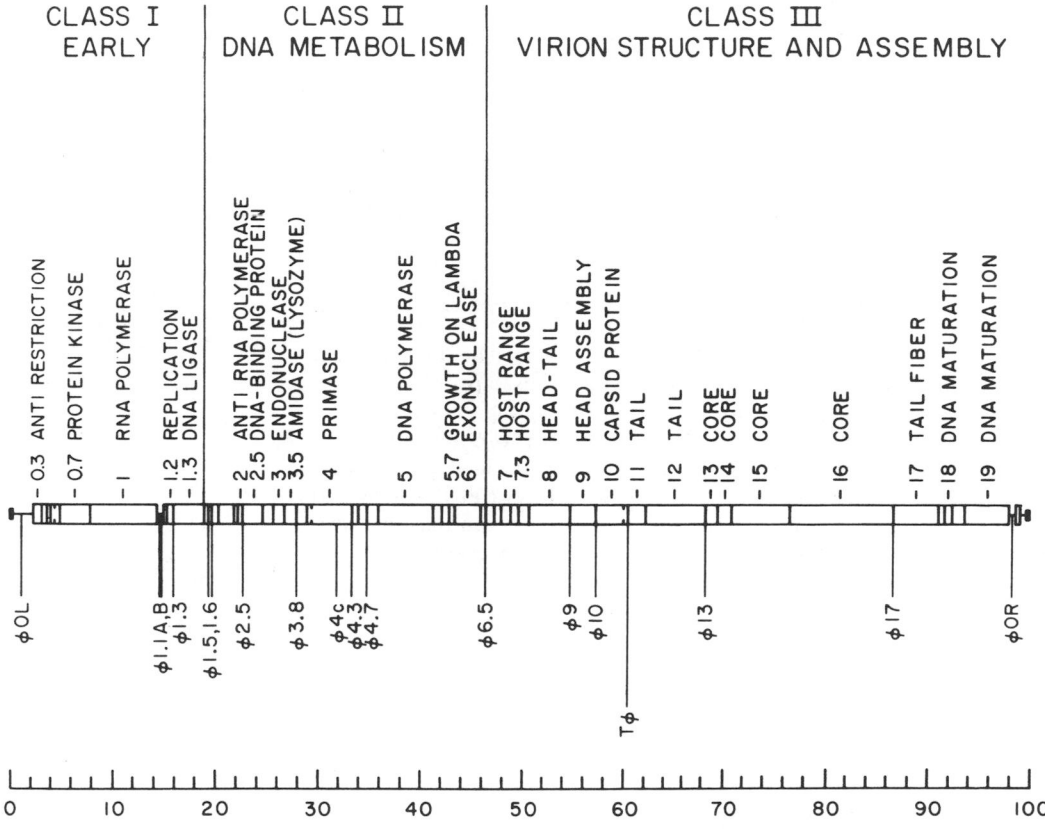

Figure 1 Genetic and physical map of T7. The positions of the terminal repetition (filled boxes), the T7 genes (open boxes), and the promoters and terminator for T7 RNA polymerase are drawn to scale, according to their positions in the nucleotide sequence (Dunn and Studier 1983). Where something is known about gene function, the gene number and function are indicated.

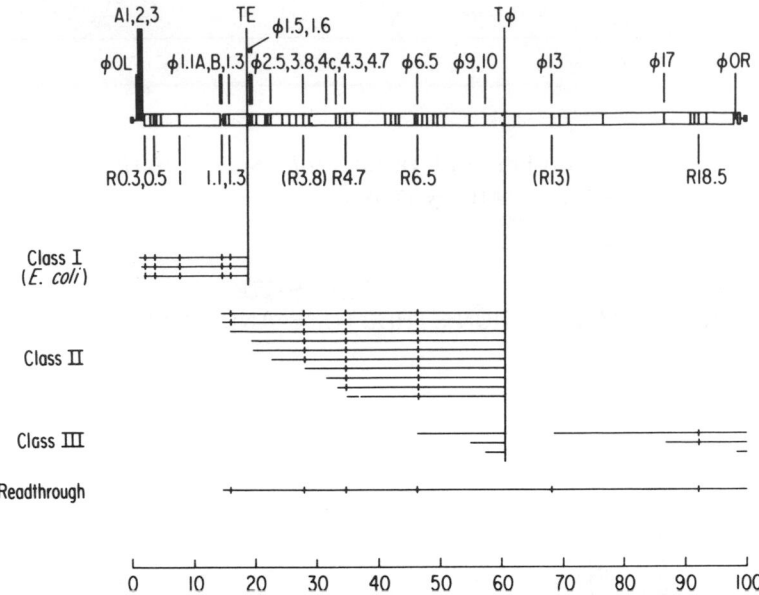

Figure 2 Synthesis and processing of T7 mRNAs. The T7 genes are represented by open boxes; the positions of transcription signals are given above the genes, and the RNase III cleavage sites are given below. The primary transcript from each promoter is represented by a horizontal line, and sites of RNase III cleavage by the short vertical lines. Apparently not all RNAs are cut at the R*3.8* and R*13* RNase III cleavage sites, which is indicated by the parentheses. RNAs produced by readthrough of Tφ are also represented.

F. RNA Coliphages

Figure 1 and Table 1 are reprinted, with permission, from van Duin (*The Bacteriophages* [ed. R. Calendar], vol. 1. Plenum Press, New York, pp. 119 and 121, respectively [1988]). The references cited in the figure caption and table notes can be found in the original paper.

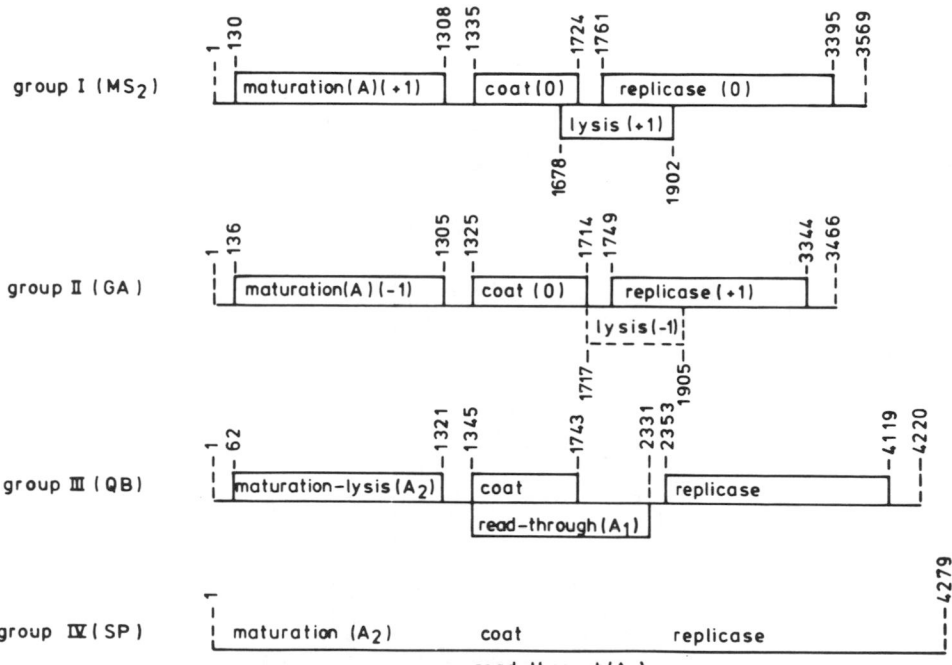

Figure 1 Genetic map of group I–IV RNA coliphages. The MS2 sequence was determined by Fiers et al. (1976); GA is from Inokuchi et al. (1986); and Qβ is from Mekler (1981). SP, which is highly homologous to Qβ, was recently sequenced and found to be 4279 nucleotides long (Hirashima, pers. comm.). For group I and II phages, the relative reading frames of the genes are indicated by −1, 0, and +1, respectively. The lysis gene in GA has been tentatively identified.

Table 1   Properties of ssRNA Phages[a]

Host	Phage	Morphology	$s_{20,w}$ (S)	Diameter (nm)	Host receptor site	Proteins Coat (daltons)	Proteins A protein (daltons)
Escherichia coli	MS2-R17-f2	Icosahedral	79–80	26.0–26.6	F pili	13,731	43,988[b]
Escherichia coli	Qβ	Icosahedral	83–84	26.0	F pili	14,125	41,000[c]
Pseudomonas aeruginosa	7S[d,e]	Icosahedral	88	25.0	Polar pili	+[h,i]	+[h,i]
	PP7[f,g]	Icosahedral		25.0	Polar pili		
Caulobacter crescentus	Cb5[i]	Icosahedral	70–71	23.0	Polar pili	12,000	40,000
	Cb12r[k,l]	Icosahedral		22.0–23.0	Polar pili		
Caulobacter bacteroides	Cb8r[k,l]	Icosahedral	—	22.0–23.0	Polar pili		
Caulobacter fusiformis	Cb23r[k,l]	Icosahedral	—	21.0–23.0	Polar pili		
RP plasmid[m]	PRR1[n]	Icosahedral	80	25.0	RP pili	+[h,o]	+[h,o]

[a] Adapted from Shapiro and Bendis (1975).
[b] Fiers et al. (1975).
[c] Weber and Konigsberg (1975).
[d] Feary et al. (1963).
[e] Feary et al. (1964).

[f] Bradley (1966).
[g] Lin and Schmidt (1972).
[h] +, Present and similar in size to the group I coliphages (MS2-R17-f2).
[i] Davis and Benike (1974).
[j] Bendis and Shapiro (1970).

[k] Schmidt and Stanier (1965).
[l] Schmidt (1966).
[m] This plasmid has a wide host range (see text).
[n] Olson and Thomas (1973).
[o] Dhaese et al. (1977).

SECTION 15

G. φX174

Table 1 and Figure 1 are reprinted, with permission, from Hayashi et al. (*The Bacteriophages* [ed. R. Calendar], vol. 2. Plenum Press, New York, pp. 3 and 4, respectively [1988]). The references cited in the table notes and figure caption can be found in the original paper.

Table 1   Genes of φX174

Gene	Gene function	DNA replication stages[a] blocked by mutation	Mol. wt. of gp in PAGE[a]	Mol. wt. of gp from sequence
A	Stg. II, stg. III	RF→RF, RF→SS	60,000	58,650
A*	Shut off host DNA repl.	Unknown (see text)	37,000	38,700
B	Morphogenesis, Ω^b	RF→SS	20,000	13,830
C	Switch stg. II to stg. III	RF→SS[c]	5,800[d]	10,050
D	Morphogenesis, $\Omega,^b$ 132S comp.	RF→SS	14,000	16,920
E	Lysis	None	10,000[e]	10,370
F	Major capsid protein, Ω^b	RF→SS	50,000	48,440
G	Major spike protein, Ω^b	RF→SS	20,000	19,020
H	Minor spike protein, $\Omega,^b$ adsorption	RF→stable SS	37,000	34,370
J	Phage internal protein, stg. III	RF→stable SS[f]	4,050[g]	4,220
K	Stimulation of phage synthesis	None	8,000[h]	6,380

[a]See Denhardt (1977) for A, A*, B, D, F, G, and H.
[b]Ω Prohead component.
[c]Aoyama and Hayashi (1986).
[d]Aoyama *et al.* (1983).
[e]Pollock *et al.* (1978).
[f]Hamatake *et al.* (1985).
[g]Freymeyer *et al.* (1977).
[h]Tessman *et al.* (1980).
Abbreviations: comp., component; repl., replication; stg. II and stg. III, stage II and stage III DNA replication.

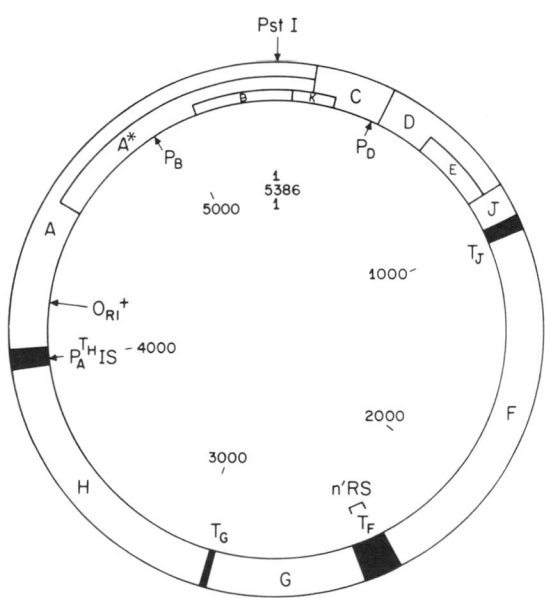

Figure 1 Genetic map of φX174. This map is constructed according to the Sanger et al. (1978) sequence data. A unique *Pst*I site is the reference point (nucleotide 1). O_{RI^+} is the origin of +-strand DNA synthesis. P_A, P_B, and P_D are promoters, and T_J, T_F, T_G, and T_H are transcription terminators (Hayashi et al. 1986). IS is the incompatibility sequence (Van der Avoort et al. 1982, 1984). n'RS is the n' protein recognition site (Shlomai and Kornberg 1980a).

Sequences in the *lac* System

A. Sequence of the *lac* operon
   1. DNA sequence from *lacI* to the beginning of *lacA*
   2. Nucleotide sequence of the *lacA* gene

B. Restriction enzyme cleavage sites in the *lac* operon

REFERENCES

Buchel, D.E., B. Gronenborn, and B. Müller-Hill. 1980. Sequence of the lactose permease gene. *Nature* **283:** 541–545.

Dickson, R.C., J. Abelson, W.M. Barnes, and W.S. Reznikoff. 1975. Genetic regulation: The *lac* control region. *Science* **187:** 27–35.

Farabaugh, P.J. 1978. Sequence of the *lacI* gene. *Nature* **274:** 765–769.

Hediger, M.A., D.F. Johnson, D.P. Nierlich, and I. Zabin. 1985. DNA sequence of the lactose operon: The *lacA* gene and the transcriptional termination region. *Proc. Natl. Acad. Sci.* **82:** 6414–6418.

Kalnins, A., K. Otto, and B. Müller-Hill. 1983. Sequence of the *lacZ* gene of *Escherichia coli. EMBO J.* **2:** 593–597.

Silhavy, T.J., M.L. Berman, and L.W. Enquist. 1984. *Experiments with Gene Fusions.* Cold Spring Harbor Labortory, Cold Spring Harbor, New York.

SECTION 16

A. Sequence of the *lac* Operon

Figure 1 is reprinted from Silhavy et al. (*Experiments with Gene Fusions.* Cold Spring Harbor Laboratory, Cold Spring Harbor, New York, pp. 273–281 [1984]). Figure 2 is reprinted, with permission, from Hediger et al. (*Proc. Natl. Acad. Sci.,* vol. 82, p. 6416 [1985]).

```
          10        20        30        40        50        60
GTTGACACCATCGAATGGCGCAAAACCTTTCGCGGTATGGCATGATAGCGCCCGGAAGAGAGTCAA          lacIP


          76        86        96        106       116       126
TTCAGGGTGGTGAATGTGAAACCAGTAACGTTATACGATGTCGCAGAGTATGCCGGTGTCTCTTAT          lacI
                Me  Ly Pr Va Th Le Ty Ap Va Al Gu Ty Al Gy Va Se Ty
                1                               10


          142       152       162       172       182       192
CAGACCGTTTCCCGCGTGGTGAACCAGGCCAGCCACGTTTCTGCGAAAACGCGGGAAAAAGTGGAA          lacI
Gn Th Va Se Ar Va Va An Gn Al Se Hi Va Se Al Ly Th Ar Gu Ly Va Gu
   20                               30


          208       218       228       238       248       258
GCGGCGATGGCGGAGCTGAATTACATTCCCAACCGCGTGGCACAACAACTGGCGGGCAAACAGTCG          lacI
Al Al Me Al Gu Le An Ty Il Pr An Ar Va Al Gn Gn Le Al Gy Ly Gn Se
40                            50                             60


          274       284       294       304       314       324
TTGCTGATTGGCGTTGCCACCTCCAGTCTGGCCCTGCACGCGCCGTCGCAAATTGTCGCGGCGATT          lacI
Le Le Il Gy Va Al Th Se Se Le Al Le Hi Al Pr Se Gn Il Va Al Al Il
                   70                                80


          340       350       360       370       380       390
AAATCTCGCGCCGATCAACTGGGTGCCAGCGTGGTGGTGTCGATGGTAGAACGAAGCGGCGTCGAA          lacI
Ly Se Ar Al Ap Gn Le Gy Al Se Va Va Va Se Me Va Gu Ar Se Gy Va Gu
                   90                              100


          406       416       426       436       446       456
GCCTGTAAAGCGGCGGTGCACAATCTTCTCGCGCAACGCGTCAGTGGGCTGATCATTAACTATCCG          lacI
Al Cy Ly Al Al Va Hi An Le Le Al Gn Ar Va Se Gy Le Il Il An Ty Pr
             110                             120


          472       482       492       502       512       522
CTGGATGACCAGGATGCCATTGCTGTGGAAGCTGCCTGCACTAATGTTCCGGCGTTATTTCTTGAT          lacI
Le Ap Ap Gn Ap Al Il Al Va Gu Al Al Cy Th An Va Pr Al Le Ph Le Ap
          130                             140


          538       548       558       568       578       588
GTCTCTGACCAGACACCCATCAACAGTATTATTTTCTCCCATGAAGACGGTACGCGACTGGGCGTG          lacI
Va Se Ap Gn Th Pr Il An Se Il Il Ph Se Hi Gu Ap Gy Th Ar Le Gy Va
150                            160                            170


          604       614       624       634       644       654
GAGCATCTGGTCGCATTGGGTCACCAGCAAATCGCGCTGTTAGCGGGCCCATTAAGTTCTGTCTCG          lacI
Gu Hi Le Va Al Le Gy Hi Gn Gn Il Al Le Le Al Gy Pr Le Se Se Va Se
                   180                             190
```

Figure 1 (*See facing page for caption.*)

16.2

```
      670       680       690       700       710       720
GCGCGTCTGCGTCTGGCTGGCTGGCATAAATATCTCACTCGCAATCAAATTCAGCCGATAGCGGAA          lacI
Al Ar Le Ar Le Al Gy Tr Hi Ly Ty Le Th Ar An Gn Il Gn Pr Il Al Gu
         200                                     210


      736       746       756       766       776       786
CGGGAAGGCGACTGGAGTGCCATGTCCGGTTTTCAACAAACCATGCAAATGCTGAATGAGGGCATC          lacI
Ar Gu Gy Ap Tr Se Al Me Se Gy Ph Gn Gn Th Me Gn Me Le An Gu Gy Il
         220                                     230


      802       812       822       832       842       852
GTTCCCACTGCGATGCTGGTTGCCAACGATCAGATGGCGCTGGGCGCAATGCGCGCCATTACCGAG          lacI
Va Pr Th Al Me Le Va Al An Ap Gn Me Al Le Gy Al Me Ar Al Il Th Gu
         240                                     250


      868       878       888       898       908       918
TCCGGGCTGCGCGTTGGTGCGGATATCTCGGTAGTGGGATACGACGATACCGAAGACAGCTCATGT          lacI
Se Gy Le Ar Va Gy Al Ap Il Se Va Va Gy Ty Ap Ap Th Gu Ap Se Se Cy
260                        270                                280


      934       944       954       964       974       984
TATATCCCGCCGTTAACCACCATCAAACAGGATTTTCGCCTGCTGGGGCAAACCAGCGTGGACCGC          lacI
Ty Il Pr Pr Le Th Th Il Ly Gn Ap Ph Ar Le Le Gy Gn Th Se Va Ap Ar
              290                                300


     1000      1010      1020      1030      1040      1050
TTGCTGCAACTCTCTCAGGGCCAGGCGGTGAAGGGCAATCAGCTGTTGCCCGTCTCACTGGTGAAA          lacI
Le Le Gn Le Se Gn Gy Gn Al Va Ly Gy An Gn Le Le Pr Va Se Le Va Ly
         310                                     320


     1066      1076      1086      1096      1106      1116
AGAAAAACCACCCTGGCGCCCAATACGCAAACCGCCTCTCCCCGCGCGTTGGCCGATTCATTAATG          lac
Ar Ly Th Th Le Al Pr An Th Gn Th Al Se Pr Ar Al Le Al Ap Se Le Me
         330                                     340


     1132      1142      1152      1162      1172      1182
CAGCTGGCACGACAGGTTTCCCGACTGGAAAGCGGGCAGTGAGCGCAACGCAATTAATGTGAGTTA          lacI/lacP
Gn Le Al Ar Gn Va Se Ar Le Gu Se Gy Gn **
    350                      360


     1198      1208      1218      1228      1238      1248
GCTCACTCATTAGGCACCCCAGGCTTTACACTTTATGCTTCCGGCTCGTATGTTGTGTGGAATTGT          lacP/lacO


     1264      1274      1284      1294      1304      1314
GAGCGGATAACAATTTCACACAGGAAACAGCTATGACCATGATTACGGATTCACTGGCCGTCGTTT          lacO/lacZ
                                    Th Me Il Th Ap Se Le Al Va Va Le
                                    1                            10
```

Figure 1 DNA sequence of the *lac* operon from *lacI* to the beginning of *lacA*. The sequence was assembled from the data of Dickson et al. (1975), Farabaugh (1978), Buchel et al. (1980), and Kalnins (1983). The first nucleotide is the first base of the *Hinc*II site in the *lacI* promoter. Genes are labeled in the right-hand margin. The sequence is translated into the corresponding protein products using a two-letter code. In this code, all abbreviations are the first two letters of the standard triplet abbreviation except Gln = Gn, Glu = Gu, Asp = Ap, and Asn = An. The first codon and amino acid of each protein is underlined, and relevant stop codons are represented by double asterisks. The amino acid residues for each protein are numbered sequentially. Only part of the *lacA* sequence is shown here. See Figure 2 for the nucleotide sequence of the *lacA* gene.

```
      1330      1340      1350      1360      1370      1380
TACAACGTCGTGACTGGGAAAACCCTGGCGTTACCCAACTTAATCGCCTTGCAGCACATCCCCCTT        lacZ
 Gn Ar Ar Ap Tr Gu An Pr Gy Va Th Gn Le An Ar Le Al Al Hi Pr Pr Ph
                     20                                 30
```

```
      1396      1406      1416      1426      1436      1446
TCGCCAGCTGGCGTAATAGCGAAGAGGCCCGCACCGATCGCCCTTCCCAACAGTTGCGCAGCCTGA        lacZ
 Al Se Tr Ar An Se Gu Gu Al Ar Th Ap Ar Pr Se Gn Gn Le Ar Se Le An
                     40                                 50
```

```
      1462      1472      1482      1492      1502      1512
ATGGCGAATGGCGCTTTGCCTGGTTTCCGGCACCAGAAGCGGTGCCGGAAAGCTGGCTGGAGTGCG        lacZ
 Gy Gu Tr Ar Ph Al Tr Ph Pr Al Pr Gu Al Va Pr Gu Se Tr Le Gu Cy Ap
                  60                                 70
```

```
      1528      1538      1548      1558      1568      1578
ATCTTCCTGAGGCCGATACTGTCGTCGTCCCCTCAAACTGGCAGATGCACGGTTACGATGCGCCCA        lacZ
 Le Pr Gu Al Ap Th Va Va Va Pr Se An Tr Gn Me Hi Gy Ty Ap Al Pr Il
               80                                 90
```

```
      1594      1604      1614      1624      1634      1644
TCTACACCAACGTAACCTATCCCATTACGGTCAATCCGCCGTTTGTTCCCACGGAGAATCCGACGG        lacZ
 Ty Th An Va Th Ty Pr Il Th Va An Pr Pr Ph Va Pr Th Gu An Pr Th Gy
100                               110                               120
```

```
      1660      1670      1680      1690      1700      1710
GTTGTTACTCGCTCACATTTAATGTTGATGAAAGCTGGCTACAGGAAGGCCAGACGCGAATTATTT        lacZ
 Cy Ty Se Le Th Ph An Va Ap Gu Se Tr Le Gn Gu Gy Gn Th Ar Il Il Ph
                     130                               140
```

```
      1726      1736      1746      1756      1766      1776
TTGATGGCGTTAACTCGGCGTTTCATCTGTGGTGCAACGGGCGCTGGGTCGGTTACGGCCAGGACA        lacZ
 Ap Gy Va An Se Al Ph Hi Le Tr Cy An Gy Ar Tr Va Gy Ty Gy Gn Ap Se
                  150                               160
```

```
      1792      1802      1812      1822      1832      1842
GTCGTTTGCCGTCTGAATTTGACCTGAGCGCATTTTTACGCGCCGGAGAAAACCGCCTCGCGGTGA        lacZ
 Ar Le Pr Se Gu Ph Ap Le Se Al Ph Le Ar Al Gy Gu An Ar Le Al Va Me
               170                               180
```

```
      1858      1868      1878      1888      1898      1908
TGGTGCTGCGTTGGAGTGACGGCAGTTATCTGGAAGATCAGGATATGTGGCGGATGAGCGGCATTT        lacZ
 Va Le Ar Tr Se Ap Gy Se Ty Le Gu Ap Gn Ap Me Tr Ar Me Se Gy Il Ph
            190                               200
```

```
      1924      1934      1944      1954      1964      1974
TCCGTGACGTCTCGTTGCTGCATAAACCGACTACACAAATCAGCGATTTCCATGTTGCCACTCGCT        lacZ
 Ar Ap Va Se Le Le Hi Ly Pr Th Th Gn Il Se Ap Ph Hi Va Al Th Ar Ph
210                               220                               230
```

Figure 1   *Continued*

```
      1990        2000        2010        2020        2030        2040
TTAATGATGATTTCAGCCGCGCTGTACTGGAGGCTGAAGTTCAGATGTGCGGCGAGTTGCGTGACT          lacZ
 An Ap Ap Ph Se Ar Al Va Le Gu Al Gu Va Gn Me Cy Gy Gu Le Ar Ap Ty
             240                                          250
```

```
      2056        2066        2076        2086        2096        2106
ACCTACGGGTAACAGTTTCTTTATGGCAGGGTGAAACGCAGGTCGCCAGCGGCACCGCGCCTTTCG         lacZ
 Le Ar Va Th Va Se Le Tr Gn Gy Gu Th Gn Va Al Se Gy Th Al Pr Ph Gy
             260                                          270
```

```
      2122        2132        2142        2152        2162        2172
GCGGTGAAATTATCGATGAGCGTGGTGGTTATGCCGATCGCGTCACACTACGTCTCAACGTCGAAA         lacZ
 Gy Gu Il Il Ap Gu Ar Gy Gy Ty Al Ap Ar Va Th Le Ar Le An Va Gu An
             280                                          290
```

```
      2188        2198        2208        2218        2228        2238
ACCCGAAACTGTGGAGCGCCGAAATCCCGAATCTCTATCGTGCGGTGGTTGAACTGCACACCGCCG         lacZ
 Pr Ly Le Tr Se Al Gu Il Pr An Le Ty Ar Al Va Va Gu Le Hi Th Al Ap
             300                                          310
```

```
      2254        2264        2274        2284        2294        2304
ACGGCACGCTGATTGAAGCAGAAGCCTGCGATGTCGGTTTCCGCGAGGTGCGGATTGAAAATGGTC         lacZ
 Gy Th Le Il Gu Al Gu Al Cy Ap Va Gy Ph Ar Gu Va Ar Il Gu An Gy Le
320                                330                          340
```

```
      2320        2330        2340        2350        2360        2370
TGCTGCTGCTGAACGGCAAGCCGTTGCTGATTCGAGGCGTTAACCGTCACGAGCATCATCCTCTGC         lacZ
 Le Le Le An Gy Ly Pr Le Le Il Ar Gy Va An Ar Hi Gu Hi Hi Pr Le Hi
             350                                          360
```

```
      2386        2396        2406        2416        2426        2436
ATGGTCAGGTCATGGATGAGCAGACGATGGTGCAGGATATCCTGCTGATGAAGCAGAACAACTTTA         lacZ
 Gy Gn Va Me Ap Gu Gn Th Me Va Gn Ap Il Le Le Me Ly Gn An An Ph An
             370                                          380
```

```
      2452        2462        2472        2482        2492        2502
ACGCCGTGCGCTGTTCGCATTATCCGAACCATCCGCTGTGGTACACGCTGTGCGACCGCTACGGCC         lacZ
 Al Va Ar Cy Se Hi Ty Pr An Hi Pr Le Tr Ty Th Le Cy Ap Ar Ty Gy Le
             390                                          400
```

```
      2518        2528        2538        2548        2558        2568
TGTATGTGGTGGATGAAGCCAATATTGAAACCCACGGCATGGTGCCAATGAATCGTCTGACCGATG         lacZ
 Ty Va Va Ap Gu Al An Il Gu Th Hi Gy Me Va Pr Me An Ar Le Th Ap Ap
             410                                          420
```

```
      2584        2594        2604        2614        2624        2634
ATCCGCGCTGGCTACCGGCGATGAGCGAACGCGTAACGCGAATGGTGCAGCGCGATCGTAATCACC         lacZ
 Pr Ar Tr Le Pr Al Me Se Gu Ar Va Th Ar Me Va Gn Ar Ap Ar An Hi Pr
430                                440                                450
```

Figure 1 *Continued*

```
      2650      2660      2670      2680      2690      2700
CGAGTGTGATCATCTGGTCGCTGGGGAATGAATCAGGCCACGGCGCTAATCACGACGCGCTGTATC      lacZ
 Se Va Il Il Tr Se Le Gy An Gu Se Gy Hi Gy Al An Hi Ap Al Le Ty Ar
              460                                 470
```

```
      2716      2726      2736      2746      2756      2766
GCTGGATCAAATCTGTCGATCCTTCCCGCCCGGTGCAGTATGAAGGCGGCGGAGCCGACACCACGG      lacZ
 Tr Il Ly Se Va Ap Pr Se Ar Pr Va Gn Ty Gu Gy Gy Gy Al Ap Th Th Al
              480                                 490
```

```
      2782      2792      2802      2812      2822      2832
CCACCGATATTATTTGCCCGATGTACGCGCGCGTGGATGAAGACCAGCCCTTCCCGGCTGTGCCGA      lacZ
 Th Ap Il Il Cy Pr Me Ty Al Ar Va Ap Gu Ap Gn Pr Ph Pr Al Va Pr Ly
              500                                 510
```

```
      2848      2858      2868      2878      2888      2898
AATGGTCCATCAAAAAATGGCTTTCGCTACCTGGAGAGACGCGCCCGCTGATCCTTTGCGAATACG      lacZ
  Tr Se Il Ly Ly Tr Le Se Le Pr Gy Gu Th Ar Pr Le Il Le Cy Gu Ty Al
     520                                 530
```

```
      2914      2924      2934      2944      2954      2964
CCCACGCGATGGGTAACAGTCTTGGCGGTTTCGCTAAATACTGGCAGGCGTTTCGTCAGTATCCCC      lacZ
  Hi Al Me Gy An Se Le Gy Gy Ph Al Ly Ty Tr Gn Al Ph Ar Gn Ty Pr Ar
540                                 550                              560
```

```
      2980      2990      3000      3010      3020      3030
GTTTACAGGGCGGCTTCGTCTGGGACTGGGTGGATCAGTCGCTGATTAAATATGATGAAAACGGCA      lacZ
 Le Gn Gy Gy Ph Va Tr Ap Tr Va Ap Gn Se Le Il Ly Ty Ap Gu An Gy An
              570                                 580
```

```
      3046      3056      3066      3076      3086      3096
ACCCGTGGTCGGCTTACGGCGGTGATTTTGGCGATACGCCGAACGATCGCCAGTTCTGTATGAACG      lacZ
 Pr Tr Se Al Ty Gy Gy Ap Ph Gy Ap Th Pr An Ap Ar Gn Ph Cy Me An Gy
              590                                 600
```

```
      3112      3122      3132      3142      3152      3162
GTCTGGTCTTTGCCGACCGCACGCCGCATCCAGCGCTGACGGAAGCAAAACACCAGCAGCAGTTTT      lacZ
 Le Va Ph Al Ap Ar Th Pr Hi Pr Al Le Th Gu Al Ly Hi Gn Gn Gn Ph Ph
              610                                 620
```

```
      3178      3188      3198      3208      3218      3228
TCCAGTTCCGTTTATCCGGGCAAACCATCGAAGTGACCAGCGAATACCTGTTCCGTCATAGCGATA      lacZ
 Gn Ph Ar Le Se Gy Gn Th Il Gu Va Th Se Gu Ty Le Ph Ar Hi Se Ap An
              630                                 640
```

```
      3244      3254      3264      3274      3284      3294
ACGAGCTCCTGCACTGGATGGTGGCGCTGGATGGTAAGCCGCTGGCAAGCGGTGAAGTGCCTCTGG      lacZ
 Gu Le Le Hi Tr Me Va Al Le Ap Gy Ly Pr Le Al Se Gy Gu Va Pr Le Ap
650                                 660                              670
```

Figure 1 *Continued*

```
         3310      3320      3330      3340      3350      3360
   ATGTCGCTCCACAAGGTAAACAGTTGATTGAACTGCCTGAACTACCGCAGCCGGAGAGCGCCGGGC          lacZ
    Va Al Pr Gn Gy Ly Gn Le Il Gu Le Pr Gu Le Pr Gn Pr Gu Se Al Gy Gn
                    680                               690
```

```
         3376      3386      3396      3406      3416      3426
   AACTCTGGCTCACAGTACGCGTAGTGCAACCGAACGCGACCGCATGGTCAGAAGCCGGGCACATCA          lacZ
    Le Tr Le Th Va Ar Va Va Gn Pr An Al Th Al Tr Se Gu Al Gy Hi Il Se
                    700                               710
```

```
         3442      3452      3462      3472      3482      3492
   GCGCCTGGCAGCAGTGGCGTCTGGCGGAAAAACCTCAGTGTGACGCTCCCCGCCGCGTCCCACGCCA          lacZ
    Al Tr Gn Gn Tr Ar Le Al Gu An Le Se Va Th Le Pr Al Al Se Hi Al Il
                    720                               730
```

```
         3508      3518      3528      3538      3548      3558
   TCCCGCATCTGACCACCAGCGAAATGGATTTTTGCATCGAGCTGGGTAATAAGCGTTGGCAATTTA          lacZ
    Pr Hi Le Th Th Se Gu Me Ap Ph Cy Il Gu Le Gy An Ly Ar Tr Gn Ph An
                    740                               750
```

```
         3574      3584      3594      3604      3614      3624
   ACCGCCAGTCAGGCTTTCTTTCACAGATGTGGATTGGCGATAAAAAACAACTGCTGACGCCGCTGC          lacZ
    Ar Gn Se Gy Ph Le Se Gn Me Tr Il Gy Ap Ly Ly Gn Le Le Th Pr Le Ar
   760                         770                             780
```

```
         3640      3650      3660      3670      3680      3690
   GCGATCAGTTCACCCGTGCACCGCTGGATAACGACATTGGCGTAAGTGAAGCGACCCGCATTGACC          lacZ
    Ap Gn Ph Th Ar Al Pr Le Ap An Ap Il Gy Va Se Gu Al Th Ar Il Ap Pr
                         790                          800
```

```
         3706      3716      3726      3736      3746      3756
   CTAACGCCTGGGTCGAACGCTGGAAGGCGGCGGGCCATTACCAGGCCGAAGCAGCGTTGTTGCAGT          lacZ
    An Al Tr Va Gu Ar Tr Ly Al Al Gy Hi Ty Gn Al Gu Al Al Le Le Gn Cy
                    810                               820
```

```
         3772      3782      3792      3802      3812      3822
   GCACGGCAGATACACTTGCTGATGCGGTGCTGATTACGACCGCTCACGCGTGGCAGCATCAGGGGA          lacZ
    Th Al Ap Th Le Al Ap Al Va Le Il Th Th Al Hi Al Tr Gn Hi Gn Gy Ly
                    830                          840
```

```
         3838      3848      3858      3868      3878      3888
   AAACCTTATTTATCAGCCGGAAAACCTACCGGATTGATGGTAGTGGTCAAATGGCGATTACCGTTG          lacZ
    Th Le Ph Il Se Ar Ly Th Ty Ar Il Ap Gy Se Gy Gn Me Al Il Th Va Ap
   850                              860
```

```
         3904      3914      3924      3934      3944      3954
   ATGTTGAAGTGGCGAGCGATACACCGCATCCGGCGCGGATTGGCCTGAACTGCCAGCTGGCGCAGG          lacZ
    Va Gu Va Al Se Ap Th Pr Hi Pr Al Ar Il Gy Le An Cy Gn Le Al Gn Va
   870                         880                             890
```

Figure 1 *Continued*

```
        3970        3980        3990        4000        4010        4020
TAGCAGAGCGGGTAAACTGGCTCGGATTAGGGCCGCAAGAAAACTATCCCGACCGCCTTACTGCCG        lacZ
Al Gu Ar Va An Tr Le Gy Le Gy Pr Gn Gu An Ty Pr Ap Ar Le Th Al Al
                        900                                 910
```

```
        4036        4046        4056        4066        4076        4086
CCTGTTTTGACCGCTGGGATCTGCCATTGTCAGACATGTATACCCCGTACGTCTTCCCGAGCGAAA        lacZ
Cy Ph Ap Ar Tr Ap Le Pr Le Se Ap Me Ty Th Pr Ty Va Ph Pr Se Gu An
                        920                                 930
```

```
        4102        4112        4122        4132        4142        4152
ACGGTCTGCGCTGCGGGACGCGCGAATTGAATTATGGCCCACACCAGTGGCGCGGCGACTTCCAGT        lacZ
Gy Le Ar Cy Gy Th Ar Gu Le An Ty Gy Pr Hi Gn Tr Ar Gy Ap Ph Gn Ph
                        940                                 950
```

```
        4168        4178        4188        4198        4208        4218
TCAACATCAGCCGCTACAGTCAACAGCAACTGATGGAAACCAGCCATCGCCATCTGCTGCACGCGG        lacZ
An Il Se Ar Ty Se Gn Gn Gn Le Me Gu Th Se Hi Ar Hi Le Le Hi Al Gu
                        960                                 970
```

```
        4234        4244        4254        4264        4274        4284
AAGAAGGCACATGGCTGAATATCGACGGTTTCCATATGGGGATTGGTGGCGACGACTCCTGGAGCC        lacZ
Gu Gy Th Tr Le An Il Ap Gy Ph Hi Me Gy Il Gy Gy Ap Ap Se Tr Se Pr
980                                     990                             1000
```

```
        4300        4310        4320        4330        4340        4350
CGTCAGTATCGGCGGAATTCCAGCTGAGCGCCGGTCGCTACCATTACCAGTTGGTCTGGTGTCAAA        lacZ
Se Va Se Al Gu Ph Gn Le Se Al Gy Ar Ty Hi Ty Gn Le Va Tr Cy Gn Ly
                        1010                                1020
```

```
        4366        4376        4386        4396        4406        4416
AATAATAATAACCGGGCAGGCCATGTCTGCCCGTATTTCGCGTAAGGAAATCCATTATGTACTATT        lacY
**                                                        Me Ty Ty Le
                                                          ‾‾
                                                           1
```

```
        4432        4442        4452        4462        4472        4482
TAAAAAACACAAACTTTTGGATGTTCGGTTTATTCTTTTTCTTTTACTTTTTTATCATGGGAGCCT        lacY
Ly An Th An Ph Tr Me Ph Gy Le Ph Ph Ph Ph Ty Ph Ph Il Me Gy Al Ty
                        10                                  20
```

```
        4498        4508        4518        4528        4538        4548
ACTTCCCGTTTTTTCCCGATTTGGCTACATGACATCAACCATATCAGCAAAAGTGATACGGGTATTA        lacY
Ph Pr Ph Ph Pr Il Tr Le Hi Ap Il An Hi Il Se Ly Se Ap Th Gy Il Il
                        30                                  40
```

```
        4564        4574        4584        4594        4604        4614
TTTTTGCCGCTATTTCTCTGTTCTCGCTATTATTCCAACCGCTGTTTGGTCTGCTTTCTGACAAAC        lacY
Ph Al Al Il Se Le Ph Se Le Le Ph Gn Pr Le Ph Gy Le Le Se Ap Ly Le
        50                              60                              70
```

Figure 1   *Continued*

```
        4630      4640      4650      4660      4670      4680
TCGGGCTGCGCAAATACCTGCTGTGGATTATTACCGGCATGTTAGTGATGTTTGCGCCGTTCTTTA          lacY
  Gy Le Ar Ly Ty Le Le Tr Il Il Th Gy Me Le Va Me Ph Al Pr Ph Ph Il
                              80                               90
```

```
        4696      4706      4716      4726      4736      4746
TTTTTATCTTCGGGCCACTGTTACAATACAACATTTTAGTAGGATCGATTGTTGGTGGTATTTATC          lacY
  Ph Il Ph Gy Pr Le Le Gn Ty An Il Le Va Gy Se Il Va Gy Gy Il Ty Le
                             100                              110
```

```
        4762      4772      4782      4792      4802      4812
TAGGCTTTTGTTTTAACGCCGGTGCGCCAGCAGTAGAGGCATTTATTGAGAAAGTCAGCCGTCGCA          lacY
  Gy Ph Cy Ph An Al Gy Al Pr Al Va Gu Al Ph Il Gu Ly Va Se Ar Ar Se
                             120                              130
```

```
        4828      4838      4848      4858      4868      4878
GTAATTTCGAATTTGGTCGCGCGCGGATGTTTGGCTGTGTTGGCTGGGCGCTGTGTGCCTCGATTG          lacY
  An Ph Gu Ph Gy Ar Al Ar Me Ph Gy Cy Va Gy Tr Al Le Cy Al Se Il Va
                             140                              150
```

```
        4894      4904      4914      4924      4934      4944
TCGGCATCATGTTCACCATCAATAATCAGTTTGTTTTCTGGCTGGGCTCTGGCTGTGCACTCATCC          lacY
  Gy Il Me Ph Th Il An An Gn Ph Va Ph Tr Le Gy Se Gy Cy Al Le Il Le
  160                        170                              180
```

```
        4960      4970      4980      4990      5000      5010
TCGCCGTTTTACTCTTTTTCGCCAAAACGGATGCGCCCTCTTCTGCCACGGTTGCCAATGCGGTAG          lacY
  Al Va Le Le Ph Ph Al Ly Th Ap Al Pr Se Se Al Th Va Al An Al Va Gy
                             190                              200
```

```
        5026      5036      5046      5056      5066      5076
GTGCCAACCATTCGGCATTTAGCCTTAAGCTGGCACTGGAACTGTTCAGACAGCCAAAACTGTGGT          lacY
  Al An Hi Se Al Ph Se Le Ly Le Al Le Gu Le Ph Ar Gn Pr Ly Le Tr Ph
                             210                              220
```

```
        5092      5102      5112      5122      5132      5142
TTTTGTCACTGTATGTTATTGGCGTTTCCTGCACCTACGATGTTTTTGACCAACAGTTTGCTAATT          lacY
  Le Se Le Ty Va Il Gy Va Se Cy Th Ty Ap Va Ph Ap Gn Gn Ph Al An Ph
                             230                              240
```

```
        5158      5168      5178      5188      5198      5208
TCTTTACTTCGTTCTTTGCTACCGGTGAACAGGGTACGCGGGTATTTGGCTACGTAACGACAATGG          lacY
  Ph Th Se Ph Ph Al Th Gy Gu Gn Gy Th Ar Va Ph Gy Ty Va Th Th Me Gy
  250                        260
```

```
        5224      5234      5244      5254      5264      5274
GCGAATTACTTAACGCCTCGATTATGTTCTTTGCGCCACTGATCATTAATCGCATCGGTGGGAAAA          lacY
  Gu Le Le An Al Se Il Me Ph Ph Al Pr Le Il Il An Ar Il Gy Gy Ly An
  270                        280                              290
```

Figure 1 *Continued*

```
      5290        5300        5310        5320        5330        5340
ACGCCCTGCTGCTGGCTGGCACTATTATGTCTGTACGTATTATTGGCTCATCGTTCGCCACCTCAG      lacY
 Al Le Le Le Al Gy Th Il Me Se Va Ar Il Il Gy Se Se Ph Al Th Se Al
                            300                             310
```

```
      5356        5366        5376        5386        5396        5406
CGCTGGAAGTGGTTATTCTGAAAACGCTGCATATGTTTGAAGTACCGTTCCTGCTGGTGGGCTGCT      lacY
 Le Gu Va Va Il Le Ly Th Le Hi Me Ph Gu Va Pr Ph Le Le Va Gy Cy Ph
                        320                             330
```

```
      5422        5432        5442        5452        5462        5472
TTAAATATATTACCAGCCAGTTTGAAGTGCGTTTTTCAGCGACGATTTATCTGGTCTGTTTCTGCT      lacY
 Ly Ty Il Th Se Gn Ph Gu Va Ar Ph Se Al Th Il Ty Le Va Cy Ph Cy Ph
                    340                             350
```

```
      5488        5498        5508        5518        5528        5538
TCTTTAAGCAACTGGCGATGATTTTTATGTCTGTACTGGCGGGCAATATGTATGAAAGCATCGGTT      lacY
 Ph Ly Gn Le Al Me Il Ph Me Se Va Le Al Gy An Me Ty Gu Se Il Gy Ph
                360                             370
```

```
      5554        5564        5574        5584        5594        5604
TCCAGGGCGCTTATCTGGTGCTGGGTCTGGTGGCGCTGGGCTTCACCTTAATTTCCGTGTTCACGC      lacY
 Gn Gy Al Ty Le Va Le Gy Le Va Al Le Gy Ph Th Le Il Se Va Ph Th Le
 380                             390                             400
```

```
      5620        5630        5640        5650        5660        5670
TTAGCGGCCCCGGCCCGCTTTCCCTGCTGCGTCGTCAGGTGAATGAAGTCGCTTAAGCAATCAATG      lacY
 Se Gy Pr Gy Pr Le Se Le Le Ar Ar Gn Va An Gu Va Al **
                            410                         417
```

```
      5686        5696        5706        5716        5726        5736
TCGGATGCGGCGCGACGCTTATCCGACCAACATATCATAACGGAGTGATCGCATTGAACATGCCAA      lacA
                                                         An Me Pr Me
                                                         ─
                                                         1
```

```
      5752        5762        5772        5782        5792        5802
TGACCGAAAGAATAAGAGCAGGCAAGCTATTTACCGATATGTGCGAAGGCTTACCGGAAAAAAGA      lacA
 Th Gu Ar Il Ar Al Gy Ly Le Ph Th Ap Me Cy Gu Gy Le Pr Gu Ly Ar
                    10                              20
```

Figure 1   *Continued*

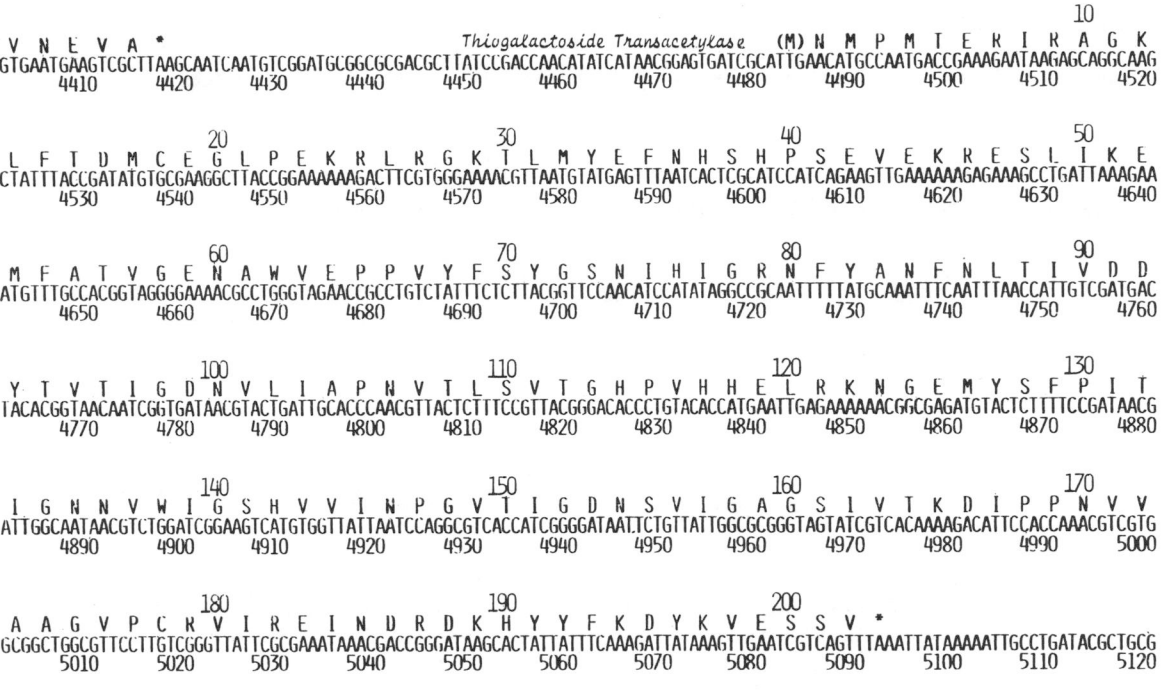

```
     V  N  E  V  A  *                          Thiogalactoside Transacetylase  (M) N  M  P  M  T  E  R  I  R  A  G  K
    GTGAATGAAGTCGCTTAAGCAATCAATGTCGGATGCGGCGCGACGCTTATCCGACCAACATATCATAACGGAGTGATCGCATTGAACATGCCAATGACCGAAAGAATAAGAGCAGGCAAG
        4410      4420      4430      4440      4450      4460      4470      4480      4490      4500      4510      4520

     L  F  T  D  M  C  E  G  L  P  E  K  R  L  R  G  K  T  L  M  Y  E  F  N  H  S  H  P  S  E  V  E  K  R  E  S  L  I  K  E
    CTATTTACCGATATGTGCGAAGGCTTACCGGAAAAAAGACTTCGTGGGAAAACGTTAATGTATGAGTTTAATCACTCGCATCCATCAGAAGTTGAAAAAAGAGAAAGCCTGATTAAAGAA
        4530      4540      4550      4560      4570      4580      4590      4600      4610      4620      4630      4640

     M  F  A  T  V  G  E  N  A  W  V  E  P  P  V  Y  F  S  Y  G  S  N  I  H  I  G  R  N  F  Y  A  N  F  N  L  T  I  V  D  D
    ATGTTTGCCACGGTAGGGGAAAACGCCTGGGTAGAACCGCCTGTCTATTTCTCTTACGGTTCCAACATCCATATAGGCCGCAATTTTTATGCAAATTTCAATTTAACCATTGTCGATGAC
        4650      4660      4670      4680      4690      4700      4710      4720      4730      4740      4750      4760

     Y  T  V  T  I  G  D  N  V  L  I  A  P  N  V  T  L  S  V  T  G  H  P  V  H  H  E  L  R  K  N  G  E  M  Y  S  F  P  I  T
    TACACGGTAACAATCGGTGATAACGTACTGATTGCACCCAACGTTACTCTTTCCGTTACGGGACACCCTGTACACCATGAATTGAGAAAAAACGGCGAGATGTACTCTTTTCCGATAACG
        4770      4780      4790      4800      4810      4820      4830      4840      4850      4860      4870      4880

     I  G  N  N  V  W  I  G  S  H  V  V  I  N  P  G  V  T  I  G  D  N  S  V  I  G  A  G  S  I  V  T  K  D  I  P  P  N  V  V
    ATTGGCAATAACGTCTGGATCGGAAGTCATGTGGTTATTAATCCAGGCGTCACCATCGGGGATAAT.TCTGTTATTGGCGCGGGTAGTATCGTCACAAAAGACATTCCACCAAACGTCGTG
        4890      4900      4910      4920      4930      4940      4950      4960      4970      4980      4990      5000

     A  A  G  V  P  C  R  V  I  R  E  I  N  D  R  D  K  H  Y  Y  F  K  D  Y  K  V  E  S  S  V  *
    GCGGCTGGCGTTCCTTGTCGGGTTATTCGCGAAATAAACGACCGGGATAAGCACTATTATTTCAAAGATTATAAAGTTGAATCGTCAGTTTAAATTATAAAAATTGCCTGATACGCTGCG
        5010      5020      5030      5040      5050      5060      5070      5080      5090      5100      5110      5120
```

Figure 2 Nucleotide sequence of the *lacA* gene. The numbers correspond to the nucleotide positions in the *lac* mRNA. The translation products of the nucleotide sequence, the end of the lactose permease, and all of thiogalactoside transacetylase are shown.

B. Restriction Enzyme Cleavage Sites in the *lac* Operon

Figure 1 is reprinted from Silhavy et al. (*Experiments with Gene Fusions.* Cold Spring Harbor Laboratory, Cold Spring Harbor, New York, p. 282 [1984]).

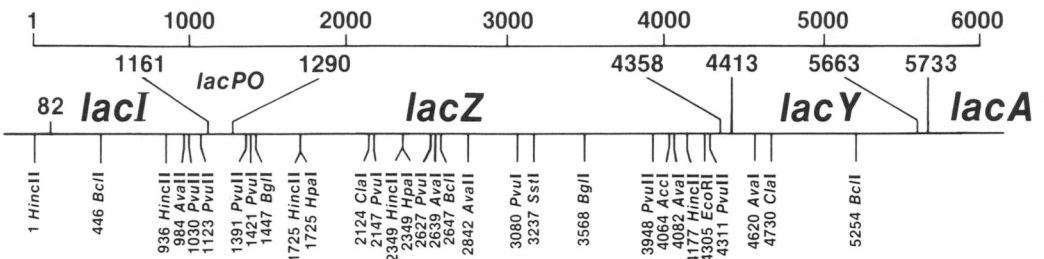

Figure 1 This map was constructed from the sequence data of Dickson et al. (1975), Farabaugh (1978), Buchel et al. (1980), and Kalnins et al. (1983). Transcription and translation is from left to right. The first nucleotide (nt) of the *Hinc*II site at the *lacI* promoter is taken as nt 1; the relative positions of other restriction enzyme recognition sites are as indicated, with the numbers referring to the positions of the first nucleotides in the sites. Also shown are the extents of the coding sequences for the products of *lacI* (nt 82–1161), *lacZ* (nt 1290–4358), *lacY* (nt 4413–5663), and the start of *lacA* (nt 5733). The first of each of these pairs of numbers refers to the position of the first nucleotide of the codon for the amino-terminal amino acid and the second refers to the last nucleotide of the codon for the carboxy-terminal residue of the indicated proteins.

SECTION · 17

Regulation

A. Promoters
1. Alignment of *E. coli* promoter sequences
2. *E. coli* σ 70 promoters
3. Compilation of DNA sequences for promoters utilizing alternative σ factors

B. SOS system
1. DNA sequences of *lexA*-repressed promoters
2. SOS genes, map positions, and activities
3. Global regulatory responses

C. Compilation of *E. coli* ribosome binding sites

D. Transcription termination

REFERENCES*

d'Aubenton Carafa, Y., E. Brody, and C. Thermes. 1990. Prediction of rho-independent *Escherichia coli* transcription terminators. *J. Mol. Biol.* **216:** 835–858.

Collado-Vides, J., B. Magasanik, and J.D. Gralla. 1991. Control site location and transcriptional regulation in *Escherichia coli*. *Microbiol. Rev.* **55:** 371–394.

Demple, B. 1991. Regulation of bacterial oxidative stress genes. *Annu. Rev. Genet.* **25:** 315–337.

Harley, C.B. and R.P. Reynolds. 1987. Analysis of *E. coli* promoter sequences. *Nucleic Acids Res.* **15:** 2343–2361.

Hoopes, B.C. and W.R. McClure. 1987. Strategies in regulation of transcription initiation. In Escherichia coli *and* Salmonella typhimurium. *Cellular and Molecular Biology* (ed. F.C. Neidhardt), pp. 1231–1240. American Society for Microbiology, Washington, D.C.

Neidhardt, F.C. 1987. Multigene systems and regulons. In Escherichia coli *and* Salmonella typhimurium. *Cellular and Molecular Biology* (ed. F.C. Neidhardt), pp. 1313–1317. American Society for Microbiology, Washington, D.C.

Neidhardt, F.C. and R.A. VanBogelen. 1987. Heat shock response. In Escherichia coli *and* Salmonella typhimurium. *Cellular and Molecular Biology* (ed. F.C. Neidhardt), pp. 1334–1345. American Society for Microbiology, Washington, D.C.

The references cited in Section 17C are listed on pp. 17.25–17.26.

SECTION 17

A. Promoters

Harley and Reynolds (1987) have compiled 263 *E. coli* promoters, using a program to aid in the alignment. This is presented in Table 1. Also, Collado-Vides et al. (1991) have tabulated the control site locations of *E. coli* promoters, as shown in Figure 1. Figure 2 shows a compilation of DNA sequences for promoters utilizing altered σ factors (Hoopes and McClure 1987).

Table 1 is reprinted, with permission, from Harley and Reynolds (*Nucleic Acids Res.*, vol. 15, pp. 2347–2351 [1987]). Figure 1 is reprinted, with permission, from Collado-Vides et al. (*Microbiol. Rev.*, vol. 55, pp. 374–377 [1991]). Figure 2 is reprinted, with permission, from Hoopes and McClure (Escherichia coli *and* Salmonella typhimurium. *Cellular and Molecular Biology* [ed. F.C. Neidhardt]. American Society for Microbiology, Washington, D.C., p. 1233 [1987]). The references cited in the table, table notes, and caption to Figure 2 can be found in the respective original papers.

Table 1 Alignment of *E. coli* Promoter Sequences

SEQUENCE (a)	TYPE (b)	-35 (c)	-10 (d)	SP (e)	PHI (f)	DISCREP. (g)	(h)	TS (i)	REF (j)
aceEF	b	ACGTAGAC<u>CTGT CTT</u>ATT GAGCTTTC	CGGC<u>GAGAG</u> TTCAAT GGGaCAGGTCCAG	17	-4.3		-4.4	4	24
ada	b	AAGATTGTTGGTTT <u>TTGCGT</u> GAT<u>GGTGA</u>	CCGGGCAGC <u>CTAAAG</u> GCTaTCCTTAACC	17	-5.5	-3.4	-4.6	4	25,26
alaS	b	AACCGCATACGGTAT TTTACC TTCCCAGTC	AAGAAAACT TATCTT ATTCCCaGTTTTCAGT	18	-3.1				9
ampC	b	TGCTATCCTGACAG TTGTCA CGCTCGATT	GGTGTCGT TACAAT CTAACGCaTCGCCAATG	16	-1.5				9
ampC/C16	b	GCTATC TTGACA GTTGTCAC	GCTGATTGG TATCGT TACAATCTaACGTATCG	17	-1.3			1,3	25
araBAD	b	TTAGCGGATCCTAC <u>CTGACG</u> CTTTTTAT	CGCAACTC <u>TCTACT</u> GTTTCTCCATaCCCGTT	16	-3.6	-3.7			9
araC	b	GCAAATAATCAATG TGGACT TTTCTGCC	GTGATTATA GACACT TTTGTTACgCGTTTTTG	17	-3.6				9
araE	b	CTGTTTCCGAC CTGACA CCTGCGTGA	GTTGTCAG TATTTT TTCACTATgTCTTACTC	19	-3.2			4	28
araI(c)	m	AGCGGATCCTAC CTGCGG CTTTTTAT	CGCAACTC TCTACT GTTTCTCCATaCCCGTT	16	-4.3			4	29
araI(c)X(c)	m	AGCGGATCCTAC CTGGGG CTTTTTATC	GCAACTCTC TACTAT TTCTCCATaCCCGTTTT	18	-3.8			4	29
argCBH	b	TTTGTTTTTCATTG <u>TTGACA</u> CACCTCTGG	TCATG<u>ATAG</u> TATCAA TATTCaTGCAGTATT	18	-2.4	-2.6			9
argCBH-P1/6-	m	TTTGTTTTTCATTG TTGACA CACCTCT	GGTCATAA TATTTAT CAATATTCaTGCAGTATT	15	-2.0				30
argCBH-P1/LL	m	TTTGTTTTTCATTG TTGACA CACCTCT	GGTCATGA TATTTAT CAATATTCaTGCAGTAT	15	-2.0				30
argE-P1	b	TTACGGCTGGTGGG TTTTAT TACGCTCA	ACGTTAGTG TATTTT TATTCaTAAATACTGCA	17	-2.6			4	31
argE-P2	b	CCGGCATCATTGCTT <u>TGCGCT</u> GAAACAGT	CAAAGCGGT <u>TATGTT</u> CATaTGCGGATGGGG	17	-3.9	-3.9		4	31
argE/LL13	m	CCGGCATCATTGCTT TGCGCT GAAACAGT	CAAAGCGGT TATATT CATaTGCGGATGGGG	17	-3.3				31
argF	b	ATTGTGAAATGGGG TTGCAA ATGAATAA	TTACACATA TAAAGT GAATTTTaATTCAATAA	17	-1.7			4	31,32
argI	b	AGAC TTGCAA ATGAATAA	TCATCCATA TAAATT GAATTTTaaTTCATTGA	17	-1.5			4	31
argR	b	TCGTCGCGCGG <u>TTGCAG</u> GAGCAAGG	CTTTGACAA <u>TATTAA</u> TCAGTCTaaaGTCTGGG	17	-3.2	-5.9		2,4	31
aroF	b	TACGGAAATATGGCA TTGAAA ACTTTACT	TTATGTGT TATCGT TACGTCaTCCTCGCTG	16	-1.9			2,4	33
aroG	b	AGTGTAAAACCCCG TTTACA CATTCTGA	CGGAAGATA TAGATT GGAAGTaTTGCATTCA	17	-1.6			2,4	33
aroH	b	GTACTAGAGAACTA GTGCAT TAGCTTAT	TTTTTTGT TATCAT GCTAaccaCCCGGCGAG	16	-3.1				9
bioA	b	GCCTTCTCCAAAAC GTGTTT <u>TTGTTGTT</u>	AATTCGGTG TAGACT TGTaaaCCTAAATCT	18	-3.8	-3.4			9
bioB	b	TTGTCATAATCGAC TTGTAA ACCAAATT	GAAAGATT TAGGTT TACAAGTctACACCGAT	17	-2.2				9
bioP98	M	TTGTTAATTCGGTG TAGACT TGTAAACC	TAAATCTTT TAAATT TGGTTTaCAAGTCGAT	17	-2.0				9
C62.5-P1	b	CACCTGCTCTCGCC <u>TTGAAA</u> TTATCTC	CCTTGT<u>CCC</u> CATCTC TCCCAcatCCTGTTTT	17	-3.3		+	4	34
carAB-P1	b	ATCCCGCCATTAAG TTGACT TTTAGCGC	CCATATCTC CAGAAT GCCGCCgTTTGCCAGA	17	-1.9			4	35
carAB-P2	b	TAAGCAGATTTGCA TTGATT TACGTCATC	ATTGTGAAT TAATAT GCAAaTAAAGTGAG	18	-2.4			4	35
cat	b	ACGTTGATCGGC A<u>CGTAA</u> GA<u>GGTCC</u>	AACTTTCAC <u>CATAAT</u> GAAATAAgATCACTACC	17	-4.2	-2.4	-5.3		9
cit.util-379	p	AAACAGGCGGGG GTCTCA GGCG<u>ACTAA</u>	CCCGCAAAC TC<u>TTAC</u> CTCTATACaTAATTCTG	18	-5.6	-5.2	‰	3,4	36-38
cit.util-431	p	GACAGGCACAGCA TTGTAC GATCAACTG	ATTTGTGCC AATAAT TAaaTGAAATCAC	18	-3.4			3,4	36-38
CloDFcloacin	p	TCATATA<u>TTGCACC</u> CTGAAA ACTGGAGG	<u>AGTAAGGT AATAAT</u> CATACTgTGTATATAT	16	-2.9	-1.5	-3.5	3	39
CloDFrnaI	p	ACACCGCGGTTGCTC TTGAAG TGTGCGCCA	AAGTCCGGC TACACT GGAACGaCAGATTTGG	18	-2.2				9
colE1-B	p	TTATAAAA<u>TCCTCT</u> TTGACT TTTAAAA	CAA<u>TACGT</u> TAAAAA TAAaTACTGTAA	15	-3.4	-4.4		1,3	40
colE1-C	p	TTATAAAATCCTCT TTGACT TTTAAAAC	AATAAGTT AAAAAT AAATACTgTACAATATAA	16	-2.4			1,3	40
ColE1-P1	p	GGAAGTCCACAGTC TTGACA GGGAAAAT	GCAGCGGCG TAGCTT TTATGCTgTATATAAA	17	-1.7				9
ColE1-P2	p	TTTTTAACTTA<u>TTG</u> TTTTAA AAGTCAAA	GAGGATTT <u>TATAAT</u> GGAAACCgCGGTAGCGT	16	-1.7	-1.9			9
colE110.13	p	GCTACAGAGTTC TTGAAG TAGTGGCCC	GACTACGGC TACACT AGAAGGACaGTATTTGG	18	-2.2			1,3	41
colicinE1 P3	p	TTTTTAACTTATTG TTTTAA AAGTCAAA	GAGGATTT TATAAT GGAAACCgCGGTAGCGT	16	-1.7				42
crp	b	AAGCGAGACCACCAG GAGACA CAAAGCGA	AAGC<u>TATGC</u> TAAAAC AGTCAGgATGCTACAG	17	-3.2		‰	2,3	43
cya	b	GTAGCGCATCTTTC TTTACG GTCAATCA	GCAAGGTGT TAAATT GATCACgTTTTAGACC	17	-1.8			1-3	44
dapD	b	AAGTGCATCAGCCG TTGACA GAGGCCCTC	AATCCAAAC GATAAA GGGTGatgTGTTTACTG	18	-2.8			4	45
deo-P1	b	CAGAAACGTTTTA TTCGAA CATCGCATCT	CGTCTGTGT TAGAAT TCTAACaTACGGTTGC	19	-3.5				9
deo-P2	b	TGATGTGTA TCGAAG TGTGTTGCG	GAGTAGATGT TAACAAT ACTAACaAACTCGCAA	19	-3.9				9
deo-P3	b	ACACCAACTGTCTA TCGCCG TATCAGCG	AATAACGG TATACT GATCTGaTCATTTAAA	16	-3.2			2,4	46
divE	b	AAACAAATTAGGGG TTTACA CGCCGCAT	CGGGATGTT TATAGT GCGGGTCaTTCCGGAAG	17	-1.2			1,2	47
dnaA-1p	b	TGCGGGGTAAAATCG <u>TGCCCG</u> CCTCGCGGC	AGGATCGTT <u>TACACT</u> TAGGCGAgTTCTGGAAA	18	-4.4	-4.9		4	48,49
dnaA-2p	b	TCTGTGAGAAACAG AAGATC TCTTCGCG	AGTTTAGGC TATGAT CCGcggtccCGATCG	17	-4.5			4	48
dnaK-P1	b	TTTGCATCTCCCC<u>C</u> TTGATG ACGTGGTTT	ACGA<u>CCCCA</u> TTTAGT AGTcaaCCGCAGTG	18	-3.2	-8.2		2,4	34
dnaK-P2	b	ATGAAATTGGGCAG <u>TTGAAA</u> CCAGACGT	TTCGCCGC TATTAC AGACTcaCAACCACA	16	-2.4	-9.3		2,4	34
dnaQ-P1	b	GCCAGCGCTAAACG TTTTCT CGCGTCCG	CGATAGCG TAAAAT AGCgccGTAACCCC	16	-2.1			2-4	50,51
Fpla-oriTpX	p	GAACCACCAACCTG TTGAGC CTTTTGT	GGAGTGGGT TAAATT ATTTaCGGATAAAG	17	-2.5			2	52
Fplas-traM	p	ATTAGGGGTGC<u>TGC</u> TAGCGG CGCGGTGT	GTTTTTTTA <u>TAGGAT</u> ACCCGTaGGGGCGGCTG	17	-4.0	-5.7		2	52
Fplas-traY/Z	p	GCGTT<u>AATAAGGT GTTAAT</u> AAAATATA	<u>GACTTTCCG TCTATT</u> TacccttttctgaTTATT	17	-3.9	-3.0	-4.1	3	53
frdABCD	b	GATCTCGTCAA A<u>TTTCA</u> GACTTATC	GATCAGAC <u>TATACT</u> GTTGTACCTATaAAGGA	16	-3.2	-3.9		4	54
fumA	b	GTACTAGTCTCAGT <u>TTTTGT</u> TAAAAAAG	TGTGTAGGA <u>TATTGT</u> TACTCGCTtttAACAGG	17	-3.5	-3.8		4	55-57
ɣ-ẟ-tnpA	p	ACACATTAACAGCA CTGTTT TTATGTGT	GCGATAATT TATAAT ATTTCCgACGGTTGCA	17	-2.4				9
ɣ-ẟ-tnpR	p	ATTCATTAACAAT <u>TTTGCA</u> ACCGTCCG	AAAATATTA <u>TAAATT</u> ATCCCACaCATAAAAAC	16	-2.4	-3.0			9
gal-P1	b	TCCATG<u>TCACAC</u>TT TTGCCA TC<u>TTTGTT</u>	<u>ATGCTATGG</u> TTATTT CATaCCATAAG	17	-3.8	-2.9	-4.0		9
gal-P2	b	CTAATTTATTCC<u>AT</u> GTCACA CTTTTCGC	ATCTTTGT <u>TATGT</u> ATGGTaTTTCATACC	16	-2.9	-3.1			9
gal-P2/mut-1	m	TAATTTATTCCAT <u>GTCACA</u> CTTTTCGC	ATCTTTGT <u>TATACT</u> <u>ATGGT</u>aTTTCATAC	16	-2.3	-4.0		3	58

Table 1   *Continued*

Name		Sequence (upstream)	Sequence (downstream)	n					Ref.
gal-P2/mut-2	m	TAATTTATTCCAT GTCACA CTTTTCGC	ATTTTTGT TATGCT ATGGTTaTTTCATAC	16	-2.9			3	58
glnL	b	CAATTCTCTGATGC TTCGCG CTTTTTATC	CGTAAAAAGC TATAAT GCACTaAATGGTGC	19	-3.2			2,4	59
glnS	b	TAAAAAACTAACAG TTGTCA GCCTGTCC	CGCTTATAA GATCAT ACGCCgttaTACGTT	17	-2.1				9
gltA-P1	b	ATTCATTGGGGACA GTTATT AGTGGTAG	ACAAGTTT AATAAT TCGGAtTGCTAAGTA	16	-4.3	-4.4		4	57,60
gltA-P2	b	AGTTGTTACAAACA TTACCA GGAAAAGCA	TATAATGCG TAAAAG TTAtGAAGTCGGT	18	-4.0	-1.8	-2.5	4	57,60
glyA	b	TCCTTTGTCAAGAC CTGTTA TCGCACAA	TGATTCGGT TATACT GTTCgCCGTTGTCC	17	-2.4			2,4	63
glyA/geneX	b	ACACCAAAGAACCA TTTACA TTGCAGGG	CTATTTTTTA TAAGAT GCATTtGAGATACAT	18	-1.9			2,4	61
gnd	b	GCATGGATAAGCTA TTTATA CTTTAATA	AGTACTTTG TATACT TATTTGCgAACATTCCA	17	-1.7			4	62
groE	b	TTTTTCCCCC TTGAAG GGGCAACG	CCATCCCCA TTTCTC TGGTCaCCAGCCGGGAA	17	-3.9	+		4	34
gyrB	b	CGGACGAAAA TTCGAA GATCGTTTACCGTGGAAAAGGG	TAAAAT AACGGATtAACCCAAGT	21	-3.2			4	63
his	b	ATATAAAAAGTTC TTGCTT TCTAACGTTG	AAAGTGGTT TAGGTT AAAAGACaTCAGTTGAA	18	-3.6				9
hisA	b	GATCTACAAACTAA TTAATA AATAGTTA	ATTAACGCT CATCAT TGTACAATGAaCTGTAC	17	-3.5	-2.7	-5.7		9
hisBp	b	CCTCCAGTGCGGTG TTTAAA TCTTTGTG	GGATCAGGG CATTAT CTTacGTGATCAG	17	-2.4			2,4	64
hisJ(St)	b	TAGAATGCTTTGCC TTGTCG GCCTGATT	AATGCAC GATAGT CGCATCGGATCTG	16	-3.0	-3.6			9
hisS	b	AAATAATAACGTGA TGGGAA GCGGCTCG	CTTCCCGTG TATGAT TGAACccgCATGGCTC	17	-2.7			4	65,66
htpR-P1	b	ACATTACGCCACTT ACGCCT GAATAATA	AAAGCGTGT TATACT CTTTCCtGCAATGGTT	17	-3.8			4	67,68
htpR-P2	b	TTCACAAGCTTGCA TTGAAC TTGTGGATA	AAATCACGG TCTGAT AAAACAgTGAATG	18	-3.7	-2.3		4	67,68
htpR-P3	b	AGCTTGCATTGAAC TTGTGG ATAAAATC	ACGGTCTGA TAAAAC AGTGAATgATAACCTCGT	17	-3.2			4	67,68
ilvGEDA	b	GCCAAAAAATATCT TGTACT ATTTACAA	AACCTATGG TAACTC TTTAGGCaTTCGTTCGA	17	-4.6	-3.9	-4.6		9
ilvIH-P1	b	CTCTGGCTGCCAA TTGCTT AAGCAAGA	TCGGACGGT TAATGT GTTttacacatttTTC	17	-3.2			2,4	69
ilvIH-P2	b	GAGGATTTTATGGT TTCTTT TCACCTTT	CCTCCTGTTT TATTCT TATtACCCCGTGT	17	-3.1	-3.1		2,4	69
ilvIH-P3	b	ATTTTAGGATTAA TTAAAA AAATAGAG	AAAATTGCTG TAAGTT GTGGGATTcAGCCGATT	17	-2.7			2,4	69
ilvIH-P4	b	TGTAGAATTTTATT CTGAAT GTCTGGGC	TCTCTATTT TAGGAT TAATTAAaAAAATAGAG	17	-2.7			2,4	69
IS1ins PL	p	CGAGGCCGGTGATG CTGCCA ACTTACTG	ATTTAGTG TATGAT GGTGtttTTGAGGTGCT	16	-2.5			1,3,4	70
IS1ins PR	p	ATATATACCTTA TGGTAA TGACTCCA	ACTTATTGA TAGTGT TTTATGTtCAGATAAT	17	-3.6	-3.3		1,3,4	70
IS2I-II	M	ATGTC TGGAAA TATAGGG	CAAATCAC TAGTAT TAAGACtaTCACTTATT	17	-2.6				9
lacI	b	GACACCATCGAATG GCCGCAA AACCTTTC	GCGGTATGG CATGAT AGCGCCCgGAAGAGAGT	17	-4.5				9
lacP1	b	TAGGCACCCCAGGC TTTACA CTTTATGCT	TCCGGCTCG TATGTT GTGTGGAaTTGTGAGC	18	-2.0				9
lacP115	M	TTTACACTTTATG CTTCCG GCTCGTATG	TTGTGTGG TATTGT GAGcggataacaATTT	17	-3.9	-2.0	-4.2		9
lacP2	b	AATGTGAGTTAGCT CACTCA TTAGGCAC	CCCAGGCTT TACACT TTATGCtTCCGGCTCG	17	-4.0	-2.6	-4.3		9
lambdac17	M	GGTGTATGCATTTA TTTGCA TACATTCA	ATCAATTGT TGTTATcTAAGGAAAT	17	-1.4				9
lambdacin	M	TAGATAACAATTGA TTGAAT GTATGCAA	ATAAATGCA TACACT ATAGGTgTGGTTTAAT	17	-1.6				9
lambdaL57	M	TGATAAGCAATGC TTTTTT ATAATGCCA	ACTTAGTA TAAAAT AGCCAACcTGTTCGACA	17	-2.4	-2.5			9
lambdaPI	f	CGGTTTTTTCTTGC GTGTAA TTGCGGAG	ACTTTCGCA TGTACT TGACACtTCAGGAGTG	17	-3.6				9
lambdaPL	f	TATCTCTGGCGGTG TTGACA TAAATACC	ACTGGCGGT GATACT GAGCACaTCAGCAGGA	17	-1.4				9
lambdaPo	f	TACCTCTGCCGAAG TTGAGT ATTTTTGC	TGTATTTGT CATAAT GACTCCtgTTGATAGAT	17	-2.1				9
lambdaPR	f	TAACACCGTGCGTG TTGACT ATTTTACC	TCTGGCGGT GATAAT GGTTGCaTGTACTAAG	17	-1.4				9
lambdaPR'	f	TTAACGGCATGATA TTGACT TATTGAAT	AAAATTGGG TAAATT TGACTCAaCGATGGGTT	17	-1.1				9
lambdaPRE	f	GAGCCTGGTTGCGT TTGTTT GCACGAACC	ATATGTAAG TATTTC CTTaGATAACAAT	18	-4.1	-5.7			9
lambdaPRM	f	AACACGCACGGTGT TAGATA TTTATCCC	TTGCGGTGA TAGATT TAACGTaTGAGCACAA	17	-2.6				9
lep	b	TCCTCGCCTCAATG TGTGTAG TGTAGAAT	GGGCGGTT TCTATT AATAcaGACGTTAAT	17	-3.4			2,4	71
leu	b	G TTGACA TCCGTTTT	TGTATCCAG TAACTC TAAAAGCATATCGCATT	17	-2.5				9
leultRNA	b	TCGATAATTAACTA TTGACG AAAAGCTG	AAAACCAC TAGAAT GGGCCTCGgTGGTAGCA	16	-1.5				9
lex	b	TGTGTCAGTTTATG TTCCAA AATCGCCT	TTTGCTGTA TATACT CACAGCaTAACTGTAT	17	-1.9				9
livJ	b	TGTCAAAATAGCTA TTCCAA TATCATAA	AAATCGGA TATGTT TTAGCaGAGTATGCT	17	-2.5			1,4	67,68
lpd	b	TGTTG TTTAAA AATTGTTA	ACAATTTTG TAAGAT ATagACAtagAACGA	17	-1.1			4	24,57
lpp	b	CCATCAAAAAAATA TTCTCA ACATAAAAA	ACTTTGTG TAATAC TTGTAACgCTACATGGA	17	-3.2	-3.3			9
lpp/P1	m	ATCAAAAAAATA TTCTCA ACATAAAAA	ACTTTGTGT TATACT TGTAACgCTACATGGA	18	-1.9				72
lpp/P2	m	ATCAAAAAAATA TTCTCA ACATAAAAA	ACTTTGTGT TATAAT TGTAACgCTACATGGA	18	-1.6				72
lpp/R1	m	ATCAAAAAAATA TTCACA ACATAAAAA	ACTTTGTG TAATAC TTGTAACgCTACATGGA	17	-2.7	-2.8			72
M1rna	b	ATCGGCAACGCGGG GTCACA AGGGCGCG	CAAACCCTC TATACT GGGGCGCgAAGCTGACC	17	-1.2				9
mac11	M	CCCCCGCAGGGAT GAGGAA GGTGGTCGA	CCGGGCTCG TATGTT GTGTGGAaTTGTGAGC	18	-4.1			4	76
mac12	M	CCCCCGCAGGGAT GAGGAA GGTCGGTCG	ACGGGCTCG TATGTT GTGTGGAaTTGTGAGC	18	-4.1			4	76
mac21	M	CCCCCGCAGGGAT GAGGAA GGTCGACCT	TCCGGCTCG TATGTT GTGTGGAaTTGTGAGC	18	-4.1			4	76
mac3	M	CCCCCGCAGGGAT GAGGAA GGTCGGTC	GACCGCTCG TATGTT GTGTGGAATTgTGAGCG	17	-3.7			4	76
mac31	M	CCCCCGCAGGGAT GAGGAA GGTCGGTC	GACCGCTCG TATATT GTGTGGAATTgTGAGCG	17	-3.1			4	76
malEFG	b	AGGGGCAAGGAGGA TGGAAA GAGGTTGC	CGTATAAA GAAACT ACAGTCCgTTTAGGTGT	16	-3.5				9
malK	b	CAGGGGGTGGAGGA TTTAAG CCATCTCC	TGATGACG CATAGT CAGCCCaTCATGAATG	16	-3.3				9
malPQ	b	ATCCCCGCAGGATG AGGAAG GTCAACAT	CGAGCCTGG CAAACT ACGCGATaACGTTGTGT	17	-4.7			2	77
malPQ/A516P1	m	ATCCCCGCAGG ATGAGG AGCCTGCC	AAACTAGC GATGAT AACGTTGTGTTgAA	16	-4.6			2,4	78
malPQ/A516P2	m	ATCCCCGCAGGAGG ATGAGG GTCCTGGCA	AACTAGCA TAACGT TGTGTTgAA	18	-4.6			2,4	78
malPQ/A517/A	m	CCCCGCAGGATGAG GTCGAG CCTGGCAA	ACTAGCGA TAACGT TGTGTTgAA	16	-4.9			2,4	78
malPQ/Ppl2	m	ATCCCCGCAGGAT GAGGAA GGTCAACAT	CGAGCCTG GAAAAC TAGCGATaaCGTTGTGT	17	-5.2	-5.2			77
malPQ/Ppl3	m	ATCCCCGCAGGAT TAGGAA GGTCAACAT	CGAGCCTGG CAAACT ACGCGATaACGTTGTGT	18	-3.9	-4.7			77
malPQ/Ppl4	m	ATCCCCGCAGGAT GAGGAA GGTCAACA	TCGAGCCTG GAAACT ACGGATaACGTTGTGT	17	-4.4				77
malPQ/Ppl5	m	ATCCCCGCAGGAT GAGAAA GGTCAACAT	CGAGCCTG CAAACT ACGGATaACGTTGTGT	18	-4.0				77
malPQ/Ppl6	m	ATCCCCGCAGGATA AGGAAG GTCAACAT	CGAGCCTGG CAAACT ACGCGATaACGTTGTGT	17	-4.7				77
malPQ/Ppl8	m	ATCCCCGCAGGATG GGGAAG GTCAACAT	CGAGCCTGG CAAACT ACGCGATaACGTTGTGT	17	-4.3				77

Table 1 *Continued*

malT	b	GTCATCGCTTGCAT TAGAAA GGTTTCTG	GCCGACCT TATAAC CATTAATTACG	16	-2.6		-3.9		9
manA	b	CGGCTCCAGGTTAC TTCCCG TAGGATTC	TTGCTTTAA TAGTGG GATTAATtCCACATTA	17	-5.0	-2.9	-2.9	4	56
metA P1	b	TTCAACATGCAGGC TGGACA TTGGCAAA	TTTTCTGGT TATCTT CAGCTaTCTGGATGT	17	-2.3			2,4	79
metA P2	b	AAGACTAATTACCA TTTTCT CTCCTTTT	AGTCATTCT TATATT CTAACGTaGTCTTTTCC	17	-1.8		-2.5	2,4	79
metBL	b	TTACCGTGACA TCGTGT AATGCACCT	GTCGGCGT GATAAT GCATATAAttTTAACCG	17	-3.9	-3.3		2,4	80
metF	b	TTTTCGG TTGACG CCGTTCGG	CTTTTCCTT CATCTT TacaTCTGGACG	17	-2.5			2-4	81
micF	b	GCGGAATGGCGAAA TAAGCA CCTAACAT	CAAGCAAT AATAAT TCAAGGTtAAAATCAAT	16	-4.6	-2.9		2,4	82,83
motA	b	GCCCCAATCGCGCG TTAACG CCTGACGAC	TGAACATCC TGTCAT GGTCAaCAGTCGA	18	-4.5				84
MuPc-1	f	AAATT TTGAAA AGTAACTTTATAGAAAAGAAT AATACT GAAAAGTCAAtttGGTG	21	-3.3	-2.0	-4.0	2,4	85
MuPc-2	f	GGAACACA TTTAAA AACCCTCC	TAAGTTTTG TAATCT ATAAAGttAGCAATTTA	17	-2.1		-4.0	2,4	85
MuPe	f	TACCAAAAAGCACC TTTACA TTAAGCTT	TTCAGTAAT TATCTT TTTAGTaAGCTAGCTA	17	-1.7			2,4	85
NRlrnaC	p	GTCACAATTCTCAA GTCGCT GATTTCAAA	AAACTGTAG TATCCT CTGCgaaacGATCCT	18	-4.1		-4.1	2-4	86
NRlrnaC/m	m	TCACAATTCTCAAG TTGCTG ATTTCAAA	AAACTGTAG TATCCT CTGCgaaacGATCCT	17	-2.8				86
NTPlrna100	p	GGAGTTTGTC TTGAAG TTATGCACC	TGTTAAGGC TAAACT GAAAGAaCAGATTTGT	18	-1.8				87
nusA	b	CAGTAT TTGCAT TTTTTACC	CAAAACGAG TAGAAT TTGCCACgTTTCAGGCG	17	-1.8			1,3	88,89
ompA	b	GCCTGACGGAG TTCACA CTTGTAAG	TTTTCAAC TACGTT GTAGACtTTAC	16	-2.7	-2.0	-7.4	3,3	90
ompC	b	GTATCATATTCGTG TTGGAT TATTCTGC	ATTTTTGGG GAGAAT GGACtTGCCGACTG	17	-2.9			3,4	92,893
ompF	b	GGTAGG TAGCGA AACGTTAG	TTTGAATGG AAAGAT GCCTGCaGACACATAAA	17	-4.6	-3.9		3,4	91
ompF/pKI217	m	GG TAGCGA AACGTTAG	TTTGCAAGC TTTAAT GGGtaGTTTATCAC	17	-3.4	-2.6		3,4	91
ompR	b	TTTCGCCGAATAAA TTGTAT ACTTAAG	CTGCGTGTT TAATAT GCTTTgTAACAATTT	15	-3.4	-2.4		4	92
pl5primer	p	ATAACGATGATCTTC TTGAGA TCGTTTTG	GTCTGCGCG TAATCT CTTGCTTgAAAACGAAA	17	-2.1			1	93
pl5rnaI	p	TAGAGGAGTTAGTC TTGAAG TCATGCGCC	GGTTAAGGC TAAACT GAAAGGaCAAGTTTTG	18	-1.8			1	93
P22ant	f	TCCAAGTTAGTGTA TTGACA TGATAGAA	GCACTCTAC TATATT CTCAATaggTCCACGG	17	-0.4				9
P22mnt	f	CCACCGTGCGACCTA TTGACA ATATAGTA	GAGTGCTTC TATCAT GTCAATAaCACTAACGTT	17	-1.5				9
P22PR	f	CATCTTAAATAAAC TTGACT AAAGATTC	CTTTAGTA GATAAT TTAAGTgTTCTTTAAT	16	-1.8				9
P22PRM	f	AAATTATC TACTAA AGGAATCT	TTAGTCAAG TTTATT TAAGATGACTTaaCTAT	17	-3.7	-3.1	-3.9		9
pBR313Htet	m	AATTCTCATGT TTGACA GCTTATCA	TCGATAAGC TAGCTT TAATGCgGTAGTTTAT	17	-1.7			1,3	94
pBR322bla	p	TTTTTCTAAATACA TTCAAA TATGTATC	CGCTCATGA GACAAT AACCCTgATAAATGCT	17	-2.6				9
pBR322P4	p	CATCTGTGCGGTAT TTCACA CCGCATATGGTGCACTCTCAG TACAAT CTGCTCTgATGCCGCAT	21	-2.7				9
pBR322primer	p	ATCAAAGGATCTTC TTGAGA TCCTTTTT	TTCTGCGCG TAATCT GCTGCTTgCAAACAAAA	17	-2.1				9
pBR322tet	p	AAGAATTCTCATGT TTGACA GCTTATCA	TCGATAAGC TTTAAT GCGGTAgTTTATCACA	17	-1.0				9
pBRH4-25	M	TCG TTTTCA AGAATTCA	TTAATGCGG TAGTTT ATCAcagTTA	17	-2.7			4	95
pBRP1	p	TTCATACACGGTGC CTGACT CGGTTAGCAATTTAACTGTGA TAAACT ACGGCAttAAAGCTTA	21	-3.3				9
pBRRNAI	p	GTGCTACAGAGTTC TTGAAG TGGTGGCCT	AACTACGGC TACACT AGAAGGacaGTATTTG	18	-2.2				9
pBRtet-10	M	AAGAATTCTCATGT TTGACA GCTTATCA	TCGATAAGC TAGTTT ATCAcagTTA	17	-1.6			4	95
pBRtet-15	M	AAGAATTCTCATGT TTGACA GCTTATCA	TCGGTAGTT TATCAC AGTTAaatTGC	17	-1.8			4	95
pBRtet-22	M	AAGAATTCTCATGT TTGACA GCTTATCAT	CGATCACAG TTAAAT TGCTAacgCAG	18	-1.8			4	95
pBRtet/TA22	M	TTCTCATGT TTGACA GCTTATCA	TCGATAAGC TAAAT TTATATAaaATTTAGCT	17	-0.7			1	96
pBRtet/TA33	M	TTCTCATGT TTGACA GCTTATCA	TCGATAAGC TAAATT TATATATAaaATTTTATAT	17	-0.7			1	96
pColViron-P1	p	TCACAATTCTCAAG TTGATA ATGAGAAT	CATTATTGA CAATACT TGTTaTTATTTTAC	17	-1.6			1,3,4	97
pColViron-P2	p	TGTTTCAACACC ATGTAT TAATTGTG	TTTATTTG TAAAAT TAATTTtctgacaATAA	16	-3.0			3,4	97
pEG3503	M	GGC TGGACT TCGAATTCA	TTAATGCGG TAGTTT ATCAcagTTA	18	-3.6			4	95
phiXA	f	AATAACCGTCAGGA TTGACA CCCTCCCA	ATTGTATGT TTTCAT GCCTCCaAATCTTGGA	17	-1.7				9
phiXB	f	GCCAGTTAAATAGC TTGCAA AATACGTGG	CCTTATGGT TACAGT ATGCCCaTCGCAGTT	18	-2.6				9
phiXD	f	TAGAGATTCTCTTG TTGACA TTTTAAAAG	ACGTGGAT TACATT CTGAGTCCgATGCTGTT	18	-1.7				9
pori-I	b	CTGTTGTTCAGTTT TTGAGT TGTGTATA	ACCCCTCAT TCTGAT CCCAGcTTATACGGT	17	-3.2				9
Pori-r	b	GATCGCACGATCTG TATACT TATTTGAGT	AAATTAACC CACGAT CCCAGCCaTTCTTCTGC	18	-4.5				9
ppc	b	CGATTTCGGCAGCAT TTGACG TCACCGCT	TTTACGTGG CTTTAT AAAAgaCGACGAAAA	17	-3.1			3,4	99
pSC101oriP1	p	T TTGTTAG AGGAGCAAACAGCGTTTGCGA CATCCT TTTGTAATACTGCGGGAA	21	-4.4		‰	2,3	102,103
pSC101oriP2	p	ATTATCA TTGACT AGCCCATC	TTTATTTG TATAAT GATTAAAATCCACATT	16	-1.4		‰	2,3	102,103
pSC101oriP3	p	ATACGCTCAGATGA TGAACA TCAGTAGG	GAAAATGCT TATGGT GTATTAGCTAAAGC	17	-3.6			2,3	102,104
pyrB1-P1	b	CTTTCACACTCCGC CCTATA AGTCGGAT	GAATGGAA TAAAT GCATAtcTGATTGCGTG	16	-4.2	-3.6		3	105
pyrB1-P2	b	TTGCATCAAATG CTTGCG CCGGCTTCT	GACGATGAG TATAAT GCCGgacAATTTGCCGG	17	-2.8			3	105
pyrD	b	TTGCCGCAGGTCAA TTCCCT TTTGGTCC	GAACTCGCA CATAAT ACgccccCGGTTTG	17	-2.6			3,4	106
pyrE-P1	b	ATGCCTTGTAAGGA TAGGAA TAACCGCC	GGAAGTCGG TAAAT GGGCAgCCACATTTG	17	-1.8			4	107,108
pyrE-P2	b	GTAGGCGGTCATA CTCGGG ATCATAGAC	GTTCCTGTT TATAAA AGGAGaGGTGGAAGG	18	-4.6			4	108
R100rna3	p	GTACCGGCTTACGC CGGGCT TCGGCGGTT	TTACTCCTG TATCAT ATGAaACAACAGAG	18	-4.3				9
R100RNAI	p	CACAGAAAGAAGTC TTGAAC TTTTTCGG	GCCATATAAC TATACT CCCCGGCaTAGCTGAAT	17	-1.6				9
R100RNAII	p	ATGGGCTTACATTC TTGAGT GTTCAGAA	GATTAGTGC TAGATT ACTGATCgTTTAAGGAA	17	-2.2				9
R1RNAII	b	ACTAAAGTAAACGC TTACT TTGTGGCG	TAGCATGC TACATT ACTGATGcGTTTAAGGAA	16	-2.4				9
recA	b	TTTCTACAAACAC TTGATA CTGTATGA	GCATACAG TATAAT TGCTTCaACAGAACAT	16	-1.1				9
rnh	b	GTAAGCGGTCATTT ATGTCA GACTGTC	GTTTTACAG TTCGAT TcaaTTACAGGA	17	-4.0		-4.5	2,3,4	50,51
rnp(RNaseP)	b	ATGCGCAACGCGGG GTGACA AGGGCGCG	CAAAACCTC TACACT GCGGCGGCgAAGCTGACC	17	-1.2			1	109
rplJ	b	TGTAAACTAATGCC TTTACG TGGGCGGT	GATTTTGTC TACAAT CTTACccCACGTATA	17	-1.8				9
rpmH1p	b	GATCCAGGACGATC CTTGCG CTTTACCC	ATCAGCCCG TAAAT CCTccacccGCGGCAG	17	-2.8		-2.9	4	48
rpmH2p	b	ATAAGGAAAGAAA TTGACT CCGGAGTC	TACAATTAA TACAAT CCGgcctcTTTAAATC	17	-1.0			4	48
rpmH3p	b	AAATTTAATGACCA TAGACA AAAATTGG	CTTAATCGA TCTAAT AAAgatcCCAGGACG	17	-2.3			4	48
rpoA	b	TTCGCATATTTTTC TTGCAA AGTTGGGT	TGAGCTGGC TAGATT AGCCAgCCAATCTTT	17	-1.8				9

Table 1 *Continued*

		sequence					
rpoB	b	CGACTTAATATACT GCGACA GGACGTCC	GTTCTGTG TAAATC GCAATGAAATGGTTTAA	16	-4.4		9
rpoD-Pa	b	CGCCCTGTTCCG CAGCTA AAACGCAC	GACCATGCG TATACT TATAgggTT	17	-3.5		2,4	110
rpoD-Pb	b	AGCCAGGT CTGACC ACCGGGCAA	CTTTTAGAG CACTAT CGTGGTACaaaT	18	-4.6	-5.9	2,4	110
rpoD-Phs	b	ATGCTGCCACCG TTGAAA AACTGTCG	ATGTGGGAC GATATA GCAGAtaaG	17	-2.9		4	110
rpoD-Phs/min	b	CCC TTGAAA AACTGTCGATGTGGGACGATA TAGCAG ATAAGAATATTgcT	21	-4.2	-2.9 -4.7	4	110
rrn4.5S	b	GGCACGGCATGGG TTGCAA TTAGCGCG	GGCACGAGT CATACT GGGCCTGCgCGTTGGTT	17	-1.9		1	111
rrnABP1	b	TTTTAAATTTCCTC TTGTCA GGCCGGAA	TAACTCCC TATAAT GGGGCACCaCTGACACG	16	-0.8			9
rrnABP2	b	GCAAAAATAAATGC TTGACT CTGTAGCG	GGAAGGCG TATTAT GCACAcccCGCGCGCGC	16	-1.4			9
rrnB-P3	b	CTATGATAAGGAT TACTCA TCTTATCCTT ATCAAACGT TAAAAT GGGCGgtgTGAGCTTG	20	-4.1		2,4	112-114
rrnB-P4	b	GCGTATCCGGTCAC CTCTCA CCTGACA	GTTCGTCG TAAAAT AGCCAAccTGTTCGACA	15	-3.8		2,4	112-114
rrnDEXP2	b	CCTGAAATTCAGGG TTGACT CTGAAAGA	GGAAACGG TAAAAT ACGGCACcTCGCGACAG	16	-1.7			9
rrnD-P1	b	GATCAAAAAAATAC TTGTGC AAAAAATT	GGGATCCC TATAAT GCGCCTCCgTTGAGACG	16	-2.7			9
rrnE-P1	b	CTGCAATTTTTCTA TTGCGG CCTGCGGA	GAACTCCC TATAAT GCGGCCTCCaTCGACACG	16	-2.3			9
rrnG-P1	b	TTTATATTTTTCGC TTGTCA GGCCGGAA	TAACTCCC TATAAT GCGGCACCACTGACACG	16	-0.8			9
rrnG-P2	b	AAGCAAAGAAATGC TTGACT CTGTAGCG	GGAAGGCG TATTAT GCACACCGCCGCGCCG	16	-1.4			9
rrnX1	b	ATGCATTTTTCCGC TTGTCT TCCTGAGC	CGACTCCC TATAAT GCGGCCTCCaTCGACACG	16	-1.2			9
RSFprimer	b	GGAATAGCGTGTTCG TTGACT TGATAGAC	CGATTGATT CATCAT CTCATaAATAAAGAA	17	-2.0			9
RSFrnaI	p	TAGAGGAGTTTGTC TTGAAG TTATGCACC	TGTTAAGGC TAAACT GAAAGAaCAGATTTTG	18	-1.8			9
S10	b	TACTAGCAATACGC TTGCGT TCGGTGGT	TAAGTATG TATAAT GGGCgggCTTGTCGT	16	-2.2			9
sdh-P1	b	ATATGTAGGTAA TTGTAA TGATTTTG	TGAACAGCC TATACT GCGGCCAGtCTCCGGAA	17	-1.0		4	57,115
sdh-P2	b	AGCTTCCGCGATTA TGGGCA GCTTCTTC	GTCAAATT TATCAT GTGGGGCaTCCTTACCG	16	-2.9		4	57,115
spc	b	CCGTTTATTTTTTC TACCCA TATCCTTG	AAGCGGTGT TAAACT GGGCgCGCCTCGATA	17	-2.2			9
spot42r	b	TTACAAAAAGTGCT TTCTGA ACTGAACA	AAAAAGAG TAAAGT TAGTCGCgtAGGGTACA	16	-3.2	-3.3		9
ssb	b	TAGTAAAAGCGGCTA TTGGTA ATGGTACAA	TCGGCGCGTT TACACT TATTCAgAACGATTTT	18	-2.9			116,117
str	b	TCGTTGTATATTTC TTGACA CCTTTTCG	GCATCGCGCC TAAAAT TCGGCgTCCTCATAT	17	-0.3			9
sucAB	b	AAATGCAGGAAATC TTTAAA AACTGCCGC	TGACACTAA GACAGT TTTAAAaGGTTCCTT	18	-3.6		4	22
supB-E	b	CCTTGAAAAAGAGG TTGACG CTGCAAGG	CTCTATACG CATAAT GGGCCCCgCAACGCCGA	17	-1.4			9
T7-A1	f	TATCAAAAAGAGTA TTGACT TAAAGTCT	AACCTATAG GATACT TACACCGCaTCGAGAGGG	17	-1.8			9
T7-A3	f	GTGAAACAAAACGG TTGACA ACATGAAG	TAAACACGG TACGAT GTACCACaTGAAACGAC	17	-1.2			9
T7-C	f	CATTGATAAGCAAC TTGACG CAATGTTA	ATGGGCTGA TAGTCT TATCTTaCAGGTCATC	17	-2.1			9
T7-D	f	CTTTAAGATAGCG TTGACT TGATGGGT	CTTTAGGTG TAGGCT TTAAGGTgTTGGCTTTA	17	-1.9			9
T7A2	f	ACGAAAAACAGGTA TTGACA ACATGAAGT	AACATGCAG TAAGAT ACAAATCgCTAGGTAAC	18	-1.3			9
T7E	p	CTTACGGATG ATGATA TTTACACA	TTACAGTGA TAGCAT GGGCCCaCTACAGATA	17	-2.4		1,3	118
TAC16	M	AATGAGCTG TTGACA ATTAATCA	TCGGCTCG TATAAT GTGTGGAaTTGTG	16	-0.4			119,120
Tn10Pin	p	TCATTAAG TTAAGG TGGATACAC	ATCTTGTCA TATGAT CAAATGGTTTCgCGAAA	18	-3.5	-5.0		9
Tn10Pout	p	AGTGTAATTGGGGG CAGAAT TGGTAAAG	ACAGTCGTG TAAAAT ATCGAGttCGCACATC	17	-2.7			9
Tn10tetA	p	ATTCCTAATTTTTTG TTGACA CTCTATCAT	TGATAGAGT TATTTT ACCACTCCCTATCAGT	18	-1.4			9
Tn10tetR *	p	TATTCATTTCACTT TTCTCT ATCACTGAT	AGGGAGTGG TAAAAT AACTCTATCAATGATA	18	-2.2			9
Tn10tetR*	p	TGATAGCGAG TGGTAA AATAACTC	TATCAATGA TACTGA CACCAAAAATTAGG	17	-3.0		4	122
Tn10xxxP1	p	TTAAAATTTTCTTG TTGATG ATTTTTAT	TTCCATGA TAGATT TAAAATaACATACC	16	-2.6		4	123
Tn10xxxP2	p	AAATGTTCTTAAGA TTGTCA CGACCACA	TCATCATGA TACCAT AAACaTACTGACGG	17	-1.8		4	123
Tn10xxxP3	p	CCATGATAGA TTTAAA ATAACATACCGTCAGTATGTT TATGGT ATCATGATGaTGTGTGTC	21	-3.3	-4.6	4	123
Tn2660bla-P3	p	TTTTTCTAAATACA TTCAAA TATGTATC	CGCTCATGA GACAAT AACCCTgATAAATGCT	17	-2.6		2,4	124
Tn2661bla-Pa	p	GGTTTATAAAATTC TTGAAG ACGAAAGG	GGCTGGTGA TAGGCT TATttttATAGGTTAA	17	-2.3		2,4	124
Tn2661bla-Pb	p	CCTC GTGATA CGCTTATT	TTTATAGGT TAATGT CATGAtaaTAAATGGTTT	17	-3.1		2,4	124
Tn501mer	p	TTTTCCATATCGC TTGACT CCGTACATG	AGTACCGAAG TAAGGT TACGGCTaTCCAATTTC	19	-3.2		3,4	125-127
Tn501merR	p	CATGCGCTTGTCCT TTCGAA TTGAAATT	GGATAGCG TAACCT TACTtCCGTACTCA	16	-3.3	-3.8	3,4	125-127
Tn5IR	p	TCCAGGATCTGATC TTCCAT GTGACCTC	CTAACATGG TAACGT TCATGATaACTTCTGCT	17	-3.4			9
Tn5neo	p	CAAGCGAACCGGAA TTGCCA GCTGGGGC	GCCCTCTGG TAAGGT TGGGAAGCCCTGCAA	17	-2.1			9
Tn7-PLE	p	ACTAGACAGAATAG TTGTAA ACTGAAAG	CAGTCGCAGT TATgct gtgagaaaaGCAT	17	-1.6		4	128
tnaA	b	AAAACAATTTCAGAA TAGACA AAAACTCT	GAGTGTAA TAATGT AGCCTCgtgTCTTGCG	16	-2.8			9
tonB	b	ATCGTCTTGCCTTA TTGAAT ATGATTGCT	ATTTGCATT TAAAAT CGAgACCTGGTTT	18	-1.3		4	129
trfA	p	AGCCGCTAAAGTTC TTGACA GCGGAACCA	ATGTTTAGC TAAACT AGAGTCtCCT	18	-1.1		4	130,131
trfB	p	AGCGGCTAAAGGTG TTGACG TGCGGAGAA	ATGTTTAGC TAAACT TCTCTCaTGT	17	-1.1		4	130
trp	b	TCTGAAATGAGCTG TTGACA ATTAATCA	TGGAACTAG TTAACT AGtacGCAATTT	17	-1.7			9
trpP2	b	ACCGGAAGAAAACC GTGACA TTTTAACA	CGTTTGTTA CAAGGT AAAGGCgACGCCGCGCC	17	-3.3			9
trpR	b	TGGGGACGTCGTTA CTGATC CGCACGTTT	ATGATATGC TATCGT ACTCTTTaGCGAGTACA	18	-4.3	-2.8		9
trpS	b	CGGCGAGCGCTATCG ATCTCA GCCACGCCT	GATGTAATT TATCAG TCtatAAATGACC	17	-4.5	-5.7		9
trxA	b	CAGCTTACTATTGC TTTACG AAAGCGTAT	CCGGTGAAA TAAAGT CAACTaGTTGGTTAA	18	-2.5		3	132
tufB	b	ATGCAATTTTTTAG TTGCAT GAACTCGC	ATGTCTCCA TAGAAT GCGGCTaCTTGATGCC	17	-1.8			9
tyrT	b	TCTCAACGTAACAC TTTACA GCGGCGCG	TCATTTGA TATGAT GCGCCCCgCTTCCCGAT	16	-1.6			9
tyrT/109	b	ACAGCGCGTCTTTG TTTACG GTAATCAA	CGATTATTC TTTAAT CGGCAGcAAAAATAA	18	-2.6		2-4	131
tyrT/140	b	TTAAGTCGTCACTA TACAAA GTACTGGCA	CAGCGGGTC TTTGTT TACggGTAATCG	18	-4.2	-5.2	2-4	131
tyrT/178	b	TGCGCGCAGGTC GTGACG TCGAGAAAA	ACGTCT TAAGTC GTGCAcTATACA	15	-5.2	-4.9	2-4	131
tyrT/212	b	G ATCATA CCTACACAG	CTGAAGA TATGAT GCGCGCAgGTCGTGACG	16	-3.6		2-4	131
tyrT/6	b	ATTTTTCTCAAC GTAACA CTTTACA GC	GGCGGGTCA TTTGAT ATGATGCgCCCCGCTTC	16	-4.1	-1.6 -1.6	2-4	131
tyrT/77	b	ATTATTCTTTAA TCGCCA GCAAAATAA	CTCGGTTACC TTTAAT CCGTTACGgATGAAAAT	19	-4.3	-4.2	2-4	131
uncI	b	TGGCTACTTATTGT TTGAAA TCACGGGG	GCGGCACCG TATAAT TTGACCgCTTTTTGAT	16	-0.6	-1.6	3,4	132,133

Table 1   *Continued*

uvrB-P1	b	TCCAGTATAATTTG TTGGCA TAATTAAG	TACGACGAG TAAAAT TACATaCCTGCCCGC	17	-1.0			9
uvrB-P2	b	TCAGAAATATTATG GTGATG AACTGTTTT	TTTATCCAG TATAAT TTGTTGgCATAATTAA	18	-2.5			9
uvrB-P3	b	ACAGTTATCCACTA TTCCTG TGGATAAC	CATGTGTAT TAGAGT TAGAAAaCACGAGGCA	17	-3.7			9
uvrC	b	GCCCATTTGCCAGT TTGTCT GAACGTGA	ATTGCAGAT TATGCT GATGatcaCCAAGG	17	-1.8		4	136
uvrD	b	TGGAAATTTCCCGC TTGGCA TCTCTGAC	CTCGCTGA TATAAT CAGCCAAaTCTGTATAT	16	-1.1		3	137
434PR	f	AAGAAAACTGTAT TTGACA AACAAGAT	ACATTGTAT GAAAAT ACAAGAAAgTTTGTTGA	17	-1.3			9
434PRM	f	ACAATGTATCTTGT TTGTCA AATACAGT	TTTTCTTGT GAAGAT TGGGGGTAAATAACAGA	17	-2.4			9

List of promoter sequences arranged alphabetically by name (a) and aligned with respect to optimal -35 (c) and -10 hexamer sequences (d) consistent with the transcriptional start. Column (b) designates promoter type: b, bacterial; p, plasmid or transposon; f, phage; M, mutation or fusion which generates a new promoter; m, point mutation in an existing promoter. The lower case base(s) downstream of the -10 region denotes experimentally determined transcriptional start point(s). Column (e) indicates spacing in base pairs between -35 and -10 hexamers. Column (f) reports relative promoter homology index (PHI) of promoter elements in columns c,d,e as described in the text. Column (g) signals discrepancies between the promoter elements consistent with transcriptional start data and the best promoter elements independent of start data (indicated by double underlines). Only discrepancies for which the PHI values of these promoters differed by at least 0.5 are shown. Column (h) signals discrepancies between the computer selected promoter elements and published -35 and -10 sequences (shown by single underlines). The figures in these columns are PHI values corresponding to the underlined promoter elements. Column (i) indicates the nature of experimental data defining the transcription start: 1, total or partial RNA sequence with identification of the 5' nucleoside triphosphate; 2, mutational or genetic identification of -35 and -10 regions; 3, high resolution sizing of in vitro transcripts; 4, high resolution S1 nuclease mapping. The 112 promoters documented by Hawley and McClure (9) are included in this compilation and can be identified by a 9 in reference column (j).

% Only one of the -35 or -10 promoter hexamers was unambiguously identified, thus no PHI value for the published promoter can be given.

+ Underlined -35 and -10 regions for these genes represent heat shock promoter elements which are apparently recognized by a distinct heat shock sigma factor (34).

Figure 1   *E. coli* σ 70 promoters. Promoters are listed in alphabetical order of regulators. Within regulons, they are listed in alphabetical order of promoters, except in cases of multiple promoters that are located one immediately after the other. All of the binding sites use the same scale relative to the site of initiation of transcription (+1) marked by a vertical line. Only sites with established regulatory effects on the respective promoter are indicated. Symbols: ▬, known or inferred binding sites of repressors; ▨, binding sites of activators; ▬, protein with a dual positive and negative effect; an abbreviation that refers to the regulatory protein is indicated above each box, except in the case of positive CRP-binding sites (▤) and positive FNR-binding sites (▨); ▨, domain defined by deletion analysis.

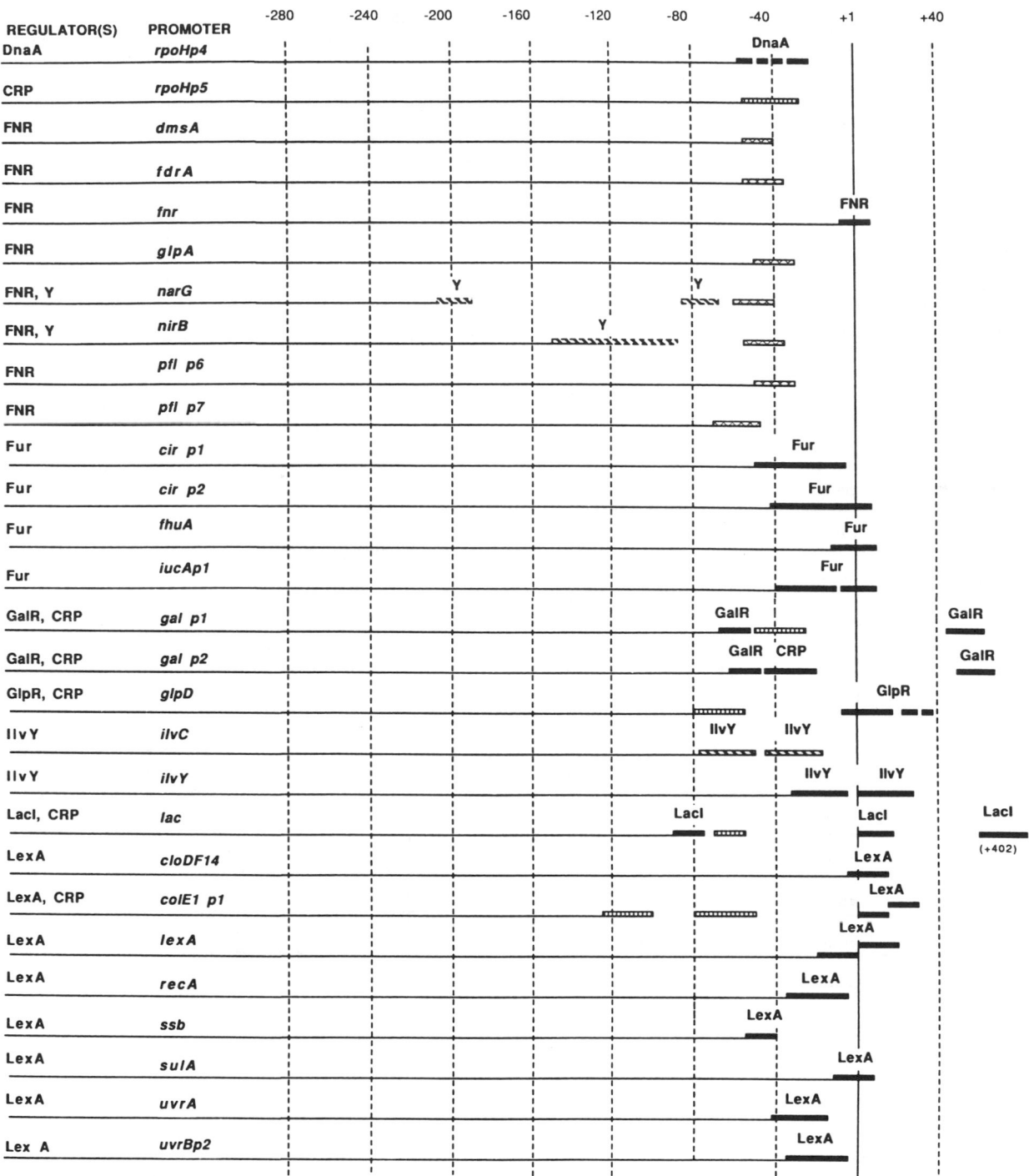

Figure 1 *Continued*

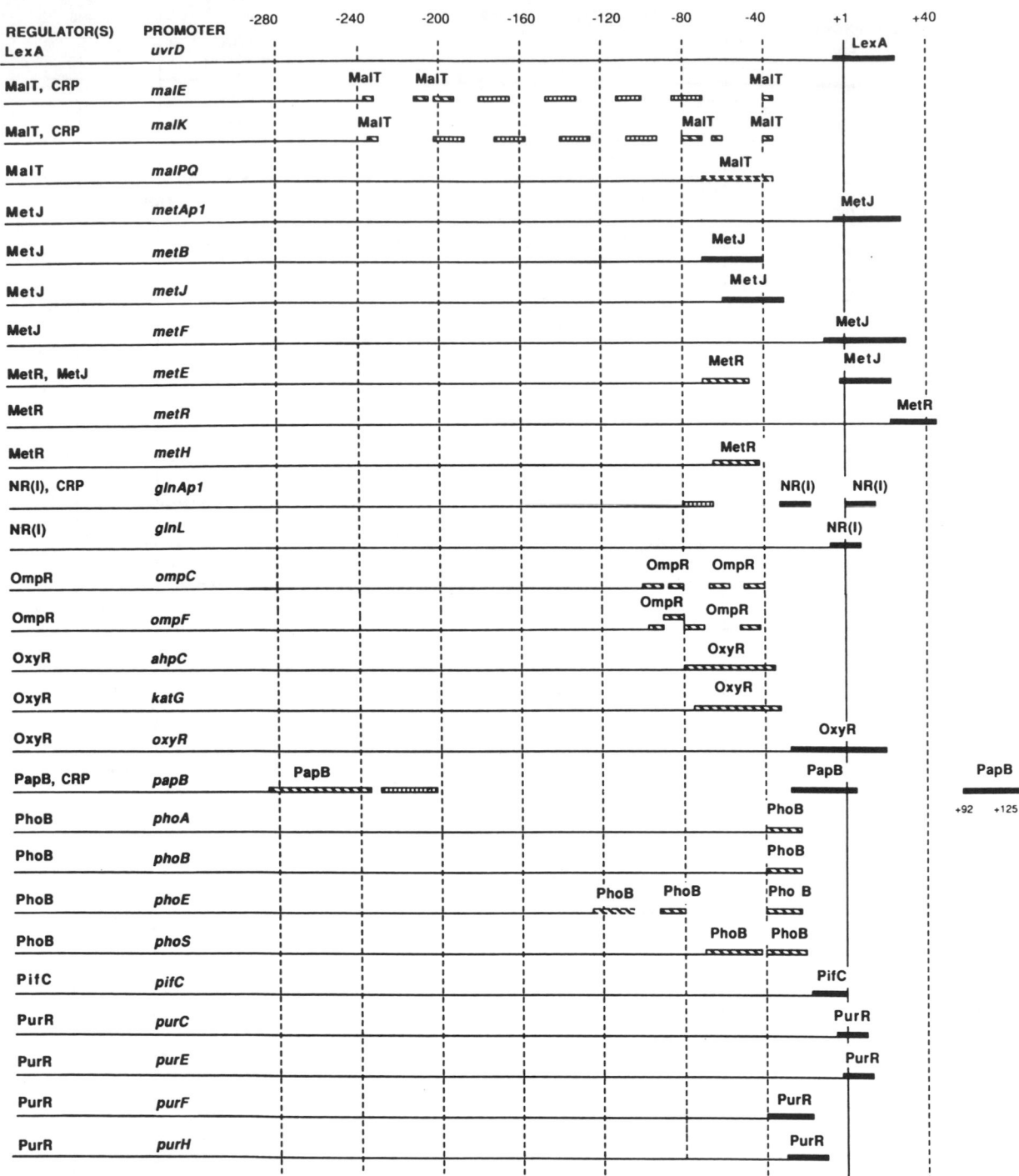

Figure 1   *Continued*

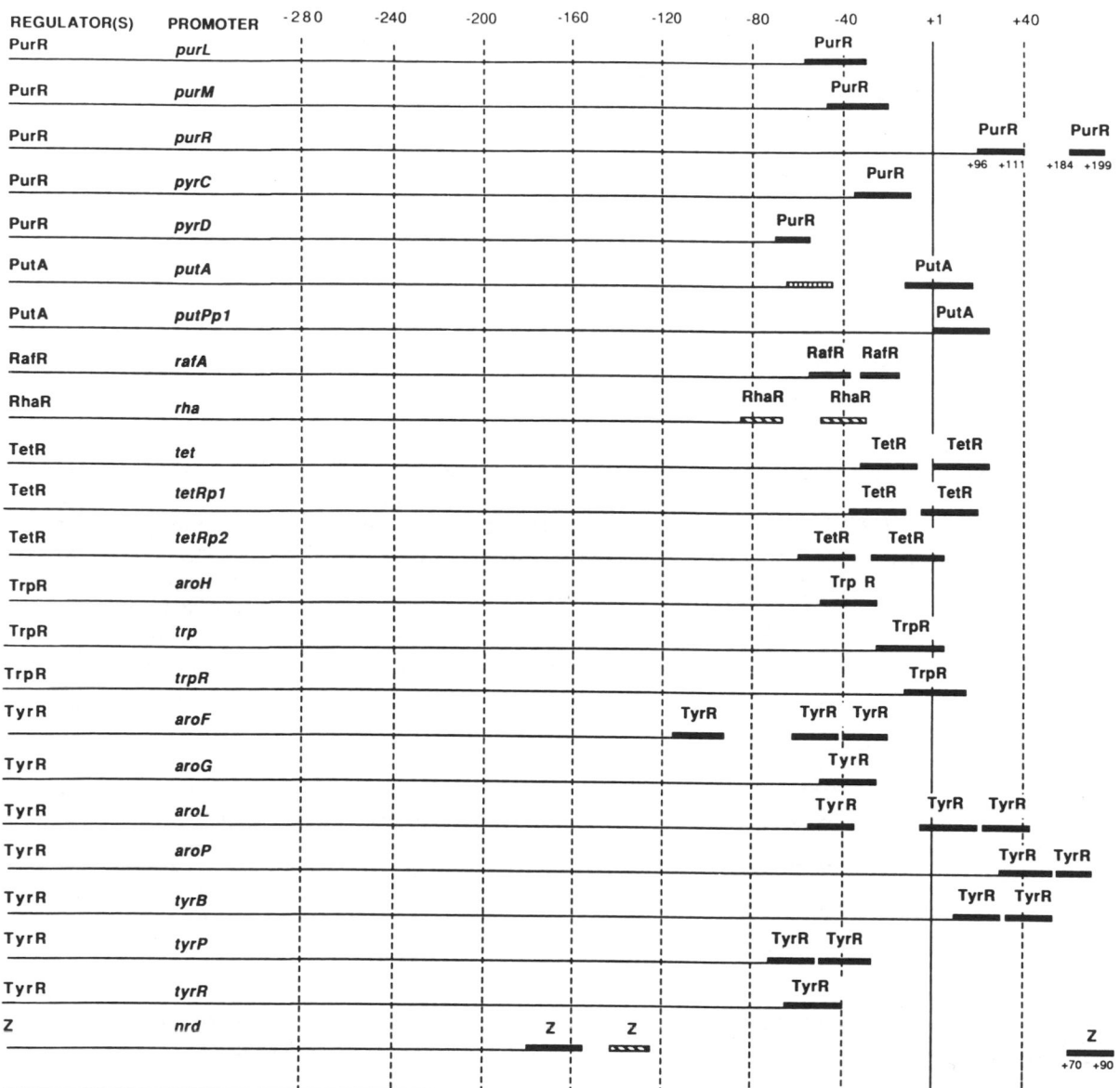

Figure 1 *Continued*

promoter	σ	sequence				
		−40	−30	−20	−10	+1
spo VG	σ^{S37}	CGAGCAGGATTTCAGAAAAAATCGTGGAATTGATACACTAA				
spo VG	σ^{S32}	TTCAGAAAAAATCGTGGAATTGATACACTAATGCTTTTATA				
spo VC	σ^{S37}	CGAGGTTTAAATCCTTATCGTTATGGGTATTGTTTGTAATA				
ctc	σ^{S32}, σ^{S37}	CGAGGTTTAAATCCTTATCGTTATGGGTATTGTTTGTAATA				
P102	σ^{S28}	AAGCTAAATGATTCTGTTTTTATGCCGATATATAATCACTA				
P201	σ^{S28}	TGTCCCTAAAGTTCCGGGCACCAAAACCGATATTAACCATA				
SP01 M25.1	σ^{gp28}	GUUAGGAGAAAACUUUUGAGGGGUUAGUUUGUUUCUUUCAA				
SP01 M25.2	σ^{gp28}	UAUAAGGUGAAGAGUUCUCCACGCAGACUUUUCUUUUGGUA				
SP01 M19.1	σ^{gp28}	CCUUAUGAGAAAAGACUAAAAAUGGGGAUUGUUUUUUGUUG				
SPO1 M12	σ^{gp28}	CAUCAGGAGAUUAUUUCAAUGAUCUCUGUUUAUUUCUUUCA				
SP01 M19.2	σ^{gp28}	CUUAGGAGACAAUUUUUGUCUCUGAGAUUUUAUUUACAAUG				
Kp nifE	σ^{60}	ATCAAGCTCCGCTTCTGGAGCGCGAATTGCATCTTCCCCCT				
Kp nifU	σ^{60}	ATTAATTTTATTCTCTGGTATCGCAATTGCTAGTTCGTTAT				
Kp nifB	σ^{60}	TGCGAAATTAACCTCTGGTACAGCATTTGCAGCAGGAAGGT				
Kp nifH	σ^{60}	TAAACAGGCACGGCTGGTATGTTCCCTGCACTTCTCTGCTG				
Kp nifM	σ^{60}	TCAGCCAGCCGTGGCTGGCCGGAAATTTGCAATACAGGGAT				
Kp nifF	σ^{60}	GTATGCAAAGCAACCTGGCACAGCCTTCGCAATACCCCTGC				
Kp nifL	σ^{60}	CATCACGCCGATAAGGGCGCACGGTTTGCATGGTTATCACC				
glnA P2	σ^{60}	GGCAATTTAAAAGTTGGCACAGATTTCGCTTTATCTTTTTT				
gro E	σ^{32}	TCCCCCTTGAAGGGGCGAAGCCATCCCCATTTCTCTGGTCA				
dnaK P1	σ^{32}	TCCCCCTTGATGACGTGGTTTACGACCCCATTTAGTAGTCA				
P2		GGGCAGTTGAAACCAGACGTTTCGCCCCTATTACAGACTCA				
C62.5	σ^{32}	TCGCTTGAAATTATTCTCCCTTGTCCCCATCTCTCCCACAT				
rpoD$_{HS}$	σ^{32}	CACCCTTGAAAAACTGTCGATGTGGGACGATATAGCAGATA				
lon	σ^{32}	CGGCGTTGAATGTGGGGGAAACATCCCCATATACTGACGTA				
total consensus		-g----ttAAAt--tt--a-t---a----tT--Tt-t---A				
		−40	−30	−20	−10	+1
σ^{70}	consensus	TTGACA			TATAAT	
σ^{32}	consensus	CCCCC		CCCC		
σ^{60}	consensus			CTGGC	TTGCA	

Figure 2 Compilation of DNA sequences for promoters utilizing alternative σ factors. The DNA sequences are shown for promoters recognized by *B. subtilis* σ factors 37, 32, and 28, *B. subtilis* phage SPO1 σ factor gp28, *E. coli* nitrogen regulation σ factor σ^{60}, and *E. coli* heat-shock σ factor σ^{32}. These sequences were aligned relative to their start points for transcription, and the distribution of base pairs was determined for each position. Those base pairs that occurred at a frequency one standard deviation above random $(6 + 2.44)$ are shown in lowercase letters; those base pairs that occurred two standard deviations above random are shown as capital letters. Base pairs that occurred above random frequency but that corresponded in position and identity with the consensus sequences previously derived as being σ^{60}- or σ^{32}-specific were not included. The DNA sequences were derived from the following promoters: *spoVG* (49); *spoVC* (72); *ctc* (80); P102 and P201 (30); SPO1 M25.1, SPO1 M25.2, and SPO1 M19.1 (54); SPO1 M19.2 and SPO1 M12 (95); *K. pneumoniae nifH* (94); *K. pneumoniae nifE, nifU, nifB,* and *nifM* (5); *K. pneumoniae nifF* and *nifL* (22); *glnAp2* (81); and σ^{32} (16).

SECTION 17

B. SOS System

Figure 1 presents DNA sequences of *lexA*-repressed promoters. SOS genes, map positions, and activities are shown in Figure 2. Tables 1 and 2 and Figures 3 and 4 focus on global regulatory responses. For an updated review of oxidative stress regulons, summarized in Figures 3 and 4, see Demple (1991).

Figure 1 and Tables 1 and 2 are reprinted, with permission, from Hoopes and McClure, Neidhardt, and Neidhardt and VanBogelen, respectively (Escherichia coli *and* Salmonella typhimurium. *Cellular and Molecular Biology*. American Society for Microbiology, Washington, D.C., pp. 1235, 1316, and 1339, respectively [1987]). Figure 2 is courtesy of David Mount, and Figures 3 and 4 are reprinted, with permission, from Demple (*Annu. Rev. Genet.*, vol. 25, pp. 321 and 330, respectively [1991]). The references cited in the tables, table notes, and caption to Figure 1 can be found in the respective original papers.

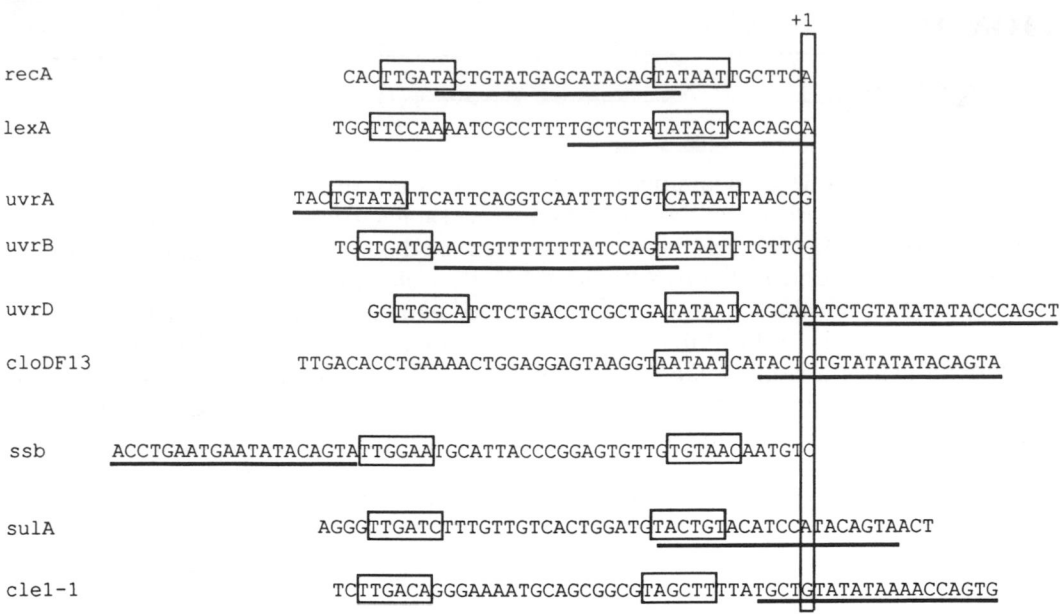

Figure 1 DNA sequences of *lexA*-repressed promoters. The DNA sequences of nine promoters repressed by *lexA* in vivo are listed. The proposed *lexA* binding sites in the promoters are underlined (see ref. 103), and the presumptive −35 and −10 regions are boxed. References: *recA* (56); *lexA* (11); *uvrA, uvrB* (85); *uvrD* (24); CloDF13 (102); *ssb* (8); *sulA* (70); and *cle1-1* (25).

DNA Repair Activities:

uvrA	(91')	Excision Repair
uvrB	(18')	Excision Repair
recN	(57')	Recombinational DNA Repair
recQ[1]	(86')	Recombinational DNA Repair
ruvAB	(41')	Recombinational DNA Repair
phr[1]	(16')	Photoreactivation of Pyrimidine Dimers

DNA replication:

ssb	(92')	DNA Replication, Recombination
uvrD	(86')	DNA Replication(Helicase II), Mismatch Repair, etc.
nrdAB[1]	(49')	Ribonucleoside Diphosphate Reductase
dnaQ(mutD)[1]	(5')	ε-subunit of Pol III (editing function)
dnaN[1]	(83')	β-subunit of Pol III (processivity)
umuDC	(26')	SOS Mutagenesis
polB(dinA)	(1.2')	DNA Polymerase II

Regulation of the SOS Response:

recA	(58')	Repressor Cleavage, Mutagenesis and Recombination
lexA/dinF[2]	(92')	Repressor of the SOS genes/?

Other activities:

sulA	(22')	Cell Division Inhibition
dinB[2]	(~8)	?
dinD[2]	(80-85)	?
dinG	(17.8')	?
dinH	(19.8')	?
dinI[3]	(23.4')	?
sosA[3]	(98.5')	?
sosB[3]	(38')	?

Figure 2 SOS genes, map positions, and activities. [1]These genes exhibit *recA*, *lexA*-dependent expression when their regulatory regions are fused to a reporter gene such as *lacZ*. Although the genes have been sequenced, they do not contain recognizable binding sites for *lexA* repressor. [2]*dinB*, *dinD*, and *dinF* were defined genetically using Mud(Amp, *lacZ*) fusions and have not yet been sequenced; the *dinF* and *lexA* genes appear to form an operon. [3]*dinI*, *sosA* (downstream from *hsdS*), and *sosB* (downstream from *ntrla*) were originally identified as specific 20-bp binding sites for *lexA* repressor on the *E. coli* chromosome; transcription from *dinI* has been shown to be *lexA*-dependent, but transcription from the other loci has not yet been investigated.

Sources include: Walker 1984 (*Microbiol. Rev. 48:* 60); Berg 1988 (*Nucleic Acids Res. 16:* 5089); Kaasch et al. 1989 (*Mol. Gen. Genet. 219:* 187); Payne and Sancar 1989 (*Mutat. Res. 218:* 207); Gilbert et al. 1990 (*Mol. Gen. Genet. 220:* 400); K. Lewis and D. Mount (unpubl.).

Table 1　Multigene Systems in *E. coli*

Multigene system	Environmental stimulus	Regulatory gene(s)	Regulated genes	Reference(s)
Nitrogen utilization	Ammonia limitation	*glnB, glnD, glnG, glnL*	*glnALG* plus others	Chapter 81
Carbon utilization	Carbon/energy limitation	*cya, crp*	*gal, deo, ara, mal, dsd, tna, lac,* plus others	Chapter 81
Phosphate utilization	Phosphate limitation	*phoB, phoM, phoR, phoU*	20+ genes	Chapter 82
Stringent response	Amino acid/energy limitation	*relA, relB, relX, spoT, gpp,* plus others	Many	Chapter 87
Heat shock response	Heat, certain toxic agents	*htpR (rpoH)*	17 genes	Chapter 83
SOS response	UV and other DNA damagers	*recA, lexA*	17 genes	Chapter 84
Adaptive response	Methylating agents	*ada*	3+ genes	10, 21, 22
Translation apparatus	Growth rate-supporting ability of medium	Many	200+ genes	Chapters 85 and 86
Osmotic stress response	High osmolarity	*envZ, ompR* *kdpD*	*ompF, ompC* *kdpABC* plus others	3, 6, 11 3, 8
Oxidative stress response	H_2O_2, other oxidants	*oxyR*	12+ genes	2
Anaerobic respiration	Presence of electron acceptors other than O_2	*fnr (=nirA, =nirR)*	20+ genes	1, 17
Anaerobic fermentation	Absence of electron acceptors	?	20–50 genes	4, 18, 20
Aerobic response	Addition of oxygen	?	20 (?) genes	19

Table 2　Promoter Regions of the Genes of the *E. coli* Heat-shock System[a]

Gene or consensus	−35 region	13- to 15-bp region	−10 region	+1
groE	TTTCCCCCTTGAA	GGGGCGAAGCCAT	CCCCATTTCTCTGGTCAC	
dnaK P1	TCTCCCCCTTGAT	GACGTGGTTTACGA	CCCCATTTAGTAG TCAA	
dnaK P2	TTGGGCAGTTGAA	ACCAGACGTTTCG	CCCCTATTACAGACTCAC	
htpG P1	GCTCTCGCTTGAA	ATTATTCTCCCTTGT	CCCCATCTCTCCCACATC	
rpoD P_{hs}	TGCCACCCTTGAA	AAACTGTCGATGTGG	GACGATATAGCAG ATAA	
lon[b]	TCTCGGCGTTGAA	TGTGGGGGAAACAT	CCCCATATACTGACGTAC	
Consensus	T tC CcCTTGAA	13–15 bp	CCCCATtTA	
E. coli[c]	tcTTGACa	16–18 bp	TATAaT	˙cat

[a] Taken from references 22 and 97.
[b] The transcriptonal start site has not yet been determined.
[c] Consensus sequence of *E. coli* promoters (54).

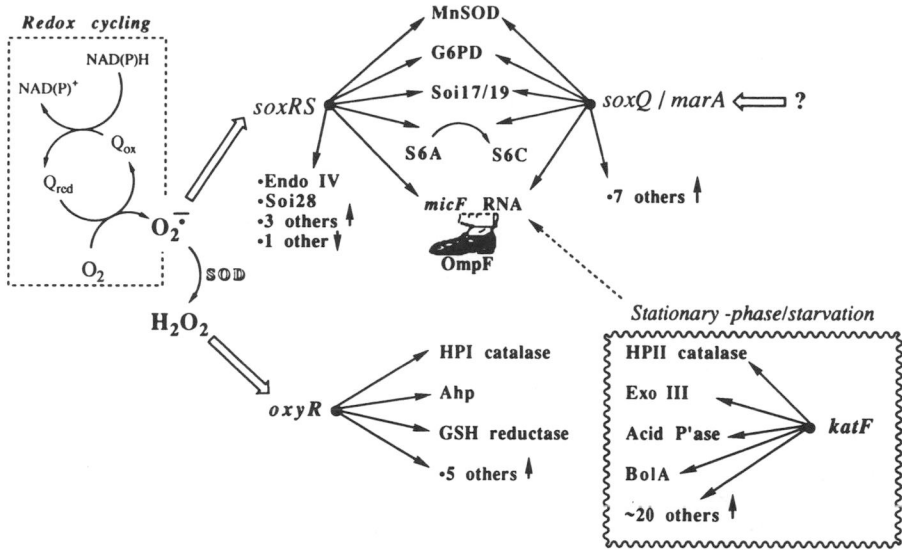

Figure 3 Regulons that defend against oxidative stress. Key agents implicated in triggering each regulon (where known) are shown. The open arrows do not necessarily imply direct interactions. In no case is the molecular activating event defined, although direct oxidation of OxyR protein is likely. Many functional overlaps are apparent for the various regulons: the large group of proteins controlled in common by *soxRS* and *soxQ/marA*; the two catalases controlled by *oxyR* and *katF*; two DNA repair enzymes, controlled by *soxRS* and *katF*; the multiregulation of OmpF via *micF* (see also Figure 4). Abbreviations: S6A and S6C, respectively, the unmodified and modified forms of ribosomal protein S6; Endo IV, endonuclease IV; Exo III, exonuclease III; Acid P'ase, acid phosphatase; GSH, glutathione; Q, a reduced or oxidized redox-cycling agent. (For other abbreviations, see Demple 1991.) Symbols: open arrows, activation of a control system; filled arrows, induction; small upward arrows, increases; small downward arrow, decrease; dashed arrow, induction suggested but not proven; foot over OmpF, repression.

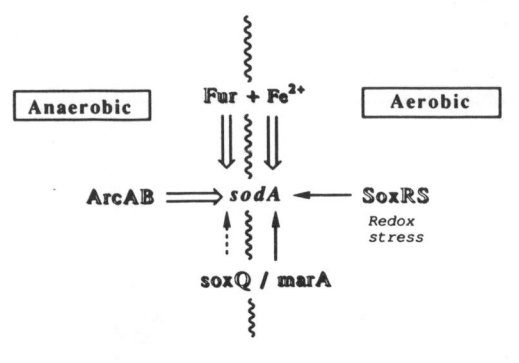

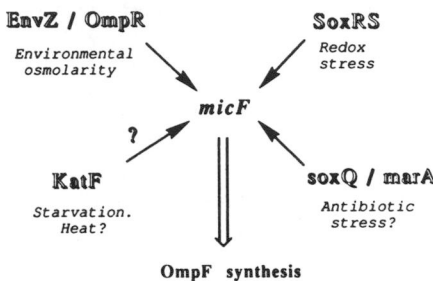

Figure 4 Examples of oxidative stress genes that are members of other global regulons. Not all of the controls on *sodA* operate simultaneously; for example, the ArcAB system is inactive under aerobic conditions. The hierarchics of control operating on either *sodA* or *micF* are not yet established. Symbols: filled arrows, positive control; open arrows, negative control; dashed arrow, possible control, but untested; wavy line, division between anaerobic and aerobic conditions.

C. Compilation of *E. coli* Ribosome Binding Sites

Contributed by

Kenneth E. Rudd[1] and Thomas D. Schneider[2]

[1]*National Center for Biotechnology Information, National Library of Medicine, National Institutes of Health, Bethesda, MD 20894;* [2]*National Cancer Institute, Frederick Cancer Research and Development Center, Laboratory of Mathematical Biology, Frederick, MD 21702-1201*

We present a compilation of 1055 *E. coli* genes that includes the ribosome binding sites and genomic map positions. The data are part of a dataset (EcoGene5) describing the chromosomal genes of *E. coli* which is derived from a nonredundant DNA sequence dataset (EcoSeq5). These DNA sequences, which comprise 38% of the *E. coli* genome, and their alignment to and integration into a genomic restriction map are described in Section 2A of this Handbook (Rudd, this volume), a mapping article that complements the compilation presented here.

The DNA sequences in Table 1 are aligned at their translational start codons, usually an ATG codon. We present DNA sequence, the primary repository of information content, but the actual binding site is of course RNA. Our dataset is composed entirely of *E. coli* chromosomal genes. We refer the reader to recent reviews by Gold (1988) and de Smit and van Duin (1990) for more comprehensive discussions of prokaryotic ribosome binding sites, particularly initiation control mechanisms. We have chosen to present the interval -21 to $+13$ on the basis of an initial information content analysis of the interval -100 to $+100$ (see below). Ribosome protection of mRNA from nuclease digestion indicated protection of the -21 to $+13$ region (Steitz 1969; Gold 1988). An earlier information analysis identified the same region, suggesting that bases out to $+13$ may play a role in recognition of this region by ribosomes (Schneider et al. 1986). A recent innovation, the toeprint (Gold 1988; Hartz et al. 1988), uses blockage of reverse transcriptase by mRNA-bound 30S particles and tRNA-fMet to identify the ribosome binding site. A toeprint on pure T4 gene-*32* mRNA (made by T7 RNA polymerase) stops reverse transcriptase at position $+15$ (Hartz et al. 1988). The correlation of information analysis and mRNA protection experiments is quite striking. Information analysis of the region downstream from the start codon in our new dataset is complicated by the fact that this is the coding region, and coding effects can be seen in the information curve for at least 100 bases downstream (Figure 1 inset). In addition, the amino-terminal residues may have a variety of constraints such as signals for export or amino peptidase recognition signals. Here, we present a preliminary information analysis of this large *E. coli* ribosome binding site dataset. We present a few observations about our new dataset, including a reevaluation of some points discussed by Gold (1988).

Information analysis of DNA sequences (Schneider et al. 1986) is considerably better than a consensus sequence identification approach. It does not round off the observed base frequencies to arbitrary values and so preserves data about which bases are favored at each base position. The dataset utilized in the present studies differs markedly from earlier datasets in that it is comprised entirely of *E. coli* chromosomal genes. Of the 124 start sites used by Stormo et al. (1982), only 41 were *E. coli* chromosomal genes. A similarly low proportion of *E. coli* genes were present among the 149 genes analyzed by Schneider et al. (1986). The preponderance of phage genes used in these studies may have complicated these studies, since these phages have been subjected to selective pressures that differ not only from phage to phage, but also from those exerted on *E. coli* chromosomal genes.

Figure 1 is a summary of our results using a sequence logo representation (Schneider and Stephens 1990) of the information content plot displayed in the Figure 1 inset. The sequence logo consists of stacks of letters. The height of the stack is proportional to the degree of conservation of the bases at that position. Conservation is measured in the information units of bits, where one bit is the choice between two equally likely possibilities. Since there are four bases in nucleic acids, the maximum possible conservation is a choice of one in four. This is the same as two choices of one in two, so the highest possible conservation is two bits, indicated by the vertical bar in Figure 1. For example, one could first ask "Is it a purine (G or A)?" and then ask "Does it form a weak base pair (A or T)?". The heights of the letters in the stacks are proportional to their frequencies at that position in the binding sites. The letters are sorted so that the most frequent one is on top.

Because information is an additive quantity (Pierce 1980), the total information (or conservation) of the sequences can be found by adding up the heights across the entire site. This value, defined as $R_{sequence}$, is 8.68 +/− 0.15 bits for this set of *E. coli* chromosomal ribosome binding sites. The previously measured value was 11.0 +/− 0.4 bits (Schneider et al. 1986), so the new dataset shows a significant drop. This may be because the old dataset was abnormally high due to the phage sequences.

As in an earlier information content analysis of *E. coli* ribosome binding sites (Schneider et al. 1986), we assume that one entire strand of the *E. coli* chromosome is transcribed and thus available for ribosome binding. However, the previous estimated length of the *E. coli* genome (3.9×10^6 bp) is supplanted by a refined estimate of 4,673,600 bp on the basis of a revised genomic restriction map partially (38%) derived from DNA sequence (Kohara et al. 1987; Rudd, Section 2A of this Handbook). Our dataset consists of 1218 sequenced *E. coli* genes, but the 164 open reading frames (orfs) in this dataset are excluded from Table 1 and the results depicted in Figure 1. A separate information content analysis of the orfs subset indicates that they are likely to be mostly real genes. The previous estimate of the total number of genes (= number of ribosome binding sites), 2574, is replaced by an extrapolation of the sequenced genes (1230 including 22 confidential gene sequences not used in this analysis) to yield an estimate of 3237 total *E. coli* protein coding genes. If we correct the earlier total gene estimate using the current chromosome length estimate, we obtain a value of 3082 genes, indicating that the earlier estimates of gene density (coding capacity) cited in Schneider et al. (1986) were quite accurate.

These numbers tell us the task facing the ribosome: It must select 3237 objects from a set of 4,673,600 possible ones. The number of binary selections required to do this, defined as $R_{frequency}$, is $= \log_2(4,673,600/3239) = 10.5$ bits. Unlike the previous dataset, this is more than $R_{sequence}$ by 1.82 +/− 0.15 bits. One explanation for this large discrepancy is that the ribosome scans only in one direction along the mRNA to find its sites.

Using a new technique (T.D. Schneider, unpubl.), we have been able to assign a numerical value that estimates the individual ribosome binding site information content. Table 2 gives these values for the 20 best and worst ribosome binding sites. It is possible that the "good" sites are highly efficient ribosome binding sites. The "poor" sites may need to be reexamined experimentally for their authenticity. Curiously, *cysX*, the only gene in EcoSeq5 that is on the complementary strand of a known gene (*cysE*), has a negative individual information value (Table 2). Despite the fact that *cysX* has been reported to be expressed (Tei et al. 1990), its unique position among *E. coli* chromosomal genes (except for the overlapping IS-element genes, which are not included in this compilation) needs to be confirmed.

Gold (1988) lists six elements within ribosome binding sites (RBS, also referred to as translational initiation regions, TIR), and we briefly address each element.

The Initiation Codon

The Table 1 dataset includes 71 genes with GTG start codons, 13 genes with TTG start codons (*betA, carA, cyaA, deoA, gshA, lacA, ndh, pnp, psd, rnd, rpsT, speD, spoT*), and 1 gene with an ATT start codon (*infC*). In addition to the genes listed in Table 1, we analyzed 162 orfs, which include 6 with GTG and 2 with TTG start codons. These alternative start codons code for fMet, not Val or Leu, and are not necessarily less efficient than ATG (Gold 1988). However, the sequence ATGTG found at the *lacI* GTG start may reduce in-frame translation efficiency due to competition from the out-of-frame ATG start (Gold 1988). Table 1 includes 5 additional genes with AT preceding the GTG start: *hdhA, kdpE, secF, tdcR,* and *ruvA*. The unusual ATT start of the *infC* gene encoding translation initiation factor IF3 is probably tolerated because of extensive homology between the 16S rRNA and the *infC* mRNA (Gold 1988). In some circumstances, AUA, AUU, and AUC can be used to initiate phage or foreign protein synthesis in *E. coli*, although not as the primary initiation codon (Kopke and Leggatt 1991; Romero and Garcia 1991).

The Shine-Dalgarno Region

The Shine-Dalgarno region (SD) consists of mRNA bases upstream of the start codon which are complementary to the nine nucleotides at the 3'-OH end of 16S rRNA. The evidence that the SD (a subset of TAAGGAGGT) is a crucial element of a ribosome binding site is very convincing (Gold 1988). The longer the region of rRNA-mRNA complementarity, the more frequently that gene is translated. We do not indicate SDs in Figure 1 but we encourage the readers to look for them. They are not always readily apparent and may be absent entirely in some cases. This could be due to other regions of the mRNA that are complementary to 16S rRNA (see below) or because a low level of expression is sufficient for that gene. It could also be a reflection of the fact that not all of the start sites in Table 1 have been experimentally confirmed.

SD/AUG Spacing

The spacing between the SD region and the start codon can vary from 5 to 13 bases with little effect on translation efficiency (Gold 1988). This affects the information content analysis since here (see below) and in earlier studies (Stormo et al. 1982; Schneider et al. 1986) the sequences are aligned at the start codon. In one earlier study (Schneider et al. 1986), an information analysis was performed with aligned SD regions. As the SD region increased, the ATG information content was lost. Adding the information content from aligned SD and separately aligned ATG gave 14 bits. This number is too high because conservation is effectively lost by the variable spacing, which is 3 bits. Interestingly, this gives an estimate of 11 bits, close to the number obtained without any special alignment. We have not yet tried to align the data in Table 1 at the SD since alignment by the SD is a difficult task.

Non-SD Sequences

This and earlier information analyses (Stormo et al. 1982; Schneider et al. 1986) indicate nonrandom nucleotide distributions at regions other than the SD and ATG positions. Various authors have suggested that regions upstream of and downstream from the start codon may be involved in complementary interactions with regions of the 16S rRNA molecule other than the 3'-OH known to interact with the SD region. Such non-SD sequences have not been demonstrated to pair with 16S rRNA and usually appear in only a small subset of ribosome binding sites, which makes us suspect that they are imaginary. The spacing between these presumptive sites and the start codon is quite variable, lessening our ability to analyze them with an information content analysis. Nonetheless, these reports are occurring with increasing frequency recently, and we point them out to the reader as they indicate ongoing research in the study of ribosome binding sites despite their generally speculative nature at present. We note that these sites are proposed to operate as translation enhancers but that they still require the presence of at least a minimally homologous SD sequence (see below for one exception). On the basis of a deletion analysis that indicated that the bases immediately upstream of the SD region of the *E. coli atpE* gene are required for maximal translation efficiency, McCarthy et al. (1985) suggested that there may be some sequence conservation in this region (but see below). This region had originally been proposed by Scherer et al. (1980) to be conserved, and this general concept is supported by this and the earlier information content studies. However, no specific base preferences could be reliably identified. Interestingly, Scherer et al. (1980) suggest (with little if any evidence) that some ribosomal protein genes may share some similarity both upstream of and downstream from the start codon. The subset of genes involved with the protein synthesizing system (PSS) of *E. coli* has also been suggested by Sprengart et al. (1990) to contain nonrandom sequences downstream from the start codon. These authors propose that these downstream sequences may base pair with bases 1471–1482 of the 16S rRNA molecule. This suggests that the PSS ribosome binding sites need to be analyzed apart from the rest of the *E. coli* ribosome binding sites. It is noteworthy that mutational analysis has implicated the bases immediately upstream of the *rplA* SD region in the translational coupling of this gene with the upstream *rplK* gene (Sor et al. 1987). A sequence immediately upstream of the phage T7 gene-*10* SD region has been shown by Olins and Rangwala (1989) to enhance translation when placed upstream of or downstream from the *E. coli lacZ* gene. These authors named this sequence (UUAACUUUA) Epsilon (enhancer of protein synthesis initiation). They suggest it is functionally complementary to bases 458–466 of 16S rRNA and indicate that the translation enhancer that McCarthy et al. (1985) identified upstream of the *atpE* start is also an Epsilon sequence. However, a recent analysis by McCarthy's group indicated that the absence of secondary structure was the only enhancer feature that could be identified by variation of the sequence immediately upstream of the SD region or downstream from the ATG (Schauder and McCarthy 1989). They suggest that the region immediately upstream of the *atpE* SD may act like an enhancer when put in front of other genes because its high AT content makes it less likely to base pair with most N-terminal coding regions. It has also been suggested that highly expressed *E. coli* genes contain an additional ribosome binding site, TP (translation-initiation promoting), upstream of the SD region (Thanaraj and Pandit 1989). This TP site (a subset of UGAUCC) has a complementary sequence immediately 5′ to the anti-SD sequence at the end of 16S rRNA (Thanaraj and Pandit 1989). Finally, the TIR of the *M. genitalium tuf* gene does not contain an SD sequence, but apparently, it can efficiently replace the *lacZ* and *cat* SD sequences in *E. coli* (Loechel et al. 1991). This sequence, UUAACAACAU, resembles the putative Epsilon sequence but is proposed to pair with nucleotides 1082–1093 of the *E. coli* 16S rRNA. Further experimental evidence seems necessary to confirm or eliminate the proposed roles of these translation enhancers and alternative SD regions, in particular the confirmation of their annealing to specific regions of 16S rRNA. However, it appears that the presence of oligo-U upstream of the SD region can serve as a binding site for the ribosomal protein S1 (Boni et al. 1991). These authors suggest that the absence of secondary structure upstream of the SD region (Schauder and McCarthy 1989) may facilitate S1 binding and that this S1-mRNA interaction is the basis of at least some of the translation enhancer effects that have been reported.

The Second Codon

The sequence logo in Figure 1 clearly shows that the five bases following the start codon (positions 3–7) constitute the third highest region of information content after the start codon and the SD region. Previous studies indicated that AAA and GCU were the most abundant second codons in the previously analyzed *E. coli* gene datasets (as reviewed by Gold 1988). Dunn and Studier (1983) reported that many T7 genes had GCU as the second codon and that these genes were highly translated. Although our full dataset (including the 162 orfs omitted from Table 1) excludes T7 genes, these two codons are still the most abundant second codons (124/1218 are AAA; 59/1218 are GCU). The ribosomal protein gene subset (52/1218 genes, 4.3%) is highly expressed and constitutes 13.6% (8/59) of the GCU second codon subset. In contrast, the AAA second codon subset contains an average percentage (4.0%) of ribosomal protein genes. This suggests that GCU in the second codon position may stimulate the translation of *E. coli* chromosomal genes just as it apparently does for phage T7 genes. The only triplets not present in the second codon position of our dataset are the three stop codons. The preponderance of A at the first position of the second codon (position 3 in Figure 1) may be influenced by the fact that a number of translation starts and stops overlap (i.e., ATGA), which may influence translational coupling (e.g., *ompR, envZ; lpxB, rnhB; motA, motB*). This arrangement, as well as the overlapping motif TGATG, e.g., *trpB trpA* (Das and Yanofsky 1989), and the presence of other closely spaced genes place many SD regions within the coding regions of upstream genes, putting additional constraints on the sequences at the *E. coli* ribosome binding sites.

mRNA Secondary Structure

Efficient translation generally requires an absence of secondary structures that would mask either the SD sequence or the start codon. Numerous translational regulatory schemes have been demonstrated to involve the unwinding of these secondary structures, often by translation of an upstream gene. On the other hand, as pointed out by Gold (1988), a stable hairpin situated between the SD sequence and the start codon could serve to shorten the SD–ATG distance to create a more optimal spacing. Recently, five different putative ATP-dependent RNA helicase genes belonging to the DEAD motif protein family have been discovered in *E. coli*: *srmB* (Nishi et al. 1988), *dbpA* (Iggo et al. 1990), *deaD* (Toone et al. 1991), *rhlE* (Kalman et al. 1991), and *rhlB* (Kalman et al. 1991). Overproduction of *srmB* suppresses an *rplX* (ribosomal protein L24) temperature-sensitive mutation (Nishi et al. 1988), and $rpsB_{ts}$ (ribosomal protein S2) mutations are suppressed by *deaD* overexpression (Toone et al. 1991). These two putative RNA helicases show specificity in that cross-suppression of the two ribosomal protein gene mutations was not observed (Toone et al. 1991). It is tempting to speculate that RNA helicases may unwind secondary structures in mRNA translation initiation regions, possibly in a sequence-dependent fashion. We eagerly await further genetic and biochemical studies on these putative *E. coli* RNA helicases. We are currently examining our ribosome binding site dataset for secondary structures that may be involved in the regulation of gene expression.

MAPPING *E. coli* GENES

Table 1 includes the genomic address and the DNA database LOCUS designation, which can be used to access additional information contained in the *E. coli* restriction map in Section 2A of this Handbook (Rudd, this volume) and in the literature citations for the DNA sequences presented here. The data in Section 2A are sorted by map position, whereas Table 1 is an alphabetical listing sorted by gene name. The genomic address in Table 1 permits the reader to correlate this data with that contained in Section 2A. The genomic addresses listed in Table 1 are presented in two different units: centisome (a unit equal to one percent of the chromosome length) and base pairs (bp) of DNA sequence.

We now describe how the chromosomal map positions in Table 1 were derived. The revised, integrated map (Ecoli5.map) is a composite of the Kohara genomic restriction map (Kohara et al. 1987) and restriction sites in the DNA sequences of EcoSeq5, which have been aligned to the Kohara map (Rudd, Section 2A of this Handbook; Rudd et al. 1991). The start codons of the protein-encoding genes (excluding orfs) in EcoGene5 are derived from GenBank and EMBL DNA sequence database entries as well as from the primary literature. Software recently developed by W. Miller of The Pennsylvania State University allows the features (including gene intervals) of individual GenBank entries (defined as local addresses) to be converted to global addresses on the E. coli integrated genomic restriction map. This new software allows one to easily derive (and update) global addresses for sequenced genes in base-pair coordinates. As with the E. coli genetic map minutes (Bachmann 1990; see Section 1 of this Handbook), the global base-pair addresses start and end at the thrABC operon. The global addresses of genes will be refined with each new EcoGene release until all of the E. coli genome is DNA sequenced. The centisome values are calculated by dividing the genomic base-pair address by 46,736, one percent of the length of the E. coli chromosome as approximated by the digital restriction map Ecoli5.map. The centisome value is usually close to the corresponding 1990 E. coli genetic map minute value (Bachmann 1990; see Section 1 of this Handbook).

Table 1 includes an EcoGene accession number for each gene, as well as a cross-referenced accession number from the SWISS-PROT protein sequence database curated by A. Bairoch (Bairoch and Boeckmann 1991). Starting with release 20, the SWISS-PROT database now includes EcoGene accession numbers for E. coli protein entries. The SWISS-PROT and GenBank entries listed in Table 1 both contain additional information about these genes, including references and gene products. GenBank entries with LOCUS names in Table 1 (usually starting with "ECO") can be retrieved from GenBank using the accession numbers in the genomic restriction map in Section 2A of this Handbook (Rudd, this volume) if the LOCUS name changes. The GenBank entries contained in EcoSeq5 are taken from GenBank Release 69 (9/15/91) and weekly updates to 12/1/91. Readers are also guided to the 1990 E. coli genetic map (Bachmann 1990; see Section 1 of this Handbook) for additional information. We used the gene designations from the 1990 genetic map where possible. For recently sequenced genes, names were taken from GenBank or from the primary literature. In a few cases, gene names were not available and provisional names are being used. For example, the gene for protease II was named tlp (trypsin-like protease) after its Salmonella typhimurium counterpart; ntrL is used for the ntr-like gene in ECONTRLA (M15328); hdh is used for the 7-α-hydroxysteroid dehydrogenase gene; and the cysT gene (see ECOCYS, M32101) that lies between cysW and cysP on the chromosome has been provisionally renamed cysU, since there is already a tRNA gene named cysT listed in the 1990 genetic map (Bachmann 1990; see Section 1 of this Handbook). Gene synonyms are not provided here, but they will be available in future releases of EcoGene. Partially sequenced genes are marked with an apostrophe. EcoGene data files and programs are available from one of us (K.E.R.) and are publicly available from the National Center for Biotechnology Information (NCBI) using the anonymous FTP computer file transfer protocol (ncbi.nlm.nih.gov, in the repository/Eco directory). The EcoGene data is also incorporated in a new Macintosh database application, GeneScape (Bouffard, Ostell, and Rudd, submitted to CABIOS).

REFERENCES

Bachmann, B.J. 1990. Linkage map of *Escherichia coli* K-12, edition 8. *Microbiol. Rev.* **54:** 130–197.

Bairoch, A. and B. Boeckmann. 1991. The SWISS-PROT protein sequence data bank. *Nucleic Acids Res.* (Suppl.) **19:** 2247–2249.

Boni, I.V., D.M. Isaeva, M.L. Musychenko, and N.V. Tzareva. 1991. Ribosome-messenger recognition: mRNA target sites for ribosomal protein S1. *Nucleic Acids Res.* **19:** 155–162.

Das, A. and C. Yanofsky. 1989. Restoration of a translational stop-start overlap reinstates translational coupling in a mutant *trpB'-trpA* gene pair of the *Escherichia coli* tryptophan operon. *Nucleic Acids Res.* **17:** 9333–9340.

de Smit, M.H. and J. van Duin. 1990. Control of prokaryotic translational initiation by mRNA secondary structure. *Prog. Nucleic Acid. Res. Mol. Biol.* **38:** 1–35.

Dunn, J.J. and F.W. Studier. 1983. Complete nucleotide sequence of bacteriophage T7 DNA and the locations of T7 genetic elements. *J. Mol. Biol.* **166:** 477–535.

Gold, L. 1988. Posttranscriptional regulatory mechanisms in *Escherichia coli*. *Annu. Rev. Biochem.* **57:** 199–233.

Hartz, D., D.S. McPheeters, R. Traut, and L. Gold. 1988. Extension inhibition analysis of translation initiation complexes. *Methods Enzymol.* **164:** 419–425.

Iggo, R., S. Picksley, J. Southgate, J. McPheat, and D.P. Lane. 1990. Identification of a putative RNA helicase in *E. coli*. *Nucleic Acids Res.* **18:** 5413–5417.

Kalman, M., H. Murphy, and M. Cashel. 1991. *rhlB*, a new *Escherichia coli* K-12 gene with an RNA helicase-like protein sequence motif, one of at least five such possible genes in a prokaryote. *New Biol.* **3:** 886–895.

Kohara, Y., K. Akiyama, and K. Isono. 1987. The physical map of the whole *E. coli* chromosome: Application of a new strategy for rapid analysis and sorting of a large genomic library. *Cell* **50:** 495–508.

Kopke, A.K. and P.A. Leggatt. 1991. Initiation of translation at an AUA codon for an archaebacterial protein gene expressed in *E. coli*. *Nucleic Acids Res.* **19:** 5169–5172.

Loechel, S., J.M. Inamine, and P. Hu. 1991. A novel translation initiation region from *Mycoplasma genitalium* that functions in *Escherichia coli*. *Nucleic Acids Res.* **19:** 6905–6911.

McCarthy, J.E., H.U. Schairer, and W. Sebald. 1985. Translational initiation frequency of *atp* genes from *Escherichia coli*: Identification of an intercistronic sequence that enhances translation. *EMBO J.* **4:** 519–526.

Nishi, K., F. Morel-Deville, J.W. Hershey, T. Leighton, and J. Schnier. 1988. An eIF-4A-like protein is a suppressor of an *Escherichia coli* mutant defective in 50S ribosomal subunit assembly. *Nature* **336:** 496–498.

Olins, P.O. and S.H. Rangwala. 1989. A novel sequence element derived from bacteriophage T7 mRNA acts as an enhancer of translation of the *lacZ* gene in *Escherichia coli*. *J. Biol. Chem.* **264:** 16973–16976.

Pierce, J.R. 1980. *An Introduction to Information Theory: Symbols, Signals and Noise*, Second Edition. Dover Publications, New York.

Romero, A. and P. Garcia. 1991. Initiation of translation at AUC, AUA and AUU codons in *Escherichia coli*. *FEMS. Microbiol. Lett.* **84:** 325–330.

Rudd, K.E., W. Miller, C. Werner, J. Ostell, C. Tolstoshev, and S.G. Satterfield. 1991. Mapping sequenced *E. coli* genes by computer: Software, strategies and examples. *Nucleic Acids Res.* **19:** 637–647.

Schauder, B. and J.E. McCarthy. 1989. The role of bases upstream of the Shine-Dalgarno region and in the coding sequence in the control of gene expression in *Escherichia coli*: Translation and stability of mRNAs in vivo. *Gene* **78:** 59–72.

Scherer, G.F., M.D. Walkinshaw, S. Arnott, and D.J. Morre. 1980. The ribosome binding sites recognized by *E. coli* ribosomes have regions with signal character in both the leader and protein coding segments. *Nucleic Acids Res.* **8:** 3895–3907.

Schneider, T.D. and R.M. Stephens. 1990. Sequence logos: A new way to display consensus sequences. *Nucleic Acids Res.* **18:** 6097–6100.

Schneider, T.D., G.D. Stormo, L. Gold, and A. Ehrenfeucht. 1986. Information content of binding sites on nucleotide sequences. *J. Mol. Biol.* **188:** 415–431.

Sor, F., M. Bolotin-Fukuhara, and M. Nomura. 1987. Mutational alterations of translational coupling in the L11 ribosomal protein operon of *Escherichia coli*. *J. Bacteriol.* **169:** 3495–3507.

Sprengart, M.L., H.P. Fatscher, and E. Fuchs. 1990. The initiation of translation in *E. coli*: Apparent base pairing between the 16S rRNA and downstream sequences of the mRNA. *Nucleic Acids Res.* **18:** 1719–1723.

Steitz, J.A. 1969. Polypeptide chain initiation: Nucleotide sequences of the three ribosomal binding sites in bacteriophage R17 RNA. *Nature* **224:** 957–964.

Stormo, G.D., T.D. Schneider, and L.M. Gold. 1982. Characterization of translational initiation sites in *E. coli*. *Nucleic Acids Res.* **10:** 2971–2996.

Tei, H., K. Murata, and A. Kimura. 1990. Structure and expression of *cysX*, the second gene in the *Escherichia coli* K-12 *cysE* locus. *Biochem. Biophys. Res. Commun.* **167:** 948–955.

Thanaraj, T.A. and M.W. Pandit. 1989. An additional ribosome-binding site on mRNA of highly expressed genes and a bifunctional site on the colicin fragment of 16S rRNA from *Escherichia coli*: Important determinants of the efficiency of translation-initiation. *Nucleic Acids Res.* **17:** 2973–2985.

Toone, W.M., K.E. Rudd, and J.D. Friesen. 1991. *deaD*, a new *Escherichia coli* gene encoding a presumed ATP-dependent RNA helicase, can suppress a mutation in *rpsB*, the gene encoding ribosomal protein S2. *J. Bacteriol.* **173:** 3291–3302.

Table 1   Compilation of *E. coli* Translation Initiation Regions

-21_to_-1	0_to_+18	Gene	Centisome	Kb	DNA_LOCUS	EG5/SP19_Acc
ATAACTATGGAGCATCTGCAC	ATGAAAACCCGTACACAAC	aceA	90.898	4247.3	ECACEB	EG10022/P05313
TCCATGACGAGGAGCTGCACG	ATGACTGAACAGGCAACAA	aceB	90.863	4245.7	ECACEB	EG10023/P08997
TAGATAGATAAGGAATAACCC	ATGTCAGAACGTTTCCCAA	aceE	2.644	123.5	ECOACE	EG10024/P06958
GGCGTAAGAGGTAAAAGAATA	ATGGCTATCGAAATCAAAG	aceF	2.701	126.2	ECOACE	EG10025/P06959
TGCTCCTGATGAGGGCGCTAA	ATGCCGCGTGGCCTGGAAT	aceK	90.930	4248.8	ECOIDHKPA	EG10026/P11071
GTATCAATTATAGGTACTTCC	ATGTCGAGTAAGTTAGTAC	ackA	51.998	2429.7	ECOACKA	EG10027/P15046
CACATCTGGAGAAACACCGCA	ATGTCAGTTATTGGTCGCA	act	20.521	958.9	ECOPFL	EG10028/P09374
TCCTTAACCAGGGAGCTGATT	ATGAAAAAAGCCACATGCT	ada	49.679	2321.3	ECOADA	EG10029/P06134
ATCCAGAAAAAGAGTGCGACC	ATGATTGATACCACCCTGC	add	36.618	1711.0	ECOADD	EG10030/P22333
TAACATTATCAGGAGAGCATT	ATGGCTGTTACTAATCGTC	adhE	27.906	1303.9	ECOADHEX	EG10031/P17547
CATTTTAAGGGGATTTTCGCA	ATGCGTATCATTCTGCTTG	adk	10.783	503.9	ECOAPTADK	EG10032/P05082
ATAACAAAGAGGTGCCAGGA	ATGAACAAAACGCTAATCG	agp	22.989	1074.2	ECOAGPA	EG10033/P19926
TATAAATCACAGGAGTCGCCC	ATGTCAGTACCCGTTCAAC	ald	32.058	1498.0	ald-eco	EG10035
AGACGTGAAACAGGAGTCATA	ATGAATTTTCATCATCTGG	aldH	29.349	1371.3	ECOALDHQ3	EG10036/P23883
GCGGATGAATAAGGAGATGCG	ATGTATACCCTGAACTGGC	alkA	46.191	2158.3	ECOALKA	EG11222/P04395
AGCTGAAAATGAGGAGAGGTA	ATGTTGGATCTGTTTGCCG	alkB	49.665	2320.7	ECOADAB	EG10037/P05050
AACACTTATCAAGGAACACAA	ATGCAAGCGGCAACTGTTG	alr'	91.911	4295.5	ECODNAB	EG10038
TGTATGGAACAGGAGACACAC	ATGAATAATAAGGGCTCCG	amn	44.248	2067.5	ECOAMN	EG10039/P15272
TTTGTATGGAAACCAGACCCT	ATGTTCAAAACGACGCTCT	ampC	94.370	4409.5	ECOAMPCFR	EG10040/P00811
CCACATATAAGGAGATCCTGC	ATGTTGTTAGAACAGGGGT	ampD	2.552	119.3	ECOAMPDE	EG10041/P13016
GCTGGTCAGCAAGGAGACAAC	ATGACGCTATTTACAACCT	ampE	2.564	119.8	ECOAMPDE	EG10042/P13017
TCACAGAGACGAGGTGGAGAA	ATGTTAGATCAAGTATGCC	amtA	95.617	4467.8	ECOAMTA	EG10043/P22255
TGATAAACAAGAGGACGGAAT	ATGAACAAGAATATGGCAG	anr	39.178	1830.6	ECANRG	EG10044
TACACGGGCGGGACCATCGGG	ATGCAGCGTTCCGAGCAGG	ansA	39.830	1861.1	ECOANSORA	EG10045/P18840
TCACGTAACTGGAGGAATGAA	ATGGAGTTTTTCAAAAAGA	ansB	66.743	3118.7	ECOANSBA	EG10046/P00805
GCCTTTGCAGGAGAGTTAACG	ATGATCAATTCGCCCCGAG	apaG	1.114	52.1	ECOAPAH	EG10047/P05636
ACACTCATTCATTAAAAGAAT	ATGGCGACATACCTTATTG	apaH	1.096	51.2	ECOAPAH	EG10048/P05637
ATGAAAAGCGGAAACATATCG	ATGAAAGCGATCTTAATCC	appA	22.444	1048.7	ECOAPPAA	EG10049/P07102
TAATTTTATGATAGGTGCAAG	ATGGATTATGTTTGCTCCG	appY	12.630	590.1	ECOAPPYAA	EG10050/P05052
GTAACAAGCAGGCATACACTT	ATGACCGCGACTGCACAGC	apt	10.660	498.1	ECOAPTADK	EG10051/P07672
CGACTCTATAAGGACACGATA	ATGACGATTTTTGATAATT	araA	1.443	67.4	ECOARAABD	EG10052/P08202
TTTTTTGGATGGAGTGAAACG	ATGGCGATTGCAATTGGCC	araB	1.475	68.9	ECOARAABD	EG10053/P08204
AATGGTGGGAGTATGAAAAGT	ATGGCTGAAGCGCAAAATG	araC	1.519	71.0	ECOARACK	EG10054/P03021
ATCCGGCACGAAGGAGTCAAC	ATGTTAGAAGATCTCAAAC	araD	1.422	66.4	ECOPOLBDA	EG10055/P08203
ACTCTCTGCTGGCAGGAAAAA	ATGGTTACTATCAATACGG	araE	64.179	2998.8	ECOARAEA	EG10056/P09830
ACACTAAAGCTGGAGAGAACC	ATGCACAAATTTACTAAAG	araF	42.780	1998.9	ECOARAFGH	EG10057/P02924
CCAGTGATTCACGGAGACGTT	ATGCAACAGTCTACCCCGT	araG	42.746	1997.3	ECOARAFGH	EG10058/P08531
TGTTGCCTGAGTAAGGAGAGT	ATGATGTCTTCTGTTTCTA	araH	42.725	1996.3	ECOARAFGH	EG10059/P08532
TCTGGTGGTTGTGATGGTGGT	ATGAAAAAAGTCATTTTAT	araJ	8.946	418.0	ECOARAJ	EG10060
TTGGCAATTTAGGTAGCAAAC	ATGCAGACCCCGCACATTC	arcA	99.973	4671.3	ECODYE	EG10061/P03026
TTGTCGTGAAGGAATTCCCTA	ATGAAGCAAATTCGTCTGC	arcB	72.091	3369.3	ECOARCB	EG10062/P22763
AATCATGCAAAGAGGTGTACC	GTGGTAAAGGAACGTAAAA	argA	63.507	2967.4	ECOARGA	EG10063/P08205
AGTCTCTTATTTAAGGGTGCA	ATGATGAATCCATTAATTA	argB	89.567	4185.1	ECOARGBCH	EG10064/P11445
ATAAGAAGGTGAATAGCCCCG	ATGTTGAATACGCTGATTG	argC	89.546	4184.1	ECOARGBCH	EG10065/P11446
GCGTAACAGGGTGATCATGAG	ATGGCAATTGAACAAACAG	argD	75.113	3509.7	ECOARGD	EG10066/P18335
CGTTCGCCATGCGAGGATAAA	ATGTCCGATTTATACAAAA	argF	6.356	297.0	ECOARGF	EG10067/P06960
AATTAAGCAGGGTGTTATTTT	ATGACGACGATTCTCAAGC	argG	71.431	3337.7	ECOARGGA	EG10068/P22767
TCAAAATAAGGAAACAGAGTT	ATGGCACTTTGGGGCGGGC	argH'	89.566	4186.0	ECOARGBCH	EG11223/P11447
CGTTAGCCACAGGAGGGATCT	ATGTCCGGGTTTTATCATA	argI	96.456	4507.0	ECOARGI	EG10069/P04391
AACCACCATATCGGGTGACTT	ATGCGAAGCTCGGCTAAGC	argR	72.849	3404.0	ECOARGR	EG10070/P15282
TCATCAATGTAAGGTATTCCG	GTGAATATTCAGGCTCTTC	argS	42.217	1972.6	ECOARGS	EG10071/P11875
ATTTTCACTTGAGGGTTATGT	ATGAAGAAGTCGATTCTCG	argT'	52.326	2445.5	ECOHISPUR2	EG10072/P09551
CTGTTGTAGAGAGTTGATTC	ATGGAATCCCTGACGTTAC	aroA	20.698	967.1	ECOAROA	EG10073/P07638
AATTAAGGTGGATGTCGCGTT	ATGGAGAGGATTGTCGTTA	aroB	75.736	3538.8	ECOAROB	EG10074/P07639
CACAACAATAACGGAGCCGTG	ATGGCTGGAAACACAATTG	aroC	52.744	2464.5	ECOAROCX	EG10075/P12008
CGCGTGCAGAAAGGGTAAAAA	ATGAAAACCGTAACTGTAA	aroD	38.188	1784.4	ECOAROD	EG10076/P05194
AGCATAAACAGGATCGCCATC	ATGCAAAAAGACGCGCTGA	aroF	58.922	2753.2	ECOPHEAB	EG10078/P00888
TATGGCAACACTGGAACAGAC	ATGAATTATCAGAACGACG	aroG	17.013	794.9	ECOAROG	EG10079/P00886
AACAAATTTCTGAGACTTGTA	ATGAACAGAACTGACGAAC	aroH	38.455	1796.8	ECOAROHA	EG10080/P00887
GTCTGACTCTCGCAATATCTT	ATGAGGTTTCAGTTCATGT	aroK	75.764	3540.1	ECOAROK	EG10081

Table 1   *Continued*

-21_to_-1	0_to_+18	Gene	Centisome	Kb	DNA_LOCUS	EG5/SP19_Acc
ACCTATTGGGGAAAACCCACG	ATGACACAACCTCTTTTTC	aroL	8.842	413.1	ECOAROLM	EG10082/P08329
GAAAAGTAAGGTGGATGAACA	ATGAGTGCGTCGTTGGCGA	aroM	8.864	414.2	ECOAROLM	EG10083/P08403
ACAACCACCGCACGAGGTTTC	ATGATGGAAGGTCAACAGC	aroP	2.583	120.7	ECOAROP	EG10084/P15993
CTTTGCAGGAAAAAAACGCTT	ATGAAAAATGTTGGTTTTA	asd	76.964	3596.2	ECOASD	EG10088/P00353
CAGGACGCAGGAGTATAAAAA	ATGAAAACCGCTTACATTG	asnA	84.632	3954.5	ECOORIASN	EG10091/P00963
ACCAGGTTAACGGAGAAGGTT	ATGTGTTCAATTTTTGGCG	asnB	15.147	707.8	ECOASNB	EG10092/P22106
CAATCCGCATAAGAAAATCCT	ATGGAAAATTATCTGATCG	asnC	84.618	3953.9	ECOORIASN	EG10093/P03809
ACTATTAACAGAGAGAATATT	ATGAGCGTTGTGCCTGTAG	asnS	21.279	994.3	ECOTGASNS	EG10094/P17242
AGCTTGAAAAAGAAGGTTCAC	ATGTCAAACAACATTCGTA	aspA	94.111	4397.4	aspAeco	EG10095/P04422
TAACCATAATGGAACCTCGTC	ATGTTTGAGAACATTACCG	aspC	21.213	991.2	ECOASPC	EG10096/P00509
AGCTAAGTTAAGGGATATCTC	ATGCGTACAGAATATTGTG	aspS	41.975	1961.3	ECOASPS	EG10097/P21889
CAGTCTTAAGGGGACTGGAGC	ATGCAACTGAATTCCACCG	atpA	84.442	3945.7	ECOUNCC	EG10098/P00822
CAACAAAGGGTAAAAGGCATC	ATGGCTTCAGAAAATATGA	atpB	84.505	3948.6	ECOUNCC	EG10099/P00855
ACGCCTTAATCGGAGGGTGAT	ATGGCAATGACTTACCACC	atpC	84.383	3942.9	ECOUNCC	EG10100/P00832
TTATTTCGTAGAGGATTTAAG	ATGGCTACTGGAAAGATTG	atpD	84.393	3943.3	ECOUNCC	EG10101/P00824
TGAAACAAACTGGAGACTGTC	ATGGAAAAACCTGAATATGG	atpE	84.499	3948.3	ECOUNCC	EG10102/P00844
ACTAAATAGAGGCATTGTGCT	GTGAATCTTAACGCAACAA	atpF	84.487	3947.8	ECOUNCC	EG10103/P00859
CAAGGCATTGAGGAGAAGCTC	ATGGCCGGCGCAAAAGAGA	atpG	84.423	3944.7	ECOUNCC	EG10104/P00837
ACTGTAAGGAGGGAGGGGCTG	ATGTCTGAATTTATTACGG	atpH	84.476	3947.2	ECOUNCC	EG10105/P00831
CATACCTCGAAGGGAGCAGGA	GTGAAAAACGTGATGTCTG	atpI	84.523	3949.4	ECOUNCC	EG10106/P03808
TTAAACCATAGCCAGAGGAAA	ATGGCTCTTCCGGTTTATG	avtA	80.546	3763.6	ECOAVT	EG10107/P09053
TGAACAGGATGGAGTTAAGTA	ATGAATCCACTGAAAGCCG	bcp	55.834	2608.9	ECOORF123	EG10108/P23480
TATTCTAACCAGGAGGTTTAT	TTGCAATTTGACTACATCA	betA	7.104	331.9	ECOBET	EG10109/P17444
CCGATTAACCGAGGAGACGTG	ATGTCCCGAATGGCAGAAC	betB	7.140	333.6	ECOBET	EG10110/P17445
CTGATGAATGGAGTGGCGAAA	ATGCCCAAATTGGGGATGC	betI	7.172	335.1	ECOBET	EG10111/P17446
CAATAACAGTGGGGATACTGG	ATGACAGACCTTTCACACA	betT	7.187	335.8	ECOBET	EG10112/P17447
TGGAAGCGAAGGAGTCAAAAA	ATGAAAGGTGATACTAAAG	bfr	74.635	3487.4	ECOBFR	EG10113/P11056
TAACGATAAAAGGAGTTAATT	ATGAAAGCATTTCCAGAAA	bglB	84.099	3929.6	ECOBGLO	EG10114/P11988
CAAAGCGGTAGAGGGCAAGTT	ATGACGGAGTTAGCCAGAA	bglF	84.130	3931.1	ECOBGLO	EG10115/P08722
TTTATAAAAAAGGTCCTTGCT	ATGAACATGCAAATCACCA	bglG	84.173	3933.1	ECOBGLO	EG10116/P11989
AATTTGGTTTACAAGTCGATT	ATGACAACGGACGATCTTG	bioA	17.489	817.2	ECOBIO	EG10117/P12995
AACACGTTTTGGAGAAGCCCC	ATGGCTCACCGCCCACGCT	bioB	17.519	818.6	ECOBIO	EG10118/P12996
ACCGTCTGCTGGAGGTGCTGC	ATGGCAACGGTTAATAAAC	bioC	17.565	820.8	ECOBIO	EG10119/P12999
TTTTTTTGGGAGTGATTGCTC	GTGAGTAAACGTTATTTTG	bioD	17.581	821.5	ECOBIO	EG10120/P13000
ATATTACAACGCGGCAGCATT	ATGAGCTGGCAGGAGAAAA	bioF	17.541	819.6	ECOBIO	EG10121/P12998
GTCAAAAGAGAACAATAGCGG	ATGAATAACATCTGGTGGC	bioH	76.322	3566.2	bioHeco	EG10122/P13001
GAGCGCAGTGGAGACAATTTC	ATGAAGGATAACACCGTGC	birA	89.935	4202.3	ECOBIRA	EG10123/P06709
CGTGGCGCGTTAGCCACAGGA	ATGGAAAACTCCTTGCAGA	bisC	79.960	3736.2	ECOBISCASD	EG10124/P20099
CGTTGTCGGAGGAGATATTTC	ATGATGATACGTGAGCGGA	bolA	9.875	461.4	ECOBOLA	EG10125/P15298
TTCTATTGTGGATGCTTTACA	ATGATTAAAAAAGCTTCCC	btuB	89.733	4192.9	ECOBTUB	EG10126/P06129
CACTTCAGCAGAACGGATACC	ATGCTGACACTTGCCCGCC	btuC	38.584	1802.9	ECOBTUCED	EG10127/P06609
TAAACTGGCGTTGGCAAAATA	ATGTCTATTGTGATGCAGT	btuD	38.555	1801.5	ECOBTUCED	EG10128/P06611
TGACCACACCCAGGAGAAACG	ATGCAAGATTCCATTCTGA	btuE	38.571	1802.3	ECOBTUCED	EG10129/P06610
CCGTAAACCAGGAATTTCCCA	ATGAGTGATGAACGCTACC	btuR	28.574	1335.2	ECOBTUR	EG10130/P13040
TTTTTACCTGGAGATATGACT	ATGAACGTTATTGCAATAT	cadA	93.900	4387.6	ECOCADABC	EG10131/P23892
AAATGAAATTAGGAGAAGAGC	ATGAGTTCTGCCAAGAAGA	cadB	93.948	4389.8	ECOCADAB	EG10132/P23891
CTTTTCGAATGAGTTTCTATT	ATGCAACAACCTGTAGTTC	cadC	93.984	4391.5	ECOCADAB	EG10133/P23890
GAATATTCTCTGGAGGGTGTT	TTGATTAAGTCAGCGCTAT	carA	0.642	30.0	ECOCARAB	EG10134/P00907
GTAATCAGGAGTAAAAGAGCC	ATGCCAAAACGTACAGATA	carB	0.667	31.1	ECOCARAB	EG10135/P00968
ATTTGACGAAGGGATGATGGC	GTGAAGAATTTATCTGGTCG	cca	68.952	3221.8	ECOCCA	EG10136/P06961
ACATGATTATGAGGCAACGCC	ATGCATCCACGTTTTCAAA	cdd	47.978	2241.8	cdd-eco	EG10137/P13652
TTGAGGTACAGGGAAAAAAAG	ATGAAAAAAGCGGGTCTTC	cdh	88.586	4139.3	ECOCDHA	EG10138/P06282
ATTGTAACGCTGGTGGTCTGC	ATGCTGGCAGCGTGGGAAT	cdsA	4.353	203.4	ECOCDS	EG10139/P06466
ATCGTCGATGAGGGCAGTTTT	ATGGAAAAGAAACACATTT	celA	39.164	1830.0	ECOCELA	EG10140/P17409
GGTTTTTAAGGGTATTTTTCT	ATGAGTAATGTTATTGCAT	celB	39.136	1828.7	ECOCELA	EG10141/P17334
AAGAAAAGAGAGGAACGATGT	ATGATGGATCTCGATAACA	celC	39.125	1828.2	ECOCELA	EG10142/P17335
AACTGAAGGCATAAGGAGTCG	ATGATGCAGCCAGTGATTA	celD	39.107	1827.3	ECOCELA	EG10143/P17410
TCTGTACTGAAGGGAGAAATT	ATGAGCCAGAAATTAAAAG	celF	39.081	1826.1	ECOCELA	EG10144/P17411
GCCATTGCAAAGGAGAAGACT	ATGTTGAAATCCCCCCTGT	cet	99.943	4669.9	ECOPHOM	EG10145/P08369
CAGCCGAACCGAGGTGACAGC	GTGAGCATGGATATAAGCC	cheA	42.511	1986.4	ECOCHEA	EG10146/P07363
TGCCGCTAAGTAAGGATTAACG	ATGAGCAAAATCAGGGTGT	cheB	42.384	1980.4	ECOCHE3	EG10147/P07330
TGAAGTGATTGAGAAGGCGCT	ATGACTTCATCTCTGCCCT	cheR	42.407	1981.5	ECOCHE3	EG10148/P07364
AATGAGTAAAAAGGTAACAAT	ATGACCGGTATGACGAATG	cheW	42.500	1985.8	ECOCHE1	EG10149/P07365
TATTTAAATCAGGAGTGTGAA	ATGGCGGATAAAGAACTTA	cheY	42.376	1980.0	ECOCHE3	EG10150/P06143

Table 1   *Continued*

-21_to_-1	0_to_+18	Gene	Centisome	Kb	DNA_LOCUS	EG5/SP19_Acc
TGGGCATGTGAGGATGCGACT	ATGATGCAACCATCAATCA	cheZ	42.362	1979.4	ECOCHE3	EG10151/P07366
TGAACGGGCGGGGCGCTAATC	ATGCTGGAACTGAATTTTT	chlD	17.280	807.4	ECOCHLJD	EG10152/P09833
TGCCTCCGCAGGAGTGTTTTC	ATGGAATTTACCACCGGAT	chlE	18.704	874.0	ECOCHLEN	EG10153/P12281
CGCGTTGTTCGGAGGCCTGTA	ATGGCGGAACTCAGCGATC	chlN	18.688	873.2	ECOCHLEN	EG10154/P12282
GTTACCTCATGGAGATATGGA	ATGTTTAGGTTGAACCCTT	cirA	48.280	2255.9	ECOCIR	EG10155/P17315
TAAAAAATTGGGGGAGGTGCCT	ATGCTCAATCAAGAACTGG	clpA	19.959	932.6	ECOCLPAA	EG10156/P15716
TGATCCGTTATGGGAGGAGTT	ATGCGTCTGGATCGTCTTA	clpB	58.691	2742.4	ECOPROT	EG10157/P03815
AATTTTATCCAGGAGACGGAA	ATGTCATACAGCGGCGAAC	clpP	9.922	463.6	ECOCLPPA	EG10158/P19245
GATGATAAAAGGATGTCCCTG	ATGATTAAGTTTAGCGCAA	cpdB	95.572	4465.7	ECOCPDB	EG10160/P08331
TGCCTGAAAAAGGGTAACGAT	ATGAAAAAATTAACCTGCT	cpsG	ND	ND	ECOPGMPMI	EG10162
CTATCTGATGGTTTCTGCTTC	ATGATAGGCAGCTTAACCG	cpxA	88.459	4133.4	ECOCPXA	EG10163/P08336
GCATCACAACAGGAGATAGCA	ATGACGTTACCGAGTGGAC	crl	5.691	265.9	ECOPHOE	EG11092
TATAACAGAGGATAACCGCGC	ATGGTGCTTGGCAAACCGC	crp	75.049	3506.8	ECOCRP	EG10164/P03020
TTTACTGCTTAGGAGAAGATC	ATGGGGTTTGTTCGATAAAC	crr	54.645	2553.4	ECOPHOSYS	EG10165/P08837
AATTATTAAAGGTAATACACT	ATGTCCGGTAAAATGACTG	cspA	80.093	3742.4	ECOCSPAA	EG10166/P15277
ACACGCACACGGATAACAACT	ATGAACAAATCAGGGAAAT	cstA	13.637	637.2	ECOCASTGE	EG10167/P15078
CGAAACTGGATAGATAACTAC	ATGGCTTTTGCCTCATTAA	cutE	14.966	699.3	ECOCUTE	EG10168
GCGTAAAATCGTGGGACACAT	ATGGTCTGGATTGATTACG	cvpA	52.400	2448.4	ECOHISPUR1	EG10169/P08550
TAGCAAATCAGGCGATACGTC	TTGTACCTCTATATTGAGA	cyaA	86.018	4019.3	ECOCYAG	EG10170/P00936
CTCATCGCCTTTCAGGTCAAA	ATGGTAGCCGCCCTGTTTG	cyaX'	86.057	4022.0	ECOCYAG	EG10171/P11291
AAAATTATGTTAAAGCAGGAA	ATGTTATGGAAAATAAATA	cybB	32.103	1500.0	ECOCYBB	EG10172/P08732
TCATGCGATGAGCAAGGAGTC	ATGATGTTAGATATAGTCG	cydA	16.734	781.9	ECOCYD	EG10173/P11026
TAAGACAGGAGTCGTCAAATG	ATGATCGATTATGAAGTAT	cydB	16.768	783.5	ECOCYD	EG10174/P11027
TAAACCATCAGGAGGTTCCACC	ATGATTCAGTCACAAATTA	cynS	7.839	366.3	M23219	EG10175/P00816
GGTCATTCCAACTGTGGCGCG	ATGACCGCCATTGCCAGCT	cynT	7.831	365.9	M23219	EG10176/P17582
AACCATCAAAAGACGTTCACG	ATGCTGCTGGTACTGGTGC	cynX	7.850	366.8	M23219	EG10177/P17583
CCGTGGAATTGAGGTCGTTAA	ATGAGACTCAGGAAATACA	cyoA	9.796	457.7	ECOCYOA	EG10178/P18400
AGGGGTTGAGGAAGAATAAAG	ATGTTCGGAAAATTATCAC	cyoB	9.753	455.7	ECOCYOA	EG10179/P18401
TTACTAAGGCAGGGCTGAAAA	ATGGCAACTGATACTTTGA	cyoC	9.740	455.1	ECOCYOA	EG10180/P18402
TTATCTGATGGGGGCGATGTA	ATGAGTCATTCTACCGATC	cyoD	9.733	454.8	ECOCYOA	EG10181/P18403
GATGCACTAAGAGCGGCGGTT	ATGATGTTTAAGCAATACC	cyoE	9.713	453.9	ECOCYOA	EG10182/P18404
GCGCACAGCAGGAGGAACATC	ATGAGCATTGAGATTGCCA	cysA	54.724	2557.0	cysPeco	EG10183/P16676
CGTCTAAGTGGATGGTTTAAC	ATGAAATTACAACAACTTC	cysB	28.712	1341.6	ECOCYSB	EG10184/P06613
CGATTTGCTGGGGGATAAATA	ATGGCGGCTGCATGACGAAA	cysC	61.889	2891.8	ECOCYSDNC	EG10185
CGGTATTGCAAAGGAACGGTT	ATGGATCAAATACGACTTA	cysD	61.932	2893.8	ECOCYSDNC	EG10186/P21156
TCTCATCGTGTGGAGTAAGCA	ATGTCGTGTGAAGAACTGG	cysE	81.441	3805.4	ECOCYSXE	EG10187/P05796
TAATTACTAAGGGGTTTTTAC	GTGGATCATTTGCCTATAT	cysG	75.304	3518.7	ECNIRBC	EG10188/P11098
AAGGCAAACAGTGAGGAATCT	ATGTCCAAACTCGATCTAA	cysH	62.212	2906.9	ECOCYSH	EG10189
TTATCAGCGAGATGTCTACTA	ATGAGCATGAGCGAAAAAC	cysI	62.229	2907.7	ECOCYSJIHA	EG10190
CTTACTGGAACATAACGACGC	ATGACGACACAGGTCCCAC	cysJ	62.266	2909.4	ECOCYSJIHA	EG10191/P14782
TCATACAGTTAAGGACAGGCC	ATGAGTAAGATTTTTGAAG	cysK	54.572	2549.9	ECOCYSK	EG10192/P11096
CACCACGTATGGATAGAGATC	GTGAGTACATTAGAACAAA	cysM	54.702	2556.0	cysPeco	EG10193/P16703
AACGTCAGGGGTATTTTTAAG	ATGAACACCGCACTTGCAC	cysN	61.902	2892.4	ECOCYSDNC	EG10194
TTAAATTTATAAGGGTGCGCA	ATGGCCGTTAACTTACTGA	cysP	54.784	2559.8	cysPeco	EG10195/P16700
CATGTCTAAACGGAATCTTCG	ATGCTAAAAATCTTCAATA	cysS	12.003	560.9	ECOCYSSG	EG10196/P21888
GTTAGCGGCGGGGCGTAACTG	ATGTTTGCTGTCTCCTCCA	cysU	54.766	2559.0	cysPeco	EG10197/P16701
TCGGCGTGTGGTAGGTCATTA	ATGGCGGAAGTTACCCAAT	cysW	54.748	2558.1	cysPeco	EG10198/P16702
CGGAACGCCAGCGGCGGTGGT	ATGCGGCGGCACCGGTTGC	cysX	81.444	3805.5	ECOCYSXE	EG10199/P20343
CGTGCCAGGCGAGGAGTGAGT	GTGAAAGCGAAGAAGCAGG	cytR	88.875	4152.8	ECOCYTR	EG10200/P06964
ACGGATGTCGTAGTTCAGACC	ATGAATACCATTTTTTCCG	dacA	14.391	672.4	ECODACA	EG10201/P04287
CCAGGTTGTTAGCGCGAGATT	ATGCGATTTTCCAGATTTA	dacB	71.656	3348.2	dacBeco	EG10202
ATGGATACTCGGGTGGCATTT	ATGACGCAATACTCCTCTC	dacC	19.044	889.8	ECODACC	EG10203/P08506
AGGTGTAATTAGTTAGTCAGC	ATGAAGAAAAATCGCGCTT	dam	75.686	3536.5	ECODAM	EG10204/P00475
TGTTTGCACAGAGGATGGCCC	ATGTTCACGGGAAGTATTG	dapA	55.800	2607.3	ECODAPA	EG10205/P05640
AATTAACAAAAGAGAATAGCT	ATGCATGATGCAAACATCC	dapB	0.614	28.7	ECODAPB	EG10206/P04036
TGATATGAAAAGAGTTTAACA	ATGCAGCAGTTACAGAACA	dapD	4.087	191.0	ECODAPD	EG10207/P03948
TTTTTCCATGAGGTGTAGTCT	ATGTCGTGCCCGGTTATTG	dapE	55.642	2599.9	ECODAPE	EG10208
GTACCCGCGTGATTGGAGTAA	ATGATGCAGTTCTCGAAAA	dapF	86.097	4023.0	ECODAPF	EG10209/P08885
AATGAGTTGGGGTTATTTAACC	ATGACGCCGGTGCAGGCCG	dbpA	30.349	1418.1	ECODEADA	EG10210/P21693
CTCAGATGGCCGGTGAAATCT	ATGCAGGAAAATATATCAG	dcm	43.747	2044.1	ECODCM	EG10211/P11876
CCTCGCGTTCAGGAGAAGAAA	ATGACAACAATGAATCCTT	dcp	29.026	1356.3	ECODCPG	EG10212
TAGACGGGATAGGTTTTTAAG	ATGGAAAAACTGCGGGTAG	ddlA	8.703	406.7	ECODDLA	EG10213
AAACTCCGGAGGAAGAACAAC	ATGACTGATAAAATCGCGG	ddlB	2.194	102.5	ECOFTSQA	EG10214/P07862
TGAAAATTGCTGATCAATTTC	ATGATGAGTTATGTAGACT	deaD	71.165	3325.3	ECODEAD	EG10215/P23304

Table 1 *Continued*

-21_to_-1	0_to_+18	Gene	Centisome	Kb	DNA_LOCUS	EG5/SP19_Acc
GAGTACAAGGCATAGGCAAAT	ATGGACCTGATTTATTTCC	dedA	52.482	2452.3	ECOHISPUR1	EG10216/P09548
GTTCAGACAGAAAGGTCCCTA	ATGAGCTGGATTGAACGAA	dedB	52.459	2451.2	ECOHISPUR1	EG10217/P08193
AGTAAGTTTCAGAATCGGTTA	GTGGGCACGATCGTGCTGG	dedD	52.416	2449.2	ECOHISPUR1	EG10218/P09549
TCTTTTCCTCGGGAGGTTACC	TTGTTTCTCGCACAAGAAA	deoA	99.515	4649.9	deoAeco	EG10219/P07650
AGCACTGAGTACGGAGAACAT	ATGAAACGTGCATTTATTA	deoB'	99.523	4651.3	deoAeco	EG10220/P07651
GCGACAAGCCAGGAGAATGAA	ATGACTGATCTGAAAGCAA	deoC	99.495	4649.0	ECODEOC	EG10221/P00882
TGTCACAAAAAGGATAAAACA	ATGGCTACCCCACACATTA	deoD	99.566	4652.3	deoDeco	EG10222/P09743
CAATAACTATTCAGAGGGATT	ATGGAAACACGTCGCGAAG	deoR	19.070	891.1	ECODEOR	EG10223/P06217
TAATGTAAAAAGGTTCTTTCA	ATGGCCAATAATACCACTG	dgkA	91.726	4286.0	ECOPLSB	EG10224/P00556
ATTTCTCACGCGGGAGAGGAC	ATGGCACAGATTGATTTCC	dgt	3.845	179.7	ECODGTP	EG10225/P15723
ACGATTTAGCATTAGCTAACT	ATGGAAACAAAAAATTTAA	dicA	35.446	1656.2	ECOICABC	EG10226/P06966
GTGAATTTTGTGATGTGGTGA	ATGCGGCTGAGCGCACGCG	dicB	35.479	1657.8	ECOICABC	EG10227/P09557
CTAACAATAAAGGTGTTTTAA	ATGCTTAAAACTGACGCTC	dicC	35.439	1655.9	ECOICABC	EG10228/P06965
TTTCAGGTATAAGGATACGTA	ATGATACAACCTATTTCCG	div'	52.553	2456.1	ECOPDXB	EG10229/P15286
TGCGTGTTAAGGAGAAGCAAC	ATGCAAGAAGGGCAAAACC	dksA	3.429	160.2	dksAeco	EG10230/P18274
CGCCACCACAAGGAGTGGAAA	ATGGTCTTCCATGACAACAA	dld	47.786	2232.8	ECODLD	EG10231/P06149
ACAGCGATCGGCGGCCTGGCA	ATGGCCAGCAGCGCATTAA	dmsA	20.329	949.9	ECODMS	EG10232/P18775
AAAAGGTGTAAGGAGTAACCG	ATGACAACCCAGTATGGAT	dmsB	20.380	952.3	ECODMS	EG10233/P18776
CAAACCCGAAGGAGGTGTGAG	ATGGGAAGTGGATGGCATG	dmsC	20.393	952.9	ECODMS	EG10234/P18777
TTTGTTCGAGTGGAGTCCGCC	GTGTCACTTTCGCTTTGGC	dnaA	83.610	3906.7	ECODNAAOP	EG10235/P03004
ATCTCGGTAACTCCATTCACT	ATGGCAGGAAATAAACCCT	dnaB	91.899	4294.1	ECODNAB	EG10236/P03005
ACCAGGATTCAGAGGGTAACG	ATGAAAAACGTTGGCGACC	dnaC	99.138	4632.3	ECODNATC	EG10237/P07905
TAAGTAAACCGGAATCTGAAG	ATGTCTGAACCACGTTTCG	dnaE	4.557	212.9	ECOLPXA	EG10238/P10443
AATTGCTAAAAATCGGGGCCT	ATGGCTGGACGAATCCCAC	dnaG	69.149	3231.1	ECORPSRPO	EG10239/P02923
TCTAGGGGCAATTTAAAAAAG	ATGGCTAAGCAAGATTATT	dnaJ	0.310	14.5	ECODNAJK	EG10240/P08622
ATATATAGTGGAGACGTTTAG	ATGGGGTAAAATAATTGGTA	dnaK	0.267	12.5	ECODNAK	EG10241/P04475
GAACATTGTCATCGTAAACCT	ATGAAATTTACCGTAGAAC	dnaN	83.586	3905.6	ECODNAAOP	EG10242/P00583
AGTCTGACATAAATGACCGCT	ATGAGCACTGCAATTACAC	dnaQ	5.234	244.6	ECORNHQ	EG10243/P03007
GTTCCATATTTGAGAAACAGT	ATGTCTTCCAGAGTTTTGA	dnaT	99.154	4633.1	ECODNATC	EG10244/P07904
CCAGCGTTTCAGAGCCTGCCA	ATGAGTTATCAGGTCTTAG	dnaX	10.675	498.8	ECOAPTADK	EG10245/P06710
GCCGCAGAAACGAAGCAGTAC	GTGCCTAAAATGCTGGCAT	dniR	5.154	240.8	ECODNIR	EG10246
GATGCTGGAAGCGCGTGGCTA	ATGGCAATATTAGGTTTAG	dpj	58.005	2710.3	ECORECOPDX	EG10247
CATCACAATTGGAGCAGAATA	ATGCGTATTTCCTTGAAAA	dppA	79.800	3728.7	ECODPPA	EG10248
CTAAGCGCAAAGAGACGTACT	ATGGAAAACGCTAAAATGA	dsdA	53.480	2498.9	ECODSDA	EG10249/P00926
GCGACAAACACGGCGATAAAC	ATGGCAGTGATAGGGTTGC	dsdC	53.453	2497.6	dsdCeco	EG10250/P11693
ACGAGATCGTGACCCGTTATG	ATGAAAAAAATCGACGTTA	dut	82.118	3837.1	ECODUTPYR	EG10251/P06968
TTCTCTCTACCGGGAGGCCCT	ATGAATCGCTGGGAAAACA	ebgA	69.405	3243.0	ECOEBGRA	EG10252/P06864
TTTGCACACGGAGAATAAGCA	ATGAGGATCATCGATAACT	ebgC	69.471	3246.1	ECOEBGRA	EG10253
GGGAAAGTAAAAAAGGTAAAC	ATGGCAACACTAAAAGACA	ebgR	69.380	3241.8	ECOEBGRA	EG10254/P06846
TAGTGACTTGAGGAAAACCTA	ATGTCCAAAATCGTAAAAA	eno'	62.614	2926.3	ECOPYRG	EG10258/P08324
ACTCTCCCGCGAGGTGAAATA	ATGGATTTCAGCGGTAAAA	entA	13.608	635.9	M24148	EG10259/P15047
AGCCTGAAGGAGAGAATCACG	ATGGCTATTCCAAAATTAC	entB	13.590	635.0	M24148	EG10260/P15048
TATCATTTTGTGGAGGATGAT	ATGGATACGTCACTGGCTG	entC	13.530	632.2	M24142	EG10261/P10377
CGTTTGTCATCAGTCTCGAAT	ATGGTCGATATGAAAACTA	entD	13.200	616.8	entDeco	EG10262/P19925
GGATTGCATTAAGGAGCGAGG	ATGAGCATTCCATTCACCC	entE	13.555	633.4	ECOENTB	EG10263/P10378
TACCCAGTTGCAGGAGGCACA	ATGAGCCAGCATTTACCTT	entF	13.300	621.5	ECOENTF	EG10264/P11454
GTAATTTGGCGAGATAATACG	ATGATCAAACAAAGGACAC	envA	2.286	106.8	ECOENVAA	EG10265/P07652
TTGGTTTTTTGAGGAATAGTA	ATGACGAAACATGCCAGGT	envC	ND	ND	envCeco	EG10266
ACGCAGGTTATCGAACAGAAT	ATGAACGGTATCGATAACC	envD	ND	ND	envCeco	EG10267
CTGATATCTGAGGAGTGCGAA	ATGCAATTGAGCAGCAGTG	envY	12.682	592.6	ECOENVY	EG10268/P10805
TGTACCGGACGGCTCTAAAGC	ATGAGGCGATTGCGCTTCT	envZ	76.112	3556.4	ECOOMPB	EG10269/P02933
GTTGAAAAAACTGGAGCTGGA	ATGACATCGATAAAAGTT	era	58.044	2712.1	ECOERA	EG10270/P06616
ATGAAGAAGGAAAGCAAAAAA	ATGAAGACCATTCTACCTG	eco	ND	ND	ECOECOA	EG10255
CTTGCGTCCTGGAGATACACA	GTGGGTAATAATTTAATGC	exbB	67.857	3170.7	ECOEXBBD	EG10271/P18783
TTACGCGCAGGATAATATCCG	ATGGCAATGCATCTTAACG	exbD	67.847	3170.2	ECOEXBBD	EG10272/P18784
AAAATAAGGCTTACAGAGAAC	ATGGTAGATAAACGCGAAT	fabA	21.904	1023.5	ECOFABAA	EG10273/P18391
TGCGACTTACAGAGGTATTGA	ATGAAACGTGCAGTGATTA	fabB	52.614	2458.5	ECOFABB	EG10274/P14926
CAAAGAGTACGGAACCCACTC	ATGGATATTCGTAAGGATTA	fabE	73.306	3425.3	ECOFABEA	EG10275/P02905
TCATCGAGTAACGAGGCGAAC	ATGCTGGATAAAATTGTTA	fabG	73.316	3425.8	ECOFABG	EG10276
CCGAAAAGTGACTGAGCGTAC	ATGTATACGAAGATTATTG	ſabH	ND	ND	ECOFABH	EG10277
AAAACGGCTTAAGGAGTCACA	ATGGAACAGGTTGTCATTG	fadA	86.823	4056.9	ECOFADAB	EG10278/P21151
TTGCGGATTCAGGAGACTGAC	ATGCTTTACAAAGGCGACA	fadB	86.849	4058.1	ECOFADAB	EG10279/P21177
TAAAAATAAATCATTGAGGTT	ATGGTCATGAGCCAGAAAA	fadL	53.064	2479.5	ECOFADLA	EG10280/P10384
CTGTGTTATGGAAATCTCACT	ATGGTCATTAAGGCGCAAA	fadR	ND	ND	fadReco	EG10281/P09371

Table 1 *Continued*

-21_to_-1	0_to_+18	Gene	Centisome	Kb	DNA_LOCUS	EG5/SP19_Acc
CGACGATACAGGACAAGAGAC	ATGTCTAAGATTTTTGATT	fba	66.084	3087.9	ECOFDAPGK	EG10282/P11604
AAACATTGCAGGGAAAGTTTT	ATGAAAACGTTAGGTGAAT	fbp	96.006	4486.0	ECOFBPASE	EG10283/P09200
GACAACGGAAAAGAAAGCTGA	ATGAGTATTCGCATAATCC	fdhE	87.964	4110.2	ECOFDHE	EG10284/P13024
TGATTAACTGGAGCGAGACCG	ATGAAAAAAGTCGTCACGG	fdhF	92.586	4326.2	fdhFeco	EG10285/P07658
CCCTGAAAAAAGAGGAAAGCA	ATGGACGTCAGTCGCAGAC	fdnG	33.306	1556.2	ECOFDNGHI	EG11227
GAAGGCGTAAGGGGCGAACAG	ATGGCTATGGAAACGCAGG	fdnH	33.371	1559.3	ECOFDNGHI	EG11228
ACGATGACGAGGAGGATCATC	ATGAGTAAGTCGAAAATGA	fdnI	33.390	1560.2	ECOFDNGHI	EG11229
AAAAATGATGATGGGGAAGGT	ATGACGCCGTTACGCGTTT	fecA	97.265	4544.8	fecAeco	EG10286/P13036
GCTTTTAGCTGGAATGTGATT	ATGTTGGCATTTATCCGTT	fecB	97.245	4543.9	M26397	EG10287/P15028
TCAGCCGCTTACCGTTGTGAA	ATGACCGCGATAAAACACC	fecC	97.223	4542.9	M26397	EG10288/P15030
CTGGCTTGTGAGGAGGCGAGG	ATGAAAATTGCGCTGGTTA	fecD	97.203	4541.9	M26397	EG10289/P15029
TTGCTTGTGAGAATGCGATAA	ATGACTTTACGAACTGAAA	fecE	97.187	4541.1	M26397	EG10290/P15031
TTAACTTTTGATGCACTCCGC	ATGTCTGACCGCGCCACTA	fecI	97.337	4548.2	ECOFECIR	EG10291/P23484
GTTCCGTCTGGAGTATGGGTT	ATGAATCCTTTGTTAACCG	fecR	97.317	4547.2	ECOFECIR	EG10292/P23485
CAAAAATGCAGGAATAAAACA	ATGAACAAGAAGATTCATT	fepA	13.217	617.6	ECOFEPAA	EG10293/P05825
TTAATAATAGGAAGCTGATTT	GTGAGACTCGCCCCCGCTCT	fepB	13.501	630.9	ECOFEPB	EG10294/P14609
AATTCAGGAGTCTCGCAAAAA	ATGACCGAATCAGTAGCCC	fepC	13.412	626.7	ECOFEPCDG	EG10295
TTCATTATCATGGAAGTTCGT	ATGTCTGGTTCTGTTGCCG	fepD	13.451	628.5	ECOFEPCDG	EG10296
CTCATTTTCAGGTTATACCCC	ATGTCATCACTGAATATTA	fepE	13.388	625.6	fepEeco	EG10297
ACGTAAAACGCGAGGTGGTGC	ATGATTTACGTCTCTCGCC	fepG	13.430	627.5	ECOFEPCDG	EG10298
TGGCAGCGTCTGAATGACGAA	ATGTTTGAGGTCACTTTCT	fes	13.272	620.1	ECOFE6	EG10299/P13039
TTCCACCCCAGGCGAGAGACA	ATGTTTGATAATTTAACCG	ffh	59.043	2758.9	ECOTRMD	EG10300/P07019
TGCTTTCCGTGGAAGAACAAA	ATGTCATATACACCGATGA	fhlA	61.470	2872.2	ECOFHLA	EG10301/P19323
CTTTACATCAGAGATATACCA	ATGGCGCGTTCCAAAACTG	fhuA	3.587	167.6	ECOFHUACD	EG10302/P06971
TAACGCCATCGGAGGTAAAGC	GTGAGTAAACGAATTGCGC	fhuB	3.673	171.6	ECOFHUB	EG10303/P06972
CCATTTCACAAGTTGGCTGTT	ATGCAGGAATACACGAATC	fhuC	3.637	169.9	ECOFHUACD	EG10304/P07821
ACCTGTGAGTTTTGTTTATTG	ATGAGCGGCTTACCTCTTA	fhuD	3.654	170.7	ECOFHUACD	EG10305/P07822
ACATAAACCAAGAGATTTCAA	ATGCTTTCAACACAATTTA	fhuE	25.026	1169.4	ECOFHUE1	EG10306/P16869
TTGAAGAGCTGAGGAGTCACT	ATGAGCGATAAATTCGGCG	fic	75.154	3511.6	ECOPABAA	EG10307/P20605
GTTTTTTGAAAGGAAAGCAGC	ATGAAAATTAAAACTCTGG	fimA	97.902	4574.6	ECOFIMA	EG10308/P04128
TGTAATTATAAGGGAAAAACG	ATGAAGAATAAGGCTGATA	fimB	97.856	4572.4	ECOFIMBE	EG10309/P04742
TCAGTTAGGAGTACTACTATT	GTGAGTAAACGTCGTTATC	fimE	97.879	4573.5	ECOFIMBE	EG10312/P04741
GAATGTCGTTAAGGGGCGTG	ATGAGAAACAAACCTTTTT	fimF	98.001	4579.2	ECOFIMFGH	EG10313/P08189
TATCAGTAACTGGAGATGCTC	ATGAAATGGTGCAAACGTG	fimG	98.012	4579.7	ECOFIMFGH	EG10314/P08190
GAACCCGGAGGGATGATTGTA	ATGAAACGAGTTATTACCC	fimH	98.024	4580.3	ECOFIMFGH	EG10315/P08191
CTGAAACAGGTTAAATAAGTA	ATGCCTTCAATTCGACTGG	firA	4.468	208.8	ECOFIRA	EG10316/P21645
AAATAAAGAGCTGACAGAACT	ATGTTCGAACAACGCGTAA	fis	73.423	3430.8	ECOFISA	EG10317/P11028
TTTCAAGAGGTTATTTCACTC	ATGGCTATCACTGGCATCT	fldA	15.439	721.4	ECOFLDA	EG10318/P23243
GCGCAAGAAAAGGGCCTGATC	ATGAGTGAAAAAAGCATTG	flhC	42.594	1990.3	ECOFLBA	EG10319/P11165
AAATAAAGTTGGTTATTCTGG	GTGGGAATAATGCATACCT	flhD	42.607	1990.8	ECOFLBA	EG10320/P11164
TGCAATATAGGATAACGAATC	ATGGCACAAGTCATTAATA	fliC	43.147	2016.1	ECOHAG	EG10321/P04949
GACACAGGAAGACCGCAACAC	ATGACTGATTACGCGATAA	fliL	43.510	2033.1	ECOFLAA	EG10322/P06973
CTTTTATTCTGCGATAACGAC	ATGGGCGATAGTATTCTTT	fliM	43.520	2033.5	ECOFLAA	EG10323/P06974
TCTGAACGAGGAACAGCCCAA	ATGAGTGACATGAATAATC	fliN	43.542	2034.5	M26294	EG10324/P15070
GCGCCGCCTGAGCCGTTAGTG	ATGAATAACCACGCTACTG	fliO'	43.541	2035.0	M26294	EG11224/P22586
AATTACGGCTTGAGCAGACCT	ATGATCCCGGAAAAGCGAA	fnr	30.126	1407.7	ECONIRR	EG10325/P03019
TTTTATATCCAGGAGAAATCA	ATGATTAGTCTGATTGCGG	folA	1.084	50.7	ECODHFOLG	EG10326/P00379
CGTAACGAATCAGCGGATACC	ATGATTATCAAACGCACTC	folC	52.430	2449.8	ECOHISPUR1	EG10327/P08192
ACAGATGGAATCCTCTCTCTG	ATGGCAGCAAAGATTATTG	folD	12.081	564.5	ECOFOLD	EG10328
CTGTTCATTCCTGGAGATGCT	ATGCCTGAATTACCCGAAG	fpg	82.040	3833.4	ECOFPG	EG10329/P05523
AAAACAATCTGGAGGAATGTC	GTGCAAACCTTTCAAGCCG	frdA	94.428	4412.2	ECOAMPCFR	EG10330/P00363
CCAATAAGAAGGAGAAGGCGA	ATGGCTGAGATGAAAAACC	frdB	94.413	4411.5	ECOAMPCFR	EG10331/P00364
AACCACGCTAAGGAGTGCAAC	ATGACGACTAAACGTAAAC	frdC	94.404	4411.1	ECOAMPCFR	EG10332/P03805
TGTACTGGTAAGGAGCCTGAG	ATGATTAATCCAAATCCAA	frdD	94.396	4410.7	ECOAMPCFR	EG10333/P03806
TCGATCCGACAGAGAAAGCGC	ATGACAACCTTAAGCTGTA	fre	86.800	4055.8	ECOFLRDA	EG10334/P23486
AAGTTTTCAAGGATTCGTAAC	GTGATTAGCGATATCAGAA	frr	4.276	199.8	ECORRFX	EG10335/P16174
ACTGACAGCAGGAGAGGCATA	ATGAAAACGCTGCTGATTA	fruA	48.604	2271.1	M23196	EG10336/P20966
TGGTCTTGGGGAGGGCGCATA	ATGAGCAGACGTGTTGCTA	fruK	48.641	2272.8	ECOFRUK	EG10337/P23539
AATGCGTTGAGAATGATCTCA	ATGCGCAATTTACAGCCCA	fruL	1.886	88.1	ECOFRURG	EG11273/P22183
CATTTTACGCAAGGGGCAATT	GTGAAACTGGATGAAATCG	fruR	1.890	88.3	ECOFRURG	EG10338/P21168
TCAGGCACAGGCAGAACAACA	ATGATCAAGGCGACGGACA	ftsA	2.231	104.2	ECOFTSQA	EG10339/P06137
TTTTGCCCGAGAGGATTAACA	ATGATTCGCTTTGAACATG	ftsE	77.567	3624.4	ECOFTSYEX	EG10340/P10115
AAATAAGGATAAACGCGACGC	ATGAAAGCAGCGGCGAAAA	ftsI	1.962	91.7	ECOPBPB	EG10341/P04286
TTCTGGAACTGGCGGACTAAT	ATGTCGCAGGCTGCTCTGA	ftsQ	2.213	103.4	ECOFTSQA	EG10342/P06136

Table 1 *Continued*

-21_to_-1	0_to_+18	Gene	Centisome	Kb	DNA_LOCUS	EG5/SP19_Acc
TGGCGGATTGAGCAATTTATC	ATGAAAAAACCGAATCATT	ftsS	77.617	3626.7	ECOFTSYEX	EG10343/P10120
TCTGGCGAAGGAGTTAGGTTG	ATGCGTTTATCTCTCCCTC	ftsW	2.112	98.7	ECOFTSW	EG10344/P16457
ACTTGCATGGAGGCGTGGGCC	ATGAATAAGCCCCATGCAA	ftsX	77.545	3623.3	ECOFTSYEX	EG10345/P10122
CAGCGAGGAGTGTAGTCGCAA	ATGGCGAAAGAAAAAAAAC	ftsY	77.581	3625.1	ECOFTSYEX	EG10346/P10121
GGCACAAATCGGAGAGAAACT	ATGTTTGAACCAATGGAAC	ftsZ	2.259	105.6	ECOFTSQAB	EG10347/P06138
TGAATCAGAGAGAGGTAGCTA	ATGGAACGAAATAAACTTG	fucA	63.173	2951.8	ECOFUCOSE	EG10348/P11550
GAATAAAGTGAGGAATCTGTA	ATGAAAAAAATCAGCTTAC	fucI	63.227	2954.4	ECOFUCOSE	EG10349/P11552
GAGTGGTTAGCCGGATAAGCA	ATGTTATCCGGCTATATTG	fucK	63.267	2956.2	ECOFUCOSE	EG10350/P11553
CGTAAAGCAACAAGGAGAAGG	ATGATGGCTAACAGAATGA	fucO	63.148	2950.6	ECOFUCOSE	EG10351/P11549
TTCTGAACCTAAGAGGATGCT	ATGGGAAACACATCAATAC	fucP	63.198	2953.0	ECOFUCOSE	EG10352/P11551
TCGAATGATGGGGTGAAAAAT	ATGAAAGCGGCACGCCAGC	fucR	63.308	2958.1	ECOFUCOSE	EG10353/P11554
ATCGGCAGTTTGAGGTTGAAA	ATGGTTTCACGGCACCAGC	fucT'	63.315	2959.1	ECOFUCOSE	EG10354/P11556
AATTTATAGAGGAAGTGTGAA	ATGCTGAAAACAATTTCGC	fucU	63.298	2957.7	ECOFUCOSE	EG10355/P11555
CAGGCAGTAAGTGAGAGAACA	ATGTCAAACAAACCCTTTC	fumA	36.284	1695.4	ECOFUMA	EG10356/P00923
CAGAGTTACAGGCTGGAAGCT	ATGTCAAACAAACCCTTTA	fumB	93.652	4376.0	ECOFUMB	EG10357/P14407
TAATCAGGTGAGGAGCAGGTC	ATGAATACAGTACGCAGCG	fumC	36.251	1693.9	ECOFUMC	EG10358/P05042
TAGTAACAGGACAGATTCCGC	ATGACTGATAACAATACCG	fur	15.423	720.7	ECOFUR	EG10359/P06975
AAAAGATAAACGAGGAAACAA	ATGGCTCGTACAACACCCA	fusA	74.738	3492.2	ECOSRTA	EG10360/P02996
GTCTTTAACTGGAGAATGCGA	ATGAACAGCAATAAAGAGT	gabT	60.161	2811.1	ECOGABT	EG10361/P22256
CACACTTTTCGCATCTTTGTT	ATGCTATGGTTATTTCATA	galE	17.121	800.0	galEeco	EG10362/P09147
CGCGAATCCGGAGTGTAAGAA	ATCAGTCTGAAAGAAAAAA	galK	17.074	797.8	galEeco	EG10363/P06976
TTTACCCACTAAGGTATTTTC	ATGGCGACCATAAAGGATG	galR	64.090	2994.7	ECOGALLYS	EG10364/P03024
TATCCCGATTAAGGAACGACC	ATGACGCAATTTAATCCCG	galT	17.099	799.0	galEeco	EG10366/P09148
AACAAATAGCTGGTGGAATAT	ATGACTATCAAAGTAGGTA	gapA	40.094	1873.4	ECOGAP	EG10367/P06977
GGAAAACCTTGCAGGAGATCT	ATGACCGTACGCGTAGCGA	gapB	66.138	3090.4	ECOFDAPGK	EG10368/P11603
GAATTGAAATGGTGTCTCTTT	ATGGCAATTAACAATACAG	gcd	2.999	140.1	ECOGCD	EG10369/P15877
ATATATAAGGGTTTTATATCT	ATGGATCAGACATATTCTC	gdhA	39.655	1852.9	ECOGDHAK	EG10372/P00370
AAGGTCAAGGAGAAGAGAGCA	ATGAAGCAGCATAAGGCGA	gef	0.365	17.1	gef-eco	EG10373/P22982
TAAAAAACAAGGAAGGCTAAT	ATGCTAGTTGTAGAACTCA	genA'	94.062	4396.1	aspAeco	EG11225/P04539
CTTAATCTCTGGAGAATAACG	ATGATAAAACCGACGTTTT	ggt	77.203	3607.4	M28722	EG10374/P18956
GCGTTATGCGAGGCAATCACC	ATGTTTTATCCGGATCCTT	gidA	84.559	3951.1	ECOUNCC	EG10375/P17112
CATCAAGAACAGGTAATCACC	GTGCTCAACAAACTCTCCT	gidB	84.544	3950.4	ECOUNCC	EG10376/P17113
AGGGCATAAACAGGAGCGATA	ATGCAGGTTTTACATGTAT	glgA	76.810	3589.0	ECOGLGA	EG10377/P08323
CAATAAAACAGGAAGACAAGC	ATGTCCGATCGTATCGATA	glgB	76.911	3593.7	ECOGLGBA	EG10378/P07762
ATGATAAAAAAGGAGTTAGTC	ATGGTTAGTTTAGAGAAGA	glgC	76.841	3590.5	ECOGLG	EG10379/P00584
TAGTTTTCAGGAAACGCCTAT	ATGAATGCTCCGTTTACAT	glgP	76.758	3586.6	ECOGLGPA	EG10380/P13031
CTGGCTGGTTCGGGAGGCAGA	ATGACACAACTCGCCATTG	glgX	76.879	3592.2	ECOGLG	EG10381/P15067
TATAAATCGGAATCAAAAACT	ATGTGTGGAATTGTTGGCG	glmS	84.304	3939.2	ECOUNCC	EG10382/P17169
CCAATCCAGGAGAGTTAAAGT	ATGTCCGCTGAACACGTAC	glnA	87.447	4086.0	ECOGLN	EG10383/P06711
GCGACCTTTTCAAGGAATAGC	ATGAAAAAGATTGATGCGA	glnB	57.711	2696.6	hmp-eco	EG10384/P05826
CAGGAAATAAAGGTGACGTTT	ATGCAACGAGGGATAGTCT	glnG	87.388	4083.3	ECOGLN	EG10385/P06713
GGATAGAAAAAAGGAAATGCT	ATGAAGTCTGTATTAAAAG	glnH	18.322	856.1	ECOGLNHPQ	EG10386/P10344
ACCTGTTCAGGAGACTGCTTT	ATGGCAACAGGCACGCAGC	glnL	87.418	4084.7	ECOGLN	EG10387/P06712
CCACGGTAACAGGAACAACAT	ATGCAGTTTGACTGGAGTG	glnP	18.305	855.3	ECOGLNHPQ	EG10388/P10345
GGAAAGAAGGATGAAAATCCT	GTGATTGAATTTAAAAACG	glnQ	18.290	854.6	ECOGLNHPQ	EG10389/P10346
TTACGCTTTGAGGAATCCACG	ATGAGTGAGGCAGAAGCCC	glnS	15.330	716.3	ECOGLNSA	EG10390/P00962
ACGAACTTCAGAGGGATAACA	ATGAAAACTCGCGACTCGC	glpA	50.661	2367.2	ECOGLPA	EG10391/P13032
GTCTGGAGAAGGAGCAGAAAG	ATGCGCTTTGATACTGTCA	glpB	50.696	2368.8	ECOGLPA	EG10392/P13033
CCAACGCGCAGGAGGCCAACA	ATGAATGACACCAGCTTCG	glpC	50.723	2370.1	ECOGLPA	EG10393/P13034
ACGAAAGTGAATGAGGGCAGC	ATGGAAACCAAAGATCTGA	glpD	76.712	3584.5	ECOSNGLPD	EG10394/P13035
TTTGTAAAGAAAGAGAGACGC	ATGGATCAGTTCGAATGTA	glpE	76.700	3583.9	ECOGLPREG	EG10395/P09390
TTAACTCTTCAGGATCCGATT	ATGAGTCAAACATCAACCT	glpF	88.741	4146.5	ECOGLPF	EG10396/P11244
GTGGAATAAGCGACAGCAACG	ATGTTGATGATTACCTCTT	glpG	76.683	3583.1	ECOGLPREG	EG10397/P09391
ATGACTACGGGACAATTAAAC	ATGACTGAAAAAAAATATA	glpK	88.708	4145.0	ECOGLYK	EG10398/P08859
AACGCAACGGAGGCTAATGGC	ATGAAATTGACGCTGAAAA	glpQ	50.603	2364.5	ECOGLPQ	EG10399/P09394
AAGTGGCGGAGCAAATCCCCA	ATGGCTCGACGCTGTTTAT	glpR	76.657	3581.9	ECOGLPREG	EG10400/P09392
CACGGGCCACGGAGGCTATCA	ATGTTGAGTATTTTTAAAC	glpT	50.626	2365.6	ECOGLPT	EG10401/P08194
AAGGCGCTAAGGAGACCTTAA	ATGGCTGATACAAAAGCAA	gltA	16.338	763.4	ECOGLTA	EG10402/P00891
GGGCGAATCGGAGGCGCGCGT	ATGACACGCAAACCCGTC	gltB	72.195	3373.4	ECOGLTB	EG10403/P09831
GCGCAGTAAGGGGTAGCAACA	ATGAGTCAGAATGTTTATC	gltD	72.292	3377.9	ECOGLTB	EG10404/P09832
TCCATTGAGGAAGTCATTCAT	ATGAAAAATATAAATTTCA	gltP	92.533	4323.7	ECOGACAR	EG10405/P21345
TGCGGATACAAAGGAGTAACT	ATGTTTCATCTCGATACTT	gltS	82.411	3850.7	ECOGLTS	EG10406/P19933
TTCTAAACGTAAGGCCATTTC	ATGAAAAATCAAAACTCGCT	gltX	54.299	2537.2	ECOGLTX	EG10407/P04805
TAGCTGAGTCAGGAGATGCGG	ATGTTAAAGCGTGAAATGA	glyA	57.650	2693.8	ECOGLYA	EG10408/P00477

Table 1   *Continued*

-21_to_-1	0_to_+18	Gene	Centisome	Kb	DNA_LOCUS	EG5/SP19_Acc
ACGTATCCAGCACGAAATAAT	ATGCAAAAGTTTGATACCA	glyQ	80.185	3746.7	ECOGLYS	EG10409/P00960
AAAGATAAGTAAGAGGCGGCT	ATGCAAGAGAAAACTTTTC	glyS	80.141	3744.7	ECOGLYS	EG10410/P00961
CACACCTGACAGGAGTATGTA	ATGTCCAAGCAACAGATCG	gnd	45.181	2111.1	ECOGND	EG10411/P00350
CAACGATAAGGACACTTTGTC	ATGACTAAACACTATGATT	gor	78.527	3669.2	ECOGOR	EG10412/P06715
AAACAGGCTGGGACACTTCAC	ATGAGCGAAAAATACATCG	gpt	5.651	264.1	ECOGPTA	EG10414/P00501
AATATTCAAGAGGTATAACAA	ATGCAAGCTATTCCGATGA	greA	71.640	3347.5	greAeco	EG10415/P21346
GAAAAAAACGCGGAGAAATTC	ATGAGTAGTAAAGAACAGA	grpE	59.111	2762.0	ECOGRPE	EG10416/P09372
AAATCTATGAGGAGAGAAATA	ATGCAAACCGTTATTTTTG	grx	18.836	880.1	ECOGRX	EG10417/P00277
TTTGACAGGCGGGAGGTCAAT	TTGATCCCGGACGTATCAC	gshA	60.639	2833.4	ECOGSHI	EG10418/P06980
TTGGGCTAACGGAGAAGAATA	ATGATCAAGCTCGGCATCG	gshB	66.575	3110.8	ECOGSHII	EG10419/P04425
TAACCACCCGTAAAAACAACC	ATGAAATTTCCCGGTAAAC	gsk'	10.844	506.8	hemHeco	EG11102/P22937
CTTGCCTCGGAATAAGCGTCA	ATGACGGAAAACATTCATA	guaA	56.476	2638.9	ECOGUABA	EG10420/P04079
ACTCTGGTCGAGATATTGCCC	ATGCTACGTATCGCTAAAG	guaB	56.511	2640.6	ECOGUABA	EG10421/P06981
ATTAACCCCAGGAATCCGCAC	ATGCGTATTGAAGAAGATC	guaC	2.434	113.7	ECOGUAC	EG10422/P15344
CAGTAGAGGGATAGCGGTTAG	ATGAGCGACCTTGCGGAGAG	gyrA	50.311	2350.8	ECOGYRAAM	EG10423/P09097
ATAAATGAGCGAGAAACGTTG	ATGTCGAATTCTTATGACT	gyrB	83.511	3902.1	ECORECFA	EG10424/P06982
TGAAGTTCACAGGAGGTTTAT	GTGTTTAATTCTGACAACC	hdhA	36.512	1706.1	hdhAeco	EG10425
TTCGTGTGGATCTACAGAGTC	ATGGAACTGAAAGCGACAA	helD	22.083	1031.9	ECOHELIV	EG10426/P15038
TTGGTATTATTTCCCGCAGAC	ATGACCCTTTTAGCACTCG	hemA	27.201	1271.0	ECOHEMA	EG10427/P13580
CTACCCTGCTTCAGGTATACT	ATGCCCCTCGATTCCACAA	hemB	8.475	396.0	M24488	EG10428/P15002
ATGATAATGACGGTAACAAGC	ATGTTAGACAATGTTTTAA	hemC	85.989	4017.9	ECOHEMCD	EG10429/P06983
CTATAACGGAGACGCCCCGGC	ATGAGTATCCTTGTCACCC	hemD	85.974	4017.2	ECOHEMCD	EG10430/P09126
TTGAATCGATAAGAGGCGGTA	ATGCGTCAGACTAAAACCG	hemH	10.802	504.7	hemHeco	EG10431
TCTACGAATCAGGAACCCTCC	ATGAGGAAGTCTGAAAATC	hemL	3.718	173.7	ECOGSAG	EG10432/P23893
CTCTCATAACAGGAAGCCATA	ATGACGGAACAAGAAAAAA	hemX	85.948	4016.0	ECOHEMCD	EG10433/P09127
CGCGCCGCAAGGAGAATAACC	ATGCTAAAAGTGTTATTGC	hemY	85.922	4014.8	ECOHEMCD	EG10434/P09128
CCAGCGTCAGGGGGAATAACG	ATGCGTAAGTCAGTTATCG	hflC	94.905	4434.5	hflXeco	EG10435
TAACAAATATGGAGCACAAAC	ATGGCGTGGAATCAGCCCG	hflK	94.878	4433.3	hflXeco	EG10436
CTATCCAAGCAAGACAAAGAT	ATGGAAGACCTCCAGGAGT	hflX	94.851	4432.0	hflXeco	EG10437
AGCGTGTGAGGAAAAGAGAGA	ATGGCTAAGGGGCAATCTT	hfq	94.841	4431.5	hfq-eco	EG10438
ATTACAACCATAGGTAGAAGT	ATGTCCGAAAAACCTTTAA	hha	10.419	486.8	ECHHAE	EG10439
CATCATTGAGGGATTGAACCT	ATGGCGCTTACAAAAGCTG	himA	38.607	1804.0	thrSeco	EG10440/P06984
CTTTAAGGAACCGGAGGAATC	ATGACCAAGTCAGAATTGA	himD	20.800	971.9	ECOHIP	EG10441/P08756
AACTTGCTGTGGACGTATGAC	ATGATGAGCTTTCAGAAGA	hipA	34.274	1601.5	hipAeco	EG10442
ACAGCAAAATCTGGAGTGGTA	ATGCCTAAACTTGTCACTT	hipB	34.246	1600.2	hipAeco	EG10443
GAAAAACTTCCTGGAGATGTG	ATGATTATTCCGGCATTAG	hisA	45.096	2107.2	ECOHISOPA	EG10444/P10371
CTTACGTGCGGAGCAAGTTTG	ATGAGTCAGAAGTATCTTT	hisB	45.060	2105.5	ECOHISOPA	EG10445/P06987
TAACGCCCTCAAGGAGCAAGC	ATGAGCACCGTGACTATTA	hisC	45.038	2104.4	ECOHISOPA	EG10446/P06986
GAAGATGATGGAGTGATCGCC	ATGAGCTTTAACACAATCA	hisD	45.010	2103.1	ECOHISOPA	EG10447/P06988
CACCGTGAAGGAGGCCATCGC	ATGCTGGCAAAACGCATAA	hisF	45.111	2107.9	ECOHISOPA	EG10448/P10373
GGTTTAAAGAGGAATAACAAA	ATGACAGACAACACTCGTT	hisG	44.990	2102.2	ECOHISOPA	EG10449/P10366
CTCGTCGAAAGGAGTGCTGTA	ATGAACGTGGTGATCCTTG	hisH	45.083	2106.6	ECOHISOPA	EG10450/P10375
ACAGGGCGTGGAGATCAGGAT	ATGTTAACAGAACAACAAC	hisI	45.128	2108.6	ECOHISOPA	EG10451/P06989
TAAACATTCACAGAGACTTTT	ATGACACGCGTTCAATTTA	hisL	44.986	2102.0	ECOHISOPA	EG11269/P03058
CTTCAACGCACTGAGAATACG	ATGTCCGAGAATAAATTAA	hisP	52.219	2440.0	ECOHISMP	EG10452/P07109
ATATTCAGAAAGAGAATAAAC	GTGGCAAAAAACATTCAAG	hisS	56.657	2647.3	hisSeco	EG10453/P04804
GGTGCAGGAGTATCTGTACTA	ATGTCCGACCAGCAACAAC	hisT	52.497	2453.0	ECOHIST1	EG10454/P07649
TGGGATGGTAAGGAGTTTATT	GTGAAAAAGTGGTTATTAG	hlpA	4.457	208.3	ompHeco	EG10455/P11457
ATGCAAAAAAAGGAAGACCAT	ATGCTTGACGCTCAAACCA	hmp	57.684	2695.3	hmp-eco	EG10456
TATAAGTTTGAGATTACTACA	ATGAGCGAAGCACTTAAAA	hns	27.842	1300.9	ECOHNS	EG10457/P08936
GCTTAATTTACGGAACTCACA	ATGAACAATAACGATCTGG	hsdM	98.731	4613.3	ECOHSDRM	EG10458/P08957
GGTACAGAAAGCTAAGATGAG	ATGTTATGGGCCTTAAATA	hsdR	98.775	4615.4	ECOHSDRM	EG10459/P08956
AGCGTTTGGTGGGGTGAAGGA	ATGAGTGCGGGGAAATTGC	hsdS	98.701	4611.9	ECOHSDSK	EG10460/P05719
GCATTATTGAGGTAGACCTAC	ATGAAAGGACAAGAAACTC	htpG	10.739	501.8	ECOAPTADK	EG10461/P10413
AACTGAGGTAAAAAGAAAATT	ATGATGCGAATCGCGCTCT	htpX	41.164	1923.4	ECOHTPX	EG10462/P23894
TGTTAATCGAGACTGAAATAC	ATGAAAAAAACCACATTAG	htrA	3.880	181.3	ECOHTRA	EG10463/P09376
AGGTTGAAAAACAGGATTGAT	ATGACGAATCTACCCAAGT	htrB	24.083	1125.3	ECOHTRB	EG10464
TTTTAGTGAGGTTTTTTTACC	ATGAATCAGACGCTACTTT	htrP	68.552	3203.2	ECOLUXH	EG10465
AACACATTGTAAGGATAACTT	ATGAACAAGACTCAACTGA	hupA	90.529	4230.1	hupAeco	EG10466/P02342
ATTATAAAGAGGAAGAGAAGA	GTGAATAAATCTCAATTGA	hupB	10.024	468.4	ECOHUPB	EG10467/P02341
CAAGGAGGAGAGACGTGCGAT	ATGAATAACGAGGAAACAT	hyaA	22.256	1039.9	ECOHYA	EG10468/P19928
AGGCAATGAGGATAAACAGGC	ATGAGCACTCAGTACGAAA	hyaB	22.280	1041.0	ECOHYA	EG10469/P19927
TAACAGCGAAGGAGAATCATC	ATGCAACAGAAAAGCGACA	hyaC	22.319	1042.9	ECOHYA	EG10470/P19929
AATAAGTAACAAGGAGCGTTC	ATGAGCGAGCAACGCGTGG	hyaD	22.334	1043.6	ECOHYA	EG10471/P19930

Table 1   *Continued*

-21_to_-1	0_to_+18	Gene	Centisome	Kb	DNA_LOCUS	EG5/SP19_Acc
GATGACACAGGAGGAGCAGGG	ATGAGCAACGACACGCCAT	hyaE	22.346	1044.1	ECOHYA	EG10472/P19931
ACCGCAGCAGGAGCGTGCCTC	ATGAGCGAAACTTTTTTCC	hyaF	22.355	1044.5	ECOHYA	EG10473/P19932
TGTTAAACGGGTAACCTGACA	ATGACTATTTGGGAAATAA	hycA	61.376	2867.9	ECOHEVOP	EG10474/P16427
CAATCACCTGAGGAATGCCTG	GTGAATCGTTTTGTAATTG	hycB	61.361	2867.1	ECOHEVOP	EG10475/P16428
GGCTCAAAGTGGAGAGGCTAA	ATGAGCGCAATTTCCCTGA	hycC	61.322	2865.3	ECOHEVOP	EG10476/P16429
TGTTTCACGAGGAGCCTGAGA	ATGAGTGTTTTATATCCGT	hycD	61.302	2864.4	ECOHEVOP	EG10477/P16430
GTGATTAAAGAGAGTTTGAGC	ATGTCTGAAGAAAAATTAG	hycE	61.265	2862.7	ECOHEVOP	EG10478/P16431
CCGCTGAAATAAGGAATCGCC	ATGTTTACCTTTATCAAAA	hycF	61.253	2862.1	ECOHEVOP	EG10479/P16432
CCATATGAAAGAGGCCATCTG	ATGACAATTTATTAGGCC	hycG	61.237	2861.3	ECOHEVOP	EG10480/P16433
TGTTGAAGAGGCGCGTATCCG	ATGAGTGAAAAGGTGGTGT	hycH	61.228	2860.9	ECOHEVOP	EG10481/P16434
TACGCGTAAGGACCCACAAGG	ATGACGCACGATAATATCG	hydG	90.603	4233.6	hydGeco	EG10482/P14375
CAATAGTTAATGGGAGGCGAT	ATGCACGAAATAACCCTCT	hypA	61.391	2868.5	ECOHYP	EG10483
GAAATAGACCAGGAGTGAGCG	ATGTGTACAACATGCGGTT	hypB	61.398	2868.9	ECOHYP	EG10484
GAACTGGCTGGAGACACAGCG	ATGTGCATAGGCGTTCCCG	hypC	61.417	2869.8	ECOHYP	EG10485
GTTGTATGGCGAGGAAAAATA	ATGCGTTTTGTTGATGAAT	hypD	61.423	2870.0	ECOHYP	EG10486
CACGGTAGCGGCGGCCAGGCG	ATGCAGCAATTAATCAACA	hypE	61.448	2871.2	ECOHYP	EG10487
CATGGTATTTAAGGACTCACT	ATGTTTTCCGCATTGCGCC	iap	61.957	2895.0	ECOIAP	EG10488/P10423
TAGCGCTCGAAGGAGAGGTGA	ATGGAAAGTAAAGTAGTTG	icdE	25.685	1200.2	ECOICLR	EG10489/P08200
TACCAGGAGCAGACAACAGCA	ATGAAACGCCCGGACTACA	iciA	65.877	3078.2	ECOICIA	EG10490
ATACAGAAAAAGGAGACTGTC	ATGGTCGCACCCATTCCCG	iclR	91.018	4252.9	ileSeco	EG10491/P16528
AAATACGGAACCGAGAATCTG	ATGAGTGACTATAAATCAA	ileS	0.486	22.7	ECOILVGMED	EG10492/P00956
ATCGAAACTGGGGGGTTAATA	ATGGCTGACTCGCAACCCC	ilvA	85.247	3983.2	ECOILVBPR	EG10493/P04968
CCAGAAGAAAAGGACTGGAGC	ATGGCAAGTTCGGGCACAA	ilvB	82.937	3875.3	ECOILVGE	EG10494/P08142
ACAACATCACGAGGAATCACC	ATGGCTAACTACTTCAATA	ilvC	85.303	3985.9	ECOILVGMED	EG10495/P05793
GACACTGGGAGTAAATAAAGT	ATGCCTAAGTACCGTTCCG	ilvD	85.207	3981.4	ECOILVGE	EG10496/P05791
GAGCGCAAAAGGAATATAAAA	ATGACCACGAAGAAAGCTG	ilvE	85.186	3980.4	ECOILVGE	EG10497/P00510
ATTCAGGACGGGGAACTAACT	ATGAATGGCGCACAGTGGG	ilvG	85.145	3978.5	ilvIeco	EG10498/P00892
CAAAACGGAGAGAACCTGATT	ATGCGCCGGATATTATCAG	ilvH	1.875	87.6	ilvIeco	EG10499/P00894
CGTCAAACAGTGAGGCAGGCC	ATGGAGATGTTGTCTGGAG	ilvI	1.838	85.9	ECOILVGE	EG10500/P00893
CAAGATGCAAGAAAAGACAAA	ATGACAGCCCTTCTACGAG	ilvL	85.140	3978.2	ECOILVGE	EG11270/P03060
AATGTTGGAGAAATTATCATG	ATGCAACATCAGGTCAATG	ilvM	85.180	3980.1	ECOILVBPR	EG10501/P13048
GAAATGGTGGGGGAATAAGCC	ATGCAAAACACAACTCATG	ilvN	82.930	3875.0	ECOILVGE	EG10502/P08143
TTGCGAGATTGAATTCACTAT	ATGACAGGAAATTTATTGC	ilvY	85.281	3984.8	ECOINFSERW	EG10503/P05827
AGTGATACCCCAGAGGATTAG	ATGGCCAAAGAAGACAATA	infA	20.022	935.6	nusAeco	EG10504/P02998
ACTGTAGCAGGAAGGAACAGC	ATGACAGATGTAACGATTA	infB	71.318	3332.4	nusAeco	EG10505/P02995
AAACAATTGGAGGAATAAGGT	ATTAAAGGCGGAAAACGAG	infC	38.702	1808.8	thrSeco	EG10506/P02999
TGGCATAACAGAGGAAAGAAA	ATGTCACTCTTCCGCAGAA	int	12.225	571.2	ECOINTDLP	EG10507
TTTTACACCGGACAATGAGTA	ATGGACTTTCCGCAGCAAC	ispA	9.576	447.5	ECOISPA	EG10508/P22939
ACAAAACCAACAAGGTCCCCA	ATGACTACTTCCATGCTCA	ivbL	82.975	3877.1	ECOILVBPR	EG11275/P03061
TAAATTAAGGAGACGAGTTCA	ATGTCGCAACATAACGAAA	katE	38.998	1822.2	ECOKATE	EG10509/P21179
GCGGAACCAGGCTTTGCTTGA	ATGTTCCGTCAAGGGATCA	katF	61.740	2884.9	ECOKATF	EG10510/P13445
ACTGTAGAGGGGACGCACATTG	ATGAGCACGTCAGACGATA	katG	89.094	4163.0	ECOKATGA	EG10511/P13029
TGCCCTGTCTGGAGAATCGCA	ATGCGTGGAGAATTTTATC	kbl	81.648	3815.1	ECOKBLTDH	EG10512/P07912
GATCAATGCGGAGGCGTTCTG	ATGGCTGCGCAAGGGTTCT	kdpA	15.787	737.7	ECOKDPABC	EG10513/P03959
GATATTGAGTGAGCACTGAAT	ATGAGTCGTAAACAACTGG	kdpB	15.743	735.6	ECOKDPABC	EG10514/P03960
CGGTCTGGTGTGAGGTTACC	ATGAGTGGATTACGTCCGG	kdpC	15.731	735.0	ECOKDPABC	EG10515/P03961
ATCTGGCGCTGGATAAACTTG	ATGAATAACGAACCCTTAC	kdpD	15.673	732.4	ECOKDPDE	EG10516/P21865
TGAAGAATTTCATGAGGATAT	GTGACAAACGTTCTGATTG	kdpE	15.659	731.7	ECOKDPDE	EG10517/P21866
ACTCAAGATTAAGGCGATCCT	ATGAAACAAAAAGTGGTTA	kdsA	27.301	1275.7	ECOKDSA	EG10518/P17579
GACTGCTGATGAGAGTAAATC	ATGAGTTTTGTGGTCATTA	kdsB	20.930	978.0	ECOKDSB	EG10519/P04951
TCCATAAGTCAGCTATTTACT	ATGCTCGAATTGCTTTACA	kdtA	82.001	3831.6	ECOKDTA	EG10520/P23282
TGGAATGGCAGGAGGCCCATC	ATGGATAGCCATACGCTGA	kefC	1.040	48.6	ECOKEFC	EG10521/P03819
TACGGCAGGAGACATAATGGC	ATGGCTGAAAACTACTGTAA	kgtP	58.538	2735.2	kgtPeco	EG10522/P17448
CAAAATGATTGTTAACACCCA	ATGAATAATCGAGTCCACC	ksgA	1.123	52.5	ECOAPAH	EG10523/P06992
TATCATAACGGAGTGATCGCA	TTGAACATGCCAATGACCG	lacA	7.876	368.0	ECOLAC	EG10524/P07464
AGTCAATTCAGGGTGGTGAAT	GTGAAACCAGTAACGTTAT	lacI	7.987	373.2	ECOLAC	EG10525/P03023
TTTCGCGTAAGGAAATCCATT	ATGTACTATTTAAAAAACA	lacY	7.891	368.7	ECOLAC	EG10526/P02920
AATTTCACACAGGAAACAGCT	ATGACCATGATTACGGATT	lacZ	7.919	370.0	ECOLAC	EG10527/P00722
AGCAATGACTCAGGAGATAGA	ATGATGATTACTCTGCGCA	lamB	91.536	4277.1	malGeco	EG10528/P02943
AAGGCACAATAATCATACTTT	ATGAAGAATATACGTAACT	lepA	58.105	2715.0	ECOLEP	EG10529/P07682
AAATAACCCTTAGGAGTTGGC	ATGGCGAATATGTTTGCCC	lepB	58.084	2714.0	ECOLEP	EG10530/P00803
AAAGAGACAAGGACCCAAACC	ATGAGCCAGCAAGTCATTA	leuA'	1.792	83.7	ECOLEUA	EG11226/P09151
CATTCATCTGGAGCTGATTTA	ATGACTCACATCGTTCGCT	leuL	1.795	83.9	ECOLEUA	EG11280/P09149
AGTGTGACAGTGGAGTTAAGT	ATGCCAGAGGTACAAACAG	leuO	1.811	84.6	ECOLEUO	EG10531/P10151

Table 1 *Continued*

-21_to_-1	0_to_+18	Gene	Centisome	Kb	DNA_LOCUS	EG5/SP19_Acc
AAAACAGGACCACTGGCTGCC	ATGCAAGAGCAATACCGCC	leuS	14.606	682.5	ECOLEUS	EG10532/P07813
TATATACACCCAGGGGGCGGA	ATGAAAGCGTTAACGGCCA	lexA	91.736	4286.5	ECOLEXA	EG10533/P03033
TTTTAGCATTGATGGTGCGAT	ATGGAATCAATCGAACAAC	lig	54.497	2546.4	ECOLIG	EG10534/P15042
CAATGTAGAGGTTAACGAAAA	ATGCGCTCACCAATTTGTC	lit	25.762	1203.7	ECOLIT	EG10535/P11072
CCTATTTAGGTGAGGCATAAG	ATGGAAAAAGTCATGTTGT	livF	77.367	3615.0	ECOLIVHMGF	EG10536/P22731
CGCAGCGAAAGGAGAGCAGGC	ATGAGTCAGCCATTATTAT	livG	77.382	3615.8	ECOLIVHMGF	EG10537/P22730
CGGGTTTAGAAAGGTTACCTT	ATGTCTGAGCAGTTTTTGT	livH	77.426	3617.8	ECOLIVHMGF	EG10538/P08340
AGAAGAATGGGGATTCTCAGG	ATGAACACAAAGGGCAAAG	livJ	77.492	3620.9	ECOLIVHMGF	EG10539/P02917
CACGAATGGGGATTTTTGACT	ATGAAACGGAATGCGAAAA	livK	77.447	3618.8	ECOLIVHMGF	EG10540/P04816
TCGCCCGGAGGTAGAGAAAGT	ATGAAACCGATGCATATTG	livM	77.399	3616.5	ECOLIVHMGF	EG10541/P22729
ATTAAACTAAGAGAGAGCTCT	ATGAATCCTGAGCGTTCTG	lon	9.969	465.8	ECOLON	EG10542/P08177
GATAAATATATAGAGGTCATG	ATGAGTACTGAAATCAAAA	lpd	2.749	128.4	ECOACE	EG10543/P00391
TCAATCTAGAGGGTATTAATA	ATGAAAGCTACTAAACTGG	lpp	37.810	1766.7	ECOLPP	EG10544/P02937
TCGTAGCCGGAGGCGTGATAC	GTGATTGATAAATCCGCCT	lpxA	4.502	210.4	ECOLPXA	EG10545/P10440
AACGCGCGGTCTGATTCGTTA	ATGACTGAACAGCGTCCAT	lpxB	4.519	211.1	ECOLPXA	EG10546/P10441
GAATACAGAGAGACAATAATA	ATGGTAGATAGCAAGAAGC	lrp	20.144	941.2	lrp-eco	EG10547/P19494
TGAAAAACGTAAGTTTGCCTG	ATGAGTCAATCGATCTGTT	lspA	0.546	25.5	ECOLSP	EG10548/P00804
AGCACTTATCTGGAGTTTGTT	ATGCCACATTCACTGTTCA	lysA	64.112	2995.7	ECOGALLYS	EG10549/P00861
CTCCCCTGACACGAGGTAGTT	ATGTCTGAAATTGTTGTCT	lysC	91.200	4261.4	ECOLYSC	EG10550/P08660
CCGGATTGCGAGAGAGCGCTA	ATGGCCGCCGTTAACTTAC	lysR	64.142	2997.1	ECOGALLYS	EG10551/P03030
GCAGGGTTATGAGGAACCAAC	ATGTCTGAACAACACGCAC	lysS	65.306	3051.5	ECOHERC	EG10552/P13030
GCTGGATTTAGAGGAACCAAA	ATGTCTGAACAAGAAACAC	lysU	93.812	4383.5	ECOLYSU	EG10553/P14825
CACCAACAAGGACCATAGATT	ATGAAAATAAAAACAGGTG	malE	91.478	4274.4	malGeco	EG10554/P02928
CCTGGAATGAGGAAGAACCCT	ATGGATGTCATTAAAAAGA	malF	91.441	4272.7	malGeco	EG10555/P02916
TGATTAAGGGAGATAACAAAA	ATGGCAATGGTCCAACCGA	malG	91.422	4271.8	ECOMALG	EG10556/P07622
ATCAGGGAGTAGGTCATCTGC	ATGGCTACCGCCAAAAAAA	malI	36.532	1707.0	ECOMALAA	EG10557/P18811
TTGTTACAAAGGGAGAAGGGC	ATGGCGAGCGTACAGCTGC	malK	91.511	4275.9	malGeco	EG10558/P02914
CCAGATTTTGAGGTGAAAACA	ATGAAAATGAATAAAAGTC	malM	91.570	4278.7	ECOMALM	EG10559/P03841
TCTAAGAAAAGTGGAACTCCT	ATGTCACAACCTATTTTTA	malP	76.452	3572.3	ECOMALP	EG10560/P00490
GCAAAACGCTAAGGAAGCTCG	ATGGAAAGCAAACGTCTGG	malQ	76.407	3570.2	ECOMALQP	EG10561/P15977
TCCACAGTGAAGTGATTAACT	ATGCTGATTCCGTCAAAAC	malT	76.516	3575.3	ECOMALT	EG10562/P06993
TTCTCTACGAGGAGTCGTTTT	ATGACGGCGAAAACAGCAC	malX	36.557	1708.2	ECOMALAA	EG10563/P19642
GTCCAGGCATAAGGATAAGAT	ATGTTCGATTTTTCAAAGG	malY	36.591	1709.8	ECOMALAA	EG10564/P23256
GATTTTCATCACAGGGGAATT	ATGATGTTAAATGCATGGC	malZ	9.185	429.2	ECOMALZ	EG10565/P21517
ACATTAAAACAGGGATTGATC	ATGCAAAAACTCATTAACT	manA	36.323	1697.2	ECOMANANA	EG10566/P00946
CGATAATAAAGGAGGTAGCAA	GTGACCATTGCTATTGTTA	manX	40.947	1913.3	ECOPTSLPM	EG10567/P08186
TACGTAAACAGGAGAAGTACA	ATGGAGATTACCACTCTTC	manY	40.969	1914.3	ECOPTSLPM	EG10568/P08187
GATAACGAACTGGACTAACAG	GTGAGCGAAATGGTTGATA	manZ	40.986	1915.1	ECOPTSLPM	EG10569/P08188
CGACGCTGATGGACAGAATTA	ATGGCTATCTCAATCAAGA	map	4.200	196.2	ECOMAP	EG10570/P07906
TAGCGGTTTGAGGAAAGGGTT	ATGATCCACAGTAGCGTAA	mazE	62.703	2929.9	ECORELA	EG10571/P18534
CCCCGCGATATACGCTTGAAG	ATGTGTGGTTATCAGGCTT	mcrA	26.041	1216.8	ECOMCRA	EG10573
AGCGATAATAGAGGCTTAGCA	ATGAGGAAGGCATATCTTA	mcrB	98.656	4609.8	ECOMCRBC	EG10574/P15005
TGACCCCTATAAACAACAGAA	ATGGACCAACAAATTATTA	mcrC	98.634	4608.8	ECOMCRBC	EG10575/P15006
TCAATATAATAAGGATTTAGG	ATGAAAGTCGCAGTCCTCG	mdh	72.820	3402.6	M24777	EG10576/P06994
CAAGCAAGCCAGGAGATCTGC	ATGATGTCTGCACCCAAAA	melA	93.567	4372.0	ECOMELA	EG10577/P06720
CGATACCCTATGAGCATTTCA	ATGACTACAAAACTCAGTT	melB	93.598	4373.5	ECOMELB	EG10578/P02921
TTCCAGGAAAGAGAGCCATCC	ATGAATACAGATACGTTTA	melR	93.541	4370.8	ECOMELOPA	EG11230/P10411
GCTCAGCAAAAATTAGCGGCA	ATGTCGTGCGATTATATGC	menD	51.204	2392.5	ECOMEND	EG10579/P17109
GATATTCCACGCTGGTAAAAA	ATGAATAAAACCGCGATTG	mepA	52.726	2463.7	ECOMEPAMR	EG10580/P14007
ATGTAGTGAGGTAATCAGGTT	ATGCCGATTCGTGTGCCGG	metA	90.837	4244.5	ECOMETAG	EG10581/P07623
AAGCCCAGGGAACTTCATCAC	ATGACGCGTAAACAGGCCA	metB	88.983	4157.8	ECOMETLB1	EG10582/P00935
TATAAAAACAGGAATCCCGAC	ATGGCGGACAAAAAGCTTG	metC	67.878	3171.7	ECOMETC	EG10583/P06721
AAGCGATTGATGAGGTAAGGT	ATGAGCTTTTTTCACGCCA	metF	89.068	4161.8	ECOMETF	EG10585/P00394
TATTAAGAAGTAATGCCTACT	ATGACTCAAGTCGCGAAGA	metG	47.189	2205.0	ECORPMET	EG10586/P00959
TTGAGCGTGTCGGGAGCAAGT	GTGAGCGACAAAGTGGAAC	metH	91.040	4253.9	ECOMTHM	EG10587/P13009
ACGAAGAGGATTAAGTATCTC	ATGGCTGAATGGAGCGGCG	metJ	88.971	4157.2	ECOMETJA	EG10588/P08338
TCTCTTTAGGTGATATTAAAT	ATGGCAAAACACCTTTTTA	metK	66.455	3105.2	ECOMETK	EG10589/P04384
GGCTGCAAACAAGGGGTAAAA	ATGAGTGTGATTGCGCAGG	metL	89.008	4159.0	ECOMETL	EG10590/P00562
ATGCCGAAGTGAAGGACTTTC	ATGATCGAAGTAAAACACC	metR	86.522	4042.8	ECOMETR	EG10591/P19797
TGGCCAACAAGGTAAAATTAT	ATGGTCAGCTCAACGACTC	mglA	48.122	2248.5	ECOMGLABCO	EG10592/P23199
ATAAGAAAAACCGGAGATACC	ATGAATAAGAAGGTGTTAA	mglB	48.156	2250.1	ECOMGLABCO	EG10593/P02927
CTTTAAGATTAGGGGCTTCCC	ATGAGTGCGTTAAATAAGA	mglC	48.100	2247.5	ECOMGLABCO	EG10594/P23200
CGGCGATAAAAGCCCTGAAAG	ATGAGTGATATCAGTAAGG	miaA	94.819	4430.5	ECOMIAA	EG10595/P16384
AGCTAATTGAGTAAGGCCAGG	ATGTCAAACACGCCAATCG	minC	26.356	1231.5	ECOMINB	EG10596/P18196

Table 1   *Continued*

-21_to_-1	0_to_+18	Gene	Centisome	Kb	DNA_LOCUS	EG5/SP19_Acc
CCTTTTTAACAAGGAATTTCT	ATGGCACGCATTATTGTTG	minD	26.338	1230.7	ECOMINB	EG10597/P18197
CGCTTGTTCGGAGGATAAGTT	ATGGCATTACTCGATTTCT	minE	26.332	1230.4	ECOMINB	EG10598/P18198
CGAATTTAAGGAATAAAGATA	ATGGCAGCTAAACACGTAA	mopA	94.214	4402.2	ECOGROESL	EG10599/P06139
TTTCTCAAAGGAGAGTTATCA	ATGAATATTCGTCCATTGC	mopB	94.207	4401.9	ECOGROESL	EG10600/P05380
CAACAGTGGAAGGATGATGTC	GTGCTTATCTTATTAGGTT	motA	42.573	1989.2	ECOMOTAB	EG10601/P09348
ACAGACGACAACCGAGGAAGC	ATGAAGAATCAAGCGCATC	motB	42.553	1988.3	ECOMOTAB	EG10602/P09349
AATTAATGAGGTCATACCCAA	ATGGATAGTTCGTTTACGC	mprA	60.538	2828.7	ECOMPRA	EG10603
CGCTTTACAGGAGAATGGGAC	ATGTTAGTTTGGCTGGCCG	mraY	2.060	96.3	ECOMUROY	EG10604/P15876
TGAATATTGCGGAGAAAAAGC	ATGGCCGGGAATGACCGCG	mrcB	3.529	164.9	ECOPONB	EG10605/P02919
TTTGAGTAGAAAACGCAGCGG	ATGAAACTACAGAACTCTT	mrdA	14.467	676.0	ECOPBPA	EG10606/P08150
CGCAGCGGAGGACCATTAATC	ATGACGGATAATCCGAATA	mrdB	14.444	674.9	ECORODA	EG10607/P15035
CTTTCAGGATTATCCCTTAGT	ATGTTGAAAAAATTTCGTG	mreB	73.186	3419.7	ECOMREB	EG10608/P13519
ATACGAGAATACGCATAACTT	ATGAAGCCAATTTTTAGCC	mreC	73.161	3418.5	ECOMERBCD	EG10609/P16926
GCGTGCGCCGGGAGGGCAATA	GTGGCGAGCTATCGTAGCC	mreD	73.151	3418.1	ECOMERBCD	EG10610/P16927
CTATACAAAAAAGGAGTCGGG	ATGAACGAACAATCCCAGG	mrp	47.163	2203.7	ECOMRPMET	EG10611/P21590
CTGATTCTGCAAGGATGTACT	ATGACGGTTCCTACCTATG	mrr	98.848	4618.8	ECOMRR	EG10612
CTTATCCGAAACTGGAAAAGC	ATGGAAACGAAAAAAAATA	msbB	41.730	1949.9	ECOMSBBA	EG10614
AACATAAGAAGGGGTGTTTTT	ATGTCATCCGATATTAAGA	mtlA	81.235	3795.8	ECOMTLA	EG10615/P00550
ACATTGATGAAGGTTAATACT	ATGAAAGCATTACATTTTG	mtlD	81.280	3797.9	ECOMTLD1	EG10616/P09424
TTTCACTGGAGAGAAGCCCTC	ATGGCAACACTAACCACCA	mtr	71.131	3323.7	ECOMTR	EG10617/P22306
AGATAGCGGAGAGGAAGAATA	ATGATTGAACGCGGTAAAT	mukB	21.039	983.1	ECOMUKB	EG10618/P22523
AAGAAGTTAATGGCGTAAAGA	ATGAATACACAACAATTGG	murC	2.162	101.0	ECOMURGC	EG10619/P17952
AACGCTGAAGGTACGTTAATC	ATGGCTGATTATCAGGGTA	murD	2.083	97.4	ECOMUROY	EG10620/P14900
CCTTCGCGAGCACTGCGAGAG	ATGACACTCGACAGCCGTG	murE	2.001	93.5	murEeco	EG10621/P22188
GCGTCTGCTGGGGGTGATTGC	ATGATTAGCGTAACCCTTA	murF	2.031	94.9	ECOMURF	EG10622/P11880
GGCGTTTGTACGAGGTTCACG	ATGAGTGGTCAAGGAAAGC	murG	2.138	99.9	ECOMURGC	EG10623/P17443
TTTTAATCAAGGTATCATGAC	ATGTCCCAACCTCGCCCAC	mutH	63.947	2988.0	ECOMUTH	EG10624/P06722
CAGGGAACCGGACATAACCCC	ATGAGTGCAATAGAAAATT	mutS	61.528	2875.0	ECOMUTS	EG10625
CTTTTTTATAGGTTTAAGACA	ATGAAAAAGCTGCAAATTG	mutT	2.382	111.3	ECOMUTT	EG10626/P08337
ACAACAGTGAATTCGGTGACC	ATGCAAGCGTCGCAATTTT	mutY	66.814	3121.9	ECOMICA	EG10627/P17802
TTGAACAGGAAAGAATATGCT	ATGAACCCTTATATTTATC	mvrC	12.300	574.7	ECOMVRC	EG10629/P23895
CGTCTAAAAGAACGTAATGCG	GTGATGGTTGCCCACTACT	nadA	16.960	792.5	ECONADA	EG10630/P11458
TTAGTAAATTAAACAAAGAAA	ATGAATACTCTCCCTGAAC	nadB	58.231	2720.9	ECONADB	EG10631/P10902
TTTTTAAAATCGGGGGTCAGA	ATGTATGCATTAACCCAGG	nagA	15.234	711.8	ECONAGACD	EG10632/P15300
TTTTACTTATTGAGGTGAATA	ATGAGACTGATCCCCCTGA	nagB	15.260	713.0	ECONAGBE	EG10633/P09375
TGTCAATATTCTGGGTAGTCA	ATGACCATTAAAAATGTAA	nagD	15.191	709.8	ECONAGACD	EG10634/P15302
GGTTCTCGTAGGGGGAATAAG	ATGAATATTTTAGGTTTTT	nagE	15.284	714.2	ECONAGBE	EG10635/P09323
CGTAACTCAATAAGAGAAAGT	ATGACACCAGGCGGACAAG	nagR	15.208	710.6	ECONAGACD	EG10636/P15301
AGCATCACTTCAGAGGTATTT	ATGGCAACGAATTTACGTG	nanA	72.582	3391.5	ECONANA	EG10637/P06995
GCGTCCCACAGGAGAAAACCG	ATGAGTAAATTCCTGGACC	narG	27.553	1287.4	narGeco	EG10638/P09152
CCAGGTACAGGAGAGCGTAAA	ATGAAAATTCGTTCACAAG	narH	27.633	1291.2	narGeco	EG10639/P11349
CACCACCGGAGGACAGCACTA	ATGCAATTCCTGAATATGT	narI	27.681	1293.4	narGeco	EG10640/P11350
CAGCAAAACGGAGCCGCATCC	ATGATCGAACTCGTGATTG	narJ	27.666	1292.7	narGeco	EG10641/P11351
AAACGAGTATCAGAGGTGTCT	ATGAGTCACTCATCCGCCC	narK	27.512	1285.5	ECONARK	EG10642/P10903
CAGACGTCCAAGGAGATACCC	ATGAGTAATCAGGAACCGG	narL	27.453	1282.8	ECONARXL	EG10643/P10957
CATCAGTGCGGGAGGTGGGAA	ATGATTCAGTATCTGAACG	narV	33.058	1544.7	ECNARZYW	EG10644/P19316
CGACAAAGCGGAGGGCGAATA	ATGCAGATCCTCAAAGTGA	narW	33.073	1545.4	ECNARZYW	EG10645/P19317
AGCCTGAAGGAAGAGGTTTAC	ATGCTTAAACGTTGTCTCT	narX	27.467	1283.4	ECONARXL	EG10646/P10956
TCAGGTACAGGAGGCGAAAAA	ATGAAAATCCGTTCACAAG	narY	33.088	1546.0	ECNARZYW	EG10647/P19318
TCCTGGAGCAGGAGTCATGTC	ATGAGTAAACTTTTGGATC	narZ	33.121	1547.6	ECNARZYW	EG10648/P19319
AATAAATTTAAGGGGGTCACG	TTGACTACGCCATTGAAAA	ndh	25.182	1176.7	ECONDH	EG10649/P00393
GAACAATTTACAGAGGTAAAA	ATGGCTATTGAACGTACTT	ndk	56.770	2652.6	ndk-eco	EG10650
GGGTTTAACAGGAGTCCTCGC	ATGAAATACATTGGAGCGC	nfo	48.415	2262.3	ECONFO	EG10651/P12638
ATCATTGCCGCTATCCTGGCG	ATGATTATGGCCAACAGCG	nhaA	0.383	17.9	ECOANTAPA	EG10652/P13738
AGAAAAGAAATCGAGGCAAAA	ATGAGCAAAGTCAGACTCG	nirB	75.223	3514.9	ECNIRBC	EG10653/P08201
GAACTGTTCACCGGACACACC	ATGTTCCTCACCTTTGGGG	nirC	75.292	3518.1	ECNIRBC	EG10654/P11097
AACTCTGGTGGAGGACAACGC	ATGACCGGTGGAAAGACA	nirD	75.277	3517.4	ECNIRBC	EG10655/P23675
TAACTTAAGGAGTGAGGAAAA	ATGGAAAGTAATTTCATTG	nlp	71.783	3354.1	nlp-eco	EG10656/P18837
CATTAATAAAAGGATAAAAAA	ATGAAACTGACAACACATC	nlpA	82.669	3862.8	nlpAeco	EG10657/P04846
TGTAAAGTTTAGGGAGATTTG	ATGGCTTACTCTGTTCAAA	nlpB	55.777	2606.3	ECONLRB34	EG10658/P21167
AAAACAGAGGAACCGTACGGA	ATGCGTTTCTGCCTTATTT	nlpC	38.543	1801.0	ECOBTUCED	EG11133
AGTTCAAAAAAAGGGCTCACG	ATGAAAAAATTAACAGTGG	nmpC	12.457	582.1	ECOEPNMPC	EG10659/P21420
CCAAAAAACAGGTACGACATAC	ATGAATCAGAATCTGCTGG	nrdA	50.495	2359.4	ECNRDAB1	EG10660/P00452
AACTCCCAACAGGACACACTC	ATGGCATATACCACCTTTT	nrdB	50.549	2361.9	ECONRDA	EG10661/P00453

Table 1   *Continued*

-21_to_-1	0_to_+18	Gene	Centisome	Kb	DNA_LOCUS	EG5/SP19_Acc
CGGTGAAACAGGGAATGTCTG	ATGAATAAAGCAAAACGCC	nth	36.826	1720.7	ECONTH	EG10662/P20625
TTTTCTGTCTGGAGGGGTTCA	ATGACATTGCAACAACAAA	ntrL	39.189	1831.2	ntrLeco	EG10663/P18843
TCACGATGTGAGGAAATTAAC	ATGAATCTTAAGCTGCAGC	nupG	66.873	3124.7	ECONUPG	EG10664/P09452
GGATGAGGTGAAAAGCCCGCG	ATGAACAAAGAAATTTTGG	nusA	71.376	3335.1	nusAeco	EG10665/P03003
GAAATTAGTAAGCGGAAATCC	GTGAAACCTGCTGCTCGTC	nusB	10.089	471.4	ECONUSAA	EG10666/P04381
TCACTGGCCTGAGGTTCTGAG	ATGTCTGAAGCTCCTAAAA	nusG	90.035	4207.0	ECOSECE	EG10667/P16921
GTTTGTCTTAAGAGAGAACGG	ATGCTGAGATTACTTGAAG	ogt	30.146	1408.6	ECOOGT	EG10668/P09168
GATGATAACGAGGCGCAAAAA	ATGAAAAAGACAGCTATCG	ompA	21.974	1026.8	ompAeco	EG10669/P02934
ATATAACAGAGGGTTAATAAC	ATGAAAGTTAAAGTACTGT	ompC	49.735	2323.9	ECOOMPC	EG10670/P06996
AAACCATGAGGGTAATAAATA	ATGATGAAGCGCAATATTC	ompF	21.242	992.6	ECOOMPF	EG10671/P02931
GAACCTTTGGGAGTACAAACA	ATGCAAGAGAACTACAAGA	ompR	76.141	3557.8	ECOOMPB	EG10672/P03025
ACGATTGAATGGAGAACTTTT	ATGCGGGCGAAACTTCTGG	ompT	12.651	591.1	ECOOMPT1	EG10673/P09169
AATGAGGGAGTCCAAAAAACA	ATGACCAACATCACCAAGA	oppA	28.003	1308.5	ECOOPPA	EG10674
ATAAATTCAGGAGAGAGTATT	ATGTTTGTAACGAGCAAAA	osmB	28.938	1352.2	ECOOSMB	EG10679/P17873
TAAGCCCACAGGAGCAACA	ATGACAATCCATAAGAAGA	osmC	33.505	1565.5	ECOOSMC	EG10680
CGTGGCGATGGAGGATGGATA	ATGAATATTCGTGATCTTG	oxyR	89.623	4187.7	ECOOXYS	EG10681/P11721
TTCTTTTGTACCGGAGCCGCC	ATGATCCTGCTTATAGATA	pabA	75.141	3511.1	ECOPABAA	EG10682/P00903
TAGCCAGTAGAGTCAGGACTG	ATGAAGACGTTATCTCCCG	pabB	40.783	1905.6	ECOPABB	EG10683/P05041
AATAGTAAAGGAATCATTGAA	ATGCAACTGAACAAAGTGC	pal	16.895	789.5	ECOPAL	EG10684/P07176
ATATCCCACTGGAGGATGACG	ATGCAGCTTGAAGTAATTC	panF	73.352	3427.4	ECOPANF	EG10685/P16256
AACGCCTACTTAAACTACTCC	ATGTACGTGATCATGGACC	parC	68.130	3183.4	ECOPARC	EG10686/P20082
CTCGAATTACTAACTTAAACC	ATGACGCAAACTTATAACG	parE	68.338	3193.2	ECOPARE	EG10687/P20083
TATTGGCTAAGGAGCAGTGAA	ATGCGCGTTAACAATGGTT	pckA	76.076	3554.7	pckAeco	EG10688/P22259
CAGCGTGGGAGTTGGCACGCA	ATGGTAAGCAGACGCGTAC	pcm	61.791	2887.3	ECOPCM	EG10689
AGTGAAAATGCCCTGAAGGTA	ATGTACAGGCTCAATAAAG	pcnB	3.377	157.8	ECOPCNB	EG10690/P13685
CGTTAAAATCCTGAGCAACTA	ATGGTTAAAAACCCAACGTG	pdxA	1.140	53.3	ECOPDXA	EG10691/P19624
TCTCATACAGGTAACACAAAC	GTGAAAATCCTTGTTGATG	pdxB	52.538	2454.9	ECOPDXB	EG10692/P05459
TTATGAATGATGAGGATTGTC	ATGGCTGAATTACTGTTAG	pdxJ	58.013	2710.7	ECORECOPDX	EG10693
TAAGATTCAGGAGCGTAGTGC	ATGGAGTTTAGTGTAAAAA	pepA	96.610	4514.2	ECOXERB	EG10694/P11648
TGATCGACAAGGAGACTTAAC	GTGTCTGAACTGTCTCAAT	pepD	5.615	262.4	ECOPEPD	EG10695/P15288
AAAAGATGCTAAAGGTTATTT	ATGACTCAACAGCCACAAG	pepN	21.344	997.3	ECOPEPN	EG10696/P04825
AAAAAACGTAAGGAGAGTGTT	ATGAGTGAGATATCCCGGC	pepP	ND	ND	ECOAPP2	EG10697/P15034
AAAAAAACAGAAGGGTAAAAA	ATGGAATCACTGGCCTCGC	pepQ	86.899	4060.5	pepQeco	EG10698/P21165
TTCCAAAGTTCAGAGGTAGTC	ATGATTAAGAAAATCGGTG	pfkA	88.535	4136.9	ECOCDHA	EG10699/P06998
ATCAGCCTATAGGAGGAAATG	ATGGTACGTATCTATACGT	pfkB	38.835	1814.6	ECOPFKBK	EG10700/P06999
ACTTAAGAAGGTAGGTGTTAC	ATGTCCGAGCTTAATGAAA	pfl	20.541	959.8	ECOPFL	EG10701/P09373
AAAATCAGAAGAGTATTGCTA	ATGAAAAACATCAATCCAA	pgi	91.240	4263.3	ECOPGI	EG10702/P11537
GAATCAACGAGAGGATTCACC	ATGTCTGTAATTAAGATGA	pgk	66.112	3089.2	ECOFDAPGK	EG10703/P11665
CAGAATACCCAGCGGCCAGTG	ATGACCGATAAAATACAGG	pgpA	ND	ND	M23546	EG10704/P18200
ACTTATCAAAAAGGAGAGGCC	ATGCGTTCGATTGCCAGAC	pgpB	ND	ND	M23628	EG10705/P18201
AACAGATAGTTACCCGTCATT	ATGCAATTTAATATCCCTA	pgsA	42.932	2006.0	ECOGLYWA	EG10706/P06978
TGATAACAAAAAGGCAACACT	ATGACATCGGAAAACCCGT	pheA	58.872	2750.9	ECOPHEAB	EG10707/P07022
AAGTCACTTAAGGAAACAAAC	ATGAAACACATACCGTTTT	pheL	58.869	2750.7	ECOPHEAB	EG11271/P03057
AAAGACACACAGGGGAAAGGC	GTGAAAAACGCGTCAACCG	pheP	13.041	609.3	ECOPHEPA	EG10708
ACCAGTGTCACCACTGACACA	ATGAGGAAAACCATGTCAC	pheS	38.665	1806.7	thrSeco	EG10709/P08312
TAAATAAGGCAGGAATAGATT	ATGAAATTCAGTGAACTGT	pheT	38.614	1804.3	thrSeco	EG10710/P07395
CTGACTTTTGAGGAAATCCAC	ATGTCATTACCACACTGCC	phnA	93.231	4356.3	ECOPHNAQ	EG10711/P16680
CTCTTAACTGAGGTCACCATC	ATGCCGTTAAGTCCCTACA	phnB	93.209	4355.3	ECOPHNAQ	EG10712/P16681
CTTTTTTAGGGAGGCTGCATC	ATGCAAACGATTATCCGTG	phnC	93.190	4354.4	ECOPHN	EG10713/P16677
CCCCATCATTGAGGAAAACGA	ATGAACGCTAAGATAATTG	phnD	93.167	4353.3	ECOPHN	EG10714/P16682
GCCTACAGACCGGAGCCAAAC	ATGCAAACCATCACCATCG	phnE	93.128	4352.5	ECOPHN	EG11283/P16683
ATAAGCGAGGCATTGATATCT	ATGCACTTGTCTACACATC	phnF	93.132	4351.7	ECOPHN	EG10715/P16684
GAATTCACTATGGAGCACTGA	ATGCACGCAGATACCGCGA	phnG	93.123	4351.3	ECOPHN	EG10716/P16685
GATGGTTCGCGGAGACAACGC	ATGACCCTGGAAACCGCTT	phnH	93.110	4350.7	ECOPHN	EG10717/P16686
CACTCATGTGGAGGTGTGCTG	ATGTACGTTGCCGTGAAAG	phnI	93.088	4349.6	ECOPHN	EG10718/P16687
GTCTGCAACAGGAGCAGAACC	ATGGCTAATCTGAGCGGCT	phnJ	93.070	4348.8	ECOPHN	EG10719/P16688
ACAGAGCGAGGCAAAAAACCA	ATGAATCAACCGTTACTTT	phnK	93.053	4348.0	ECOPHN	EG10720/P16678
ACCCGAACCGAGGAAAAAGCA	ATGATTAACGTACAAAACG	phnL	93.036	4347.2	ECOPHN	EG10721/P16679
GCACCCAATGGGAGCCTCTTC	ATGATTATCAATAACGTTA	phnM	93.012	4346.1	ECOPHN	EG10722/P16689
TCAGGGTAAAAGGGTGTTCTG	ATGATGGGAAAACTGATTT	phnN	93.000	4345.5	ECOPHN	EG10723/P16690
TCCATCAGAAGGAGAAACACC	ATGCCTGCTTGTGAGCTTC	phnO	92.991	4345.1	ECOPHN	EG10724/P16691
GCTTCACCAAGGCGCTGTAAC	ATGAGCCTGACCCTCACGC	phnP	92.975	4344.4	ECOPHN	EG10725/P16692
TCCCCATTGGGGTGAGGGGCG	ATGCCTGCTCCATACCCAA	phnQ	92.964	4343.8	ECOPHN	EG10726
TTTGTACATGGAGAAAATAAA	GTGAAACAAAGCACTATTG	phoA	8.744	408.6	ECOPHOAA	EG10727/P00634

Table 1 *Continued*

-21_to_-1	0_to_+18	Gene	Centisome	Kb	DNA_LOCUS	EG5/SP19_Acc
TTTATTACAACAGGGCAAATC	ATGGCGAGACGTATTCTGG	phoB	9.071	423.8	ECOPHOB	EG10728/P08402
TAATTAAAATCAGGAATGAAA	ATGAAAAAGAGCACTCTGG	phoE	5.701	266.4	ECOPHOE	EG10729/P02932
ATATAGCCTGAGGGGCCTGTA	ATGCGTATCGGCATGCGGT	phoM	99.911	4668.5	ECOPHOM	EG10730/P08401
TTAAGACAGGGAGAAATAAAA	ATGCGCGTACTGGTTGTTG	phoP	25.569	1194.7	ECOPHOPQ	EG10731
CTATCTGTTCGAATTGCGCTG	ATGAAAAAATTACTGCGTC	phoQ	25.538	1193.3	ECOPHOPQ	EG10732
TTCTTAACTGGAGTATCTTAC	GTGCTGGAACGGCTGTCGT	phoR	9.087	424.6	ECOPHORG	EG10733/P08400
TGAATCCTCCCAGGAGACATT	ATGAAAGTTATGCGTACCA	phoS	84.275	3937.8	ECOPHOS	EG10734/P06128
CGGTTGATTCAGGAGTGCGTT	ATGGACAGTCTCAATCTTA	phoU	84.197	3934.2	ECOPHOS	EG10735/P07656
CTTGCGCCATTCAGGAGTTTT	ATGACTACCCATCTGGTCT	phr	16.057	750.3	ECOPHRORF	EG10736/P00914
TACCAAAAGGTAACGCAAGCA	ATGAATGCTGCTATTTTCC	phtL	38.692	1807.9	thrSeco	EG11272/P06985
CTTTTACCAGGGGGATTTAAC	ATGCTTATTGGCTATGTAC	pin	26.013	1215.5	ECOPINP	EG10737/P03014
TTGTTTATTGCACCATAATAT	ATGACTGATTTGCCTGTGG	pinO	74.360	3474.6	ECORPSJ	EG11263/P03825
TCGTTACGGAGAACACGACCG	ATGCGGACTCTGCAGGGCT	pldA	86.320	4033.4	ECOPLDAA	EG10738/P00631
GTTTAAGAGGCAGATTACCCG	ATGTTTCAGCAGCAAAAAG	pldB	86.407	4037.5	ECOPLDB	EG10739/P07000
ACCAGAGGCTTTACATCGTTT	ATGTCCGGCTGGCCACGAA	plsB	91.670	4283.4	ECOPLSB	EG10740/P00482
TTCAGAGAAACTCTCTACATT	ATGGCACTTGCAATGAAAG	pmbA	96.063	4488.6	ECOPMBA	EG10741
CGAGGACGCTACTGCGCACCT	ATGACACAATTCGCTTCTC	pncB	21.312	995.8	ECOPNCB	EG10742/P18133
AGAAAGAGAAAGGATATTACA	TTGCTTAATCCGATCGTTC	pnp	71.226	3328.1	ECORPSOP	EG10743/P05055
AAACCGATGGAAGGGAATATC	ATGCGAATTGGCATACCAA	pntA	36.073	1685.6	ECOPNTAB	EG10744/P07001
GCAAAAATTAAGGGGTAACAT	ATGTCTGGAGGATTAGTTA	pntB	36.043	1684.1	ECOPNTAB	EG10745/P07002
TGATAAACAGGCACGGACATT	ATGGTTCAGATCCCCCAAA	polA	87.236	4076.2	ECOPOLA	EG10746/P00582
GTTTTTTGATGGATTTTCAGC	GTGGCGCAGGCAGGTTTTA	polB	1.370	64.0	ECPOLB	EG10747/P21189
AAACTAAATGGGAAATTTCCA	GTGAAGTTCGTAAAGTATT	ponA	75.860	3544.6	ECOPONA	EG10748/P02918
TAAGCCATCCGTTGCGTTTAC	ATGGGACAGAGTAAAAAAT	potA	ND	ND	ECOPOTABCD	EG10749
ACAAGTAAGTTCCAGAATGTA	GTGATTGTCACTATTGTCG	potB	ND	ND	ECOPOTABCD	EG10750
GAATAAGAAGGTGGAACTCGA	ATGATCGGTCGACTGCTTC	potC	ND	ND	ECOPOTABCD	EG10751
AGGTAACACAGGGGACGTTAA	ATGAAAAAATGGTCACGCC	potD	ND	ND	ECOPOTABCD	EG10752
CCTGTTGAAAGGGGAAAAATT	ATGAGTCAGGCTAAATCGA	potE	15.571	727.6	ECOPOTESPE	EG10753
ACATTCAGGAGATGGAGAACC	ATGAAACAAACGGTTGCAG	poxB	19.444	908.5	ECOPOXB	EG10754/P07003
ATATTTTTAAAGGAAACAGAC	ATGAGCTTACTCAACGTCC	ppa	95.892	4480.6	M23550	EG10755/P17288
AGATGGGGTGTCTGGGGTAAT	ATGAACGAACAATATTCCG	ppc	89.454	4179.8	ECOPPCG	EG10756/P00864
TCTGTTTGTTAAGGAAATCTC	ATGTTCAAATCGACCCTGG	ppiA	75.172	3512.5	ECOFIC1	EG10757/P20752
TCAACGGAACAGGATGCAAAA	ATGGTTACTTTCCACACCA	ppiB	11.989	560.2	ECOAPPIB	EG10758
TTATCACAAAAGGATTGTTCG	ATGTCCAACAATGGCTCGT	ppsA	38.376	1793.1	ECOPEPSYN	EG10759/P23538
GAAACGGAGGCCGGGCCAGGC	ATGAACATGTTTTTTAGGC	prc	41.187	1924.5	ECOPRC	EG10760
TTACAGGGTGCATTTACGCCT	ATGAAGCCTTCTATCGTTG	prfA	27.229	1272.3	ECORF1X	EG10761/P07011
CTCAATAAAAGAAATCAGACC	ATGTTTGAAATTAATCCGG	prfB	65.320	3053.0	ECORF2X	EG10762/P07012
ATTTCAAGTCAGGATGATGCT	ATGCCCGTTGCCCACGTTG	priA	88.901	4154.0	ECOPRIA	EG10763/P17888
GATTCTGAAGAGTAATTTCTG	ATGACCAACCGTCTGGTGT	priB	95.380	4456.7	ECORPSFRI	EG10764/P07013
CATTTTCATTGAGGTCTTATC	GTGAAAACCGCCCTGCTGC	priC	10.636	497.0	ECOPRIC	EG10765
AAATCGAGGAATAAGTAGCAG	ATGGCTAAACAACCGGGAT	prlA	74.130	3463.8	ECORPLN	EG10766/P03844
TACCCGTTAAGGAGCAGGCTG	ATGCTGGAACAAATGGGCA	proA	5.753	268.8	ECOPHOEA	EG10767/P07004
AAAATTGAATGGCAGAGAATC	ATGAGTGACAGCCAGACGC	proB	5.729	267.7	ECOPHOEA	EG10768/P07005
TTTCACGGCAGGAGTGAGGCA	ATGGAAAAGAAAATCGGTT	proC	8.810	411.6	ECOPROC	EG10769/P00373
TCCAACTGGAACCGTAACAAC	ATGCGTACTAGCCAATACC	proS	4.815	225.0	ECODRPA	EG10770/P16659
ATAAAGGAATCTTTCTATTGC	ATGGCAATTAAATTAGAAA	proV	60.411	2822.8	ECOPROU	EG10771/P14175
TAGATCGTGAGGGGGTAAATA	ATGGCTGATCAAAATAATC	proW	60.437	2824.0	ECOPROU	EG10772/P14176
CTGCCAAAAAAAGGAATAACA	ATGCGACATAGCGTACTTT	proX	60.461	2825.1	ECOPROU	EG10773/P14177
CGCATGCCTGAGGTTCTTCTC	GTGCCTGATATGAAGCTTT	prs	27.142	1268.2	ECOPRS	EG10774/P08330
AACAATGGCCTGGAGGCTACC	TTGTTAAAATTCATTTAAAC	psd	94.615	4421.0	ECOPSD	EG10775/P10740
AGAAATCAAGAGGACAACATT	ATGGGTATTTTTTCTCGCT	pspA	29.464	1376.7	ECOPSP	EG10776
CTCATCGTCTAAGGAGTACTT	ATGAGCGCGCTATTTCTGG	pspB	29.480	1377.5	ECOPSP	EG10777
TCCGAACTGGAGGGATCGCTA	ATGGCGGGCATTAATCTCA	pspC	29.484	1377.7	ECOPSP	EG10778
CCGTCAACTGTGAGGAAAGTT	ATGAATACTCGCTGGCAAC	pspD	29.492	1378.1	ECOPSP	EG10779
CATCCATAGAAGGACGCTTAC	ATGTTTAAAAAAGGCTTAC	pspE	29.499	1378.3	ECOPSP	EG10780
ACAGAGAAGAAATGCACTGTG	ATGTTGTCAAAATTTAAGC	pssA	58.492	2733.1	pssAeco	EG10781/P23830
TAAGAATGAGGGGGCACGCTA	ATGGCTATGGTTGAAATGC	pstA	84.233	3935.9	ECOPHOS	EG10782/P07654
CACGATGAGGAAAAGATTGCA	ATGAGTATGGTTGAAACTG	pstB	84.213	3934.9	ECOPHOS	EG10783/P07655
CGTTTAACTGAACAGTAACTT	ATGGCTGCAACCAAGCCTG	pstC	84.252	3936.8	ECOPHOS	EG10784/P07653
ACACACTCAGGACAAAAAAAC	GTGACGATTAAATTGATTG	pth	27.074	1265.3	pth-eco	EG10785
AGTTGATTATGAGGTCCGTGA	ATGCCCCGCAGCACCTGGT	ptr	63.652	2974.2	ECOPTR	EG10786/P05458
TACTCAGGAGCACTCTCAATT	GTGTTTAAGAATGCATTTG	ptsG	24.994	1167.9	ECOPTSG	EG10787/P05053
CCTATAAGTTGGGGAAATACA	ATGTTCCAGCAAGAAGTTA	ptsH	54.601	2551.3	ECOPHOSYS	EG10788/P07006
AGTCACAAGTAAGGTAGGGTT	ATGATTTCAGGCATTTTAG	ptsI	54.608	2551.6	ECOPHOSYS	EG10789/P08839

Table 1 *Continued*

-21_to_-1	0_to_+18	Gene	Centisome	Kb	DNA_LOCUS	EG5/SP19_Acc
AGCAAACGGTGATTTTGAAAA	ATGGGTAACAACGTCGTCG	purA	94.935	4435.9	ECOPURAA	EG10790/P12283
TCACACCCAGGAGTGATAAAG	ATGCAAAAGCAAGCTGAGT	purC	55.758	2605.3	ECOPURCA	EG10791/P21155
CCGCCATTAATGGAGCAATAG	ATGAAAGTATTAGTGATTG	purD	90.631	4234.8	ECOPURHD	EG10792/P15640
ACCACAGGAGTTTTAAGACGC	ATGTCTTCCCGCAATAATC	purE	11.962	558.9	ECOPUREK	EG10793/P09028
GCTTAACGAGGAAAAAGACGT	ATGTGCGGTATTGTCGGTA	purF	52.367	2446.9	ECOPURF1	EG10794/P00496
GTCAAATCCAGGGGATTTACC	ATGCAACAACGTCGTCCAG	purH	90.659	4236.1	ECOPURHD	EG10795/P15639
CCCGGACCCGCGAGGTGCGGC	ATGAAACAGGTTTGCGTCC	purK	11.939	557.9	ECOPUREK	EG10796/P09029
AGCTTAGAAGACGAGAGACTT	ATGATGGAAATTCTGCGTG	purL	57.818	2701.6	ECOPURLA	EG10797/P15254
TAACGCGTGGGGACCCAAGCA	GTGACCGATAAAACCTCTC	purM	56.268	2629.2	ECOPURMN	EG10798/P08178
ACAACGCGTGGTTATCGAATA	ATGAATATTGTGGTGCTTA	purN	56.290	2630.2	ECOPURMN	EG10799/P08179
TAGGGTCTGGAGTGAAATGGA	ATGGCAACAATAAAAGATG	purR	37.403	1747.7	ECOPURRRP	EG10800/P15039
CTGAACAACAGGAGTAATGGC	ATGGGAACCACCACCATGG	putA'	23.267	1087.4	ECPUTP	EG10801/P09546
ATAATTTTTTGGAGACTTTAG	ATGGCTATTAGCACACCGA	putP	23.285	1088.0	ECPUTP	EG10802/P07117
GACAATTTGCCGGGAGGATGT	ATGGTTCAGTGTGTTCGAC	pyrL	96.356	4502.3	ECOPYRBI	EG11279/P09150
TTTCAGTCAACGGAGTATTAC	ATGTCCAGAAGGCTTCGCA	pykA	41.696	1948.3	ECOPYKAA	EG10803/P21599
AACTTAAAGACTAAGACTGTC	ATGAAAAAGACCAAAATTG	pykF	37.773	1765.0	ECOPK1	EG10804/P14178
CCAGGCGTCAGGAGATAAAAG	ATGGCTAATCCGCTATATC	pyrB	96.335	4501.4	ECOPYRBIA	EG10805/P00479
TCAGCCGGAGCATAGAGATTA	ATGACTGCACCATCCCAGG	pyrC	24.201	1130.8	ECOPYRC	EG10806/P05020
ACCGGGAATCCAGGAGAGTTC	ATGTACTACCCCTTCGTTC	pyrD	21.659	1012.0	ECOPYRD	EG10807/P05021
AGTAAGATGAGGAGCGAAGGC	ATGAAACCATATCAGCGCC	pyrE	82.144	3838.3	ECPYRE	EG10808/P00495
CATCATCAAGAAGGTCTGGTC	ATGACGTTAACTGCTTCAT	pyrF	28.912	1351.0	ECOPYRF	EG10809/P08244
ACCTAACTTCTCAGGTTCAGC	ATGACAACGAACTATATTT	pyrG	62.637	2926.8	ECOPYRG	EG10810/P08398
GTACTGTAAGGGGAAATAGAG	ATGACACACGATAATAAAT	pyrI	96.325	4500.9	ECOPYRBIA	EG10811/P00478
ATTTTAGTCTGAGTCAGTGTC	ATGCGCGTTACCGATTTCT	queA	9.238	431.7	ECOQBIO	EG10812/P21516
ACCTTAATTAAGGAGAAAAAC	ATGATTACCAATTATGAAG	racC	30.509	1425.6	M24905	EG10813/P15033
GCGTGACGTTCTGAGGCCGTC	ATGGAAGCATTACTTCAGC	rbsA	84.772	3961.1	ECORBS	EG10814/P04983
CGACTACAGGACATCTTGAAT	ATGAACATGAAAAAACTGG	rbsB	84.826	3963.6	ECORBS	EG10815/P02925
GCGTGAATCAGGAGTAAAAAA	ATGACAACCCAGACTGTCT	rbsC	84.804	3962.6	ECORBS	EG10816/P04984
GTTTCGCTGATGGAGAAAAAA	ATGAAAAAAGGCACCGTTC	rbsD	84.763	3960.6	ECORBS	EG10817/P04982
CATTTTATGGACATCCCGAAT	ATGCAAAACGCAGGCAGCC	rbsK	84.847	3964.6	ECORBS	EG10818/P05054
CGCATTGAGTGAGGGTATGCC	ATGTCAACGATTATTATGG	rcsA	43.589	2036.8	ECORCSA	EG10820
TGTAGCAAGGTAGCCTATTAC	ATGAACAATATGAACGTAA	rcsB	49.830	2328.4	ECORCSBC	EG10821/P14374
ACCCTGAAAGCCTCGCGCTAC	ATGTTCAGAGCATTGGCGT	rcsC	49.848	2329.2	ECORCSBC	EG10822/P14376
CCCGGCATGACAGGAGTAAAA	ATGGCTATCGACGAAAACA	recA	60.815	2841.6	ECORECE	EG10823/P03017
TGCCCCTGATGAGTGAAAAGA	ATGAGTGATGTCGCCGAGA	recB	63.576	2970.7	ECORECB	EG10824/P08394
CGTCAGTAGTCAGGAGCCGCT	ATGTTAAGGGTCTACCATT	recC	63.717	2977.3	ECORECC	EG10825/P07648
TATGACCCTGGAGGAGGCGTA	ATGAAATTGCAAAAGCAAT	recD	63.537	2968.8	ECORECD	EG10826/P04993
CTGAGTTGAGGTTAAAAAACA	ATGAGCACAAAACCACTCT	recE	30.458	1423.5	M24905	EG10827/P15032
TGTCATGCCAATGAGACTGTA	ATGTCCCTCACCCGCTTGT	recF	83.563	3904.6	ECORECFA	EG10828/P03016
GCAGGCTGCAGGGTAAGTGCC	ATGAAAGGTCGCCTGTTAG	recG	82.363	3848.5	ECORCG	EG10829
GACCAGCGGTAAATAATTCGC	GTGAAACAACAGATACACC	recJ	65.364	3054.2	ECORECJXPR	EG10830/P21893
TTTTTCATACAGGAAACGACT	ATGTTGGCACAACTGACCA	recN	59.153	2764.0	ECORECN	EG10831/P05824
CGATCTTTAAGAGTAACTCCG	ATGGAAGGCTGGCAGCGCG	recO	58.029	2711.4	ECORECO	EG10832/P15027
TTTTTATTTCAGGCAATCGGG	GTGAATGTGGCGCAGGCGG	recQ	86.341	4034.4	ECORECQ	EG10833/P15043
TGGCTTTAAGATGCCGTTCTG	ATGCAAACCAGCCCGCTGT	recR	10.724	501.1	ECOAPTADK	EG10834/P12727
TGGTCCCTAAAGGAGAGGACG	ATGGTTGCGGTAAGAAGTG	relA	62.710	2930.2	ECORELA	EG10835/P11585
TAATTACAAGAGGTGTAAGAC	ATGGGTAGCATTAACCTGC	relB	35.393	1653.8	ECORELB	EG10836/P07007
TGTGACGCTGGATGAACTCTG	ATGGCGTATTTTCTGGATT	relE	35.387	1653.5	ECORELB	EG11131/P07008
CGTTGACAGGAGAAGCAGGCT	ATGAAGCAGCAAAAGGCGA	relF	35.382	1653.3	ECORELB	EG11130/P07009
TAATTATCCGGAAGGATTCTG	ATGATGAACATCGAAGAAC	rem	35.372	1652.8	ECORELB	EG11129/P07010
CGAAGATTGAGCAATACACCT	ATGCGTCTAAACCCCGGCC	rep	85.356	3988.4	ECOREPHEL	EG10837/P09980
CATAATTCGAAGGTTACAGTT	ATGATCATCGTTACCGGCG	rfaD	81.704	3817.7	rfaDeco	EG10838/P17963
CAGGCCGCTGTTGGCATTGTT	ATGATGGTGTTCGGCAAGC	rfe	85.518	3995.9	rfe-eco	EG10840
CGAAAGATAAAAGGAAATCGC	ATGGCAAGTATTTCATCGC	rfs'	43.176	2017.9	ECOHAGFLG	EG10841
GTGATATTCGCCAGGGACGGG	ATGGCTTTCTGCAATAACG	rhaR	88.360	4128.7	ECORHAC	EG10842/P09378
TGACGACATCAGGAGGCCAGT	ATGACCGTATTACATAGTG	rhaS	88.342	4127.9	ECORHAC	EG10843/P09377
AGTTTAAGAACTCACACCACT	ATGAATCTTACCGAATTAA	rho	85.479	3994.1	ECORHO	EG10845/P03002
CCGATAAAACAAGGATGAGAA	ATGAGCGGAAAACCGGCAG	rhsA	81.015	3785.5	ECORHSA	EG10846/P16916
ATAATAGAAAAGGATTTTACG	ATGAGCGGAAAACCGGCAG	rhsB	77.944	3642.0	ECORHSBA	EG10847/P16917
CGATAAAACAAGGATGAGCAG	ATGAGCGGAAAACCGGCGG	rhsC	15.843	740.3	ECORHSC	EG10848/P16918
ATAACTAAAAGGGCACTTTAT	ATGAGCGGAAAACCAGCGG	rhsD	11.343	530.0	ECORHSDG	EG10849/P16919
TATGGCAACAAATTTGCACAT	ATGAACACGATTTCTTCCC	rimI	99.316	4640.6	ECORIMI	EG10850/P09453
ACGGGAAGGAGTAGGTATAGA	ATGTTTGGCTATCGCAGTA	rimJ	24.300	1135.4	ECORIMJ	EG10851/P09454
AAATTGCCATATTGTCCCGGG	ATGGAACGCTCTATTCGTG	rimK	19.284	901.1	ECORIMK	EG10852/P17116

Table 1 *Continued*

-21_to_-1	0_to_+18	Gene	Centisome	Kb	DNA_LOCUS	EG5/SP19_Acc
CATGGATTGACTGGAGATAAG	ATGACTGAAACGATAAAAG	rimL	32.267	1507.7	ECORIML	EG10853/P13857
AAAGCGTGTAAGAGGTGCGCA	ATGCGTAAGCAGTGGCTCG	rlpA	14.420	673.8	ECORLPA	EG10854/P10100
TGGCTAAGCGCGGGAGGAAGC	GTGCGATATCTGGCAACAT	rlpB	14.593	681.9	ECORLPB	EG10855/P10101
TGTATGAGTTCCACACCCATT	ATGAAAGCATTCTGGCGTA	rna	13.949	651.8	ECORIBI34	EG10856/P21338
GCATTTATTTATTGGTATCGC	ATGAACCCCATCGTAATTA	rnc	58.064	2713.1	ECORNC1	EG10857/P05797
GCCCACTAAAGAGAAAACAAT	TTGAATTACCAAATGATTA	rnd	40.617	1897.9	ECORND	EG10858/P09155
AAGAATAATGAGTAAGTTACG	ATGAAAAGAATGTTAATCA	rne	24.635	1151.1	ECOAMSG	EG10859/P21513
ATTACAGGAAGTCTACCAGAG	ATGCTTAAACAGGTAGAAA	rnhA	5.223	244.0	ECORNHQ	EG10860/P00647
AGCCGTTCTGGAGTTAGCACA	ATGATCGAATTTGTTTATC	rnhB	4.543	212.3	ECOLPXA	EG10861/P10442
AGTAATAAAGCTAACCCCTGA	GTGGTTAAGCTCGCATTTC	rnpA	83.656	3908.9	ECORNPA	EG10862/P06277
TTTCAAGCCGGAGATTTCAAT	ATGCGTCCAGCAGGCCGTA	rph	82.159	3839.0	ECPYRE	EG10863/P03842
CTGGTAGTGGAGGACTAAGAA	ATGGCTAAACTGACCAAGC	rplA	90.059	4208.1	ECORPLRPO	EG10864/P02384
GTAAGTCGGAGGAGTAATACA	ATGGCAGTTGTTAAATGTA	rplB	74.297	3471.6	ECORPOS10	EG10865/P02387
AGCGATTGAGAGGTTGAAACA	ATGATTGGTTTAGTCGGTA	rplC	74.334	3473.3	ECORPOS10	EG10866/P02386
TGAAGGCGTAAGGAGATAGCA	ATGGAATTAGTATTGAAAG	rplD	74.321	3472.7	ECORPOS10	EG10867/P02388
CAAGTAATTTGGAGTAGTACG	ATGGCGAAACTGCATGATT	rplE	74.219	3467.9	ECORPLN	EG10868/P02389
GTAGCCTAATCGGACGAAAAA	ATGTCTCGTGTTGCTAAAG	rplF	74.191	3466.7	ECORPLN	EG10869/P02390
CGACTTTGAGAGGATAAGGTA	ATGCAAGTTATTCTGCTTG	rplI	95.393	4457.3	ECORPSFRI	EG10870/P02418
AAACATCCAGGAGCAAAGCTA	ATGGCTTTAAATCTTCAAG	rplJ	90.083	4209.2	ECORPLRPO	EG10871/P02408
ACCCAACTTGAGGAATTTATA	ATGGCTAAGAAAGTACAAG	rplK	90.050	4207.7	ECORPLRPO	EG10872/P02409
TGATATTCAGGAACAATTTAA	ATGTCTATCACTAAAGATC	rplL	90.095	4209.8	ECORPLRPO	EG10873/P02392
TATTTATTGGGTAAGCTTTTA	ATGAAAACTTTTACAGCTA	rplM	72.710	3397.5	ECORPSI	EG10874/P02410
TTGACATTAGCGGAGCCTAAA	ATGATCCAAGAACAGACTA	rplN	74.238	3468.8	ECORPLN	EG10875/P02411
GTTAAAGTTGAGGAGTAAGAG	ATGCGTTTAAATACTCTGT	rplO	74.159	3465.2	ECORPLN	EG10876/P02413
CGTAAATAAGGAGCGTCGCTG	ATGTTACAACCAAAGCGTA	rplP	74.259	3469.8	ECORPOS10	EG10877/P02414
TTTACTGAGAAGGATAAGGTC	ATGCGCCATCGTAAGAGTG	rplQ	74.063	3460.7	ECORPA	EG10878/P02416
AAGAAGAAGTAAGGTAACACT	ATGGATAAGAAAATCTGCTC	rplR	74.183	3466.3	ECORPLN	EG10879/P02419
CCCAGGATAAGAGATTAAATT	ATGAGCAACATTATTAAGC	rplS	58.995	2756.6	ECOTRMD	EG10880/P02420
TATAGATACAGGAGAGCACAT	ATGGCTCGCGTAAAACGTG	rplT	38.696	1808.1	thrSeco	EG10881/P02421
GAAATAAGGTAGGAGGAAGAG	ATGGAAACTATCGCTAAAC	rplV	74.283	3471.0	ECORPOS10	EG10882/P02423
GCAAGTTGAGGAGATGCTGGC	ATGATTCGTGAAGAACGTC	rplW	74.315	3472.4	ECORPOS10	EG10883/P02424
AAGTACTCTAAGGAGCGAATC	ATGGCAGCGAAAATCCGTC	rplX	74.231	3468.5	ECORPLN	EG10884/P02425
ATTAATTTAAGAGAGAAAGAA	ATGTTTACTATCAACGCAG	rplY	49.098	2294.1	rplYeco	EG10885/P02426
TTTGATTTTTGGAGAATAGAC	ATGTCCCGAGTCTGCCAAG	rpmB	82.063	3834.5	ECORPMBG	EG10886/P02428
TGTAACTAAGACGGTGATGTA	ATGAAAGCAAAAGAGCTGC	rpmC	74.255	3469.6	ECORPOS10	EG10887/P02429
GAAATTCTGGGGAAATAAACC	ATGGCAAAGACTATTAAAA	rpmD	74.168	3465.6	ECORPLN	EG10888/P02430
GCCTTAAACCGAGGTTTTCCC	ATGAAAAAAGATATTCACC	rpmE'	88.933	4156.4	ECOPRIAY	EG10889/P02432
CGTAATTGAGGAGTAAGGTCC	ATGGCCGTACAACAGAATA	rpmF	24.767	1157.3	ECORPMFA	EG10890/P02435
AGTACTTAGAGGAAATAAATC	ATGGCTAAAGGTATTCGTG	rpmG	82.059	3834.3	ECORPMBG	EG10891/P02436
CAAGTTTAGGTAGAAATCGCC	ATGAAACGCACTTTTCAAC	rpmH	83.653	3908.7	ECORNPA	EG10892/P02437
TGCGAAGTGGAAGTTATTAAA	ATGCCAAAATTAAAGACCG	rpmI	38.704	1808.5	thrSeco	EG11231/P07085
GAGAAGTTACGGAGAGTAAAA	ATGAAAGTTCGTGCTTCCG	rpmJ	74.127	3463.7	ECORPLN	EG11232/P21194
TAGTACCAAAGAGAGGACACA	ATGCAGGGTTCTGTGACAG	rpoA	74.072	3461.1	ECORPA	EG10893/P00574
GTCAGCGAGCTGAGGAACCCT	ATGGTTTACTCCTATACCG	rpoB	90.110	4210.5	ECORPLRPO	EG10894/P00575
AACTCCGACGGGAGCAAATCC	GTGAAAGATTTATTAAAGT	rpoC	90.197	4214.6	ECORPLRPO	EG10895/P00577
AAGTGTGTGGATTACCGTCTT	ATGGAGCAAAACCCGCAGT	rpoD	69.191	3233.0	ECORPSRPO	EG10896/P00579
ATATCGATTGAGAGGATTTGA	ATGACTGACAAAATGCAAA	rpoH	77.521	3622.3	ECOLIVHMGF	EG10897/P00580
CACCTGTGGAGCTTTTTAAGT	ATGGCACGCGTAACTGTTC	rpoZ	82.296	3845.4	ECOSPOT	EG10899/P08374
ATATAAACCTGAAGATTAAAC	ATGACTGAATCTTTTGCTC	rpsA	20.761	970.1	ECORPSA	EG10900/P02349
TTTTATATAGAGGTTTTAATC	ATGGCAACTGTTTCCATGC	rpsB	4.225	197.4	ECORPSBTS	EG10901/P02351
CTGAGACTCTGGAGACTAGCA	ATGGGTCAGAAAGTACATC	rpsC	74.268	3470.2	ECORPOS10	EG10902/P02352
CAGGTTTGTTGGAGAAAGAAA	ATGGCAAGATATTTGGGTC	rpsD	74.094	3462.1	ECORPA	EG10903/P02354
GTTCTAAGGTAGAGGTGTAAG	ATGGCTCACATCGAAAAAC	rpsE	74.172	3465.8	ECORPLN	EG10904/P02356
TAATCCGTAAGGAGCAATTCG	ATGCGTCATTACGAAATCG	rpsF	95.372	4456.3	ECORPSFRI	EG10905/P02358
AATTAACAACGGAGTATTTCC	ATGCCACGTCGTCGCGTCA	rpsG	74.784	3494.3	rpsGeco	EG10906/P02359
AATCACGGGAGGTAAAGACAG	ATGAGCATGCAAGATCCGA	rpsH	74.203	3467.2	ECORPLN	EG10907/P02361
CATCTAATCGGATTATAGGCA	ATGGCTGAAAATCAATACT	rpsI	72.702	3397.1	ECORPSI	EG10908/P02363
AATAATTGGAGCTCTGGTCTC	ATGCAGAACCAAAGAATCC	rpsJ	74.348	3474.0	ECORPSJ	EG10909/P02364
AATAATCGGGGTGATTGAATA	ATGGCAAAGGCACCAATTC	rpsK	74.108	3462.8	ECORPA	EG10910/P02366
GCTAAAACCAGGAGCTATTTA	ATGGCAACAGTTAACCAGC	rpsL	74.797	3495.0	ECOSTR1	EG10911/P02367
ATTAAATAGTAGGAGTGCATA	GTGGCCCGTATAGCAGGCA	rpsM	74.116	3463.2	ECORPA	EG10912/P02369
CAAGTAAGGTAGGGTTACTAA	ATGGCTAAGCAATCAATGA	rpsN	74.212	3467.6	ECORPLN	EG10913/P02370
TATATACTTTGGAGTTTTAAA	ATGTCTCTAAGTACTGAAG	rpsO	71.277	3330.5	ECORPSOP	EG10914/P02371
TTATTCACACAAGAGGATGTT	ATGGTAACTATTCGTTTAG	rpsP	59.033	2758.4	ECOTRMD	EG10915/P02372

Table 1 *Continued*

-21_to_-1	0_to_+18	Gene	Centisome	Kb	DNA_LOCUS	EG5/SP19_Acc
GAACGAGAAGGCGGGTGCGTA	ATGACCGATAAAATCCGTA	rpsQ	74.249	3469.4	ECORPOS10	EG10916/P02373
TAGATTCTGGAGACTAGCCAT	ATGGCACGTTATTTCCGTC	rpsR	95.387	4457.1	ECORPSFRI	EG10917/P02374
AATAATTTTAGAGGATAAGCC	ATGCCACGTTCTCTCAAGA	rpsS	74.290	3471.3	ECORPOS10	EG10918/P02375
AACACATTTGGGAGTTGGACC	TTGGCTAATATCAAATCAG	rpsT	0.452	21.1	ECORPST	EG10919/P02378
TTAATCAAAGGTGAGAGGCAC	ATGCCGGTAATTAAAGTAC	rpsU	69.142	3230.7	ECORPSRPO	EG10920/P02379
TAACCTTTGTGGAGCACTATC	ATGCTGAAACCAGAAATGA	rsgA	42.841	2001.8	ECORSGA	EG10921/P23887
ATTAACCCATTCAACAGAACT	GTGACGCGCCATGGCAAAT	rts	89.956	4203.3	ECORTSA	EG10922/P15044
TTCATTACGCAGGAGCGTCAT	GTGATAGGCAGACTCAGAG	ruvA	41.903	1957.9	ECORUVABA	EG10923/P08576
CGCCGCGTTATGAGGTAAAGG	ATGATTGAAGCAGACCGTC	ruvB	41.881	1956.9	ECORUVABA	EG10924/P08577
ACAGCAAAACGGAGACGCGTG	ATGGCTATTATTCTCGGCA	ruvC	41.935	1959.4	ruvCeco	EG10925
AGAAATAACGGATTTAACCTA	ATGATGAATGACGGTAAGC	sbcB	44.833	2094.9	ECOSBCB	EG10926/P04995
CCTCGCCGGAGAACACGAAGC	ATGAAAATTCTCAGCCTGC	sbcC	8.974	419.3	ECOSBCC	EG10927/P13458
AAGAAGTTAGCAGGAGTGCAT	ATGTTTAAGTCTTTTTTCC	sbmA	8.633	403.4	ECOSBMA	EG10928
ACAACATAAGAGAGTCGGGCG	ATGAACAAGTGGGGCGTAG	sbp	88.562	4138.2	ECOCDHA	EG10929/P06997
ATCGTGATTAGTCTATTCGAC	ATGTTTAAGGTGGGGATTG	sdaA	40.829	1907.8	ECOSDAA	EG10930/P16095
CGTTGTGGTGTGGGGTGTGTG	ATGAAATTGCCAGTCAGAG	sdhA	16.396	766.1	ECOGLTA	EG10931/P10444
TACTAATGCGGAGACAGGAAA	ATGAGACTCGAGTTTTCAA	sdhB	16.434	767.9	ECOGLTA	EG10932/P07014
AACAGCATGTGGGCGTTATTC	ATGATAAGAAATGTGAAAA	sdhC	16.380	765.4	ECOGLTA	EG10933/P10446
ACTTCTCGCAGGAGTCCTCGT	ATGGTAAGCAACGCCTCCG	sdhD	16.388	765.8	ECOGLTA	EG10934/P10445
TCAGGGGCGTTGCGGTTTACT	ATGCAGGATAAGGATTTTT	sdiA	43.016	2010.0	ECOUVRC	EG10935/P07026
CGGGGCGTTTGAGATTTTATT	ATGCTAATCAAATTGTTAA	secA	2.323	108.5	ECOSECA	EG10936/P10408
ACACTTAAGGGTTTTCTACAC	ATGTCAGAACAAAACAACA	secB	81.479	3807.2	ECOSECB	EG10937/P15040
TTTTTCCCTAAGGGAATTGCC	GTGTTAAACCGTTATCCTT	secD	9.295	434.3	ECOSECDF	EG10938/P19673
AAACTTCTGACAGGTTGGTTT	ATGAGTGCGAATACCGAAG	secE	90.026	4206.6	ECOSECE	EG10939/P16920
TGTCAATCTGAGGAGTGCGAT	GTGGCACAGGAATATACTG	secF	9.335	436.2	ECOSECDF	EG10940/P19674
CCTTGATCAGCCAGGTTTCCT	ATGACAACCGAAACGCGTT	selA	80.966	3783.2	ECOSELA	EG10941/P23328
GTTTTTGGAGATGTTGTTGAA	ATGATTATTGCGACTGCCG	selB	80.927	3781.4	ECOSELB	EG10942/P14081
TTGCATTGACAGGAGATGTCC	ATGAGCGAGAACTCGATTC	selD	39.745	1857.1	ECOSELD	EG10943/P16456
TCAAAAGACAGGATTGGGTAA	ATGGCAAAGGTATCGCTGG	serA	65.811	3075.1	ECOSERA	EG10944/P08328
AACGATTTTACAGGAGCCTTA	ATGCCTAACATTACCTGGT	serB	99.661	4656.7	ECOSERB	EG10945/P06862
AACGCAACGTGGTGAGGGGAA	ATGGCTCAAATCTTCAATT	serC	20.673	966.0	ECOAROA	EG10946/P23721
TTCGATAAGCACAGGATAAGC	ATGCTCGATCCCAATCTGC	serS	20.294	948.3	ECOSERS	EG10947/P09156
GACGATAAAAGCCCCCCAGGG	ATGGATATTCAAAAAAGAG	sfcA'	33.449	1563.3	sfcAeco	EG10948
GCAATAACAAGGATTGTCGCA	ATGGAATTTTCTCCCCCTC	sfsA	3.443	160.9	ECOSFS1A	EG10949/P18273
TTACACTTAGAGGATGCGCTT	GTGGAAAAAGCCAAACAAG	slt	99.787	4662.6	ECOSLTY	EG10950/P03810
CAAATTTTCAAAGGGTGGAAG	ATGGCTCGCACAAAACTGA	smp	99.645	4656.0	ECOSERB	EG10951/P18838
TACGCATTGGGTACACCGCTG	ATGTCCGATCCATTTGGTA	smpA	59.196	2766.0	smpAeco	EG10952/P23089
CCGACAATACTGGAGATGAAT	ATGAGCTATACCCTGCCAT	sodA	88.405	4130.8	ECOSOD	EG10953/P00448
TAATAATAAAGGAGAGTAGCA	ATGTCATTCGAATTACCTG	sodB	37.349	1745.2	ECOSODB	EG10954/P09157
GCTGTAAAAGGACAGTGAATC	ATGCCCGCTAATGCTCGCT	sohA	70.541	3296.1	ECOSOHA	EG10955/P15373
GTTTAACCAAGGTGGGGACTC	GTGGAATTGTTGTCTGAAT	sohB	28.616	1337.1	ECOSOHB	EG10956
TAACTTGAGGTAAAGCGATTT	ATGGAAAAGAAATTACCCC	soxR	92.178	4307.1	ECOSOXRS	EG10957/P22538
ACACTGAAAAGAGGCAGATTT	ATGTCCCATCAGAAAATTA	soxS	92.169	4306.7	ECOSOXRS	EG10958/P22539
TAAAATAATTTGAGGTTCGCT	ATGTCTGACGACATGTCTA	speA	66.396	3102.4	ECOSPEA	EG10959/P21170
CTTTGTAATAGGAGTCCATCC	ATGAGCACCTTAGGTCATC	speB	66.373	3101.4	ECOSPEAA	EG10960/P16936
GACCAGTTTGACCCATATCTC	ATGGGGCAGGGTTTTCCAC	speC	66.901	3126.0	ECOSPEC	EG10961/P21169
TCCTAAGGAGAAGATAAGAAA	TTGAAAAAACTGAAACTGC	speD	2.901	135.6	ECOSPDE	EG10962/P09159
TAACAAAGGAGGTATCAACCC	ATGGCCGAAAAAAAAACAGT	speE	2.919	136.4	ECOSPDE	EG10963/P09158
CCTAAAATAAAGAGATGAAAA	ATGTCAAAATTAAAAATTG	speF	15.599	728.9	ECOPOTESPE	EG10964
TAATCACAAAGCGGGTCGCCC	TTGTATCTGTTTGAAAGCC	spoT	82.303	3845.7	ECOSPOT	EG10966/P17580
CGAAACCGAAATTAATGTTTT	ATGAACCCAACACGTTATG	spoU	82.348	3847.8	ECOSPOT	EG10967/P19396
GACCTTAAGTTGGGAGAATAC	ATGCGAACCCTTTGGCGAT	sppA	39.785	1859.0	ECOSPPA	EG10968/P08395
TAAATCCTGAAGGAGAGAACA	ATGATAGAAACCATTACTC	srlA	60.866	2844.0	ECOGUT	EG10969/P05705
TCTATCAATAGAGGCTGAAAC	ATGACCGTTATTTATCAGA	srlB	60.899	2845.6	ECOGUT	EG10970/P05706
GAATCTGTTAAGGAGTAAAAA	ATGAATCAGGTTGCCGTTG	srlD	60.907	2845.9	ECOGUT	EG10971/P05707
CCACTACATTAAGGAAAAGTT	ATGGTATCCGCACTCATCA	srlM	60.926	2846.8	ECOGUT	EG10972/P15081
AACGCGGGACGTCAGACGTTA	ATGCTGGAGTTGCAGGAAG	srlQ	60.952	2848.1	ECOGUTQ	EG10973/P17115
CTAAATCAGGTAATCACGCCC	ATGAAACCTCGTCAGCGTC	srlR	60.935	2847.3	ECOGUT	EG10974/P15082
CGCCCCACACAGAGGTAGAAC	ATGACTGTAACGACTTTTT	srmB	44.441	2076.6	ECOSRMB	EG10975/P21507
TTTTTTTCAGGAGCACGAAC	ATGGCCAGCAGAGGCGTAA	ssb	92.101	4303.5	ECOSSB	EG10976/P02339
TGACTATACCTGGAGGTTTTC	ATGGCTGTCGCTGCCAACA	sspA	72.680	3396.0	ECOSSPG	EG10977/P05838
TCTGGGCCGGAGTTAATCTGT	ATGGATTTGTCACAGCTAA	sspB	72.669	3395.5	sspBeco	EG10978
AAAGATGCTTAAGGGATCACG	ATGCAGAACAGCGCTTTGA	sucA	16.456	768.9	ECOGLTA	EG10979/P07015
CGAATAAATAAAGGATACACA	ATGAGTAGCGTAGATATTC	sucB	16.516	771.7	ECOGLTA	EG10980/P07016

Table 1   *Continued*

-21_to_-1	0_to_+18	Gene	Centisome	Kb	DNA_LOCUS	EG5/SP19_Acc
ACTGAAGGATGGACAGAACAC	ATGAACTTACATGAATATC	sucC	16.548	773.2	ECOGLTA	EG10981/P07460
TGCCGCAGTGGAGGGGAAATA	ATGTCCATTTTAATCGATA	sucD	16.573	774.4	ECOGLTA	EG10982/P07459
AACATCCAGTGAGAGAGACCG	ATGCATCCGATGCTGAACA	suhB	57.189	2672.2	ECOSUHBA	EG10983/P22783
CTCACAGGGGCTGGATTGATT	ATGTACACTTCAGGCTATG	sulA	22.004	1028.2	ompAeco	EG10984/P08846
GTTAATTGAAATGGAAAAAGT	ATGAAGAACTGGAAAACGC	surA	1.161	54.3	surAeco	EG10985/P21202
GGGTACATAGCGAGGGAAAGT	ATGGAACGTTGCGGCTGGG	tag	79.938	3735.2	ECOTAG	EG10986/P05100
GCCGCCTGATGGGGAGCGTTG	ATGTTTAATCGTATTCGAA	tap	42.426	1982.4	ECOCHE2	EG10987/P07018
TTGTTTTCAGGAAGGTGCCTT	ATGATTAACCGTATCCGCG	tar	42.461	1984.0	ECOCHE2	EG10988/P07017
AGGTAATTAACGTAGGTCGTT	ATGAGCACTATTCTTCTTC	tdcA	70.317	3285.7	ECTDCRAB	EG10989/P11036
TTCAGGCGAAGAGGTTTTATA	ATGCATATTACATACGATC	tdcB	70.294	3284.6	ECTDCRAB	EG10990/P05792
TTAATTCGTTGAGGATAGGAT	ATGAGTACTTCAGATAGCA	tdcC	70.266	3283.2	ECTDCRAB	EG10991/P11867
GGCGATAACATCATTCGTTAT	GTGGTTAATACAAAAAAGG	tdcR	70.343	3286.8	ECTDCRAB	EG10992/P11866
GTTATCGCCTGAGGATGTGAG	ATGAAAGCGTTATCCAAAC	tdh	81.626	3814.0	ECOKBLTDH	EG10993/P07913
ATCATTGAGGGCCTGTGGCTG	ATGGCACAGCTATATTTCT	tdk	27.864	1302.0	ECOTDKG	EG10994/P23331
GCTTTGTTACTGGAGAGTTAT	ATGAGTCAGGCGCTAAAAA	tesB	10.303	481.4	ECOTESB	EG10995
TTCTGACGTAGTGGAGAAAAA	ATGAAATTTGAACTGGACA	tgt	9.262	432.8	ECOTGT	EG10996/P19676
CTTCATTTAAGGGGCCACTAA	ATGGAATTGCAAGACGACA	thdF	83.708	3911.4	thdFeco	EG10997
AAAGGTAACGAGGTAACAACC	ATGCGAGTGTTGAAGTTCG	thrA	0.009	0.4	ECOTHR	EG10998/P00561
CATGGAAGTTAGGAGTCTGAC	ATGGTTAAAGTTTATGCCC	thrB	0.061	2.9	ECOTHR	EG10999/P00547
GCACGAGTACTGGAAAACTAA	ATGAAACTCTACAATCTGA	thrC	0.081	3.8	ECOTHR	EG11000/P00934
ATTACAGAGTACACAACATCC	ATGAAACGCATTAGCACCA	thrL	0.006	0.3	ECOTHR	EG11277/P03059
CACTGCAAATAAGGATATAAA	ATGCCTGTTATAACTCTTC	thrS	38.722	1809.3	thrSeco	EG11001/P00955
CAACACGTTTCCTGAGGAACC	ATGAAACAGTATTTAGAAC	thyA	63.831	2982.6	ECOTHYA	EG11002/P00470
GTGATTTTTTGAGGTAACAAG	ATGCAAGTTTCAGTTGAAA	tig	9.889	462.1	tig-eco	EG11003/P22257
TCAACCAGAAAGAACAATAAC	ATGCTACCAAAAGCCGCCC	tlp	ND	ND	tlp-eco	EG11004
AAATAAATGAAGGATTATGTA	ATGGAAAACTTTAAACATC	tnaA	83.749	3913.3	ECOTNAA	EG11005/P00913
CCTTCCTCTAAAGGTGGCATC	ATGACTGATCAAGCTGAAA	tnaB	83.782	3914.8	ECOTNAB	EG11006/P23173
TCTTGCGAGGATAAGTGCATT	ATGAATATCTTACATATAT	tnaL	83.743	3913.0	ECOTNAA	EG11276/P09408
AACAGTTTTTGGAAACCGAGA	GTGTCAAAGGCAACCGAAC	tolA	16.837	786.7	ECOTOLAB	EG11007/P19934
ATCGGTCCAGATAAGGGAGAT	ATGATGAAGCAGGCATTAC	tolB	16.867	788.1	ECOTOLAB	EG11008/P19935
AATACTGCTTCACCACAAGGA	ATGCAAATGAAGAAATTGC	tolC	68.430	3197.5	ECOTOLCMP	EG11009/P02930
TATTGTCGCGGAGTTTAACCA	GTGACTGACATGAAATATCC	tolQ	16.812	785.5	ECOTOLQRA	EG11010/P05828
GAGAGCAACAAGGGGTAAGCC	ATGGCCAGAGCGCGTGGAC	tolR	16.826	786.2	ECOTOLQRA	EG11011/P05829
AGACCTGGTTTTTCTACTGAA	ATGATTATGACTTCAATGA	tonB	28.219	1318.6	ECOTONB	EG11012/P02929
ATCAAATTAGGTAAGGTGAAT	ATGGGTAAAGCTCTTGTCA	topA	28.652	1338.8	ECTOPA	EG11013/P06612
TGGTTGAGATTCGTTAATTCA	ATGCGGTTGTTTATTGCCG	topB	39.703	1855.2	ECOTOPB	EG11014/P14294
TTATAAGCGTGGAGAATTAAA	ATGCGACATCCTTTAGTGA	tpiA	88.603	4140.1	ECOTPIA	EG11015/P04790
CCGCAATTAAATATTCTGCCC	ATGCGGGGAAGGATGAGAA	tpr	27.708	1294.7	ECOTGY1	EG11016/P02338
TTCGCCAAAGGAGAATGATTG	ATGAAATCCCCCGCACCTT	treA	26.795	1252.0	ECOTREA	EG11017/P13482
CCTGGACGAGAGACAACGGTA	ATGAATACAACTCCCTCAC	trg	32.132	1501.4	ECOTRG	EG11018/P05704
TGAGATAACGGGTCGCGACTG	ATGAAAATTATCATTCTGG	trkA	73.973	3456.5	ECOTRKAG	EG11019
TAATTGAAAGAATATTTAGAT	ATGAATACATCTCATGTAA	trkG	30.651	1432.5	ECTRKG	EG11020
AGAACTAAGGAAGCGGCAGAG	ATGCATTTTCGCCGCCATTA	trkH	86.942	4062.4	pepQeco	EG11021/P21166
TTTAATTTTTCAGGATACATC	ATGACCCCCGAACACCTTC	trmA	89.701	4191.4	ECOTRMA	EG11022/P23003
TAAACGGTAAAAGACGGCGCT	ATGTGGATTGGCATAATTA	trmD	59.004	2757.0	ECOTRMD	EG11023/P07020
GAAAGCACGAGGGGAAATCTG	ATGGAACGCTACGAATCTC	trpA	28.334	1323.9	ECOTGP	EG11024/P00928
CGCATATTAAGGAAAGGAACA	ATGACAACATTACTTAACC	trpB	28.351	1324.7	ECOTGP	EG11025/P00932
CTGGCGGCACGAGGGTAAATG	ATGCAAACCGTTTTAGCGA	trpC	28.377	1325.9	ECOTGP	EG11026/P00909
TCATGCACAGGAGACTTTCTG	ATGGCTGACATTCTGCTGC	trpD	28.406	1327.3	ECOTGP	EG11027/P00904
GAACAAAATTAGAGAATAACA	ATGCAAACACAAAAACCGA	trpE	28.440	1328.9	ECOTGP	EG11028/P00895
CACGTAAAAAGGGTATCGACA	ATGAAAGCAATTTTCGTAC	trpL	28.475	1330.5	ECOTGP	EG11274/P03053
CGCTAACAATGGCGACATATT	ATGGCCCAACAATCACCCT	trpR	99.830	4664.7	ECOTRPR	EG11029/P03032
GCGAAAATCAGGAATCGAAAA	ATGACTAAGCCCATCGTTT	trpS	75.635	3534.1	ECOTRPS	EG11030/P00954
AAGTCAACCTTTAGTTGGTTA	ATGTTACACCAACAACGAA	trxA	85.464	3993.4	ECORHOB	EG11031/P00274
TCACATCTCCGAGGATTTTAG	ATGGCTGAAATTACCGCAT	tsf	4.246	198.4	ECORPSBTS	EG11033/P02997
GGTCCACAGGAAAGAGAAACC	ATGTTAAAACGTATCAAAA	tsr	98.950	4623.5	ECOTSR	EG11034/P02942
TTTTCAAACAGTGGCATACAT	ATGAAAAAAACATTACTGG	tsx	9.372	437.9	ECOTSX	EG11035/P22786
ACAATAGTAAGGAATATAGCC	GTGTCTAAAGAAAAATTTG	tufA	74.711	3490.9	ECOSTR3	EG11036/P02990
CGTGTCTTAGAGGGACAATCG	ATGTCTAAAGAAAAGTTTG	tufB	89.996	4205.2	ECOGTUFB	EG11037/P02990
GTAACTAAAGTGGTTAATATT	ATGGCGCGTTACGATCTCG	tus	36.231	1692.9	tus-eco	EG11038/P16525
GCGTGGCTTAAGAGGTTTATT	ATGGTTGCTGAATTGACCG	tyrA	58.898	2752.1	ECOPHEAB	EG11039/P07023
GTAAACCTGGAGAACCATCGC	GTGTTTCAAAAAGTTGACG	tyrB	91.959	4296.9	tyrBeco	EG11040/P04693
ATCGTCAGGACAGAAGAAAGC	GTGAAAAACAGAACCCTGG	tyrP	42.878	2003.5	ECOTYRPA	EG11041/P18199
TTTTTTCAGGTGAAGGTTCCC	ATGCGTCTGGAAGTCTTTT	tyrR	ND	ND	ECOTYRR	EG11042/P07604

Table 1   *Continued*

-21_to_-1	0_to_+18	Gene	Centisome	Kb	DNA_LOCUS	EG5/SP19_Acc
AGTCTGACAGGGCAACTATTT	ATGAAACGACTCATTGTAG	ubiX	52.353	2446.2	ECOHISPUR2	EG11044/P09550
CGATTCACAGAGGAGTTGTAT	ATGTCCAAGTCTGATGTTT	udp	86.585	4045.8	ECOUDP	EG11045/P12758
ACAGTTCAGGAATTAACCGTA	ATGTCATCATCCCGTCCGG	ugpA	77.309	3612.3	ECOUGP	EG11046/P10905
GTACAACAAGAGAGATAAACG	ATGAAACCGTTACATTATA	ugpB	77.330	3613.3	ECOUGP	EG11047/P10904
TGGTCGATAGTGAGAAATAAG	ATGGCAGGACTGAAATTAC	ugpC	77.268	3610.4	ECOUGP	EG11048/P10907
TGAAAGCAAGGTGCGTTACCA	ATGATTGAGAACCGTCCGT	ugpE	77.291	3611.5	ECOUGP	EG11049/P10906
TGGTGAAACAGGACAACGAGT	ATGAGTAACTGGCCTTATC	ugpQ	77.252	3609.7	ECOUGPQQ	EG11050/P10908
TTTACCCGCCAGGACAAGACC	ATGATCACCGTTGCCCTTA	uhpA	82.916	3874.3	ECOUHP	EG11051/P10940
CCGCATGTTTGATGGCTGGTG	ATGAAGACGTTGTTCTCCC	uhpB	82.884	3872.8	ECOUHP	EG11052/P09835
CAACGCTATGTCTAAGGTTTG	ATGATGTTGCCGTTTCTGA	uhpC	82.855	3871.5	ECOUHP	EG11053/P09836
TATTCATTTCAGGAGTAACCC	ATGCTGGCTTTCTTAAACC	uhpT	82.823	3870.0	ECOUHP	EG11054/P13408
TATCTTAATGAGGAGTCCCTT	ATGTTACGTCCTGTAGAAA	uidA	36.455	1703.4	ECOUIDAA	EG11055/P05804
CGTCGTTAAGGCGATGCGCTG	ATGTTTGCCCTCTGTGATG	umuC	26.483	1237.5	ECOUMUCD	EG11056/P04152
TATAACTTCAGGCAGATTATT	ATGTTGTTTATCAAGCCTG	umuD	26.474	1237.0	ECOUMUCD	EG11057/P04153
GAAGATTCGCAGGAGAGCGAG	ATGGCTAACGAATTAACCT	ung	58.368	2727.3	ECOUNG	EG11058/P12295
CTATTTCTGGAGTAAACCACC	ATGTCTGAAGGCTGGAACA	usg	52.515	2453.8	ECOHIST1	EG11059/P08390
AATCAGGTCAGGGAGAGAAGT	ATGAAATTATTGCAGCGGG	ushA	10.946	511.5	ECOUSHA	EG11060/P07024
CGTTTAATCCGGGAAAGGTGA	ATGGATAAGATCGAAGTTC	uvrA	92.036	4300.5	ECOUVRAA	EG11061/P07671
AACTCCTTCAGGTAGCGACTC	ATGAGTAAACCGTTCAAAC	uvrB	17.608	822.8	ECOUVRB	EG11062/P07025
AGCCAGCCAGGCGTTTATCGC	ATGTACGATGCTGGTGGTA	uvrC	42.947	2006.7	ECOUVRC	EG11063/P07028
TATTTTTACGCGGCGGTGCCA	ATGGACGTTTCTTACCTGC	uvrD	86.166	4026.2	ECOUVRD	EG11064/P03018
CTAATCAGAAGGGAACCCATT	GTGAAAACACTAAATCGTC	uxaB'	56.132	2623.4	ECOUXAB	EG11065
CTATGTGAAAGAGGAAAAATC	ATGGAACAGACCTGCGCTG	uxuA'	ND	ND	ECOUXU1	EG11066
TGAATACGGCAACCTGGAAAA	ATGGAAAAGACATATAACC	valS	96.536	4510.7	ECOVALS	EG11067/P07118
TGCGTCAGCAAGAGGCACAAC	ATGGCCGACGTTCACGATA	vsr	43.738	2043.7	ECODCM	EG11068/P09184
GGAGCGTTGGATTGAACGCGT	ATGACCGATTTACACACCG	xerC	86.130	4024.5	ECOXERC	EG11069/P22885
CTGAAAAGAACGGGAAGATTT	ATGAAGAAAGGTTTTATGT	xprA	65.401	3055.9	ECORECJXPR	EG11070/P21892
GAGTGACCATAAGGGGCGCAA	GTGAAACAGGATCTGGCAC	xprB	65.417	3056.7	ECORECJXPR	EG11071/P21891
TTAGAATTTGATCTCGCTCAC	ATGTTACCTTCTCAATCCC	xse	56.546	2642.2	ECOXSEA	EG11072/P04994
TCTGAATAAATGGCAGCGACT	ATGAAATTTGTCTCTTTTA	xthA	39.421	1842.0	ECOXTHA	EG11073/P09030
TACCTGATTATGGAGTTCAAT	ATGCAAGCCTATTTTGACC	xylA	80.316	3752.8	ECOXYLABA	EG11074/P00944
TTTTTTTAAGGAACGATCGAT	ATGTATATCGGGATAGATC	xylB	80.283	3751.3	ECOXYLABA	EG11075/P09099
AATGGTCTAAGGCAGGTCTGA	ATGAATACCCAGTATAATT	xylE	91.382	4269.9	ECOXYLE	EG11076/P09098
AGAGATACCACTCTTCACCTG	ATGCAGCCCGCTTACTGCT	xylU	ND	ND	ECOXYLUP1	EG11077/P05056
AGTTAACTTAAGGAGAATGAC	ATGGCGGTAACGCAAACAG	zwf	41.636	1945.5	ECOZWF	EG11221/P22992

Table 2   Individual Information Content of Selected Translation Initiation Regions Given in Bits per Site

xylU	-5.05	rts	-1.05	racC	15.48	fepA	16.17
plsB	-3.71	cdsA	-0.79	srlA	15.68	fimB	16.42
rcsC	-3.05	dicB	-0.48	dapB	15.73	nlpA	16.74
katF	-2.40	mcrA	-0.21	rpmD	15.89	proW	16.83
cysX	-1.66	speC	0.00	tnaA	15.93	speF	17.10

Figure 1   *E. coli* ribosome binding sites

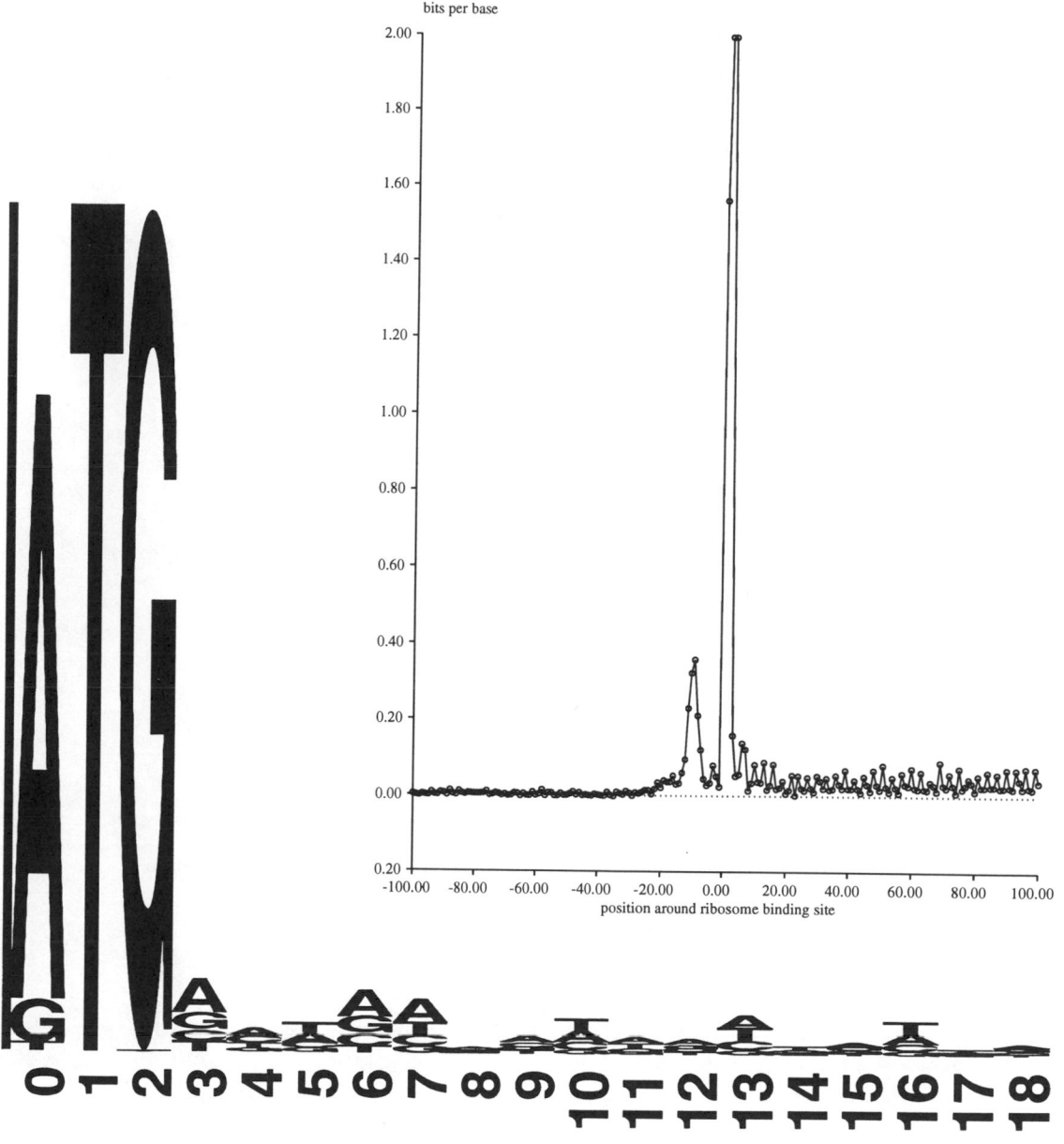

Figure 1 *Continued*

D. Transcription Termination

d'Aubenton Carafa and co-workers have compiled and aligned 148 rho-independent transcription termination sequences. Table 1 is reprinted, with permission, from d'Aubenton Carafa et al. (*J. Mol. Biol.*, vol. 216, pp. 839–843 [1990]). The references cited in the table in parentheses can be found in the original paper.

Table 1   *E. coli* **Rho-independent Terminator Sequences**

A. *Sequences found in the literature*

Terminators with UUCG and variant loops

```
ECOPONA      2671   GCACAGGAATTGTTCTGATTAAAAAAGGCGCTTCGGCGCCTTTTCAGTTTGCTGAC        (1)
                             ---->    <----
ECORELB      1293   GCATCCTCAACGCACCCGCACTTAACCCGCTTCGGCGGGTTTTTGTTTTTATTTT        (2)
                             ---->    <----
ECORNPB       482   CAGTTTCACCTGATTTACGTAAAAACCCGCTTCGGCGGGTTTTTGCTTTTGGAGG        (3)
                             ---->    <----
ECORPLRPO   11299   CGTTAATCCGCAAATAACGTAAAAAACCCGCTTCGGCGGGTTTTTTTATGGGGGGA       (4)
                             ---->    <----
ECORPSI      1065   CGGCAGAAAACAATTTTCGAAAAAAACCCGCTTCGGCGGGTTTTTTTTATAGGGAAG      (5)
                             ---->    <----
ECOTGGA       434   CAAAGCAATAAGCAGTGTCGTGAAACCACCTTCGGGTGGTTTTTTTGTGCCTGCA        (6)
                             ---->    <----
ECOTGRHLP     553   CTCTCTCGCCGACCAATTTTGAACCCCGCTTCGGCGGGGTTTTTTGTTTTCTGTG        (7)
                            ----->    <-----
ECOTCY1 †   *1655   AAGCGAATTTTAGTTCACATAGACCCTGCTTCGGCGGGGTTTTTTTATGGGCACG        (8)
                            ----->    <-----
ECOPHEAB      380   CTTTATATTGAGTGTATCGCCAACGCGCCTTCGGGCGCGTTTTTTGTTGACAGCG        (9)
                            ----->    <-----
ECOTGOP       875   CGTACCCCAGCCACATTAAAAAAAGCTCGCTTCGGCGAGCTTTTTGCTTTTCTGCG       (10)
                            ----->    <-----
ECOGLYA      1751   TTTTTCGTTTATGATCATCAAGGCTTCCTTCGGGAAGCCTTTCTACGTTATCGCG        (11)
                            ------>    <------
ECORGNG       307   GCCGTCCAGTAAATGATAAAACGAGCCCTTCGGGGCTCGTTTTTGTCTATAAGTT        (12)
                            ------>    <------
ECOHIMA       814   TTCAAGTAGCACTGTCTGTGGAGCCTTCGGGCTCACAGCTTTGTTCGCCTTTGTT        (13)
                           ---- ---->     <--------
ECOPROC       908   CGGACGTCAGGCCGCCACTTCGGTGCGGTTACGTCCGGCTTTCTTTGCTTTGTAA        (14)
                   -------  ------ ->     <----------------


ECOSERS      1629   ATACCCAATTTTTCTGAATCTAAAAACGCCTGCGGGCGCTTTTTTTTGTCTCCCTT       (15)
                             --->    <---
ECOOMPC      1590   TCTAATCTCGATTGATATCGAACAAGGGCCTGCGGGCCCTTTTTTTCATTGTTTTC       (16)
                             ---->    <----
ECOPRS       1456   ACACTAATCGAACCCGGCTCAAAGACCCGCTGCGGCGGGTTTTTTTGTCTGTAAT        (17)
                             ---->    <----
ECORPA       3071   AGTAATCTGAAGCAACGTAAAAAAAACCCGCCCCGGCGGGTTTTTTTATACCCGTA       (18)
                             ---->    <----
ECORPMBG      650   TCTCGCTTTGATGTAACAAAAAAACCCCGCCCCGGCGGGGTTTTTTGTTATCTGCT       (19)
                            ----->    <-----
ECOFPG         31   CTCGCTTTGATGTAACAAAAAAAACCTCGCTCCGGCGGGGTTTTTGTTATCTGCTT       (20)
                            ----->    <-----
ECOGLNSA     2152   AGAAACAGCAAACAATCCAAAACGCCGCGTTCGGCGGCGTTTTTCTGCTTTTCTT        (21)
                            ----->    <-----
ECOFUR        713   AAGCCAGCCTGAACGAGAAAAGCCAACCTGCGGGTTGGCTTTTTTTATGCAAGGGA       (22)
                           ------>    <------
ECORPSOP     2794   GAGTAAGGTTGCCATTTGCCCTCCGCTGCGGCGGGGGGCTTTTAACCGGGCAGGA        (23)
                          -------->     <--------
```

Table 1   *Continued*

Terminators with GAAA and variant loops

ECOCYSB   518   TGTCAGGTTTTTATAAACAAAGGGTCGCGAAAGCGGCCCTTTTTTATTCGCATAT   (24)
                                    ------>   <-------

ECORPLN   84   CACTCTCTCAATACGAATAAACGGCTCAGAAATGAGCCGTTTATTTTTTCTACCC   (25)
                                    ------>   <-------

ECORGNX3   547   CCGCCCCTGCCAGAAATCATCCTTAGCGAAAGCTAAGGATTTTTTTTATCTGAAA   (26)
                                    ------->   <-------

ECORGNB   6613   CATCAAATAAAACGAAAGGCTCAGTCGAAAGACTGGGCCTTTCGTTTTATCTGTT   (27)
                                    -------->   <--------

ECOUNCC   10490   GCTTGAAAAGCACAAAAGCCAGTCTGGAAACAGGCTGGCTTTTTTTTGCGCGTGT   (28)
                                    -------->   <--------

ECOILVGE   231   CTTAACGAACTAAGACCCCCGCACCGAAAGGTCCGGGGGTTTTTTTTGACCTTAA   (29)
                                    ------- --> <-- -------

ECORPSRPO   966   TGTAGTTGTAAGGCCGTGCTTCCGAAAGGAATGCGCGGCTTATTTTCGTTTATGA   (30)
                                    ----------->   <--- -------

ECORPST   542   AATTTGCTGAAGCTTTGTGAAAAAGCCCGCGCAAGCGGGTTTTTTTTATGCCTGCT   (31)
                                    ---->   <----

ECOHTPR   989   TCCGCTATTAAGCAGAGAACCCTGGATGAGAGTCCGGGGTTTTTGTTTTTTGGGC   (32)
                                    ------->   <-------

ECORGNDIS   588   CATCAAATAAAACAAAAGGCTCAGTCGGAAGACTGGGCCTTTGTTTTATCTGTT   (33)
                                    -------->   <--------

ECODEOD3   67   CGATTGCCTTGTGAAGCCGGAGCGGGAGACTGCTCCGGCTTTTTAGTATCTATTC   (34)
                                    --------->   <---------

Other terminators

ECOACE   5664   TCTGGTGATGTAAGTAAAAGAGCCGGCCCAACGGCCGGCTTTTTTTCTGGTAATCT   (35)
                                    ------>   <-------

ECOACE   7439   GGAACATCCGGCAATTAAAAAAGCGGCTAACCACGCCGCTTTTTTTTACGTCTGCA   (35)
                                    ---->   <----

ECOAMPCFR   4095   GCATCGCCAATGTAAATCCGGCCCGCCTATGGCGGGCCGTTTTGTATGGAAACCA   (36,37)
                                    -------->   <-------

ECOAMPCFR   5279   CTCTACAGTAAAATTCCATCGGGTCCGAATTTTCGGACCTTTTCTCCGCTTTTCC   (36,37)
                                    ------>   <-------

ECOARAABD   4328   ACAGCCAATCAAACGAAACCAGGCTATAATCAAGCCTGGTTTTTTTGATGGAATTA   (38)
                                    ------>   <-------

ECOAROG   1654   ATTGTTAGCAACAAAAAAGCCGACTCACTTGCAGTCGGCTTTCTCATTTTAAACG   (39)
                                    ------>   <-------

ECOASPAW   1556   GGGTAGTACAAATAAAGAAGGCACGTCAGATGACGTGCCTTTTTTTCTTGTGAGCA   (40)
                                    -------->   <--------

ECOBGLO   514   CAAAACCTGACATAACCAGAGAATACTGGTGAAGTCGGGTTTTTTTGTTTATAAA   (41)
                                    ------ ----> <---- -------

ECOBGLO   1471   GCAAACCTGAAAAAAAATTGCTTGATTCACGTCAGGCCGTTTTTTTCAGGTTTTT   (41)
                                    ------->   <------

ECOBIRA   164   GGCATCAAATTAAGCAGAAGGCCATCCTGACGGATGGCCTTTTTGCGTTTCTACA   (42)
                                    -------->   <--------

ECOCARAB   54   TTTAATTTATTGATTATAAAGGGCTTTAATTTTTGGCCCTTTTATTTTTGGTGTT   (43,44)
                                    ---->   <------

ECOCARAB   4896   TTTGATCGAATAACTAATACGGTTCTCTGATGAGGACCGTTTTTTTTTGCCCATT   (43,44)
                                    -------->   <--------

ECOCDH   299   CAGGATAATTATCAAACCCGGTGGTTTCTCGCGACCGGGTTTTTTATTTGTCACG   (45)
                                    ------ ->   <- ------

ECOCDHA   1220   GCGCACTTTGTCAGCAATATGAGGCGGATTTCTTCCGCCTTTTTTAATCCCTCAAC   (46)
                                    ----->   <-----

ECOCHE2   17   TTCCTCAATTGAAATGAACCCGATGATCTGCGCATCGGGTTTTTTATTTCAATTT   (47)
                                    ------>   <------

ECOCRP   999   TAATCCCGTCGGAGTGGCGCGTTACCTGGTAGCGCGCCATTTTGTTTCCCCCGAT   (48)
                                    ----------->   <-----------

ECOCYSB   1705   TTTTCGGCACCTTTTATGTAGCGAAGGTGCCGGAATATATTCTCTTTTGTTACTT   (24)
                                    --------------> <------------- ---

ECODAPD   1035   GACTAAAAGTATGCACACGGGCAGCACGACGCTGCCCGATTTTTTTGCAGGGATG   (49)
                                    ------->   <-------

Table 1 *Continued*

```
ECODLD      2003   GTTCTGTGCCGTCTGCCCCGCCGCCGCCATTTGGGCGGCTTTTTGTTTTTTATAG    (50)
                        ----->        <-----
ECODYE       819   CTTTACCACCGTCAAAAAAAACGGCGCTTTTTAGCGCCGTTTTTATTTTTCAACC    (51)
                        ------>       <------
ECOENVAA    1074   GGCAAATCTGGCACTCTCTCCGGCCAGGTGAACCAGTCGTTTTTTTTTGAATTTT    (52)
                     --   ------>       <------    -------
ECOFDHF     2194   CGTAATACCGTCCTTTCTACAGCCTCCTTTCGGAGGCTGTTTTTTTATCCATTCG    (53)
                        ------->      <------
ECOFOLC     1056   GCCTGATAACTGATAAGGGCAGGGCCACTGGCTCTGCCCTTTTGCTATTCTCACC    (54)
                        --------->    <---------
ECOFUMA     1984   GCACCCGCTGTGTGAAATAAACAGAGCGCCTTCGGGGCGTTTTTTTACATGGCAC    (55)
                          ---->     <-----
ECOGLGPA    2488   TAAGTTCACCAATAAATAGAACGGGGCCAAAGGGTCCCGTTTTTTTCCGCCATCA    (56)
                         ------>      <------
ECOGLNAL     478   GAAGTTTGTCAGCTATCTGTAGCCCATCTCTGCATGGGCTTTTTTCTCCGTCAAT    (57)
                        ----->        <-----
ECOGLNSA     142   CATCCAACAATGACAAGCGGTGGAGATCTTCTCTGCCGCTTTTTTTTTCATCAAT    (21)
                        --------->    <---------
ECOGLTA    *1072   TGTAAATATTTTAACCCGCCGTTCATATGGCGGGCTTGATTTTTATATGCCTAAA    (58)
                      ----------->   <------ ----
ECOGLTA    13006   GCTGTCGGTTTCGACATGGTTGGCCATCGTATGATGGCCTTTTTTGTGCTTATCG    (59)
                        ------>       <------
ECOGLTX     1421   CGTGTAAAAAGATAAACGGCAGGAGATAATATCCTGCCGTTTTTTATTTATGCTG    (60)
                        ------->      <-------
ECOGLYS     3169   TAATAACGCCGTTATTAAATAGCCTGCCATCTGGCAGGCTTTTTTTATCGCTAAA    (61)
                        ------>       <------
ECOGPT       642   CCGGTCGCTAATCTTTTCAACGCCTGGCACTGCCGGGCGTTGTTCTTTTTAACTT    (62)
                        ------->      <-------
ECOGUABA    1768   GGGCTCCTGATTCTCTTCGCCCGACTTCATGTCGGGCGATTTATATTATCTGTTT    (63)
                        --------->    <--------
ECOGUABA    3433   GCACTATGAATGAACAAAACCCTCTGTTACTACAGAGGGTTTTTTATCTTCAAGA    (64)
                        --------->    <--------
ECOHIMA     1502   CTGGCATAAGCCAGTTGAAAGAGGGAGCTAGTCTCCCTCTTTTCGTTTCAACGCC    (65)
                        ------>       <------
ECOHIMA     1705   GAAACGGAAAACAGCGCCTGAAAGCCTCCCAGTGGAGGCTTTTTTTGTATGCGCG    (65)
                        ----->        <-----
ECOHIMA     5646   CGAGTAATCTGATCTAACTAAAAAGGCCGCTCTGCGGCCTTTTTTCTTTTCACTG    (66)
                        ----->        <-----
ECOHIP        23   CGAGTAATTCTCTGACTCTTCGGGATTTTTATTCCGAAGTTTGTTGAGTTTACTT    (67)
                        --------->    <---------
ECOHIP       478   TACTCAAACTTGAACGAGAGAAAAGCACCTGTCGGGTGCTTTTTTTCATTTCTCTA    (67)
                        ----->        <-----
ECOHISOP     581   ACGCATGAGAAAGCCCCCGGAAGATCACCTTCCGGGGGCTTTTTATTGCGCGGTT    (68)
                        ---------->   <----------
ECOHISPUR   3974   CATAAAAAGTCGATGGCGTTGAATATTTTTTCAGCGCCATTTTTATTGATGCGCG    (69)
                        --------->    <----------
ECOHISPUR   5457   CAGGGCGCATAACTCCGCTGTTGCCCTGTTTCAGGGCAATTTTGCAACCGCGATC    (69)
                        -------->     <--------
ECOHLY      4407   AATAGCAATCTTACTGGGCTGTGCCACATAAGATTGCTATTTTTTTGGAGTCATA    (70)
                     -----------  -->  <--  -----------
ECOHU2       449   TGGCAGTGAACAGTTTTAACGAAGGGGTGGTTTCACCCCTTTTGTCTTTCTGGCG    (71)
                        ----->        <-----
ECOILVBPR    261   ACACGATTCCAAAACCCCGCCGGCGCAAACCGGGCGGGGTTTTTCGTTTAAGCAC    (72)
                        -------->     <--- -----
ECOILVBPR   2346   CTTGAACAACATCGCGCTTATCGTTAAGGTAAGCGCGTATTTTTTTTACCCGCCA    (73)
                        ---------->   <----------
ECOINFA      278   CGAAGAGAAAGAACGAGTAAAAGGTCGGTTTAACCGGCCTTTTTATTTTAT       (74)
                        ------>       <------
ECOLEUA      155   AGCACGCAGTCAAACAAAAAAACCCGCGCCATTGCGCGGGTTTTTTTATGCCCGAA    (75)
                        ------>       <------
ECOLEXA      709   AACATATCTCTGAGACCGCGATGCCGCCTGGCGTCGCGGTTTGTTTTTCATCTCT    (76)
                        --------->    <---------
ECOMALM     1177   GCTGATTTTGCAACAACTGGTGCGGTCTCTTGGCGCACCTTTTTTTATGCTTCCT    (77)
                        ----->        <-----
ECOMAP       997   TCTCGCACGACGAATAAGATGAAGCCGGCGAATGCCGGCTTTTTTTAATGCGATAA    (78)
                        ----->        <-----
```

Table 1 *Continued*

ECOMETLB2	135	TAGTTTATATTTGCAGTCCGGTTTGCTTTGCATACCGGATTTTCTTTTTCTTACC ------- --> <-- ------	(79)
ECOMOTAB	62	AGGCGCAACATTCCAGCAGCGGTAACGACGTACCGCTGCTTTTTTTTGCCCCAAT --------> <---------	(80)
ECOMUTD	1536	AAATGAGTTCAGAGAGCCGCAAGATTTTTAATTTTGCGGTTTTTTTGTATTTGAA ---------> <---------	(81)
ECONUSA	270	TTATAGGGTTCAGTTATATAAAGCCCCGATTTATCGGGGTTTTTTGTTATCTGAC -----> <-----	(82)
ECONUSA	311	TTTTGTTATCTGACTACAGAATAACTGGGCTTTAGGCCCTTTTTTTAAGTCTTGG ----> <-----	(82)
ECONUSA	5137	TCCACCTGAAATTAATTTTAAAAAGGGGCTAACAGCCCCTTTTTTGTCAGGAGAA -----> <-----	(83)
ECONUSB	566	GTTCCATGATCTGGCAGCGATGGGGGCCGATCGCCGGCCTTTTCTTTTTACCTGC ----> <----	(84)
ECOOMPB	2308	AAAGAAGGGTAAATAAACGGGAGGCGAAGGTGCCTCCCGTTTTGCTTTCTATAAG --------> <---------	(85)
ECOOMPF	1	AAAACTAATCCGCATTCTTATTGCGGATTAGTTTTTTCTTAGCTAAT ---------> <---------	(86)
ECOOMPF	1554	CTCTTTGTTAAATGCCGAAAAAACAGGACTTTGGTCCTGTTTTTTTTATACCTTC -----> <-----	(86)
ECOPAPA	2044	CTGATAATCCGGTCGGTAAACAGCGGAAATATTCCGCTGTTTATTTCTCAGGGTA --------> <--------	(87)
ECOPHEA	101	GCGAAGACGAACAATAAAGGCCTCCCAAATCGGGGGGCCTTTTTTTATTGATAACA --------> <--------	(88)
ECOPHEAB	*1831	TAATAATCCAGTGCCGGATGATTCACATCATCCGGCACCTTTTCATCAGGTTGGA -----------> <-----------	(9)
ECOPHEV	372	GGGCACCACTAATTCTTAAGAACCCGCCCACAAGGCGGGTTTTTGCTTTTGGATC -----> <-----	(89)
ECOPHOAA	1696	TAAAACCGCGCCCGGCAGTGAATTTTCGCTGCCGGGTGGTTTTTTTGCTGTTAGC -------------> <-------------	(90)
ECOPHOM	4585	AATAAGGTTGAAAAATAAAAACGGCGCTAAAAAGCGCCGTTTTTTTTGACGGTGG ------> <------	(91)
ECOPHORG	1363	TTTGTCATCTTTTATTGCCATAAGCCAGTCGATGCTGGCTTATTTTCTTTGCAGT -----> <-----	(92)
ECOPOLA	3297	TGAATATTTTAGCCGCCCCAGTCAGTAATGACTGGGGCGTTTTTTATTGGGCGAA -----------> <-----------	(93)
ECOPTSHI	2791	TAATTCTTGCCGCAGTGAAAAATGGCGCCATCGGCGCCATTTTTTATGCTTCCG ------> <------	(94)
ECOPTSLPM	3069	CTGTAAGACTGTTGTACACTACCGGGGCCTTTTGGCCCCTTTTTTTATCTGGAGG -----> <-----	(95)
ECOPYRBI	549	CCCGTTGATCACCCATTCCCAGCCCCTCAATCGAGGGGCTTTTTTTTGCCCAGGC ------> <------	(96)
ECOPYRC	759	TTCCATTTACTGATTAATCACGAGGGCGCATTCGCGCCCTTTATTTTTCGTGCAA -----> <-----	(97)
ECOPYRE	1072	CGCTGGCAAACTGATTTTTAAGGCGACTGATGAGTCGCCTTTTTTTTGTCTGTAG ------> <------	(98)
ECOR1PARB	451	TCCGGTAAGTAGCAACCTAGAGGCGGGCGCAGGCCCGCCTTTTCAGGACTGATGC ------> <------	(99)
ECORECA	1299	ATCGTCTTGTTTGATACACAAGGGTCGCATCTGCGGCCCTTTTGCTTTTTTAAGT ------> <------	(100)
ECORECC	612	ACGAAACATCCTGCCAGAGCCGACGCCAGTGTGCGTCGGTTTTTTTTACCCTCCGT ------> <------	(101)
ECORECFA	3713	CAAGAAGCTATAAGAAAAGGGCGGAGATCATCTCCGCCCTTTTTATTTCTGCAAT --------> <---------	(102)
ECORGNF	371	TCCGCCACTTATTAAGAAGCCTCGAGTTAACGCTCGAGGTTTTTTTTCGTCTGTA -------> <-------	(103)
ECORHO	1728	ATTTGTCTTATGCCAAAAACGCCACGTGTTTACGTGGCGTTTTGCTTTTATATCT -------> <-------	(104)
ECORPLRPO	43	CGTTGCACAAGGCGTGAGATTGGAATACAATTTCGCGCCTTTTGTTTTTATGGGC -----------> <------------	(105)
ECORPLRPO	2677	AAATCAGGCTGATGGCTGGTGACTTTTTAGTCACCAGCCTTTTGCGCTGTAAGGC -----------> <----------	(105)
ECORPSBTS	1166	GTAGTACACGTTTGGTTAGGGGGCCTGCATATGGCCCCCTTTTTCACTTTTATAT ------> <------	(106)
ECORPSBTS	2107	AGCAGTCTTAATTATCAAAAAGGAGCCGCCTGAGGGCGCTTTTTGTGCCCATCTT ---> <---	(106)

Table 1 *Continued*

ECORPSFRI	1896	TTCAACGAGACGTAAAAAGCGCCCGACCATTGGTCGGCGTTTTGCTTTCTATTTT -----> <-----	(107)
ECORPSO	417	ATTCTTGCGAGTTTCAGAAAAGGGGGCCTGAGTGGCCCCTTTTTTCAAGCTGACG -----> <-----	(108)
ECORPSRPO	4861	CACTAGGCCCTCTGCACAAACGCCACCTTTTCGGTGGCGTTTTTTATCGCCCACG ------> <-------	(109)
ECORRNE	116	GGCATCAAATTAGAAAAACCCCGGTCCATAAGGCCGGGGTTTTTTGCATATCAAT -------> <-------	(110)
ECOSOD	799	ATATCAGCTTAAAAAATGAACCATCGCCAACGGCGGTGGTTTTTTTGTGATCAAT -------> <-------	(111)
ECOSODB	870	GTTGTCAGCAACTGTAACGCAGAAGGTTATCCTTCTGCGTTTTTGTTTAATTAGC --------> <-------	(112)
ECOSPDE	1797	TGTTTAACGGCTCTGGCGGAGCTCCCAGGCTCCGCCAGATTTATTTACTTCTGCT -----------> <------------	(113)
ECOSPPA	2197	TAAGTCTTGTACTGAGTGGCCGACAGATCGTCGGCCACATTATTTTTTACGTCGA --------> <--------	(114)
ECOTGLSP	340	TCGGCACCAAAAGTATGTAAATAGACCTCAACTGAGGTCTTTTTTTTATGCCTGAA -----> <-----	(115)
ECOTGP	211	GTAAAGCAATCAGATACCCAGCCCGCCTAATGAGCGGGCTTTTTTTTGAACAAAA -----> <-----	(116)
ECOTGP	6800	GCGCAGTTAATCCCACAGCCGCCAGTTCCGCTGGCGGCATTTTAACTTTCTTTAA --------> <--------	(116)
ECOTGPRO	315	TCACCGACCAAATTCGAAAAGCCTGCTCAACGAGCAGGCTTTTTTGCATCTGCAG -------> <-------	(117)
ECOTGS	214	CCGCCAAATAAGATAAGGGGTTAGCTAAATGCTAACCCCTTTTTCTTTTGCCTGT --------> <--------	(118)
ECOTGS	1089	CACCCGCTAACACTTTCATAGCCTCGCTTTATGCGGGGCTTTGTTTTTTGTTACA ------> <------	(118)
ECOTGTUFB	1918	TTTGACGCAATGCGCACTAAAGGGCATCATTTGATGCCCTTTTTGCACGCTTTCG ------> <------	(119)
ECOTHR	123	AGGAAACACAGAAAAAAGCCCGCACCTGACAGTGCGGGCTTTTTTTTTCGACCAA --------> <--------	(120)
ECOTHR	4889	TATCTCAATCAGGCCGGGTTTGCTTTTATGCAGCCCGGCTTTTTTATGAAGAAAT ------ ---> <----------	(121)
ECOTONB	1048	ATTCAGTAAGCAGAAAGTCAAAAGCCTCCGACCGGAGGCTTTTGACTATTACTCA -----> <-----	(122)
ECOTONB	*1108	TACCTGTTGAGTAATAGTCAAAAGCCTCCGGTCGGAGGCTTTTGACTTTCTGCTT -----> <-----	(122)
ECOTRMD	1706	TTAATATGACACCGGACTCCGTTCCTCGATGGGGTCCGGTTGTTTTATTCACACA ------------> <------------	(123)
ECOTRMD	3787	CGACATCCTGTTAAGAAGGGCTGGCCAATTGGCTGGCCCTTTTTTTATCTGTTTGC --------> <--------	(123)
ECOTRXB	1069	AAAAGTAAAGAAGGCGACACCATGCGACTATGGGTCGCCTTTATTTTTTCCCCGT ------ ----> <----------	(124)
ECOUNCC	14188	CGGTTGAGTAATAAATGGATGCCCTGCGTAAGCGGGGCATTTTTCTTCCTGTTAT -------> <--------	(28)
ECOUVRAA	3066	GCCGATGCTGTAATCGTTAAGGCCGCTTTCTGAGCGGCCTTTTCCTTTCAGAGTT -------> <--------	(125)
ECOUVRC	1205	CAGATGCAAAAACCGGCTGAAAGGCACGCTATCAGCCGGTTTTATATTACTGACG -------- -> <- ---------	(126)
ECOUVRD	2567	GGAGTCGGTGTAACGTTCCCGGATGCGGTGCTGCGCACCTTATTTGGCCTAAAAA ----> <----	(127)
ECOVALS	2999	AAAACACAGTGATGAAAACGAAGGCCGGAGCATGCTCCGTTTTTTATCTCTTACA ----> <----	(128)
ECOXYLABA	1547	GCAGTCCGTTGGCCCGGTTATCGGTAGCGATACCGGGCATTTTTTTAAGGAACGA ------ ----> <----------	(129)
ECOXYLE	1881	CTTCCTGTCCAGCACGCCGCGCCATTTCGGCGTGCTGACTTTTTACTCCCGCTTC ---------> <---------	(130)

We show all the sequences of the terminators (sample A). Each line contains the name of the file in GenBank® containing the sequence, the position in the sequence of the 1st base (5' end), and the reference number (in general from GenBank®); the arrows indicate the complementary sequences (not including the poly(T) stretch).

Restriction Enzyme Specificities

Roberts and Macelis (1991) have compiled all of the restriction enzymes known (see also Kessler and Manta 1990). Table 1, which was provided by R.J. Roberts, lists restriction enzymes commercially available as of January 1992. The database is also available as a computer file, as are monthly updates. Requests for the entire database and updates should be sent by electronic mail to: roberts@cshl.org.

REFERENCES

Kessler, C. and V. Manta. 1990. Specificity of restriction endonucleases and DNA modification methyltransferases: A review. *Gene* **92:** 1–248.

Roberts, R.J. and D. Macelis. 1991. Restriction enzymes and their isoschizomers. *Nucleic Acids Res.* (Suppl.) **19:** 2077–2109.

Table 1   Commercially Available Restriction Enzymes and Prototypes

	Type I enzymes	
Enzyme	Recognition sequence[2]	Me site[3]
_Cfr_AI	GCANNNNNNNNGTGG	
_Eco_AI	GAGNNNNNNNGTCA	2(6),-3(6)
_Eco_BI	TGANNNNNNNNTGCT	3(6),-4(6)
_Eco_DI	TTANNNNNNNGTCY	
_Eco_DXXI	TCANNNNNNNATTC	
_Eco_EI	GAGNNNNNNNATGC	
_Eco_KI	AACNNNNNNGTGC	2(6),-3(6)
_Eco_R124I	GAANNNNNNRTCG	
_Eco_R124/3I	GAANNNNNNNRTCG	-3(6)
_Sty_SBI	GAGNNNNNNRTAYG	2(6),-4(6)
_Sty_SJI	GAGNNNNNGTRC	
_Sty_SPI	AACNNNNNNGTRC	2(6),-3(6)
_Sty_SQI	AACNNNNNNRTAYG	

	Type III enzymes	
Enzyme	Recognition sequence[2]	Me site[3]
_Eco_PI	AGACC	3(6)
_Eco_P15I	CAGCAG(25/27)	5(6)
_Hin_fIII	CGAAT	
_Sty_LTI	CAGAG	4(6)

Table 1 *Continued*

Type II enzymes

Enzyme[1]	Isoschizomers	Recognition[2] Sequence	Me[3] site	Commercial[4] source
*Aat*II		GACGT↑C		EGJLMNOPRSUVX
*Acc*I		GT↑MKAC		ABDEGIJKLMNOPQRSUVX
*Aci*I		CCGC(-3/-1)		N
*Acy*I		GR↑CGYC		EMRV
	*Aha*II	GR↑CGYC		G
	*Bbi*II	GR↑CGYC		AK
	*Bsa*HI	GR↑CGYC		N
	*Hin*1I	GR↑CGYC		DFOU
*Afl*II		C↑TTAAG		ABGJKNU
	*Bfr*I	C↑TTAAG		M
	*Esp*4I	C↑TTAAG		F
*Afl*III		A↑CRYGT		BGJMNU
*Age*I		A↑CCGGT		N
*Aha*III		TTT↑AAA		
	*Dra*I	TTT↑AAA		ABDEFGIJKLMNOPQRSUVX
*Alu*I		AG↑CT	3(5)	ABDEFGHIJKLMNOPQRSUVX
*Alw*NI		CAGNNN↑CTG		NU
*Apa*I		GGGCC↑C	4(5)	BDEGIJKLMNOPRUVX
	*Bsp*120I	G↑GGCCC		DF
*Apa*BI		GCANNNNN↑TGC		
*Apa*LI		G↑TGCAC		EGJKNUX
	*Alw*44I	G↑TGCAC		DFORU
	*Sno*I	G↑TGCAC		BJLMV
	*Vne*I	G↑TGCAC		D
*Asc*I		GG↑CGCGCC		N
*Asu*I		G↑GNCC		R
	*Bsi*ZI	G↑GNCC		W
	*Cfr*13I	G↑GNCC	4(5)	DFKOU
	*Nsp*IV	G↑GNCC		J
	*Sau*96I	G↑GNCC		BDEGJLMNORVX
*Asu*II		TT↑CGAA		J
	*Bpu*14I	TT↑CGAA		D
	*Bsi*CI	TT↑CGAA		UW
	*Bsp*119I	TT↑CGAA		DF

Table 1   *Continued*

	*Bst*BI	TT↑CGAA		N
	*Csp*45I	TT↑CGAA		ORV
	*Lsp*I	TT↑CGAA		JL
	*Nsp*V	TT↑CGAA		ABGKOP
	*Sfu*I	TT↑CGAA		M
*Ava*I		C↑YCGRG		ABEGIJKLMNOPRSUVX
	*Ama*87I	C↑YCGRG		D
	*Bco*I	C↑YCGRG		W
	*Eco*88I	C↑YCGRG		DF
	*Nsp*III	C↑YCGRG		JP
*Ava*II		G↑GWCC		ABDEGIJKMNPRSX
	*Bme*18I	G↑GWCC		D
	*Eco*47I	G↑GWCC		DFOU
	*Nsp*HIII	GGWCC		J
	*Sin*I	G↑GWCC	4(5)	JLRSV
*Ava*III		ATGCAT		GJ
	*Eco*T22I	ATGCA↑T		KOU
	*Mph*1103I	ATGCA↑T		F
	*Nsi*I	ATGCA↑T		BELMNRVX
	*Ppu*10I	A↑TGCAT		F
	*Zsp*2I	ATGCA↑T		D
*Avr*II		C↑CTAGG		N
	*Bln*I	C↑CTAGG		K
*Bae*I		ACNNNNGTAYC		
*Bal*I		TGG↑CCA	4(5)	AIJKRSVX
	*Msc*I	TGG↑CCA		BNU
*Bam*HI		G↑GATCC	5(4)	ABDEFGHIJKLMNOPQRSUVWX
	*Bst*I	G↑GATCC		JP
*Bbv*I		GCAGC(8/12)	2(5),-2(5)	EIJNX
	*Bst*71I	GCAGC(8/12)		R
*Bbv*II		GAAGAC(2/6)		
	*Bbs*I	GAAGAC(2/6)		N
*Bcc*I		CCATC		
*Bcef*I		ACGGC(12/13)		
*Bcg*I[5]		GCANNNNNTCG(12/10)		N
*Bcl*I		T↑GATCA		BDFGIJLMNOPRSUVX
	*Bsi*QI	T↑GATCA		W
	*Fba*I	TGATCA		K
*Bet*I		W↑CCGGW		

Table 1 *Continued*

*Bgl*I		GCCNNNN↑NGGC	ABDEFGHIJLMNOPQRSUVWX
*Bgl*II		A↑GATCT	ABDEFGHIJKLMNOPQRSUVWX
*Bin*I		GGATC(4/5)	
	*Alw*I	GGATC(4/5)	NU
*Bpu*10I		CCTNAGC(-5/-2)	
*Bsa*AI		YAC↑GTR	N
*Bsa*BI		GATNN↑NNATC	N
	*Bsi*BI	GATNN↑NNATC	W
	*Mam*I	GATNN↑NNATC	M
*Bse*PI		GCGCGC	
	*Bss*HII	G↑CGCGC	BDEGJLMNOQRUVX
*Bsg*I		GTGCAG(16/14)	N
*Bsi*I		CTCGTG(-5/-1)	
*Bsi*YI		CCNNNNN↑NNGG	MUW
	*Bsl*I	CCNNNNN↑NNGG	N
*Bsm*I		GAATGC(1/-1)	DEJLMNOUVX
	*Bsc*CI	GAATGC	W
*Bsm*AI		GTCTC(1/5)	NU
	*Alw*26I	GTCTC(1/5)	DFR
*Bsp*GI		CTGGAC	
*Bsp*HI		T↑CATGA	NU
	*Rsp*XI	T↑CATGA	G
*Bsp*MI		ACCTGC(4/8)	NU
*Bsp*MII		T↑CCGGA	
	*Acc*III	T↑CCGGA	DEGJKQRV
	*Bse*AI	T↑CCGGA	M
	*Bsi*MI	T↑CCGGA	W
	*Bsp*13I	T↑CCGGA	D
	*Bsp*EI	T↑CCGGA	N
	*Kpn*2I	T↑CCGGA	F
	*Mro*I	T↑CCGGA	MOU
*Bsr*I		ACTGG(1/-1)	N
*Bst*EII		G↑GTNACC	BEGJLMNOPRSUVX
	*Bst*PI	G↑GTNACC	K
	*Eco*91I	G↑GTNACC	DF
	*Eco*O65I	G↑GTNACC	GK
*Bst*XI		CCANNNNN↑NTGG	BEJKLMNOQRUVX
*Cac*8I		GCN↑NGC	
*Cau*II		CC↑SGG	

Table 1 *Continued*

	*Bcn*I	CC↑SGG	2(4)	DFK
	*Nci*I	CC↑SGG		BEGJLMNOUVWX
*Cfr*I		Y↑GGCCR	4(5)	
	*Eae*I	Y↑GGCCR	4(5)	EGJKLMNUVX
*Cfr*10I		R↑CCGGY	2(5)	ADFKMNOU
*Cla*I		AT↑CGAT	5(6)	ABDGJKMNPQRSVX
	*Ban*III	AT↑CGAT		OU
	*Bsc*I	AT↑CGAT		JL
	*Bsi*XI	AT↑CGAT		W
	*Bsp*106I	AT↑CGAT		E
	*Bsp*DI	AT↑CGAT		N
	*Bsu*15I	AT↑CGAT		DF
*Cvi*JI		RG↑CY	3(5)	
*Cvi*RI		TG↑CA	4(6)	
*Dde*I		C↑TNAG	1(5)	BEGIJLMNOPRUVX
*Dpn*I*		GA↑TC		ABEIJLMNRSUVX
*Dra*II		RG↑GNCCY		EGJM
	*Eco*O109I	RG↑GNCCY		DFJKLNOUVX
	*Pss*I	RGGNC↑CY		I
*Dra*III		CACNNN↑GTG		EMNUX
*Drd*I		GACNNNN↑NNGTC		N
*Drd*II		GAACCA		
*Dsa*I		C↑CRYGG		M
*Eam*1105I		GACNNN↑NNGTC		DFN
*Eci*I		TCCGCC		
*Eco*31I		GGTCTC(1/5)		DF
	*Bsa*I	GGTCTC(1/5)		N
*Eco*47III		AGC↑GCT		DFKLMNORU
*Eco*57I		CTGAAG(16/14)	5(6),5(6)	DFN
*Eco*NI		CCTNN↑NNNAGG		NU
*Eco*RI		G↑AATTC	3(6)	ABDEFGHIJKLMNOPQRSUVWX
*Eco*RII[6]		↑CCWGG	2(5)	BEGJOUV
+	*Apy*I	CC↑WGG		M
+	*Bsi*LI	CC↑WGG		W
+	*Bst*NI	CC↑WGG	2(4)	EJNX
+	*Bst*OI	CC↑WGG		R
	*Ecl*136I	CCWGG		D
+	*Mva*I	CC↑WGG	2(4)	ADFKMOU
*Eco*RV		GAT↑ATC	2(6)	ABDEGIJKLMNOPQRSUVWX

Table 1   *Continued*

	*Eco*32I	GAT↑ATC	DF
*Esp*I		GC↑TNAGC	GJU
	*Bpu*1102I	GC↑TNAGC	DFN
	*Cel*II	GC↑TNAGC	LM
*Esp*3I		CGTCTC(1/5)	DFN
*Fau*I		CCCGC(4/6)	
*Fin*I		GTCCC	
*Fnu*DII		CG↑CG	
	*Acc*II	CG↑CG	DEGJKQVX
	*Bsp*50I	CG↑CG	DF
	*Bst*UI	CG↑CG	NU
	*Mvn*I	CG↑CG	DM
	*Tha*I	CG↑CG	BI
*Fnu*4HI		GC↑NGC	N
*Fok*I		GGATG(9/13)	3(6),-2(6) DEFGIKMNRUVX
*Fse*I		GGCCGG↑CC	
*Fsi*I		R↑AATTY	
*Gdi*II		YGGCCG(-5/-1)	
*Gsu*I		CTGGAG(16/14)	DFN
	*Bpm*I	CTGGAG(16/14)	N
*Hae*I		WGG↑CCW	
*Hae*II		RGCGC↑Y	BDEIJKLMNOPRSUVX
	*Bsp*143II	RGCGC↑Y	DF
*Hae*III		GG↑CC	3(5) ABDGHIJKLMNOPQRSUVX
	*Bsh*I	GGCC	W
	*Bsu*RI	GG↑CC	3(5) DFGJ
	*Pal*I	GG↑CC	EJPV
*Hga*I		GACGC(5/10)	DNUX
*Hgi*AI		GWGCW↑C	NX
	*Alw*21I	GWGCW↑C	DF
	*Asp*HI	GWGCW↑C	M
*Hgi*CI		G↑GYRCC	
	*Acc*B1I	G↑GYRCC	D
	*Ban*I	G↑GYRCC	DEGIJMNOPRUVX
	*Eco*64I	G↑GYRCC	DF
*Hgi*EII		ACCNNNNNNGGT	
*Hgi*JII		GRGCY↑C	
	*Ban*II	GRGCY↑C	BEGIJKLMNOPRSUVX
	*Eco*24I	GRGCY↑C	DF

Table 1   *Continued*

*Hha*I		GCG↑C	2(5)	BEGJKNOPRSUX
	*Cfo*I	GCG↑C		BIJLMRV
	*Hin*6I	G↑CGC		DF
	*Hin*P1I	G↑CGC		NX
*Hin*dII		GTY↑RAC	5(6)	EM
	*Hinc*II	GTY↑RAC		ABDEFGHIJKLNOPQRSUVX
*Hin*dIII		A↑AGCTT	1(6)	ABDEFGHIJKLMNOPQRSUVWX
*Hin*fI		G↑ANTC		ABDEFGIJKLMNOPQRSUVWX
*Hpa*I		GTT↑AAC	5(6)	ABDEFGIJKLMNOPQRSUVX
*Hpa*II		C↑CGG	2(5)	ABDEFGJLMNOPQRSUVX
	*Hap*II	C↑CGG		GIK
	*Msp*I	C↑CGG	1(5)	ABDEFGIJKLMNOPQRSUVWX
*Hph*I		GGTGA(8/7)	-2(5)	NUVX
*Kpn*I		GGTAC↑C	4(6)	ABDEFGIJKLMNOPQRSUVX
	*Acc*65I	G↑GTACC		DFN
	*Asp*718I	G↑GTACC		M
*Ksp*632I		CTCTTC(1/4)		M
	*Eam*1104I	CTCTTC(1/4)		DF
	*Ear*I	CTCTTC(1/4)		N
*Mae*I		C↑TAG		M
	*Rma*I	C↑TAG		N
*Mae*II		A↑CGT		M
*Mae*III		↑GTNAC		M
*Mbo*I[7]		↑GATC		BDEFGIJKNPQRSVX
	*Bsp*143I	GATC		DF
+	*Bsp*AI	↑GATC		J
	*Dpn*II	↑GATC	2(6)	NU
	*Nde*II	↑GATC		BGM
+	*Sau*3AI	↑GATC	4(5)	ABDEGIJKLMNOPQRSUVX
*Mbo*II		GAAGA(8/7)	5(6)	BDFGIJKNOPQRSUVX
*Mcr*I		CGRY↑CG		M
	*Bsi*EI	CGRY↑CG		NW
*Mfe*I		C↑AATTG		
	*Mun*I	C↑AATTG	3(6)	FN
*Mlu*I		A↑CGCGT		ABDEFGIJKLMNOPQRSUVX
*Mly*I		GACTC(5/5)		
*Mme*I		TCCRAC(20/18)		
*Mnl*I		CCTC(7/6)		EGJNUX
*Mse*I		T↑TAA		NU

Table 1 *Continued*

	*Tru*9I	T↑TAA		D
*Msl*I		CAYNNNNRTG		
*Mst*I		TGC↑GCA		X
	*Aos*I	TGC↑GCA		J
	*Avi*II	TGC↑GCA		M
	*Fdi*II	TGC↑GCA		U
	*Fsp*I	TGC↑GCA		NSU
*Mwo*I		GCNNNNN↑NNGC		
	*Bsp*WI	GCNNNNN↑NNGC		M
*Nae*I		GCC↑GGC		EKLMNOUVX
*Nar*I		GG↑CGCC		BDEJMNOPRUVX
	*Bbe*I	GGCGC↑C		K
	*Ehe*I	GGC↑GCC		FOU
	*Kas*I	G↑GCGCC		N
	*Nun*II	GG↑CGCC		GJ
*Nco*I		C↑CATGG		ABDEFGHIJKLMNOPQRSUVWX
*Nde*I		CA↑TATG	4(6)	BEFGKLMNPSUVX
*Nhe*I		G↑CTAGC		BEGJKLMNOPRUVX
*Nla*III		CATG↑		NU
*Nla*IV		GGN↑NCC		NU
	*Bsc*BI	GGN↑NCC		I
*Not*I		GC↑GGCCGC		ABDEFGIJKLMNOPQRSUVWX
*Nru*I		TCG↑CGA		BDEGIJKLMNOPQUVWX
	*Bsp*68I	TCG↑CGA		DF
	*Spo*I	TCG↑CGA		R
*Nsp*I		RCATG↑Y		AKMU
	*Nsp*HI	RCATG↑Y		J
*Nsp*BII		CMG↑CKG		JU
*Pac*I		TTAAT↑TAA		N
*Pfl*1108I		TCGTAG		
*Pfl*MI		CCANNNN↑NTGG		NU
	*Acc*B7I	CCANNNN↑NTGG		D
	*Van*91I	CCANNNN↑NTGG		FM
*Ple*I		GAGTC(4/5)		NU
*Pma*CI		CAC↑GTG		AK
	*Bbr*PI	CAC↑GTG		M
	*Eco*72I	CAC↑GTG		DF
	*Pml*I	CAC↑GTG		NU
*Pme*I		GTTT↑AAAC		N

Table 1 *Continued*

*Ppu*MI		RG↑GWCCY		NU
	*Psp*5II	RGGWCCY		F
*Psh*AI		GACNN↑NNGTC		K
*Pst*I		CTGCA↑G	5(6)	ABDEFGHIJKLMNOPQRSUVWX
*Pvu*I		CGAT↑CG		ABDEFGJKLMNOPQRSUVX
	*Bsp*CI	CGAT↑CG		E
	*Xor*II	CGAT↑CG		B
*Pvu*II		CAG↑CTG	4(4)	ABDEFGIJKLMNOPQRSUVWX
	*Psp*5I	CAGCTG		D
*Rle*AI		CCCACA(12/9)		
*Rsa*I		GT↑AC		ABDEGIJLMNOPQRSUVWX
	*Afa*I	GT↑AC		K
	*Csp*6I	G↑TAC		DF
*Rsr*II		CG↑GWCCG		BEJMNUX
	*Cpo*I	CG↑GWCCG		DFK
	*Csp*I	CG↑GWCCG		ORV
*Sac*I		GAGCT↑C		ADEGIJKLMNOPQRSUVWX
	*Ecl*136II	GAG↑CTC		FN
	*Sst*I	GAGCT↑C		B
*Sac*II		CCGC↑GG		EIJLNOPRUVX
	*Cfr*42I	CCGC↑GG		DF
	*Kpn*378I	CCGC↑GG		D
	*Ksp*I	CCGC↑GG		EM
	*Mra*I	CCGCGG		GJ
	*Sfr*303I	CCGC↑GG		D
	*Sst*II	CCGC↑GG		B
*Sal*I		G↑TCGAC		ABDEFGHIJKLMNOPQRSUVX
	*Rtr*I	G↑TCGAC		D
*Sap*I		GCTCTTC(1/4)		
*Sau*I		CC↑TNAGG		M
	*Aoc*I	CC↑TNAGG		E
	*Axy*I	CC↑TNAGG		GJV
	*Bse*21I	CC↑TNAGG		D
	*Bsu*36I	CC↑TNAGG		NR
	*Cvn*I	CC↑TNAGG		B
	*Eco*81I	CC↑TNAGG		ADFKOU
	*Mst*II	CC↑TNAGG		EX
*Sca*I		AGT↑ACT		ABDEFGIJKLMNOPRSUVX
*Scr*FI		CC↑NGG		EGMNOSUVX

Table 1 *Continued*

	*Dsa*V	↑CCNGG		M
*Sdu*I		GDGCH↑C		DFJ
	*Bmy*I	GDGCH↑C		M
	*Bsp*1286I	GDGCH↑C		EGKNRUX
	*Nsp*II	GDGCH↑C		J
*Sec*I		C↑CNNGG		
	*Bsa*JI	C↑CNNGG		N
*Sfa*NI		GCATC(5/9)		DNUX
*Sfe*I		C↑TRYAG		
	*Sfc*I	C↑TRYAG		N
*Sfi*I		GGCCNNNN↑NGGCC		ABDEGIJLMNOPQRSUVX
*Sgr*AI		CR↑CCGGYG		M
*Sma*I		CCC↑GGG	2(4)	ABDEFGIJKLMNOPQRSUVWX
	*Cfr*9I	C↑CCGGG	2(4)	DFOU
	*Psp*AI	C↑CCGGG		E
	*Xma*I	C↑CCGGG		DEINRUVX
*Sna*I		GTATAC		
	*Bst*1107I	GTA↑TAC		DFN
*Sna*BI		TAC↑GTA		EGJKLMNRVX
	*Eco*105I	TAC↑GTA		DFOU
*Spe*I		A↑CTAGT		BEGKLMNORSUVWX
*Sph*I		GCATG↑C		ABDEGIJKLMNOPQRSUX
	*Bbu*I	GCATG↑C		RV
	*Pae*I	GCATG↑C		DF
*Spl*I		C↑GTACG		AK
	*Bsi*WI	C↑GTACG		MNUW
	*Pfl*23II	C↑GTACG		F
*Sse*8387I		CCTGCA↑GG		AK
*Ssp*I		AAT↑ATT		BDEFGKLMNORUVX
*Stu*I		AGG↑CCT		ABEGIJKLMNPRVX
	*Aat*I	AGG↑CCT		OU
	*Eco*147I	AGG↑CCT		DF
*Sty*I		C↑CWWGG		BDEGJMNRUVX
	*Bss*T1I	C↑CWWGG		D
	*Eco*130I	C↑CWWGG		DFU
	*Eco*T14I	C↑CWWGG		AK
*Swa*I		ATTT↑AAAT		M
*Taq*I		T↑CGA	4(6)	ABDEFGIJLMNOPQRSUVWX
	*Tth*HB8I	T↑CGA	4(6)	K

Table 1   *Continued*

*Taq*II[8]		GACCGA(11/9)		
		CACCCA(11/9)		
*Tfi*I		G↑AWTC		N
*Tsp*45I		GTSAC		
*Tsp*EI		AATT		
*Tth*111I		GACN↑NNGTC		EGIJKNPRUVX
	*Asp*I	GACN↑NNGTC		M
*Tth*111II		CAARCA(11/9)		
*Vsp*I		AT↑TAAT		DFKR
	*Ase*I	AT↑TAAT		NU
	*Asn*I	AT↑TAAT		M
*Xba*I		T↑CTAGA	6(6)	ABDEFGHIJKLMNOPQRSUVWX
*Xcm*I		CCANNNNN↑NNNNTGG		NU
*Xho*I		C↑TCGAG		ABDEFGHIJKLMNOPQRSUVX
	*Ccr*I	C↑TCGAG		X
	*Pae*R7I	C↑TCGAG	5(6)	NX
	*Sla*I	C↑TCGAG		D
*Xho*II		R↑GATCY		EGMRVX
	*Bst*YI	R↑GATCY		NU
	*Mfl*I	R↑GATCY		AK
*Xma*III		C↑GGCCG	4(5)	B
	*Bst*ZI	C↑GGCCG		R
	*Eag*I	C↑GGCCG		N
	*Ecl*XI	C↑GGCCG		M
	*Eco*52I	C↑GGCCG		DEFKORU
*Xmn*I		GAANN↑NNTTC		DEGJNUX
	*Asp*700I	GAANN↑NNTTC		M

Notes to Table 1

1. * signifies that *Dpn*I and its isoschizomers require the presence of 6-methyladenosine within the recognition sequence GATC.

2. Recognition sequences are given using the standard abbreviations (Eur. J. Biochem. 150: 1-5, 1985) to represent ambiguity:

R	=	G or A
Y	=	C or T
M	=	A or C
K	=	G or T
S	=	G or C
W	=	A or T
H	=	A or C or T
B	=	G or T or C
V	=	G or C or A
D	=	G or A or T
N	=	A or C or G or T

3. The site of methylation by the cognate methylase when known is indicated as follows. The first number shows the base within the recognition sequence that is modified. A negative number indicates the complementary strand, numbered from the 5' base of that strand. The number in parentheses indicates the specific methylation involved. (6) = N6-methyladenosine; (5) = 5-methylcytosine; (4) = N4-methylcytosine.

4. Commercial sources of restriction enzymes are abbreviated as follows:

A	Amersham (12/91)
B	BRL (6/91)
D	Palliard Chemical (5/91)
E	Stratagene (9/91)
F	Fermentas (6/91)
G	BioExcellence (formerly Anglian) (10/91)
H	American Allied (6/91)
I	IBI (6/91)
J	Janssen Biochimica (2/91)
K	Takara (10/91)
L	Northumbria Biologicals Ltd. (10/91)
M	Boehringer Mannheim (9/91)
N	New England Biolabs (7/91)
O	Toyobo (6/91)
P	PL-Pharmacia-LKB (9/91)
Q	Molecular Biology Resources (10/91)
R	Promega Corporation (10/91)
S	Sigma (6/91)
U	USB (10/91)
V	Serva (2/91)
W	ILS (5/91)
X	New York Biolabs (4/91)
Y	P.C. Bio (9/91)

5. *Bcg*I cleaves on both sides of the recognition sequence: 10 bases 5' to the recognition sequence and 12 bases 3' to it on both strands. Thus the recognition site is excised in a fragment, 34 base pairs long, with 2-base 3'-extensions at each end.

6. *Eco*RII isoschizomers fall into two classes based upon their sensitivity to methylation. *Eco*RII will not cleave when the second cytosine in the recognition sequence is methylated to 5-methylcytosine whereas *Mva*I will cleave such a sequence. Isoschizomers of *Eco*RII that are like *Mva*I are indicated by +.

7. *Mbo*I isoschizomers fall into two classes based upon their sensitivity to methylation. *Mbo*I will not cleave when the recognition sequence contains 6-methyladenosine whereas *Sau*3AI will not cleave when its recognition sequence contains 5-methylcytosine. Isoschizomers of *Mbo*I that are like *Sau*3AI are indicated by +.

8. *Taq*II recognizes two distinct sequences: GACCG and CACCCA.

SECTION · 19

The Genetic Code and Codon Usage in Selected Organisms

Figure 1, which is redrawn, with permission, from Suzuki et al. (1989), shows the genetic code. Some rare exceptions to the code in different organisms have been reviewed by Fox (1987) and Jukes and Osawa (1990). In addition, recent evidence for the evolution of the genetic code is reviewed by Osawa et al. (1992).

Table 1, which is excerpted from Wada et al. (1991), depicts their compilation of codon usage for 1187 genes in *E. coli*, 130 genes in *S. typhimurium*, 206 genes in *B. subtilis*, 65 genes in phage λ, and 101 genes in phage T4. Wada et al. (1991) have compiled the codon usage from over 15,000 genes in GenBank. A tape or disk listing the codon usage can be obtained by writing to:

Dr. T. Ikemura
National Institute of Genetics
Mishima, Shizuoka-ken 411
Japan

REFERENCES

Fox, T.D. 1987. Natural variation in the genetic code. *Annu. Rev. Genet.* **21:** 67–91.

Jukes, T.H. and S. Osawa. 1990. The genetic code in mitochondria and chloroplasts. *Experientia* **46:** 1117–1126.

Osawa, S., T.H. Jukes, K. Watanabe, and A. Muto. 1992. Recent evidence for evolution of the genetic code. *Microbiol. Rev.* **56:** 229–264.

Suzuki, T., A.J.F. Griffits, J.H. Miller, and R.C. Lewontin. 1989. *Introduction to Genetic Analysis*, Fourth Edition. W.H. Freeman, New York.

Wada, K., Y. Wada, H. Doi, F. Ishibashi, T. Gojobori, and T. Ikemura. 1991. Codon usage tabulated from the GenBank genetic sequence data. *Nucleic Acids Res.* (Suppl.) **19:** 1981–1986.

Second letter

First letter		U	C	A	G	Third letter
U		UUU ⎫ Phe UUC ⎭ UUA ⎫ Leu UUG ⎭	UCU ⎫ UCC ⎬ Ser UCA ⎪ UCG ⎭	UAU ⎫ Tyr UAC ⎭ UAA Stop UAG Stop	UGU ⎫ Cys UGC ⎭ UGA Stop UGG Trp	U C A G
C		CUU ⎫ CUC ⎬ Leu CUA ⎪ CUG ⎭	CCU ⎫ CCC ⎬ Pro CCA ⎪ CCG ⎭	CAU ⎫ His CAC ⎭ CAA ⎫ Gln CAG ⎭	CGU ⎫ CGC ⎬ Arg CGA ⎪ CGG ⎭	U C A G
A		AUU ⎫ AUC ⎬ Ile AUA ⎭ AUG Met	ACU ⎫ ACC ⎬ Thr ACA ⎪ ACG ⎭	AAU ⎫ Asn AAC ⎭ AAA ⎫ Lys AAG ⎭	AGU ⎫ Ser AGC ⎭ AGA ⎫ Arg AGG ⎭	U C A G
G		GUU ⎫ GUC ⎬ Val GUA ⎪ GUG ⎭	GCU ⎫ GCC ⎬ Ala GCA ⎪ GCG ⎭	GAU ⎫ Asp GAC ⎭ GAA ⎫ Glu GAG ⎭	GGU ⎫ GGC ⎬ Gly GGA ⎪ GGG ⎭	U C A G

Figure 1   The genetic code.

Table 1   Codon Usage in Selected Organisms

Source:		E. coli	S. typh.	B. subt.	λ	T4
No. genes:		1,187	130	206	65	101
ARG	CGA	3.1	3.3	3.7	7.1	5.7
	CGC	22.0	24.3	8.6	16.6	6.0
	CGG	4.6	5.3	5.5	10.0	1.1
	CGU	24.7	22.1	9.0	16.8	19.3
	AFA	2.0	2.3	11.7	9.3	8.7
	AGG	1.3	1.6	3.7	4.7	1.7
LEU	CUA	3.0	4.0	5.2	3.2	6.8
	CUC	9.8	10.0	9.2	9.2	4.1
	CUG	54.8	54.6	20.7	36.3	6.5
	CUU	9.9	9.8	23.2	13.6	20.1
	UUA	10.3	12.0	19.4	8.7	26.5
	UUG	11.2	11.9	12.9	5.7	10.5
SER	UCA	6.3	6.2	14.7	13.3	17.6
	UCC	9.6	11.4	8.2	11.2	3.3
	UCG	7.9	8.8	6.1	8.8	4.0
	UCU	10.4	8.5	15.2	6.8	24.5
	AGC	15.0	16.6	13.8	17.1	5.6
	AGU	7.1	6.6	6.2	11.4	10.9
THR	ACA	6.4	5.0	23.4	13.2	17.3
	ACC	24.6	25.1	7.6	21.1	6.5
	ACG	12.5	16.4	13.3	19.2	5.2
	ACU	10.5	7.7	9.7	9.1	27.3
PRO	CCA	8.1	5.6	6.9	8.8	13.3
	CCC	4.2	5.4	2.6	4.8	1.3
	CCG	24.2	25.3	14.1	17.2	5.7
	CCU	6.5	7.6	11.0	7.4	14.9
ALA	GCA	20.6	12.1	21.7	28.8	20.5
	GCC	23.7	27.2	12.3	27.7	5.5
	GCG	33.3	40.3	18.5	25.5	6.5
	GCU	17.8	14.1	19.4	17.0	33.3
GLY	GGA	6.7	6.3	22.0	13.1	20.3
	GGC	30.7	36.0	22.4	21.4	9.3
	GGG	9.6	10.8	9.6	14.5	4.1
	GGU	28.0	19.7	14.8	19.7	31.5
VAL	GUA	11.8	11.9	15.6	10.0	18.9
	GUC	14.3	18.4	16.4	11.4	5.6
	GUG	25.3	25.9	16.8	24.0	5.2
	GUU	20.4	14.9	20.0	18.9	32.4
LYS	AAA	36.9	35.7	55.3	37.1	63.8
	AAG	11.9	11.8	20.5	20.0	16.9
ASN	AAC	24.2	23.2	20.3	21.3	15.8
	AAU	15.9	17.8	23.4	18.7	42.2
GLN	CAA	13.0	12.3	21.7	9.8	23.5
	CAG	30.1	32.5	18.2	33.1	11.4
HIS	CAC	11.0	9.2	7.3	6.9	4.6
	CAU	11.5	11.2	14.6	10.7	12.9
GLU	GAA	43.7	40.7	53.8	37.2	57.6
	GAG	19.3	21.5	24.1	27.5	10.4

Table 1   *Continued*

Source:		E. coli	S. typh.	B. subt.	λ	T4
No. genes:		1,187	130	206	65	101
ASP	GAC	22.3	22.8	20.5	24.8	14.9
	GAU	32.0	33.6	34.4	32.3	47.6
TYR	UAC	13.4	12.4	11.1	12.4	9.6
	UAU	14.9	15.5	21.5	17.6	31.0
CYS	UGC	6.2	5.7	3.8	8.6	3.2
	UGU	4.7	4.1	3.1	3.7	6.5
PHE	UUC	18.2	15.9	13.4	14.8	11.0
	UUU	18.5	20.5	26.8	19.2	31.5
ILE	AUA	3.8	4.5	8.7	8.7	11.6
	AUC	27.1	26.9	26.9	20.7	11.4
	AUU	27.0	27.5	37.0	23.0	50.7
MET	AUG	26.5	25.7	26.3	27.7	26.0
TRP	UGG	12.8	10.9	8.9	16.7	14.4
TER	UAA	2.0	2.0	2.0	1.8	2.8
	UAG	0.2	0.1	0.4	0.4	0.3
	UGA	0.8	0.8	0.9	2.5	1.1
TOTAL		395,727	45,068	62,101	14,149	24,026

Frequency (per one thousand) of codon usages summed up for individual organisms and phages. The name of each species is listed at the top of the column in abbreviated form. The number of genes summed for each species is listed on the line specified as "No. genes," and the total codon numbers thus summed is listed on the bottom line.

Natural and Synthetic Nonsense Suppressors

Table 1 shows nonsense suppressors derived by single base substitutions in tRNA genes of *E. coli;* Table 2 shows the suppressors derived by double and triple base changes; and Table 3 shows the nonsense suppressors in phage T4. Consult Eggertsson and Söll (1988) for a recent review on nonsense suppressors. Table 4 compares the efficiencies of synthetic nonsense suppressors with those generated in vivo (see also Kleina and Miller 1990; Kleina et al. 1990).

Tables 1–3 are reprinted, with permission, from Eggertsson and Söll (*Microbiol. Rev.*, vol. 52, pp. 356, 358, and 360, respectively [1988]). The references cited in the tables and table notes can be found in the original paper. Table 4 is reprinted, with permission, from Kleina and Miller (*J. Mol. Biol.*, vol. 212, p. 300 [1990]). The references cited in the table and table notes can be found in the Introduction to Unit 4 of the Laboratory Manual portion of this volume and in the original paper.

REFERENCES

Eggertsson, G. and D. Söll. 1988. Transfer ribonucleic acid-mediated suppression of termination codons in *Escherichia coli. Microbiol. Rev.* **52:** 354–374.

Kleina, L.G. and J.H. Miller. 1990. Genetic studies of the *lac* repressor. XIII. Extensive amino acid replacements generated by the use of natural and synthetic nonsense suppressors. *J. Mol. Biol.* **212:** 295–318.

Kleina, L.G., J.-M. Masson, J. Normanly, J. Abelson, and J.H. Miller. 1990. Construction of *Escherichia coli* amber suppressor tRNA genes. II. Synthesis of additional tRNA genes and improvement of suppressor efficiency. *J. Mol. Biol.* **213:** 705–717.

Table 1   Termination Suppressors Derived by Single Base Substitutions in tRNA Genes of *E. coli*

tRNA[a]	Gene	Suppressor symbol	Map position (min)[b]	Anticodon change[c]	Suppressor type	Alternate symbols and comments	Reference(s)[d]
$tRNA^{Gln}_{CUG}$	*glnV*(α,β)	*supE*	15.5	CUG-CUA	Amber	Su_{II}, Su-2; in the *metT* operon	95
$tRNA^{Leu}_{CAA}$	*leuX*	*supP*	97	CAA-CUA[e]	Amber	Su-6	203, 229
$tRNA^{Ser}_{CGA}$	*serU*	*supD*	43	CGA-CUA	Amber	su_I, Su-1	193, 194, 204
$tRNA^{Trp}_{CCA}$	*trpT*	*supU*	84.5	CCA-CUA	Amber	Su-7; in the *rrnC* operon; inserts glutamine	159, 189
$tRNA^{Tyr}_{GUA1}$	*tyrT* (*tyrV*)[f]	*supF*	27	GUA-CUA	Amber	su_{III}, Su-3, s_{Yme1}	76
$tRNA^{Tyr}_{GUA2}$	*tyrU*	*supZ*	90	GUA-CUA	Amber	In the *thrU*(*tufB*) operon	Pétursdóttir et al. unpublished data
$tRNA^{Gln}_{UUG}$	*glnU*(α,β)	*supB*	15.5	UUG-UUA	Ochre	su_B; in the *metB* operon	147
$tRNA^{Glu}_{UUC}$	*gltT*	*glT*(SuUAA/G)	90	UUC-UUA	Ochre	$tRNA^{Glu}$-Su_{oc}; in the *rrnB* operon; obtained in vitro	160
$tRNA^{Lys}_{UUU}$	*lysT*(α,β)	*supL*	16.5	UUU-UUA	Ochre	*supG*, suβ, Su-5; in an operon with *valT*	156, 230
$tRNA^{Lys}_{UUU}$	*lysV*	*supN*	52	UUU-UUA	Ochre	Coding sequences of *lysT* and *lysV* genes are identical	207
$tRNA^{Tyr}_{GUA1}$	*tyrT* (*tyrV*)	*supC*	27	GUA-UUA	Ochre	su_C, Su-4, *supO*	5
$tRNA^{Tyr}_{GUA2}$	*tyrU*	*supM*	90	GUA-UUA	Ochre	*sup15B*	52, 54, 146
$tRNA^{Gly}_{UCC}$	*glyT*	*glyT*(SuUGA)	90	UCC-UCA	Opal	In the *thrU*(*tufB*) operon; also reads UGG	157
$tRNA^{Trp}_{CCA}$	*trpT*	*trpT*(SuUGA)	84.5	CCA-UCA	Opal	Su-7-UGA; inserts tryptophan; also reads UGG	159, 189, 219
$tRNA^{Trp}_{CCA}$	*trpT*	*Su-9*	84.5	None	Opal	G-A change in position 24; inserts tryptophan	88, 159

[a] For sequences of tRNAs, see reference 192.

[b] From reference 12.

[c] The changed nucleotide is underlined.

[d] Sequences of tRNA genes are given in reference 192. Organization of tRNA operons is described in reference 64.

[e] A second mutational change (A to G in position 27) which was found in one analysis of the *supP* gene (229) is apparently not important for suppressor function (203).

[f] *tyrT* and *tyrV* are tandemly duplicated tRNA genes in the same transcription unit. Other tandemly duplicated genes are denoted by Greek letters, e.g., *glnU*(α,β).

Table 2   Termination Suppressors Derived by Double or Triple Base Substitutions

tRNA[a]	Gene	Suppressor symbol	Map position (min)[b]	Anticodon change	Suppressor type	Alternate symbols and comments	Reference(s)[c]
$tRNA^{Gly}_{UCC}$	glyT	glyT(SuUAG-8)	90	UCC-UCCA-UCUA	Amber	Suppressor tRNA has 8 base pairs in anticodon loop	134
$tRNA^{Gly}_{CCC}$	glyU	glyU(SuUAG)	61.5	CCC-CUA	Amber		128
$tRNA^{Ser}_{VGA}$	serT	serT(SuUAG)	22	VGA-CUA	Amber	Obtained by directed mutagenesis in vitro	J. Rogers and D. Söll, unpublished data
$tRNA^{Gln}_{CUG}$	glnV(α,β)	glnV(SuUAA/G)	15.5	CUG-UUA	Ochre	$Su\text{-}2_{oc}$; derived from supE	94, 144
$tRNA^{Gly}_{UCC}$	glyT	glyT(SuUAA/G)	90	UCUA-UUA	Ochre	Derived from glyT(SuUAG-8)	135
$tRNA^{Gly}_{CCC}$	glyU	glyU(SuUAA/G)	61.5	CCC-UUA	Ochre		128
$tRNA^{Gly}_{GCC}$	glyV(α,β,γ)	glyV(SuUAA/G)	95	GCC-UUA	Ochre	Derived in three steps	130, 133
$tRNA^{Leu}_{CAA}$	leuX	leuX(SuUAA/G)	97	CAA-UUA	Ochre	Derived from supP in vitro	Rogers and Söll, unpublished data
$tRNA^{Ser}_{CGA}$	serU	serU(SuUAA/G)	43	CUA-UUA	Ochre	supDoc; derived from supD in vivo and in vitro	130, 150; Rogers and Söll, unpublished data
$tRNA^{Trp}_{CCA}$	trpT	supV	84.5	CCA-UUA	Ochre	Su-8; inserts glutamine	159, 189, 190
$tRNA^{Gly}_{CCC}$	glyU	glyU(SuUGA)	61.5	CCC-UCA	Opal	Also reads UGG	128
$tRNA^{Gly}_{GCC}$	glyV(α,β,γ)	glyV(SuUGA)	95	GCC-UCA	Opal	Also reads UGG	127, 128

[a] For sequences of tRNAs, see reference 192.
[b] From reference 12.
[c] For sequences of tRNA genes, see reference 192. Organization of tRNA genes is described in reference 64.

Table 3   Termination Suppressors in Phage T4

tRNA[a]	Suppressor symbol	Anticodon change	Suppressor type	Comments	Reference(s)
$tRNA^{Arg}_{UCU}$	$psu^+_4 op$	UCU-UCA	Opal	A weak suppressor	100
$tRNA^{Gln}_{UUG}$	psu^+_2	UUG-UUA	Ochre		40, 180
$tRNA^{Gln}_{UUG}$	$psu^+_2 am$ $(psu^+_2\text{-}C34)$	UUG-CUA	Amber	Derived from psu^+_2	39
$tRNA^{Gly}_{UCC}$	$psu^+_5 op$	UCC-UCA	Opal	A weak suppressor	120
$tRNA^{Leu}_{UAA}$	$psu^+_3 am$	UAA-CUA	Amber	Double substitution in one step	63
$tRNA^{Leu}_{UAA}$	$psu^+_3 oc$	UAA-UUA	Amber	Derived from $psu^+_3 am$	63
$tRNA^{Leu}_{UAA}$	$psu^+ op$	UAA-UCA	Opal	Derived from $psu^+_3 oc$	63
$tRNA^{Ser}_{UGA}$	$psu^+_1 (psu^+_a)$	UGA-CUA	Amber	Obtained in one step	122, 216
$tRNA^{Ser}_{UGA}$	psu^+_b	UGA-UUA?	Ochre	Sequencing not reported	216
$tRNA^{Ser}_{UGA}$	$psu^+_1 op$	UGA-UCA?	Opal	Sequencing not reported	100

[a] All U nucleotides at 5' end of anticodon (wobble position) are modified.

Table 4   Nonsense Suppressors Employed to Generate Altered Repressors

Suppressor	Codons recognized	Amino acid inserted	Efficiency (%)	Reference
A. *Synthetic*				
Phe	UAG	Phenylalanine	48–100	g
GluA	UAG	85% Glutamic acid / 15% glutamine	8–100	h, i
Cys	UAG	Cysteine	17–51	g
HisA	UAG	Histidine	16–100	h, i
ProH	UAG	Proline	9–60	h, i
Lys	UAG	Lysine	9–29	h, i
Ala	UAG	Alanine	8–83	h, i
Gly1	UAG	Glycine	39–67	h, i
FTORI 26	UAG	Arginine		j
B. *Natural*				
Su1(*supD*)	UAG	Serine	6–54	a, b
Su2*(*supE*)	UAG	Glutamine	41–61	a, b, c, h*
Su3(*supF*)	UAG	Tyrosine	11–100	a, b
Su5(*supG*)	UAA, UAG	Lysine	6–30**	a, b, d, h**
Su6(*supP*)	UAG	Leucine	30–100	a, e
Su9	UGA	Tryptophan	0·1–30	a, f

Both naturally occurring and synthetic suppressors have been employed to create amino acid substitutions in the *lac* repressor. The efficiency of each suppressor is given as a range of values, since the mRNA sequence following the nonsense codon influences suppression efficiency (Bossi, 1983; Miller & Albertini, 1983). The glutamine-inserting derivative of Su2 (*) is the improved suppressor described by Bradley *et al.* (1981). (**) The measurements of the lysine-inserting Su5 were made in a strain containing the *uar-1* mutation (Ryden & Isaksson, 1984). a, Eggertsson & Soll (1988); b, Gorini (1970); c, Bradley *et al.* (1981); d, Ryden & Isaksson (1984); e, Yoshimura *et al.* (1984); f, Hirsch (1971); g, Normanly *et al.* (1986); h, Kleina *et al.* (unpublished results); i, Normanly *et al.* (unpublished results); j, McClain & Foss (1988).

Mutagen and Mutator Specificities

Table 1 lists mutagens commonly used in *E. coli* and Table 2 gives mutators in *E. coli*. The references cited in Table 2 can be found in the Introduction to Unit 4 of the Laboratory Manual portion of this volume. See also Unit 4 for specific protocols.

Table 1 Mutagens Commonly Used in *E. coli*

Mutagen	Specificity	Mechanism	Advantages	Disadvantages
MNNG (*N*-methyl-*N'*-nitro-*N*-nitrosoguanidine)	Principally G:C→A:T transitions	Generates O^6-methylguanine	Very powerful mutagen	Dangerous to handle; frequent secondary mutations
EMS (ethylmethane sulfonate)	Principally G:C→A:T transitions	Generates O^6-ethylguanine	Powerful mutagen	Dangerous to handle; some secondary mutations
UV (ultraviolet) irradiation	All base substitutions, although favors G:C→A:T transitions; frequent hot spots; also induces frameshifts, deletions, and rearrangements	Generates photoproducts that require SOS bypass		High amount of killing required (relative to EMS) for mutagenesis; not a powerful mutagen; certain strains too sensitive
BPDE (benzo[a]pyrene diolepoxide)	Principally G:C→T:A transversions; frameshifts	Generates adducts that require SOS bypass; may stimulate depurination		Extremely dangerous to handle and difficult to obtain
2AP (2-aminopurine)	A:T→G:C and G:C→A:T transitions	Acts as a base analog	Safe and easy to use; works well on *recA* strains	Relatively weak mutagen
ICR 191	Frameshifts, mainly additions and deletions at monotonous runs of G (or C)	Probably stabilizes looped out bases by stacking between them	Causes only frameshifts, which are usually nonleaky	Some strains too sensitive
5AZ (5-azacytidine)	G:C→C:G transversions			Weak mutagen
NH₂OH (hydroxylamine)	G:C→A:T transitions when used in vitro	Reacts with cytosine to generate N^4-hydroxycytosine	Useful for treatment of phage or plasmid DNA in vitro; can be powerful mutagen under these conditions	Causes only one type of base change; more laborious to use than many mutagens
Nitrous acid	Principally transitions, deletions			High amount of killing required for good mutagenesis
Sodium bisulfite	G:C→A:T transitions		Can be used in vitro	Weak mutagen
NQO (4-nitroquinoline-1-oxide)	G:C→A:T transitions, and to a lesser extent G:C→T:A transversions; some frameshifts	Makes adducts that require SOS bypass		Extremely dangerous to handle
Mutator genes (see also Table 2) Nonspecific *mutD*	All base substitutions, frameshifts	Lacks editing function for DNA replication	No treatment required; convenient for phage and plasmids	Genetic construction required for chromosomal mutations; must move mutator out after use or move phage or plasmid
Specific *mutT* *mutY, mutM* *mutH, mutL, mutS, uvrD (mutU)*	A:T→C:G transversions G:C→T:A transversions A:T→G:C and G:C→A:T transitions; frameshifts	Lack different repair systems (see Table 2)	No treatment required	Not as strong as *mutD*; requires strain construction
mutY, mutM (double)	G:C→T:A transversions	Inability to repair 8-oxodG lesions and mispairs	Very powerful (as strong as *mutD*)	Requires strain construction
Transposable elements	Insertions; can be used for deletions and other rearrangements		Generate nonleaky mutations; mutations are often associated with antibiotic resistance markers to facilitate mapping and cloning	Will not result in missense changes; some inserts are lethal; requires some genetic expertise
Spontaneous (no mutagen)	All base substitutions, frameshifts, deletions, insertions		Wide spectrum of mutations; ease of application; no secondary mutations	Low levels of mutants; many siblings in each culture

Table 2 Mutators in *E. coli*

Locus	Map position (min)	Specificity	Strength	Defect (if known)	References
mutT	2	A:T→C:G transversions	Moderate	Prevents incorporation of A:8-oxodG mispairs by hydrolyzing 8-oxodGTP	a, b, c, d
mutH	61	G:C→A:T and A:T→G:C transitions; frameshifts	Strong	Lacks methyl-directed mismatch repair system	e, f, g, h
mutL	95	G:C→A:T and A:T→G:C transitions; frameshifts	Strong	Lacks methyl-directed mismatch repair system	e, f
mutS	59	G:C→A:T and A:T→G:C transitions; frameshifts	Strong	Lacks methyl-directed mismatch repair system	e, f
uvrD (*mutU*)	86	G:C→A:T and A:T→G:C transitions; frameshifts	Strong	Lacks helicase II and the methyl-directed mismatch repair system	e, f, i, j
mutD	5	All base substitutions; frameshifts	Very strong	Altered ε subunit of DNA polymerase III	k, l
mutY	64	G:C→T:A transversions	Moderate	Lacks glycosylase that corrects G:A and 8-oxodG:A mispairs	m, n, o
mutM	82	G:C→T:A transversions	Moderate	Fapy glycosylase (8-oxodG glycosylase)	p, z
mutA	95	A:T→T:A, G:C→T:A and A:T→C:G transversions	Moderate		q
mutC	42	A:T→T:A, G:C→T:A and A:T→C:G transversions	Weak/moderate		q
dam	74	G:C→A:T and A:T→G:C transitions; frameshifts	Moderate	Lacks DNA adenine methylase	e, f, r
ung	56	G:C→A:T transitions	Weak/moderate	Lacks uracil-DNA glycosylase	s
sodA	88		Weak	Lacks superoxide dismutase, manganese	t, u
oxyR	89		Weak	Lacks positive regulator of oxidative damage genes	v, w
polA	87	Frameshifts; deletions	Weak/moderate	Lacks DNA polymerase I	x, y

References: (a) Cox 1976; (b) Bhatnagar and Bessman 1988; (c) Akiyama et al. 1989; (d) Maki and Sekiguchi 1992; (e) Radman et al. 1980; (f) Lu et al. 1984; (g) Grafstrom and Hoess 1983; (h) Grafstrom and Hoess 1987; (i) Finch and Emmerson 1984; (j) Hickson et al. 1983; (k) Scheuermann et al. 1983; (l) Cox and Horner 1986; (m) Nghiem et al. 1988; (n) Au et al. 1989; (o) Michaels et al. 1990b, 1992; (p) Cabrera et al. 1988; (q) Michaels et al. 1990a; (r) Brooks et al. 1983; (s) Duncan and Miller 1980; (t) Touati 1983; (u) Carlioz and Touati 1986; (v) Christman et al. 1985; (w) Christman et al. 1989; (x) Coukell and Yanofsky 1970; (y) Vaccaro and Siegel 1975; (z) Michaels et al. 1991 and references therein.

SECTION·22

Properties of Amino Acids

Table 1 is reprinted from Sambrook et al. (Book 3, pp. D.2–D.5 [1989]).

REFERENCE

Sambrook, J., E.F. Fritsch, and T. Maniatis. 1989. *Molecular Cloning. A Laboratory Manual*, Second Edition. Cold Spring Harbor Laboratory Press, Cold Spring Harbor, New York.

Table 1 Properties of Amino Acids

Amino acid	Three-letter symbol	One-letter symbol	Mass[a] (daltons)	pK_a of ionizing side chain	Structure
Alanine	Ala	A	89.09		(structure)
Arginine	Arg	R	174.2	12.48	(structure)
Asparagine	Asn	N	132.1		(structure)
Aspartic acid	Asp	D	133.1	3.86	(structure)
Cysteine	Cys	C	121.12	8.33	(structure)
Glutamine	Gln	Q	146.15		(structure)

Name			MW	pKa
Glutamic acid	Glu	E	147.13	4.25
Glycine	Gly	G	75.07	
Histidine	His	H	155.16	6.0
Isoleucine	Ile	I	131.17	
Leucine	Leu	L	131.17	
Lysine	Lys	K	146.19	10.53

Table 1 *Continued*

Amino acid	Three-letter symbol	One-letter symbol	Mass[a] (daltons)	pK$_a$ of ionizing side chain	Structure
Methionine	Met	M	149.21		
Phenylalanine	Phe	F	165.19		
Proline	Pro	P	115.13		
Serine	Ser	S	105.09		

Threonine	Thr	T	119.12	
Tryptophan	Trp	W	204.22	
Tyrosine	Tyr	Y	181.19	10.07
Valine	Val	V	117.15	

Weighted mean = 126.7

[a] The polymerization of amino acids into a polypeptide chain results in a net loss of 18 daltons per peptide bond due to elimination of water during condensation.

Atomic Weights

Table 1 is reprinted from Sambrook et al. (Book 3, pp. B.6–B.7 [1989]).

REFERENCE

Sambrook, J., E.F. Fritsch, and T. Maniatis. 1989. *Molecular Cloning. A Laboratory Manual*, Second Edition. Cold Spring Harbor Laboratory Press, Cold Spring Harbor, New York.

Table 1   Atomic Weights

Element	Symbol	Atomic number	Atomic weight[a]
Actinium	Ac	89	227.02
Aluminum	Al	13	26.98
Americium	Am	95	(243)
Antimony	Sb	51	121.75
Argon	Ar	18	39.94
Arsenic	As	33	74.92
Astatine	At	85	(210)
Barium	Ba	56	137.33
Berkelium	Bk	97	(247)
Beryllium	Be	4	9.01
Bismuth	Bi	83	208.98
Boron	B	5	10.81
Bromine	Br	35	79.90
Cadmium	Cd	48	112.41
Calcium	Ca	20	40.08
Californium	Cf	98	(251)
Carbon	C	6	12.01
Cerium	Ce	58	140.12
Cesium	Cs	55	132.90
Chlorine	Cl	17	35.45
Chromium	Cr	24	51.99
Cobalt	Co	27	58.93
Copper	Cu	29	63.54
Curium	Cm	96	(247)
Dysprosium	Dy	66	162.50
Einsteinium	Es	99	(252)
Erbium	Er	68	167.26
Europium	Eu	63	151.96
Fermium	Fm	100	(257)
Fluorine	F	9	18.99
Francium	Fr	87	(223)
Gadolinium	Gd	64	157.25
Gallium	Ga	31	69.72
Germanium	Ge	32	72.59
Gold	Au	79	196.96
Hafnium	Hf	72	178.49
Helium	He	2	4.00
Holmium	Ho	67	164.93
Hydrogen	H	1	1.00
Indium	In	49	114.82
Iodine	I	53	126.90
Iridium	Ir	77	192.22
Iron	Fe	26	55.84
Krypton	Kr	36	83.80
Lanthanum	La	57	138.90
Lawrencium	Lr	103	(260)
Lead	Pb	82	207.2
Lithium	Li	3	6.94
Lutetium	Lu	71	174.96
Magnesium	Mg	12	24.30
Manganese	Mn	25	54.93
Mendelevium	Md	101	(258)
Mercury	Hg	80	200.59
Molybdenum	Mo	42	95.94

Table 1 *Continued*

Element	Symbol	Atomic number	Atomic weight[a]
Neodymium	Nd	60	144.24
Neon	Ne	10	20.17
Neptunium	Np	93	237.04
Nickel	Ni	28	58.69
Niobium	Nb	41	92.90
Nitrogen	N	7	14.00
Nobelium	No	102	(259)
Osmium	Os	76	190.2
Oxygen	O	8	15.99
Palladium	Pd	46	106.42
Phosphorus	P	15	30.97
Platinum	Pt	78	195.08
Plutonium	Pu	94	(244)
Polonium	Po	84	(209)
Potassium	K	19	39.09
Praseodymium	Pr	59	140.90
Promethium	Pm	61	(145)
Protactinium	Pa	91	231.03
Radium	Ra	88	226.02
Radon	Rn	86	(222)
Rhenium	Re	75	186.20
Rhodium	Rh	45	102.90
Rubidium	Rb	37	85.46
Ruthenium	Ru	44	101.07
Samarium	Sm	62	150.36
Scandium	Sc	21	44.95
Selenium	Se	34	78.96
Silicon	Si	14	28.08
Silver	Ag	47	107.86
Sodium	Na	11	22.98
Strontium	Sr	38	87.62
Sulfur	S	16	32.06
Tantalum	Ta	73	180.94
Technetium	Tc	43	(98)
Tellurium	Te	52	127.60
Terbium	Tb	65	158.92
Thallium	Tl	81	204.38
Thorium	Th	90	232.03
Thulium	Tm	69	168.93
Tin	Sn	50	118.69
Titanium	Ti	22	47.88
Tungsten	W	74	183.85
Unnilhexium	(Unh)	106	(263)
Unnilpentium	(Unp)	105	(262)
Unnilquadium	(Unq)	104	(261)
Unnilseptium	(Uns)	107	(262)
Uranium	U	92	238.02
Vanadium	V	23	50.94
Xenon	Xe	54	131.29
Ytterbium	Yb	70	173.04
Yttrium	Y	39	88.90
Zinc	Zn	30	65.38
Zirconium	Zr	40	91.22

[a]Numbers in parentheses are the mass numbers of the most stable isotope of that element.

SECTION·24

Additional Procedures

A. Transformation
  1. TFB-based chemical transformation protocol
  2. FSB-based frozen storage of competent cells
  3. PEG/DMSO one-step transformation procedure
  4. Rapid colony transformation
  5. Electroshock transformation of *E. coli*
  6. General protocol for electroshock transformation of gram-negative microorganisms

B. Mutagenesis
  1. Mutagenesis with aflatoxin B_1
  2. Mutagenesis with 1,2-dibromoethane

REFERENCES

Foster, P.L. 1991. In vivo mutagenesis. *Methods Enzymol.* **204:** 114–125.

Hanahan, D., J. Jessee, and F.R. Bloom. 1991. Plasmid transformation of *Escherichia coli* and other bacteria. *Methods Enzymol.* **204:** 63–113.

SECTION 24

A. Transformation

Experiment 23 in the Laboratory Manual portion of this volume details a number of transformation protocols from Hanahan and co-workers (Hanahan et al. 1991). Some of these protocols and some additional procedures are reprinted here, with permission, from Hanahan et al. (*Methods Enzymol.*, vol. 204, pp. 75–80 and 85–86 [1991]).

TFB-BASED CHEMICAL TRANSFORMATION PROTOCOL

Materials

SOB and SOC media (See Section 25C of this Handbook for preparation of these media.)

Table 1   Standard Transformation Buffer (TFB)

Compound	Amount/liter	Final concentration
KCl (ultrapure)	7.4 g	100 mM
MnCl · 4H$_2$O	8.9 g	45 mM
CaCl$_2$ · 2H$_2$O	1.5 g	10 mM
HACoCl$_3$	0.8 g	3 mM
Potassium MES (final pH 6.20 ± 0.10)	20 ml of 0.5 M stock (pH 6.3)	10 mM

Preparation: Equilibrate a 0.5 M solution of MES [2(N-morpholino) ethane sulfonic acid] to pH 6.3 using concentrated KOH. Then sterilize by filtration through a 0.2-μm membrane and store in aliquots at −20°C. Make a solution of 10 mM potassium MES, using the 0.5 M MES stock and the purest water available. Add the salts as solids, and then filter the solution through a 0.22-μm prerinsed membrane. Aliquot into sterile flasks and store at 4°C. HACoCl$_3$ is hexamminecobalt trichloride. TFB is stable for over 1 year.

Table 2   DMSO and DTT Solution (DnD)

Compound	Amount/10 ml final volume	Final concentration
DTT	1.53 g	1 M
DMSO (spectroscopy grade)	9 ml	90% (v/v)
Potassium acetate	100 μl of a 1 M stock (pH 7.5)	10 mM

Procedure

1. Pick several 2–3-mm diameter colonies from a freshly streaked SOB agar plate and disperse in 1 ml of SOB medium by vortexing. Use one colony per 10 ml of culture medium. The cells are best streaked from a frozen stock or fresh stab about 16–20 hours prior to initiating liquid growth.

2. Inoculate the cells into an Erlenmeyer flask containing SOB medium. Use a culture volume to flask volume ratio between 1:10 and 1:30 (e.g., 30–100 ml in a 1-liter flask).

3. Incubate at 37°C with moderate agitation until the cell density is 4–7 × 10^7 viable cells/ml (OD$_{550}$ = 0.4 for DH5, 0.5 for DH5α and DH5αF′).

4. Collect the culture into 50-ml polypropylene centrifuge tubes (such as Falcon 2070 tubes), and chill on ice for 10–15 minutes. (Take a 10-μl aliquot of cells to determine viable cell density by plating a 10^6 dilution on an SOB agar plate.)

5. Pellet the cells by centrifugation at 750–1000 g (2000–3000 rpm in a clinical centrifuge) for 12–15 minutes at 4°C. Drain the pelleted cells thoroughly by inverting the tubes on paper towels and rapping sharply to remove any liquid. A micropipette can be used to draw off recalcitrant drops.

6. Resuspend the cells in one-third culture volume TFB by vortexing moderately. Incubate on ice for 10–15 minutes.

7. Pellet the cells and drain thoroughly as in Step 5.

8. Resuspend the cells in TFB to 1/12.5 of the original volume. (Each 2.5 ml of culture is concentrated into 200 μl of TFB.)

9. Add DMSO and DTT solution (DnD) to 3.5% (v/v) (7 μl per 200 μl of cell suspension). Squirt the DnD into the center of the cell suspension and immediately swirl the tube for several seconds. Incubate the tubes on ice for 10 minutes.

10. Add a second, equal aliquot of DnD as in Step 9 to give a 7% final concentration. Incubate the tubes on ice for 10–20 minutes.

11. Pipette 210-μl aliquots into chilled 17 mm × 100 mm polypropylene tubes (Falcon 2059 or equivalent).

12. Add the DNA solution in a volume of less than 20 μl, swirling to mix. Incubate the tubes on ice for 20–40 minutes. (Ligations should be diluted or precipitated in ethanol.)

13. Heat-shock the cells by placing the tubes in a 42°C water bath for 90 seconds. Return the tubes to ice to quench the heat shock; allow 2 minutes for cooling.

14. Add 800 μl of SOC medium to each tube. Incubate at 37°C with moderate agitation for 30–60 minutes. Spread the cells on agar plates containing appropriate antibiotics (or other conditions) to select for transformants. The incubation period should be omitted for M13 transfections.

FSB-BASED FROZEN STORAGE OF COMPETENT CELLS

Materials

SOB medium (See Section 25C of this Handbook for preparation of this medium.)

Table 1   Frozen Storage Buffer (FSB)

Compound	Amount/liter	Final concentration
KCl	7.4 g	100 mM
MnCl · 4H$_2$O	8.9 g	45 mM
CaCl$_2$ · 2H$_2$O	1.5 g	10 mM
HACoCl$_3$	0.8 g	3 mM
Potassium acetate	10 ml of 1 M stock (pH 7.5)	10 mM
Redistilled glycerol (final pH 6.20 ± 0.10)	100 g	10% (w/v)

Preparation: Equilibrate a 1 M solution of potassium acetate to pH 7.5 using KOH. Then sterilize by filtration through a 0.2-μm membrane and store frozen. Prepare a 10 M potassium acetate, 10% glycerol solution using this stock and the purest water available. Add the salts as solids, and adjust the pH (if necessary) to 6.4 using 0.1 N HCl. Do not adjust the pH upward with base. (The pH may drift for 1–2 days before settling at 6.1–6.2.) Sterilize the solution by filtration through a prerinsed 0.2-μm filter and store at 4°C. HACoCl$_3$ is hexamminecobalt trichloride.

Procedure

1. Pick several 2–3-mm diameter colonies from a freshly streaked SOB agar plate and disperse in 1 ml of SOB medium by vortexing. Use one colony per 10 ml of culture medium. The cells are best streaked from a frozen stock or fresh stab about 16–20 hours prior to initiating liquid growth.

2. Inoculate the cells into SOB medium in an Erlenmeyer flask. Use a culture volume to flask volume ratio between 1:10 and 1:30 (e.g., 30–100 ml in a 1-liter flask).

3. Incubate at 30°C with moderate agitation until the cell density is 6–9 × 10^7 viable cells/ml (OD$_{550}$ = 0.4 for DH5, 0.5 for DH5α and DH5αF′).

4. Collect the culture into 50-ml polypropylene centrifuge tubes (such as Falcon 2070 tubes), and chill on ice for 10–15 minutes. (Take a 10-μl aliquot of cells to determine viable cell density by plating a 10^6 dilution on an SOB agar plate.)

5. Pellet the cells by centrifugation at 750–1000 g (2000–3000 rpm in a clinical centrifuge) for 12–15 minutes at 4°C. Drain the pelleted cells thoroughly by inverting the tubes on paper towels and rapping sharply to remove any liquid. A micropipette can be used to draw off recalcitrant drops.

6. Resuspend the cells in one-third culture volume FSB by vortexing moderately. Incubate on ice for 10–15 minutes.

7. Pellet the cells and drain thoroughly as in Step 5.

8. Resuspend the cells in FSB to 1/12.5 of the original volume. (Each 2.5 ml of culture is concentrated into 200 μl of FSB.)

9. Add DMSO to 3.5% (v/v) (7 μl per 200 μl of cell suspension). Squirt the DMSO into the center of the cell suspension and immediately swirl the tube for several seconds. Incubate the tubes on ice for 5 minutes. (DTT is not used here.)

10. Add a second, equal aliquot of DMSO as in Step 9 to give a 7% final concentration. Incubate the tubes on ice for 10–15 minutes.

11. Pipette 210-μl aliquots into chilled screw-cap polypropylene tubes, 1.5-ml microcentrifuge tubes, or snap-cap polypropylene tubes.

12. Flash freeze by placing the tubes in a dry ice/ethanol bath or liquid nitrogen for several minutes. (Be careful that the ethanol does not get inside the tubes. An option to avoid total immersion is to set just the bottom half of the tube in the bath.)

13. Transfer the tubes to a −80°C freezer.

PEG/DMSO ONE-STEP TRANSFORMATION PROCEDURE

Materials

Table 1   Transformation and Storage Solution (TSS)

Compound	Amount/liter	Final concentration in TSS
a. LB broth		
Bacto tryptone	10 g	~0.85%
Bacto yeast extract	5 g	~0.4%
NaCl	5 g	~8 mM
Dissolve in 1 liter ultrapure water and autoclave		
b. 2.0 M Mg^{++} solution		
1 M MgCl$_2$ · 6H$_2$O	203 g	10 mM
1 M MgSO$_4$ · 7H$_2$O	247 g	10 mM
Dissolve in 1 liter ultrapure water and filter sterilize		
c. 2 M glucose solution	360 g	20 mM
Prepare in 1 liter ultrapure water and sterilize by filtration		
d. PEG (polyethylene glycol)		10%
Use either molecular weight 3350 or 8000; add as a solid directly into TSS		
e. DMSO (ultrapure spectroscopy grade)		5%

Preparation: Add solid PEG to LB to make a 10% (w/v) solution. Add an aliquot of the 2.0 M Mg^{++} solution to achieve a final concentration of 20 mM. Measure the pH, which should be about 6.8. If the pH is higher, the solution can be titrated with 1.0 N HCl. (The final pH of the TSS should be between 6.5 and 6.8.) Filter sterilize the solution through a standard 0.45-μm filter unit. Add DMSO to the filtered solution to a final concentration of 5% (v/v). Store TSS at 4°C or on ice until ready to use. Alternatively, the PEG, Mg^{++}, and DMSO can be added to the LB, the pH adjusted, and the solution filter sterilized through a 0.2-μm nylon filter unit.

Procedure

1. Incubate *E. coli* at 37°C in LB broth to an OD$_{550}$ of 0.4–0.5, corresponding to a cell density of 5×10^7 cells/ml.
2. Pellet the cells by centrifugation at 1000 g for 10 minutes at 4°C. Resuspend in one-tenth volume cold TSS.
3. Incubate the cells on ice for 20 minutes.
4. Transfer 100-μl aliquots of the cell suspension into chilled polypropylene tubes and mix with 100 pg–1 ng of DNA. Incubate the tubes on ice for 30 minutes. (Any remaining cells can be frozen in dry ice/ethanol and stored at −80°C for subsequent use.)
5. Add 0.9 ml of LB medium containing 20 mM glucose (or use SOC, see Section 25C of this Handbook), and incubate at 37°C with moderate agitation (225 rpm) for 1 hour.
6. Plate cells, diluting into LB (or SOC) if necessary.

RAPID COLONY TRANSFORMATION

Materials

Standard transformation buffer (TFB) (See Table 1 on page 24.2 of this Handbook.)
SOC medium (See Section 25C of this Handbook for preparation of this medium.)

Procedure

1. Pick several colonies (or a clump of cells) from a plate using a tungsten inoculating loop or a wooden applicator stick, being careful to take no agar along with the cells.
2. Disperse the colonies in 200 μl of chilled TFB by vigorous vortexing or by repeated pipetting. (CCMB 80 [see Hanahan et al. 1991] or FSB [see Table 1 on page 24.4 of this Handbook] can be used instead of TFB.)
3. Incubate the cells on ice for 10 minutes.
4. Add DNA solution (10–1000 ng) in less than 20 μl, swirl to mix, and incubate on ice for 10 minutes.
5. Heat-shock the cells at 37–42°C for 90 seconds. (This step is optional if >100 ng of DNA is used.)
6. Add 400–800 μl of SOC medium, and incubate at 37°C for 20–60 minutes. (This step is also optional and unnecessary if >100 ng of DNA is used in the transformation.)
7. Plate several fractions (e.g., 1%, 5%, 25%) on appropriate selective media. Incubate to establish colonies.

ELECTROSHOCK TRANSFORMATION OF *E. COLI*

Materials

DNA
  DNA used for electroporation should be free of phenol, ethanol, and detergents, as with the high-efficiency chemical transformation protocols. In addition, it is very important that the DNA solution being used for electroshock transformation has a very low ionic strength and thus, a high resistance. Protocols that achieve these conditions can be found in Hanahan et al. (1991).
SOB and SOC media (See Section 25C of this Handbook for preparation of these media.)
Electroshock buffer (EWB; 10% redistilled glycerol in ultrapure water)
Tris · EDTA (TE)

Procedure

1. Pick a single 2–3-mm colony from a freshly streaked SOB − Mg (SOB without Mg) agar plate and disperse it in 1 ml of SOB − Mg by vortexing. The cells are best streaked from a frozen stock or a fresh stab about 16–20 hours prior to initiating liquid growth.
2. Inoculate the cells into a 500-ml Erlenmeyer flask containing 50 ml of SOB − Mg.
3. Incubate at 37°C overnight (preferably <12 hours) at 275 rpm.
4. The following day, use 7.5 ml of this fresh overnight culture to inoculate 750 ml of SOB − Mg in a 2.8-liter Fernbach flask (nonbaffled).
5. Incubate at 37°C with moderate agitation until an OD_{550} of 0.75 is reached ($3–6 \times 10^8$ cells/ml).
6. Collect the cell suspension into chilled polypropylene centrifuge tubes.
7. Pellet the cells by centrifugation at 2600 g for 12–15 minutes at 4°C. Carefully decant the supernatant.
8. Resuspend the cell pellet in an equal volume (to the original) of chilled 4°C EWB. Resuspension requires vigorous agitation by vortexing or rapping against a solid object.

9. Centrifuge again to pellet the cells, and immediately decant the supernatant. (Cell loss is difficult to avoid but should be minimized.)

10. Resuspend the cell pellet as in Step 8. Centrifuge the cells, again carefully decanting the supernatant.

11. Resuspend the cell pellet with the few drops of excess liquid left in tube, and measure the volume of the cell suspension.

12. Determine the OD_{550} by diluting a small portion (e.g., 1%) of the cell suspension 300×. Then adjust the volume of the concentrated cell slurry with cold EWB to produce a final OD_{550} of 200–250/ml.

13. Dispense the cell suspension in 120-μl aliquots into cold cryotubes, freeze in a dry ice/ethanol bath, and store at −80°C.

Electroshock Transformation

1. Thaw the cells on ice, and aliquot 20-μl volumes into chilled polypropylene tubes. Add 1 μl of DNA solution in 0.5× TE.

2. Transfer individual aliquots of cells plus DNA into chilled electroporation chambers and electroshock under optimal conditions for the strain. For DH10B, these values are 16.7 kV/cm, a resistance of 4000 ohms, and a capacitance of 2 μF. (An alternative, if 16.7 kV/cm cannot be achieved, is 12.5 kV/cm, 25 μF, and 200 ohms.)

3. Immediately remove the electroshocked cells from the chamber, place in a chilled Falcon 2059 tube, and add 1 ml of SOC recovery medium. Incubate at 37°C for 1 hour at 225 rpm.

4. Plate on selective medium.

GENERAL PROTOCOL FOR ELECTROSHOCK TRANSFORMATION OF GRAM-NEGATIVE MICROORGANISMS

Materials

Growth medium rich in nutrients and low in ionic strength (i.e., do not add salts unless necessary for growth)

Electroshock buffer (EWB; 10% redistilled glycerol in ultrapure water)

Procedure

1. Incubate a liquid culture to mid-log phase in sufficient volume to give at least 0.5–1.0 ml of cells at a density of 5×10^{10} to 10^{11} cells/ml. (The *E. coli* protocol on pages 24.6–24.7 may prove to be a useful example.)

2. Chill the cells on ice and wash twice in an equal volume of growth medium by pelleting cells with low-speed centrifugation, decanting the supernatant, adding medium, and resuspending the cell pellet by vortexing.

3. Wash the cells twice in an equal volume of chilled EWB as in Step 2. (EWB can be formulated as described in Materials above or it can be buffered with 0.2 M phosphate or 1 mM HEPES, both at pH 7.0 [see Dower et al. 1988, *Nucleic Acids Res. 16:* 6127]. It is advisable to try both the buffered and nonbuffered versions initially.)

4. Following the second wash in EWB, pellet the cells by centrifugation, drain the supernatant, and resuspend the pellet in the residual buffer on the wall of the centrifuge tube. Adjust the concentration to approximately 10^{10}–10^{11} cells/ml by determining the optical density of a dilution of the cell suspension, as is elaborated in the *E. coli* protocol on pages 24.6–24.7.

5. Cells can be used immediately for electroshock transformation or aliquots can be frozen in a dry ice/ethanol bath and stored at −80°C for subsequent use.

SECTION 24

B. Mutagenesis

Unit 4 of the Laboratory Manual portion of this volume details protocols for mutagenesis. The following additional procedures are reprinted, with permission, from Foster (*Methods Enzymol.*, vol. 204, pp. 121–122 and 123 [1991]).

MUTAGENESIS WITH AFLATOXIN B_1

Aflatoxin B_1 (AFB) is inactivated by water and light. The stock is prepared by adding the appropriate volume of dichloromethane (CH_2Cl_2) to the vial of AFB as received to give a 10 mM (3.12 mg/ml) solution (this is safer than weighing out an aliquot). Twenty-microliter aliquots are dispensed, evaporated under inert gas (N_2 or Ar) or vacuum, and stored at −20°C under dry N_2 or Ar in tightly capped tubes. Immediately before use, 40 μl of DMSO is added to an AFB aliquot, resulting in a 5 mM solution. This working stock is further diluted in DMSO to the appropriate concentrations. If immediately stored at −20°C under dry gas, the 5 mM stock can be used several times, but more dilute stocks are discarded after one use. Gas is dried by passing it through a column of $CaSO_4$; dichloromethane and DMSO are purged with and stored under the dry gas. Although extremely light-sensitive, AFB can be handled under yellow light. It fluoresces blue under UV light, which property can be used to check for contamination.

Mid-log-phase cells are centrifuged, washed with E salts, and resuspended at 10× in cold E salts. (E salts = 57 mM K_2HPO_4, 9.5 mM citric acid, 17 mM $NaNH_4HPO_4$, 0.8 mM $MgSO_4$ [pH 7].) Cells are diluted 1:5 into S9 mixture and a 0.5-ml aliquot dispensed for each dose to be used. (S9 mixture = 0.1 M sodium phosphate buffer [pH 7.4] with 8 mM $MgCl_2$, 33 mM KCl, 5 mM glucose-6-phosphate, 4 mM NADP, and 20–80 μl/ml microsomes.) Five microliters of an appropriate AFB dilution in DMSO is added to give a 5–100 μM final concentration. For each treatment, including the control, the same amount of DMSO (1% of the total volume) is added to the cells. The cells are then incubated at 37°C for 60 minutes, diluted with 0.5 ml of cold E salts, washed twice, and resuspended in 0.5 ml of cold E salts. Plating for survival and outgrowth are as described by Foster (1991). Mutation rates vary greatly among different targets, but a frequency of 10^{-5} for Rif^r at 50% killing is typical for a uvr^- strain carrying a $mucAB^+$ plasmid (see Foster 1991). AFB induces primarily G:C → T:A transversions (Foster et al. 1983, *Proc. Natl. Acad. Sci. 80:* 2695).

MUTAGENESIS WITH 1,2-DIBROMOETHANE

1,2-Dibromoethane (EDB) is an extremely volatile, light-sensitive liquid. It should be stored at room temperature in the dark and used only in a chemical hood.

The highest levels of mutation are obtained by mutating the cells with gaseous EDB. The following method is adapted from Rosenkranz (1977, *Environ. Health Perspect. 21:* 79). Using a reversion assay, 10^8 cells are plated in top agar on minimum medium containing a limiting amount of the required nutrient (see Foster 1991). A filter paper disk is fixed to the top of the inverted petri dish with a small amount of top agar. Working in a chemical hood, 1–20 μl of EDB is added to the disk, and the plate is closed and immediately sealed with Parafilm. The time elapsed between adding the EDB and sealing the plate is a critical variable and should be as short as possible. Mutants are scored after 2 days of incubation at 37°C.

With this method, Foster obtained up to 6000 EDB-induced mutants per plate, which was 10- to 100-fold higher than levels achieved with more conventional treatment methods (Foster et al. 1988, *Mutat. Res. 194:* 171). EDB is poorly soluble in aqueous solutions unless dispersed in a solvent such as DMSO. Because of its lipid solubility, it disrupts cell membranes (Brem et al. 1974, *Biochem. Pharmacol. 23:* 2345). Thus, its acute toxicity to cells may be unrelated to DNA damage, and the gas-phase treatment may be successful because it allows long exposures to subtoxic doses. Although this method is not widely applicable, it may be useful for other lipid-soluble volatile mutagens. EDB induces predominantly transitions at G:C and A:T sites, most of which are SOS-independent (Foster et al. 1988, *Mutat. Res. 194:* 171).

SECTION · 25

Formulas and Recipes

A. Minimal salts
   1. M63 medium
   2. M9 medium
   3. Minimal A medium (1× A)
   4. Minimal A medium (10× A)

B. Minimal agar plates

C. Rich media
   1. LB medium
   2. YT medium
   3. 2× YT medium (2YT)
   4. Tryptone broth (H medium for plate lysates)
   5. R medium for phage lysates
   6. Terrific broth
   7. SOB medium
   8. SOC medium

D. Indicator plates
   1. Xgal glucose plates
   2. Tetrazolium plates
   3. EMB plates
   4. MacConkey plates

E. Antibiotics
   1. Tetracycline
   2. Streptomycin
   3. Kanamycin
   4. Ampicillin
   5. Chloramphenicol
   6. Rifampicin
   7. Nalidixic acid
   8. Trimethoprim

F. Buffers
   1. Phosphate buffers
   2. Citrate buffer
   3. Saline (for dilutions)
   4. Z buffer for β-galactosidase assays
   5. Tris buffers

G. Soft agar (top agar)
  1. Minimal "F-top" agar
  2. H-top agar
  3. R-top agar

H. Media for storage of strains
  1. Stabs
  2. Storage in glycerol
  3. Freeze-drying

REFERENCES

Green, A.A. 1933. The preparation of acetate and phosphate buffer solutions of known pH and ionic strength. *J. Am. Chem. Soc.* **55:** 2331–2336.

ISCO. 1982. *ISCOTABLES. A Handbook of Data for Biological and Physical Sciences*, Eighth Edition. ISCO, Inc., Lincoln, Nebraska.

Pardee, A.B., F. Jacob, and J. Monod. 1959. The genetic control and cytoplasmic expression of "inducibility" in the synthesis of β-galactosidase by *E. coli. J. Mol. Biol.* **1:** 165–178.

Sambrook, J., E.F. Fritsch, and T. Maniatis. 1989. *Molecular Cloning. A Laboratory Manual*, Second Edition. Cold Spring Harbor Laboratory Press, Cold Spring Harbor, New York.

Tartof, K.D. and C.A. Hobbs. 1987. Improved media for growing plasmid and cosmid clones. *Bethesda Res. Lab. Focus* **9:** 12.

SECTION 25

A. Minimal Salts

In order to use each of the following as growth media, 1 ml of a 1 M solution of $MgSO_4 \cdot 7H_2O$ should be added per liter after autoclaving. In addition, 10 ml of a 20% solution of a carbon source (either a sugar or glycerol) should be added per liter. Vitamins, such as B1, are added to a final concentration of 1 μg/ml and amino acids at 40 μg/ml in the L form. Use deionized water for all recipes.

1. M63 Medium (Pardee et al. 1959)

per liter*

KH_2PO_4	13.6 g
$(NH_4)_2SO_4$	2 g
$FeSO_4 \cdot 7H_2O$	0.5 mg

Adjust pH to 7.0 with KOH.

*The original M63 formula of Pardee et al. (1959) calls for 0.2 g of $MgSO_4 \cdot 7H_2O$/liter.

2. M9 Medium

per liter

Na_2HPO_4	6 g
KH_2PO_4	3 g
NaCl	0.5 g
NH_4Cl	1 g

After autoclaving, add 10 ml of a 0.01 M solution of $CaCl_2$.

3. Minimal A Medium (1× A)

per liter

K_2HPO_4	10.5 g
KH_2PO_4	4.5 g
$(NH_4)_2SO_4$	1 g
sodium citrate $\cdot 2H_2O$	0.5 g

4. Minimal A Medium (10× A)

per liter

K_2HPO_4	105 g
KH_2PO_4	45 g
$(NH_4)_2SO_4$	10 g
sodium citrate $\cdot 2H_2O$	5 g

B. Minimal Agar Plates

The salt solution and the agar should be prepared and autoclaved separately at 15 psi for 15 minutes. Therefore, the salts are usually prepared in more concentrated form. It is convenient to autoclave either 15 g of agar in 900 ml of H_2O and the salts (at 10× normal strength) in 100 ml of H_2O or the agar in 500 ml of H_2O and the salts (at 2× normal strength) in 500 ml of H_2O. The salts used for the plates required for the experiments in the Laboratory Manual portion of this volume is minimal A medium, although any of the minimal media listed in Section 25A above will suffice. Mg^{++} and nutrients are prepared and autoclaved separately. Normally, 40 μg/ml of the L-amino acid are sufficient. Vitamins are required in smaller amounts (1 μg/ml). These are then added to the salts after autoclaving. Use deionized water in all cases.

All plates are prepared with the final concentration per liter.

Difco agar	15 g
salts: (in this case 1× A)	
K_2HPO_4	10.5 g
KH_2PO_4	4.5 g
$(NH_4)_2SO_4$	1 g
sodium citrate · $2H_2O$	0.5 g
$MgSO_4 \cdot 7H_2O$	1 ml from a 1 M stock solution after autoclaving
B1 (thiamine hydrochloride)	0.5 ml from a 1% stock solution (excess)
amino acids as required	4 ml from a 10 mg/ml stock solution
sugar	10 ml from a 20% stock solution
antibiotics as required	see Section 25E below

C. Rich Media

If these media are used in plates, add 15 g of Difco agar per liter (except in tryptone broth and R medium). Autoclave 1 liter for 20–30 minutes at 15 psi on liquid cycle. Use deionized water for all recipes.

Terrific broth, SOB medium, and SOC medium are reprinted from Sambrook et al. (*Molecular Cloning. A Laboratory Manual*, Second Edition. Cold Spring Harbor Laboratory Press, Cold Spring Harbor, New York, Book 3, p. A.2 [1989]).

1. LB Medium (Luria-Bertani Medium)

per liter

Bacto tryptone	10 g
Bacto yeast extract	5 g
NaCl	10 g

Dissolve. Adjust pH to 7.0 with 5 N NaOH (several drops).

2. YT Medium

per liter

Bacto tryptone	8 g
Bacto yeast extract	5 g
NaCl	5 g

Dissolve. Adjust pH to 7.0 with 5 N NaOH (several drops).

3. 2× YT Medium (2YT)

per liter

Bacto tryptone	16 g
Bacto yeast extract	10 g
NaCl	5 g

Dissolve. Adjust pH to 7.0 with 5 N NaOH (several drops).

4. Tryptone Broth (H Medium for Plate Lysates)

per liter

Bacto tryptone	10 g
NaCl	8 g

Use 12 g of agar per liter for plates.

5. R Medium for Phage Lysates

per liter

Bacto tryptone	10 g
Bacto yeast extract	1 g
NaCl	8 g

After autoclaving, add 2 ml of 1 M $CaCl_2$ + 5 ml of 20% glucose. Use 12 g of agar per liter for plates.

6. Terrific Broth (Tartof and Hobbs 1987)

per liter

To 900 ml of H_2O, add	
Bacto tryptone	12 g
Bacto yeast extract	24 g
glycerol	4 ml

Shake until the solutes have dissolved, and sterilize by autoclaving for 20 minutes at 15 psi on liquid cycle.

Allow the solution to cool to 60°C or less, and then add 100 ml of a sterile solution of 0.17 M KH_2PO_4, 0.72 M K_2HPO_4. (This solution is made by dissolving 2.31 g of KH_2PO_4 and 12.54 g of K_2HPO_4 in 90 ml of H_2O. After the salts have dissolved, adjust the volume of the solution to 100 ml with H_2O and sterilize by autoclaving for 20 minutes at 15 psi on liquid cycle.)

7. SOB Medium

per liter

To 950 ml of H_2O, add	
Bacto tryptone	20 g
Bacto yeast extract	5 g
NaCl	0.5 g

Shake until the solutes have dissolved. Add 10 ml of a 250 mM solution of KCl. (This solution is made by dissolving 1.86 g of KCl in 100 ml of H_2O.) Adjust the pH to 7.0 with 5 N NaOH (~0.2 ml). Adjust the volume of the solution to 1 liter with H_2O. Sterilize by autoclaving for 20 minutes at 15 psi on liquid cycle.

Just before use, add 5 ml of a sterile solution of 2 M $MgCl_2$. (This solution is made by dissolving 19 g of $MgCl_2$ in 90 ml of H_2O. Adjust the volume of the solution to 100 ml with H_2O and sterilize by autoclaving for 20 minutes at 15 psi on liquid cycle.)

8. SOC Medium

SOC medium is identical to SOB medium except that it contains 20 mM glucose. After the SOB medium has been autoclaved, allow it to cool to 60°C or less and then add 20 ml of a sterile 1 M solution of glucose. (This solution is made by dissolving 18 g of glucose in 90 ml of H_2O. After the sugar has dissolved, adjust the volume of the solution to 100 ml with H_2O and sterilize by filtration through a 0.22-micron filter.)

D. Indicator Plates

1. Xgal Glucose Plates

These plates are prepared in the same manner as glucose minimal plates (see Section 25B above). After autoclaving, add 1 ml of a 40 μg/ml solution of Xgal in N,N-dimethylformamide to the salts immediately before mixing.

2. Tetrazolium Plates

These plates work best with Difco antibiotic medium No. 2. Dissolve 25.5 grams of this medium in 950 ml of distilled water. Add 50 mg of 2,3,5-triphenyltetrazolium chloride and continue to heat until it is completely dissolved. Autoclave, and then add 50 ml of a 20% solution of the desired sugar (e.g., lactose). It is extremely important that the tetrazolium be added before autoclaving.

In cases where Difco antibiotic medium No. 2 is not available, or if colonies do not grow well on this medium, the two media below have given satisfactory results.

To 950 ml of distilled water, add either a or b:

a.	Bacto beef extract	1.5 g
	Bacto yeast extract	3 g
	peptone	6 g
	Difco agar	15 g

b.	Difco nutrient agar	23 g
	NaCl	1 g

3. EMB Plates

Prepared medium can be purchased from Difco Laboratories. This comes with and without added sugar (lactose). The latter is preferred. In the case of EMB lactose medium, dissolve and autoclave in accordance with the manufacturer's instructions. For EMB base or EMBO, add 50 ml of a 20% solution of the desired sugar after autoclaving.

EMB plates can also be prepared using the following formula:

To 930 ml of distilled water, add	
Bacto tryptone	10 g
Bacto yeast extract	1 g
NaCl	5 g
Difco agar	15 g
KH_2PO_4	2 g

Autoclave, and then add 10 ml each of sterile solutions of 4% eosin yellow and 0.65% methylene blue. These solutions should be autoclaved separately. Also add 50 ml of a 20% solution of the desired sugar.

4. MacConkey Plates

These plates are made with prepared media that can be purchased from Difco Laboratories. MacConkey medium with lactose added is also available and is preferable to the base agar. Some *E. coli* strains that grow well on this medium do not grow as well on reconstructed lactose MacConkey medium made from MacConkey base. Any sugar (50 ml of a 20% solution) can be added to MacConkey base after autoclaving.

E. Antibiotics

Make fresh stock solution for each use and use distilled water in all recipes where H_2O is indicated.

1. Tetracycline (15 mg/ml stock solution)

Weigh out 15 mg of tetracycline and dissolve in 0.5 ml of sterile H_2O and 0.5 ml of ethanol (under fume hood). Wrap the test tube in foil. Add 1 ml of the stock solution directly to 1 liter of cooling agar. (Final concentration: 15 μg/ml)

2. Streptomycin (100 mg/ml stock solution)

Make more stock solution than will be needed since some volume is lost in filter sterilization. Weight out 200 mg of streptomycin and add 2 ml of H_2O. Filter sterilize. Add 1 ml of the stock solution directly to 1 liter of cooling agar. (Final concentration: 100 μg/ml)

3. Kanamycin (30 mg/ml or 50 mg/ml stock solution)

Depending on the type of plates, prepare a 30 mg/ml or a 50 mg/ml stock solution. Make more stock solution than will be needed since some volume is lost in filter sterilization. Weigh out 60 or 100 mg of kanamycin and dissolve in 2 ml of H_2O. Filter sterilize. Add 1 ml of the stock solution directly to 1 liter of cooling agar. (Final concentration: 30 or 50 μg/ml depending on type of medium)

4. Ampicillin (100 mg/ml stock solution)

Make more stock solution than will be needed since some volume is lost in filter sterilization. Weigh out 100 mg of ampicillin and add 1 ml of H_2O. Filter sterilize. Add 1 ml of the stock solution directly to 1 liter of cooling agar. (Final concentration: 100 μg/ml)

5. Chloramphenicol (20 mg/ml stock solution)

Weight out 20 mg of chloramphenicol and add 1 ml of ethanol (under fume hood). Add 1 ml of the stock solution directly to 1 liter of cooling agar. (Final concentration: 20 μg/ml)

6. Rifampicin (50 mg/ml stock solution)

Weigh out 100 mg of rifampicin and dissolve in 2 ml of methanol (under fume hood). Vortex immediately to prevent rifampicin from sticking to the bottom of the tube. Add ~5 drops of 10 N NaOH to facilitate dissolving the rifampicin. Add 2 ml of the stock solution directly to 1 liter of cooling agar. (Final concentration: 100 μg/ml)

Rifampicin is light-sensitive, so plates should be stored in the dark or wrapped in aluminum foil. The lifetime of plates can sometimes be increased by adding minimal A salts (100 ml of 10× minimal A salts per liter of medium).

7. Nalidixic Acid (100 mg/ml stock solution)

Weigh out 100 mg of nalidixic acid and dissolve in 1 N NaOH. Add 0.3 ml of the stock solution directly to 1 liter of cooling agar. (Final concentration: 30 μg/ml)

8. Trimethoprim

When used in media, weigh out under sterile conditions 10 mg of trimethoprim and add the powder directly to 1 liter of cooling agar. (Final concentration: 10 μg/ml)

F. Buffers

Tables 1–3 are reprinted from Sambrook et al. (*Molecular Cloning. A Laboratory Manual*, Second Edition. Cold Spring Harbor Laboratory Press, Cold Spring Harbor, New York, Book 3, pp. B.21, B.21, and B.1, respectively [1989]).

1. Phosphate Buffers

Table 1   Preparation of 0.1 M Potassium Phosphate Buffer at 25°C

pH	Volume of 1 M K_2HPO_4 (ml)	Volume of 1 M KH_2PO_4 (ml)
5.8	8.5	91.5
6.0	13.2	86.8
6.2	19.2	80.8
6.4	27.8	72.2
6.6	38.1	61.9
6.8	49.7	50.3
7.0	61.5	38.5
7.2	71.7	28.3
7.4	80.2	19.8
7.6	86.6	13.4
7.8	90.8	9.2
8.0	94.0	6.0

Data from Green (1933).

Table 2   Preparation of 0.1 M Sodium Phosphate Buffer at 25°C

pH	Volume of 1 M Na_2HPO_4 (ml)	Volume of 1 M NaH_2PO_4 (ml)
5.8	7.9	92.1
6.0	12.0	88.0
6.2	17.8	82.2
6.4	25.5	74.5
6.6	35.2	64.8
6.8	46.3	53.7
7.0	57.7	42.3
7.2	68.4	31.6
7.4	77.4	22.6
7.6	84.5	15.5
7.8	89.6	10.4
8.0	93.2	6.8

Data from ISCO (1982).

Dilute the combined 1 M stock solution to 1000 ml with distilled water. pH is calculated according to the Henderson-Hasselbalch equation:

$$pH = pK' + \log\left[\frac{(\text{proton acceptor})}{\text{proton donor}}\right]$$

where $pK' = 6.86$ at 25°C.

2. Citrate Buffer, pH 5.5 (0.1 M)

citric acid (anhydrous)	9.6 g
NaOH	4.4 g

Add distilled water to a final volume of 500 ml, and adjust pH with 10 N NaOH. Autoclave.

or

0.1 M citric acid	4.7 volumes
0.1 M Na$_3$ citrate	15.4 volumes

3. Saline (for dilutions)

NaCl	8.5 g
H$_2$O	1 liter

4. Z Buffer for β-galactosidase Assays

per liter

Na$_2$HPO$_4 \cdot 7$H$_2$O	16.1 g
NaH$_2$PO$_4 \cdot$ H$_2$O	5.5 g
KCl	0.75 g
MgSO$_4 \cdot 7$H$_2$O	0.246 g
β-mercaptoethanol	2.7 ml

Do not autoclave! Adjust pH to 7.0. Store in refrigerator.

5. Tris Buffers

Table 3   Preparation of Tris Buffers of Various Desired pH Values

Desired pH (25°C)	Volume of 0.1 N HCl
7.10	45.7
7.20	44.7
7.30	43.4
7.40	42.0
7.50	40.3
7.60	38.5
7.70	36.6
7.80	34.5
7.90	32.0
8.00	29.2
8.10	26.2
8.20	22.9
8.30	19.9
8.40	17.2
8.50	14.7
8.60	12.4
8.70	10.3
8.80	8.5
8.90	7.0

Tris buffers (0.05 M) of the desired pH can be made by mixing 50 ml of 0.1 M Tris base with the indicated volume of 0.1 N HCl and then adjusting the volume of the mixture to 100 ml with H$_2$O.

G. Soft Agar (Top Agar)

Kept molten at 45°C.

1. Minimal "F-Top" Agar

per liter
Difco agar	8 g
NaCl	8 g

2. H-Top Agar

per liter
Bacto tryptone	10 g
Difco agar	8 g
NaCl	8 g

3. R-Top Agar

per liter
Bacto tryptone	10 g
Bacto yeast extract	1 g
Difco agar	8 g
NaCl	8 g

After autoclaving, add 2 ml of 1 M $CaCl_2$ + 5 ml of 20% glucose.

SECTION 25

H. Media for Storage of Strains

See Unit 2 in the Laboratory Manual portion of this volume for additional information.

1. Stabs

A variety of media are used for stabs, including LB and minimal media. I recommend the following:

per liter

Bacto tryptone	10 g
Bacto yeast extract	5 g
NaCl	10 g
Difco agar	6 g

Some investigators use this medium with 2 g of glucose added per liter.

A second widely used medium is

per liter

Difco nutrient broth (powder)	10 g
NaCl	5 g
Difco agar	6 g

Supplements are generally not added to rich medium except for Thy$^-$ strains, in which case thymine at 50 μg/ml is included in the medium.

2. Storage in Glycerol

A fresh overnight culture, grown in either minimal or rich medium, is used. Spin down the culture and resuspend in buffer (e.g., 0.1 M phosphate). For liquid glycerol cultures, add 2 ml of sterile 80% glycerol to 5 ml of the culture and place in the freezer at $-10°C$ to $-15°C$. Cells are recovered from storage by inoculating fresh medium with a few drops of the glycerol culture. For cultures at $-70°C$, add 0.5 ml of sterile 80% glycerol to 2 ml of culture, vortex, and store at $-70°C$.

3. Freeze-drying

Lyophilization, or freeze-drying, is the surest way to store *E. coli* for a long period of time. See Unit 2 in the Laboratory Manual portion of this volume for procedures.

Photographing Bacterial Colonies

The following article is reprinted, with permission, from Shapiro (1985).

REFERENCE

Shapiro, J.A. 1985. Photographing bacterial colonies. *ASM News* **51:** 62–69.

Photographing Bacterial Colonies

James A. Shapiro

Photography can serve as a valuable tool for research in microbiology in several ways. It is a superb means of documenting routine results and avoiding subjective judgments in scoring semiquantitative phenomena, such as the relative growth of different cultures on a particular solid medium. A photograph of the petri dish is much clearer and more definitive than a table with results recorded as "+++" or "weak." Some phenomena are very difficult to reduce to statistics but are readily visible in a picture, such as different kinds of sectoring patterns (Fig. 1). Certain types of observations can only be presented photographically, particularly those involving the morphology of microbial growth (plaques, colonies, lawns, etc.). The cost of a 35-millimeter black and white negative is only a few cents for film and chemicals, and the negative constitutes a permanent record of an experiment at a particular time point.

When the same sample must be observed repeatedly (for example, to record the emergence of colonies over time), my own experience has shown that photography is by far the most convenient method of data collection (J. A. Shapiro, "Observations on the formation of clones containing *araB-lacZ* cistron fusions," Mol. Gen. Genet. **194**:79–90, 1984). My work on fusion colony emergence also showed that a series of pictures can yield unexpected dividends by revealing features of the data that were not anticipated when the experiment began, such as clustering of colonies in a particular area of the growth substrate. Another example of the value of sequential photos is shown in Fig. 2. In reviewing the bottom panel of this series (taken 15 days after plating), I

realized that I could not tell whether this growth had arisen from 3 or 4 initial CFU. Fortunately, routine photography 2 days after plating (top panel) showed that there were only 3 CFU plated in this area and that a small "bud" on one colony later developed into a region with its own concentric and sectorial structure. Because of the extraordinary amount of information that a picture can contain, photography can record data that involves too many parameters to be abstracted into meaningful tables and graphs.

The usefulness of photography extends beyond mere recording of data. The mechanics of taking pictures enhances your powers of observation. In choosing what subjects to photograph and focusing the camera (see Fig. 3), you must concentrate on your experiments with great intensity, and this process often results in your noticing details and patterns that would normally be overlooked. Making prints from the negatives has the same effect, both because it too requires great concentration and because the photographic process generally enlarges the subject and enhances subtle contrasts at each step.

This lesson was brought home to me very sharply about 2 years ago. At that time, I was taking daily photographs to document the process of genetic fusions in *Escherichia coli* and was also trying to use fusions to study hydrocarbon oxidation in *Pseudomonas putida*. Our photographer, Gerry Grofman, suggested that I try using a new film, and so I tested it out on some of my genetically manipulated *P. putida* strains that had

Dr. Shapiro is a member of the Department of Molecular Genetics and Cell Biology at the University of Chicago. His photographs have appeared several times on the cover of ASM News *(March and July 1984 issues and this month), and others are scheduled for future issues.*

FIG. 1. *(Facing page, left-hand column)* Four sectored colonies of an *E. coli* strain carrying a plasmid with a Mu d*lac* element. These colonies all originated from single colony-forming units from a culture dilution prepared by resuspending a single-transformant colony. The expression of beta-galactosidase encoded by the plasmid results in hydrolysis of a chromogenic substrate (commonly called XGal) incorporated into the agar and consequent staining of the bacteria which produced the enzyme.

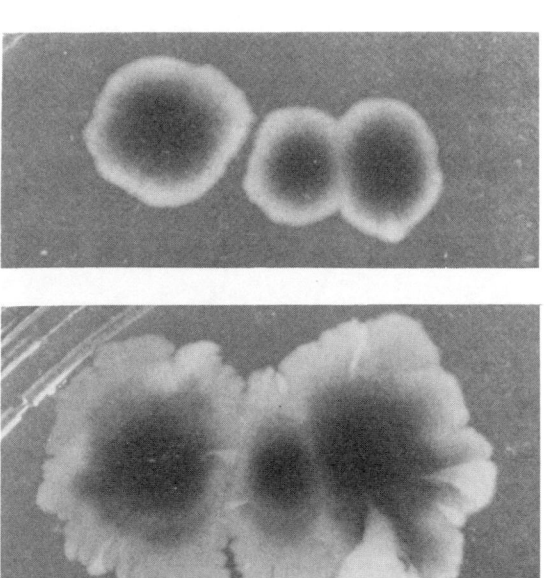

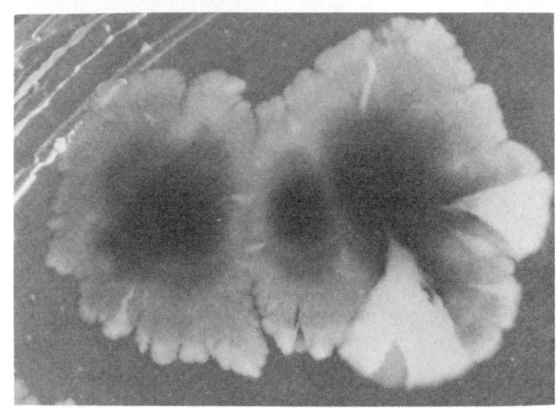

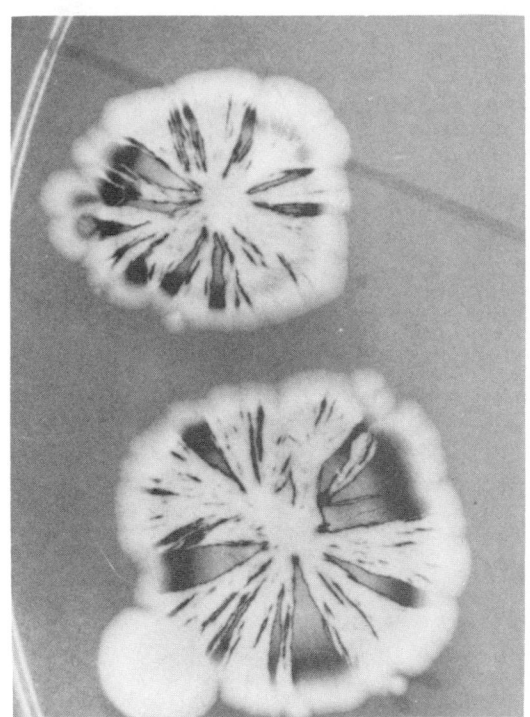

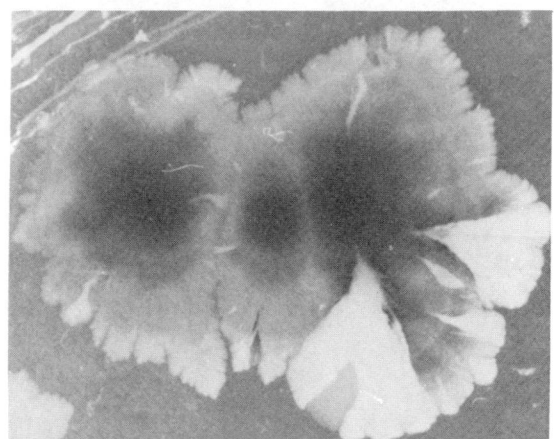

FIG. 2. *(Above)* Development of three *P. putida* (Mu d*lac*) colonies from single colony-forming units on XGal agar, photographed (top to bottom) 2, 5, 9, and 15 days after plating.

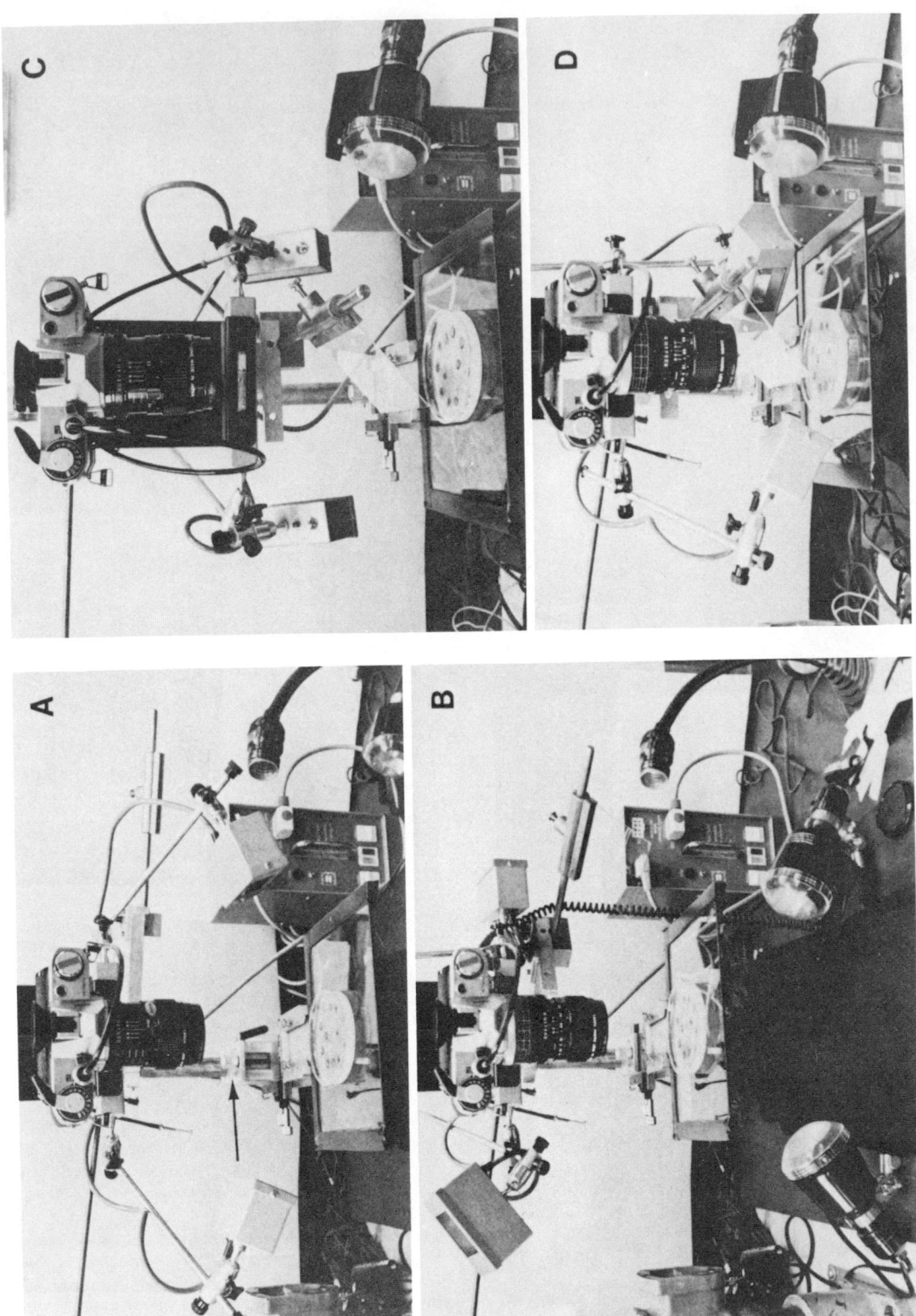

FIG. 3. Camera and lighting setups for colony photography. The panels show how various electronic flash units and the beam splitter are arranged for different types of illumination. (A) Lateral illumination from the two flashes on the Bowens Texturelite (no longer commercially available). Note the fittings on the glass plate supporting the petri dish. There is a screw control knob (arrow) for vertical adjustment. (I almost never use this latter feature but simply position the dish on the glass support by hand.) The entire support unit can be moved up and down on the copy stand. (B) Dark-field illumination from the two flash units in screw sockets on the base of the copy stand. One of these units (the master unit) is triggered by a synchronization cable from the camera, and the other is a slave unit activated by the flash from the first unit. (C) Axial illumination with the beam splitter plate. The beam splitter is also attached to the copy stand and can be turned out of the way for other kinds of illumination, as shown in panels A and B. The light comes from the flash unit to the right, mounted in a gooseneck lamp, and can be directed to a specific area in the subject by changing the position of the plate, the position of the light source, and the angle of the plate (there is an adjustment for this). The area of axial illumination on the subject can be determined by focusing the lens on the light source rather than on the petri dish and making adjustments to place the bulb in the appropriate position (usually the center) of the visual field. (D) Combined lateral and axial illumination. The lateral flash units are positioned so that they do not block the light to the beam splitter and also so that they do not create reflections on the beam splitter plate. In this illustration, both light sources are electronic flash (the beam splitter unit operating as a slave to the lateral units), facilitating color photography as shown on the cover. However, the relative intensity of the lateral and axial illumination recorded on film can be controlled by using an incandescent light source on the beam splitter and adjusting the exposure duration; longer exposures do not change the degree of lateral flash illumination but do increase the amount of axial illumination. It is sometimes necessary to mask the bright metal components of the camera setup because they create reflections (for example, the metal bars holding the Texturelite flash units). In all the panels, an additional stabilizing bar can be seen coming from the upper left to give greater rigidity to the camera support.

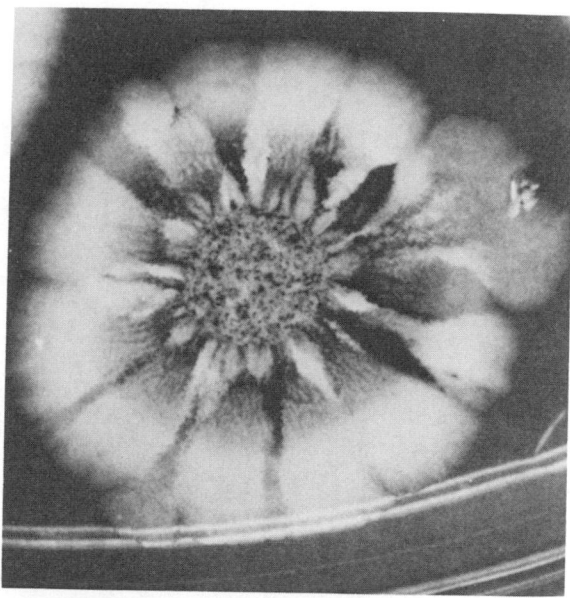

FIG. 4. *P. putida* colony produced by spotting about 10^4 cells of a culture carrying a Mu d*lac* element in the CAM-OCT plasmid on XGal agar. The site of original inoculation is the salt-and-pepper circle in the center. Note where the emerging sector at the right has been sampled with a sterile toothpick.

been stained on agar for beta-galactosidase activity. The final 8 × 10 prints showed elaborate patterns of enzyme activity that I had never noticed before, and after a little practice with different inoculation procedures a few selections eventually resulted in some very striking colonies. Figure 4 shows one example of an elaborate flowerlike pattern of beta-galactosidase expression. This unexpected outcome to some "photographic" experimentation has opened up a whole new field of bacteriological research for me, because these patterns have made it possible to study the internal organization and growth of bacterial colonies (J. A. Shapiro, "The use of Mu*dlac* transposons as tools for vital staining to visualize clonal and non-clonal patterns of organization in bacterial growth on agar surfaces," J. Gen. Microbiol. 130:1169–1181, 1984).

There are two further aspects of photography which are less obviously utilitarian but deserve mention nonetheless. The first is that the habit of recording your work photographically leads you to begin designing your experiments so that the results can be understood from the data in a picture. This often stimulates new experimental methods and makes it easier to present your work to other scientists (who, incidentally, have to accept less on faith when they see results in a photo rather than in a statistical or graphic summary). The second aspect of using photography is that it introduces an esthetic element into the research which can serve as a source of pleasure to the investigator and can also stimulate outside interest in the work.

Because taking and printing pictures can literally be such eye-opening experiences, I highly recommend

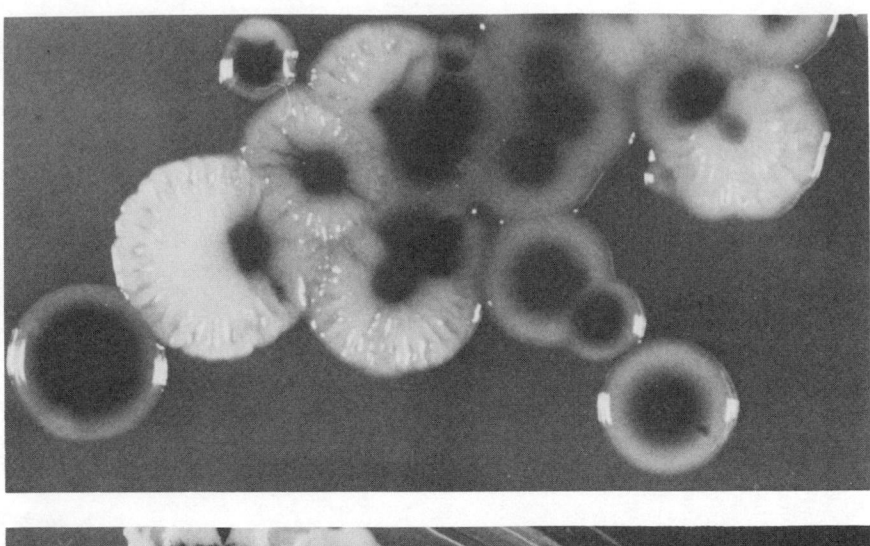

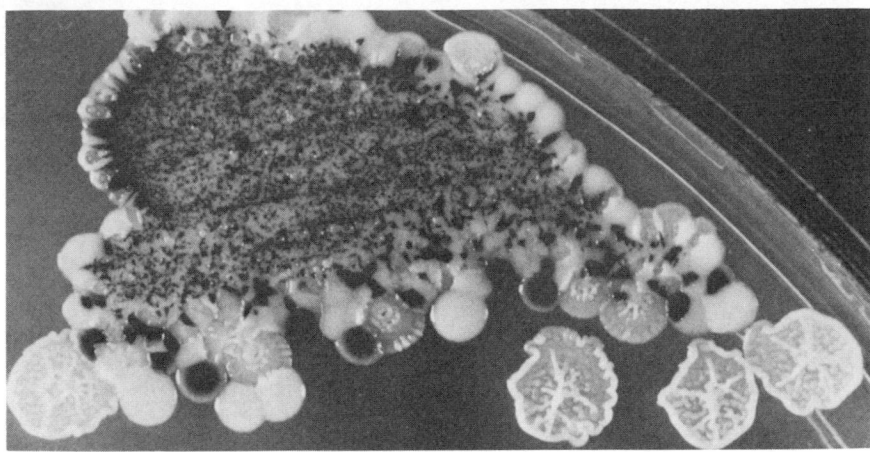

FIG. 5. Two streaks showing *P. putida* colonies with prominent three-dimensional structures on XGal agar. (Top) Colonies produced by streaking out a dark sector of a single colony. (Bottom) Colonies produced by streaking a mating mixture in which the recipient culture contained bacteria that produced the unpigmented wheellike colonies. Lighting was with two flash units about 180 degrees apart, angled down onto the bacteria at about 45 degrees.

that researchers do all their own photography rather than take their samples to a staff photographer. By research standards, the basic financial investment needed is rather small. Most scientists have access to an equipped darkroom, which is only essential for printing. A camera, macro lens, lights, copy stand, and developing accessories can be obtained for several hundred dollars (about the cost of a balance or pH meter). There are many good books available for learning basic photographic techniques and several specialist texts for scientists. One which has proved particularly useful to me is A. A. Blaker's *Handbook for Scientific Photography* (Freeman Publications, San Francisco; $31.75 in 1983).

My own photographic setup has grown in bits and pieces and now includes several standard items as well as some accessories built in our shop (Fig. 3). The camera, fitted with a 50-millimeter macro lens, cost about $300. I use a motor drive for convenience, but that is not essential. The camera is mounted on a standard copy stand to which two special fittings were added: one to hold an adjustable horizontal glass plate for supporting petri dishes and other specimens, and one to hold an angled glass plate between the camera and specimen to serve as a beam splitter for axial illumination (described in detail below). A piece of black velvet or construction paper is usually placed at the base of the stand to provide a dark background, but sometimes one or more light sources are placed there for special needs, such as for dark-field illumination or for copying negatives to make slides. Although I use the special Bowens Texturelite double flash unit shown in Fig. 3 for macrophotography because it was given to me by a colleague who no longer used it, two gooseneck lamps fitted with screw-in flash units (one master and one slave, about $35 to $50) are just as good and

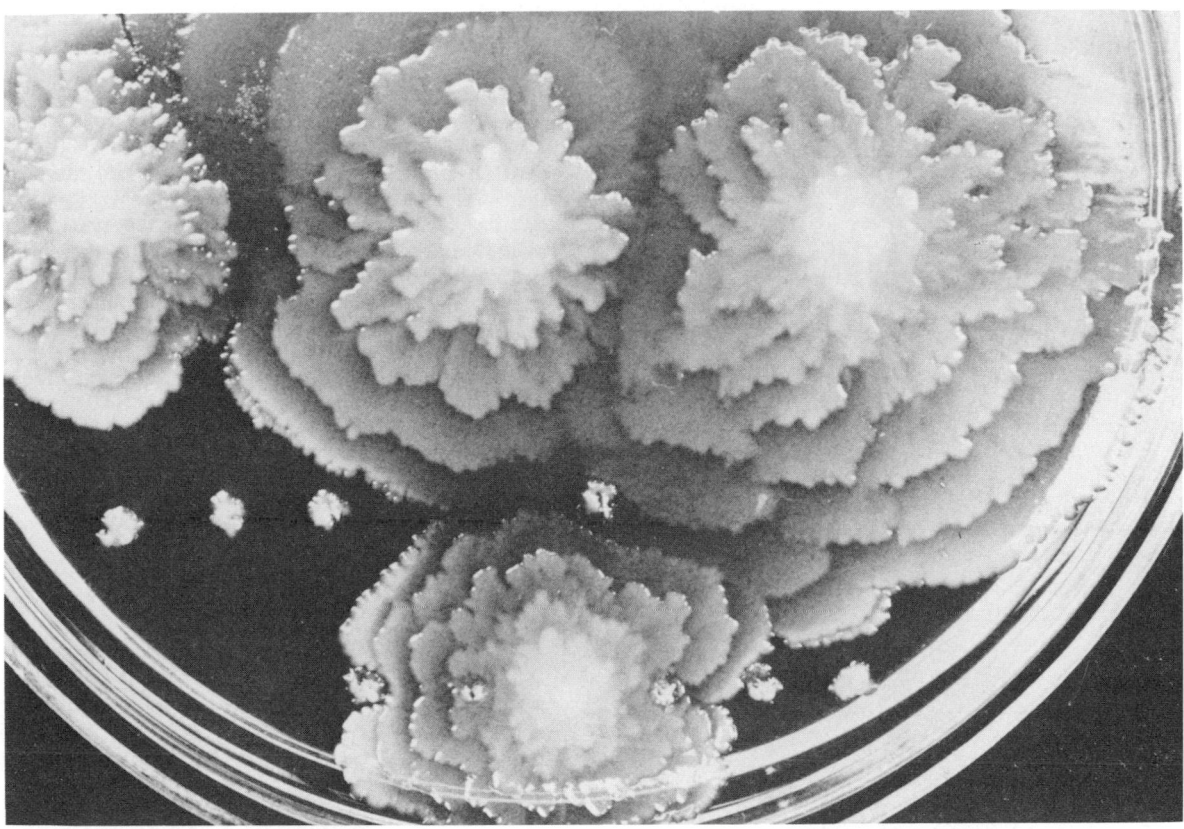

FIG. 6. Swarms of *Proteus mirabilis* on kanamycin agar after receiving a resistance plasmid from *E. coli*. Each inoculation point was stabbed with a mating mixture. When no transfer took place, the donor cells grew without swarming. When the *P. mirabilis* cells received the plasmid, they were able to swarm and form the terraced flowerlike structures shown here. Lighting was with fluorescent lamps about 6 feet above the petri dish.

sometimes even more flexible for special lighting set-ups. As shown in the figure, I also use an extra gooseneck lamp to hold a third light source for use with the beam splitter.

The main elements in a good setup for colony photography include stability (because very slight vibrations will blur close-ups), fine focus control through a screw mounting to raise and lower either the camera or (in my case) the glass plate holding the petri dish (thereby adjusting the focus without altering the magnification), and a flexible system for positioning lights easily for different illumination needs. Many useful examples are provided in L. Lefkowitz's *Manual of Close-Up Photography* (Watson-Guptill Publications, Inc., New York, 1979; $12.95), and a large photo supply store should have a variety of camera stands and lamps. I load my own black and white film (Kodak Technical Pan 2415, which has an extremely fine grain and a wide range of contrasts with different developers) from 100- or 150-foot rolls into reusable cassettes, so that my film cost is less than 4 cents per frame. For processing I use a light-tight changing bag for loading exposed film into small stainless steel developing

tanks, and developing is done at a sink in my laboratory.

One of the key variables in producing good negatives is lighting, and here your best guide is experience. Figure 3 illustrates four different arrangements for illumination. The normal lighting setup, balanced lamps angled onto the subject from above and to the side (panel A) often produces very beautiful pictures which pick up details of three-dimensional structure in the colonies (Fig. 5). Dark-field illumination, in which the light reaches the camera because its path has been altered by passage through the subject (Fig. 3B), is very good for photographing whole petri dishes with no shadows and also for some close-up work to reveal internal colony details. Dark-field illumination is especially useful for making repeated photographs of the same dish because it can be done without removing the lid, thus avoiding the risk of contamination. However, dark-field illumination produces a very flat image. Sometimes just the normal room lighting will produce the best results. For example, some of my most interesting pictures were made by using the fluorescent lights in the ceiling of my laboratory because they provided

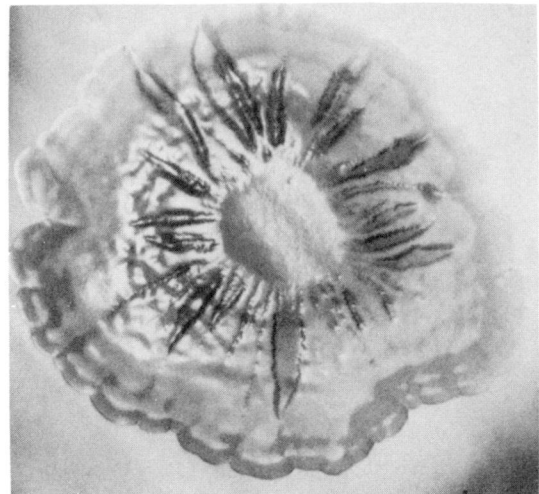

FIG. 7. Two views of the same *P. putida* colony grown on XGal agar from an inoculum of about 10^4 cells. The bottom panel was photographed with ordinary side lighting; the top panel was photographed with axial illumination.

3C). To illuminate along the axis of the lens, I shine light horizontally onto a glass plate at about 45 degrees; this bounces the light directly onto the specimen, and I can then photograph it through the glass. The difference in the results obtained with side lighting and those obtained with axial lighting from the beam splitter can be seen in two pictures of the same colony (Fig. 7). The lateral illumination shows the interior details of the colony but produces a very two-dimensional image. Axial illumination produces a striking picture of the colony surface but loses some of the interior details. By combining both light sources (Fig. 3D), it is possible to obtain satisfactory images which depict both internal details and overall colony structure, and two examples are presented in Fig. 8. These pictures are especially informative because they demonstrate the sectorial (i.e., clonal) nature of some

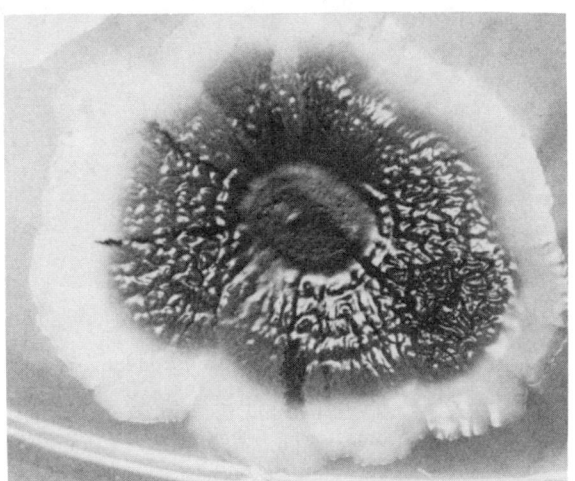

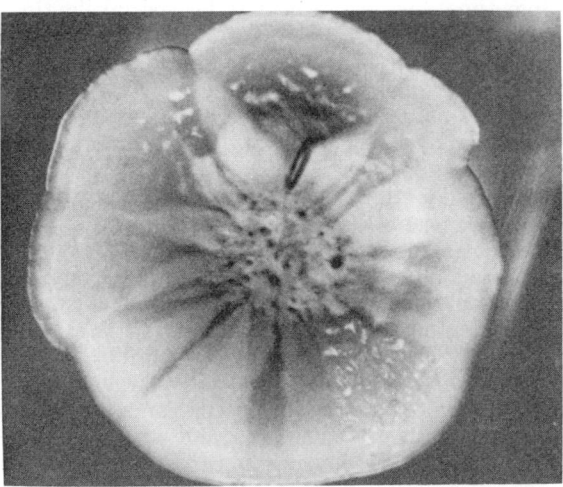

FIG. 8. Two *P. putida* colonies on XGal agar photographed with both lateral and axial illumination. Note the coincidence of sectorial changes in both beta-galactosidase expression and surface texture.

almost vertical illumination over a large part of the surface of a petri dish, highlighting the surface texture of extensive bacterial growth, such as terraced swarms of *Proteus mirabilis* (Fig. 6 and the cover photo of the March 1984 *ASM News*). The direction of illumination is critical in revealing very subtle features of three-dimensional structure. So-called axial illumination (along the axis of the lens) enhances the appearance of surface texture, because flat surfaces bounce the light directly into the lens, whereas even a slightly inclined surface will deflect it away from the lens. For very close-up work, overhead lights are not best because the specimen is in the shadow of the camera, and so other methods are needed. After reading about it in Blaker's book (p. 59), I adopted the use of a beam splitter (Fig.

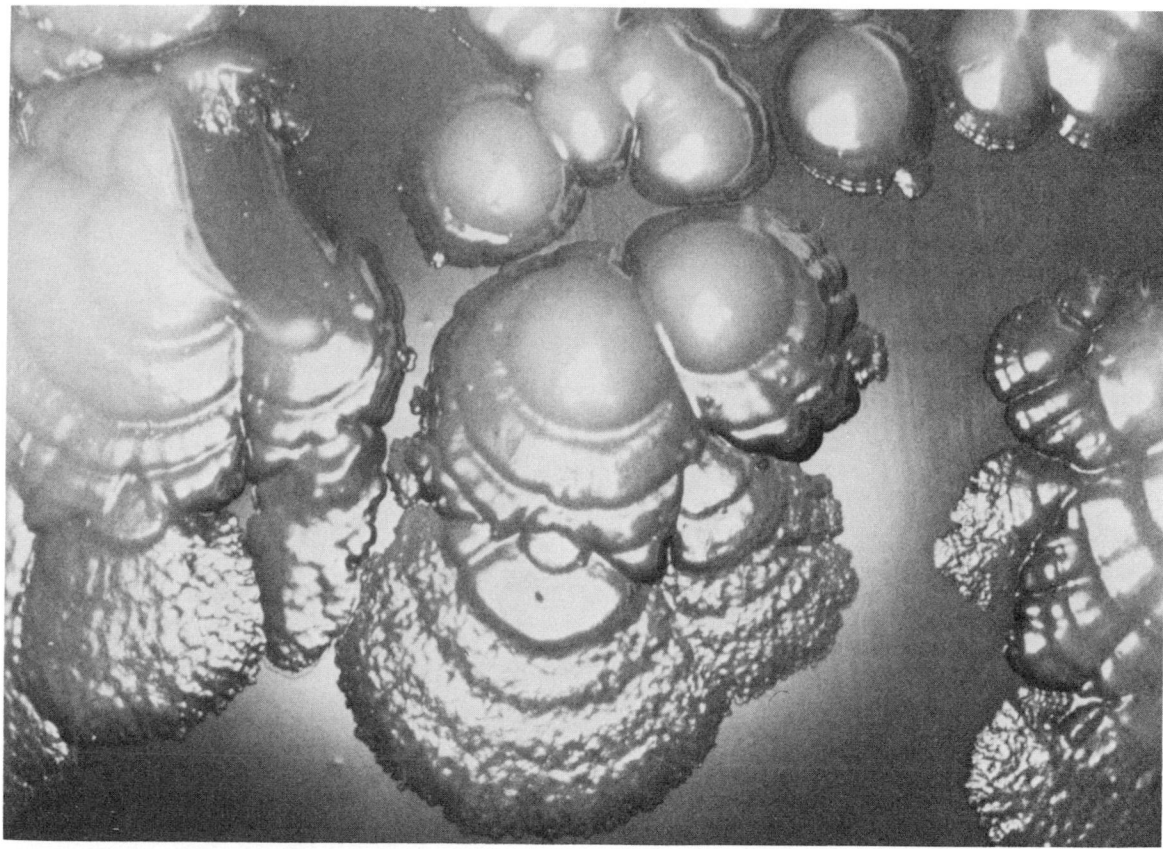

FIG. 9. *E. coli* colonies on MacConkey agar. These bacteria were isolated from a urine sample in the clinical microbiology laboratory at the University of Chicago Clinics. Axial illumination. Note the circumferential terracing (cf. Fig. 6) and the boundaries between adjacent clones.

changes in colony architecture. (Another example was published as the cover photo for the July 1984 *ASM News*).

Once the basics have been worked out, photographing colonies and other microbiological subjects can become as effortless as picking and streaking—and photography certainly offers many more opportunities for imagination and creativity. One of the most gratifying aspects of the technique is that so much information can be obtained from sources outside planned research protocols, such as the MacConkey agar plates of *E. coli* isolates from urine that were given me by Josephine Morello, who runs our clinical lab (Fig. 9), and particularly attractive microorganisms, such as the *Chromobacterium violaceum* cultures which a Belgian col-

league, Max Mergeay, sent me (cover photo). One of the main sources of satisfaction to many biologists is the appreciation of natural beauty in their research material, and a close look at colonies shows that microorganisms have as much artistic potential as plants and animals.

Acknowledgments

I am grateful to G. Gibson and A. Kittler for making the custom attachments to my photo stand, to G. Grofman and B. McClintock for advice on photographic techniques, and to J. Morello and M. Mergeay for supplying the bacteria illustrated in Fig. 6 and 8 and on the cover.

Research utilizing colony photography was supported by National Science Foundation grant PCM-8200971 and by Public Health Service grants GM-24960 and CA-19265 from the National Institutes of Health.

SECTION · 27

Commercial Suppliers

Listed below are the major commercial suppliers of materials commonly used in bacterial genetics. The information in this list, which is modified from Sambrook et al. 1989, is subject to change, but at the time of preparation (March 1992), it was correct.

REFERENCE

Sambrook, J., E.F. Fritsch, and T. Maniatis. 1989. *Molecular Cloning. A Laboratory Manual*, Second Edition. Cold Spring Harbor Laboratory Press, Cold Spring Harbor, New York.

Abbott Diagnostics, 850 Maude Avenue, Mountainview, CA 94043, USA. Telephone 415-969-1200 or 800-933-5535. Fax 415-969-0131.

Air Products and Chemicals, Inc., 7201 Hamilton Boulevard, Allentown, PA 18195, USA. Telephone 215-481-8257 or 800-345-3148.

Aldrich Chemical Co., Inc., 1001 West Saint Paul Avenue, Milwaukee, WI 53233, USA. Telephone 414-273-3850 or 800-558-9160. Fax 414-273-4979 or 800-962-9591.

American Bioanalytical, 10 Huron Drive, Natick, MA 01760, USA. Telephone 800-443-0600. Fax 508-655-2754.

American Hoechst Corporation, Route 202-206 North, P.O. Box 2500, Somerville, NJ 08876, USA. Telephone 201-231-2000 or 800-235-2637. Fax 201-231-3225. In Europe: Hoechst AG., Postfach 800320, 6320-Frankfurt am Main 80, Germany. Telephone (49)-(0)69-3050. In UK: Hoechst UK Ltd., Hoechst House, Salisbury Road, Hounslow, Middlesex TW4 6JH, UK. Telephone (44)-(0)1-5 70-7712. Fax (44)-(0)1-236-6336.

American Scientific Products. *See* Baxter Healthcare Corporation.

Amersham Corporation, 2636 South Clearbrook Drive, Arlington Heights, IL 60005, USA. Telephone 708-593-6300 or 800-323-9750. Fax 708-593-8236. In UK: Amersham International plc, Research Products Division, White Lion Road, Amersham, Buckinghamshire HP7 9LL, UK. Telephone (44)-(0)2404-4444. Fax (44)-(0)296-85190.

Amicon Corporation, 24 Cherry Hill Drive, Danvers, MA 01923, USA. Telephone 508-777-3622 or 800-343-0696. Fax 508-777-6204.

Applied Biosystems Inc., 850 Lincoln Centre Drive, Foster City, CA 94404, USA. Telephone 415-570-6667. Fax 415-572-2743.

BACHEM, Inc., 3132 Kashiwa Street, Torrance, CA 90505, USA. Telephone 310-539-4171. Fax 310-530-1571.

Baker. *See* J.T. Baker Inc.

Baxter Healthcare Corporation, Scientific Products Division, 1430 Waukegan Road, McGaw Park, IL 60085, USA. Telephone 708-689-8410 or 800-633-7370. Fax 708-473-2114.

Baxter Hospital Supply Corp. *See* Baxter Healthcare Corp.

BDH Diagnostics, Ltd., Broom Road, Poole, Dorset BH124 NN, England. Telephone 0202-745520.

Beckman Instruments Inc., 45 Belmont Drive, P.O. Box 6764, Somerset, NJ 08875-6764, USA. Telephone 201-560-0076 or 800-742-2345. Fax 201-560-1448. In Europe: Beckman Instruments International SA., 22 Rue Juste-Olivier, CH-1260 Nyon, Switzerland. Telephone (41)-(0)22-631181. Fax (41)-(0)22-621810.

Becton Dickinson, Immunocytometry Systems, 2375 Garcia Avenue, Mountain View, CA 94043, USA. Telephone 415-968-7744. Fax 415-966-8614. In Europe: P.O. Box 13, Erembodegem 9440, Belgium. Telephone (32)-5378-7830.

Becton Dickinson, Labware, 2 Bridgewater Lane, Lincoln Park, NJ 07035, USA. Telephone 201-628-1144 or 800-235-5953. Fax 201-628-1533.

Becton Dickinson, Primary Care Diagnostics, 1 Becton Drive, Franklin Lakes, NJ 07417, USA. Telephone 201-848-6500. Fax 201-848-6475.

Bellco Biotechnology, P.O. Box B, 340 Erudo Road, Vineland, NJ 08360, USA. Telephone 609-691-1075 or 800-257-7043. Fax 609-691-3240.

Bethesda Research Laboratories Inc. (BRL), P.O. Box 6009, 8717 Grovemont Circle, Gaithersburg, MD 20877, USA. Telephone 301-840-8000 or 800-828-6686. Fax 301-258-8238.

Biolog, Inc., 3447 Investment Blvd., Suite 3, Hayward, CA 94545. Telephone 510-785-2585 or 800-284-4949. Fax 510-782-4639.

Bio-Rad Laboratories, 1414 Harbor Way, Richmond, CA 94804, USA. Telephone 415-234-4130 or 800-227-5589. Fax 415-232-4257. In UK: Bio-Rad Laboratories Ltd., Caxton Way, Watford Business Park, Watford, Hertfordshire WD1 8RP UK. Telephone (44)-(0)923-240322. Fax (44)-(0)923-247825.

Boehringer Mannheim Biochemicals, Orders: P.O. Box 50414, Indianapolis, IN 46250, USA. Telephone 317-849-9350 or 800-262-1640. Fax 317-576-2754. Other Information: 9115 Hague Road, Indianapolis, IN 46256, USA. In Europe: Sandhoferstrasse 116, Postfach 310120, D6800 Mannheim, Germany. Telephone (49)-(0)621-7591. Fax (49)-(0)621-7592890. In UK: BCL-Boehringer, Boehringer-Mannheim House, Bell Lane, Lewes East Sussex BN7 1LG, UK. Telephone (44)-(0)273-480444. Fax (44)-(0)273-480266.

Brinkmann Instruments Inc., 1 Cantiague Road, Westbury, 11590, USA. Telephone 516-334-7500. Fax 516-334-7506. In Europe: Eppendorf-Netheler-Hinz GmbH., P.O. Box 650670, D-2000 Hamburg 65, Germany. Telephone (49)-(0)40-538-01-0. Fax (49)-(0)40-538-01-556. In UK: BDH Ltd., Apparatus Division, P.O. Box 8, Dagenham, Essex RM8 1RY, UK. Telephone (44)-(0)1-597-8821. Fax (44)-(0)1-597-8300.

BRL. *See* Bethesda Research Laboratories Inc.

Calbiochem, 10933 North Torrey Pines Road, La Jolla, CA 92037, USA. Telephone 619-450-9600 or 800-854-3417. Fax 619-453-3552. In UK: Cambridge Bioscience, 42 Devonshire Road, Cambridge CB1 2BL, UK. Telephone (44)-(0)223-316855. Fax (44)-(0)223-460396.

Cetus. *See* Perkin Elmer Cetus.

Ciba-Corning Diagnostics, 333 Coney Street, East Walpole, MA 02032, USA. Telephone 508-668-5000 or 800-343-0893. Fax 508-660-4591.

Clay Adams. *See* Becton Dickinson, Primary Care Diagnostics.

Collaborative Biomedical Products, 2 Oak Park, Bedford, MA 01730, USA. Telephone 617-275-0004. Fax 617-275-0043.

Corning Glass Works, Science Products Division, MP-21-5-8, Corning, NY 14831, USA. Telephone 607-974-4667. Fax 607-974-7919.

Curtin Matheson Scientific Inc., 9999 Veterans Memorial Drive, Houston, TX 77038, USA. Telephone 713-820-9898 or 713-820-1661. Fax 713-878-2221.

Difco Laboratories, P.O. Box 331058, Detroit, MI 48232-7058, USA. Telephone 313-961-0800 or 800-521-0851. Fax 313-961-1501. In UK: Difco Laboratories, P.O. Box 14B, Central Avenue, East Molesey, Surrey KT8 0SE, UK. Telephone (44)(0)1-979-9951. Fax (44)-(0)1-979-2506.

Dow Chemical Company, AgOrganics Department, Building 9001-9008, Midland, MI 48641-1706, USA. Telephone 517-636-1000. Fax 517-636-3373.

Dow Corning Corporation, Dow Corning Center, Midland, MI 48686-0994, USA. Telephone 517-496-4000. Fax 517-496-6974.

Drummond Scientific Corporation, 500 Parkway, Broomall, PA 19008, USA. Telephone 215-353-0200 or 800-523-7480. Fax 215-353-6204.

du Pont. *See* E.I. du Pont de Nemours & Company Inc.

Eastman Kodak Special Products Division, 2400 Mt. Read Blvd., Rochester, NY 14650, USA.

Telephone 716-588-3124 or 800-242 2424. In UK: Kodak Ltd., Laboratory and Special Products Division, Acornfield Road, Knowsley Industrial Park North, Liverpool L33 7UF, UK. Telephone (44)-(0)51-548-6560. Fax (44)-(0)51-547-2404.

E.I. du Pont de Nemours & Company Inc., Medical Department, Biotechnology Division, P.O. Box 80024, Barley Mill Plaza, Wilmington, DE 19880-0024, USA. Telephone 302-992-3416 or 800-551-2121. Fax 302-992-3474. In UK: Dupont (UK) Ltd., Industrial Products Division, Wedgewood Way, Stevenage, Hertfordshire SG1 4QN, UK. Telephone (44)-(0)438-734000. Fax (44)-(0)-438-734154.

ElectroNucleonics Laboratories Inc. *See* Pharmacia ENI Diagnostics Inc.

Eppendorf-Netheler-Hinz GmbH. *See* Brinkmann Instruments Inc.

Ericomp, 10055 Barnes Canyon Road, Suite G, San Diego, CA 92121, USA. Telephone 619-457-1888 or 800-541-8471. Fax 619-457-2937.

Falcon. *See*, e.g., Baxter Healthcare Corporation; Becton Dickinson, Labware.

Fisher Scientific Co., 52 Fadem Road, Springfield, NJ 07081, USA. Telephone 201-467-6400 or 800-766-7000. Fax 201-379-7638.

Flow Laboratories Inc. *See*, for USA, ICN Biomedicals. In UK: Flow Laboratories Ltd., Woodcock Hill, Harefield Road, Rickmansworth, Hertfordshire WD3 1 PQ, UK. Telephone (44)-(0)923-774666. Fax (44)-(0)-923-777005.

Fluka Chemical Corporation, 980 South 2nd Street, Ronkonkoma, NY 11779-7204, USA. Telephone 516-467-0980. Fax 516-467-0663. In Europe: Fluka Chemie AG., Industriestrasse 25, CH-9470 Buchs Switzerland. Telephone (41)-(0)85-69511. Fax (41)-(0)85-65449. In UK: Fluka Chemicals Ltd., Peakdale Road, Glossop, Derbyshire, UK. Telephone (44)-(0)4574-62518. Fax (44)-(0)4574-4307.

FMC Marine Colloids Division, Bioproducts Department, 5 Maple Street, Rockland, ME 04841, USA. Telephone 207-594-3360 or 800-341-1574. Fax 207-594-3391.

Fuji Medical Systems, 90 Viaduct Road, Stamford, CT 06907, USA. Telephone 203-353-0300 or 800-431-1850. Fax 203-353-0926.

Gelman Sciences Inc., 600 South Wagner Road, Ann Arbor, MI 48106, USA. Telephone 313-665-0651. Fax 313-761-1208.

GIBCO, 3175 Staley Road, Grand Island, NY 14072, USA. Telephone 716-773-0790 or 800-828-6686. Fax 800-331-2286. In UK: GIBCO-BRL Ltd., Unit 4, Cowley Mill Trading Estate, Longbridge Way, Uxbridge UB8 2YG, UK. Telephone (44)-(0)895-36355. Fax (44)-(0)895-53159.

Gilson Medical Electronics Inc., 3000 West Beltline Highway, P.O. Box 27, Middleton, WI 53562, USA. Telephone 608-836-1551 or 800-445-7661. Fax 608-831-4451. In Europe: Gilson Medical Electronics SA., 72 Rue Gambetta, BP 45, Villiers-le-Bel, 95400 France. Telephone (33)-139-905441.

Glass Vials, Inc., 1352 James Street, Baltimore, MD 21223, USA. Telephone 410-685-6910. Fax 410-625-7939.

Hamilton Company, 4970 Energy Way, Reno, NV 89502, USA. Telephone 702-786-7077 or 800-648-5950. Fax 702-323-7259.

Health Products. *See* Pierce Chemical Co.

Heat Systems Ultrasonics Inc., 1938 New Highway, Farmingdale, NY 11735, USA. Telephone 516-694-9555 or 800-645-9846. Fax 516-694-9412.

Hoechst. *See* American Hoechst Corporation.

Hoefer Scientific Instruments, P.O. Box 77387, 654 Minnesota Street, San Francisco, CA 94107, USA. Telephone 415-282-2307 or 800-227-4750. Fax 415-821-1081.

IBI. *See* International Biotechnologies Inc.

ICN Biochemicals, P.O. Box 28050, Cleveland, OH 44128, USA. Telephone 216-831-3000 or 800-321-6842. Fax 216-831-2569.

ICN Biomedicals Inc, Flow Laboratories, 3300 Hyland Avenue, Costa Mesa, CA 92713, USA. Telephone 714-545-0100 or 800-854-0530. Fax 714-557-4872. In UK: ICN Biomedicals Ltd., Lincoln Road, High Wycombe, Bucks HP12 3XJ, UK. Telephone (44)-(0)494-443826. Fax (44)-(0)494-436048.

ICN Radiochemicals, P.O. Box 19536, Irvine, CA 92713, USA. Telephone 714-545-0113 or 800-854-0530. Fax 714-557-4872.

IEC. *See* International Equipment Company.

Intermec Corporation, 14100 Laurel Park Drive, Suite C., Laurel, MD 20707, USA. Telephone 410-792-2133. Fax 301-498-0754.

International Biotechnologies Inc. (IBI), P.O. Box 9558, 25 Science Park, New Haven, CT 06535, USA. Telephone 203-786-5600 or 800-243-2555. Fax 203-786-5694.

International Equipment Company (IEC), Division of Damon Corporation, 300 Second

Avenue, Needham Heights, MA 02194, USA. Telephone 617-449-0800 or 800-225-8856. Fax 617-444-6743. In UK: Damon/ IEC UK Ltd., Unit 7, Lawrence Way, Brewers Hill Road, Dunstable, Bedfordshire LU6 1BD, UK. Telephone 44)-(0)582-604669. Fax (44)-(0)582-609257.

ISCO, Instrument Division, P.O. Box 5347, Lincoln, NB 68505-9987, USA. Telephone 402-464-0231 or 800-228-4250. Fax 402-464-4543.

J.T. Baker Inc., 222 Red School Lane, Phillipsburg, NJ 08865, USA. Telephone 908-859-2151 or 800-582-2537. Fax 908-859-9318.

Kodak. *See* Eastman Kodak Special Products Division.

Kontes, Spruce Street, P.O. Box 729, Vineland, NJ 08360, USA. Telephone 609-692-8500 or 800-323-7150. Fax 609-692-3242.

Life Sciences Inc., 2900 72nd Street North, St. Petersburg, FL 33710, USA. Telephone 813-345-9371 or 800-237-4323. Fax 813-347-2957.

LKB. *See* Pharmacia LKB Biotechnology Inc.

Lumigen, Inc., P.O. Box 07339, Detroit, MI 48207, USA. Telephone 313-577-6012. Fax 313-577-2299.

Mallinckrodt. *See* Baxter Healthcare Corporation.

Markem Corp., 40-T Putnam Street, Keene, NH 03431, USA. Telephone 603-352-1130.

Miles Inc., 1127 Myrtle Street, Elkart, IN 46515, USA. Telephone 219-264-8111 or 800-348-7414. Fax 219-262-6747. In UK: Miles Laboratories Ltd., Miles Scientific, Stoke Court, Stoke Poges, Slough SL2 4LY, UK. Telephone (44)-(0)2814-5151. Fax (44)-(0)2814-3893.

Millipore Corporation, 80 Ashby Road, Bedford, MA 01730, USA. Telephone 617-275-9200 or 800-225-1380. In UK: Millipore (UK) Ltd., Millipore House, The Boulevard, Ascot Road, Croxley Green, Watford WD1 8YW, UK. Telephone (44)-(0)923-816375. Fax (44)-(0)923-818297.

Nalge. *See*, e.g., Baxter Healthcare Corporation; Fisher Scientific.

NEN. *See* New England Nuclear.

New Brunswick Science Co., Inc., 44 Talmadge Road, Edison, NJ 08818-4005, USA. Telephone 201-287-1200 or 800-631-5417. Fax 201-287-4222.

New England Biolabs Inc., 32 Tozer Road, Beverly, MA 01915, USA. Telephone 508-927-5054 or 800-632-5227. Fax 508-921-1350.

New England Nuclear (NEN), 549 Albany Street, Boston, MA 02118, USA. Telephone 617-482-9595 or 800-551-2121. In Europe: du Pont de Nemours (Deutschland) GmbH., Biotechnology Systems Division, NEN Research Products, Postfach 40 12 40, 6072 Dreieich 4, Germany. Telephone (49)-(0)6103-803-155. Fax (49)-(0)6103-897.

Nunc Inc., 2000 North Aurora Road, Naperville, IL 60566, USA. Telephone 312-983-5700 or 800-288-6863. Fax 312-416-2519. In UK: GIBCO Ltd., P.O. Box 35, Trident House, Renfrew Road, Paisley, Scotland PA3 4EF. Telephone (44)-(0)41-889-6100. Fax (44)-(0)41-887-1167. In Europe: Nunc A/S., Postbox 280, Kamstrup, DK 4000, Roskilde, Denmark. Telephone (45)-(0)2-359065. Fax (45)-(0)2-350105.

Perkin Elmer Cetus, 761 Main Avenue, Norwalk, CT 06859, USA. Telephone 800-762-4003. Fax 203-761-9645.

Pharmacia ENI Diagnostics Inc., 9033 Red Branch Road, Columbia, MD 21045, USA. Telephone 410-381-4800. Fax 410-381-9164.

Pharmacia LKB Biotechnology Inc., 800 Centennial Avenue, Piscataway, NJ 08854, USA. Telephone 201-457-8000 or 800-526-3593. Fax 201-457-0557. In UK: Pharmacia LTB Biotechnology Ltd., Pharmacia House, Midsummer Boulevard, Milton Keynes, Buckinghamshire MK9 3HP UK. Telephone (44)-(0)908-661101. Fax (44)-(0)908-690091. In Europe: Pharmacia LKB Biotechnology AB., P.O. Box 175, Bjorkgatan 30, 751 82 Uppsala, Sweden. Telephone (46)-(0)18-163000. Fax (46)-(0)18-143820.

Pierce Chemical Company, P.O. Box 117, Rockford, IL 61105, USA. Telephone 815-968-0747 or 800-874-3723. Fax 815-968-7316.

Promega Biotec, 2800 South Fish Hatchery Road, Madison WI 53711, USA. Telephone 608-274-4330 or 800-356-9526. Fax 608-273-6967.

Rainin Instrument Company, Mack Road, Woburn, MA 01801, USA. Telephone 617-935-3050. Fax 617-938-8157.

Raylo Chemicals, 8045 Argyll Road, Edmonton, Alberta, Canada T6C 4A9. Telephone 403-468-6060. Fax 403-468-4784.

Research Organics, Inc., 4353 East 49th. Street, Cleveland, OH 44125-1083, USA. Telephone 216-883-8025 or 800-321-0570. Fax. 216-883-1576.

Revco Scientific Inc., 275 Aiken Road, Asheville, NC 28804, USA. Telephone 704-658-2711 or 800-252-7100.

Rohm & Haas Company, Independence Mall West, Philadelphia, PA 19105, USA. Telephone 215-592-3000. Fax 215-592-3377. In UK: Rohm & Haas (UK) Ltd., Lennig House, 2 Mason's Avenue, Croydon CR9 3NB, UK. Telephone (44)-(0)1-686-8844. Fax (44)-(0)1-681-3207.

Sarstedt Inc., Research Products Division, Box 468, Newton, NC 28658-0468, USA. Telephone 704-465-4000 or 800-257-5101. Fax 704-465-4003.

Sartorius Corporation, 140 Wilbur Place, Bohemia, NY 11716, USA. Telephone 800-368-7178 or 800-227-2842. Fax 516-563-5065.

Savant Instruments Inc., 110-103 Bi-county Blvd., Farmingdale, NY 11735, USA. Telephone 516-249-4600 or 800-634-8886. Fax 516-249-4639.

Schleicher & Schuell Inc., 10 Optical Avenue Keene, NH 03431, USA. Telephone 603-352-3810 or 800-245-4024. Fax 603-357-3627.

Scientific Manufacturing Industries. *See* Baxter Healthcare Corporation.

Scientific Products. *See* Baxter Healthcare Corporation.

Serva Biochemicals Inc., 50 A and S Drive, Paramus, NJ 07652. Telephone 201-967-5900 or 800-645-3412. Fax 201-967-8858.

Sigma Chemical Company, P.O. Box 14509, St. Louis, MO 63178-9916, USA. Telephone 314-771-5750 or 800-325-3010. Fax 314-771-5757. In UK: Sigma Chemical Company Ltd., Fancy Road, Poole, Dorset BH17 7NH, UK. Telephone (44)-(0)202-733114. Fax (44)-(0)202-715460.

Sorvall. *See* E.I. du Pont de Nemours & Company Inc.

Sterilin Instruments Ltd., Lampton House, Lampton Road, Hounslow, Middlesex TW3 4EE, UK. Telephone (44)-(0)1-572-2468. Fax (44)-(0)1-572-7301.

Stratagene, 11099 North Torrey Pines Road, La Jolla, CA 92037, USA. Telephone 619-535-5400 or 800-424-5444. Fax 619-535-5430. In Europe: Stratagene GmbH., Postfach 105466, D-6900, Heidelberg, Germany. Telephone (49)-(0)6221-40-06-34. Fax (49)-(0)6221-40-06-39.

Takara Biochemical Inc., 719-Allston Way, Berkeley, CA 94710. Telephone 510-649-9859 or 800-544-9899. Fax 510-649-8933. In Japan: Takara Shuzo Co., Ltd., Shijo-Higasinotoin, Shimogya-ku, Kyoto, 600-91, Japan. (81)75-241-5167. Fax (81)75-241-5208.

Thomas Scientific Company, 99 High Hill Road, Swedesboro, NJ 08085, USA. Telephone 609-467-2000 or 800-345-2100. Fax 609-467-3087.

3M Company, 3M Center, St. Paul, MN 55144, USA. Telephone 612-733-5454 or 800-362-3456. Fax 612-736-3090.

Ultraviolet Products. *See* UVP Inc.

United States Biochemical Corporation (USB), P.O. Box 22400, Cleveland, OH 44122, USA. Telephone 216-765-5000 or 800-321-9322. Fax 216-464-5075.

USB. *See* United States Biochemical Corporation.

UVP Inc. (Ultraviolet Products), P.O. Box 1501, 5100 Walnut Grove Avenue, San Gabriel, CA 91776, USA. Telephone 818-285-3123 or 800-452-6788. Fax 818-285-2940.

Value Plastics Inc., 3350 Eastbrook Drive, Fort Collins, CO 80525, USA. Telephone 303-223-8306. Fax 303-223-0953.

Vangard International, 1111A Green Grove Road, Neptune, NJ 07753, USA. Telephone 908-922-4900. Fax 908-922-0557.

Van Waters and Rogers, P.O. Box 34325, Seattle, WA 98124-1325, USA. Telephone 206-889-3400. Fax 206-889-4100.

VirTis Company, Route 208, Gardiner, NY 12525, USA. Telephone 914-255-5000 or 800-431-8232. Fax 914-255-5338.

VWR Scientific Corp., P.O. Box 13645, Philadelphia, PA 19101-3645. Telephone 215-891-2770 or 800-777-8977.

Waters, Division of Millipore, 34 Maple Street, Milford, MA 01757, USA. Telephone 508-478-2000 or 800-252-4752. Fax 508-872-1990 or 508-478-2000.

Whatman Laboratory Products Inc., 9 Bridewell Place, Clifton, NJ 07014, USA. Telephone 201-773-5800 or 800-631-7290. Fax 201-472-6949. In UK: Whatman Ltd., Springfield Mill, Maidstone, Kent ME14 2LE, UK. Telephone (44)-(0)622-692022. Fax (44)-(0)622-691425.

Wheaton Scientific, Division of Wheaton Industries, 1000 N. Tenth Street, Millville, NJ 08332, USA. Telephone 609-825-1100 or 800-225-1437. Fax 609-825-1368.

Worthington Biochemical, Halls Mill Road,

Treehold, NJ 07728, USA. Telephone 201-462-3838 or 800-445-9603. Fax 201-308-4453. In UK: Cambridge Bioscience, 42 Devonshire Road, Cambridge CB1 2BL, UK. Telephone (44)-(0)223-316855. Fax (44)-(0)223-460396.

Index*

*Decimal numbers (e.g., 1.1) are used for pages in the Handbook.

I.1